# Essential UNIVERSITY PHYSICS

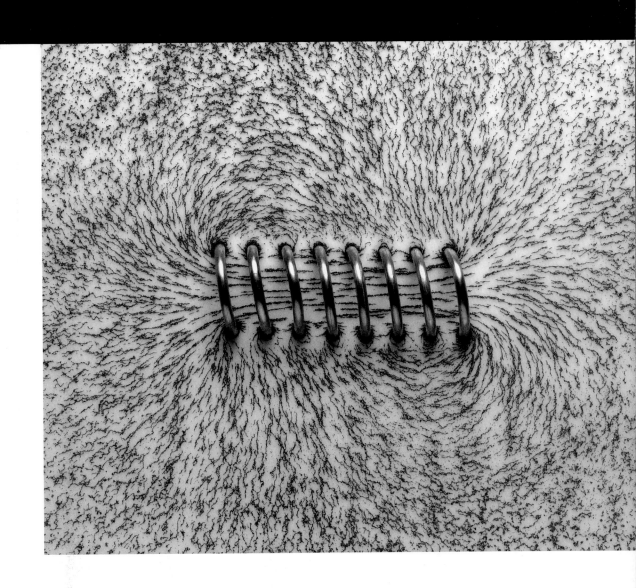

## Problem Solving Strategies

Volume 1 (pp. 1–326) includes chapters 1–19.
Volume 2 (pp. 327–727) includes chapters 20–39.

Essential **UNIVERSITY PHYSICS**

FIRST EDITION

Volume 2 **Richard Wolfson**
Chapters 20–39 *Middlebury College*

**PEARSON**

Addison
Wesley

San Francisco Boston New York
Cape Town Hong Kong London Madrid Mexico City
Montreal Munich Paris Singapore Tokyo Toronto

*Editor-in-Chief:* Adam Black, Ph.D.
*Development Editor:* Brad Patterson
*Project Manager:* Laura Kenney
*Project Editor:* Martha Steele
*Managing Producer:* Claire Masson
*Director of Marketing:* Christy Lawrence
*Managing Editor:* Corinne Benson
*Production Supervisor:* Nancy Tabor
*Production Service:* Elm Street Publishing Services, Inc.
*Composition:* Preparé, Inc.
*Text Design:* Elm Street Publishing Services, Inc.
*Cover Design:* Hespenheide Design
*Illustrations:* Rolin Graphics
*Photo Research:* Cypress Integrated Systems
*Director, Image Resource Center*: Melinda Patelli
*Rights and Permissions Manager*: Zina Arabia
*Cover Printer:* Phoenix Color Corporation
*Printer and Binder:* Von Hoffmann Graphics Owensville

*Cover Image*: John Eisele, Colorado State University

*Photo Credits*: See page C-1.

ISBN 0-8053-3838-1 Volume 2 without MasteringPhysics

ISBN 0-8053-4004-1 Volume 2 with MasteringPhysics

ISBN 0-8053-3998-1 Volume 2 Professional copy

Library of Congress Cataloging-in-Publication Data
Wolfson, Richard.
    Essential university physics
        p. cm.
    Includes index.
    ISBN 0-8053-9212-2
    1. Physics—Textbooks.   I. Title.

QC21.3.W65 2007
530—dc22                              2007040286

5  6  7  8  9  10  —VON4 — 11  10

www.aw.com/physics

PEARSON
Addison
Wesley

# Brief Contents

# About the Author

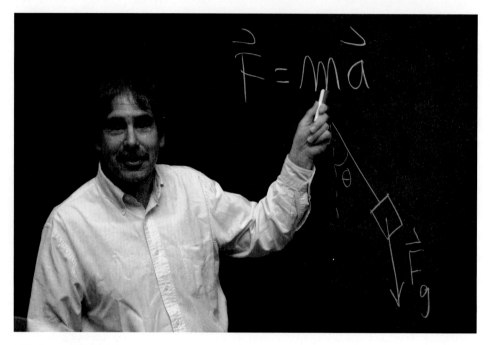

## Richard Wolfson

Richard Wolfson is the Benjamin F. Wissler Professor of Physics at Middlebury College, where he has taught since 1976. He did undergraduate work at MIT and Swarthmore College, and he holds an M.S. degree from the University of Michigan and Ph.D. from Dartmouth. His ongoing research on the Sun's corona and climate change has taken him to sabbaticals at the National Center for Atmospheric Research in Boulder, Colorado; St. Andrews University in Scotland; and Stanford University.

Rich is a committed and passionate teacher. This is reflected in his many publications for students and the general public, including the video series *Einstein's Relativity and the Quantum Revolution: Modern Physics for Nonscientists* (The Teaching Company, 1999) and *Physics in Your Life* (The Teaching Company, 2004); books *Nuclear Choices: A Citizen's Guide to Nuclear Technology* (MIT Press, 1993), *Simply Einstein: Relativity Demystified* (W. W. Norton, 2003), and *Energy, Environment, and Climate* (W. W. Norton, 2007); and articles for *Scientific American* and the *World Book Encyclopedia*.

Outside of his research and teaching, Rich enjoys hiking, canoeing, gardening, cooking, and watercolor painting.

Introductory physics texts have grown ever larger, more massive, more encyclopedic, more colorful, and more expensive. *Essential University Physics* bucks that trend—without compromising coverage, pedagogy, or quality. The text benefits from the author's three decades of teaching introductory physics, seeing firsthand the difficulties and misconceptions that students face as well as the "Got It!" moments when big ideas become clear. It also builds of the author's honing multiple editions of a previous calculus-based textbook and on feedback from hundreds of instructors and students.

## Goals of This Book

Physics is the fundamental science, at once fascinating, challenging, and subtle—and yet simple in a way that reflects the few basic principles that govern the physical universe. My goal is to bring this sense of physics alive for students in a range of academic disciplines who need a solid calculus-based physics course—whether they're engineers, physics majors, premeds, biologists, chemists, geologists, mathematicians, computer scientists, or other majors. My own courses are populated by just such a variety of students, and among my greatest joys as a teacher is having students who took a course only because it was required say afterwards that they really enjoyed their exposure to the ideas of physics. More specifically, my goals include:

- Helping students build the analytical and quantitative skills and confidence needed to apply physics in problem-solving for science and engineering.

- Adressing key misconceptions and helping students build a stronger conceptual understanding.

- Helping students see the relevance and excitement of the physics they're studying with wide-ranging and contemporary applications in science, technology, and everyday life.

- Helping students develop an appreciation of the physical universe at its most fundamental level.

- Engaging students with an informal, conversational writing style that balances precision with approachability.

## Pedagogical Innovations

This book is *concise*, but it's also *progressive* in its embrace of proven techniques from physics education research and *strategic* in its approach to learning physics. Chapter 1 introduces the IDEA framework for problem solving, and every one of the book's subsequent **worked examples** employs this framework. IDEA—an acronym for Identify, Develop, Evaluate, Assess—is not a "cookbook" method for students to apply mindlessly, but rather a tool for organizing students' thinking and discouraging equation hunting. It begins with an interpretation of the problem and an identification of the key physics concepts involved; develops a plan for reaching the solution; carries out the mathematical evaluation; and assesses the solution to see that it makes sense, to compare the example with others, and to mine additional insights into physics. In nearly all of the text's worked examples, the Develop phase includes making a drawing, and most of these use a hand-drawn style to encourage students to make their own drawings—a step that research suggests they often skip. IDEA provides a common approach to all physics problem solving, an approach that emphasizes the conceptual unity of physics and helps break the

- Complete edition Volumes 1–2 (shrinkwrapped) (ISBN 0-8053-9212-2): Chapters 1–39
- Volume 1 (ISBN 0-8053-3829-2): Chapters 1–19
- Volume 2 (ISBN 0-8053-3838-1): Chapters 20–39
- Complete edition Volumes 1–2 (shrinkwrapped) with MasteringPhysics (ISBN 0-8053-0296-4): Chapters 1–39
- Volume 1 with Mastering-Physics™ (ISBN 0-8053-3978-7): Chapters 1–19
- Volume 2 with Mastering-Physics™ (ISBN 0-8053-4004-1): Chapters 20–39

typical student view of physics as a hodgepodge of equations and unrelated ideas. In addition to IDEA-based worked examples, other pedagogical features include:

- **Problem-Solving Strategy boxes** that follow the IDEA framework to provide detailed guidance for specific classes of physics problems, such as Newton's second law, conservation of energy, thermal-energy balance, Gauss's law, or multiloop circuits.

- **Tactics boxes** that reinforce specific essential skills such as differentiation, setting up integrals, vector products, drawing free-body diagrams, simplifying series and parallel circuits, or ray tracing.

- **Got It? boxes** that provide quick checks for students to test their conceptual understanding. Many of these use a multiple-choice or quantitative ranking format to probe student misconceptions and facilitate their use with classroom-response systems.

- **Tips** that provide helpful problem-solving hints or warn against common pitfalls and misconceptions.

- **Chapter openers** that include a forward-looking **To Learn** list of new concepts and skills in the chapter ahead, and a backward-looking **To Know** list of important ideas on which the chapter builds. Both lists reference specific chapter sections by number.

- **Applications**, self-contained presentations typically shorter than half a page, provide interesting and contemporary instances of physics in the real world, such as: bicycle stability; flywheel energy storage; laser vision correction; ultracapacitors; wind energy; magnetic resonance imaging; global climate change; combined-cycle power generation; circuit models of the cell membrane; CD, DVD, and "Blu-Ray" technologies; and radiocarbon dating.

- **For Thought and Discussion** questions at the end of each chapter designed for peer learning or for self-study to enhance students' conceptual understanding of physics.

- **Annotated figures** that adopt the research-based approach of including simple "instructor's voice" commentary to help students read and interpret pictorial and graphical information.

- **End-of-chapter** problems that begin with simpler exercises keyed to individual chapter sections and ramp up to more challenging and often multistep problems that synthesize chapter material. At the end are context-rich problems focusing on real-world situations.

- **Chapter summaries** that combine text, art, and equations to provide a synthesized overview of each chapter. Each summary is hierarchical, beginning with the chapter's "big picture" ideas, then focusing on key concepts and equations, and ending with a list of "Cases and Uses"—specific instances or applications of the physics presented in the chapter.

## Organization

This contemporary book is *concise, strategic,* and *progressive,* but it's *traditional* in its organization. Following the introductory Chapter 1, the book is divided into six parts. Part One (Chapters 2–12) develops the basic concepts of mechanics, including Newton's laws and conservation principles as applied to single particles and multiparticle systems. Part Two (Chapters 13–15) extends mechanics to oscillations, waves, and fluids. Part Three (Chapters 16–19) covers thermodynamics. Part Four (Chapters 20–29) deals with electricity and magnetism. Part Five (Chapters 30–32) treats optics, first in the geometrical optics approximation and then including wave phenomena. Part Six (Chapters 33–39) introduces relativity and quantum physics. Each part begins with a brief description of its coverage, and ends with a conceptual summary and a challenge problem that synthesizes ideas from several chapters.

*Essential University Physics* is available in two paperback volumes, so students can purchase only what they need—making the low-cost aspect of this text even more attractive. Volume 1 includes Parts One, Two, and Three, mechanics through thermodynamics. Volume 2 contains Parts Four, and Five, and Six, electricity and magnetism along with optics and modern physics.

## Instructor Supplements

The **Instructor Solutions Manual, Chapters 1–39** (ISBN 0-8053-0342-1), written by Edw. S. Ginsberg, University of Massachusetts, with contributing author Sen-Ben Liao, University of California, provides complete and detailed solutions to all end-of-chapter problems. Solutions to odd-numbered problems follow the Interpret/Develop/Evaluate/Assess problem-solving framework used in the textbook.

The cross-platform **Instructor Resource CD-ROMs** (ISBN 0-8053-4008-4) consist of the **Simulation and Image Presentation CD-ROM** and the **Instructor Supplement CD-ROM**. The *Simulation and Image Presentation CD-ROM* provides all line figures and tables from the textbook in JPEG format. In addition, all the key equations, problem-solving strategies, and chapter summaries are provided in multiple editable Word formats. In-class weekly multiple-choice questions for use with various Classroom Response Systems (CRS) are also provided, based on the Got It? questions in the text. The *Instructor Supplement CD-ROM* provides the Instructor Solutions Manual in convenient editable multiple Word formats and as PDFs.

**MasteringPhysics™** (www.masteringphysics.com) is the most advanced, educationally effective, and widely used physics homework and tutorial system in the world. It provides instructors with a library of extensively pretested end-of-chapter problems and rich, Socratic tutorials that incorporate a wide variety of answer types, wrong-answer feedback, and adaptive help (comprising hints or simpler subproblems upon request). MasteringPhysics™ allows instructors to quickly build wide-ranging homework assignments of just the right difficulty and duration and provides them with efficient tools to analyze class trends—or the work of any student—in unprecedented detail and to compare the results either with the national average or with the performance of previous classes.

The **Transparency Acetates** (ISBN 0-8053-0404-5) provide more than 200 key figures from *Essential University Physics,* 1st edition.

The **Printed Test Bank** (ISBN 0-8053-0404-5) provides more than 2000 multiple-choice, true-false, and conceptual questions.

The **Computerized Test Bank** (ISBN 0-8053-0401-0) provides all of the questions from the Printed Test Bank on a cross-platform CD-ROM. More than half of the questions can be assigned with randomized numerical values.

## Student Supplements

**Student Solutions Manuals, Chapters 1–19** (ISBN 0-8053-4009-2) and **Chapters 20–39** (ISBN 0-8053-0405-3), written by Edw. S. Ginsberg, University of Massachusetts, with contributing author Sen-Ben Liao, University of California, provide detailed, step-by-step solutions to all of the odd-numbered and selected other end-of-chapter problems from the textbook. These solutions consistently follow the same Interpret/Develop/Evaluate/Assess problem-solving framework used in the textbook, reinforcing good problem-solving techniques.

**MasteringPhysics™** (www.masteringphysics.com) is the most advanced, widely used, and educationally proven physics tutorial system in the world. It is the result of eight years of detailed studies of how real students work physics problems and of precisely where they need help. Studies show that students who use MasteringPhysics™ significantly improve their scores on final exams and conceptual tests such as the Force Concept Inventory. MasteringPhysics™ achieves this improvement by providing students with instantaneous feedback specific to their wrong answers, simpler subproblems upon request when they get stuck, and partial credit for their method(s) used. This individualized, 24/7 tutor system is recommended by 9 out of 10 students to their peers as the most effective and time-efficient way to study.

## Acknowledgments

A project of this magnitude isn't the work of its author alone. First and foremost among those I thank for their contributions are the now several thousand students I've taught in calculus-based introductory physics courses at Middlebury College. Over the years your questions have taught me how to convey physics ideas in many different ways appropriate to your diverse learning styles. You've helped identify the "sticking points" that challenge introductory physics students, and you've showed me ways to help you avoid and "unlearn" the misconceptions that many students bring to introductory physics.

Thanks also go to my Middlebury faculty colleagues and to numerous instructors and students from around the world who have contributed valuable suggestions that were incorporated in the revisions of my earlier introductory physics text, *Physics for Scientists and Engineers* (Wolfson and Pasachoff, third edition: Addison-Wesley, 1999). I've heard you, and you'll find still more of your suggestions implemented in *Essential University Physics*.

Experienced physics instructors thoroughly reviewed every chapter of this book, and reviewers' comments resulted in substantive changes—and sometimes in major rewrites—to the first drafts of the manuscript. We list all these reviewers below. But first, special thanks are due to six individuals who made exceptional contributions to the quality and in some cases the very existence of this book. First is Professor Jay Pasachoff of Williams College, whose willingness more than two decades ago to take a chance on an inexperienced coauthor has made writing introductory physics a large part of my professional career. Dr. Adam Black, physics editor and Ph.D. physicist at Addison-Wesley, had the vision to see promise in a new introductory text that would respond to the rising chorus of complaints about massive, encyclopedic, and expensive physics texts. Brad Patterson, the book's development editor, brought his graduate-level knowledge of physics to a role that made him a real collaborator and the closest this book has to a coauthor. Brad is responsible for many of the book's innovative features, and it was a pleasure to work with him. Mark Hollabaugh of Normandale Community College is responsible for most of the new context-rich problems. We've gone to great lengths to make this book as error-free as possible, and much of the credit for that happy situation goes to Linda Barton of the Rochester Institute of Technology. Not only did she check the numbers for every worked example, but she also read the entire book in page proof with a professionally critical eye, and she's responsible for many improvements to both text and art made even at that late stage. Very special thanks are due to Edw. Ginsberg, of the University of Massachusetts, with whom I've collaborated since the first edition of *Physics for Scientists and Engineers*. Ed developed new calculus-based problems for this book and, most important, provided valuable suggestions for the entire book when it was in the final manuscript phase. Ed Ginsberg's stamp is on every chapter, and readers will be better served because of his contributions.

Finally, I thank Martha Steele, Nancy Tabor, and Laura Kenney at Addison-Wesley, and Heather Johnson of Elm Street Publishing Services, for their highly professional efforts in shepherding this book through its vigorous production schedule. And, as always, I thank my family, my colleagues, and my students for the patience they showed during the intensive process of writing this book.

## Reviewers

John R. Albright, *Purdue University–Calumet*
Rama Bansil, *Boston University*
Richard Barber, *Santa Clara University*

Linda S. Barton, *Rochester Institute of Technology*
Rasheed Bashirov, *Albertson College of Idaho*
Chris Berven, *University of Idaho*
David Bixler, *Angelo State University*
Ben Bromley, *University of Utah*
Charles Burkhardt, *St. Louis Community College*
Susan Cable, *Central Florida Community College*
George T. Carlson, Jr., *West Virginia Institute of Technology–West Virginia University*
Catherine Check, *Rock Valley College*
Norbert Chencinski, *College of Staten Island*
Carl Covatto, *Arizona State University*
David Donnelly, *Texas State University–San Marcos*
David G. Ellis, *University of Toledo*
Tim Farris, *Volunteer State Community College*
Paula Fekete, *Hunter College of The City University of New York*
Idan Ginsburg, *Harvard University*
James Goff, *Pima Community College*
Mark Hollabaugh, *Normandale Community College*
Rex W. Joyner, *Indiana Institute of Technology*
Nikos Kalogeropoulos, *Borough of Manhattan Community College–The City University of New York*
Kevin T. Kilty, *Laramie County Community College*
Duane Larson, *Bevill State Community College*
Kenneth W. McLaughlin, *Loras College*
Tom Marvin, *Southern Oregon University*
Perry S. Mason, *Lubbock Christian University*
Mark Masters, *Indiana University–Purdue University Fort Wayne*
Jonathan Mitschele, *Saint Joseph's College*
Gregor Novak, *United States Air Force Academy*
Richard Olenick, *University of Dallas*
Robert Philbin, *Trinidad State Junior College*
Russell Poch, *Howard Community College*
Steven Pollock, *Colorado University–Boulder*
James Rabchuk, *Western Illinois University*
George Schmiedeshoff, *Occidental College*
Natalia Semushkina, *Shippensburg University of Pennsylvania*
Anwar Shiekh, *Dine College*
David Slimmer, *Lander University*
Chris Sorensen, *Kansas State University*
Ronald G. Tabak, *Youngstown State University*
Gajendra Tulsian, *Daytona Beach Community College*
Henry Weigel, *Arapahoe Community College*
Arthur W. Wiggins, *Oakland Community College*
Fredy Zypman, *Yeshiva University*

Welcome to physics! Maybe you're taking introductory physics because you're majoring in a field of science or engineering that requires a semester or two of physics. Maybe you're premed, and you know that medical schools are increasingly interested in seeing calculus-based physics on your transcript. Perhaps you're really gung-ho and plan to major in physics. Or maybe you want to study physics further as a minor associated with related fields like math or chemistry or to complement a discipline like economics, environmental studies, or even music. Perhaps you had a great high-school physics course, and you're eager to continue. Maybe high-school physics was an academic disaster for you, and you're approaching this course with trepidation. Or perhaps this is your first experience with physics. Whatever your reason for taking introductory physics, welcome!

And whatever your reason, my goals for you are similar: I'd like to help you develop an understanding and appreciation of the physical universe at a deep and fundamental level; I'd like you to become aware of the broad range of natural and technological phenomena that physics can explain; and I'd like to help you strengthen your analytic and quantitative problem-solving skills. Even if you're studying physics only because it's a requirement, I want to help you engage the subject and come away with an appreciation for this fundamental science and its wide applicability. One of my greatest joys as a physics teacher is having students tell me after the course that they had taken it only because it was required, but found they really enjoyed their exposure to the ideas of physics.

Physics is fundamental. To understand physics is to understand how the world works, both in everyday life and on scales of time and space so small and so large as to defy intuition. For that reason I hope you'll find physics fascinating. But you'll also find it challenging. Learning physics will challenge you with the need for precise thinking and language; with subtle interpretations of even commonplace phenomena; and with the need for skillful application of mathematics. But there's also a simplicity to physics, a simplicity that results because there are in physics only a very few really basic principles to learn. Those succinct principles encompass a universe of natural phenomena and technological applications.

I've been teaching introductory physics for decades, and this book distills everything my students have taught me about the many different ways to approach physics; about the subtle misconceptions students often bring to physics; about the ideas and types of problems that present the greatest challenges; and about ways to make physics engaging, exciting, and relevant to your life and interests.

I have some specific advice for you that grows out of my long experience teaching introductory physics. Keeping this advice in mind will make physics easier (but not necessarily easy!), more interesting, and, I hope, more fun:

- *Read* each chapter thoroughly and carefully before you attempt to work any problem assignments. I've written this text with an informal, conversational style to make it engaging. It's not a reference work to be left alone until you need some specific piece of information; rather, it's an unfolding "story" of physics—its big ideas and their applications in quantitative problem solving. You may think physics is hard because it's mathematical, but in my long experience I've found that failure to *read* thoroughly is the biggest single reason for difficulties in introductory physics.

- *Look for the big ideas*. Physics isn't a hodge-podge of different phenomena, laws, and equations to memorize. Rather, it's a few big ideas from which flow myriad applications, examples, and special cases. In particular, don't think of physics as a jumble of equations that you choose among when solving a problem. Rather, identify those few big ideas and the equations that represent them, and try to see how seemingly distinct examples and special cases relate to the big ideas.

- *When working problems*, *re-read* the appropriate sections of the text, paying particular attention to the worked examples. Follow the IDEA strategy described in Chapter 1 and used in every subsequent worked example. Don't skimp on the final Assess step. Always ask: Does this answer make sense? How can I understand my answer in relation to the big principles of physics? How was this problem like others I've worked, or like examples in the text?

- *Don't confuse physics with math.* Mathematics is a tool, not an end in itself. Equations in physics aren't abstract math, but statements about the physical world. Be sure you understand each equation for what it says about physics, not just as an equality between mathematical terms.

- *Work with others.* Getting together informally in a room with a blackboard is a great way to explore physics, to clarify your ideas and help others clarify theirs, and to learn from your peers. I urge you to discuss physics problems together with your classmates, to contemplate together the "For Thought and Discussion" questions at the end of each chapter, and to engage one another in lively dialog as you grow your understanding of physics, the fundamental science.

# Detailed Contents

# Electromagnetism

Electromagnetism is one of the fundamental forces, and it governs the behavior of matter from the atomic scale to the macroscopic world. Electromagnetic technology, from computer microchips to cell phones and on to large electric motors and generators, is essential to modern society. Even our bodies rely heavily on electromagnetism, as electric signals pace our heartbeat, electrochemical processes transmit nerve impulses, and the electric structure of cell membranes mediates the flow of materials into and out of the cell.

■ A significant portion of humankind's energy is in the form of electricity, as evidenced by this composite satellite image of Earth at night. Nearly all that electrical energy is produced by generators, devices that exploit an intimate relation between electricity and magnetism.

Four fundamental laws describe electricity and magnetism. Two deal separately with the two phenomena, while the other two reveal profound connections that make electricity and magnetism aspects of a single phenomenon—electromagnetism. The 10 chapters of Part 4 will give you a thorough grounding in electromagnetism. You'll come to understand those fundamental laws, learn how electromagnetism determines the structure and behavior of nearly all matter, and explore the electromagnetic technologies that play so important a role in your life. Finally, you'll see how the laws of electromagnetism lead to electromagnetic waves and thus help us understand the nature of light.

# 20

# Electric Charge, Force, and Field

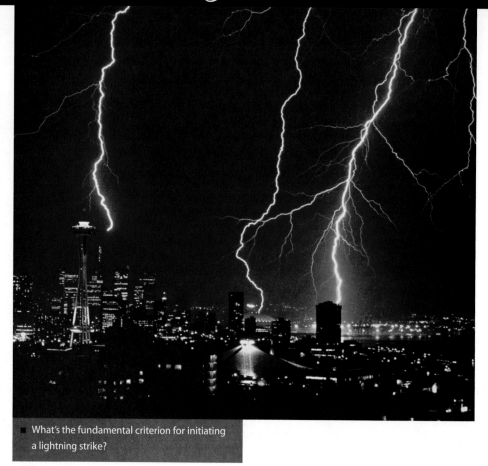

■ What's the fundamental criterion for initiating a lightning strike?

What holds your body together? What keeps a skyscraper standing? What holds your car on the road as you round a turn? What governs the electronic circuitry in your computer or MP3 player, or provides the tension in your climbing rope? What enables a plant to make sugar from sunlight and simple chemicals? What underlies the awesome beauty of a thunderstorm? The answer, in all cases, is the **electric force**. With the exception of gravity, all the forces we've encountered in mechanics—including tension forces, normal forces, compression forces, and friction—are based on electric interactions; so are the forces responsible for all of chemistry and biology. The electric force, in turn, involves a fundamental property of matter—namely, electric charge.

## 20.1 Electric Charge

**Electric charge** is an intrinsic property of the electrons and protons that, along with uncharged neutrons, make up ordinary matter. What is electric charge? At the most fundamental level we don't know. We don't know what mass "really" is either, but we're familiar with it because we've spent our lives pushing objects around. Similarly, our knowledge of electric charge results from observing the behavior of charged objects.

Charge comes in two varieties, which Benjamin Franklin designated *positive* and *negative*. Those names are useful because the total charge on an object—the object's **net charge**—is the algebraic sum of its constituent charges. Like charges repel, and opposites attract, a fact that constitutes a qualitative description of the electric force.

## Quantities of Charge

All electrons carry the same charge, and all protons carry the same charge. The proton's charge has *exactly* the same magnitude as the electron's, but with opposite sign. Given that electrons and protons differ substantially in other properties—like mass—this electric relation is remarkable. Exercise 14 shows how dramatically different our world would be if there were even a slight difference between the magnitudes of the electron and proton charges.

The magnitude of the electron or proton charge is the **elementary charge** $e$. Electric charge is **quantized**; that is, it comes only in discrete amounts. In a famous experiment in 1909, the American physicist R. A. Millikan measured the charge on small oil drops and found it was always a multiple of a basic value we now know as the elementary charge.

Elementary particle theories show that the basic unit of charge is actually $\frac{1}{3}e$. Such "fractional charges" reside on quarks, the building blocks of protons, neutrons, and many other particles. Quarks always join to produce particles with integer multiples of the full elementary charge, and it seems impossible to isolate individual quarks.

The SI unit of charge is the **coulomb** (C), named for the French physicist Charles Augustin de Coulomb (1736–1806). Although the coulomb's formal definition is in terms of electric current, it's convenient to describe 1 coulomb as about $6.25 \times 10^{18}$ elementary charges, making the elementary charge approximately $1.60 \times 10^{-19}$ C.

## Charge Conservation

Electric charge is a conserved quantity, meaning that the net charge in a closed region remains constant. Charged particles may be created or annihilated, but always in pairs of equal and opposite charge. The net charge always remains the same.

## 20.2 Coulomb's Law

Rub a balloon; it gets charged and sticks to your clothing. Charge another balloon, and the two repel (Fig. 20.1). Socks cling to your clothes as they come from the dryer, and bits of Styrofoam cling annoyingly to your hands. Walk across a carpet, and you'll feel a shock when you touch a doorknob. All these are common examples where you're directly aware of electric charge.

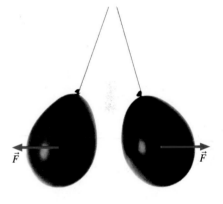

**FIGURE 20.1** Two balloons carrying similar electric charges repel each other.

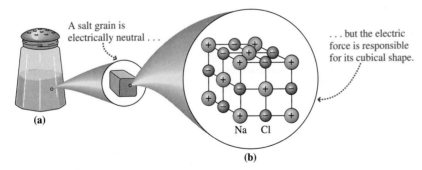

**FIGURE 20.2** (a) A single salt grain is electrically neutral, so the electric force isn't obvious. (b) Actually, the electric force determines the structure of salt.

Electricity would be unimportant if the only significant electric interactions were these obvious ones. In fact, the electric force dominates all interactions of everyday matter, from the motion of a car to the movement of a muscle. It's just that matter on a large scale is almost perfectly neutral, meaning it carries zero net charge. Therefore electric effects aren't obvious. But at the molecular level, the electric nature of matter is immediately evident (Fig. 20.2).

The unit vector $\hat{r}$ always points *away* from $q_1$.

$$\vec{F}_{12} = \frac{kq_1q_2}{r^2}\hat{r}$$

Here the product $q_1q_2$ is positive, so $\vec{F}_{12}$ is in the same direction as $\hat{r}$.

$q_1$ ⊕     $q_2$ ⊕ $\hat{r}$ → $\vec{F}_{12}$

(a)

$q_1$ ⊕ ← $\vec{F}_{12}$   $q_2$ ⊖ $\hat{r}$

Here the charges have opposite signs, so $q_1q_2 < 0$ and $\vec{F}_{12}$ points opposite $\hat{r}$.

(b)

**FIGURE 20.3** Quantities in Coulomb's law for calculating the force $\vec{F}_{12}$ that $q_1$ exerts on $q_2$.

Attraction and repulsion of electric charges imply a force. Joseph Priestley and Charles Augustin de Coulomb investigated this force in the late 1700s. They found that the force between two charges acts along the line joining them, with the magnitude proportional to the product of the charges and inversely proportional to the square of the distance between them. **Coulomb's law** summarizes these results:

$$\vec{F}_{12} = \frac{kq_1q_2}{r^2}\hat{r} \qquad \text{(Coulomb's law)} \qquad (20.1)$$

where $\vec{F}_{12}$ is the force charge $q_1$ exerts on $q_2$ and $r$ is the distance between the charges. The proportionality constant $k$ has SI value $9.0 \times 10^9$ N·m²/C². Force is a vector, and $\hat{r}$ is a unit vector that helps determine its direction. Figure 20.3 shows that $\hat{r}$ lies on a line passing through the two charges and points in the direction *from $q_1$ toward $q_2$*. Reverse the roles of $q_1$ and $q_2$, and you'll see that $\vec{F}_{21}$ has the same magnitude as $\vec{F}_{12}$ but the opposite direction; thus Coulomb's law obeys Newton's third law. Figure 20.3 also shows that the force is in the same direction as the unit vector when the charges have the same sign, but opposite the unit vector when the charges have different signs. Thus Coulomb's law accounts for the fact that like charges repel and opposites attract.

---

**PROBLEM SOLVING STRATEGY 20.1** **Coulomb's Law**

The key to using Coulomb's law is to remember that force is a vector, and to realize that Coulomb's law in the form 20.1 gives both the magnitude and direction of the electric force. Dealing carefully with vector directions is especially important in situations with more than two charges.

**INTERPRET** First, make sure you're dealing with the electric force alone. Identify the charge or charges on which you want to calculate the force. Next, identify the charge or charges producing the force. This is the **source charge**.

**DEVELOP** Begin with a drawing that shows the charges, as in Fig. 20.4. If you're given charge coordinates, place the charges on the coordinate system; if not, choose a suitable coordinate system. Determine the unit vector(s) in Equation 20.1. If the charges lie along or parallel to a coordinate axis, then the unit vector will be one of the unit vectors $\hat{i}$, $\hat{j}$, or $\hat{k}$, perhaps with a minus sign. In Fig. 20.4, the force on $q_3$ due to $q_1$ is such a case. When the two charges don't lie on a coordinate axis, like $q_1$ and $q_2$ in Fig. 20.4, you can find the unit vector by noting that the displacement vector $\vec{r}_{12}$ points in the desired direction, from the source charge to the charge experiencing the force. Dividing $\vec{r}_{12}$ by its own magnitude then gives the unit vector in the direction of $\vec{r}_{12}$: $\hat{r} = \vec{r}_{12}/r_{12}$.

**EVALUATE** Evaluate the electric force using Equation 20.1, $\vec{F}_{12} = (kq_1q_2/r^2)\hat{r}$, with $\hat{r}$ the unit vector you've just found.

**ASSESS** As always, assess your answer to see that it makes sense. Is the direction of the force you found consistent with the signs and placements of the charges giving rise to the force?

Here $\hat{r} = \dfrac{\vec{r}_{12}}{r_{12}} = \frac{4}{5}\hat{i} + \frac{3}{5}\hat{j}$.

$\vec{r}_{12} = 4\hat{i} + 3\hat{j}$ m

$r_{12} = \sqrt{4^2 + 3^2}$ m $= 5$ m

$\hat{r}$ points away from $q_1$, so here $\hat{r} = \hat{i}$.

**FIGURE 20.4** Finding unit vectors.

**GOT IT? 20.1**   Charge $q_1$ is located at $x = 1$ m, $y = 0$. What should you use for the unit vector $\hat{r}$ in Coulomb's law if you're calculating the force $q_1$ exerts on a charge $q_2$ located at (a) the origin and (b) the point $x = 0$, $y = 1$ m? Explain why you can answer without knowing the sign of either charge.

---

**EXAMPLE 20.1**   **Finding the Force: Two Charges**

A 1.0-$\mu$C charge is at $x = 1.0$ cm, and a $-1.5$-$\mu$C charge is at $x = 3.0$ cm. What force does the positive charge exert on the negative one? How would the force change if the distance between the charges tripled?

**INTERPRET** Following our strategy, we identify the $-1.5$-$\mu$C charge as the one on which we want to find the force and the 1-$\mu$C charge as the source charge.

**DEVELOP** We're given the coordinates $x_1 = 1.0$ cm and $x_2 = 3.0$ cm. Our drawing, Fig. 20.5, shows both charges at their positions on the $x$ axis. With the source charge $q_1$ to the left, the unit vector in the direction from $q_1$ toward $q_2$ is $\hat{\imath}$.

**EVALUATE** Now we use Coulomb's law to evaluate the force:

$$\vec{F}_{12} = \frac{kq_1q_2}{r^2}\hat{r}$$

$$= \frac{(9.0 \times 10^9 \, \text{N} \cdot \text{m}^2/\text{C}^2)(1.0 \times 10^{-6} \, \text{C})(-1.5 \times 10^{-6} \, \text{C})}{(0.020 \, \text{m})^2}\hat{\imath}$$

$$= -34\hat{\imath} \, \text{N}$$

This force is for a separation of 2 cm; if that distance tripled, the force would drop by a factor of $1/3^2$, to $-3.8\hat{\imath}$ N.

**FIGURE 20.5** Sketch for Example 20.1.

The charges have opposite signs, so $q_1q_2$ is negative, and $\vec{F}$ is opposite the direction of $\hat{r}$.

The unit vector is in the direction from $q_1$ to $q_2$, so here it's $\hat{\imath}$.

**ASSESS** Make sense? Although the unit vector $\hat{\imath}$ points in the $+x$ direction, the charges have opposite signs and that makes the force direction opposite the unit vector, as shown in Fig. 20.5. In simpler terms, we've got two opposite charges, so they attract. That means the force on a charge at $x = 3$ cm *from* an opposite charge at $x = 1$ cm had better be in the $-x$ direction. ∎

---

Coulomb's law for the electric force is similar to Newton's law for the gravitational force (see Section 8.2). Both show the same inverse-square decrease with distance, and both are proportional to the product of the interacting charges or masses. But there's an important difference: There's only one kind of mass, and the gravitational force is always attractive. That means large concentrations of mass result in large gravitational forces. But charge comes in two varieties that attract each other, so large concentrations of charge tend to be electrically neutral and therefore result in weak electric forces. The electric force between individual particles is vastly stronger than the gravitational force between the same particles (see Exercise 17), and it's only because of the nearly complete cancellation of positive and negative charge in bulk matter that gravity becomes important on the macroscopic scale.

## Point Charges and the Superposition Principle

Coulomb's law is strictly true only for **point charges**—charged objects of negligible size. Electrons and protons can usually be treated as point charges; so, approximately, can any two charged objects if their separation is large compared with their size. But often we're interested in the electric effects of **charge distributions**—arrangements of charge spread over space. Charge distributions are present in molecules, a memory cell in a computer chip, your heart, and a thundercloud. We need to combine the effects of two or more charges to find the electric effects of such charge distributions.

Figure 20.6 shows two charges $q_1$ and $q_2$ that constitute a simple charge distribution. We want to know the net force these exert on a third charge $q_3$. To find that net force, you might calculate the forces $\vec{F}_{13}$ and $\vec{F}_{23}$ from Equation 20.1, and then vectorially add them. And you'd be right: The force that $q_1$ exerts on $q_3$ is unaffected by the presence of $q_2$, and vice versa, so you can apply Coulomb's law separately to the pairs $q_1q_3$ and $q_2q_3$ and then combine the results. That may seem obvious, but nature needn't have been so simple.

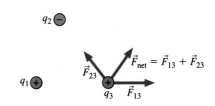

**FIGURE 20.6** The superposition principle lets us add vectorially the forces from two or more charges.

The fact that electric forces add vectorially is called the **superposition principle**. Our confidence in this principle is ultimately based on experiments that show electric and indeed electromagnetic phenomena behave according to the principle. With superposition we can solve relatively complicated problems by breaking them into simpler parts. If the superposition principle did not hold, the mathematical description of electromagnetism would be far more complicated than it is.

Although the force that one point charge exerts on another decreases with the inverse square of the distance between them, the same is not necessarily true of the force resulting from a charge distribution. The next example provides a case in point.

---

**EXAMPLE 20.2** **Finding the Force: Raindrops**

Charged raindrops are ultimately responsible for thunderstorms, as they result in substantial electric charge within specific regions of a thundercloud. Suppose two drops with equal charge $q$ are on the $x$ axis at $x = \pm a$. Find the electric force on a third drop with charge $Q$ at an arbitrary point on the $y$ axis.

**INTERPRET** Coulomb's law applies, and we identify $Q$ as the charge for which we want the force. The two charges $q$ are the source charges.

**DEVELOP** Figure 20.7 is our drawing, showing the charges, the individual force vectors, and their sum. The drawing shows that the distance $r$ in Coulomb's law is the hypotenuse $\sqrt{a^2 + y^2}$. It's clear from the symmetry that the net force is in the $y$ direction, so we need to find only the $y$ components of the unit vectors. The $y$ components are clearly the same for each, and the drawing shows that they're given by $\hat{r}_y = y/\sqrt{a^2 + y^2}$.

**EVALUATE** From Coulomb's law, the $y$ component of the force from each $q$ is $F_y = (kqQ/r^2)\hat{r}_y$, and the net force on $Q$ becomes

$$\vec{F} = 2\left(\frac{kqQ}{a^2 + y^2}\right)\left(\frac{y}{\sqrt{a^2 + y^2}}\right)\hat{j} = \frac{2kqQy}{(a^2 + y^2)^{3/2}}\hat{j}$$

The factor of 2 comes from the two charges $q$, which contribute equally to the net force.

**ASSESS** Make sense? Evaluating $\vec{F}$ at $y = 0$ gives zero force. Here, midway between the two charges, $Q$ experiences equal but opposite forces and the net force is zero. At large distances $y \gg a$, on the other hand, we can neglect $a^2$ compared with $y^2$, and the force becomes $\vec{F} = k(2q)Q\hat{j}/y^2$. This is just what we would expect from a single

**FIGURE 20.7** The force on $Q$ is the vector sum of the forces from the individual charges.

charge $2q$ a distance $y$ from $Q$—showing that the system of two charges acts like a single charge $2q$ at distances large compared with the charge separation. In between our two extremes the behavior of force with distance is more complicated; in fact, its magnitude initially increases as $Q$ moves away from the origin and then begins to decline.

In drawing Fig. 20.7, we tacitly assumed that $q$ and $Q$ have the same signs. But our analysis holds even if they don't; then the product $qQ$ is negative, and the forces actually point opposite the directions shown in Fig. 20.7. ∎

---

## 20.3 The Electric Field

In Chapter 8 we defined the gravitational field at a point as the gravitational force per unit mass that an object at that point would experience. In that context, we can think of $\vec{g}$ as the *force per unit mass* that any object would experience due to Earth's gravity. So we can picture the gravitational field as a set of vectors giving the magnitude and direction of the gravitational force per unit mass at each point, as shown in Fig. 20.8a.

We can do the same thing with the electric force, defining the **electric field** as the force per unit charge:

> The electric field at any point is the force per unit charge that would be experienced by a charge at that point. Mathematically,
>
> $$\vec{E} = \frac{\vec{F}}{q} \qquad \text{(electric field)} \qquad (20.2a)$$

Right at this point the electric field is described by the vector $\vec{E}_1$. That means a point charge $q$ placed here would experience an electric force $q\vec{E}_1$.

Right at this point the gravitational field is described by the vector $\vec{g}$. That means a mass $m$ placed here would experience a gravitational force $m\vec{g}$.

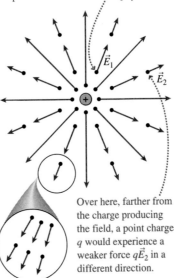

Over here, farther from the charge producing the field, a point charge $q$ would experience a weaker force $q\vec{E}_2$ in a different direction.

The gravitational field is a continuous entity, so there are field vectors everywhere. We just can't draw them all. **(a)**

The electric field is a continuous entity, so there are field vectors everywhere. We just can't draw them all.

**(b)**

**FIGURE 20.8** (a) Gravitational and (b) electric fields, here represented as sets of vectors.

The electric field exists at every point in space. When we represent the field by vectors, we can't draw one everywhere, but that doesn't mean there isn't a field at all points. Furthermore, we draw vectors as extended arrows, but each vector represents the field at only one point—namely, the tail end of the vector. Figure 20.8b illustrates this for the electric field of a point charge.

The field concept leads to a shift in our thinking about forces. Instead of the action-at-a-distance idea that Earth reaches across empty space to pull on the Moon, the field concept says that Earth creates a gravitational field and the Moon responds to the field at its location. Similarly, a charge creates an electric field throughout the space surrounding it. A second charge then responds to the field at its immediate location. Although the field reveals itself only through its effect on a charge, the field nevertheless exists at all points, whether or not there are charges present. Right now you probably find the field concept a bit abstract, but as you advance in your study of electromagnetism you'll come to appreciate that fields are an essential feature of our universe, every bit as real as matter itself.

We can use Equation 20.2a as a prescription for measuring electric fields. Place a point charge at some point, measure the electric force it experiences, and divide by the charge to get the field. In practice, we need to be careful because the field generally arises from some distribution of source charges. If the charge we're using to probe the field—the **test charge**—is large, the field it creates may disturb the source charges, altering their configuration and thus the field they create. For that reason, it's important to use a very small test charge.

If we know the electric field $\vec{E}$ at a point, we can rearrange Equation 20.2a to find the force on any point charge $q$ placed at that point:

$$\vec{F} = q\vec{E} \qquad \text{(electric force and field)} \qquad (20.2b)$$

If the charge $q$ is positive, then this force is in the same direction as the field, but if $q$ is negative, then the force is opposite to the field direction.

Equations 20.2 show that the units of electric field are newtons per coulomb. Fields of hundreds to thousands of N/C are commonplace, while fields of 3 MN/C will tear electrons from air molecules.

**EXAMPLE 20.3** **Force and Field: A Thunderstorm**

A charged raindrop carrying $10\ \mu C$ experiences an electric force of $0.30\ N$ in the $+x$ direction. What's the electric field at its location? What would the force be on a $-5.0$-$\mu C$ drop at the same point?

**INTERPRET** In this problem we distinguish between an electric force and an electric field. The electric field exists with or without the charged raindrop present, and the electric force arises when the charged raindrop is in the electric field.

**DEVELOP** Knowing the electric force and the charge on the raindrop, we can use Equation 20.2a, $\vec{E} = \vec{F}/q$, to get the electric field. Once we know the field, we can use Equation 20.2b, $\vec{F} = q\vec{E}$, to calculate the force that would act if a different charge were at the same point.

**EVALUATE** Equation 20.2a gives the electric field:

$$\vec{E} = \frac{\vec{F}}{q} = \frac{0.30\hat{\imath}\ N}{10\ \mu C} = 30\hat{\imath}\ kN/C$$

Acting on a $-5.0$-$\mu C$ charge, this field would result in a force

$$\vec{F} = q\vec{E} = (-5.0\ \mu C)(30\hat{\imath}\ kN/C) = -0.15\hat{\imath}\ N/C$$

**ASSESS** Make sense? The force on the second charge is opposite the direction of the field because now we've got a negative charge in the same field.

✓**TIP** The Field Is Independent of the Test Charge

Does the electric field in this example point in the $-x$ direction when the charge is negative? No. The field is independent of the particular charge experiencing that field. Here the electric field points in the $+x$ direction *no matter what charge* you put in the field. For a positive charge, the force $q\vec{E}$ points in the *same* direction as the field; for a negative charge, $q < 0$, the force is *opposite* the field.

■

## The Field of a Point Charge

Once we know the field of a charge distribution, we can calculate its effect on other charges. The simplest charge distribution is a single point charge. Coulomb's law gives the force on a test charge $q_{test}$ located a distance $r$ from a point charge $q$: $\vec{F} = (kqq_{test}/r^2)\hat{r}$, where $\hat{r}$ is a unit vector pointing *away* from $q$. The electric field arising from $q$ is the force per unit charge, or

$$\vec{E} = \frac{\vec{F}}{q_{test}} = \frac{kq}{r^2}\hat{r} \qquad \text{(field of a point charge)} \qquad (20.3)$$

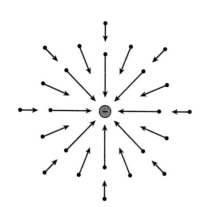

**FIGURE 20.9** Field vectors for a negative point charge.

Since it's so closely related to Coulomb's law for the electric force, we also refer to Equation 20.3 as Coulomb's law. The equation contains no reference to the test charge $q_{test}$ because the field of $q$ exists independently of any other charge. Since $\hat{r}$ always points *away* from $q$, the direction of $\vec{E}$ is radially outward if $q$ is positive and radially inward if $q$ is negative. Figure 20.9 shows some field vectors for a negative point charge, analogous to those of the positive point charge in Fig. 20.8b.

## 20.4 Fields of Charge Distributions

Since the electric force obeys the superposition principle, so does the electric field. That means the field of a charge distribution is the vector sum of the fields of the individual point charges making up the distribution:

$$\vec{E} = \vec{E}_1 + \vec{E}_2 + \vec{E}_3 + \cdots = \sum_i \vec{E}_i = \sum_i \frac{kq_i}{r_i^2}\hat{r}_i \qquad (20.4)$$

Here the $\vec{E}_i$'s are the fields of the point charges $q_i$ located at distances $r_i$ from the point where we're evaluating the field—called, appropriately, the **field point**. The $\hat{r}_i$'s are unit vectors pointing *from* each point charge *toward* the field point. In principle, Equation 20.4 gives the electric field of *any* charge distribution. In practice, the process of summing the individual field vectors is often complicated unless the charge distribution contains relatively few charges arranged in a symmetric way.

Finding electric fields using Equation 20.4 involves the same strategy we introduced for finding the electric force; the only difference is that there's no charge to experience the force. The first step then involves identifying not a charge but the field point. We still need to find the appropriate unit vectors and form the vector sum in Equation 20.4.

EXAMPLE 20.4 **Finding the Field: Two Protons**

Two protons are 3.6 nm apart. Find the electric field at a point between them, 1.2 nm from one of the protons. Then find the force on an electron at this point.

**INTERPRET** We follow our electric-force strategy, except that instead of identifying the charge experiencing the force, we identify the field point as being 1.2 nm from one proton. The source charges are the two protons; they produce the field we're interested in.

**DEVELOP** Let's have the protons define the $x$ axis, as drawn in Fig. 20.10. Then the unit vector $\hat{r}_1$ from the left-hand proton toward the field point is $+\hat{i}$, while $\hat{r}_2$ from the right-hand proton toward $P$ is $-\hat{i}$.

**EVALUATE** We now evaluate the field at $P$ using Equation 20.4:

$$\vec{E} = \vec{E}_1 + \vec{E}_2 = \frac{ke}{r_1^2}\hat{i} + \frac{ke}{r_2^2}(-\hat{i}) = ke\left(\frac{1}{r_1^2} - \frac{1}{r_2^2}\right)\hat{i}$$

We wrote $e$ for $q$ here because the protons' charge is the elementary charge.

Using $e = 1.6 \times 10^{-19}$ C, $r_1 = 1.2$ nm, and $r_2 = 2.4$ nm gives $\vec{E} = 750\hat{i}$ MN/C. An electron in this field will then experience a force $\vec{F} = q\vec{E} = -e\vec{E} = -0.12\hat{i}$ nN.

**ASSESS** Make sense? The field points in the positive $x$ direction, reflecting the fact that $P$ is closer to the left-hand proton with its stronger field at $P$. The force on the electron, on the other hand, is in the $-x$ direction; that's because the electron is negative (we used $q = -e$ for its charge), so the force it experiences is opposite the field. That field of almost 1 GN/C sounds huge—but that's not unusual at the microscopic scale, where we're close to individual elementary particles. ■

Unit vectors point *from* the source charges *toward* the field point $P$.

**FIGURE 20.10** Finding the electric field at $P$.

## The Electric Dipole

One of the most important charge distributions is the **electric dipole**, consisting of two point charges of equal magnitude but opposite sign. Many molecules are essentially dipoles, so understanding the dipole helps explain molecular behavior (Fig. 20.11). During contraction the heart muscle becomes essentially a dipole, and physicians performing electrocardiography are measuring, among other things, the strength and orientation of that dipole. Technological devices, including radio and TV antennas, also make use of the dipole configuration.

**FIGURE 20.11** A water molecule behaves like an electric dipole.

EXAMPLE 20.5 **The Electric Dipole: Modeling a Molecule**

A molecule may be modeled approximately as a positive charge $q$ at $x = a$ and a negative charge $-q$ at $x = -a$. Evaluate the electric field on the $y$ axis, and find an approximate expression valid at large distances $(y \gg a)$.

**INTERPRET** Here's another example where we'll use our strategy in applying Equation 20.4 to calculate the field of a charge distribution. We identify the field point as being anywhere on the $y$ axis and the source charges as being $\pm q$.

**DEVELOP** Figure 20.12 is our drawing. The individual unit vectors point from the two charges toward the field point, but the *negative* charge contributes a field *opposite* its unit vector; we've indicated the individual fields in Fig. 20.12. Here symmetry makes the $y$ components cancel, giving a net field in the $-x$ direction. So we need only the $x$ components of the unit vectors, which Fig. 20.12 shows are $\hat{r}_{x-} = a/r$ for the negative charge at $-a$ and $\hat{r}_{x+} = -a/r$ for the positive charge at $a$.

**EVALUATE** We then evaluate the field using Equation 20.4:

$$\vec{E} = \frac{k(-q)}{r^2}\left(\frac{a}{r}\right)\hat{i} + \frac{kq}{r^2}\left(-\frac{a}{r}\right)\hat{i} = -\frac{2kqa}{(a^2 + y^2)^{3/2}}\hat{i}$$

where in the last step we used $r = \sqrt{a^2 + y^2}$. For $y \gg a$ we can neglect $a^2$ compared with $y^2$, giving

$$\vec{E} \simeq -\frac{2kqa}{y^3}\hat{i} \qquad (y \gg a)$$

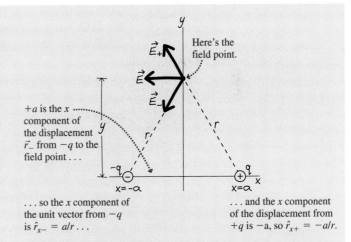

$+a$ is the $x$ component of the displacement $\vec{r}_-$ from $-q$ to the field point . . .

. . . so the $x$ component of the unit vector from $-q$ is $\hat{r}_{x-} = a/r$ . . .

. . . and the $x$ component of the displacement from $+q$ is $-a$, so $\hat{r}_{x+} = -a/r$.

**FIGURE 20.12** Finding the field of an electric dipole.

**ASSESS** Make sense? The dipole has no net charge, so at large distances its field can't have the inverse-square drop-off of a point-charge field. Instead the dipole field falls faster, here as $1/y^3$.

✓**TIP** Approximations

Making approximations requires care. Here we're basically asking for the field when $y$ is so large that $a$ is negligible *compared with* $y$. So we neglect $a^2$ compared with $y^2$ when the two are summed, but we *don't* neglect $a$ when it appears in the numerator.

■

Example 20.5 shows that the dipole field at large distances decreases as the inverse *cube* of distance. Physically, that's because the dipole has zero *net* charge. Its field arises entirely from the slight separation of two opposite charges. Because of this separation the dipole field isn't exactly zero, but it's weaker and more localized than the field of a point charge. Many complicated charge distributions exhibit the essential characteristic of a dipole—namely, they're neutral but consist of separated regions of positive and negative charge—and at large distances such distributions all have essentially the same field configuration.

At large distances the dipole's physical characteristics $q$ and $a$ enter the equation for the electric field only through the product $qa$. We could double $q$ and halve $a$, and the dipole's electric field would remain unchanged. In general, a dipole's electric properties are characterized completely by its **electric dipole moment** $p$, defined as the product of the charge $q$ and the separation $d$ between the two charges making up the dipole:

$$p = qd \quad \text{(dipole moment)} \tag{20.5}$$

In Example 20.5 the charge separation was $d = 2a$, so there the dipole moment was $p = 2aq$. In terms of the dipole moment, the field in Example 20.5 can then be written

$$\vec{E} = -\frac{kp}{y^3}\hat{\imath} \quad \left(\begin{array}{c}\text{dipole field for } y \gg a, \\ \text{on perpendicular bisector}\end{array}\right) \tag{20.6a}$$

You can show in Problem 51 that the field on the dipole axis is given by

$$\vec{E} = \frac{2kp}{x^3}\hat{\imath} \quad \left(\begin{array}{c}\text{dipole field} \\ \text{for } x \gg a, \text{ on axis}\end{array}\right) \tag{20.6b}$$

Because the dipole isn't spherically symmetric, its field depends not only on distance but also on orientation; for instance, Equations 20.6 show that the field along the dipole axis at a given distance is twice as strong as along the bisector. So it's important to know the orientation of a dipole in space, and therefore we generalize our definition of the dipole moment to make it a vector of magnitude $p = qd$ in the direction from the negative toward the positive charge (Fig. 20.13).

**FIGURE 20.13** The dipole moment vector has magnitude $p = qd$ and points from the negative toward the positive charge.

---

**GOT IT? 20.2** Far from a charge distribution, you measure an electric field strength of 800 N/C. What will the field strength be if you double your distance from the charge distribution, if the distribution consists of (a) a point charge or (b) a dipole?

---

## Continuous Charge Distributions

Although any charge distribution ultimately consists of pointlike electrons and protons, it would be impossible to sum all the field vectors from the $10^{23}$ or so particles in a typical piece of matter. Instead, it's convenient to make the approximation that charge is spread continuously over the distribution. If the charge distribution extends throughout a volume, we describe it in terms of the **volume charge density** $\rho$, with units of C/m$^3$. For charge distributions spread over surfaces or lines, the corresponding quantities are the **surface charge density** $\sigma$ (C/m$^2$) and the **line charge density** $\lambda$ (C/m).

To calculate the field of a continuous charge distribution, we divide the charged region into very many small charge elements $dq$, each small enough that it's essentially a point charge. Each $dq$ then produces an electric field $d\vec{E}$ given by Equation 20.3: $d\vec{E} = (k\,dq/r^2)\hat{r}$. We then form the vector sum of all the $d\vec{E}$'s (Fig. 20.14). In the limit of infinitely many infinitesimally small $dq$'s and their corresponding $d\vec{E}$'s, that sum becomes an integral and we have

$$\vec{E} = \int d\vec{E} = \int \frac{k\,dq}{r^2}\hat{r} \quad \left(\begin{array}{c}\text{field of a continuous} \\ \text{charge distribution}\end{array}\right) \tag{20.7}$$

The limits of this integral include the entire charge distribution.

Calculating the field of a continuous charge distribution involves the same strategy we've already used: We identify the field point and the source charges—although now the

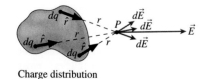

Charge distribution

**FIGURE 20.14** The electric field at $P$ is the vector sum of the fields $d\vec{E}$ arising from the individual charge elements $dq$, each calculated using the appropriate distance $r$ and unit vector $\hat{r}$.

source is a continuous charge distribution. Summing the individual field contributions now presents us with an integral, and that means writing the unit vectors $\hat{r}$ and distances $r$ in terms of coordinates over which we can integrate. Setting up the integral involves the same strategy we outlined in Chapter 9 to find the center of mass of a continuous distribution of matter. We used that strategy again in Chapter 10 to calculate rotational inertias, and you might want to review it.

---

**EXAMPLE 20.6** **Evaluating the Field: A Charged Ring**

A ring of radius $a$ carries a charge $Q$ distributed evenly over the ring. Find an expression for the electric field at any point on the axis of the ring.

**INTERPRET** We identify the field point as lying anywhere on the ring's axis, and the source charge as the entire ring.

**DEVELOP** Let's take the $x$ axis to coincide with the ring axis, with the center of the ring at $x = 0$ (Fig. 20.15). Figure 20.15 shows that the $y$ components of the field contributions from pairs of charge elements on opposite sides of the ring cancel; therefore the net field points in the $x$ direction and we need only the $x$ components of the unit vectors. Those are the same for all unit vectors—namely, $\hat{r}_x = x/r$.

**FIGURE 20.15** The electric field of a charged ring points along the ring axis, since field components perpendicular to the axis cancel in pairs.

**EVALUATE** We're now ready to set up the integral of Equation 20.7. Here each charge element contributes the same amount $dE_x = (k\,dq/r^2)\hat{r}_x = (k\,dq/r^2)(x/r)$ to the field. Figure 20.15 shows that $r = \sqrt{x^2 + a^2} = (x^2 + a^2)^{1/2}$, so the integral becomes

$$E = \int_{ring} dE_x = \int_{ring} \frac{kx\,dq}{(x^2 + a^2)^{3/2}} = \frac{kx}{(x^2 + a^2)^{3/2}} \int_{ring} dq$$

The last step follows because we have a fixed field point $P$, so its co-ordinate $x$ is a constant for the integration. But the remaining integral is just the sum of all the charge elements on the ring—namely, the total charge $Q$. So our result becomes

$$E = \frac{kQx}{(x^2 + a^2)^{3/2}} \qquad \text{(on-axis field, charged ring)}$$

This is the magnitude; the direction is along the $x$ axis, away from the ring if $Q$ is positive and toward it if $Q$ is negative.

**ASSESS** Make sense? At $x = 0$ the field is zero. Of course: A charge placed at the ring center is pulled (or pushed) equally in all directions—no net force, so no electric field. But for $x \gg a$, we get $E = kQ/x^2$—just what we expect for a point charge $Q$. As always, a finite-size charge distribution looks like a point charge at large distances. ∎

---

**EXAMPLE 20.7** **Line Charge: A Power Line's Field**

A long, straight electric power line coincides with the $x$ axis and carries a uniform line charge density $\lambda$ (unit: C/m). Find the electric field on the $y$ axis using the approximation that the wire is infinitely long.

**INTERPRET** We identify the field point as being a distance $y$ from the wire, and the source charge as the whole wire.

**DEVELOP** Figure 20.16 is our drawing, showing a coordinate system with the field point $P$ along the $y$ axis. We divide the rod into small charge elements $dq$ and note that field contributions from two such elements $dq$ on opposite sides of the $y$ axis contribute fields $d\vec{E}$ whose $x$ components cancel. Then we need only the $y$ component of each unit vector, and Figure 20.16 shows that's $\hat{r}_y = y/r$.

**EVALUATE** We're now ready to set up the integral of Equation 20.7. As described in Chapter 9's integral strategy, we need to relate $dq$ to a geometric variable so we can do the integral. Here our wire has charge density $\lambda$ C/m, so if a charge element has length $dx$, then its charge is

This is the $y$ component of the displacement $\vec{r}$ from $dq$ to $P$ . . .

. . . so the $y$ component of the unit vector $\hat{r}$ is $y/r$.

**FIGURE 20.16** The field of a charged line is the vector sum of the fields $d\vec{E}$ from all the individual charge elements $dq$ along the line.

$dq = \lambda\,dx$. Putting all this together gives the $y$ component of the field from an arbitrary $dq$ anywhere on the wire:

$$dE_y = \frac{k\,dq}{r^2}\hat{r}_y = \frac{k\lambda\,dx}{r^2}\frac{y}{r} = \frac{k\lambda y}{(x^2 + y^2)^{3/2}}dx$$

*cont'd.*

where we used $r = \sqrt{x^2 + y^2}$. Since the $x$ components cancel, we can sum—that is, integrate—the $y$ components to get the net field:

$$E = E_y = \int_{-\infty}^{+\infty} \frac{k\lambda y \, dx}{(x^2 + y^2)^{3/2}} = k\lambda y \int_{-\infty}^{+\infty} \frac{dx}{(x^2 + y^2)^{3/2}}$$

$$= k\lambda y \left[ \frac{x}{y^2 \sqrt{x^2 + y^2}} \right]_{-\infty}^{+\infty} = k\lambda y \left[ \frac{1}{y^2} - \left( -\frac{1}{y^2} \right) \right] = \frac{2k\lambda}{y}$$

Here we used the integral table in Appendix A and applied the limits $x = \pm\infty$. Our result is the field's magnitude; the direction is away from the line for positive $\lambda$ and toward the line for negative $\lambda$.

**ASSESS** Make sense? For an infinite line there's nothing to favor one direction along the line over another, so the only way the field can point is radially, away from or toward the line (Fig. 20.17). And because the line is infinite, it never resembles a point no matter how far away we are. As a result the field falls more slowly than the field of a point charge—in this case, as $1/y$. More generally, if we let $r$ designate the radial distance from the line rather than the diagonal in Fig. 20.16, then the field

**FIGURE 20.17** Field vectors for an infinite line of positive charge point radially outward, with magnitude decreasing inversely with distance.

decreases as $1/r$. An infinite line is impossible, but our result holds approximately for finite lines of charge as long as we're much closer to the line than its length, and not near an end. Far from a *finite* line, on the other hand, its field will resemble that of a point charge. ∎

## 20.5 Matter in Electric Fields

Electric fields give rise to forces on charged particles. Because matter consists of such particles, much of the behavior of matter is fundamentally determined by electric fields.

### Point Charges in Electric Fields

The motion of a single charge in an electric field is governed by the definition of the electric field, $\vec{F} = q\vec{E}$, and Newton's law, $\vec{F} = m\vec{a}$. Combining these equations gives the acceleration of a particle with charge $q$ and mass $m$ in an electric field $\vec{E}$:

$$\vec{a} = \frac{q}{m}\vec{E} \tag{20.8}$$

This equation shows that it's the charge-to-mass ratio, $q/m$, that determines a particle's response to an electric field. Electrons, nearly 2000 times less massive than protons but carrying the same charge, are readily accelerated by electric fields. Many practical devices, from X-ray machines to TV tubes, make use of electrons accelerated in electric fields.

When the electric field is uniform, problems involving the motion of charged particles reduce to the constant-acceleration problems of Chapter 2. A pair of charged plates provides a uniform field and is used to deflect charged particles in devices ranging from video displays to inkjet printers (Fig. 20.18).

When the field isn't uniform, it's generally more difficult to calculate particle trajectories. An important exception is a particle moving perpendicular to a field that points radially. The result is uniform circular motion (see Section 5.3), as shown in the next example.

**FIGURE 20.18** A pair of parallel charged plates creates a uniform electric field that deflects a charged particle.

**EXAMPLE 20.8** Particle Motion: An Electrostatic Analyzer

Two curved metal plates establish an electric field given by $E = E_0(b/r)$, where $E_0$ and $b$ are constants with the units of electric field and length, respectively. The field points toward the center of curvature, and $r$ is the distance from the center. Find an expression for the speed $v$ with which a proton entering vertically from below in Fig. 20.19 will leave the device moving horizontally.

**INTERPRET** This problem is about charged-particle motion in an electric field that points radially. We're asked for the condition that will have a proton exiting the field region moving horizontally. Figure 20.19 shows that this requires its trajectory to be a circular arc.

**FIGURE 20.19** An electrostatic analyzer.                    *cont'd.*

**DEVELOP** Equation 20.8, $\vec{a} = (q/m)\vec{E}$, determines the acceleration of a charged particle in an electric field. Here we want uniform circular motion, so our plan is to write this equation with the given field and the acceleration $v^2/r$ that we know applies in circular motion. Then we'll solve for $v$.

**EVALUATE** Under these conditions, Equation 20.8 becomes

$$a = \frac{v^2}{r} = \frac{eE}{m} = \frac{e}{m}E_0\frac{b}{r}$$

We then solve to get $v = \sqrt{eE_0 b/m}$.

**ASSESS** Make sense? Strengthen the field by increasing $E_0$ or $b$, and the electric force becomes greater. For a given speed, that would result in more bending of the trajectory; to maintain the desired trajectory, speed must therefore increase. Note that the radius $r$ canceled from our equations, showing that it doesn't matter where the protons enter the device. That's because the $1/r$ decrease in field strength matches the $1/r$ dependence of the acceleration. This device is called an electrostatic analyzer because it can sort charged particles by speed and charge-to-mass ratio. Spacecraft use such analyzers to characterize charged particles in interplanetary space. ∎

---

**GOT IT? 20.3** An electron, a proton, a deuteron (a neutron combined with a proton), a helium-3 nucleus (2 protons, 1 neutron), a helium-4 nucleus (2 protons, 2 neutrons), a carbon-13 nucleus (6 protons, 7 neutrons), and an oxygen-16 nucleus (8 protons, 8 neutrons) all find themselves in the same electric field. Rank order their accelerations from lowest to highest under the assumption (only approximately correct) that the neutron and proton have the same mass and that the mass of a composite particle is the sum of the masses of its constituent neutrons and protons. Note any that have the same acceleration.

## Dipoles in Electric Fields

Earlier in this chapter we calculated the field of an electric dipole, which consists of two opposite charges of equal magnitude. Here we study a dipole's response to electric fields. Since the dipole provides a model for many molecules, our results help explain molecular behavior.

Figure 20.20 shows a dipole with charges $\pm q$ separated a distance $d$, located in a uniform electric field. The dipole moment vector $\vec{p}$ has magnitude $qd$ and points from the negative to the positive charge (recall Fig. 20.13). Since the field is uniform, it's the same at both ends of the dipole. Since the dipole charges are equal in magnitude but opposite in sign, they experience equal but opposite forces $\pm q\vec{E}$—and therefore there's no net force on the dipole.

However, Fig. 20.20 shows that the dipole does experience a torque that tends to align it with the field. In Chapter 11 we described torque as the cross product of the position vector with the force: $\vec{\tau} = \vec{r} \times \vec{F}$, where the magnitude of the torque vector is $rF\sin\theta$ and its direction is given by the right-hand rule. Figure 20.20 thus shows that the torque about the center of the dipole due to the force on the positive charge has magnitude $\tau_+ = rF\sin\theta = \left(\frac{1}{2}d\right)(qE)\sin\theta$. The torque associated with the negative charge has the same magnitude, and both torques are in the same direction since both tend to rotate the dipole clockwise. Thus the net torque has magnitude $\tau = qdE\sin\theta$; applying the right-hand rule shows that this torque is into the page. But $qd$ is the magnitude of the dipole moment $\vec{p}$, and Fig. 20.20 shows that $\theta$ is the angle between the dipole moment vector and the electric field $\vec{E}$; therefore we can write the torque vectorially as

$$\vec{\tau} = \vec{p} \times \vec{E} \qquad \text{(torque on a dipole)} \qquad (20.9)$$

Because of this torque, it takes work to rotate a dipole in an electric field. If we start with the dipole oriented at right angles to the field $(\theta = \pi/2)$, then Equation 10.18 gives the work required to rotate it until it makes an angle $\theta$ with the field:

$$W = \int_{\pi/2}^{\theta} \tau\, d\theta = \int_{\pi/2}^{\theta} pE\sin\theta\, d\theta = pE\left[-\cos\theta\right]_{\pi/2}^{\theta} = -pE\cos\theta$$

This work ends up as stored potential energy $U$. Since the product of two vector magnitudes with the cosine of the angle between them defines the dot product, we can write the potential energy as

$$U = -\vec{p} \cdot \vec{E} \qquad (20.10)$$

where $U = 0$ corresponds to the dipole at right angles to the field.

Torque rotates dipole clockwise.

**FIGURE 20.20** A dipole in a uniform electric field experiences a torque, but no net force.

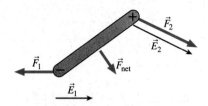

**FIGURE 20.21** When the electric field differs in magnitude or direction at the two ends of the dipole, the dipole experiences a nonzero net force as well as a torque.

**FIGURE 20.22** Dipole *B* aligns with the field of dipole *A* and then experiences a net force toward *A*.

When the electric field isn't uniform, the charges at opposite ends of the dipole experience forces that differ in magnitude and/or aren't exactly opposite in direction. Then the dipole experiences a net force as well as a torque (Fig. 20.21). An important instance of this effect is the force on a dipole in the field of another dipole (Fig. 20.22). Because the dipole field falls off rapidly with distance and because the dipole responding to the field has closely spaced charges of equal magnitude but opposite sign, the dipole-dipole force is quite weak and falls extremely rapidly with distance. This weak force, which Fig. 20.22 shows to be attractive, is partly responsible for the van der Waals interaction between gas molecules that we mentioned in Chapter 17.

## Conductors, Insulators, and Dielectrics

Bulk matter contains vast numbers of point charges—namely, electrons and protons. In some matter—notably metals, ionic solutions, and ionized gases—individual charges are free to move throughout the material. In these **conductors**, the application of an electric field results in the ordered motion of electric charge that we call **electric current**. We'll consider conductors and current in later chapters.

Materials in which charge is not free to move are **insulators**, since they don't support electric current. Insulators, however, still contain charges—it's just that their charges are bound into neutral molecules. Some molecules, like water, have intrinsic dipole moments and therefore rotate in response to an applied electric field. Even if they don't have dipole moments, molecules may respond to an electric field by stretching and acquiring **induced**

---

**APPLICATION** **Microwave Cooking and Liquid Crystals**

The torque on dipoles in electric fields forms the basis of two widespread contemporary technologies: the microwave oven and the liquid crystal display (LCD).

A microwave oven works by generating an electric field whose direction changes several billion times per second. Water molecules, whose dipole moment is much greater than most others, attempt to align with the field. But the field is changing, so the molecules swing rapidly back and forth. As they jostle against each other, the energy they gain from the field is dissipated as heat that cooks the food.

Liquid crystals consist of dipole-like molecules that all align in the same direction.

Applying an electric field changes that orientation.

$\vec{E}$

Normal liquid    Liquid crystal    Aligned by external field

Computer displays, flat-panel TVs, digital cameras, cell phones, watches, and many other devices display visual images using liquid crystals. These unique materials combine the fluidity of a liquid with the order of a solid. The liquid crystal consists of long molecules whose chemical structure results in a dipole-like charge separation. In response to each others' electric fields, the molecules tend to align. As the figure shows, an external electric field can rotate the liquid crystal dipoles, altering the material's optical properties. With optical components we'll

study in Chapter 29, different sections of a liquid crystal display can then be made to appear visible or invisible. Liquid crystal displays consume very little power, but they generate no light of their own and therefore most have a built-in light source. The photo shows an iPod—a popular device that sports a liquid-crystal display; also shown is a microphoto of the liquid crystals.

**dipole moments** (Fig. 20.23). In either case, the application of an electric field results in the alignment of molecular dipoles with the field (Fig. 20.24). The fields of the dipoles, pointing from their positive to their negative charges, then reduce the applied electric field within the material. We'll explore the consequences of this effect further in Chapter 23. Materials in which molecules either have intrinsic dipole moments or acquire induced moments are called **dielectrics**.

If the electric field applied to a dielectric becomes too great, individual charges are ripped free, and the material then acts like a conductor. Such **dielectric breakdown** can cause severe damage in electric equipment. On a larger scale, lightning results from dielectric breakdown in air.

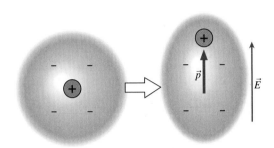

**FIGURE 20.23**  A molecule stretches in response to an electric field, acquiring a dipole moment.

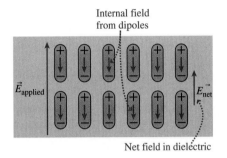

**FIGURE 20.24**  Alignment of molecular dipoles in a dielectric reduces the electric field within the dielectric.

## Big Picture

This chapter introduces several big ideas. First is **electric charge,** a fundamental property of matter that comes in positive and negative forms. Like charges repel and opposites attract; this is the **electric force.** It's convenient to define the **electric field** as the force per unit charge that a charge would experience if placed in the vicinity of other charges. Both force and field obey the **superposition principle**, meaning that the effects of several charges add vectorially.

## Key Concepts and Equations

**Coulomb's law** describes the electric force between point charges:

$$\vec{F}_{12} = \frac{kq_1 q_2}{r^2}\,\hat{r}$$

The electric field is the force per unit charge, $\vec{E} = \vec{F}/q$, and therefore the force a given charge $q$ experiences in a field is $\vec{F} = q\vec{E}$.

Electric field at $P$.

Put a charge $q$ at $P$, and the force on $q$ is $\vec{F} = q\vec{E}$.

The field of a point charge follows from Coulomb's law:

$$\vec{E} = \frac{kq}{r^2}\,\hat{r}$$

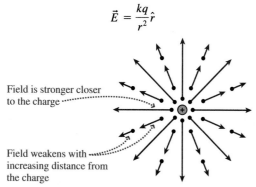

Field is stronger closer to the charge

Field weakens with increasing distance from the charge

Fields of charge distributions are found by summing fields of individual point charges, or by integrating in the case of continuously distributed charge:

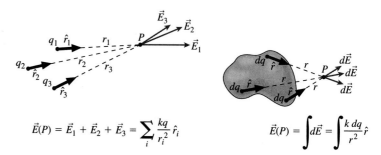

$$\vec{E}(P) = \vec{E}_1 + \vec{E}_2 + \vec{E}_3 = \sum_i \frac{kq}{r_i^2}\,\hat{r}_i$$

$$\vec{E}(P) = \int d\vec{E} = \int \frac{k\,dq}{r^2}\,\hat{r}$$

## Cases and Uses

A **dipole** consists of equal but opposite charges $\pm q$ a distance $d$ apart. For distances large compared with $d$, the dipole field drops as $1/r^3$, and the dipole is completely characterized by its **dipole moment** $qd$.

The field of an infinite line drops as $1/r$: $E = 2k\lambda/r$, with $\lambda$ the charge per unit length. This is a good approximation to the field near an elongated structure like a wire.

Point charges respond to electric fields with acceleration proportional to the charge-to-mass ratio $q/m$.

A dipole in an electric field experiences a torque that tends to align it with the field: $\vec{\tau} = \vec{p} \times \vec{E}$.

Torque rotates dipole clockwise.

If the field is nonuniform, there's also a net force on the dipole.

**Dielectrics** are insulating materials whose molecules act like electric dipoles.

# For Thought and Discussion

1. Discuss this statement: It is precisely because the electric force is so strong that the electric nature of most everyday interactions isn't obvious.

2. The gravitational force between an electron and a proton is about $10^{-40}$ times weaker than the electric force between the two. Since matter consists largely of electrons and protons, why is the gravitational force ever important?

3. A free neutron is unstable and soon decays to other particles. One of the decay products is a proton. Must there be others? If so, what electric properties must they have?

4. Where in Fig. 20.5 could you put a third charge so it would experience no net force? Would it be in stable or unstable equilibrium?

5. Why should the test charge used to measure an electric field be small?

6. Equation 20.3 gives the electric field of a point charge. Does the direction of $\hat{r}$ depend on whether the charge is positive or negative? Does the direction of $\vec{E}$ depend on the sign of the charge?

7. Is the electric force on a charged particle always in the direction of the field? Explain.

8. Why does a dipole produce an electric field at all? After all, the dipole has no net charge.

9. The ring in Example 20.6 carries total charge $Q$, and the point $P$ is the same distance $r = \sqrt{x^2 + a^2}$ from all parts of the ring. So why isn't the electric field of the ring just $kQ/r^2$?

10. A spherical balloon is initially uncharged. If you spread positive charge uniformly over the balloon's surface, would it expand or contract? What would happen if you spread negative charge instead?

11. Under what circumstances is the path of a charged particle a parabola? A circle?

12. Why should there be a force between two dipoles? After all, each has zero net charge.

13. Dipoles $A$ and $B$ are both located in the field of a point charge $Q$, as shown in Fig. 20.25. Does either experience a net torque? A net force? If each dipole is released from rest, describe qualitatively its subsequent motion.

**FIGURE 20.25** For Thought and Discussion 13

# Exercises and Problems

## Exercises

### Section 20.1 Electric Charge

14. Suppose the electron and proton charges differed by one part in one billion. Estimate the net charge you would carry.

15. A typical lightning flash delivers about 25 C of negative charge from cloud to ground. How many electrons are involved?

16. Protons and neutrons are made from combinations of the two most common quarks, the $u$ quark and the $d$ quark. The $u$ quark's charge is $+\frac{2}{3}e$, while the $d$ quark carries $-\frac{1}{3}e$. How could three of these quarks combine to make (a) a proton and (b) a neutron?

### Section 20.2 Coulomb's Law

17. Find the ratio of the electric force between a proton and an electron to the gravitational force between the two. Why doesn't it matter that you aren't told the distance between them?

18. The electron and proton in a hydrogen atom are 52.9 pm apart. What is the magnitude of the electric force between them?

19. How far apart should an electron and a proton be so the force of Earth's gravity on the electron is equal to the electric force arising from the proton? *Note:* Your answer shows why gravity is unimportant on the molecular scale!

20. Earth carries a net charge of $-4.3 \times 10^5$ C. The force due to this charge is the same as if it were concentrated at Earth's center. How much charge would you have to place on a 1.0-g mass in order for the electric and gravitational forces on it to balance?

21. A charge $q$ is at the point $x = 1$ m, $y = 0$. Write expressions for the unit vectors you would use in Coulomb's law if you were finding the force that $q$ exerts on other charges located at (a) $x = 1$ m, $y = 1$ m; (b) the origin; and (c) $x = 2$ m, $y = 3$ m. Note that you don't know the sign of $q$. Why doesn't this matter?

22. A proton is at the origin and an electron is at the point $x = 0.41$ nm, $y = 0.36$ nm. Find the electric force on the proton.

### Section 20.3 The Electric Field

23. An electron placed in an electric field experiences an electric force of $6.1 \times 10^{-10}$ N. What is the field strength?

24. What is the magnitude of the force on a 2.0-$\mu$C charge in a 100-N/C electric field?

25. A 68-nC charge experiences a 150-mN force in a certain electric field. Find (a) the field strength and (b) the force that a 35-$\mu$C charge would experience in the same field.

26. A $-1.0$-$\mu$C charge experiences a $10\hat{i}$-N electric force in a certain electric field. What force would a proton experience in the same field?

27. The electron in a hydrogen atom is 52.9 pm from the proton. What is the proton's electric field strength at this distance?

### Section 20.4 Fields of Charge Distributions

28. In Fig. 20.26, point $P$ is midway between the two charges. Find the electric field in the plane of the page (a) 5.0 cm to the left of $P$, (b) 5.0 cm directly above $P$, and (c) at $P$.

**FIGURE 20.26** Exercise 28

29. The dipole moment of the water molecule is $6.2 \times 10^{-30}$ C·m. What would be the separation distance if the molecule consisted of charges $\pm e$? *Note:* The effective charge is actually less because electrons are shared by the oxygen and hydrogen atoms.

30. The electric field 22 cm from a long wire carrying a uniform line charge density is 1.9 kN/C. What will be the field strength 38 cm from the wire?

31. What is the line charge density on a long wire if the electric field 45 cm from the wire has magnitude 260 kN/C and points toward the wire?

32. Find the magnitude of the electric field due to a charged ring of radius $a$ and total charge $Q$, at a point on the ring axis a distance $a$ from the ring's center.

### Section 20.5 Matter in Electric Fields

33. In his famous 1909 experiment that demonstrated quantization of electric charge, R. A. Millikan suspended small oil drops in an electric field. With a field strength of 20 MN/C, what mass drop can be suspended when the drop carries a net charge of 10 elementary charges?

34. How strong an electric field is needed to accelerate electrons in a TV tube from rest to one-tenth the speed of light in a distance of 5.0 cm?

35. A proton moving to the right at $3.8 \times 10^5$ m/s enters a region where a 56-kN/C electric field points to the left. (a) How far will the proton get before its speed reaches zero? (b) Describe its subsequent motion.

36. An electrostatic analyzer like that of Example 20.8 has $b = 7.5$ cm. What should be the value of $E_0$ if the device is to select protons moving at 84 km/s?

## Problems

37. A 2-g ping-pong ball rubbed against a wool jacket acquires a net positive charge of 1 $\mu$C. Estimate the fraction of the ball's electrons that have been removed.

38. Two charges, one twice as large as the other, are located 15 cm apart and experience a repulsive force of 95 N. What is the magnitude of the larger charge?

39. A proton is on the $x$ axis at $x = 1.6$ nm. An electron is on the $y$ axis at $y = 0.85$ nm. Find the net force the two exert on a helium nucleus (charge $+2e$) at the origin.

40. A 9.5-$\mu$C charge is at $x = 16$ cm, $y = 5.0$ cm, and a $-3.2$-$\mu$C charge is at $x = 4.4$ cm, $y = 11$ cm. Find the force on the negative charge.

41. A charge $3q$ is at the origin, and a charge $-2q$ is on the positive $x$ axis at $x = a$. Where would you place a third charge so it would experience no net electric force?

42. You have two charges $+4q$ and one charge $-q$. How would you place them along a line so there's no net force on any of the three?

43. In Fig. 20.27 take $q_1 = 68$ $\mu$C, $q_2 = -34$ $\mu$C, and $q_3 = 15$ $\mu$C. Find the electric force on $q_3$.

**FIGURE 20.27** Problems 43 and 44

44. In Fig. 20.27 take $q_1 = 25$ $\mu$C and $q_2 = 20$ $\mu$C. If the force on $q_1$ points in the $-x$ direction, (a) what is $q_3$ and (b) what is the magnitude of the force on $q_1$?

45. Three identical charges $+q$ and a fourth charge $-q$ form a square of side $a$. (a) Find the magnitude of the electric force on a charge $Q$ placed at the center of the square. (b) Describe the direction of this force.

46. A 65-$\mu$C point charge is at the origin. Find the electric field at the points (a) $x = 50$ cm, $y = 0$; (b) $x = 50$ cm, $y = 50$ cm; and (c) $x = -25$ cm, $y = 75$ cm.

47. A 1.0-$\mu$C charge and a 2.0-$\mu$C charge are 10 cm apart. Find a point where the electric field is zero.

48. A proton is at the origin and an ion is at $x = 5.0$ nm. If the electric field is zero at $x = -5$ nm, what is the charge on the ion?

49. (a) Find an expression for the electric field on the $y$ axis due to the two charges $q$ in Fig. 20.7. (b) At what point is the field on the $y$ axis a maximum?

50. A dipole lies on the $y$ axis and consists of an electron at $y = 0.60$ nm and a proton at $y = -0.60$ nm. Find the electric field (a) midway between the two charges; (b) at the point $x = 2.0$ nm, $y = 0$; and (c) at the point $x = -20$ nm, $y = 0$.

51. Show that the field on the $x$ axis for the dipole of Example 20.5 is given by Equation 20.6b, for $x \gg a$.

52. You're 1.5 m from a charge distribution whose size is much less than 1 m. You measure an electric field strength of 282 N/C. You move to a distance of 2.0 m, and the field strength becomes 119 N/C. What is the net charge of the distribution? *Hint:* Don't try to calculate the charge. Determine instead how the field decreases with distance, and from that infer the charge.

53. Three identical charges $q$ form an equilateral triangle of side $a$, with two charges on the $x$ axis and one on the positive $y$ axis. (a) Find an expression for the electric field at points on the $y$ axis above the uppermost charge. (b) Show that your result reduces to the field of a point charge $3q$ for $x \gg a$.

54. Two identical small metal spheres initially carry charges $q_1$ and $q_2$. When they're 1.0 m apart, they experience a 2.5-N attractive force. Then they're brought together so charge moves from one to the other until they have the same net charge. They're again placed 1.0 m apart, and now they repel with a 2.5-N force. What were the original values of $q_1$ and $q_2$?

55. Two 34-$\mu$C charges are attached to the opposite ends of a spring of spring constant 150 N/m and equilibrium length 50 cm. By how much does the spring stretch?

56. A thin rod carries a total charge $Q$ distributed uniformly over its length $L$. Use integration to show that the electric field strength a distance $x$ along the rod's direction from either end of the rod is $E = kQ/[x(x+L)]$.

57. An electron is moving in a circular path around a long, uniformly charged wire carrying 2.5 nC/m. What is the electron's speed?

58. Figure 20.28 shows a device its inventor claims will separate isotopes of a particular element. (Isotopes of the same element have nuclei with the same charge but different masses.) Atoms of the element are first stripped completely of their electrons, then accelerated from rest through an electric field chosen to give the desired isotope exactly the right speed to pass through the electrostatic analyzer (see Example 20.8). Prove that the device won't work—that is, that it won't separate different isotopes.

**FIGURE 20.28** An isotope separator that won't work (Problem 58).

59. What is the line charge density on a long wire if a 6.8-$\mu$g particle carrying 2.1 nC describes a circular orbit about the wire with speed 280 m/s?

60. A dipole with dipole moment 1.5 nC·m is oriented at 30° to a 4.0-MN/C electric field. (a) What is the magnitude of the torque on the dipole? (b) How much work is required to rotate the dipole until it's antiparallel to the field?

61. A molecule has its dipole moment aligned with a 1.2-kN/C electric field. If it takes $3.1 \times 10^{-27}$ J to reverse the molecule's orientation, what is its dipole moment?

62. Two identical dipoles, each of charge $q$ and separation $a$, are a distance $x$ apart, as shown in Fig. 20.29. By considering forces between pairs of charges in the different dipoles, calculate the net force between the dipoles. (a) Show that, in the limit $a \ll x$, the force has magnitude $6kp^2/x^4$, where $p = qa$ is the dipole moment. (b) Is the force attractive or repulsive?

**FIGURE 20.29** Problem 62

63. A dipole with charges $\pm q$ and separation $2a$ is located a distance $x$ from a point charge $+Q$, with its dipole moment vector perpendicular to the $x$ axis, as shown in Fig. 20.30. Find expressions for the magnitude of (a) the net torque and (b) the net force on the dipole, both in the limit $x \gg a$. (c) What is the direction of the net force?

**FIGURE 20.30** Problem 63

64. An electron is at the origin, and an ion with charge $+5e$ is at $x = 10$ nm. Find a point where the electric field is zero.

65. The electric field on the axis of a uniformly charged ring has magnitude 380 kN/C at a point 5.0 cm from the ring center. The magnitude 15 cm from the center is 160 kN/C; in both cases the field points away from the ring. Find the radius and charge of the ring.

66. An *electric quadrupole* consists of two oppositely directed dipoles in close proximity. (a) Calculate the field of the quadrupole shown in Fig. 20.31 for points to the right of $x = a$, and (b) show that for $x \gg a$ the quadrupole field falls off as $1/x^4$.

**FIGURE 20.31** Problem 66

67. A straight wire 10 m long carries 25 $\mu$C distributed uniformly over its length. (a) What is the line charge density on the wire? Find the electric field strength (b) 15 cm from the wire axis, not near either end, and (c) 350 m from the wire. Make suitable approximations in both cases.

68. Figure 20.32 shows a thin rod of length $L$ carrying charge $Q$ distributed uniformly over its length. (a) What is the line charge density on the rod? (b) What must be the electric field direction on the rod's perpendicular bisector (taken to be the $y$ axis)? (c) Modify

the calculation of Example 20.7 to find an expression for the electric field at a point $P$ a distance $y$ along the perpendicular bisector. (d) Show that your result for part (c) reduces to the field of a point charge $Q$ for $y \gg L$.

**FIGURE 20.32** Problem 68

69. Figure 20.33 shows a thin, uniformly charged disk of radius $R$. Imagine the disk divided into rings of varying radii $r$, as suggested in the figure. (a) Show that the area of such a ring is very nearly $2\pi r\, dr$. (b) If the surface charge density on the disk is $\sigma$ C/m$^2$, use the result of part (a) to write an expression for the charge $dq$ on an infinitesimal ring. (c) Use the result of part (b) along with the result of Example 20.6 to write the infinitesimal electric field $dE$ of this ring at a point on the disk axis, taken to be the positive $x$ axis. (d) Integrate over all such rings (that is, from $r = 0$ to $r = R$) to show that the net electric field on the disk axis has magnitude

$$E = 2\pi k\sigma \left( 1 - \frac{x}{\sqrt{x^2 + R^2}} \right)$$

**FIGURE 20.33** Problem 69

70. Use the result of Problem 69 to show that the field of an *infinite*, uniformly charged flat sheet is $2\pi k\sigma$, where $\sigma$ is the surface charge density. Note that this result is independent of distance from the sheet.

71. Use the binomial theorem to show that, for $x \gg R$, the result of Problem 69 reduces to the field of a point charge whose total charge is the charge density $\sigma$ times the disk area.

72. A semicircular loop of radius $a$ carries positive charge $Q$ distributed uniformly over its length. Find the electric field at the center of the loop (point $P$ in Fig. 20.34). *Hint:* Divide the loop into charge elements $dq$ as shown in Fig. 20.34, and write $dq$ in terms of the angle $d\theta$. Then integrate over $\theta$ to get the net field at $P$.

**FIGURE 20.34** Problem 72

73. In Example 20.2, find the position for $Q$ where it will experience the greatest force.

74. A thin rod carries a charge $Q$ distributed uniformly over its length $L$, and is situated on the $x$ axis between $x = \pm L/2$. (a) Find the electric field at an arbitrary point $(x, y)$. (You will have to do separate integrals for the $x$ and $y$ components.) (b) Show that your result reduces to the result of Problem 68, when $x = 0$, and to the result of Problem 56, when $y = 0$ and $x > L/2$.

75. A thin rod extends along the $x$ axis from $x=0$ to $x=L$, and carries a line charge density $\lambda = \lambda_0(x/L)^2$, where $\lambda_0$ is a constant. Find the electric field at the position $x=-L$.

76. A rod of length $2L$ lies on the $x$ axis, centered at the origin, and carries a line charge density given by $\lambda = \lambda_0(x/L)$, where $\lambda_0$ is a constant. (a) Find an expression for the electric field strength at points on the $x$ axis, for $x>L$. (b) Show that for $x \gg L$, your result has the $1/x^3$ dependence of a dipole field, and by comparison, determine the dipole moment of the rod.

77. Your friend who likes to write science fiction proposes a world in which the electron and proton charges differ by a factor of one part in one billion (see Exercise 14). Your friend thinks this difference won't matter in the macroscopic world of people. An average-sized 65-kg person would weigh about 650 N on Earth. Is the electrostatic force between two such people, 10 m apart, on the imaginary world negligible? How do you advise your friend?

78. Suppose you have a friend who has a part-time job translating computer maintenance training materials from Japanese into English. A section of a manual for an inkjet printer is missing, and an important derivation is needed. The printer works by steering drops with charge $q$ and mass $m$ between two parallel surfaces where there is a uniform electric field $E$ that deflects them the appropriate amount (Figure 20.35). The field extends over the length $L$ and width $d$. The reader of the manual must know the minimum speed of the drop if it is to make it through the electric field region without hitting either of the plates.

**FIGURE 20.35** Problem 78

79. A magician friend is planning to use an electric field to levitate some small beads. They have a mass of 10 g each, and he will give them 5.0 $\mu$C of charge. In order for the trick to work, the beads must remain stationary under the influence of the electric field. What should the strength and direction of the electric field be?

80. You have a summer job working for a company that manufactures telecommunications equipment. One day, when a proposal for a huge contract is being completed, the lead engineer calls in sick. (The fact that this also is the seasonal opening of the local golf course is coincidental.) One item missing from the proposal is an equation for an electric field of an antenna element. A thin rod of length $L$ has a line charge density given by $\lambda = \lambda_0(x^2/L^2)\sin(\pi x/L)$, where $\lambda_0$ is a constant. The rod attaches horizontally to a vertical, nonconductive mast, and you must find an expression for the field at the end nearest the mast.

# Answers to Chapter Questions

## Answer to Chapter Opening Question

The electric field in the atmosphere must be strong enough to rip electrons from air molecules, making the air an electrical conductor. This typically happens when $E$ exceeds about 3 MV/m.

## Answers to GOT IT? Questions

20.1 (a) $-\hat{\imath}$; (b) $-\frac{\sqrt{2}}{2}\hat{\imath} + \frac{\sqrt{2}}{2}\hat{\jmath}$; the unit vector always points away from the source charge regardless of the sign.

20.2 (a) drops by $1/2^2$, to 200 N/C; (b) drops by $1/2^3$, to 100 N/C.

20.3 carbon-13, (oxygen-16, helium-4, deuteron—all the same), helium-3, proton, electron

- Huge sparks jump to the operator's cage in the Hall of Electricity at the Boston Museum of Science, but the operator is unharmed. Why?

▶ **To Learn**

By the end of this chapter you should be able to

- Represent electric fields using field lines and interpret field-line diagrams (21.1).
- Explain Gauss's law for electric fields and relate it to Coulomb's law (21.3).
- Calculate electric fields of symmetric charge distributions using Gauss's law (21.4).
- Give the electric fields of basic charge distributions, and estimate the fields of arbitrary distributions under appropriate approximations (21.5).
- Explain the concept of electrostatic equilibrium, and why in equilibrium Gauss's law requires excess electric charge to reside only on the surface of a conductor (21.6).

◀ **To Know**

- This chapter expands on the concepts of electric force and, especially, electric field (20.2, 20.3).
- Integration will become more central—though also easier to apply—than in Chapter 20. For reviews of integration, see (9.1) and Appendix A.2.

In this chapter we introduce an elegant way of describing electric fields that makes it much easier to calculate the fields of certain charge distributions. In the process we will formulate one of the four fundamental laws of electromagnetism—a statement that's equivalent to Coulomb's law but that gives deeper insights into the electric field.

## 21.1 Electric Field Lines

We can visualize electric fields by drawing **electric field lines**, continuous lines whose direction is everywhere the same as that of the electric field. To draw a field line, determine the field direction at some point. Move a small distance in the direction of the field, and evaluate the field direction at the new point. Extending the process in both directions from your starting point traces out an electric field line. You'll find that field lines begin on positive charges and either end on negative charges or extend to infinity. Drawing many field lines gives a picture of the overall field.

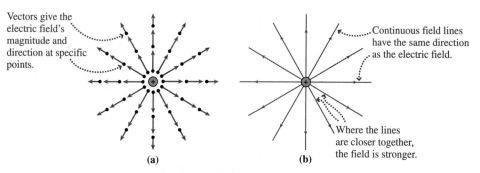

Vectors give the electric field's magnitude and direction at specific points.

Continuous field lines have the same direction as the electric field.

Where the lines are closer together, the field is stronger.

**(a)**  **(b)**

**FIGURE 21.1** Vectors (a) and field lines (b) provide two ways to visualize the electric field.

To explore the field of a positive point charge, as shown in Fig. 21.1a, start near the charge. The field points radially outward, so move a little way outward. The field is still radial. Repeat the process, and you'll trace a straight line that extends indefinitely. So the field lines of a positive point charge are straight lines that begin on the charge and extend radially to infinity (Fig. 21.1b).

In Fig. 21.1b the field lines spread apart as they extend farther from the point charge. Coulomb's law shows that the field weakens farther from the charge, so in Fig. 21.1b the field is stronger where the lines are closer and weaker when they're farther apart. This is generally true, and allows us to infer the field's relative magnitude as well as direction from field-line pictures.

To trace the field lines of a charge distribution, follow the net field—the vector sum of the field contributions from all charges in the distribution. Usually the field direction varies, so the line is curved. Figure 21.2a shows the details for one field line of a dipole. Figure 21.2b shows a number of dipole field lines; here you can see that the field is strongest near the individual charges and in the region between them. The electric field exists everywhere, so there are really infinitely many field lines. We obviously can't draw them all. To make field-line pictures precise, we associate a fixed number of field lines with a charge of a given magnitude. In Fig. 21.3, for example, eight field lines correspond to a charge of magnitude $q$. Study the figure, and you'll see how all the fields are consistent with this convention.

Direction of net field is tangent to field line.

$\vec{E}_+$

$\vec{E}$

$\vec{E}_-$

Field line

**(a)**

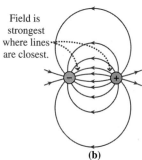

Field is strongest where lines are closest.

**(b)**

**FIGURE 21.2** Field of an electric dipole. (a) At each point, the field-line direction is that of the *net* electric field, $\vec{E} = \vec{E}_+ + \vec{E}_-$. (b) Tracing several field lines shows the overall dipole field.

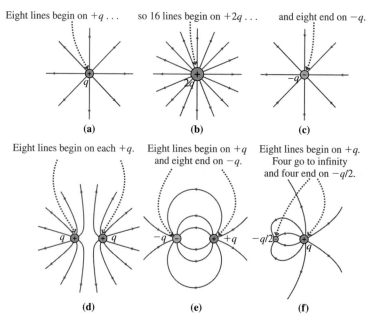

Eight lines begin on $+q$ . . .   so 16 lines begin on $+2q$ . . .   and eight end on $-q$.

**(a)**  **(b)**  **(c)**

Eight lines begin on each $+q$.   Eight lines begin on $+q$ and eight end on $-q$.   Eight lines begin on $+q$. Four go to infinity and four end on $-q/2$.

**(d)**  **(e)**  **(f)**

**FIGURE 21.3** Field lines for six charge distributions, using the convention that eight lines correspond to a charge of magnitude $q$.

## 21.2 Electric Flux and Field

In Fig. 21.4 we've surrounded each charge distribution from Fig. 21.3 with several surfaces. Each surface is closed, so it's impossible to get from inside to outside without crossing the surface. (Figure 21.4 shows only the two-dimensional cross section of each surface.) How many field lines emerge from within each surface?

In Fig. 21.4a the answer for surfaces 1 and 2 is obvious: eight. For surface 3 one field line crosses twice going out and once going in; if we count the inward-going line as negative, then there's still a net of eight lines going out. *Any* closed surface surrounding $+q$ will have eight lines emerging from it, for the simple reason that eight lines begin on the charge and extend to infinity; to get there they all have to cross the closed surface.

What about surface 4? Two lines cross going outward and two inward, for a net of zero lines emerging. What's different is that surface 4 doesn't enclose the charge. You can convince yourself that *any* surface not enclosing the charge will have as many lines going in as out, for zero net field lines emerging.

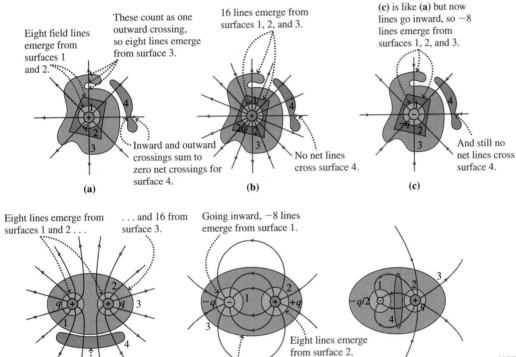

**FIGURE 21.4** The number of field lines emerging from a closed surface depends only on the net charge enclosed.

Figure 21.4b is identical except that now 16 field lines emerge from any surface surrounding the charge: $+2q$ enclosed, so 16 field lines emerge. And Fig. 21.4c is similar, too, but with the negative charge, the field lines go inward and we count them as negative: $-q$ enclosed, so $-8$ field lines emerge. And, in all three cases, zero net lines emerge from surface 4, the one that doesn't enclose the charge: zero charge enclosed, so zero field lines emerging. The same is true even for surface 3 surrounding the dipole in Fig. 21.4e; here there are two charges within the surface, but the *net charge enclosed* is zero, and sure enough, there are zero net field lines emerging. Study the rest of Fig. 21.4 and you can convince yourself that in all cases **the number of field lines emerging from any closed surface is proportional to the net charge enclosed.**

This statement is very general. It doesn't matter what shape the surface is or whether the enclosed charge is a single point charge or a lot of charges carrying the same net charge. Nor does it matter how the charges are arranged, as long as they're *enclosed* by the surface in question. The presence of charges *outside* the surface doesn't alter our conclusion about the number of field lines emerging—although it may alter the shape of individual lines.

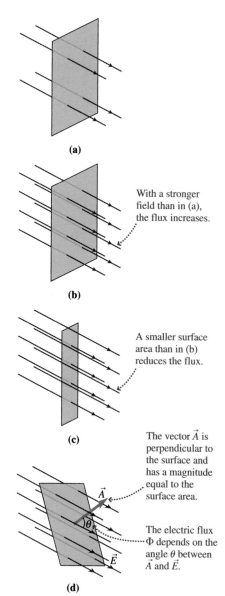

**(a)**

With a stronger field than in (a), the flux increases.

**(b)**

A smaller surface area than in (b) reduces the flux.

**(c)**

The vector $\vec{A}$ is perpendicular to the surface and has a magnitude equal to the surface area.

The electric flux $\Phi$ depends on the angle $\theta$ between $\vec{A}$ and $\vec{E}$.

**(d)**

**FIGURE 21.5** Electric flux through flat surfaces.

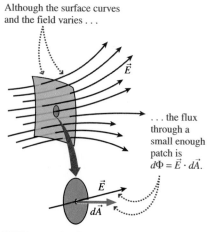

Although the surface curves and the field varies . . .

. . . the flux through a small enough patch is $d\Phi = \vec{E} \cdot d\vec{A}$.

**FIGURE 21.6** The flux through a small area $dA$.

We'll now rephrase our statement mathematically to obtain one of the four fundamental laws of electromagnetism. As we proceed, remember that the mathematics just reflects what's clear from Fig. 21.4: The number of field lines emerging from a closed surface depends only on the net charge enclosed.

## Electric Flux

Figure 21.5 shows that the number of field lines crossing a flat surface depends on three factors: the field strength $E$, the surface area $A$, and the orientation of the surface relative to the field. We get the most field-line crossings when the surface is perpendicular to the field and none when it's parallel; Fig. 21.5d shows an intermediate case. If we define a vector $\vec{A}$ normal to the surface, then the number of field-line crossings is proportional to $\cos\theta$, where $\theta$ is the angle between the normal vector $\vec{A}$ and the field $\vec{E}$. Putting all this together, we find that the number of field lines crossing the surface is proportional to $EA\cos\theta$. This quantity has a definite value that captures the spirit of the more vague "number of field lines crossing a surface." We call $EA\cos\theta$ the **electric flux** $\Phi$ through the surface. If we make the magnitude of the surface normal vector $\vec{A}$ equal to the surface area $A$, then we can define the flux using the vector dot product:

$$\Phi = \vec{E} \cdot \vec{A} \qquad (21.1)$$

where the dot product, defined in Chapter 6, is the product of the two vector magnitudes with the cosine of the angle between them. Since the units of $\vec{E}$ are N/C, flux is measured in N·m²/C.

The surfaces in Fig. 21.5 are *open* surfaces, meaning it's possible to get from one side to the other without passing through the surface. For open surfaces there's an ambiguity in the sign of $\Phi$, since we could have taken $\vec{A}$ in either of the two directions along the perpendicular to the surface. But for *closed* surfaces, we unambiguously define the direction of $\vec{A}$ as the direction of the outward-pointing normal to the surface.

✓**TIP** The Flux Is Not the Field

The flux $\Phi$ and field $\vec{E}$ are related but distinct quantities. The field is a vector defined at each point in space. The flux is a scalar and a global property, depending on how the field behaves over an extended surface rather than at a single point; it's a quantification of the number of field lines crossing a surface.

What if a surface is curved and/or the field varies with position? Then we divide the surface into patches, each small enough that it's essentially flat and that the field is essentially uniform over each (Fig. 21.6). If a patch has area $dA$, then Equation 21.1 gives the flux through it: $d\Phi = \vec{E} \cdot d\vec{A}$, where the vector $d\vec{A}$ is normal to the patch. The total flux through the surface is then the sum over all the patches. If we make the patches arbitrarily small, that sum becomes an integral, and the flux is

$$\Phi = \int_{\text{surface}} \vec{E} \cdot d\vec{A} \qquad (21.2)$$

The limits of the integral range over the entire surface, picking up contributions from all the patches $d\vec{A}$. Although the integral can be difficult to evaluate, we'll find it most useful in cases where it's almost trivial.

**GOT IT? 21.1** The figure shows a cube of side $s$ in a uniform electric field $\vec{E}$. What is the flux through each of the three cube faces $A$, $B$, and $C$ with the cube oriented as in (a)? Repeat for the orientation in (b), with the cube rotated 45°.

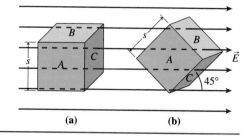

**(a)**          **(b)**

## 21.3 Gauss's Law

We've seen that the number of field lines emerging from a closed surface is proportional to the charge enclosed. Now that we've developed electric flux to express more rigorously the notion "number of field lines," we can state the following: **The electric flux through any closed surface is proportional to the net charge enclosed by that surface.** Writing the same thing mathematically gives $\Phi = \oint \vec{E} \cdot d\vec{A} \propto q_{\text{enclosed}}$, where the circle indicates that the integral is over a *closed* surface.

To evaluate the proportionality between flux and charge, consider a positive point charge $q$ and a spherical surface of radius $r$ centered on the charge (Fig. 21.7). The flux through this surface is given by Equation 21.2:

$$\Phi = \oint \vec{E} \cdot d\vec{A} = \oint E \, dA \cos\theta$$

But Fig. 21.7 shows that the surface normal $d\vec{A}$ and the electric field $\vec{E}$ are parallel at any point on the sphere, so $\cos\theta = 1$. Since the electric field varies as $1/r^2$, its magnitude is the same everywhere at the fixed radius $r$ of our sphere. Thus, we can take $E$ outside the integral, giving

$$\Phi = \oint_{\text{sphere}} E \, dA = E \oint_{\text{sphere}} dA = E(4\pi r^2)$$

where the last step follows because $\oint dA$ is just the surface area of the sphere. Now, the electric field of a point charge is given by Equation 20.3: $E = kq/r^2$. So we have $\Phi = E(4\pi r^2) = (kq/r^2)(4\pi r^2) = 4\pi kq$. Since the point charge $q$ is the only charge inside our spherical surface, we see that the proportionality constant between flux and enclosed charge is $4\pi k$.

Before proceeding, we introduce the so-called permittivity constant $\epsilon_0$, defined as $\epsilon_0 = 1/4\pi k$, where $k$ is the Coulomb constant. The value of $\epsilon_0$ is $8.85 \times 10^{-12} \text{ C}^2/\text{N} \cdot \text{m}^2$. There's no physics here, just a new constant that conveys the same information as $k$. That there are two redundant constants is a historical artifact, and we switch now from $k$ to $\epsilon_0$ because doing so makes subsequent formulas simpler. In terms of $\epsilon_0$, the proportionality $4\pi k$ between flux and enclosed charge becomes $1/\epsilon_0$. So our statement that the flux through any closed surface is proportional to the net charge enclosed becomes

$$\oint \vec{E} \cdot d\vec{A} = \frac{q_{\text{enclosed}}}{\epsilon_0} \qquad \text{(Gauss's law)} \qquad (21.3)$$

Here the integral is taken over *any closed surface*, and $q_{\text{enclosed}}$ is the charge enclosed *by that surface.*

Equation 21.3 is **Gauss's law,** one of four fundamental relations that govern the behavior of electromagnetic fields throughout the universe. Whether you journey into a star in some remote galaxy, down among the strands of a DNA molecule, or into the microprocessor chip at the heart of your computer, you'll find that the flux of the electric field through any closed surface depends only on the enclosed charge. In nearly 200 years of experiments, no electric field has ever been found to violate Gauss's law.

Gauss's law, though clothed in the mathematical finery of a surface integral, is just a more rigorous way of saying what's clear in Fig. 21.4: The number of field lines emerging from a closed surface is proportional to the net charge enclosed.

### Gauss and Coulomb

Gauss's law and Coulomb's law look completely different, but they're equivalent. Figure 21.8 shows that the heart of this equivalence is the inverse-square law. Gauss's law tells us that the flux through the two surfaces in the figure is the same and is equal to $q/\epsilon_0$. But why? Because, as our arguments leading to Gauss's law show, the flux through a spherical surface of radius $r$ centered on a point charge is the product of the surface area $4\pi r^2$ and the electric field $E$ at the surface. But Coulomb's law says that the electric field drops off as $1/r^2$. As $r$ increases, the surface area grows as $r^2$, but the $1/r^2$ decrease in field strength

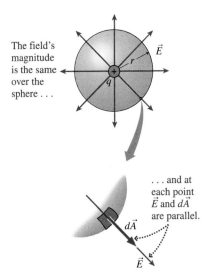

The field's magnitude is the same over the sphere . . .

. . . and at each point $\vec{E}$ and $d\vec{A}$ are parallel.

**FIGURE 21.7** The electric field of a point charge, shown with a spherical surface centered on the charge.

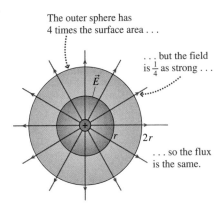

The outer sphere has 4 times the surface area . . .

. . . but the field is $\frac{1}{4}$ as strong . . .

. . . so the flux is the same.

**FIGURE 21.8** Gauss's law follows from the inverse-square aspect of Coulomb's law.

just compensates, giving a constant value for the flux. If the inverse-square law (e.g., Coulomb's law) didn't hold, then the flux wouldn't be constant and Gauss's law wouldn't hold either.

It's also the inverse-square law that makes electric field-line pictures useful for visualizing fields. Field lines begin or end only on charges; otherwise, they go off to infinity. As the field lines of a point charge spread in three dimensions, the number crossing any spherical surface (or any closed surface) remains the same. But larger spheres have larger surface areas, in proportion to $r^2$—and that means the density of field lines drops as $1/r^2$, accurately reflecting the field strength. Once again, the inverse-square law (Coulomb) and the relation between flux and enclosed charge (Gauss) are intimately connected. Incidentally, field-line pictures printed in a book or drawn on a blackboard generally can't be quantitatively correct because they don't show the spreading of field lines in all three dimensions.

We've been talking here only about isolated point charges, but we emphasize that Gauss's law applies to *all* electric fields, no matter how complicated the charge distributions that produce them. That's because the superposition principle allows us to add vectorially the electric fields described individually by Coulomb's law for the point-charge field. Since Gauss's and Coulomb's law are equivalent, our argument leading to Gauss's law still applies when the field $\vec{E}$ is a superposition of point-charge fields.

---

**GOT IT? 21.2** A spherical surface surrounds an isolated positive charge, as shown. If a second charge is placed outside the surface, which of the following will be true of the total flux through the surface? (a) it doesn't change; (b) it increases; (c) it decreases; (d) it increases or decreases depending on the sign of the second charge. Repeat for the electric field on the surface at the point between the charges.

A spherical surface surrounds a point charge.

A second charge is placed outside the surface. What happens to the total flux through the surface?

. . . and to the electric field at this point?

---

## 21.4 Using Gauss's Law

Gauss's law is a universal statement about electric fields; it's true for *any* surface enclosing *any* charge distribution. For charge distributions with sufficient symmetry—symmetric about a point, a line, or a plane—Gauss's law also provides a powerful alternative to Coulomb's law that makes electric-field calculations much easier. For such distributions it's possible to evaluate the flux integral on the left-hand side of Gauss's law (Equation 21.3) without actually knowing the field. We can then solve for $E$ in terms of the enclosed charge on the right-hand side. We begin with a general strategy for applying Gauss's law to symmetric charge distributions, followed by discussion and examples of the three symmetries.

---

**PROBLEM SOLVING STRATEGY 21.1** **Gauss's Law**

**INTERPRET** Check that your charge distribution has sufficient symmetry to use Gauss's law to find the electric field, and identify the symmetry. Is it spherical, line, or plane? If the charge distribution doesn't exhibit one of these symmetries, then Gauss's law—though always true—won't help you find the field.

**DEVELOP** Draw a diagram showing the charge distribution, and use the symmetry to infer the direction of the electric field. Then draw an appropriate **gaussian surface**—an imaginary surface that will let you evaluate the flux integral in Gauss's law. The field should have constant magnitude over the surface and should be perpendicular to the surface. Sketch some field lines; the symmetry should indicate their direction. With line and plane symmetry, there may be parts of the surface where the field is perpendicular and parts where it's parallel; we'll show in examples how to handle these situations. If you can't find a suitable gaussian surface, there probably isn't sufficient symmetry to use Gauss's law to calculate the field.

*cont'd.*

**EVALUATE**

- Evaluate the flux $\Phi = \oint \vec{E} \cdot d\vec{A}$ over your gaussian surface. Since you've found a surface to which the field is perpendicular, $\vec{E}$ and the surface normal vector $d\vec{A}$ are parallel, so $\cos\theta = 1$ and the dot product becomes $E\,dA$. With the field strength $E$ constant over the surface, it can come out of the integral, leaving $\oint dA$. And that's the surface area $A$. So the flux will be $EA$. If $\vec{E}$ is parallel to some parts of the area—as happens in line and plane symmetry—then $\vec{E} \perp d\vec{A}$, so $\vec{E} \cdot d\vec{A} = 0$ and there's no contribution to the flux from those areas.
- Evaluate the *enclosed* charge. This may or may not be the same as the total charge, depending on whether the position at which you're evaluating the field lies outside or inside the charge distribution.
- Evaluate the field $E$ by invoking Gauss's law, equating the flux to $q_{enclosed}/\epsilon_0$, and solving for $E$. This is the field magnitude; the direction should be evident from symmetry.

**ASSESS** Does your answer make sense? Does the field behave as you would expect given what you know of simpler charge distributions—point charges, line charges, or charged sheets, depending on the symmetry?

You'll quickly get the hang of this strategy because Gauss's law is useful for finding the field only in the three common symmetries. That means you need to evaluate the integral for the flux just once for a given symmetry.

## Spherical Symmetry

A charge distribution has spherical symmetry when the charge density depends only on the radial distance $r$ from the center of the distribution—also called the *point of symmetry*. A point charge is one example; so is the uniformly charged sphere that we'll treat in the next example. So, in fact, is a nonuniform spherical charge, provided the charge density varies solely with distance from the center. The only electric field consistent with spherical symmetry is a field that points in the radial direction, either away from or toward the point of symmetry. The next example shows the application of Gauss's law to spherical symmetry.

---

**EXAMPLE 21.1**   **Gauss's Law: A Uniformly Charged Sphere**

A charge $Q$ is spread uniformly throughout a sphere of radius $R$. Find the electric field at all points, first outside and then inside the sphere.

**INTERPRET** The charge distribution has spherical symmetry, so we can use Gauss's law to find the field.

**DEVELOP** Figure 21.9 shows the spherical charge distribution. Given the symmetry, the field has to point radially, outward for a positive charge or inward for a negative one. We've shown some field lines on our drawing. So an appropriate gaussian surface is itself a sphere, centered on the center of the charge distribution. Since we're asked for the field both outside and inside the charge distribution, we've drawn two such gaussian surfaces in Fig. 21.9, one outside and one inside the distribution.

Surface 1 encloses all of the charge $Q$ . . .

. . . while surface 2 encloses only some of it.

**FIGURE 21.9** Finding the field of a uniformly charged sphere.

**EVALUATE**

- We begin with the flux integral $\Phi = \oint \vec{E} \cdot d\vec{A}$. Our strategy shows that with $\vec{E}$ perpendicular to the area, $\cos\theta = 1$ and the dot product becomes $E\,dA$. So the flux integral simplifies to $\oint E\,dA$. With the gaussian surface centered on the point of symmetry, all points on the surface are the same distance from the center, so $E$ has the same value over our surface. $E$ comes outside the integral, leaving $\oint dA$, with the integral over the entire spherical surface. That's just the surface area, $4\pi r^2$, so the flux is $\Phi = 4\pi r^2 E$. Which surface were we using in doing this calculation? It doesn't matter; the result holds for *any* spherical surface centered on the symmetry point at the center of the charge distribution. So our expression $\Phi = 4\pi r^2 E$ is valid for both our gaussian surfaces.
- Now we evaluate the *enclosed* charge. That *is* different for the two surfaces, so we'll first deal with surface 1, which lies outside the charge distribution. This gaussian surface encloses the entire charge $Q$, so $q_{enclosed} = Q$ for $r > R$. The interior is a bit trickier. Here, gaussian surface 2 has radius $r$, so it encloses volume $\frac{4}{3}\pi r^3$. The volume of the entire sphere is $\frac{4}{3}\pi R^3$, so gaussian surface 2 encloses a fraction $r^3/R^3$ of the total volume. We're told that the sphere is uniformly charged, so that's also the fraction of the total charge $Q$ enclosed by the gaussian surface. So for $r < R$, the enclosed charge is $q_{enclosed} = Q(r^3/R^3)$.

*cont'd.*

■ Next we apply Gauss's law, evaluating the field by equating the flux $\Phi = 4\pi r^2 E$ to the enclosed charge divided by $\epsilon_0$. When we're outside the sphere, $q_{enclosed} = Q$, so we have $4\pi r^2 E = Q/\epsilon_0$, or

$$E = \frac{Q}{4\pi\epsilon_0 r^2} \quad \text{(field outside spherical charge distribution)} \quad (21.4)$$

Inside the sphere, $q_{enclosed} = Q(r^3/R^3)$; equating $q_{enclosed}/\epsilon_0$ to the flux $\Phi = 4\pi r^2 E$ gives $4\pi r^2 E = Qr^3/\epsilon_0 R^3$. Solving for $E$, we get

$$E = \frac{Qr}{4\pi\epsilon_0 R^3} \quad \text{(field inside uniformly charged sphere)} \quad (21.5)$$

**ASSESS** Make sense? Consider first the field outside the spherical charge. Since $1/4\pi\epsilon_0$ is the Coulomb constant $k$, this field can be written as $E = kQ/r^2$—precisely the field of a point charge! Inside, in contrast, the field increases linearly with distance from the center, reflecting two opposite effects. First, we're getting farther from the center, which causes a $1/r^2$ decrease in the field. At the same time, the enclosed charge grows as the enclosed volume—that is, as $r^3$. The result is a linear increase in field strength. Figure 21.10 shows the combined results for the field both inside and outside the sphere. As a check on our results, the figure shows that our two expressions agree at $r = R$. ■

**FIGURE 21.10** Field strength versus radial distance for a uniformly charged sphere of radius $R$.

Our result for the field outside a spherical charge distribution (Equation 21.4) is *not* an approximation but an exact result that holds right up to the sphere's surface. So a spherical charge distribution produces exactly the same field as a point charge located at its center—at least for points outside the charge. Imagine how hard it would have been to calculate this using the superposition principle! Yet somehow all the charge elements produce $d\vec{E}$ vectors that add to give the same field as a single point charge. That result doesn't even require the sphere to be charged uniformly, as long as the distribution of charge is spherically symmetric. Outside *any* spherically symmetric charge distribution, the electric field is exactly the same as that of a point charge at the center.

Incidentally, this result also holds for gravity because it, too, obeys an inverse-square law. That's why we can treat planets as though they're point masses located at their centers. And our result for the field inside the charged sphere (Equation 21.5) shows why the gravitational acceleration $g$ decreases approximately linearly as one descends into Earth's interior.

What if our charged sphere wasn't uniformly charged throughout its interior? The next example considers the extreme case when there's no charge in the interior.

**EXAMPLE 21.2** **Gauss's Law: A Hollow Spherical Shell**

A thin, hollow spherical shell of radius $R$ contains a total charge $Q$ distributed uniformly over its surface. Find the electric field both inside and outside the shell.

**INTERPRET** This is a spherically symmetric charge distribution, so we can apply Gauss's law just as we did in Example 21.1. There we found that the field outside any spherically symmetric charge distribution is that of a point charge: $E = kQ/r^2$, pointing radially. So we need to find the field only inside the spherical shell.

**DEVELOP** Figure 21.11 is a sketch of the hollow shell. With spherical symmetry, the appropriate gaussian surface is itself a sphere, which we've drawn inside the shell.

**EVALUATE**

■ The flux $\Phi = 4\pi r^2 E$ from Example 21.1 holds for *any* spherical gaussian surface with spherical symmetry, so we've already got the flux.

■ Next we need the enclosed charge. But there's no charge inside the hollow shell and therefore none inside our gaussian surface. So $q_{enclosed} = 0$.

**FIGURE 21.11** Sketch for Example 21.2.

■ Equating the flux $4\pi r^2 E$ to $q_{enclosed}/\epsilon_0$—zero in this case—gives $4\pi r^2 E = 0$. Therefore $E = 0$ inside the shell.

*cont'd.*

**ASSESS** Make sense? The radius $r$ of our gaussian sphere is arbitrary as long as $r < R$, so we're *inside* the shell. Thus this example shows that the electric field is exactly zero *everywhere* inside the shell. Figure 21.12 shows how this remarkable result is a consequence of the inverse-square law. ∎

Charges at $B$ contribute $\vec{E}_B$ to the field at $P$ . . .

... and charges at $A$ contribute $\vec{E}_A$. The two fields cancel.

**FIGURE 21.12** At any point $P$ inside a charged shell, the field from the relatively few nearby charges at $A$ is exactly canceled by the field from the more numerous but more distant charges at $B$. The result is zero field everywhere inside the shell.

**GOT IT? 21.3** A spherical shell carries charge $Q$ uniformly distributed over its surface. If the charge on the shell doubles, what happens to the electric field (a) inside and (b) outside the shell?

---

**EXAMPLE 21.3**  **Gauss's Law: A Point Charge Within a Shell**

A positive point charge $+q$ is at the center of a spherical shell of radius $R$ carrying charge $-2q$, distributed uniformly over its surface. Find expressions for the field strength inside and outside the shell.

**INTERPRET** This problem is about a charge distribution with spherical symmetry.

**DEVELOP** We want to find the fields both inside and outside the distribution, so we show two spherical gaussian surfaces in our sketch, Fig. 21.13. We've also drawn some field lines, which will help in assessing our answer.

Eight lines start on $+q$.

Any gaussian surface outside the shell has eight lines going inward.

Sixteen lines end on $-2q$.

Any gaussian surface inside the shell has eight lines going outward.

**FIGURE 21.13** A shell carrying charge $-2q$ surrounds a point charge $+q$.

**EVALUATE**

- We already know that the flux through a spherical gaussian surface is $\Phi = 4\pi r^2 E$ when we have spherical symmetry, so that's the flux for both surfaces.

- The outer surface encloses the charge $+q$ at the center and $-2q$ on the shell, so the net charge enclosed is $q_{enclosed} = -q$ for $r > R$. The inner surface encloses only the point charge $+q$, so $q_{enclosed} = +q$ for $r < R$.

- Now we find the field by equating the flux $\Phi = 4\pi r^2 E$ to $q_{enclosed}/\epsilon_0$. The result for $r > R$ is $E = -q/4\pi\epsilon_0 r^2$, where the minus sign appears because the enclosed charge is $-q$. For $r < R$ with enclosed charge $+q$, the result is $E = q/4\pi\epsilon_0 r^2$.

**ASSESS** Make sense? We've seen that the field outside a spherically symmetric charge distribution is that of an equivalent point charge at the center. Here the net charge is $-q$, and our result for $r > R$ is indeed the field of a point charge $-q$. Another way to see this is to apply superposition: The field outside due to the shell alone is that of a point charge $-2q$; that adds to the field of the central point charge $+q$ to give, again, the field of a point charge $-q$ outside the shell. In Example 21.2 we found that the shell produces no field in its interior, so here superposition leaves us with the field of the central point charge alone, just as our result shows. The field lines in Fig. 21.13 also show how our results make sense. ∎

---

✓ **TIP**  Symmetry Matters!

We used the fact that there's no charge inside an empty spherical shell to conclude that the field inside the shell is zero. But be careful: That conclusion follows only when there's enough symmetry, as Fig. 21.14 makes obvious. Here, the gaussian surface encloses zero net charge, so the flux through the surface is zero. But the electric field both on and inside the surface isn't zero.

**FIGURE 21.14** A spherical gaussian surface surrounds a dipole. We have $q_{enclosed} = 0$, but $E \neq 0$ inside the surface because the dipole isn't spherically symmetric.

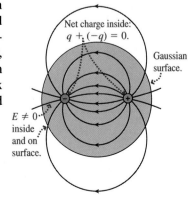

Net charge inside: $q + (-q) = 0$.

Gaussian surface.

$E \neq 0$ inside and on surface.

## Line Symmetry

A charge distribution has line symmetry when its charge density depends only on the perpendicular distance $r$ from a line, called the *symmetry axis*. Symmetry then requires that the field point radially and that the field magnitude depend only on distance from the axis. It also requires the charge distribution to be infinitely long, so there's no variation parallel to the line. That's impossible, of course, but nevertheless the infinite line-symmetric charge distribution is a reasonable approximation to elongated structures like wires. The next two examples explore the application of Gauss's law to line symmetry.

---

**EXAMPLE 21.4**    **Gauss's Law: An Infinite Line of Charge**

Use Gauss's law to find the electric field of an infinite line charge carrying charge density $\lambda$ in coulombs per meter.

**INTERPRET** An infinite line has line symmetry, so we can apply Gauss's law to find the electric field.

**DEVELOP** Symmetry requires that the field point radially and that its magnitude be the same at a given distance $r$ from the line charge. So an appropriate gaussian surface is a cylinder coaxial with the line, as we've drawn in Fig. 21.15.

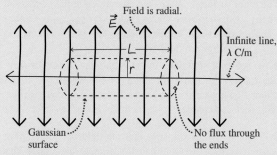

Field is radial.

$\vec{E}$

Infinite line, $\lambda$ C/m

$L$

$r$

Gaussian surface

No flux through the ends

**FIGURE 21.15** A cylindrical gaussian surface surrounds a line charge.

**EVALUATE**
- We begin with the flux $\Phi = \oint \vec{E} \cdot d\vec{A}$. $\vec{E}$ is everywhere perpendicular to the curved part of the cylinder, so here the dot product $\vec{E} \cdot d\vec{A}$ becomes $E\, dA$, and this part of the flux integral simpli-

fies to $\int_{\text{curved part}} E\, dA$. But the cylinder has ends, and Gauss's law requires that we consider the flux through those ends, too. No field lines emerge from the ends, so there's no flux. Mathematically, $\vec{E}$ is perpendicular to the normal vector $d\vec{A}$, so $\vec{E} \cdot d\vec{A} = 0$ on the ends. All we need is the flux through the curved part. Since the cylindrical gaussian surface is concentric with the line of symmetry, all points on it are the same distance $r$ from the line, and thus the magnitude $E$ is the same over the curved part of the cylinder. So we can take $E$ outside the integral, giving $E \int_{\text{curved part}} dA$. The integral is just the surface area of the curved part, which, if you unwound it, would be a rectangle of length $L$ and width equal to the circumference $2\pi r$. So the surface area is $2\pi rL$, and the flux becomes $\Phi = 2\pi rLE$. This result depends only on the symmetry, so it applies in all problems involving line symmetry.
- Next we need the enclosed charge. The line carries $\lambda$ C/m and our gaussian cylinder encloses $L$ meters of the line, so $q_{\text{enclosed}} = \lambda L$.
- Finally, we invoke Gauss's law by equating the flux $\Phi$ to $q_{\text{enclosed}}/\epsilon_0$. The result, $2\pi rLE = \lambda L/\epsilon_0$, solves to give

$$E = \frac{\lambda}{2\pi\epsilon_0 r} \qquad \text{(field of a line charge)} \qquad (21.6)$$

**ASSESS** Make sense? We worked this same problem in Example 20.7 in a much more difficult calculation involving a complicated integral. Since $1/2\pi\epsilon_0 = 2k$, our result here is the same. But Gauss's law makes the problem much easier! ∎

---

Although Example 21.4 involved an infinitesimally thin line of charge, you can see from Fig. 21.16 that our result must hold *outside* any charge with line symmetry. And, as we argued in Example 20.7, it's a good approximation for the field of any long, cylindrical structure as long as we're not too near its ends.

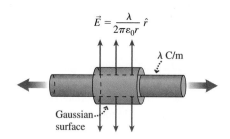

$\vec{E} = \frac{\lambda}{2\pi\varepsilon_0 r}\, \hat{r}$

$\lambda$ C/m

**FIGURE 21.16** The arguments of Example 21.4 apply outside *any* cylindrical charge distribution.

Gaussian surface

**EXAMPLE 21.5** **Gauss's Law: A Hollow Pipe**

A thin-walled pipe 3.0 m long and 2.0 cm in radius carries a net charge $q = 5.7\ \mu C$ distributed uniformly over its surface. Find the electric field both 1.0 cm and 3.0 cm from the pipe axis, far from either end.

**INTERPRET** Although the pipe has finite length, both distances are small compared with the length, so we can approximate the pipe as an infinitely long structure with line symmetry.

**DEVELOP** With line symmetry the appropriate gaussian surface is a cylinder coaxial with the pipe. We've drawn two such cylinders in Fig. 21.17, one for each radius where we're asked to evaluate the field.

**FIGURE 21.17** Gaussian surfaces for Example 21.5.

**EVALUATE**

- We showed in Example 21.4 that the flux integral in line symmetry gives $\Phi = 2\pi r L E$.
- Next we need the enclosed charge. At 3 cm we're outside the pipe, so the gaussian surface with this radius encloses all the charge: $q_{enclosed} = 5.7\ \mu C$. The pipe is hollow, so at 1 cm the enclosed charge is zero.
- Equating the flux to $q_{enclosed}/\epsilon_0$ and solving for $E$ give

$$E = \frac{q_{enclosed}}{2\pi\epsilon_0 rL} = \frac{5.7\ \mu C}{(2\pi\epsilon_0)(3.0 \times 10^{-2}\ \text{m})(3.0\ \text{m})} = 1.1\ \text{MN/C}$$

for the field at 3 cm and $E = 0$ for the field inside the pipe.

**ASSESS** Make sense? *Inside* the pipe, there's no field, and for the same reason as inside a uniformly charged hollow sphere—namely, that fields from near and far parts of the pipe cancel due, ultimately, to the inverse-square law. Again, be careful: That result follows because of the symmetry, although we'll soon see that with *conducting* pipes and shells there's no interior field even without that symmetry. We argued earlier that the field *outside* any line-symmetric distribution should be given by Equation 21.6, $E = \lambda/2\pi\epsilon_0 r$. In our result, the quantity $q_{enclosed}/L$ is the line charge density $\lambda$, so our result is indeed consistent with that equation. ∎

## Plane Symmetry

A charge distribution has plane symmetry when its charge density depends only on the perpendicular distance from a plane. The only electric-field direction consistent with this symmetry is perpendicular to the plane of symmetry. As with line symmetry, true plane symmetry implies a charge distribution that's infinite in extent. That's impossible—but plane symmetry remains a good approximation when charge is spread uniformly over large, flat surfaces or slabs. The next example illustrates the application of Gauss's law to plane symmetry.

**EXAMPLE 21.6** **Gauss's Law: A Sheet of Charge**

An infinite sheet of charge carries uniform surface charge density $\sigma$ in coulombs per square meter. Find the resulting electric field.

**INTERPRET** Since the sheet is infinite, we have plane symmetry with the sheet itself as the symmetry plane.

**DEVELOP** Figure 21.18 is our sketch. We choose as our gaussian surface a cylinder that straddles the charged sheet, extending equally on either side. Then the electric field is perpendicular to the ends and parallel to the sides; symmetry requires that it be uniform over the ends.

A gaussian surface has sides parallel and ends perpendicular to $\vec{E}$.

The distribution extends infinitely in two dimensions.

The sheet carries a uniform charge of $\sigma$ C/m².

Symmetry plane

The field is perpendicular to the plane.

Each end has area $A$.

**FIGURE 21.18** An infinite sheet of charge, with a gaussian surface straddling the sheet.

*cont'd.*

**EVALUATE**

- First we evaluate the flux through our gaussian cylinder. The field is parallel to the sides, so there's no flux contribution here. Then the total flux is through the ends, each of which has area $A$. Since the field is uniform over each end and perpendicular to the ends, the flux through each end is $EA$. So the total flux through both ends is $\Phi = 2EA$. This result is independent of the details of the charge distribution, so it holds in all cases of plane symmetry.

- Next we need the enclosed charge. The gaussian surface encloses an area of the sheet equal to the area $A$ of its ends. With surface charge density $\sigma$ on the sheet, the enclosed charge is $q_{enclosed} = \sigma A$.

- Now we apply Gauss's law, equating the flux to $q_{enclosed}/\epsilon_0$. Thus $2EA = \sigma A/\epsilon_0$, so

$$E = \frac{\sigma}{2\epsilon_0} \quad \text{(field of a charged sheet)} \quad (21.7)$$

The direction of this field on either side of the sheet is outward from the sheet if it's positively charged and inward if negative.

**ASSESS** Make sense? With an infinite plane, symmetry requires that the field lines be perpendicular to the plane. So they don't spread out, and that means the field doesn't vary with distance—as Equation 21.7 shows because it doesn't involve the distance from the sheet. Although our result is exact only for a truly infinite sheet, it's a good approximation near any large, flat, uniformly charged surface as long as we're not close to an edge. ∎

## 21.5 Fields of Arbitrary Charge Distributions

Although Gauss's law is always true, most charge distributions lack the symmetry needed to apply Gauss's law to find the field. The alternative, Coulomb's law, is hard to use in all but the simplest cases. But we can often learn a lot by considering the distributions whose fields we calculated here and in Chapter 20. Figure 21.19 summarizes four of these fields. For the last three, note the simple relation between the number of dimensions and the behavior of the field. The plane has two dimensions, and its field doesn't decrease with distance. The line has one dimension, and its field decreases as $1/r$. The point has no

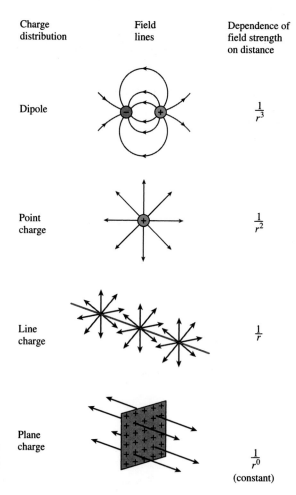

| Charge distribution | Field lines | Dependence of field strength on distance |
|---|---|---|
| Dipole | | $\frac{1}{r^3}$ |
| Point charge | | $\frac{1}{r^2}$ |
| Line charge | | $\frac{1}{r}$ |
| Plane charge | | $\frac{1}{r^0}$ (constant) |

**FIGURE 21.19** Fields of a dipole, a point charge, a charged line, and a charged plane.

dimensions, and its field falls as $1/r^2$. In a sense, the dipole continues this progression. It consists of opposite point charges whose effects nearly cancel; no wonder its field decreases faster still, as $1/r^3$. In fact, there's a hierarchy of charge distributions whose fields decrease ever faster, as dipoles nearly cancel dipoles, and so on. Scientists and engineers use this hierarchy in modeling charged structures ranging from molecules to radio antennas.

Good thinking coupled with knowledge of simpler charge distributions can provide a reasonable approximation to the field of an arbitrary distribution. Consider the uniformly charged disk in Fig. 21.20. For points much closer to the disk surface than to the edge, the disk looks almost like an infinite plane of charge. For these points the field is approximately that of an infinite plane—a field that points away from the plane and doesn't vary with distance. Far from the disk, meanwhile, its exact size and shape are unimportant. Its field resembles that of a point charge; it's radially outward and falls as the inverse square of the distance from the disk. The field at intermediate distances is harder to determine. But the infinite-plane field lines close to the disk must connect smoothly to the point-charge field lines far away. If we sketch these, as in Fig. 21.20, we have a rough picture of the field everywhere. Don't underestimate the value of a simple approximation like this! It can often tell all we need to know about a situation and may provide a clearer understanding than would a detailed calculation.

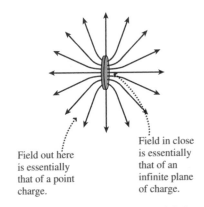

Field out here is essentially that of a point charge.

Field in close is essentially that of an infinite plane of charge.

**FIGURE 21.20** The field of a charged disk.

---

**GOT IT? 21.4**    A uniformly charged sheet measures 1 m on each side, and you're told the total charge $Q$. What expressions would you use to get approximate values for the field magnitude (a) 1 cm from the sheet (but not near an edge) and (b) 1 km from the sheet?

---

## 21.6 Gauss's Law and Conductors

### Electrostatic Equilibrium

We've defined conductors as materials that contain free charges, like the free electrons in metals. Figure 21.21 shows what happens when an electric field is applied to a conductor. Free charges respond to the electric force $q\vec{E}$ by moving—in the direction of the field if they're positive, opposite the field if negative. The resulting charge separation gives rise to an electric field within the material that's opposite to the applied field. As more charge moves, this internal field becomes stronger until its magnitude eventually equals that of the applied field. At that point free charges within the conductor experience zero net force, and the conductor is in **electrostatic equilibrium**. Although individual charges continue to move about in random thermal motion, there's no longer any net charge motion. Once equilibrium is reached, the internal and applied fields are equal but opposite, and therefore:

> The electric field is zero inside a conductor in electrostatic equilibrium.

It could not be otherwise: Since a conductor contains free charges, the presence of any internal electric field would result in bulk charge motion, and we wouldn't have equilibrium. This result doesn't depend on the size or shape of the conductor, the magnitude or direction of the applied field, or even the nature of the material as long as it's a conductor. This is a macroscopic view; it considers only average fields within the material. At the atomic and molecular level, there are still strong electric fields near individual electrons and positive ions. But the *average* field, taken over larger distances, is zero inside a conductor in electrostatic equilibrium.

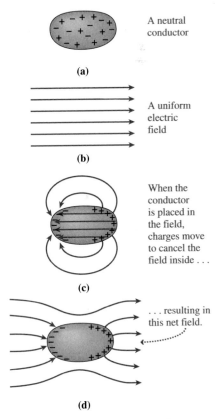

A neutral conductor

**(a)**

A uniform electric field

**(b)**

When the conductor is placed in the field, charges move to cancel the field inside . . .

**(c)**

. . . resulting in this net field.

**(d)**

**FIGURE 21.21** A conductor in a uniform electric field.

### Charged Conductors

Although they contain free charges, conductors are normally electrically neutral because they include equal numbers of electrons and protons. But suppose we give a conductor a nonzero net charge, for example, by injecting excess electrons into its interior. There's a mutual repulsion among the electrons and, because these are *excess* electrons, there's no

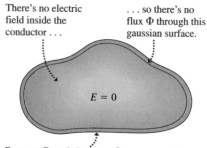

There's no electric field inside the conductor . . .

. . . so there's no flux Φ through this gaussian surface.

$E = 0$

Because Gauss's law says $\Phi \propto q_{enclosed}$, all excess charge resides on the conductor surface.

**FIGURE 21.22** Gauss's law implies that any net charge resides on the surface of a conductor in electrostatic equilibrium.

compensating attraction from positive charges. We might expect, therefore, that the electrons will move as far apart as possible—namely, to the surface of the conductor.

We now use Gauss's law to prove rigorously that excess charge *must* be at the surface of a conductor in electrostatic equilibrium. Figure 21.22 shows a piece of conducting material with a gaussian surface drawn just below the material surface. In equilibrium there's no electric field inside the conductor, and thus the field is zero everywhere on the gaussian surface. The flux, $\oint \vec{E} \cdot d\vec{A}$, through the gaussian surface is therefore also zero. But Gauss's law says that the flux through a closed surface is proportional to the net charge enclosed, and therefore the net charge inside our gaussian surface must be zero. This is true no matter where the gaussian surface is as long as it's *inside* the conductor. We can move it arbitrarily close to the conductor surface and it still encloses no net charge. If there is a net charge on the conductor, it lies outside the gaussian surface, and therefore we conclude: **If a conductor in electrostatic equilibrium carries a net charge, it must reside on the conductor surface.**

**EXAMPLE 21.7** **Gauss's Law: A Hollow Conductor**

An irregularly shaped conductor has a hollow cavity. The conductor itself carries a net charge of 1 $\mu$C, and there's a 2-$\mu$C point charge inside the cavity. Find the net charge on the cavity wall and on the outer surface of the conductor, assuming electrostatic equilibrium.

**INTERPRET** This problem involves a conductor in electrostatic equilibrium, which means (1) there's no electric field inside the conducting material and (2) the net charge resides on the conductor surface—which in this case includes both the inner and outer surfaces.

**DEVELOP** We draw the situation in Fig. 21.23. Our plan is to apply Gauss's law to find the charges. We consider a gaussian surface inside the conductor and enclosing the cavity, as shown.

$E = \vec{0}$

$+3 \mu C$

$+2 \mu C$

$-2 \mu C$

**FIGURE 21.23** The gaussian surface encloses zero net charge, so there must be $-2$ $\mu$C on the cavity wall.

**EVALUATE** Since there's no electric field inside the conductor, the flux through the gaussian surface is zero, and therefore the net charge enclosed is also zero. But there's that +2-$\mu$C point charge in the cavity. For the gaussian surface to enclose zero net charge, there must be $-2$ $\mu$C somewhere else—and the only place it can be is on the cavity wall. However, the entire conductor carries $+1$ $\mu$C. With $-2$ $\mu$C on the inside wall, that leaves $+3$ $\mu$C on the outer surface.

**ASSESS** Make sense? Yes: This distribution of charge is the only one possible that's consistent with both Gauss's law and the requirement that $E$ is zero inside a conductor in electrostatic equilibrium. As another check, think about what this charge distribution must look like from far away—namely, a point charge with its net charge of 3 $\mu$C. Since the fields of the cavity wall and the inner point charge don't penetrate the conductor, the only field lines that reach out beyond the conductor are those from the charge on its outer surface. So that charge must be 3 $\mu$C, as we've found. ∎

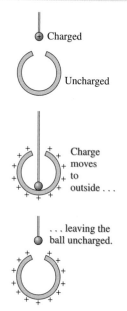

Charged

Uncharged

Charge moves to outside . . .

. . . leaving the ball uncharged.

**FIGURE 21.24** Experimental test of Gauss's law.

**GOT IT? 21.5** A conductor carries a net charge $+Q$. There's a hollow cavity inside the conductor that contains a point charge $-Q$. In electrostatic equilibrium, is the charge on the outer surface of the conductor (a) $-2Q$, (b) $-Q$, (c) 0, (d) $Q$, or (e) $2Q$?

## Experimental Tests of Gauss's Law

That net charge moves to a conductor surface provides a sensitive test of Gauss's law, and thus—through the arguments of pp. 351–352—a test of the inverse-square law for the electric force. Figure 21.24 shows a charged conducting ball touched to the inside of a hollow, initially neutral conductor. Charge flows to the outer surface of the hollow conductor, leaving the ball uncharged. Measuring zero charge on the ball confirms the inverse-square law; such experiments show that the exponent in $1/r^2$ is indeed 2 to some 16 decimal places!

## The Field at a Conductor Surface

There can't be an electric field *within* a conductor in electrostatic equilibrium, but there *may* be a field right at the conductor surface (Fig. 21.25a). Such a field must be perpendicular to the surface; otherwise, charge would move along the surface in response to the field's parallel component, and we wouldn't have equilibrium.

We can compute the strength of this surface field by considering a small gaussian surface that straddles the conductor surface, as shown in Fig. 21.25b. There's no flux through the sides, and because the field is zero inside the conductor, there's no flux through the inner end either. So the only flux is through the outer end, with area $A$. Since the end is perpendicular to the field, the flux is $EA$. The gaussian surface encloses charge $\sigma A$, where $\sigma$ is the surface charge density. Gauss's law equates the flux with $q_{enclosed}/\epsilon_0$, so we have $EA = \sigma A/\epsilon_0$, or

$$E = \frac{\sigma}{\epsilon_0} \quad \text{(field at conductor surface)} \tag{21.8}$$

This result shows that large fields develop where the charge density on a conductor is high. Engineers who design electrical devices must avoid high charge densities whose associated fields lead to sparks, arcing, and breakdown of electric insulation.

Equation 21.8 gives a field that depends only on the local charge density. Does that mean the field at a conductor surface arises only from the local charge? No! As always, the field is the vector sum of contributions from all charges. Remarkably, Gauss's law requires that charges on a conductor arrange themselves in such a way that the field at any point on the conductor surface depends only on the surface charge density right at that point—even though that field arises from *all* the charges on the surface (as well as from charges elsewhere if there are such)!

Consider a thin, flat, isolated, conducting sheet that has charge density $\sigma$ on one of its two faces (Fig. 21.26a). Equation 21.8 shows that the field at the surface of this plate is $\sigma/\epsilon_0$. But if the plate is large and flat, we can approximate it as an infinite sheet of charge—for which we found earlier (Equation 21.7) that the field should be $\sigma/2\epsilon_0$. Is there a contradiction here? No! If the plate is isolated, then symmetry requires that the charge spread

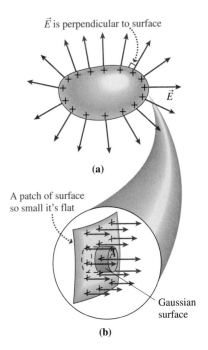

**FIGURE 21.25** (a) The electric field at the surface of a charged conductor is perpendicular to the conductor surface. (b) A gaussian surface straddles the conductor surface.

---

**APPLICATION** **Shielding and Lightning Safety**

We've seen that charge moves to the outside of a conductor surface, leaving the interior free of charge and electric field—even if the interior is hollow. This is the basis of electric shielding, in which a conducting enclosure keeps out external electric fields. A common example is the coaxial cable that connects your MP3 player to a stereo system or carries TV signals from a cable box to your video display or DVD recorder. Coaxial cables also connect electronic instruments in scientific and medical research, and they carry radio and TV signals to and from antennas. The photo shows a typical coaxial cable, consisting of an inner copper wire, insulation, and an outer braided copper shield. External electric fields from radio, TV, cell phone, Bluetooth, and other wireless technologies, or the random electrical "noise" that permeates our environment, all cause charges to move in the shield so as to cancel these external fields inside the cable. That prevents interference with the signals carried in the cable. In another application of this shielding principle, researchers doing experiments with very weak electric signals often construct entire rooms with conducting walls to minimize electric interference.

Shielding is also the reason a car is a relatively safe place to be in a thunderstorm. A lightning strike dumps charge on the car's metal body, but the charge distributes itself on the outside so as to prevent any electric fields from developing inside the car. That, in turn,

prevents harmful currents from flowing through the cars' occupants. The operator's cage in this chapter's opening photo has the same effect, harmlessly deflecting charge from the artificial lightning and keeping the interior free of electric fields.

Strictly speaking, charge resides on the outside of a conductor only in equilibrium. But electrons in metals respond so quickly that equilibrium results almost instantaneously—meaning that metallic shielding is effective even against the rapidly varying electric fields of high-frequency radio, TV, and microwave signals.

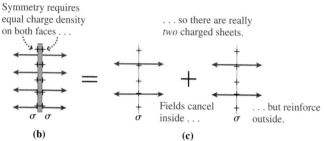

**FIGURE 21.26** (a) An isolated, charged conducting plate. Its field points outward from both faces. (b) Edge-on view of the plate. (c) The field anywhere is the sum of the fields of the two faces, each treated as a single charged sheet.

Charge resides on inner faces, giving two oppositely charged sheets.

**FIGURE 21.27** Edge-on view of two parallel conducting plates carrying opposite charges.

itself evenly over *both* faces. If one face has charge density $\sigma$, so must the other—so we really have *two* charge sheets, each with density $\sigma$ (Fig 21.26b). Each gives a field of magnitude $\sigma/2\epsilon_0$, and *outside* the conductor those fields superpose to give the net field $\sigma/\epsilon_0$ (Fig. 21.26c). *Inside* the conductor their directions are opposite, and the result is zero field inside the conductor. Applying Equation 21.8 skips these details. But because Equation 21.8 was derived on the assumption that the field inside the conductor is zero, it "knows" about charges everywhere on the conductor—and in this case that means on the second face.

Equation 21.8 also applies to a pair of oppositely charged conducting plates (Fig. 21.27); the result, for the field between the plates, is $\sigma/\epsilon_0$, where $\sigma$ is the surface charge density on either plate. Why not $2\sigma/\epsilon_0$? Again, Equation 21.8 gives the field at a conductor surface— and it takes into account other charges that may be present. Here each plate's charge attracts the other's opposite charge, all to the *inner* face. Each plate is thus a single charge layer, giving a field $\sigma/2\epsilon_0$, and between the plates the fields sum to Equation 21.8's result, $\sigma/\epsilon_0$. Beyond the plates the fields sum to zero—a result that also follows from Equation 21.8 because now there's zero surface charge on the outer faces.

## Big Picture

The big idea here is **Gauss's law**—a universal statement about electric fields that's equivalent to Coulomb's inverse-square law but is expressed in terms of the global behavior of the field over any closed surface. Using the **electric field-line** picture, Gauss's law says that the number of field lines emerging from a closed surface depends only on the net charge enclosed; more rigorously, it says that the **electric flux** through the surface is proportional to the enclosed charge.

Eight lines pass through any closed surface surrounding $q$.

Point charge $q$

The number of lines through a surface depends on the net charge enclosed.

Point charges $+q$, $-q/2$

## Key Concepts and Equations

**Electric flux** $\Phi$ describes the amount of electric field crossing an area.

$\Phi = EA$ for a flat area perpendicular to a uniform field

$$\Phi = \int \vec{E} \cdot d\vec{A}, \text{ in general}$$

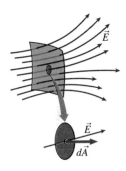

In terms of flux, Gauss's law reads $\oint \vec{E} \cdot d\vec{A} = q_{enclosed}/\epsilon_0$.

Here $\epsilon_0 = 1/4\pi k = 8.85 \times 10^{-12} \, \text{C}^2/\text{N} \cdot \text{m}^2$ is another way of expressing the Coulomb constant $k = 9.0 \times 10^9 \, \text{N} \cdot \text{m}^2/\text{C}^2$.

## Cases and Uses

Gauss's law gives the fields of symmetric charge distributions:

Spherical symmetry:

Outside: $E = \dfrac{Q}{4\pi\epsilon_0 r} = \dfrac{kQ}{r^2}$

Inside uniformly charged sphere:

$$E = \frac{kQr}{R^3}$$

Inside hollow sphere: $E = 0$

Line symmetry:

$\lambda$ C/m

Outside: $E = \dfrac{\lambda}{2\pi\epsilon_0 r}$

Inside hollow pipe:

$$E = 0$$

Plane symmetry:

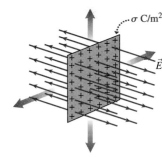

$\sigma$ C/m$^2$

$\vec{E}$

Outside: $E = \dfrac{\sigma}{2\epsilon_0}$

Gauss's law and conductors:

- The field is zero inside a conductor in electrostatic equilibrium.
- Any net charge resides on a conductor's surface.
- The field at the surface is perpendicular to the surface and has magnitude $\sigma/\epsilon_0$.

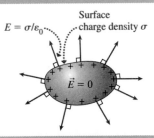

$E = \sigma/\varepsilon_0$ ···· Surface charge density $\sigma$

$\vec{E} = 0$

# For Thought and Discussion

1. Can electric field lines ever cross? Why or why not?
2. If identical charged particles are placed at points $A$ and $B$ in Fig. 21.28, which will experience the greater force?

**FIGURE 21.28** For Thought and Discussion 2

3. The electric flux through a closed surface is zero. Must the electric field be zero on that surface? If not, give an example.
4. If the flux of the gravitational field through a closed surface is zero, what can you conclude about the region interior to the surface?
5. Under what conditions can the electric flux through a surface be written as $EA$, where $A$ is the surface area?
6. Eight field lines emerge from a closed surface surrounding an isolated point charge. Would this fact change if a second identical charge were brought to a point just *outside* the surface? If not, would anything change? Explain.
7. If a charged particle were released from rest on a curved field line, would its subsequent motion follow the field line? Explain.
8. Gauss's law describes the flux of the electric field through a surface that may enclose charge. Must the field in Gauss's law arise only from charges within the closed surface?
9. In a certain region the electric field points to the right and its magnitude increases as you move to the right, as shown in Fig. 21.29. Does the region contain net positive charge, net negative charge, or zero net charge?

**FIGURE 21.29** For Thought and Discussion 9 (Left ends mark the beginnings of field lines, which extend indefinitely to the right.)

10. A point charge is located a fixed distance from a uniformly charged sphere, outside the sphere. If the sphere shrinks in size without losing any charge, what happens to the force on the point charge?
11. The field of an infinite line of charge decreases as $1/r$. Why isn't this a violation of the inverse-square law?
12. Why can't you use Gauss's law to determine the field of a uniformly charged cube? Why wouldn't it work to draw a cubical gaussian surface?
13. You're sitting inside an uncharged, hollow spherical shell. Suddenly someone dumps a billion coulombs of charge on the shell, distributed uniformly. What happens to the electric field at your location?
14. Does Gauss's law apply to a spherical gaussian surface not centered on a point charge, as shown in Fig. 21.30? Would this be a useful surface to use in calculating the electric field?

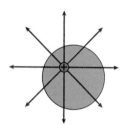

**FIGURE 21.30** For Thought and Discussion 14

15. An insulating sphere carries charge spread uniformly throughout its volume. A conducting sphere has the same radius and net charge, but of course the charge is spread over its surface only. How do the electric fields outside these two charge distributions compare?
16. Why must the electric field be zero inside a conductor in electrostatic equilibrium?
17. The electric field of a flat sheet of charge is $\sigma/2\epsilon_0$. Yet the field of a flat conducting sheet—even a thin one, like a piece of aluminum foil—is $\sigma/\epsilon_0$. Explain this apparent discrepancy.

# Exercises and Problems

## Exercises

### Section 21.1 Electric Field Lines
18. What is the net charge shown in Fig. 21.31? The magnitude of the middle charge is 3 $\mu$C.

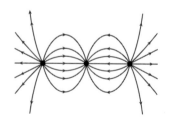

**FIGURE 21.31** Exercise 18

19. A charge $+2q$ and a charge $-q$ are near each other. Sketch some field lines for this charge distribution, using the convention of eight lines for a charge of magnitude $q$.
20. Two charges $+q$ and a charge $-q$ are at the vertices of an equilateral triangle. Sketch some field lines for this charge distribution.

21. The net charge shown in Fig. 21.32 is $+Q$. Identify each of the charges $A$, $B$, and $C$ shown.

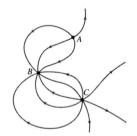

**FIGURE 21.32** Exercise 21

### Section 21.2 Electric Flux and Field
22. A flat surface with area 2.0 m$^2$ is in a uniform electric field of 850 N/C. What is the electric flux through the surface when it is (a) at right angles to the field, (b) at 45° to the field, and (c) parallel to the field?
23. What is the electric field strength in a region where the flux through a 1.0 cm $\times$ 1.0 cm flat surface is 65 N·m$^2$/C, if the field is uniform and the surface is at right angles to the field?

24. A flat surface with area $0.14 \text{ m}^2$ lies in the $x$-$y$ plane, in a uniform electric field given by $\vec{E} = 5.1\hat{\imath} + 2.1\hat{\jmath} + 3.5\hat{k}$ kN/C. Find the flux through this surface.

25. The electric field on the surface of a 10-cm-diameter sphere is perpendicular to the sphere and has magnitude 47 kN/C. What is the electric flux through the sphere?

### Section 21.3 Gauss's Law

26. A sock comes out of the dryer with 1 trillion $(10^{12})$ excess electrons. What's the electric flux through a surface surrounding the sock?

27. What is the electric flux through each closed surface shown in Fig. 21.33?

**FIGURE 21.33** Exercise 27

28. A 6.8-$\mu$C charge and a $-4.7$-$\mu$C charge are inside an uncharged sphere. What is the electric flux through the sphere?

29. A 2.6-$\mu$C charge is at the center of a cube 7.5 cm on each side. What is the electric flux through one face of the cube? *Hint:* Think about symmetry, and don't do an integral.

### Section 21.4 Using Gauss's Law

30. The electric field at the surface of a uniformly charged sphere of radius 5.0 cm is 90 kN/C. What would be the field strength 10 cm from the surface?

31. A solid sphere 25 cm in radius carries $14 \, \mu$C, distributed uniformly throughout its volume. Find the electric field strength (a) 15 cm, (b) 25 cm, and (c) 50 cm from the sphere's center.

32. A 10-nC point charge is located at the center of a thin spherical shell of radius 8.0 cm carrying $-20$ nC distributed uniformly over its surface. What are the magnitude and direction of the electric field (a) 2.0 cm, (b) 6.0 cm, and (c) 15 cm from the point charge?

33. The electric field strength outside a charge distribution and 18 cm from its center has magnitude 55 kN/C. At 23 cm the field strength is 43 kN/C. Does the distribution have spherical or line symmetry?

34. An electron close to a large, flat sheet of charge is repelled from the sheet with a $1.8 \times 10^{-12}$ N force. Find the surface charge density on the sheet.

35. Find the field produced by a uniformly charged sheet carrying 87 pC/cm².

36. What surface charge density on an infinite sheet will produce an electric field of 1.4 kN/C?

### Section 21.5 Fields of Arbitrary Charge Distributions

37. A rod 50 cm long and 1.0 cm in radius carries a 2.0-$\mu$C charge distributed uniformly over its length. What is the approximate magnitude of the electric field (a) 4.0 mm from the rod surface, not near either end, and (b) 23 m from the rod?

38. What is the approximate field strength 1 cm above an ordinary sheet of paper carrying a uniform surface charge density $\sigma = 45 \text{ nC/m}^2$?

39. The disk in Fig. 21.20 is uniformly charged to 5.0 $\mu$C and has area $0.14 \text{ m}^2$. Find the approximate field strength (a) 1 mm from the disk, not near an edge, and (b) 2.5 m from the disk.

### Section 21.6 Gauss's Law and Conductors

40. What is the electric field strength just outside the surface of a conducting sphere carrying surface charge density 1.4 $\mu$C/m²?

41. A net charge of 5.0 $\mu$C is applied on one side of a solid metal sphere 2.0 cm in diameter. After electrostatic equilibrium is reached, what are (a) the volume charge density inside the sphere and (b) the surface charge density on the sphere? Assume there are no other charges or conductors nearby. (c) Which of your answers depends on this assumption, and why?

42. A point charge $+q$ lies at the center of a spherical conducting shell carrying a net charge $\frac{3}{2}q$. Sketch the field lines both inside and outside the shell, using eight field lines to represent a charge of magnitude $q$.

43. A total charge of 18 $\mu$C is applied to a thin, square metal plate 75 cm on a side. Find the electric field strength near the plate's surface.

## Problems

44. What is the flux through the hemispherical open surface of radius $R$ shown in Fig. 21.34? The uniform field has magnitude $E$. *Hint:* Don't use a messy integral!

**FIGURE 21.34** Problem 44

45. An electric field is given by $\vec{E} = E_0(y/a)\hat{k}$, where $E_0$ and $a$ are constants. Find the flux through the square that lies in the $x$-$y$ plane and is bounded by the points $(0, 0)$, $(0, a)$, $(a, a)$, $(a, 0)$.

46. The electric field in a certain region is given by $\vec{E} = ax\hat{\imath}$, where $a = 40$ N/C·m and $x$ is the position in meters. Find the volume charge density in the region. *Hint:* Apply Gauss's law to a cube 1 m on a side.

47. Positive charge is spread uniformly over the surface of a spherical balloon 70 cm in radius, resulting in an electric field of 26 kN/C at the balloon's surface. Find the field strength (a) 50 cm from the balloon's center and (b) 190 cm from the center, (c) What is the net charge on the balloon?

48. A solid sphere 2.0 cm in radius carries a uniform volume charge density. The electric field 1.0 cm from the sphere's center has magnitude 39 kN/C. (a) At what other distance does the field have this magnitude? (b) What is the net charge on the sphere?

49. A point charge $-2Q$ is at the center of a spherical shell of radius $R$ carrying charge $Q$ spread uniformly over its surface. What is the electric field at (a) $r = \frac{1}{2}R$ and (b) $r = 2R$? (c) How would your answers change if the charge on the shell were doubled?

50. A spherical shell of radius 15 cm carries 4.8 $\mu$C, distributed uniformly over its surface. At the center of the shell is a point charge. (a) If the electric field at the surface of the sphere is 750 kN/C and points outward, what is the charge of the point charge? (b) What is the field just inside the shell?

51. A spherical shell 30 cm in diameter carries a total charge of 85 $\mu$C distributed uniformly over its surface. A 1.0-$\mu$C point charge is located at the center of the shell. What is the electric field strength (a) 5.0 cm from the center and (b) 45 cm from the center? (c) How would your answers change if the charge on the shell were doubled?

52. A thick, spherical shell of inner radius $a$ and outer radius $b$ carries a uniform volume charge density $\rho$. Find an expression for the electric field strength in the region $a < r < b$, and show that your result is consistent with Equation 21.5 when $a = 0$.

53. A long, thin wire carrying 5.6 nC/m runs down the center of a long, thin-walled, hollow pipe with radius 1.0 cm carrying $-4.2$ nC/m spread uniformly over its surface. Find the electric field (a) 0.50 cm from the wire and (b) 1.5 cm from the wire.

54. An infinitely long rod of radius $R$ carries a uniform volume charge density $\rho$. Show that the electric field strengths outside and inside the rod are given, respectively, by $E = \rho R^2/2\epsilon_0 r$ and $E = \rho r/2\epsilon_0$, where $r$ is the distance from the rod axis.

55. A long, solid rod 4.5 cm in radius carries a uniform volume charge density. If the electric field strength at the surface of the rod (not near either end) is 16 kN/C, what is the volume charge density?

56. If you "painted" positive charge on the floor, what surface charge density would be necessary to suspend a 15-$\mu$C, 5.0-g particle above the floor?

57. A slab of charge extends infinitely in two dimensions and has thickness $d$ in the third dimension, as shown in Fig. 21.35. The slab carries a uniform volume charge density $\rho$. Find expressions for the electric field strength (a) inside and (b) outside the slab, as functions of the distance $x$ from the center plane.

**FIGURE 21.35** Problem 57

58. A solid sphere 10 cm in radius carries a uniform 40-$\mu$C charge distribution throughout its volume. It is surrounded by a concentric shell 20 cm in radius, also uniformly charged with 40 $\mu$C. Find the electric field (a) 5.0 cm, (b) 15 cm, and (c) 30 cm from the center.

59. A nonconducting square plate 75 cm on a side carries a uniform surface charge density. The electric field strength 1 cm from the plate, not near an edge, is 45 kN/C. What is the approximate field strength 15 m from the plate?

60. A 250-nC point charge is placed at the center of an uncharged spherical conducting shell 20 cm in radius. (a) What is the surface charge density on the outer surface of the shell? (b) What is the electric field strength at the shell's outer surface?

61. An irregular conductor containing an irregular, empty cavity carries a net charge $Q$. (a) Show that the electric field inside the cavity must be zero. (b) If you put a point charge inside the cavity, what value must it have in order to make the surface charge density on the outer surface of the conductor everywhere zero?

62. A coaxial cable consists of an inner wire and a concentric cylindrical outer conductor (Fig. 21.36). If the conductors carry equal but opposite charges, show that there is no net charge on the *outside* of the outer conductor.

**FIGURE 21.36** Problem 62

63. A point charge $-q$ is at the center of a spherical shell carrying charge $+2q$. That shell, in turn, is concentric with a larger shell carrying charge $-\frac{3}{2}q$. Draw a cross section of this structure, and sketch the electric field lines using the convention that eight lines correspond to a charge of magnitude $q$.

64. A point charge $q$ is at the center of a spherical shell of radius $R$ carrying charge $2q$ spread uniformly over its surface. Write expressions for the electric field strength at (a) $\frac{1}{2}R$ and (b) $2R$.

65. The volume charge density inside a solid sphere of radius $a$ is given by $\rho = \rho_0 r/a$, where $\rho_0$ is a constant. Find (a) the total charge and (b) the electric field strength within the sphere, as a function of distance $r$ from the center.

66. Figure 21.37 shows a rectangular box with sides $2a$ and length $L$ surrounding a line of charge with uniform line charge density $\lambda$. The line passes directly through the center of the box faces. Using an expression for the field of a line charge, integrate over strips of width $dx$ as shown to find the electric flux through one face of the box. Multiply by 4 to get the total flux through the box, and show that your result is consistent with Gauss's law.

**FIGURE 21.37** Problem 66

67. The charge density within a uniformly charged sphere of radius $R$ is given by $\rho = \rho_0 - ar^2$, where $\rho_0$ and $a$ are constants, and $r$ is the distance from the center. Find an expression for $a$ such that the electric field outside the sphere is zero.

68. Calculate the electric fields in Example 21.2 directly, using the superposition principle and integration. Consider the shell to be composed of charge elements that are coaxial rings, whose axis passes through the field point, which is a distance $r$ from the center. *Hint*: Consult Example 20.6. You'll have to evaluate the cases $r < R$ and $r > R$ separately.

69. A solid sphere of radius $R$ carries a volume charge density $\rho = \rho_0 e^{r/R}$, where $\rho_0$ is a constant and $r$ is the distance from the center. Find an expression for the electric field strength at the sphere's surface.

70. A friend is working on a biology experiment and needs to create an electric field of magnitude 430 N/C at a distance of 10 cm from the central portion of a large nonconducting square plate 4.5 m on each side. She needs to know how much charge to put on the plate. What do you tell her?

71. Your roommate is taking a history of science course and has an assignment to write a paper on the remarkable similarities between the gravitational and electric forces. The history professor said something to the effect that Gauss's law could be used to find the gravitational field *inside* the Earth, but didn't explain how to do it. Looking at your roommate's notes, you find this: $\oint \vec{g} \cdot d\vec{A} = -4\pi G M_{\text{enclosed}}$. Can you help?

# Answers to Chapter Questions

## Answer to Chapter Opening Question

Gauss's law requires that electric charge remain on the outside of the metal cage, arranging itself so there's no electric field within the cage.

## Answers to GOT IT? Questions

21.1 (a) $\Phi_A = 0$, $\Phi_B = 0$, $\Phi_C = s^2 E$;
(b) $\Phi_A = 0$, $\Phi_B = \Phi_C = s^2 E \cos 45° = s^2 E/\sqrt{2}$.
21.2 Flux: (a); doesn't change. Field: (d); increase if charges are opposite, decrease if same
21.3 (a) The field stays zero; (b) the field $kQ/r^2$ doubles.
21.4 (a) $E = \sigma/2\epsilon_0$, with $\sigma = Q/A$ and $A = 1$ m$^2$;
(b) $E = kQ/r^2$, with $r = 1000$ m.
21.5 (c).

■ This parasailer landed on a 138,000-volt power line. Why didn't he get electrocuted?

▶ **To Learn**

By the end of this chapter you should be able to

■ Explain the meaning of electric potential difference, including the familiar term "volt" (22.1).

■ Calculate the potential difference between two points in simple electric fields using a line integral (22.1).

■ Describe the potential in the field of a point charge, and use superposition to calculate potential differences by summing or integrating over point charges (22.2).

■ Explain the concept of equipotentials, and determine electric fields from potential differences (22.3).

■ Describe qualitatively how charge distributes itself on conductors (22.4).

◀ **To Know**

■ This chapter merges the concepts of work (6.1, 6.2), conservative forces (7.1), and potential energy (7.2) with the electric field (20.3).

■ We build on results from Chapters 20 and 21, including the fields of a dipole (20.4), a spherical charge distribution (21.4), a line charge (20.4, 21.4), and a charged sheet (21.4).

■ We further our understanding of charged conductors (21.6).

■ We use Tactics 9.1 for handling integrals (9.1).

Like gravity, the electric force is conservative. That means the work done in moving a charge against an electric force results in stored potential energy. It's convenient to consider the energy per unit charge, a measure that defines the concept of electric potential. Here we'll see how potential provides a simpler approach to calculating electric fields and helps characterize everyday devices like batteries.

## 22.1 Electric Potential Difference

In Chapter 7 we defined the potential-energy difference $\Delta U_{AB}$ as the negative of the work $W_{AB}$ done by a conservative force $\vec{F}$ on an object moved from point $A$ to point $B$ (Equation 7.2):

$$\Delta U_{AB} = U_B - U_A = -W_{AB} = -\int_A^B \vec{F} \cdot d\hat{r}$$

where $d\hat{r}$ is an element of the path from $A$ to $B$, and $\Delta U_{AB}$ is independent of the path taken from $A$ to $B$. When the force doesn't vary, we can calculate the work more easily using Equation 6.5, $W = \vec{F} \cdot \Delta \vec{r}$.

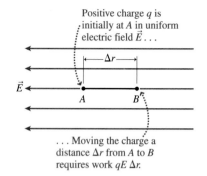

Positive charge $q$ is initially at $A$ in uniform electric field $\vec{E}$ ...

... Moving the charge a distance $\Delta r$ from $A$ to $B$ requires work $qE\,\Delta r$.

**FIGURE 22.1** Work done in moving a charge $q$ against an electric field $\vec{E}$.

**TABLE 22.1** Force and Field, Potential Energy and Electric Potential

| Quantity | Symbol/Equation | Units |
|---|---|---|
| Force | $\vec{F}$ | N |
| Electric field | $\vec{E} = \dfrac{\vec{F}}{q}$ | N/C or V/m |
| Potential-energy difference | $\Delta U = -\displaystyle\int_A^B \vec{F} \cdot d\vec{r}$ | J |
| Electric potential difference | $\Delta V = \dfrac{\Delta U}{q}$ or $\Delta V = -\displaystyle\int_A^B \vec{E} \cdot d\vec{r}$ | J/C or V |

Potential difference $\Delta V_{AB}$ depends only on points $A$ and $B$.

Calculating potential difference along any path (1, 2, or 3) gives $\Delta V_{AB} = E\Delta r$.

**FIGURE 22.2** Potential difference is path independent.

Consider a positive charge $q$ moved between points $A$ and $B$ a distance $\Delta r$ apart in a uniform electric field $\vec{E}$, as shown in Fig. 22.1. Since the field is uniform, a constant electric force $\vec{F} = q\vec{E}$ acts on the charge, so we use Equation 6.5 to evaluate the work done by the field, and the resulting potential energy change:

$$\Delta U_{AB} = -W_{AB} = -q\vec{E} \cdot \Delta\vec{r} = -qE\,\Delta r \cos 180° = qE\,\Delta r$$

where the factor $\cos 180° = -1$ appears because $\vec{E}$ and $\Delta\vec{r}$ have opposite directions. Make sense? Pushing a positive charge from $A$ to $B$ against the electric field is like pushing a car up a hill: Potential energy increases in both cases. Let go of the charge, and the field accelerates it back to $A$, just as gravity would accelerate the car back down the hill.

If we had moved a charge $2q$ in Fig. 22.1, the potential-energy change $\Delta U$ would have been twice as great; a charge $\frac{1}{2}q$ would have cut $\Delta U$ in half. Since $\Delta U$ is proportional to charge, it's convenient to consider the *potential-energy change per unit charge* involved in moving charge between two points. Mathematically, we write $\vec{F} = q\vec{E}$ in our general expression for $\Delta U_{AB}$ and divide by $q$. The result defines the **electric potential difference $\Delta V$**:

> The electric potential difference from point $A$ to point $B$ is the potential-energy change per unit charge in moving from $A$ to $B$:
>
> $$\Delta V_{AB} = \frac{\Delta U_{AB}}{q} = -\int_A^B \vec{E} \cdot d\vec{r} \qquad \text{(electric potential difference)} \qquad (22.1a)$$

Here $\Delta$ and the subscripts $AB$ show explicitly that we're talking about a *change* or *difference* from one point to another. We'll sometimes use just the symbol $V$ for potential difference, in cases where the starting point $A$ is understood. Note that potential difference, although computed from vectors, is itself a scalar quantity.

The change from potential energy to electric potential is analogous to the change we made in Chapter 20 when we introduced the electric field as the electric force *per unit charge*; similarly, the electric potential difference is the change in potential energy *per unit charge*. The reason is the same: We want to express electric properties in terms that don't involve specific charges. Table 22.1 summarizes the relations among force and field, potential energy and electric potential.

In the special case of a uniform field, Equation 22.1a reduces to

$$\Delta V_{AB} = -\vec{E} \cdot \Delta\vec{r} \qquad \text{(uniform field)} \qquad (22.1b)$$

where $\Delta\vec{r}$ is a vector from $A$ to $B$. Figure 22.1 shows the special case when the field $\vec{E}$ and path $\Delta\vec{r}$ are in opposite directions; here, Equation 22.1b gives $\Delta V_{AB} = E\,\Delta r$.

Potential difference can be positive or negative, depending on whether the path goes against or with the field. Moving a positive charge through a positive potential difference is like going uphill: Potential energy increases. Moving a positive charge through a negative potential difference is like going downhill: Potential energy decreases. The converse is true for a negative charge; even though the potential difference remains the same, the force is opposite and so the potential energy reverses sign.

We emphasize that potential difference is a property *of two points*; it doesn't depend on the path between those points. In Fig. 22.1, considering a straight path from $A$ to $B$ made the calculation of potential difference easy, but we would have found the same result—albeit with much more effort—using any path (Fig. 22.2).

---

**GOT IT? 22.1** What would happen to the potential difference $V_{AB}$ in Fig. 22.1 if (a) the electric field strength were doubled; (b) the distance $\Delta r$ were doubled; (c) the points were moved so the path lay at right angles to the field; and (d) the positions of $A$ and $B$ were interchanged?

---

## The Volt and the Electronvolt

Potential difference measures work or energy per unit charge, so its units are joules per coulomb, as shown in Table 22.1. Potential difference is important enough that this unit has a special name, the **volt** (V). To say that a car has a 12-V battery, for example, means the battery does 12 J of work on every coulomb of charge that moves between its terminals. That's generally true; multiplying the first equality in Equation 22.1a by $q$ shows that the change in potential energy of a charge $q$ as it moves through a potential difference $\Delta V$ is $\Delta U = q\,\Delta V$. If a charge $q$ "falls" freely through a potential difference $\Delta V$, it therefore gains kinetic energy given by $|q\,\Delta V|$.

We often use the term **voltage** to mean potential difference, especially in electric circuits. The two are subtly different, however, when changing magnetic fields are present; more on this in Chapter 27. Table 22.2 lists some typical potential differences in technological and natural systems.

✓**TIP** Potential Difference Involves Two Points

Potential difference is the energy per unit charge involved in moving *between those points*. This is ultimately a practical matter; if you forget it, you won't be able to hook up a voltmeter properly, or connect jumper cables safely to your car battery! This chapter's opening photo provides a dramatic illustration of this point.

Sometimes we say "the potential (or voltage) at point $P$." This is *always* a shorthand way of talking, and we *must* have in mind some other point. What we mean is the potential difference going from that other point to $P$.

In molecular, atomic, and nuclear systems it's often convenient to measure energy in **electronvolts** (eV), defined as **the energy gained by a particle carrying one elementary charge when it moves through a potential difference of 1 volt**. Since one elementary charge is $1.6 \times 10^{-19}$ C, 1 eV is $1.6 \times 10^{-19}$ J. Energy in eV is particularly easy to calculate when charge is given in elementary charges; then, with $\Delta V$ in volts, $q\,\Delta V$ gives the energy in eV. However, the eV is *not* an SI unit and should be converted to joules before calculating other quantities, like velocity.

**GOT IT? 22.2** (a) A proton (charge $e$), (b) an alpha particle (charge $2e$), and (c) a singly ionized oxygen atom each moves through a 10-V potential difference. What's the work in eV done on each?

**TABLE 22.2** Typical Potential Differences

| | |
|---|---|
| Between human arm and leg due to heart's electrical activity | 1 mV |
| Across biological cell membrane | 80 mV |
| Between terminals of flashlight battery | 1.5 V |
| Car battery | 12 V |
| Electric outlet (depends on country) | 100–240 V |
| Between long-distance electric transmission line and ground | 365 kV |
| Between base of thunderstorm cloud and ground | 100 MV |

---

**EXAMPLE 22.1** Potential Difference, Work, and Energy: X Rays

In an X-ray tube, a uniform electric field of 300 kN/C extends over a distance of 10 cm, from an electron source to a target; the field points from the target toward the source. Find the potential difference between source and target and the energy gained by an electron as it accelerates from source to target (where its abrupt deceleration produces X rays). Express the energy in both electronvolts and joules.

**INTERPRET** This problem requires first calculating the potential difference from the field and then the energy from the potential difference.

**DEVELOP** Figure 22.3 is our drawing with point $A$ the source and point $B$ the target. Equation 22.1b, $\Delta V_{AB} = -\vec{E}\cdot\Delta\vec{r}$, determines the potential difference in this uniform field. Given the potential difference, or energy per unit charge, we can find the energy gain from the magnitude of the product $q\,\Delta V$.

**EVALUATE** With the field and path in opposite directions, $\cos\theta = -1$ in the dot product, so Equation 22.1b gives

$$\Delta V_{AB} = E\,\Delta r = (300\text{ kN/C})(0.10\text{ m}) = 30\text{ kV}.$$

Although this difference is positive, a *negative* electron moves "downhill" from source to target, and thus gains kinetic energy as work gets done on it. With the charge measured in elementary charges, the product $|q\,\Delta V|$ gives this energy directly in electronvolts: $(1\text{ elementary charge } e)(30\text{ kV}) = 30\text{ keV}$. With $1.6 \times 10^{-19}$ J/eV, this is 4.8 fJ.

**ASSESS** Make sense? An electronvolt is a lot smaller than a joule, so the SI answer (in fJ $= 10^{-15}$ J) is numerically much smaller. ∎

**FIGURE 22.3** Sketch for Example 22.1.

**EXAMPLE 22.2** **Potential of a Charged Sheet**

An isolated, infinite charged sheet carries a uniform surface charge density $\sigma$. Find an expression for the potential difference from the sheet to a point a perpendicular distance $x$ from the sheet.

**INTERPRET** This is a question about calculating the potential difference from the field.

**DEVELOP** The result of Example 21.6 gives the field of a charged sheet: It's uniform, with magnitude $E = \sigma/2\epsilon_0$ and direction perpendicular to the sheet. We've drawn the sheet and a few of its field lines in Fig. 22.4. Since the field is uniform, Equation 21.1b, $\Delta V_{AB} = -\vec{E} \cdot \Delta \vec{r}$, determines the potential difference.

**EVALUATE** Moving away from the sheet means going in the direction of the field (assuming positive $\sigma$), so $\cos\theta = 1$ in the dot product, and we evaluate to get

$$V_{0x} = -Ex = -\frac{\sigma x}{2\epsilon_0}$$

Here we've used $x$ for the displacement $\Delta r$ and $V_{0x}$ for the potential difference because we're measuring from the sheet $(x = 0)$ to the point $x$.

**ASSESS** Make sense? Our result shows that the potential difference in a *uniform* field varies *linearly* with distance. Moving a positive charge away from the sheet is like going "downhill," in this case with a con-

**FIGURE 22.4** Sketch for Example 22.2. The field also extends to the left from the sheet, but we haven't drawn that.

stant slope. Give the sheet a negative charge $(\sigma < 0)$ and the potential difference changes sign; now moving a positive charge away from the sheet is going "uphill." ∎

## Curved Paths and Nonuniform Fields

If the electric field isn't uniform or the path isn't straight, then we need to use the integral in Equation 22.1a to find the potential difference. With that equation we're dividing the path into segments $d\vec{r}$, each so short that it's essentially straight with a uniform field over its length (Fig. 22.5). Then Equation 22.1b gives the potential difference $dV = -\vec{E} \cdot d\vec{r}$ across the segment, and in integrating we're summing infinitely many infinitesimal $dV$ values to get the potential difference between two points $A$ and $B$. We'll see some examples in the next section.

**GOT IT? 22.3** The figure shows three straight paths $AB$ of the same length, each in a different electric field. The field at $A$ is the same in each. Rank the potential differences $\Delta V_{AB}$.

(a)　　　　(b)　　　　(c)

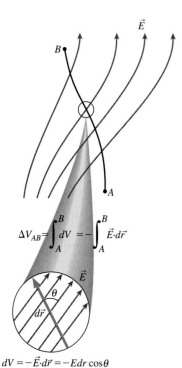

$$\Delta V_{AB} = \int_A^B dV = -\int_A^B \vec{E} \cdot d\vec{r}$$

$$dV = -\vec{E} \cdot d\vec{r} = -E\,dr\cos\theta$$

**FIGURE 22.5** The integral in Equation 22.1a is the sum of infinitely many infinitesimally small potential differences $dV$.

## 22.2 Calculating Potential Difference

Here we use Equation 22.1a to calculate the potential differences for several charge distributions. Most important is the point charge, which then provides an easy way to find the potential differences for more complicated charge distributions.

## The Potential of a Point Charge

Equation 20.3 gives the electric field of a point charge: $\vec{E} = (kq/r^2)\hat{r}$. Let's find the potential difference between two points $A$ and $B$ at distances $r_A$ and $r_B$ from a positive point charge, as shown in Fig. 22.6. We can't just multiply the distance $r_B - r_A$ by $E$ because the field varies with position. Instead we integrate, using Equation 22.1a:

$$\Delta V_{AB} = V(B) - V(A) = -\int_{r_A}^{r_B} \vec{E} \cdot d\vec{r} = -\int_{r_A}^{r_B} \frac{kq}{r^2}\hat{r} \cdot d\vec{r}$$

As we move from $A$ to $B$, the path elements are increments $dr$ in the radial direction, so we write $d\vec{r} = \hat{r}\, dr$. Then the potential difference is

$$\Delta V_{AB} = -\int_{r_A}^{r_B} \frac{kq}{r^2}\hat{r} \cdot \hat{r}\, dr = -kq \int_{r_A}^{r_B} r^{-2}\, dr$$

since the dot product of the unit vector $\hat{r}$ with itself is 1. Evaluating the integral gives

$$\Delta V_{AB} = -kq\left[-\frac{1}{r}\right]_{r_A}^{r_B} = kq\left(\frac{1}{r_B} - \frac{1}{r_A}\right) \tag{22.2}$$

Make sense? For $r_B > r_A$, the potential difference is negative, showing that a positive test charge at $r_A$ would "fall" toward $r_B$. Going the other way would require that work be done on a positive charge, as it's pushed "up" against the repulsive force of the charge $q$. Our result holds as well for $q < 0$, in which case the sign of the potential difference changes.

Although we derived Equation 22.2 for two points on the same radial line, Fig. 22.7 shows that the result holds for *any* two points in the field of a charge $q$. It doesn't matter which point is at the greater distance either; if $r_B < r_A$, Equation 22.2 still gives the correct potential difference, which then becomes positive, showing that we do work to move a positive test charge *toward* a positive $q$.

## The Zero of Potential

Only potential differences have physical significance. But it's often convenient to define a point of zero potential; we can then speak of the potential $V$ at some other point $P$, meaning the potential difference between our zero point and $P$. In this context the expression $\Delta V_{AB}$ can be written $V(B) - V(A)$. The choice for the zero of potential is usually based on mathematical or physical convenience. In electric power systems, Earth, called "ground," is usually taken as the zero of potential; in automobile electric systems, the car's body is a convenient zero point.

When we deal with isolated charges, it's convenient to take the zero of potential at infinity. Then $r_A \to \infty$ in Equation 22.2 and $1/r_A$ becomes zero. We'll omit the subscript on $r_B$ because it can be any point; then Equation 22.2 becomes

$$V_{\infty r} = V(r) = \frac{kq}{r} \quad \text{(point-charge potential)} \tag{22.3}$$

When we call this expression $V(r)$ "the potential of a point charge," we really mean that $V(r)$ is the potential difference going from a point very far from a charge $q$ to a point a distance $r$ from the charge—an interpretation that's consistent with our definition of potential difference as depending on *two* points. Because the field outside any spherically symmetric charge distribution is that of a point charge, Equation 22.3 also gives the potential outside a spherically symmetric charge distribution.

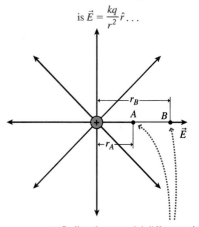

The electric field of a point charge

is $\vec{E} = \dfrac{kq}{r^2}\hat{r}$ ...

... so finding the potential difference $\Delta V_{AB}$ between $A$ and $B$ requires integration because $\vec{E}$ varies with position.

**FIGURE 22.6** Potential difference in the field of a point charge.

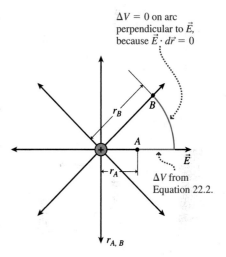

$\Delta V = 0$ on arc perpendicular to $\vec{E}$, because $\vec{E} \cdot d\vec{r} = 0$

$\Delta V$ from Equation 22.2.

**FIGURE 22.7** Potential difference is path independent, so $\Delta V_{AB}$ here still follows from Equation 22.2.

Does it bother you that potential difference can be finite over an infinite distance? The reason lies in the inverse-square dependence of the field, which drops so rapidly that the work done in moving a charge from infinity to the vicinity of a point charge remains finite. We found an analogous result in Chapter 8, where it took only a finite amount of energy to escape completely from a planet's gravitational attraction. As long as a charge distribution is finite in size—so its field at large distances falls at least as fast as $1/r^2$—it makes sense to take the zero of potential at infinity.

---

**GOT IT? 22.4** You measure a potential difference of 50 V between two points a distance 10 cm apart in the field of a point charge. If you move closer to the charge and measure the potential difference over another 10-cm interval, will it be (a) greater, (b) less, or (c) the same?

---

**EXAMPLE 22.3** **Potential and Work: At the Science Museum**

The Hall of Electricity at the Boston Museum of Science contains a large Van de Graaff generator, a device that builds up charge on a metal sphere (see Chapter 21's opening photo). The sphere has radius $R = 2.30$ m and develops a charge $Q = 640$ $\mu$C. Considering this to be a single isolated sphere, find (a) the potential at its surface, (b) the work needed to bring a proton from infinity to the sphere's surface, and (c) the potential difference between the sphere's surface and a point $2R$ from its center.

**INTERPRET** This problem is about potential differences in the field of a spherically symmetric charge distribution. In Chapter 21 we found that the field outside such a distribution is identical to that of a point charge. The term "potential" is meaningless unless we're talking about two points, so here, with a point-charge field, we interpret the question as asking us to take the zero of potential at infinity.

**DEVELOP** Because the field outside the spherical charge distribution is the same as that of a point charge, Equation 22.3, $V(r) = kQ/r$, determines the potential for $r \geq R$. We've sketched this $1/r$ potential curve in Fig. 22.8. Because the zero of potential is at infinity, we can multiply the potential at the surface by the proton's charge to get the work required to bring a proton from infinity. Finally, we can evaluate the potential difference $\Delta V_{R2R}$ from the potentials at $R$ and $2R$.

**EVALUATE** (a) Equation 22.3 gives

$$V(R) = \frac{kQ}{R} = 2.50 \text{ MV}$$

using $Q$ and $R$ given for the museum's device. (b) This 2.5-MV result is the potential difference between infinity and the sphere's surface.

Then the work needed to move a proton—1 elementary charge $e$—from infinity is 2.5 MeV or $4.0 \times 10^{-13}$ J. (c) We find the potential difference from the surface to $2R$ by subtracting the potentials at the two points:

$$\Delta V_{R2R} = V(2R) - V(R) = \frac{kQ}{2R} - \frac{kQ}{R} = -\frac{kQ}{2R} = -1.25 \text{ MV}$$

**ASSESS** Make sense? The potential difference $\Delta V_{R2R}$ is negative because we're moving away from the positively charged sphere. Our result also shows that fully half the potential difference between the sphere and infinity occurs within one radius of the sphere's surface—a consequence of the rapid $1/r^2$ decrease in the field. ∎

**FIGURE 22.8** Sketch for Example 22.3.

**Potential Difference: A High-Voltage Power Line**

A long, straight power-line wire has radius 1.0 cm and carries line charge density $\lambda = 2.6\ \mu C/m$. Assuming no other charges are present, what's the potential difference between the wire and the ground, 22 m below?

**INTERPRET** We can interpret the long, straight wire as essentially an infinitely long charge distribution with line symmetry.

**DEVELOP** In Chapter 21 we found that the field outside any line-symmetric distribution is that of a line charge, $\vec{E} = (\lambda/2\pi\epsilon_0 r)\hat{r}$, so this equation determines the power line's field. We haven't been given any explicit expression for potential differences in this field, so because the field varies with position, our plan is to apply Equation 22.1a, $\Delta V = -\int \vec{E}\cdot d\vec{r}$. We've drawn the situation in Fig. 22.9.

**EVALUATE** We evaluate the integral in Equation 22.1a over a straight path perpendicular to the wire, from its surface at $r_A$ to the ground at $r_B$:

$$\Delta V_{AB} = -\int_{r_A}^{r_B} \vec{E}\cdot d\vec{r} = -\int_{r_A}^{r_B} \frac{\lambda}{2\pi\epsilon_0 r}\hat{r}\cdot\hat{r}\,dr$$

$$= -\frac{\lambda}{2\pi\epsilon_0}\int_{r_A}^{r_B} \frac{dr}{r} = -\frac{\lambda}{2\pi\epsilon_0}\ln r\Big|_{r_A}^{r_B} \qquad (22.4)$$

$$= \frac{\lambda}{2\pi\epsilon_0}\ln\left(\frac{r_A}{r_B}\right)$$

where the last step follows because $\ln x - \ln y = \ln(x/y)$. The numbers of this example give $\Delta V = -360$ kV, a value typical of long-distance electric power transmission lines.

We want the potential difference from the wire surface, a distance $r_A$ from the axis, to the ground, a distance $r_B$ below.

**FIGURE 22.9** A long, straight power line approximated as an infinite charged rod whose field is that of a line charge.

**ASSESS** Make sense? Our result is negative because the path $AB$ goes away from a positive charge. (Mathematically, $r_A < r_B$ so the logarithm is negative.) The symbolic form of our answer shows that we can't let $r_B$ go to infinity. Physically, that's because we're assuming the charge distribution has infinite extent, so it never resembles a point charge no matter how far away we get; mathematically, it's because the field falls off more slowly than a point-charge field—namely, as $1/r$. In practice, our answer here should be modified to account for the presence of other wires and of charges drawn to the ground surface. ∎

## Finding Potential Differences Using Superposition

When we don't know the field of a charge distribution, or when the field is too complicated to integrate easily, we can find the potential using superposition. This often provides an easier approach to calculating the field as well.

Consider a charge $q$ brought from infinity to a point $P$ in the vicinity of some other charges. The superposition principle states that the electric field of a charge distribution is the sum of the fields of the individual charges that make up the distribution. Therefore the work per unit charge—that is, the potential difference—between infinity and $P$ is the sum of the potential differences associated with the individual point charges. Mathematically, we find $V(P)$ by summing Equation 22.3 over the individual point charges $q_i$:

$$V(P) = \sum_i \frac{kq_i}{r_i} \qquad (22.5)$$

where the $r_i$'s are the distances from each of the charges to the point $P$. Equation 22.5 has one enormous advantage over its counterpart for the electric field, Equation 20.4. Electric potential is a *scalar*, so the sum in Equation 22.5 is a scalar sum, with no angles, vector components, or unit vectors.

**EXAMPLE 22.5** **Discrete Charges: The Dipole Potential**

An electric dipole consists of point charges $\pm q$ a distance $2a$ apart. Find the potential at an arbitrary point $P$, and approximate for the case where the distance to $P$ is large compared with the charge separation.

**INTERPRET** We have two point charges, so this problem is based on the point-charge potential, and therefore we'll take the zero of potential at infinity.

**DEVELOP** Figure 22.10 is our drawing, showing the distances from the two charges to a point $P$. Our plan is to apply superposition, summing the potentials of the individual point charges at $P$ as determined in Equation 22.5, $V(P) = \Sigma(kq/r)$.

This is approximately the difference $r_2 - r_1$

**FIGURE 22.10** Finding the dipole potential.

**EVALUATE** Applying Equation 22.5 gives

$$V(P) = \frac{kq}{r_1} + \frac{k(-q)}{r_2} = \frac{kq(r_2 - r_1)}{r_1 r_2}$$

This is an exact result valid for any $P$. We're also asked for an approximation for large distances. If $r$ is the distance to the dipole center, as shown in Fig. 22.10, then for $r \gg a$, the quantities $r_1$, $r_2$, and $r$ are nearly the same and the term $r_1 r_2$ is very nearly $r^2$. We have to be a little more careful with the term $r_1 - r_2$ because here we're comparing

nearly equal quantities. Figure 22.10 shows that this term—the difference between the distances from the two charges to $P$—is approximately $2a \cos\theta$. So the dipole potential for $r \gg a$ becomes

$$V(r, \theta) = \frac{k(2aq)\cos\theta}{r^2} = \frac{kp\cos\theta}{r^2} \qquad \text{(dipole potential)} \qquad (22.6)$$

with $p = 2aq$ the dipole moment.

**ASSESS** Make sense? The dipole *potential* drops as $1/r^2$; earlier, we found the dipole *field* dropping as $1/r^3$. The difference of one power in $r$ occurs because the potential results from integrating the field over distance. The same is true for the point charge, whose *field* drops as $1/r^2$ while its *potential* drops as $1/r$. Note also that Equation 22.6 gives $V = 0$ when $\theta = 90°$. There, on the dipole's perpendicular bisector, a charge brought from infinity is always moving at right angles to the dipole field (recall Fig. 21.2), so no work is involved (Fig. 22.11). ∎

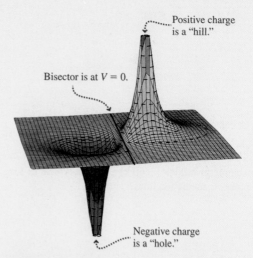

**FIGURE 22.11** 3-D plot of the dipole potential in the plane of Fig. 22.10.

**GOT IT? 22.5** The figure shows three paths from infinity to a point $P$ on a dipole's perpendicular bisector. Compare the work done in moving a charge to $P$ on each of the paths.

## Continuous Charge Distributions

We can calculate the potential of a continuous charge distribution by considering it to be made up of infinitely many infinitesimal charge elements $dq$. Each acts like a point charge and therefore contributes to the potential at some point $P$ an amount $dV$ given by $dV = k\, dq/r$, where the zero of potential is at infinity. The potential at $P$ is the sum—that is, the integral—of the contributions $dV$ from all the charge elements:

$$V = \int dV = \int \frac{k\, dq}{r} \qquad \left( \begin{array}{c} \text{potential of a continuous} \\ \text{charge distribution} \end{array} \right) \qquad (22.7)$$

where the integration is over the entire charge distribution.

**EXAMPLE 22.6**   **Potential of a Continuous Distribution:**
                   **A Charged Ring**

A total charge $Q$ is distributed uniformly around a thin ring of radius $a$. Find the potential on the ring's axis.

**INTERPRET** We interpret the ring as a continuous charge distribution.

**DEVELOP** Equation 22.7, $V = \int k\,dq/r$, gives the potential for continuous charge distributions. Figure 22.12 is our drawing, showing an $x$ axis coincident with the ring axis, with $x = 0$ at the ring center. Charge elements $dq$ in this case are small segments of the ring, and Fig. 22.12 shows that the distance $r = \sqrt{x^2 + a^2}$ is the same for all charge elements.

**EVALUATE** Equation 22.7 becomes

$$V(x) = \int \frac{k\,dq}{r} = \frac{k}{r}\int dq = \frac{kQ}{r} = \frac{kQ}{\sqrt{x^2 + a^2}} \qquad (22.8)$$

The integration here simplified because $r$ is the same for all charge elements within the ring and so comes outside the integral; the remaining integral, $\int dq$, is the total charge $Q$. Example 22.7 will present a more typical—and more challenging—integral.

**ASSESS** Make sense? At large distances $(x \gg a)$, $a^2$ is negligible and our result becomes $V(x) = kQ/x$. This is the potential of a point charge $Q$—just as we'd expect when we're so far from the ring that its size isn't significant. At the ring's center, on the other hand, $V(0) = kQ/a$. Here we're a distance $a$ from the ring, and since potential is a *scalar*, the different directions don't matter. The result is therefore the same as being a distance $a$ from a point charge $Q$.   ∎

**FIGURE 22.12** A charged ring.

**EXAMPLE 22.7**   **Potential of a Continuous Distribution:**
                   **A Charged Disk**

A charged disk of radius $a$ carries a charge $Q$ distributed uniformly over its surface. Find the potential at a point $P$ on the disk axis, a distance $x$ from the disk.

**INTERPRET** This problem, too, involves a continuous charge distribution.

**DEVELOP** Equation 22.7, $V = \int k\,dq/r$, determines the potential. But now all parts of the charge distribution aren't the same distance from $P$, so we have to set up the integral using the procedure first outlined in Chapter 9 and used more recently in Chapter 20 when we calculated the fields of continuous charge distributions. We've drawn the disk and its axis in Fig. 22.13. The preceding example suggests that we divide the disk into ring-shaped charge elements, as we've drawn in the figure. Each ring contributes a potential $dV$ given by Equation 22.8: $dV = k\,dq/\sqrt{x^2 + r^2}$, where $r$ is the ring radius. We get the potential of the entire disk by integrating over all the rings that make up the disk:

$$V(x) = \int_{\text{disk}} dV = \int_{r=0}^{r=a} \frac{k\,dq}{\sqrt{x^2 + r^2}}$$

Before we can evaluate this integral, we need to relate the charge element $dq$ and the geometric variable $r$. Here the relation involves area:

The ratio of ring area to disk area is the same as the ratio of $dq$ to the total charge $Q$. "Unwinding" a ring gives a rectangle of length $2\pi r$ and width $dr$, so the ring area is $2\pi r\,dr$. The disk area is $\pi a^2$, so $dq/Q = 2\pi r\,dr/\pi a^2$, giving $dq = (2Q/a^2)r\,dr$.

**EVALUATE** Using this result in our integral for the potential $V(x)$ gives

$$V(x) = \int_0^a \frac{2kQ}{a^2}\frac{r\,dr}{\sqrt{x^2 + r^2}} = \frac{kQ}{a^2}\int_0^a \frac{2r\,dr}{\sqrt{x^2 + r^2}}$$

Now, $2r\,dr = d(r^2) = d(x^2 + r^2)$ since $x$ is a constant with respect to the integration. The integral therefore has the form $u^{-1/2}\,du$, where $u = x^2 + r^2$, and the result is $2u^{1/2}$ or

$$V(x) = \frac{2kQ}{a^2}\sqrt{x^2 + r^2}\,\Big|_{r=0}^{r=a} = \frac{2kQ}{a^2}\left(\sqrt{x^2 + a^2} - |x|\right)$$

**ASSESS** Make sense? Figure 22.14 shows that it does. Close to the disk, the potential resembles that of an infinite sheet, changing linearly with distance (recall Example 22.2); far away, it has the $1/r$ behavior of a point-charge potential. It's only at intermediate distances—on the order of the disk radius $a$—that we really need our exact expression.   ∎

**FIGURE 22.13** A charged disk, showing a ring-shaped charge element $dq$ of radius $r$ and width $dr$.

**FIGURE 22.14** Charged-disk potential approaches that of an infinite sheet for points close to the disk, and that of a point charge far from the disk.

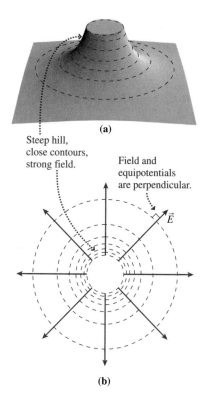

**(a)**

Steep hill,
close contours,
strong field.

Field and
equipotentials
are perpendicular.

$\vec{E}$

**(b)**

**FIGURE 22.15** A flat-topped hill (a) and its contour map (b) represent equipotentials (dashed curves) for a charged spherical shell, in a plane through the shell's center.

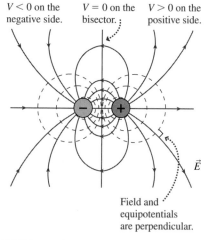

$V < 0$ on the negative side.  $V = 0$ on the bisector.  $V > 0$ on the positive side.

$\vec{E}$

Field and equipotentials are perpendicular.

**FIGURE 22.16** Equipotentials (dashed curves) and field lines for a dipole, in a plane containing the dipole. You should convince yourself that the equipotentials provide a contour map of Fig. 22.11.

## 22.3 Potential Difference and the Electric Field

It takes no work to move a charge at right angles to an electric field, so there's no potential difference between two points on a surface perpendicular to the field. Such surfaces are called **equipotentials**. Equipotentials are like contour lines on a map (Fig. 22.15). A contour is a line of constant elevation, so it takes no work to move along it. Where contours are close, elevation changes rapidly. Similarly, closely spaced equipotentials indicate a large potential difference between nearby points. That means there must be a large electric field present. Figure 22.15 might just as well represent electric potential, with closely spaced equipotentials—steep slopes on the "potential hill"—representing strong electric fields. Similarly, the equipotentials for a dipole (Fig. 22.16) describe the steep "hill" of the positive charge and the "hole" of the negative charge that we showed in Fig. 22.11. There is one difference, though: Equipotentials are surfaces in three dimensions, and when we draw them as contour lines, we're showing only the surfaces' intersections with a plane.

**GOT IT? 22.6** The figure shows cross sections through two equipotential surfaces. In both diagrams the potential difference between adjacent equipotentials is the same. Which could represent the field of a point charge? Explain.

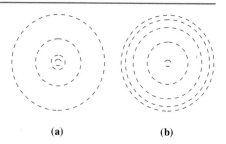

**(a)**          **(b)**

## Calculating Field from Potential

Given electric field lines, we can construct equipotentials. Conversely, given equipotentials, we can reconstruct the field by sketching field lines at right angles to the equipotentials. Specifying the potential at each point thus conveys all the information needed to determine the field.

We can quantify the relation between potential and field by considering the potential difference $dV$ between two nearby points. Suppose they're separated by a small displacement $dx$ in the $x$ direction. Then Equation 22.1b becomes $dV = -E_x \, dx$, where we handled the dot product by considering only the component of $\vec{E}$ along the displacement. Dividing through by $dx$ shows that we can write the electric-field component in the $x$ direction as $E_x = -dV/dx$. We can write similar expressions for the $y$ and $z$ components. When a function depends on more than one variable, as the potential generally does, we write derivatives with the partial derivative symbol $\partial$ instead of $d$ to indicate the rate of change with respect to only one variable. Thus we have $E_x = -\partial V/\partial x$, $E_y = -\partial V/\partial y$, and $E_z = -\partial V/\partial z$. Putting together these three components lets us write the entire electric-field vector:

$$\vec{E} = -\left(\frac{\partial V}{\partial x}\hat{\imath} + \frac{\partial V}{\partial y}\hat{\jmath} + \frac{\partial V}{\partial z}\hat{k}\right) \qquad (22.9)$$

Equation 22.9 confirms that the electric field is strong where the potential changes rapidly. The minus sign here is the same as in Equation 22.1: It says that if we move in the direction of *increasing* potential, then we're moving *against* the electric field. Equation 22.9 also shows that the units of electric field, N/C, can be written as volts per meter—a unit widely used in both science and engineering.

Because potential is a scalar, it's often easier to calculate the potential and then use Equation 22.9 to get the field. The next example shows how.

**EXAMPLE 22.8** **Field from Potential: A Charged Disk**

Use the result of Example 22.7 to find the electric field on the axis of a charged disk.

**INTERPRET** Example 22.7 gives the potential of a charged disk, so this problem is about calculating electric field from potential.

**DEVELOP** The result of Example 22.7 is a rather complicated expression for the potential on the axis of a charged disk: $V(x) = (2kQ/a^2)(\sqrt{x^2 + a^2} - |x|)$. Equation 22.9 shows that we can get the electric field by differentiating the potential with respect to all three coordinates. But here the potential depends only on $x$, so $\vec{E}$ has only an $x$ component—a fact that should also be evident from symmetry. Our plan is to apply Equation 22.9, differentiating $V(x)$ to get the field component $E_x$.

**EVALUATE** We apply Equation 22.9 to $V(x)$ to get

$$E_x = -\frac{dV}{dx} = -\frac{d}{dx}\left[\frac{2kQ}{a^2}\left(\sqrt{x^2 + a^2} - |x|\right)\right]$$

$$= \frac{2kQ}{a^2}\left(\pm 1 - \frac{x}{\sqrt{x^2 + a^2}}\right)$$

where the $+/-$ signs apply for $x > 0$ and $x < 0$, respectively. Because $V$ depends only on $x$, we wrote the total derivative $dV/dx$ rather than the partial derivative.

**ASSESS** Make sense? In Fig. 22.14 we showed that the field of a disk ought to look like that of a charged sheet close up and like a point charge far away. For $|x| \ll a$, our result gives $|E_x| = 2kQ/a^2$. With $k = 1/4\pi\epsilon_0$ and $Q/\pi a^2 = \sigma$, this is indeed the field $E = \sigma/2\epsilon_0$ of a charged sheet. You can show in Problem 71 that the case $|x| \gg a$ reduces to the field of a point charge $Q$, as expected. ∎

✓**TIP** Field and Potential

Note that the *values* of field and potential at a single point aren't related; rather, as Equation 22.9 shows, field measures the *rate of change* of potential. Field and potential are related in the same way as acceleration and velocity; their *values* are independent, but one is the rate of change of the other—although with a negative sign in the case of field and potential. In particular, the field can be zero where the potential isn't, and vice versa, as Figure 22.17 shows.

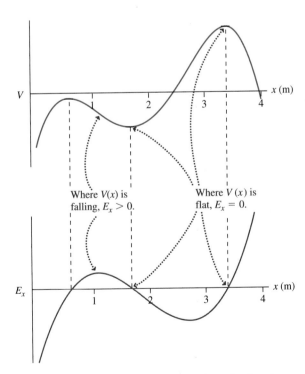

Where $V(x)$ is falling, $E_x > 0$.

Where $V(x)$ is flat, $E_x = 0$.

**FIGURE 22.17** *Values* of field and potential aren't related; rather, field is the negative rate of change of potential with position.

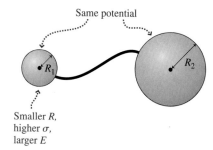

Same potential

Smaller $R$,
higher $\sigma$,
larger $E$

**FIGURE 22.18** Two conducting spheres connected by a conducting wire.

## 22.4 Charged Conductors

There's no electric field inside a conducting material in electrostatic equilibrium, and at the conductor surface there's no field component parallel to the surface. Therefore it takes no work to move a test charge on or inside a conductor—and that means **a conductor in electrostatic equilibrium is an equipotential.**

Consider an isolated, spherical conductor of radius $R$ carrying charge $Q$. Charge is distributed uniformly over its surface, so the field outside the sphere is that of a point charge. Then the potential at its surface is $V(R) = kQ/R$, as we found in Example 22.3. Now consider two widely separated spheres of different sizes. If we connect them by a thin conducting wire (Fig. 22.18), then the system constitutes a single conductor, and charge will move through the wire until both spheres are at the same potential. But since the spheres are widely separated, each still has an essentially spherical charge distribution, so $V(R) = kQ/R$ gives each sphere's potential. Because the spheres have the same potential, $kQ_1/R_1 = kQ_2/R_2$. We can write each charge as the surface area multiplied by the surface charge density: $Q = 4\pi R^2 \sigma$. Substituting for the $Q$'s in the above equation and solving for the ratio of surface charge densities then give

$$\frac{\sigma_1}{\sigma_2} = \frac{R_2}{R_1}$$

Thus the *smaller* sphere has the *higher* surface charge density. Since the electric field at a conductor surface has magnitude $E = \sigma/\epsilon_0$, the field must also be stronger at the smaller sphere.

This discussion of spherical conductors provides a qualitative description of nonspherical conductors as well. All parts of an irregularly shaped conductor must be at the same potential. Where the conductor surface curves sharply, it's like a small sphere and therefore has a higher surface charge density and a stronger electric field. In general, the field is strongest where the surface curves most sharply.

Because a conductor is an equipotential and the electric field is perpendicular to the conductor surface, equipotentials just above the surface have approximately the same shape as the surface. Because the electric field is stronger where the surface curves sharply, there must be more field lines emerging from such regions. Far from a charged conductor, on the other hand, its field resembles that of a point charge, with radial field lines and circular equipotentials. Using these limiting cases, we can sketch the approximate form for the field of an arbitrary conductor (Fig. 22.19). But this conclusion about surface charge

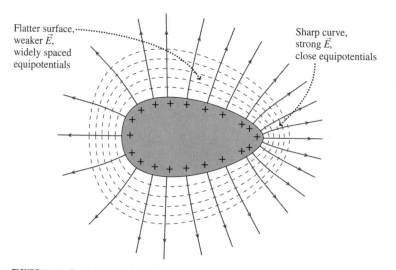

Flatter surface,
weaker $\vec{E}$,
widely spaced
equipotentials

Sharp curve,
strong $\vec{E}$,
close equipotentials

**FIGURE 22.19** Equipotentials and field of a charged conductor.

density and curvature applies strictly only to *isolated* conductors—those far from any other charges. The field of a nearby charge will modify the charge distribution on a conductor, altering its surface charge distribution (Fig. 22.20).

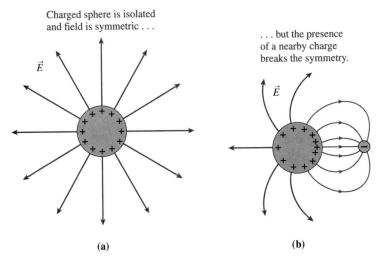

**FIGURE 22.20** Distribution of charge on a conductor changes in the presence of another charge.

APPLICATION **Corona Discharge, Pollution Control, and Xerography**

The large electric fields that develop at sharply curved conductors can cause serious problems in electric equipment. Fields stronger than 3 MN/C strip electrons from air molecules, making air a conductor. The result is a blue glow, called **corona discharge**, resulting from electrons recombining with atoms. Corona discharge causes energy loss from high-voltage transmission lines, so engineers try to avoid sharp edges on conducting structures. The photo shows corona discharge leaking current across a power-line insulator.

Corona discharge can also be useful. Pollution-control devices called **electrostatic precipitators** use a thin wire at a high negative potential to produce a strong field that ionizes gas molecules. Ions adhere to pollutant particles, which are then attracted to positively charged collecting plates. Such devices remove up to 99% of particulate pollutants from power plants and factories.

You're using corona discharge when you make a photocopy or use a laser printer. The "ink" in these devices consists of tiny plastic "toner" particles that adhere to charged regions on a special light-sensitive drum. The drum is initially charged uniformly by a corona discharge from a thin wire at high voltage; light then neutralizes all but the dark regions. Toner particles are electrically attracted to those regions and then transferred to paper, where they're melted in to make a permanent image.

The big idea here is **electric potential difference**—a measure of the energy per unit charge involved in moving charge between two points in an electric field. Because the electric field is conservative, potential difference is path independent and thus depends only on the two points in question.

## Key Concepts and Equations

Electric potential difference between points $A$ and $B$ is the negative of the line integral of the electric field over any path from $A$ to $B$:

$$\Delta V_{AB} = \frac{\Delta U_{AB}}{q} = -\int_A^B \vec{E} \cdot d\vec{r}$$

When a charge "falls" through a potential difference $\Delta V$, it gains energy $q\Delta V$.

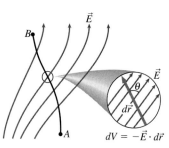

$$dV = -\vec{E} \cdot d\vec{r}$$

Potentials of charge distributions follow by summing or integrating the fields of pointlike charge elements:

$$V = \sum \frac{kq_i}{r_i} \quad \text{(discrete charges)}$$

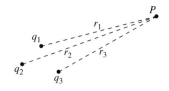

$$V = \int \frac{k\,dq}{r} \quad \text{(continuous charge distribution)}$$

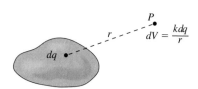

$$dV = \frac{kdq}{r}$$

In a uniform field, the potential difference becomes

$$\Delta V_{AB} = -\vec{E} \cdot \Delta \vec{r}$$

The potential in the field of a point charge is $V(r) = kq/r$, where the zero of potential is taken at infinity and $r$ is the distance from the point charge.

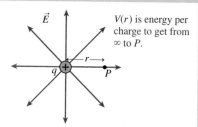

$V(r)$ is energy per charge to get from $\infty$ to $P$.

**Equipotentials** are surfaces of constant potential perpendicular to the electric field. Where equipotentials are close, the field is strong. The field component in a given direction depends on the rate at which potential changes with position; thus,

$$E_x = -\frac{dV}{dx}$$

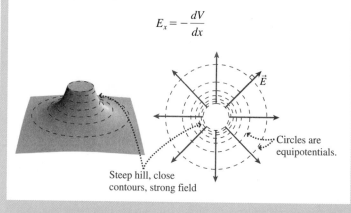

Steep hill, close contours, strong field

Circles are equipotentials.

## Cases and Uses

The dipole potential is

$$V = \frac{kp\cos\theta}{r^2}$$

where $p = qd$ is the dipole moment and the angle $\theta$ is measured from the dipole axis

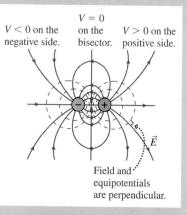

$V = 0$ on the bisector. $V < 0$ on the negative side. $V > 0$ on the positive side.

Field and equipotentials are perpendicular.

In charged conductors, the charge density is generally highest, and the field strongest, where a conductor curves sharply.

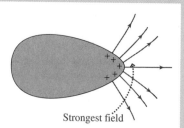

Strongest field

# For Thought and Discussion

1. Why can a bird perch on a high-voltage power line without getting electrocuted?
2. One proton is accelerated from rest by a uniform electric field, another proton by a nonuniform electric field. If they move through the same potential difference, how do their final speeds compare?
3. Would a free electron move toward higher or lower potential?
4. A proton and a positron (a particle with the electron's mass carrying charge $+e$) are accelerated through the same potential difference. How do their final energies compare? Their final speeds?
5. The electric field at the center of a uniformly charged ring is obviously zero, yet Example 22.6 shows that the potential at the center is not zero. How is this possible?
6. Must the potential be zero at any point where the electric field is zero? Explain.
7. Must the electric field be zero at any point where the potential is zero? Explain.
8. The potential is constant throughout an entire volume. What must be true of the electric field within that volume?
9. In considering the potential of an infinite flat sheet, why isn't it useful to take the zero of potential at infinity?
10. "Cherry picker" trucks for working in trees or power lines often carry electrocution hazard signs. Explain how this hazard arises and why it might be more of a danger to someone on the ground than to a worker on the truck.
11. Is it possible for equipotential surfaces to intersect? Explain.
12. Is the potential at the center of a hollow, uniformly charged spherical shell higher, lower, or the same as at the surface?
13. A solid sphere contains positive charge uniformly distributed throughout its volume. Is the potential at its center higher, lower, or the same as at the surface?
14. Two equal but opposite charges form a dipole. Describe the equipotential surface on which $V = 0$.
15. The electric potential in a region increases linearly with distance. What can you conclude about the electric field in this region?

# Exercises and Problems

## Exercises

### Section 22.1 Electric Potential Difference

16. How much work does it take to move a 50-$\mu$C charge against a 12-V potential difference?
17. The potential difference between the two sides of an ordinary electric outlet is 120 V. How much energy does an electron gain when it moves from one side to the other?
18. It takes 45 J to move a 15-mC charge from point $A$ to point $B$. What's the potential difference $\Delta V_{AB}$?
19. Show that 1 V/m is the same as 1 N/C.
20. Find the magnitude of the potential difference between two points located 1.4 m apart in a uniform 650-N/C electric field, if a line between the points is parallel to the field.
21. A charge of 3.1 C moves from the positive to the negative terminal of a 9.0-V battery. How much energy does the battery impart to the charge?
22. A proton, an alpha particle (a bare helium nucleus), and a singly ionized helium atom are accelerated through a potential difference of 100 V. Find the energy each gains.

### Section 22.2 Calculating Potential Difference

23. An electric field is given by $\vec{E} = E_0 \hat{j}$, where $E_0$ is a constant. Find the potential as a function of position, taking $V = 0$ at $y = 0$.
24. The classical picture of the hydrogen atom has a single electron in orbit a distance 0.0529 nm from the proton. Calculate the electric potential associated with the proton's electric field at this distance.
25. The potential at the surface of a 10-cm-radius sphere is 4.8 kV. What is the charge on the sphere, assuming it's distributed in a spherically symmetric way?
26. What is the maximum potential allowable on a 5.0-cm-diameter metal sphere if the electric field at the sphere's surface is not to exceed the 3-MV/m breakdown field of air?
27. A 3.5-cm-diameter isolated metal sphere carries a net charge of 0.86 $\mu$C. (a) What is the potential at the sphere's surface? (b) If a proton were released from rest at the sphere's surface, what would be its speed far from the sphere?

### Section 22.3 Potential Difference and the Electric Field

28. In a uniform electric field, equipotential planes that differ by 1.0 V are 2.5 cm apart. What is the field strength?
29. Figure 22.21 shows a plot of potential versus position along the $x$ axis. Make a plot of the $x$ component of the electric field for this situation.

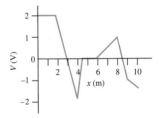

**FIGURE 22.21** Exercise 29

30. Figure 22.22 shows some equipotentials in the $x$-$y$ plane. (a) In what region is the electric field strongest? What are (b) the direction and (c) the magnitude of the field in this region?

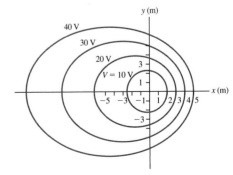

**FIGURE 22.22** Exercise 30

31. The electric potential in a region of space is given by $V = 2xy - 3zx + 5y^2$, with $V$ in volts and the coordinates in meters. If point $P$ is at $x = 1$ m, $y = 1$ m, $z = 1$ m, find (a) the potential at $P$ and (b) the $x$, $y$, and $z$ components of the electric field at $P$.

**Section 22.4 Charged Conductors**

32. (a) What is the maximum potential (measured from infinity) for the sphere of Example 22.3 before dielectric breakdown of air occurs at the sphere's surface? *Note:* Breakdown of air occurs at a field strength of 3 MV/m. (b) What is the charge on the sphere when it's at this potential?

33. The spark plug in an automobile engine has a center electrode made from wire 2.0 mm in diameter. The electrode is worn to a hemispherical shape, so it behaves approximately like a charged sphere. What is the minimum potential on this electrode that will ensure the plug sparks in air? Neglect the presence of the second electrode.

34. A large metal sphere has three times the diameter of a smaller sphere and carries three times as much charge. Both spheres are isolated, so their surface charge densities are uniform. Compare (a) the potentials and (b) the electric field strengths at their surfaces.

# Problems

35. Two points $A$ and $B$ lie 15 cm apart in a uniform electric field, with the path $AB$ parallel to the field. If the potential difference $\Delta V_{AB}$ is 840 V, what is the field strength?

36. Figure 22.23 shows a uniform electric field of magnitude $E$. Find expressions for the potential differences (a) $\Delta V_{AB}$ and (b) $\Delta V_{BC}$. (c) Use your results to determine $\Delta V_{AC}$.

**FIGURE 22.23** Problem 36

37. The electric field within the membrane separating the inside and outside of a biological cell is approximately 8.0 MV/m and is essentially uniform. If the membrane is 10 nm thick, what is the potential difference across the membrane?

38. What is the potential difference between the terminals of a battery that can impart $7.2 \times 10^{-19}$ J to each electron that moves between the terminals?

39. Electrons in a TV tube are accelerated from rest through a 25-kV potential difference. With what speed do they hit the TV screen?

40. What is the charge on an ion that gains $1.6 \times 10^{-15}$ J when it moves through a potential difference of 2500 V?

41. Two large, flat metal plates are a distance $d$ apart, where $d$ is small compared with the plate size. If the plates carry surface charge densities $\pm\sigma$, show that the magnitude of the potential difference between them is $V = \sigma d/\epsilon_0$.

42. An electron passes point $A$ moving at 6.5 Mm/s. At point $B$ the electron has come to a complete stop. Find the potential difference $\Delta V_{AB}$.

43. A 5.0-g object carries a net charge of 3.8 μC. It acquires a speed $v$ when accelerated from rest through a potential difference $V$. A 2.0-g object acquires twice the speed under the same circumstances. What is its charge?

44. Points $A$ and $B$ lie 20 cm apart on a line extending radially from a point charge $Q$, and the potentials at these points are $V_A = 280$ V and $V_B = 130$ V. Find $Q$ and the distance $r$ between $A$ and the charge.

45. A sphere of radius $R$ carries a negative charge of magnitude $Q$, distributed in a spherically symmetric way. Find the "escape speed" for a proton at the sphere's surface—that is, the speed that would enable the proton to escape to arbitrarily large distances.

46. A thin spherical shell of charge has radius $R$ and total charge $Q$ distributed uniformly over its surface. What is the potential at its center?

47. A solid sphere of radius $R$ carries a net charge $Q$ distributed uniformly throughout its volume. Find the potential difference from the sphere's surface to its center. *Hint:* Consult Example 21.1.

48. Find the potential as a function of position in an electric field given by $\vec{E} = ax\hat{i}$, where $a$ is a constant and $V = 0$ at $x = 0$.

49. A coaxial cable consists of a 2.0-mm-diameter inner conductor and an outer conductor of diameter 1.6 cm and negligible thickness (Fig. 22.24). If the conductors carry line charge densities of $\pm 0.56$ nC/m, what is the magnitude of the potential difference between them?

**FIGURE 22.24** Problem 49

50. The potential difference between the surface of a 3.0-cm-diameter power line and a point 1.0 m distant is 3.9 kV. What is the line charge density on the power line?

51. Three equal charges $q$ form an equilateral triangle of side $a$. Find the potential at the center of the triangle.

52. A charge $+Q$ lies at the origin and $-3Q$ at $x = a$. Find two points on the $x$ axis where $V = 0$.

53. Two identical charges $q$ lie on the $x$ axis at $\pm a$. (a) Find an expression for the potential at all points in the $x$-$y$ plane. (b) Show that your result reduces to the potential of a point charge for distances large compared with $a$.

54. Find the potential 10 cm from a dipole of moment $p = 2.9$ nC·m (a) on the dipole axis, (b) at 45° to the axis, and (c) on the perpendicular bisector. The dipole separation is much less than 10 cm.

55. A thin plastic rod 20 cm long carries 3.2 nC distributed uniformly over its length. (a) If the rod is bent into a ring, find the potential at its center. (b) If the rod is bent into a semicircle, find the potential at the center (i.e., at the center of the circle of which the semicircle is part).

56. A thin ring of radius $R$ carries a charge $3Q$ distributed uniformly over three-fourths of its circumference, and $-Q$ over the rest. What is the potential at the center of the ring?

57. The potential at the center of a uniformly charged ring is 45 kV, and 15 cm along the ring axis the potential is 33 kV. Find the ring's radius and its total charge.

58. The annulus shown in Fig. 22.25 carries a uniform surface charge density $\sigma$. Find an expression for the potential at an arbitrary point $P$ on its axis.

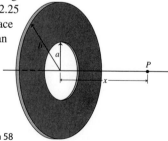

**FIGURE 22.25** Problem 58

59. (a) Find the potential as a function of position in the electric field $\vec{E} = E_0(\hat{i} + \hat{j})$, where $E_0 = 150$ V/m. Take the zero of potential at the origin. (b) Find the potential difference from the point $x = 2.0$ m, $y = 1.0$ m to the point $x = 3.5$ m, $y = -1.5$ m.

60. The potential in a certain region is given by $V = axy$, where $a$ is a constant. (a) Determine the electric field in the region. (b) Sketch some equipotentials and field lines.

61. Use Equation 22.6 to calculate the electric field on the perpendicular bisector of a dipole, and show that your result is equivalent to Equation 20.6a.

62. Use the result of Example 22.6 to determine the on-axis field of a charged ring, and verify that your answer agrees with the result of Example 20.6.

63. The electric potential in a region is given by $V = -V_0(r/R)$, where $V_0$ and $R$ are constants, $r$ is the radial distance from the origin, and the zero of potential is taken at $r = 0$. Find the magnitude and direction of the electric field in this region.

64. Two metal spheres each 1.0 cm in radius are far apart. One sphere carries 38 nC of charge, the other $-10$ nC. (a) What is the potential on each? (b) If the spheres are connected by a thin wire, what will be the potential on each once equilibrium is reached? (c) How much charge must move between the spheres in order to achieve equilibrium?

65. Two conducting spheres are each 5.0 cm in diameter and each carries 0.12 $\mu$C. They are 8.0 m apart. Determine (a) the potential on each sphere, (b) the field strength at the surface of each sphere, (c) the potential midway between the spheres, and (d) the potential difference between the spheres.

66. A 2.0-cm-radius metal sphere carries 75 nC and is surrounded by a concentric spherical conducting shell of radius 10 cm carrying $-75$ nC. (a) Find the potential difference between the shell and the sphere. (b) How would your answer change if the shell charge were changed to $+150$ nC?

67. A coaxial cable consists of a 2.0-mm-radius central wire carrying 75 nC/m and a concentric outer conductor of radius 10 mm carrying $-75$ nC/m. (a) Find the potential difference between the outer and inner conductors. (b) How would your answer change if the outer conductor were charged to $+150$ nC/m?

68. A sphere of radius $R$ carries a nonuniform but spherically symmetric volume charge density that results in an electric field in the sphere given by $\vec{E} = E_0(r/R)^2\hat{r}$, where $E_0$ is a constant. Find the potential difference from the sphere's surface to its center.

69. The potential as a function of position in a region is given by $V(x) = 3x - 2x^2 - x^3$, with $x$ in meters and $V$ in volts. Find (a) all points on the $x$ axis where $V = 0$, (b) an expression for the electric field, and (c) all points on the $x$ axis where $E = 0$.

70. A conducting sphere 5.0 cm in radius carries 60 nC. It's surrounded by a concentric spherical conducting shell of radius 15 cm carrying $-60$ nC. (a) Find the potential at the sphere's surface, taking the zero of potential at infinity. (b) Repeat for the case when the shell also carries $+60$ nC.

71. Show that the result of Example 22.8 approaches the field of a point charge for $x \gg a$. *Hint:* You'll need to apply the binomial approximation (Appendix A) to the quantity $1/\sqrt{x^2 + a^2}$.

72. The potential on the axis of a uniformly charged disk at 5.0 cm from the disk center is 150 V; the potential 10 cm from the disk center is 110 V. Find the disk radius and its total charge.

73. A uranium nucleus (mass 238 u, charge 92$e$) decays, emitting an alpha particle (mass 4 u, charge 2$e$) and leaving a thorium nucleus (mass 234 u, charge 90$e$). At the instant the alpha particle leaves the nucleus, the centers of the two are 7.4 fm apart and essentially at rest. Find their speeds when they're a great distance apart. Treat each particle as a spherical charge distribution.

74. A disk of radius $a$ carries a nonuniform surface charge density given by $\sigma = \sigma_0(r/a)$, where $\sigma_0$ is a constant. (a) Find the potential at an arbitrary point on the disk axis, a distance $x$ from the disk center. (b) Use the result of part (a) to find the electric field on the disk axis, and (c) show that the field reduces to an expected form for $x \gg a$.

75. An open-ended cylinder of radius $a$ and length $2a$ carries charge $q$ spread uniformly over its surface. Find the potential on the cylinder axis at its center. *Hint:* Treat the cylinder as a stack of charged rings, and integrate.

76. The electric field in a certain device is $\vec{E} = (a/x^4)\hat{i}$, where $a = 55$ V·m³. The device is grounded (i.e., $V = 0$) at $x = 1.3$ m. Find the potential at $x = 2.8$ m.

77. A line charge extends along the $x$ axis from $-L/2$ to $L/2$. Its line charge density is $\lambda = \lambda_0(x/L)^2$, where $\lambda_0$ is a constant. Find an expression for the potential on the $x$ axis for $x > L/2$. Check that your expression reduces to an expected result for $x \gg L$.

78. Repeat Problem 77 for the charge distribution $\lambda = \lambda_0 x/L$. *Hint:* What does this charge distribution resemble at large distances?

79. You're an engineer for a telecommunications company and you're assigned to a team designing a new antenna. The antenna is a thin rod of length $L$, and your job is to verify the lead engineer's calculation for the electric potential in a plane that perpendicularly bisects the rod when the rod carries a charge $Q$ uniformly distributed over its length. You find this equation in the lead engineer's notes:

$$V(r) = \frac{2kQ}{L} \ln\left(\frac{L}{2r} + \sqrt{1 + \frac{L^2}{4r^2}}\right)$$

along with a drawing suggesting that $r$ is the distance from the rod in the bisecting plane. Is the calculation correct?

80. You're part of a team investigating lightning. To calibrate your instruments, you need to find the electric field at the base of a vertical tower of height $h$ that carries a nonuniform line charge density $\lambda(y) = by$, where $b$ is a constant with the units C/m². You're asked for an expression giving the electric field as a function of distance $x$ from the base of the tower. You don't like to integrate vectors, so you decide to find first the potential and then the field. What expression do you get for the field $E(x)$?

# Answers to Chapter Questions

## Answer to Chapter Opening Question

That 138,000 volts is a measure of electric potential difference—the energy per unit charge involved in moving electric charge between two points. Luckily, the parasailer is in contact with only one wire, so he doesn't experience that lethal potential difference.

## Answers to GOT IT? Questions

22.1 (a) doubles; (b) doubles; (c) becomes zero; (d) reverses sign.

22.2 (a) 10 eV; (b) 20 eV; (c) 10 eV.

22.3 (c) has the highest $\Delta V_{AB}$; (b) has the lowest.

22.4 (a), because the field is stronger.

22.5 They're all equal to zero, because the potential anywhere on the perpendicular bisector of a dipole is zero, and they're all the same, because potential difference is path independent.

22.6 (a), because the equipotentials are closer nearer the center, indicating a stronger field. In (b) the field actually gets stronger farther from the center.

# 23 Electrostatic Energy and Capacitors

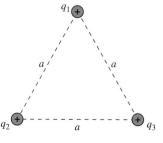

■ The lifesaving jolt of a defibrillator requires a large amount of energy delivered in a short time. Where does that energy come from?

Figure 23.1 shows three positive charges arranged to form a triangle. Stored in this charge distribution is energy, **electrostatic energy** representing the work done against the repulsive electric forces as the charges were brought into proximity. Although this example may seem trivial, its implications are not. Energy storage in configurations of electric charge is a vital aspect of natural and technological systems. The energy of chemical reactions—including metabolizing food and burning fuels—is ultimately electric energy released in the rearrangement of molecular charge distributions. Energy storage using charged conductors is essential in technologies ranging from computer memories to camera flashes to high-powered lasers.

**FIGURE 23.1** Electrostatic energy is stored in this configuration of three point charges.

## 23.1 Electrostatic Energy

Let's find the energy stored in the configuration of Fig. 23.1, assuming we start with widely separated charges and bring them in sequentially. It takes no work to bring in charge $q_1$, since there's initially no electric field. But with $q_1$ in place, bringing in $q_2$ means doing work against $q_1$'s electric field. In Chapter 22 we found that the potential of a point charge $q$ is $V = kq/r$. So the potential $V_1$ *due to* $q_1$ at the eventual location of $q_2$ is $kq_1/a$, where $a$ is the side of the triangle. That potential is the energy per unit charge; given $q_2$'s charge, the work needed to bring in $q_2$ is $W_2 = q_2V_1 = kq_1q_2/a$. Then we bring in $q_3$, which experiences the electric fields of both $q_1$ and $q_2$. This requires work done against both fields; following our reasoning for $q_2$, that work is $W_3 = kq_1q_3/a + kq_2q_3/a$. The denominator is the same in both terms because the charges form an equilateral triangle. So the total work done to assemble this charge distribution is

$$W_2 + W_3 = kq_1q_2/a + kq_1q_3/a + kq_2q_3/a.$$

Because the electric field is conservative, this work becomes the stored electrostatic energy, $U$.

Although we considered the three charges in Fig. 23.1 to be positive, our expression for work holds no matter what the signs. That means electrostatic energy can be positive or negative, depending on the sign of the work done in assembling a charge distribution. If it's negative, then it takes work to separate the charges. Although we considered assembling our charges in the order 1, 2, 3, the expression for work would have been the same no matter what the order—showing that electrostatic energy is a property of a charge distribution, not of how that distribution is assembled. Figure 23.1 is a simple metaphor for a molecule. A water molecule, for example, consists of a negatively charged oxygen atom and two positively charged hydrogen atoms. The electrostatic energy is negative and represents the energy it would take to dissociate the molecule; equivalently, it's the energy released when the water forms from individual atoms.

## 23.2 Capacitors

In technological applications, we often store energy using a pair of electrical conductors that carry equal but opposite charges. Such a system constitutes a **capacitor**. Although capacitors come in many configurations, it's easiest to analyze the **parallel-plate capacitor** consisting of two closely spaced conducting plates as shown in Fig. 23.2a. Understanding this device not only is technologically valuable, but will also provide us with deep insights into the electric field and electrostatic energy.

Initially both capacitor plates are electrically neutral. We charge the capacitor by transferring charge between the plates, building up positive charge on one plate and equal negative charge on the other. In practice, we accomplish this by connecting the capacitor to a battery, but here it's easier to imagine grabbing charge from one plate and physically moving it to the other. Charge on the plates produces an electric field between them, as shown in Fig. 23.2b. With closely spaced plates that field is essentially uniform in the region between the plates, except right near the edges. Outside, the field is so small as to be negligible. So we can approximate the parallel-plate capacitor as having a uniform field confined entirely to the region between its plates.

In Chapter 21 we showed that the electric field at the surface of a conductor is $E = \sigma/\epsilon_0$, with $\sigma$ the charge per unit area. Here we've got charge spread uniformly over the capacitor plates, so if there's charge $Q$ on a plate, then $\sigma = Q/A$, and the uniform field between the plates is $E = Q/\epsilon_0 A$. (If you think this should be doubled because there are two plates, reread the discussion around Figs. 21.26 and 21.27 to see why not.) In this uniform field, the potential difference between the plates is the product of the field and the plate separation: $V = Ed = Qd/\epsilon_0 A$.

### Capacitance

We can rewrite our expression for the potential difference between the plates of the capacitor in the form $Q = (\epsilon_0 A/d)V$. We added the parentheses to emphasize two things: First, charge is linearly proportional to potential difference and, second, the proportionality factor depends only on the constant $\epsilon_0$ and on the geometry—here the plate area and

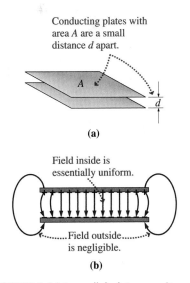

**FIGURE 23.2** (a) A parallel-plate capacitor consists of closely spaced conducting plates with area $A$ and spacing $d$. (b) Edge-on view, showing the electric field.

spacing—of the two charged conductors. This geometric factor, which gives the ratio of charge to potential difference, defines the **capacitance** of a configuration of two conductors:

$$C = \frac{Q}{V} \quad \text{(capacitance)} \tag{23.1}$$

Capacitance depends on the physical arrangement of the conductors that make up a capacitor, and it's a constant for a given capacitor. Our expression $Q = (\epsilon_0 A/d)V$ shows that the capacitance of a parallel-plate capacitor is

$$C = \frac{\epsilon_0 A}{d} \quad \text{(parallel-plate capacitor)} \tag{23.2}$$

Problems 41 and 42 show how to calculate capacitance for other configurations.

Equation 23.1 shows that the units of capacitance are coulombs/volt. This unit has its own name, the **farad** (F), in honor of the 19th-century scientist Michael Faraday. One farad is a large capacitance; practical capacitors are often measured in $\mu$F $(10^{-6}$ F$)$ or pF $(10^{-12}$ F$)$. Incidentally, Equation 23.2 shows that the units of $\epsilon_0$ may be expressed as F/m.

## Energy Storage in Capacitors

Imagine moving a small charge $dQ$ from the negative to the positive plate of a capacitor when there's a potential difference $V$ between the plates. Since potential difference is work per unit charge, this takes work $dW = V\,dQ$. The additional charge increases the electric field in the capacitor, resulting in an increase $dV$ in the potential difference. Equation 23.1 shows that the increases $dQ$ and $dV$ are related by $dQ = C\,dV$. So the work involved in moving the charge $dQ$ between the plates becomes $dW = V\,dQ = CV\,dV$.

If we start with the capacitor uncharged and then begin transferring charge between the plates, we'll need to do increasing amounts of work because the electric field and the potential difference increase continuously with the charge we've already transferred. The total work involved will be the sum of all the $dW$ values. In this case, where the potential difference increases continuously, that sum becomes an integral:

$$W = \int dW = \int_0^V CV\,dV = \tfrac{1}{2}CV^2$$

where the last step follows because the integral has the familiar form $\int x\,dx = \tfrac{1}{2}x^2$. The work we do in charging the capacitor is stored as potential energy $U$, so

$$U = \tfrac{1}{2}CV^2 \quad \text{(energy in a capacitor)} \tag{23.3}$$

We can measure potential difference $V$ directly, with a voltmeter, so it's more useful to express the energy in terms of voltage rather than charge.

✓ **TIP** Charged but Neutral

A "charged capacitor" means a capacitor with one plate positive and the other negative; overall, the capacitor remains neutral (Fig. 23.3). The charge $Q$ refers to the *magnitude* of the charge on either plate—*not* to the capacitor's net charge, which is zero.

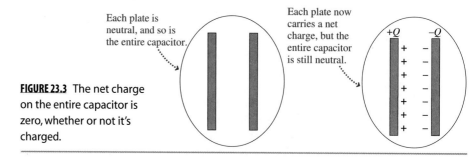

Each plate is neutral, and so is the entire capacitor.

Each plate now carries a net charge, but the entire capacitor is still neutral.

**FIGURE 23.3** The net charge on the entire capacitor is zero, whether or not it's charged.

**EXAMPLE 23.1**  **Capacitance, Charge, and Energy:
A Parallel-Plate Capacitor**

A capacitor consists of two circular metal plates of radius $R = 12$ cm, separated by $d = 5.0$ mm. (a) Find its capacitance. Find (b) the charge on the plates and (c) the stored energy when the capacitor is connected to a 12-V battery.

**INTERPRET** Because the two circular metal plates have an area much larger than their separation distance, we can treat the field between them as uniform. So we identify the configuration as a parallel-plate capacitor.

**DEVELOP** We've sketched the capacitor in Fig. 23.4. Equation 23.2, $C = \epsilon_0 A/d$, determines the capacitance for part (a) from the separation distance and plate area $(A = \pi R^2)$. For parts (b) and (c) the 12-V battery maintains a 12-V potential difference across the capacitor. Knowing that voltage and the capacitance, we can find the capacitor's charge from Equation 23.1, $C = Q/V$, and the stored energy from Equation 23.3, $U = \frac{1}{2}CV^2$.

**EVALUATE** We first solve part (a) for the capacitance:

$$C = \frac{\epsilon_0 A}{d} = \frac{\epsilon_0 \pi R^2}{d} = 80 \text{ pF}$$

**FIGURE 23.4** Sketch for Example 23.1.

For part (b) the definition of capacitance then gives

$$Q = CV = (80 \text{ pF})(12 \text{ V}) = 960 \text{ pC},$$

or just under 1 nC. The corresponding stored energy follows for part (c) from

$$U = \tfrac{1}{2}CV^2 = \tfrac{1}{2}(80 \text{ pF})(12 \text{ V})^2 = 5760 \text{ pJ},$$

or about 5.8 nJ.

**ASSESS** Make sense? At 80 pF, this is a pretty small capacitor, so no wonder the charge and energy are measured in nano-units (nC and nJ). ∎

## 23.3 Using Capacitors

Capacitors are essential in modern technology. They range from the billions of 25-fF $(10^{-15}$ F$)$ capacitors that store individual bits of information in your computer's memory, to millifarad-range capacitors that smooth the 60-Hz AC power to provide steady current to your stereo amplifier, to ultracapacitors measuring hundreds of farads that store electric energy for short bursts of power in some gas–electric hybrid and fuel-cell vehicles (Fig. 23.5).

**FIGURE 23.5** Typical capacitors. The large unit at left is a 220-mF electrolytic capacitor, while the flat disk-shaped capacitor at right has the greatest capacitance, 1 F. The smaller capacitors range from 43 pF to 2.2 mF.

### Practical Capacitors

Equation 23.2 shows that the way to achieve a large capacitance is with large plate area and small spacing. That's true in general, whether or not a capacitor has the parallel-plate geometry of Fig. 23.2a. Inexpensive capacitors are often made from two long strips of aluminum foil separated by thin plastic insulation. This foil "sandwich" is rolled into a compact cylinder, wires are attached, and the whole thing is dipped in a protective coating. Very large capacitances are achieved with electrolytic capacitors, in which a thin insulating layer develops chemically under the influence of the applied voltage. Capacitors are among the hardest components to fabricate on integrated-circuit chips, but small-capacitance units can be made by alternating conductive material with an insulating layer.

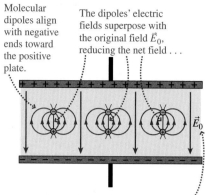

Molecular dipoles align with negative ends toward the positive plate.

The dipoles' electric fields superpose with the original field $\vec{E}_0$, reducing the net field . . .

$\vec{E}_0$

. . . charge $Q$ stays the same, so the reduced field $\vec{E} = \vec{E}_0/\kappa$ results in a lower potential $V = V_0/\kappa$ and therefore larger capacitance $C = \kappa C_0$.

**FIGURE 23.6** A capacitor with a dielectric.

Our analysis of the parallel-plate capacitor assumed air between the plates. But most capacitors have solid insulating materials, or **dielectrics**, that contain molecular dipoles but no free charge. In Section 20.5 we showed how the alignment of molecular dipoles in a dielectric reduces the field in the material. In a capacitor, the effect is to reduce also the potential difference $V$ between the plates (Fig. 23.6). The factor by which the field and potential difference decrease is the **dielectric constant $\kappa$** (Greek kappa). For a given charge $Q$, decreased potential difference means a larger capacitance $C = Q/V$. Thus a parallel-plate capacitor with a dielectric between its plates has capacitance

$$C = \kappa \frac{\epsilon_0 A}{d} \quad \text{(parallel-plate capacitor with dielectric)} \quad (23.4)$$

Most materials have dielectric constants between about 2 and 10; see Table 23.1. Some tantalum compounds have much higher values of $\kappa$, making this rare element a crucial material in today's electronic age.

Another practical consideration is a capacitor's **working voltage**, the maximum safe potential difference beyond which there's a risk of dielectric breakdown. For a given material, breakdown occurs at a fixed electric field; for air it's 3 MV/m, while polyethylene breaks down at 50 MV/m. In a parallel-plate capacitor the field is $E = V/d$, so the smaller the spacing, the lower the allowed voltage before breakdown. Thus there's a trade-off between large capacitance (small $d$) and high working voltage (large $d$). Large-capacitance, high-voltage capacitors are expensive!

**TABLE 23.1** Properties of Some Common Dielectrics

| Dielectric Material | Dielectric Constant | Breakdown Field (MV/m) |
|---|---|---|
| Air | 1.0006 | 3 |
| Aluminum oxide | 8.4 | 670 |
| Glass (Pyrex) | 5.6 | 14 |
| Paper | 3.5 | 14 |
| Plexiglas | 3.4 | 40 |
| Polyethylene | 2.3 | 50 |
| Polystyrene | 2.6 | 25 |
| Quartz | 3.8 | 8 |
| Tantalum oxide | 26 | 500 |
| Teflon | 2.1 | 60 |
| Water | 80 | depends on time and purity |

**EXAMPLE 23.2  Finding Charge and Energy: Which Capacitor?**

A 100-$\mu$F capacitor has a working voltage of 20 V, while a 1.0-$\mu$F capacitor is rated at 300 V. Which can store more charge? More energy?

**INTERPRET** This problem involves the charge and energy stored in capacitors, now constrained by the working voltage.

**DEVELOP** Equation 23.1, in the form $Q = CV$, determines the charge, and Equation 23.3, $U = \frac{1}{2}CV^2$, determines the stored energy. Setting the voltage to the working voltage will give the maximum charge and energy.

**EVALUATE** For the charges on the two capacitors, we get from Equation 23.1,

$$Q_{100\,\mu F} = CV = (100\ \mu F)(20\ V) = 2.0\ mC$$

and, similarly, $Q_{1\,\mu F} = 0.30$ mC. The energies follow from Equation 23.3:

$$U_{100\,\mu F} = \tfrac{1}{2}CV^2 = \tfrac{1}{2}(100\ \mu F)(20\,V)^2 = 20\ mJ$$

and, similarly, $U_{1\,\mu F} = \frac{1}{2}(1.0\ \mu F)(300\ V)^2 = 45$ mJ. So the 100-$\mu$F capacitor stores more charge, but the 1-$\mu$F capacitor holds more energy.

**ASSESS** Make sense? The larger capacitor holds more *charge*, despite its lower working voltage. But the *energy* depends on $V^2$, so the smaller capacitor wins out because of its much higher working voltage. ∎

**GOT IT? 23.1** You need to replace a capacitor with one that can store more energy. Which will give you greater energy increase: (a) a capacitor with twice the capacitance and the same working voltage as the old one, or (b) a capacitor with the same capacitance but twice the working voltage?

## Connecting Capacitors: Parallel

Connecting capacitors together lets us achieve capacitance or working voltage that might not be available in a single capacitor. There are two simple ways to connect capacitors and other electronic components: **parallel** and **series** (Fig. 23.7).

With capacitors in parallel, a conducting wire connects together the top plates of each capacitor and another connects the bottom plates. Therefore the top plates are both at the same potential and so are the bottom plates. That means **two capacitors in parallel have the same potential difference between their plates.** We'll find that's always true for electric components in parallel. We want the equivalent capacitance of the parallel combination, meaning the ratio of the total charge on both capacitors to their common voltage $V$. Given the definition $C = Q/V$, we can write $Q_1 = C_1 V$ and $Q_2 = C_2 V$. So the total charge is $Q = Q_1 + Q_2 = C_1 V + C_2 V$. The equivalent capacitance is then $C = Q/V$ or $C = C_1 + C_2$, a result that generalizes to any number of parallel capacitors:

$$C = C_1 + C_2 + C_3 + \cdots \quad \text{(parallel capacitors)} \quad (23.5)$$

In other words, capacitors in parallel add. You can understand this result physically by considering two parallel-plate capacitors with equal spacing. Connecting them in parallel amounts to adding their plate areas, giving a larger capacitance. Because parallel capacitors experience the same voltage, the working voltage of a parallel combination is the voltage of whichever capacitor has the lowest working voltage.

## Connecting Capacitors: Series

Figure 23.8 is a closer look at the series combination of Fig. 23.7b, showing what happens if we put charge $+Q$ on the top plate of $C_1$ and charge $-Q$ on the lower plate of $C_2$. Each of these charged plates pulls the opposite charge to the other plate of the individual capacitors—and that means **two capacitors in series carry the same charge.** But now the voltages can be different; they're given by Equation 23.1 as $V_1 = Q/C_1$ and $V_2 = Q/C_2$, where $Q$ is the common charge. Since the electric fields in the two capacitors point the same way, the voltage across the series combination is $V = V_1 + V_2 = Q/C_1 + Q/C_2$. Dividing through by $Q$ gives $V/Q$, which is the inverse of the equivalent capacitance $Q/V$. Thus

$$\frac{1}{C} = \frac{1}{C_1} + \frac{1}{C_2}$$

This result generalizes to any number of capacitors, showing that capacitors in series add reciprocally:

$$\frac{1}{C} = \frac{1}{C_1} + \frac{1}{C_2} + \frac{1}{C_3} + \cdots \quad \text{(series capacitors)} \quad (23.6a)$$

With two capacitors it's straightforward to invert Equation 23.6a to get

$$C = \frac{C_1 C_2}{C_1 + C_2} \quad (23.6b)$$

Either way, the combined capacitance is less than the individual capacitances. You can make sense of this by considering parallel-plate capacitors with equal plate areas. Put them in series, and you're increasing the effective spacing and thus decreasing the capacitance. In series, each capacitor gets less than the full voltage $V$; just how much depends on the ratio of the capacitances. That means the working voltage of the series combination is higher than that of any individual capacitor.

---

**GOT IT? 23.2** You have two identical capacitors with capacitance $C$. How would you connect them to get equivalent capacitances (a) $2C$ and (b) $\frac{1}{2}C$? Which combination would have the higher working voltage?

---

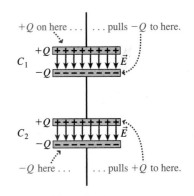

The two top plates are at the same potential . . .
. . . as are the two bottom plates.

So two capacitors in parallel have the same potential difference.

**(a)**

$C_1$

$C_2$

**(b)**

**FIGURE 23.7** Connecting capacitors: (a) parallel and (b) series. ‡ is the standard circuit symbol for a capacitor.

$+Q$ on here . . .  . . . pulls $-Q$ to here.

$C_1$  $+Q$  $\vec{E}$  $-Q$

$C_2$  $+Q$  $\vec{E}$  $-Q$

$-Q$ here . . .  . . . pulls $+Q$ to here.

**FIGURE 23.8** Capacitors in series carry the same charge.

**EXAMPLE 23.3** Equivalent Capacitance: Connecting Capacitors

Find the equivalent capacitance of the combination shown in Fig. 23.9a. If the maximum voltage to be applied between points $A$ and $B$ is 100 V, what should be the working voltage of $C_1$?

(a)

$C_2$ and $C_3$ form the parallel equivalent $C_{23}$.

(b)

$C_1$ and $C_{23}$ form the series equivalent $C_{123}$.

(c)

**FIGURE 23.9** Finding the equivalent capacitance.

**INTERPRET** This problem is about an electric circuit—in this case, an assemblage of three capacitors.

**DEVELOP** To handle such circuit problems, we find combinations of series and parallel components, and then simplify the circuit by treating each combination as a single component. Here all components are capacitors, and each time we compute an equivalent capacitance for two capacitors, we'll redraw the circuit with the new equivalent capacitance. We begin by noting that $C_2$ and $C_3$ are in parallel, so the equivalent capacitance is given by Equation 23.5: $C_{23} = C_2 + C_3 = 4.0\ \mu\text{F}$. In Fig. 22.9b we've redrawn the original circuit showing this combina-

tion of the two individual capacitors. Next we see that $C_1$ is in series with $C_{23}$, so their equivalent capacitance follows from Equation 23.6b:

$$C_{123} = \frac{C_1 C_{23}}{C_1 + C_{23}} = \frac{(12\ \mu\text{F})(4.0\ \mu\text{F})}{12\ \mu\text{F} + 4.0\ \mu\text{F}} = 3.0\ \mu\text{F}$$

We've redrawn the circuit again with this equivalent capacitance (Fig. 23.9c).

Now we want the working voltage of $C_1$. Our plan is to go backward from the simple circuit of Fig. 23.9c until we have enough information to find the voltage on $C_1$. With the voltage $V_{AB}$ across $A$ and $B$ known, we can calculate the charge on $C_{123}$ using Equation 23.1: $Q_{123} = C_{123} V_{AB}$. But $C_{123}$ is the series combination of $C_1$ and $C_{23}$, and we know that series capacitors carry the same charge—and that's the charge of their equivalent capacitance. So $Q_1 = Q_{123}$, and we can apply Equation 23.1 again, this time to $C_1$ alone, to get $V_1 = Q_1/C_1$.

**EVALUATE** With $V_{AB} = 100$ V across the combination $C_{123}$, the corresponding charge is $Q_{123} = C_{123} V_{AB} = (3.0\ \mu\text{F})(100\ \text{V}) = 300\ \mu\text{C}$. Because $Q_1 = Q_{123}$, the charge on $C_1$ is also $300\ \mu\text{C}$. We then substitute this into $V_1 = Q_1/C_1$ to get $V_1 = (300\ \mu\text{C})/(12.0\ \mu\text{F}) = 25$ V, the minimum working voltage for $C_1$.

**ASSESS** Make sense? Since $C_1$ is in series with $C_{23}$, it doesn't "feel" the full 100 V applied across $AB$, so its working voltage can be lower. And because its capacitance is *larger*, its share of the voltage is *smaller*, thanks to the relation $V = Q/C$ and the fact that series capacitors carry the *same charge*.

✓**TIP** Series and Parallel

Parallel components have their two ends connected directly together; series components are connected in such a way that if you move through one component, the only place you can go is into the next. In Fig. 23.9a, $C_2$ and $C_3$ are definitely in parallel. But $C_1$ isn't in series with either of the other single capacitors because after $C_1$, the circuit splits and you could go into either $C_2$ or $C_3$. Equations 23.5 and 23.6 apply *only* to true parallel and series combinations. $C_1$ *is* in series with the combination $C_{23}$, so we could apply Equation 23.6b in analyzing Fig. 23.9b.

∎

---

**APPLICATION** Bursts of Power

As San Francisco's BART trains decelerate, their kinetic energy is stored as electric energy in an ultracapacitor. The stored energy is then used to accelerate the train. This system saves BART some 320 megawatt-hours of energy each year.

Capacitors are excellent devices for short-term storage and supply of electric energy because they can deliver their stored energy very quickly—much more quickly than a battery that might contain a lot more total energy.

When you use a flash camera, you have to wait a few seconds before the flash is ready to fire again. That's because the flash requires a power—energy per time—far greater than the camera's battery could supply. So the battery's energy gradually charges a capacitor, whose energy is then dumped abruptly to power the short burst of the flash lamp. It takes a while to recharge the capacitor before it's ready again. Much the same thing happens in a defibrillator, which delivers several hundred joules to restore a heart's normal beating. Again, the energy is stored in capacitors, which can discharge in a few milliseconds. On a much larger scale, whole rooms full of capacitors store the energy that drives nanosecond laser pulses pouring millions of joules into tiny targets in experiments aimed at making nuclear fusion a viable energy source. And increasingly, so-called ultracapacitors supply extra energy for bursts of power in machinery from amusement park rides to mass-transit trains to gas–electric hybrid cars.

## 23.4 Energy in the Electric Field

What's the difference between a charged and an uncharged capacitor? Not the total charge, which is zero, but the arrangement of charge. And with the charge arrangement comes stored energy. Where, exactly, is this energy? We can ask the same question for the triangular charge distribution we assembled in Fig. 23.1. The individual charges didn't change, but their arrangement did. With the new arrangement came energy, but where is that energy?

In both cases something else has changed: the electric field. There's no electric field in the uncharged capacitor, but once charged, there's a field between the plates. The triangular distribution started with three isolated point-charge fields and ended with a more complex field. So where's the stored energy? The answer is the same in both cases: The energy is in the electric field. In fact, *every* electric field represents stored energy. Rearrange the charges to their original state—by discharging the capacitor or letting the three point charges fly apart—and you get back that energy. Because electric forces govern much of the behavior of everyday matter, many seemingly different forms of energy are actually electric. Burn gasoline or metabolize food, and you're rearranging the charge distributions we call molecules into new configurations whose electric fields contain less energy.

If electric fields store energy, then the amount of stored energy should depend on the field. Since field strength can vary with position, we specify the **energy density**, or energy stored per unit volume. For a capacitor, we can use Equation 23.1 in the form $V = Q/C$ to write the stored energy $U = \frac{1}{2}CV^2$ as $U = Q^2/2C$. For a parallel-plate capacitor, Equation 23.2 gives $C = \epsilon_0 A/d$, so the stored energy can be written as $U = Q^2 d/2\epsilon_0 A$. This energy is associated with the uniform electric field inside the capacitor, where it occupies a volume $Ad$. So the energy density is $U/Ad = Q^2/2\epsilon_0 A^2$. We can rewrite this in terms of the field, which we found to be $E = Q/\epsilon_0 A$. Then $Q = \epsilon_0 AE$, and the energy density becomes

$$u_E = \tfrac{1}{2}\epsilon_0 E^2 \qquad \text{(electric energy density)} \qquad (23.7)$$

Although we derived Equation 23.7 for the uniform field of a parallel-plate capacitor, it is in fact universal. Anywhere there's an electric field, there's also stored energy with density, in J/m$^3$, given by $\frac{1}{2}\epsilon_0 E^2$. The deep significance of Equation 23.7 lies in this statement: *Every* electric field represents stored energy. The energy that drives much of the physical universe, from everyday events here on Earth to happenings in distant galaxies, results from the release of energy stored in electric fields.

---

**EXAMPLE 23.4** **Electric Energy: A Thunderstorm**

Typical electric fields in thunderstorms average around $10^5$ V/m. Consider a cylindrical thundercloud with height 10 km and diameter 20 km, and assume a uniform electric field of $1 \times 10^5$ V/m. Find the electric energy contained in this storm.

**INTERPRET** This problem is about stored electric energy.

**DEVELOP** Since the field and hence the energy density are uniform, our plan is to find the energy density and then multiply it by the storm's cylindrical volume to calculate the total electric energy. We'll use Equation 23.7, $u_E = \frac{1}{2}\epsilon_0 E^2$, for the energy density.

**EVALUATE** The energy density is
$$u_E = \tfrac{1}{2}\epsilon_0 E^2 = \tfrac{1}{2}\epsilon_0 (1 \times 10^5 \text{V/m})^2 = 4.4 \times 10^{-2} \text{J/m}^3.$$

The cylindrical storm has volume

$$V = \pi r^2 h = \pi(10 \text{ km})(10\text{km})^2 = 3.1 \times 10^{12} \text{m}^3.$$

Multiplying energy density by volume gives the total stored energy:

$$U = u_E V = (4.4 \times 10^{-2} \text{J/m}^3)(3.1 \times 10^{12} \text{ m}^3) = 140 \text{ GJ}$$

**ASSESS** Make sense? A gallon of gasoline contains about 0.1 GJ (see Appendix C), so the thunderstorm stores the energy equivalent of about 1400 gallons of gasoline. That's not a whole lot for such a vast volume, showing that the energy density of macroscopic electric fields can't compare with the electric energy density locked into the molecular structure of a fuel. You'll never see cars running on the energy stored in atmospheric electric fields! ∎

---

When the electric field is uniform, as in our thunderstorm, the total energy is the product of energy density and volume. But when the field changes with position, we call on the power of calculus. Consider a small volume $dV$, so small that the electric field is essentially uniform over this volume. The stored energy is then $dU = u_E dV = \frac{1}{2}\epsilon_0 E^2 \, dV$. The total energy in the field is the sum—here the integral—of all the $dU$ values:

$$U = \tfrac{1}{2}\epsilon_0 \int E^2 \, dV \qquad (23.8)$$

Because Equation 23.8 gives the energy stored in an electric field, it also represents the work done in assembling the charge distribution resulting in that field. The next example illustrates this point.

---

**EXAMPLE 23.5** **Work and Energy: A Shrinking Sphere**

A sphere of radius $R_1$ carries charge $Q$ distributed uniformly over its surface. How much work does it take to compress the sphere to a smaller radius $R_2$?

**INTERPRET** This problem involves the difference in stored electric energy of a charge distribution, which we know is equal to the work done in arranging the distribution. Here we start with a charged sphere already assembled, and rearrange the charge by shrinking the sphere to a smaller radius.

**DEVELOP** We have spherical symmetry, so the field and thus the stored energy outside the original radius $R_1$ don't change. Therefore we need to find the energy stored in the new field created when the sphere shrinks. Figure 23.10 is our sketch of the situation before and after the sphere shrinks. Here the field varies with position, so Equation 23.8,

$$U = \tfrac{1}{2}\epsilon_0 \int E^2 \, dV,$$

gives the stored energy. Our plan is to evaluate the field in the region $R_2 < r < R_1$ and use the result in Equation 23.8. Given the spherical symmetry, the new field is a

**(a)**

The work involved in shrinking the sphere ends up as energy in the electric field here.

**FIGURE 23.10** (a) A charged sphere and its electric field. (b) Shrinking the sphere creates field and energy in the region $R_2 < r < R_1$.

**(b)**

point-charge field: $E = kQ/r^2$. To use Equation 23.8 we need an appropriate volume element $dV$. With spherical symmetry, Fig. 23.11 shows that we can use a thin spherical shell of volume $dV = 4\pi r^2 \, dr$. Then Equation 23.8 becomes

$$U = \tfrac{1}{2}\epsilon_0 \int E^2 \, dV = \tfrac{1}{2}\epsilon_0 \int_{R_2}^{R_1} \left(\frac{kQ}{r^2}\right)^2 4\pi r^2 \, dr = \frac{kQ^2}{2} \int_{R_2}^{R_1} r^{-2} \, dr$$

where we substituted $1/4\pi k$ for $\epsilon_0$.

**FIGURE 23.11** A thin spherical shell has volume $dV = 4\pi r^2 \, dr$.

Surface area is $4\pi r^2$ ...

... and thickness is $dr$

**EVALUATE** The integral is $\int r^{-2} \, dr = \dfrac{r^{-1}}{-1} = -\dfrac{1}{r}$, so

$$U = \frac{kQ^2}{2}\left(-\frac{1}{r}\right)\Bigg|_{R_2}^{R_1} = \frac{kQ^2}{2}\left(\frac{1}{R_2} - \frac{1}{R_1}\right)$$

**ASSESS** Make sense? Here $R_2 < R_1$, so the stored energy is positive and indicates that this much work had to be done to shrink the sphere. Of course: The sphere carries charge of the same sign, and shrinking it moves that charge closer together, against the repulsive electric force. Letting $R_1$ go to infinity gives the work needed to assemble a spherical surface charge distribution. Putting $R_2 = 0$ makes the work and therefore the stored energy infinite—suggesting that the notion of a point charge is an impossible idealization. Problem 71 explores some implications of this result in the theory of elementary particles. ∎

---

**GOT IT? 23.3** You're at a point $P$ a distance $a$ from a point charge $+q$. You then place a point charge $-q$ a distance $a$ on the opposite side of $P$ as shown. What happens to (a) the electric field strength and (b) the electric energy density at $P$? (c) Does the total electric energy $U = \int u_E dV$ of the entire field increase, decrease, or remain the same?

## Big Picture

The big idea here came at the end: *All* electric fields represent stored energy. This energy is associated with the work needed to assemble a distribution of electric charge, and may be negative or positive.

You do positive work to assemble this charge distribution ...

... and therefore the stored electric energy $U$ is positive.

$U > 0$

You do negative work to assemble this charge distribution ...

... and therefore the stored electric energy $U$ is negative.

$U < 0$

## Key Concepts and Equations

The **energy density** in an electric field $E$ is $u_E = \frac{1}{2}\epsilon_0 E^2$.

Integrating over volume gives the total electric energy $U$ stored in a field:

$$U = \int u_E \, dV.$$

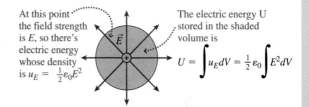

At this point the field strength is $E$, so there's electric energy whose density is $u_E = \frac{1}{2}\varepsilon_0 E^2$

The electric energy U stored in the shaded volume is

$$U = \int u_E dV = \frac{1}{2}\varepsilon_0 \int E^2 dV$$

## Cases and Uses

A **capacitor** is a pair of insulated conductors used to store electric energy. **Capacitance** is the ratio of charge to potential difference:

$$C = Q/V$$

For a parallel-plate capacitor:

$$C = \epsilon_0 A/d$$

Potential difference $V$

$Q$   $A$    Spacing $d$

$-Q$

The energy stored in a capacitor is $U = \frac{1}{2}CV^2$.

Capacitors in parallel add: $C = C_1 + C_2$.

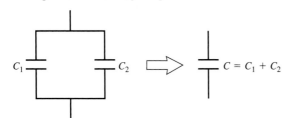

$C = C_1 + C_2$

Capacitors in parallel have the same voltage.

Capacitors in series add reciprocally: $\dfrac{1}{C} = \dfrac{1}{C_1} + \dfrac{1}{C_2}$.

$C_1$

$C_2$

$\dfrac{1}{C} = \dfrac{1}{C_1} + \dfrac{1}{C_2}$

Capacitors in series have the same charge.

A **dielectric** between capacitor plates increases the capacitance, as determined by the **dielectric constant** $\kappa$ of the material: $C \rightarrow \kappa C_0$.

# For Thought and Discussion

1. Two positive point charges are initially infinitely far apart. Is it possible, using only a finite amount of work, to move them until they're located a small distance $d$ apart?

2. How does the energy density a certain distance from a negative point charge compare with the energy density the same distance from a positive point charge of equal magnitude?

3. A dipole consists of two equal but opposite charges. Is the total energy stored in the field of the dipole zero? Why or why not?

4. Charge is spread over the surface of a balloon. The balloon is then allowed to expand. What happens to the energy of the electric field?

5. Why doesn't the superposition principle hold for electric-field energy densities? That is, if you double the field strength at some point, why don't you simply double the energy density as well?

6. A student argues that the total energy associated with the electric field of a charged sphere must be infinite because its field extends throughout an infinite volume. Critique this argument.

7. A capacitor is said to carry a charge $Q$. What's the net charge on the entire capacitor?

8. Does the capacitance of a capacitor describe the maximum amount of charge it can hold, in the same way that the capacity of a bucket describes the maximum amount of water it can hold? Explain.

9. A capacitor of capacitance $C$ is charged to a potential difference $V$ and carries charge $Q$. Why isn't the stored energy given simply by $CV^2$? After all, the work required to move a charge $Q$ against a potential difference $V$ is $QV$, and $C = Q/V$, so $Q = CV$.

10. Is a force needed to hold the plates of a charged capacitor in place? Explain.

11. A solid conducting slab is inserted between the plates of a capacitor, not touching either plate. Does the capacitance increase, decrease, or remain the same?

12. Two capacitors are storing equal amounts of energy, yet one has twice the capacitance of the other. How do their voltages compare?

13. An air-insulated parallel-plate capacitor is connected to a battery that imposes a potential difference $V$ across the capacitor. If a dielectric slab is inserted between the capacitor plates, what happens to (a) the potential difference, (b) the capacitor charge, and (c) the capacitance?

# Exercises and Problems

## Exercises

### Section 23.1 Electrostatic Energy

14. Four 50-$\mu$C charges are brought from far apart onto a line where they're spaced at 2.0-cm intervals. How much work does it take to assemble this charge distribution?

15. Three point charges $+q$ and a fourth, $-\frac{1}{2}q$, are assembled to form a square of side $a$. Find an expression for the electrostatic energy of this charge distribution.

16. Repeat Exercise 15 for the case when the fourth charge is $-q$.

17. If the three particles in Fig. 23.1 have identical charge $q$ and mass $m$, and if they're released from their positions on the triangle, what speed $v$ will they have when they're far away?

18. To a very crude approximation, a water molecule consists of a negatively charged oxygen atom and two "bare" protons, as shown in Fig. 23.12. Calculate the electrostatic energy of this configuration, which is therefore the magnitude of the energy released in forming this molecule from widely separated atoms. *Note:* Your answer is an overestimate because electrons are actually "shared" among the three atoms, spending more time near the oxygen.

### Section 23.2 Capacitors

19. A capacitor consists of square conducting plates 25 cm on a side and 5.0 mm apart, carrying charges $\pm 1.1$ $\mu$C. Find (a) the electric field between the plates, (b) the potential difference between the plates, and (c) the stored energy.

20. An uncharged capacitor has parallel plates 5.0 cm on a side, spaced 1.2 mm apart. (a) How much work is required to transfer 7.2 $\mu$C from one plate to the other? (b) How much work is required to transfer a second 7.2 $\mu$C?

21. (a) How much charge must be transferred between the initially uncharged plates of Exercise 20 in order to store 15 mJ of energy? (b) What will be the potential difference between the plates?

22. A capacitor's plates hold 1.3 $\mu$C when charged to 60 V. What is its capacitance?

23. The "memory" capacitor in a VCR has a capacitance of 4.0 F and is charged to 3.5 V. What is the charge on its plates?

24. What voltage is needed to put 1.6 mC on a 100-$\mu$F capacitor?

25. Show that the units of $\varepsilon_0$ may be written as F/m.

26. Find the capacitance of a parallel-plate capacitor consisting of circular plates 20 cm in radius separated by 1.5 mm.

27. A parallel-plate capacitor with 1.1-mm plate spacing has $\pm 2.3$ $\mu$C on its plates when charged to 150 V. What is the plate area?

28. The power supply of a stereo receiver contains a 2500-$\mu$F capacitor charged to 35 V. How much energy does it store?

29. Find the capacitance of a capacitor that stores 350 $\mu$J when the potential difference across its plates is 100 V.

**FIGURE 23.12** Exercise 18

## Section 23.3 Using Capacitors

30. You have a 1.0-$\mu$F and a 2.0-$\mu$F capacitor. What values of capacitance could you get by connecting them in series or parallel?

31. Two capacitors are connected in series and the combination is charged to 100 V. If the voltage across each capacitor is 50 V, how do their capacitances compare?

32. (a) What is the equivalent capacitance of the combination shown in Fig. 23.13? (b) If a 100-V battery is connected across the combination, what is the charge on each capacitor? (c) What is the voltage across each?

**FIGURE 23.13** Problem 32

33. You're given three capacitors: 1.0 $\mu$F, 2.0 $\mu$F, and 3.0 $\mu$F. Find (a) the maximum, (b) the minimum, and (c) two intermediate values of capacitance you could achieve with various combinations of all three capacitors.

## Section 23.4 Energy in the Electric Field

34. The energy density in a uniform electric field is 3.0 J/m$^3$. What is the field strength?

35. A car battery stores about 4 MJ of energy. If all this energy were used to create a uniform electric field of 30 kV/m, what volume would it occupy?

36. Air undergoes dielectric breakdown at a field strength of 3 MV/m. Could you store energy in a uniform electric field in air with the same energy density as that of liquid gasoline? *Hint:* See Appendix C.

37. Find the electric-field energy density at the surface of a proton, taken to be a uniformly charged sphere 1 fm in radius.

## Problems

38. A charge $Q_0$ is at the origin. A second charge, $Q_x = 2Q_0$, is brought to the point $x = a$, $y = 0$. Then a third charge $Q_y$ is brought to the point $x = 0$, $y = a$. If it takes twice as much work to bring in $Q_y$ as it did $Q_x$, what is $Q_y$ in terms of $Q_0$?

39. A conducting sphere of radius $a$ is surrounded by a concentric spherical shell of radius $b$. Both are initially uncharged. How much work does it take to transfer charge from one to the other until they carry charges $\pm Q$?

40. A pair of closely spaced square conducting plates measure 10 cm on a side. The electric-field energy density between the plates is 4.5 kJ/m$^3$. What is the charge on the plates?

41. A capacitor consists of two long concentric metal cylinders (Fig. 23.14). Find an expression for its capacitance in terms of the dimensions shown.

42. A capacitor consists of a conducting sphere of radius $a$ surrounded by a concentric conducting shell of radius $b$. Show that its capacitance is $C = [ab/k(b - a)]$.

43. Figure 23.15 shows a capacitor consisting of two electrically connected plates with a third plate between them, spaced so its surfaces are a distance $d$ from the other plates. The plates have area $A$. Neglecting edge effects, show that the capacitance is $2\epsilon_0 A/d$.

**FIGURE 23.15** Problem 43

44. A certain capacitor stores 40 mJ of energy when charged to 100 V. (a) How much would it store when charged to 25 V? (b) What is its capacitance?

45. Which can store more energy: a 1-$\mu$F capacitor rated at 250 V or a 470-pF capacitor rated at 3 kV?

46. A circuit application calls for a 10-$\mu$F capacitor that can store 12 mJ. What should be its voltage rating? The capacitors are available with voltage ratings that are multiples of 25 V.

47. A 0.01-$\mu$F, 300-V capacitor costs 25¢; a 0.1-$\mu$F, 100-V capacitor costs 35¢; and a 30-$\mu$F, 5-V capacitor costs 88¢. (a) Which can store the most charge? (b) Which can store the most energy? (c) Which is the most cost-effective energy-storage device, as measured by energy stored per unit cost?

48. A medical defibrillator stores 950 J of energy in a 100-$\mu$F capacitor. (a) What is the voltage across the capacitor? (b) If the capacitor discharges 300 J of its stored energy in 2.5 ms, what power is delivered during this time?

49. A camera flashtube requires 5.0 J of energy per flash. The flash duration is 1.0 ms. (a) What power does the flashtube use *while it is actually flashing*? (b) If the flashtube operates at 200 V, what size capacitor is needed to supply the flash energy? (c) If the flashtube is fired once every 10 s, what is its *average* power consumption?

50. The NOVA laser fusion experiment at Lawrence Livermore Laboratory in California can deliver $10^{14}$ W (roughly one hundred times the output of all the world's power plants) of light energy when its lasers are on. But the laser pulse lasts only $10^{-9}$ s. (a) How much energy is delivered in one pulse? (b) The capacitor bank supplying this energy has a total capacitance of 0.26 F. Only about 0.17% (i.e., 0.0017) of the capacitor energy actually appears as light. To what voltage must the capacitor bank be charged?

51. A solid conducting slab is inserted between the plates of a charged capacitor, as shown in Fig. 23.16. The slab thickness is 60% of the plate spacing, and its area is the same as the plates'. (a) What happens to the capacitance? (b) What happens to the stored energy, assuming the capacitor is not connected to anything?

**FIGURE 23.14** Problems 41 and 79

**FIGURE 23.16** Problem 51

52. What is the equivalent capacitance of the four identical capacitors in Fig. 23.17, measured between $A$ and $B$?

**FIGURE 23.17** Problem 52

53. You have an unlimited supply of 2.0-$\mu$F, 50-V capacitors. Describe combinations that would be equivalent to (a) a 2.0-$\mu$F, 100-V capacitor and (b) a 0.50-$\mu$F, 200-V capacitor.

54. What is the equivalent capacitance in Fig. 23.18?

**FIGURE 23.18** Problems 54 and 55

55. In Fig. 23.18, find the energy stored in the 1-$\mu$F capacitor when a 50-V battery is connected between points $A$ and $B$.

56. Two capacitors $C_1$ and $C_2$ are in series, with a voltage $V$ across the combination. Show that the voltages across the individual capacitors are

$$V_1 = \frac{C_2 V}{C_1 + C_2} \quad \text{and} \quad V_2 = \frac{C_1 V}{C_1 + C_2}$$

57. A 0.10-$\mu$F capacitor rated at 50 V is in series with a 0.20-$\mu$F capacitor rated at 200 V. What is the maximum voltage that should be applied across the series combination?

58. A parallel-plate capacitor has plates with 50-cm$^2$ area separated by a 25-$\mu$m layer of polyethylene. Find (a) its capacitance and (b) its working voltage.

59. A 470-pF capacitor consists of two circular plates 15 cm in radius, separated by a sheet of polystyrene (dielectric constant 2.3, breakdown field 50 MV/m). (a) What is the thickness of the sheet? (b) What is the working voltage?

60. The first accurate estimate of the thickness of biological cell membranes used a capacitive technique, in which the capacitance per unit area of cell membrane was determined through a macroscopic measurement of the electric properties of a suspension of cells; the result was a value of about 1 $\mu$F/cm$^2$ for a wide range of cells. Assuming a dielectric constant of about 3 for the membrane material, find the membrane thickness. *Note:* Your answer is the thickness of the bipolar lipid layer alone, and is lower by a factor of about 3 than values based on X-ray techniques; the full membrane may be thicker still.

61. A variable "trimmer" capacitor used to make fine adjustments has a capacitance range of 10–30 pF. The trimmer is in parallel with a capacitor of about 0.001 $\mu$F. Over what percentage range can the capacitance of the combination vary?

62. The electric-field strength as a function of position $x$ in a certain region is given by $E = E_0(x/x_0)$, where $E_0 = 24$ kV/m and $x_0 = 6.0$ m. Find the total energy stored in a cube 1.0 m on a side located between $x = 0$ and $x = 1.0$ m. *Note:* The field strength is independent of $y$ and $z$.

63. A sphere of radius $R$ contains charge $Q$ spread uniformly throughout its volume. Find an expression for the electrostatic energy contained within the sphere itself. *Hint:* Consult Example 21.1.

64. A sphere of radius $R$ carries a total charge $Q$ distributed over its surface. Show that the total energy stored in its electric field is $U = kQ^2/2R$.

65. A uranium-235 nucleus contains 92 protons and 143 neutrons and has a diameter of 6.6 fm. Assuming that the proton charge is distributed uniformly throughout the nucleus, calculate the total electrostatic energy of this configuration. *Hint:* See Problems 63 and 64.

66. Two 4.0-mm-diameter water drops each carry 15 nC. They're initially separated by a great distance. Find the change in the electrostatic potential energy if they're brought together to form a single spherical drop. Assume all charge resides on the drops' surfaces.

67. A 2.1-mm-diameter wire carries a uniform line charge density $\lambda = 28$ $\mu$C/m. How much energy is contained in a space 1.0 m long within one wire diameter of the wire surface?

68. A typical lightning flash transfers 30 C across a potential difference of 30 MV. Assuming such flashes occur every 5 s in the thunderstorm of Example 23.4, roughly how long could the storm continue if its electric energy were not replenished?

69. Show that the result of Problem 42 reduces to that of a parallel-plate capacitor when the separation $b - a$ is much less than the radius $a$.

70. A solid sphere contains a uniform volume charge density. What fraction of the total electrostatic energy of this configuration is contained *within* the sphere?

71. A classical view of the electron pictures it as a purely electric entity, whose Einstein rest mass energy, $E = mc^2$ is the energy stored in its electric field. If the electron were a sphere with charge distributed uniformly over its surface, what radius would it have in order to satisfy this condition? *Note:* Your answer for the electron's "size" isn't consistent with modern quantum mechanics or with experiments that suggest the electron is a true point particle.

72. An air-insulated parallel-plate capacitor of capacitance $C_0$ is charged to voltage $V_0$ and then disconnected from the charging battery. A slab of material with dielectric constant $\kappa$, whose thickness is essentially equal to the capacitor spacing, is then inserted halfway into the capacitor (Fig. 23.19). Determine (a) the new capacitance, (b) the stored energy, and (c) the force on the slab in terms of $C_0$, $V_0$, $\kappa$, and the capacitor-plate length $L$.

**FIGURE 23.19** Problem 72

73. Repeat parts (b) and (c) of Problem 72, now assuming the battery remains connected while the slab is inserted.

74. A transmission line consists of two parallel wires, of radius $a$ and separation $b$, carrying uniform linear charge densities, $\pm\lambda$ respectively. In the approximation $a \ll b$, their electric field is the superposition of the fields from two long straight lines of charge. Find the capacitance per unit length for this transmission line.

75. An infinitely long rod of radius $R$ carries a uniform volume charge density $\rho$. Find an expression for the electrostatic energy per unit length contained within the rod. *Hint:* See Problem 21.61.

76. An electric field $\vec{E} = E_0\hat{i}/x^3$ exists in a cylindrical volume of radius $R$ and length $L$, extending along the $x$ axis from $x = L$ to $x = 2L$. Find the total electrostatic energy contained in this cylinder.

77. (a) Write the electrostatic potential energy of a pair of oppositely charged, closely spaced parallel plates as a function of $x$, their separation, $A$, their area, and $Q$, the magnitude of their charges. (b) Differentiate this with respect to $x$, to find the magnitude of the attractive force between the plates. Why is the force not equal to the charge on one plate times the electric field between the plates?

78. An unknown capacitor $C$ is connected in series with a 3.0-$\mu$F capacitor, this pair is placed in parallel with a 1.0-$\mu$F capacitor, and the entire combination is put in series with a 2.0-$\mu$F capacitor. (a) Make a circuit diagram of this network. (b) When a potential difference of 100 V is applied across the open ends of this network, the total energy stored in all the capacitors is 5.8 mJ. Find $C$.

79. A proposal for a big telecommunications contract continues to haunt you. Another team member rushes into your office and says, "I need to know the energy density of a cylindrical capacitor as a function of radial position within the capacitor!" From Gauss's law, you know that $\vec{E} = (\lambda/2\pi\epsilon_0 r)\hat{r}$. You tell your colleague not to worry and not only calculate the energy density but also integrate to find the stored energy $U$. What do you get? As a check on your result, show that $U = \frac{1}{2}CV^2$. *Note:* Figure 23.14 showed such a capacitor.

80. You are planning to install a home video studio. A very important piece of information is missing from the specifications. You must find the capacitance per unit length for a coaxial cable. The cable has an inner radius of 1.0 mm and an outer conductor 3.0 mm in radius separated by polyethylene insulation that acts as a dielectric. What's the capacitance per unit length?

81. Your home video studio will use a TV antenna cable often known as "twin lead." Two 0.50-mm-diameter wires are spaced 12 mm apart. Your equipment requires that the total capacitance of the cable not exceed 1000 pF. The cable is 30 m long. A technical paper on the cable gives the potential difference between the two conductors as

$$\Delta V = \frac{\lambda}{\pi\epsilon_0} \ln\left(\frac{b - a}{a}\right)$$

where $a$ is the radius of the wire and $b$ is the separation. Is your cable within the specifications?

# Answers to Chapter Questions

## Answer to Chapter Opening Question

The energy is stored in the electric field of a pair of charged conductors—a capacitor—and dumped quickly to the defibrillator when it's needed.

## Answers to GOT IT? Questions

23.1 (b), because $U$ depends on $V^2$.

23.2 (a) Parallel; (b) series. The working voltage of the series combination is twice that of the parallel combination, which is the same as that of the individual capacitors.

23.3 (a) $E(P)$ doubles; (b) $u_E(P)$ quadruples; (c) $U$ decreases, since the charges are attracted and therefore you do negative work to bring in the negative charge.

# 24 Electric Current

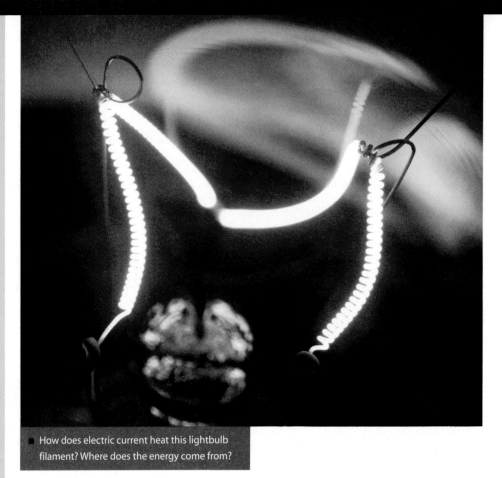

- How does electric current heat this lightbulb filament? Where does the energy come from?

We now relax the assumption of electrostatic equilibrium, and consider situations in which charges are moving. The flow of charge constitutes **electric current**, and it generally occurs in materials containing free charges—that is, in conductors.

Electric current is essential in many technological and natural processes. Currents in lightbulbs, toasters, and stoves produce heat and light. Currents in electric motors run hard disks, refrigerators, and subway trains. In computers, currents move and process data. In your body, they regulate heartbeat and control your muscles. Currents in Earth's liquid outer core generate the planet's magnetism, protecting us from cosmic radiation. And currents in the Sun are responsible for giant eruptions that can spew high-energy particles toward Earth.

## 24.1 Electric Current

Quantitatively, current is the net rate of charge crossing an area. Its units are coulombs per second, which is given the special name **ampere** (A) after the French physicist André Marie Ampère (1775–1836). In electronics and biomedical applications, currents are small enough that milliamperes (mA) and microamperes ($\mu$A) are widely used. When current $I$ is steady or a time average will do, we write

$$I = \frac{\Delta Q}{\Delta t} \quad \text{(steady current)} \qquad (24.1a)$$

where $\Delta Q$ is the charge crossing the given area in time $\Delta t$. For time-varying currents we take the limit of small time intervals:

$$I = \frac{dQ}{dt} \quad \text{(instantaneous current)} \qquad (24.1b)$$

Current is in the direction in which *positive* charge flows. If the moving charge is negative, as with electrons in a metal, then the current is opposite the charge motion.

A current may consist of one kind of moving charge, or both. If it's both, then the net current is the sum of the currents carried by positive and negative charges (Fig. 24.1a). That's why the bulk motion of a neutral object—even though it contains lots of positive and negative charge—doesn't constitute a current (Fig. 24.1b).

**GOT IT? 24.1** Which of the following represents a nonzero current? What's its direction? (a) a beam of electrons moves from left to right; (b) a beam of protons moves upward; (c) in a solution, positive ions move left and negative ions move right; (d) blood, carrying positive and negative ions at the same speed, moves up through a vein; (e) a metal car with no net charge speeds westward.

## Current: A Microscopic Look

Current depends on the speed of the charge carriers, their density, and their charge. In some cases, like a beam of electrons in vacuum, "speed" here means the actual speed of the charges. But in typical conductors, charges are moving about at high speed with random thermal velocities that don't result in a net flow of charge. When a current is present, there's an additional and usually very small **drift velocity** superposed on the charges' random motion, and it's this drift velocity that determines the current. We'll see this in more detail when we consider metallic conductors.

Figure 24.2 shows a conductor that contains $n$ charges per unit volume, each with charge $q$ and drift speed $v_d$. We want to express the current in terms of these microscopic properties and macroscopic properties like length and area. With $A$ the conductor's cross-sectional area, a length $L$ of conductor has volume $AL$ and contains $nAL$ individual charges for a total charge $\Delta Q = nALq$. Moving at $v_d$, this charge takes time $\Delta t = L/v_d$ to pass a given point. Then the current is

$$I = \frac{\Delta Q}{\Delta t} = \frac{nALq}{L/v_d} = nAqv_d \qquad (24.2)$$

Protons moving right: $I$ is to right.

Net current

*Negative* electrons moving left: $I$ is to *right*.

**(a)**

Zero net current

$\vec{v}$

Both charges moving right; no net current.

**(b)**

**FIGURE 24.1** The net current is the sum of the currents carried by both positive and negative charges.

$\vec{v}_d \longrightarrow$

$n$ charges/unit volume, each charge $q$

$A$

$L$

This volume contains charge $\Delta Q = nALq$.

**FIGURE 24.2** A conductor of cross-sectional area $A$ containing $n$ charges per unit volume.

---

**EXAMPLE 24.1** **Finding Current: A Copper Wire**

A 5.0-A current flows in a copper wire with cross-sectional area 1.0 mm², carried by electrons with number density $n = 1.1 \times 10^{29}$ m⁻³. Find the electrons' drift speed.

**INTERPRET** We're given microscopic parameters, so this is a problem about the relation between current and the parameters $n$, $q$, and $v_d$.

**DEVELOP** Figure 24.3 is our sketch. Equation 24.2, $I = nAqv_d$, relates current to the macroscopic parameters, so our plan is to solve for $v_d$.

$I = 5.0$ A

$n = 1.1 \times 10^{29}$ m⁻³

$A = 1.0$ mm²

$v_d = ?$

**FIGURE 24.3** Sketch for Example 24.1.

**EVALUATE** Solving, we get

$$v_d = \frac{I}{nAq}$$

$$= \frac{5.0 \text{ A}}{(1.1 \times 10^{29} \text{ m}^{-3})(10^{-6} \text{ m}^2)(1.6 \times 10^{-19} \text{ C})} = 0.28 \text{ mm/s}$$

**ASSESS** Make sense? Our answer seems awfully small. After all, when you flip a light switch, the light comes on immediately, not several thousand seconds later as our answer might imply. But the answer is right. Electrons in the wire all get their "marching orders" from the electric field, and that's established almost instantaneously. As a result electrons throughout the wire start moving almost simultaneously. That's why the light comes on immediately. ∎

✓ **TIP** Drift Speed and Signal Speed

Example 24.1 points to an important distinction between the drift speeds of electrons and speed of electric signals. The former is typically about 1 mm/s, but the latter is close to the speed of light. When you connect a wire across a battery, for example, an electric field develops, starting at both battery terminals and moving along the wire at nearly the speed of light. As soon as electrons experience the field, they start to move. So there's almost no time delay before the start of the current.

### Current Density

Currents aren't always confined to wires. Currents in the Earth, in chemical solutions, in your body, and in ionized gases flow in ill-defined paths, and their magnitude and direction may vary with position. We characterize such diffuse currents in terms of **current density**, $\vec{J}$, a vector whose direction at each point is that of the local current and whose magnitude is the current per unit area. Dividing Equation 24.2 by area and using the drift velocity vector $\vec{v}_d$ instead of speed $v_d$ gives the current density:

$$\vec{J} = nq\vec{v}_d \quad \text{(current density)} \qquad (24.3)$$

When the current density is uniform, as in a wire, the total current is just the product of the current density and the wire's cross-sectional area. But when the current density varies, it's necessary to integrate to find the total current; see Problem 58.

---

**EXAMPLE 24.2** **Current and Current Density: Through the Cell Membrane**

Ion channels are narrow pores that allow ions to pass through cell membranes (Fig. 24.4). A particular channel has a circular cross section 0.15 nm in radius; it opens for 1 ms and passes $1.1 \times 10^4$ singly ionized potassium ions. Find both the current and the current density in the channel.

Ion channels

Lipid molecules

~0.3 nm

**FIGURE 24.4** Diagram of a cell membrane, showing ions passing through an ion channel.

**INTERPRET** This problem describes a flow of individual ions and asks for two distinct but related quantities: current and current density.

**DEVELOP** Current is the *rate* of charge passing through a given area, here the opening of an ion channel. Equation 24.1a, $I = \Delta Q/\Delta t$, determines the current. Current density, however, is *current per unit area*, which we can compute from $J = I/A$.

**EVALUATE** With each ion carrying charge $e$, a total charge $\Delta Q = 1.1 \times 10^4 e = 1.8 \times 10^{-15}$ C flows through the channel in $\Delta t = 10^{-3}$ s, giving a current $I = \Delta Q/\Delta t = 1.8$ pA. For current density we then find

$$J = \frac{I}{A} = \frac{1.8 \times 10^{-12}\,\text{A}}{\pi(0.15 \times 10^{-9}\,\text{m})^2} = 2.5 \times 10^7\,\text{A/m}^2$$

**ASSESS** Make sense? How can something so tiny as a cell have a current density of 25 million amperes per square meter? No problem: Current density measures current *per unit area*. The ion channel is so small that the total current—1.8 picoamperes—is tiny. But that channel is impressive in its own right; its 25 MA/m² is about four times the maximum safe current density in typical household wiring. ∎

---

## 24.2 Conduction Mechanisms

Electric fields exert forces on charges, so it's the presence of electric fields in conductors that results in electric current. Fields in conductors? Yes. With moving charge we no longer have electrostatic equilibrium, so the field inside a conductor does *not* have to be zero. Newton's law suggests that an electric field should *accelerate* free charges in a conductor, resulting in an ever-increasing current. But in most conductors charges collide with things—usually ions—and lose the energy they've gained from the field. These collisions provide an effective force that counters the electric force, and the end result is that it takes an electric field to sustain a steady current. In most materials the field and current are in the same direction, and we can therefore express the relation between the two with the equation

$$\vec{J} = \sigma\vec{E} \quad \text{(Ohm's law, microscopic version)} \quad (24.4a)$$

where the quantity $\sigma$ is called the **conductivity** of the material.

## Ohm's Law: A Microscopic View

For many common conductors, including metals, conductivity $\sigma$ is independent of electric field. Such materials are called **ohmic**, and for them Equation 24.4a states that current density and electric field are linearly proportional. In **nonohmic** materials, conductivity does depend on field, and thus the relationship between $\vec{J}$ and $\vec{E}$ isn't linear.

You may be familiar with the *macroscopic* version of **Ohm's law**, which relates electric current and voltage in a piece of conducting material. Equation 24.4a is the *microscopic* version of Ohm's law, describing the relation between electric field and current density *at each point* within a conductor. The macroscopic version is helpful in analyzing electric circuits, and we'll derive it in the next section. But the microscopic version is important in biophysics, geophysics, astrophysics, semiconductor engineering, and other areas where electric fields vary with position and we want to know what's going on at each point.

The conductivity $\sigma$ tells how large a current density will result from a given electric field; it's a measure of how easily charges in the material can move. A perfect conductor would have $\sigma = \infty$; a perfect insulator, $\sigma = 0$. A related quantity is **resistivity**, $\rho$, defined as the inverse of conductivity: $\rho = 1/\sigma$. Then Equation 24.4a can be written as

$$\vec{J} = \frac{\vec{E}}{\rho} \quad (24.4b)$$

Resistivity tells how hard it is for charge to move in a material; the higher the resistivity, the stronger the electric field needed to produce a given current density. You may be familiar with electrical *resistance*, and we'll soon see how resistance and resistivity are related.

Equation 24.4b shows that the units of resistivity are V·m/A. One V/A is given the name **ohm**, $\Omega$, after the German physicist Georg Ohm (1789–1854), who explored the relation between voltage and current. Thus the SI units of resistivity are $\Omega$·m; reciprocally, those of conductivity are $(\Omega\text{·m})^{-1}$. Conductivity and resistivity vary dramatically; their range is among the broadest of any physical quantity, spanning some 24 orders of magnitude. Table 24.1 lists the resistivities of some typical materials. Measurement of electrical resistivity provides information on the composition of materials in fields ranging from medicine to geophysics.

**TABLE 24.1** Resistivities

| Material | Resistivity ($\Omega$ · m) |
|---|---|
| **Metallic conductors (20°C)** | |
| Aluminum | $2.65 \times 10^{-8}$ |
| Copper | $1.68 \times 10^{-8}$ |
| Gold | $2.24 \times 10^{-8}$ |
| Iron | $9.71 \times 10^{-8}$ |
| Mercury | $9.84 \times 10^{-7}$ |
| Silver | $1.59 \times 10^{-8}$ |
| **Ionic solutions (in water, 18°C)** | |
| 1-molar $CuSO_4$ | $3.9 \times 10^{-4}$ |
| 1-molar HCl | $1.7 \times 10^{-2}$ |
| 1-molar NaCl | $1.4 \times 10^{-4}$ |
| $H_2O$ | $2.6 \times 10^{5}$ |
| Blood, human | 0.70 |
| Seawater (typical) | 0.22 |
| **Semiconductors (pure, 20°C)** | |
| Germanium | 0.47 |
| Silicon | 23.0 |
| **Insulators** | |
| Ceramics | $10^{11}$–$10^{14}$ |
| Glass | $10^{10}$–$10^{14}$ |
| Polystyrene | $10^{15}$–$10^{17}$ |
| Rubber | $10^{13}$–$10^{16}$ |
| Wood (dry) | $10^{8}$–$10^{14}$ |

---

**EXAMPLE 24.3**  **Finding the Electric Field: Household Wiring**

A 1.8-mm-diameter copper wire carries 15 A to a household appliance. Find the magnitude of the electric field in the wire.

**INTERPRET** This problem asks us to calculate the electric field within a conductor carrying an electric current.

**DEVELOP** Equation 24.4b, $\vec{J} = \vec{E}/\rho$, relates electric field and current density. Here we're given the total current $I$ and the wire diameter, so we can write the current density as $J = I/A$, with $A$ the wire's cross-sectional area. We also need the resistivity of copper from Table 24.1. Then we can solve Equation 24.4b for the electric field.

**EVALUATE** Solving for the field magnitude, we have

$$E = J\rho = \frac{I\rho}{A} = \frac{(15\text{ A})(1.68 \times 10^{-8}\ \Omega\text{·m})}{\pi(0.90 \times 10^{-3}\text{ m})^2} = 99\text{ mV/m}$$

**ASSESS** Make sense? This number is a lot smaller than the electric fields we discussed in electrostatic situations. Because copper is such a good conductor, a weak field can drive a substantial current. In well-engineered circuits, the field inside conducting wires is often so small as to be negligible, even when the current is large. ∎

---

**GOT IT? 24.2** Two wires carry the same current $I$. Wire A has a larger diameter, a higher density of current-carrying electrons, and a lower resistivity than wire B. Rank order (a) the current densities, (b) the electric fields, and (c) the drift speeds in the two wires.

**FIGURE 24.5** Atoms of a metal form a regular crystal lattice.

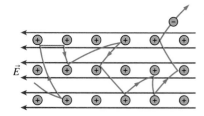

**FIGURE 24.6** An electron's path in a metal is almost completely random, but in the presence of an electric field there's a slight drift antiparallel to the field.

## Conduction in Metals

Metals are good conductors because they contain abundant free electrons, which respond readily to electric fields. Each atom in a metal typically contributes one or more electrons to this "sea" of free electrons. The remaining ions form a regular structure called the crystal lattice (Fig. 24.5). Electrons move through the lattice at speeds of about $10^6$ m/s, colliding frequently with ions and bouncing off in random directions. In the absence of an electric field, there's no net flow of electrons in any particular direction, and so no current.

We'll now consider what happens when an electric field is applied to a metal, and we'll show why metals obey Ohm's law. However, our explanation is necessarily incomplete because a full description of metallic conduction involves quantum mechanics.

An electric field accelerates negative electrons in the direction opposite the field. But like a car in stop-and-go traffic, the electron soon gives up the energy and speed it gained from the field. For the car, that happens at the next stoplight; for the electron, it's at the next collision with an ion, where it rebounds in a random direction (Fig. 24.6). Like the car, the electron thus acquires an average velocity that's proportional to the acceleration it experiences between collisions—that is, proportional to the electric field. There's one difference, though, between the electron and the car: The electron has also a high random thermal velocity, so the average velocity is a tiny effect superposed on the electron's random thermal motion. That average velocity is the drift velocity, $\vec{v}_d$. All electrons share this common drift velocity, so their motion constitutes a current proportional to $v_d$.

The drift velocity depends on two things: the electrons' acceleration and the rate at which they undergo collisions. The electric field provides the acceleration, so $v_d$ is proportional to $E$. The collision rate depends on how fast the electrons are moving, and here comes the important point: Because thermal motions are so fast, the additional drift velocity makes essentially no difference in the collision rate, so the latter is constant. Therefore the drift velocity and hence the current are directly proportional to the electric field—and that's what makes Equation 24.4a a direct proportionality between current density and field. That's why metals are ohmic.

Although a metal's conductivity is independent of the applied field, it does depend on temperature $T$. That's because the thermal speed and hence the collision rate increase with temperature, decreasing conductivity and increasing resistivity. Classical physics gives thermal speed proportional to $\sqrt{T}$, as we saw in Section 17.1, so we might expect resistivity to depend similarly on temperature. Experiment, however, shows that resistivity is nearly linear with temperature (Fig. 24.7)—a result that can be explained only using quantum mechanics.

**FIGURE 24.7** Resistivity of copper has a nearly linear dependence on temperature, in contrast to the classical prediction of a dependence on $\sqrt{T}$.

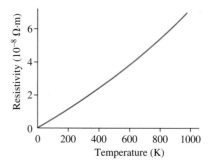

Although the current associated with random thermal motions averages to zero, at any given instant short-term fluctuations can result in more electrons moving in a particular direction. The result is a very small current whose direction and magnitude fluctuate randomly. This **thermal noise** can overwhelm currents of interest in sensitive electronic equipment. Circuits like the amplifiers in radio telescopes are often cooled to decrease thermal noise.

## Ionic Solutions

Liquid solutions contain positive and negative ions that respond to an electric field by moving in opposite directions, resulting in a net current. Conductivity is limited by collisions between ions and neutral atoms and, as Table 24.1 suggests, ionic solutions are poorer conductors than metals. Ionic conduction is essential to life, as the transport of ions through cell membranes in Example 24.2 suggests. Electric eels use ionic conduction to sense and kill their prey. Some technologies depend on ionic conduction, including batteries and fuel cells, electroplating, and the hydrolysis of water to make hydrogen and oxygen. Ionic conduction plays a role in the corrosion of metals, and an ionic solution—sweat—increases our vulnerability to electric shock.

## Plasmas

**Plasma** is ionized gas that conducts because it contains free electrons and ions. It takes substantial energy to ionize atoms, so plasmas usually exist only at high temperatures. Plasmas are rare on Earth; they're found in fluorescent lamps, plasma TVs, neon signs, the ionosphere, flames, and lightning flashes. Yet much of the universe's ordinary matter is in the plasma state; stars, in particular, are mostly plasma.

The electric properties of plasma make it so different from ordinary gas that plasma is often called "the fourth state of matter." Some plasmas—like the Sun's corona—are so diffuse and therefore collisions so rare as to make them far better conductors than metals. These "collisionless" plasmas can sustain large currents with minimal electric fields.

## Semiconductors

Even in insulators, random thermal motions dislodge a few electrons, giving these materials very modest conductivity. In a few materials—notably the element silicon—this effect is significant at room temperature. Such materials have conductivities between those of good insulators and metals, so they're called **semiconductors**. Semiconductors make possible the microelectronic technology so pervasive in modern civilization. Here we give a qualitative description of semiconductors based on classical physics; we'll revisit semiconductors from a quantum-mechanical viewpoint in Chapter 37.

A dislodged electron leaves behind a "hole," into which an adjacent electron, nudged by the electric field, can "fall" (Fig. 24.8) The result is a movement of holes in the direction of the field. Thus holes act as positive charges, so a pure semiconductor contains equal numbers of negative charge carriers (electrons) and positive carriers (holes).

The key to semiconductor technology lies in **doping** with impurities that greatly alter a semiconductor's intrinsic conductivity. Figure 24.9 shows how a single phosphorus atom, with five valence electrons, fits into silicon's crystal structure but leaves a free electron. It doesn't take much phosphorus for these extra electrons to constitute the vast majority of charge carriers. Since the charge carriers are negative, the material is an **N-type semiconductor**. Doping with trivalent atoms like boron, in contrast, leaves extra holes and makes a **P-type** semiconductor.

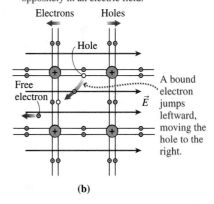

**FIGURE 24.8** Structure of a silicon crystal, showing each atom bound to each of its neighbors by two shared electrons. (a) Thermal motion dislodges electrons, creating electron-hole pairs. (b) An electric field drives electrons and holes in opposite directions, creating an electric current.

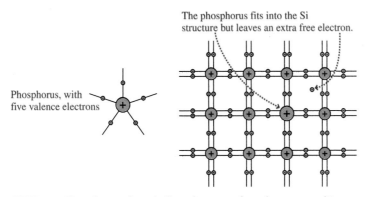

**FIGURE 24.9** Phosphorus-doped silicon has extra free electrons, making it an *N*-type semiconductor.

The essential element of nearly every semiconductor device is the **PN junction**. Electrons and holes diffuse across such a junction and recombine, depleting the junction region of charge carriers and making it a poor conductor. Applying a voltage from the *P* to the *N* region—but not the other way—lets charge flow through the junction. So the *PN* junction conducts in one direction but not the other (Fig. 24.10). The wide range of semiconductor devices in use today results largely from carefully engineered combinations of *PN* junctions.

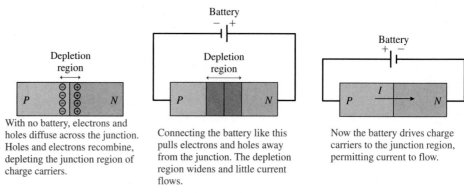

With no battery, electrons and holes diffuse across the junction. Holes and electrons recombine, depleting the junction region of charge carriers.

Connecting the battery like this pulls electrons and holes away from the junction. The depletion region widens and little current flows.

Now the battery drives charge carriers to the junction region, permitting current to flow.

**FIGURE 24.10** A *PN* junction conducts in only one direction.

---

**APPLICATION** **The Transistor**

Few inventions have revolutionized society as much as the transistor, the semiconductor device at the root of all modern electronics. The figure shows one type, the field-effect transistor, or FET. This particular FET is a slab of *P*-type semiconductor with two embedded *N*-type regions. Normally no current can flow through the transistor because one of its two *PN* junctions is backward, as shown in part (a) of the figure. But atop the so-called *channel* between the *N*-type regions is a thin insulating layer and over it a metal layer called the *gate*. Make the gate positive, and it pulls electrons into the channel, making it

temporarily *N*-type, as in part (b). That eliminates the *PN* junctions, and now current flows through the transistor.

Varying the gate voltage continuously makes the transistor an *amplifier*, in which a weak gate signal controls a larger current. Swinging between fully on and off makes the transistor a digital switch, providing the binary 1 and 0 from which all digital information is built. Today, transistors by the billions are fabricated on single chips of silicon, making the powerful microprocessors that are the "brains" of computers.

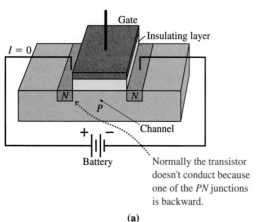

Normally the transistor doesn't conduct because one of the *PN* junctions is backward.

**(a)**

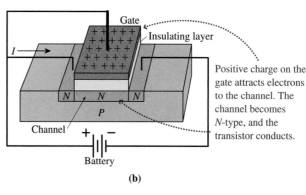

Positive charge on the gate attracts electrons to the channel. The channel becomes *N*-type, and the transistor conducts.

**(b)**

## Superconductors

In 1911 the Dutch physicist H. Kamerlingh Onnes found that the resistivity of mercury dropped to zero at a temperature of 4.2 K. Today we know thousands of substances that become **superconductors** at sufficiently low temperatures. Currents in superconductors persist for years without measurable decrease, suggesting the resistivity is truly zero (Fig. 24.11). For decades the known superconductors were metals and metal alloys that required cooling with liquid helium. Then, in 1986, physicists at IBM's Zurich laboratory stunned the scientific community with the discovery of ceramic materials that become superconducting at around 100 K—high enough to cool with inexpensive liquid nitrogen. The search for higher-temperature superconductors continues, with the highest temperature reported now over 160 K. Development of a room-temperature superconductor could revolutionize much of electrical technology.

Superconductors offer loss-free flow of electric power. Today, superconductors are widely used in high-strength electromagnets, including those in MRI scanners; in devices that measure weak magnetic fields in biomedical, geophysical, and other applications; for electric-power transmission in high-density urban applications; and in motors for ship propulsion. Expect more applications in the near future.

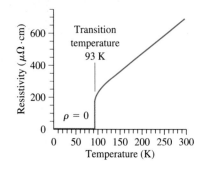

**FIGURE 24.11** Resistivity versus temperature for a thin film of yttrium–barium–copper-oxide superconductor.

## 24.3 Resistance and Ohm's Law

How much current does it take to run this hair dryer? Do I risk a fatal shock if I touch this wire? How long an extension cord can I use with this electric saw? How long will it take to recharge my cell phone? Is the wiring in my house safe? All these questions are ultimately about the electric current flowing in wires, bodies, and batteries. The answer in each case depends on two things: the voltage $V$ applied across the object, and the **resistance** $R$ that the object offers to the flow of electric current.

The macroscopic version of Ohm's law relates voltage, current, and resistance. Ohm's law states that the current through an object is proportional to the voltage across it and inversely proportional to the object's resistance:

$$I = \frac{V}{R} \qquad \text{(Ohm's law, macroscopic version)} \qquad (24.5)$$

Ohm's law shows that a given voltage can push more current through a lower resistance. It's worth noting two extreme cases: An **open circuit** is a nonconducting gap with infinite resistance. No matter what the voltage is across an open circuit, Equation 24.5 shows that no current can flow. One example of an open circuit is a switch in its "off" position. A **short circuit**, in contrast, has zero resistance. In a short circuit, current of any magnitude is possible without any voltage or electric field. A switch in its "on" position approximates a short circuit. An unintentional short circuit is often dangerous; short circuits in household wiring, for example, are a leading cause of fires because they allow large currents to flow, resulting in excessive heating. All real situations, with the exception of superconductors, lie between the extremes of short and open circuits.

We can understand how the macroscopic version of Ohm's law follows from the microscopic version by considering the conductor shown in Fig. 24.12. Suppose there's a uniform electric field $\vec{E}$ within the conductor. Then there must be a uniform current density given by Equation 24.4b: $\vec{J} = \vec{E}/\rho$, where $\rho$ is the resistivity of the material. Then the total current is $I = JA = EA/\rho$, where $A$ is the conductor's cross-sectional area. If the conductor has length $L$, then the potential difference between its ends is $V = EL$, since the electric field is uniform. Solving to get $E = V/L$ and using the result in our expression for $I$ give

**FIGURE 24.12** A cylinder of conducting material with resistivity $\rho$.

$$I = \frac{VA}{L\rho} = \frac{V}{\rho L/A}$$

Comparison with the macroscopic Ohm's law, Equation 24.5, lets us identify the resistance with the term $\rho L/A$. Thus resistance depends on the resistivity $\rho$ and the geometry—length and area—of the particular piece of material:

$$R = \frac{\rho L}{A} \tag{24.6}$$

We derived this expression for a conductor of uniform cross section; although Ohm's law still holds for a nonuniform conductor, integration is required to calculate the resistance (see Problem 62). Equations 24.5 and 24.6 both show that the units of resistance are ohms ($\Omega$).

We emphasize that neither version of Ohm's law is fundamental; rather, each is an empirical law that provides a good description of electrical conduction in some materials. Table 24.2 summarizes the relation between microscopic and macroscopic quantities in Ohm's law.

**TABLE 24.2** Microscopic and Macroscopic Quantities and Ohm's Law

| Microscopic | Macroscopic | Relation |
|---|---|---|
| Electric field, $\vec{E}$ | Voltage, $V$ | $\vec{E}$ is defined at each point in a material; $V$ is the integral of $\vec{E}$ over a path. In a uniform field, $V = EL$. |
| Current density, $\vec{J}$ | Current, $I$ | $\vec{J}$ is defined at each point in a material; $I$ is the integral of $\vec{J}$ over an area. With uniform current density, $I = JA$. |
| Resistivity, $\rho$ | Resistance, $R$ | $\rho$ is a property of a given type of material; $R$ is a property of a particular piece of that material. In a piece with uniform cross section, $R = \rho L/A$. |
| Ohm's law $\vec{J} = \dfrac{\vec{E}}{\rho}$ | Ohm's law $I = \dfrac{V}{R}$ | Microscopic version relates current density to electric field at a point in a material. Macroscopic version relates current through to voltage across a given piece of material. |

---

**EXAMPLE 24.4** **Resistance and Ohm's Law: Starting Your Car**

A copper wire 0.50 cm in diameter and 70 cm long connects your car's battery to the starter motor. What's the wire's resistance? If the starter motor draws a current of 170 A, what's the potential difference across the wire?

**INTERPRET** This problem involves Ohm's law, and we identify the wire as the object in which we want to relate current, voltage, and resistance.

**DEVELOP** Figure 24.13 shows the wire. Equation 24.6, $R = \rho L/A$, determines the resistance, so our plan is first to use the resistivity of copper from Table 24.1 in Equation 24.6 and then to use the resulting resistance in Ohm's law (Equation 24.5), $I = V/R$, to find the potential difference.

$$I = 170 \text{ A}$$
$$d = 0.50 \times 10^{-2} \text{ m}$$
$$\rho = 1.68 \times 10^{-8} \ \Omega \cdot \text{m}$$
$$R = ?$$
$$A = \pi \left(\frac{d}{2}\right)^2$$
$$L = 0.70 \text{ m}$$

**FIGURE 24.13** Sketch for Example 24.4.

**EVALUATE** Table 24.1 gives $\rho = 1.68 \times 10^{-8} \ \Omega \cdot \text{m}$ for copper, so for the resistance we get

$$R = \frac{\rho L}{A} = \frac{(1.68 \times 10^{-8} \ \Omega \cdot \text{m})(0.70 \text{ m})}{\pi (0.25 \times 10^{-2} \text{ m})^2} = 0.60 \text{ m}\Omega$$

Then Ohm's law gives the voltage: $V = IR = (170 \text{ A})(0.60 \text{ m}\Omega) = 0.10 \text{ V}$.

**ASSESS** Make sense? These numbers seem awfully small. They should be! A wire carrying a large amount of current needs to have a very low resistance so the voltage across the wire remains low. We want that 12-V potential difference from the battery to appear across the starter motor, not the connecting wires. A thinner, higher-resistance wire would mean lower voltage across the starter and a significant reduction in current. ∎

A resistor is a piece of conductor made to have a specific resistance. Heating elements used in electric stoves, hair dryers, irons, space heaters, and the like are all essentially resistors; so are the filaments of ordinary incandescent lightbulbs. In all these cases the resistance—ultimately resulting from collisions between conduction electrons and lattice ions—provides a means of turning electric energy into heat. Resistors also set appropriate values of current and voltage in electronic circuits; for this purpose, they're made in a wide range of resistances. Resistors are rated not only by their resistance but also by the maximum power they can dissipate without overheating.

**GOT IT? 24.3** The figure shows three pieces of wire. (1) and (2) are made from the same material, while (3) is made from a material with twice the resistivity. (1) and (3) have twice the diameter of (2). (a) Which has the highest resistance? (b) If the same voltage is applied across each, which will pass the largest current?

(1)                              (2)                              (3)

## 24.4 Electric Power

Impose a potential difference $V$ across a resistor, and a current $I$ flows through it. The quantity $V$ is the energy gained per unit charge as charge "falls" through the potential difference. In a resistor, that energy is dissipated through collisions, heating the material. So $V$ is also the energy per unit charge going into heating. Meanwhile, the current $I$ is the rate at which charge flows through the resistor. Then the energy per unit time—that is, the power dissipated in heating the resistor—is the product of the energy per unit charge and the rate at which charge moves through the conductor:

$$P = IV \quad \text{(electric power)} \tag{24.7}$$

Although we developed Equation 24.7 for power dissipated as heat in a resistor, it holds any time electrical energy is being converted to some other form. If we measure 5 V across an electric motor and 2 A through the motor, we can conclude that the motor is converting electrical to mechanical energy at the rate of 10 W (actually less because some of the power goes into heating).

Solving Ohm's law for $V$ and putting the result in Equation 24.7 give

$$P = I^2 R \tag{24.8a}$$

Solving instead for $I$ gives

$$P = \frac{V^2}{R} \tag{24.8b}$$

These are useful forms when we know the resistance and either the voltage or the current.

✓**TIP** What's Constant?

Equation 24.8a seems to imply that power increases with increasing resistance, while Equation 24.8b seems to suggest the opposite. Both implications are correct—*if* $I$ in Equation 24.8a and $V$ in Equation 24.8b are constants. But there's no contradiction because $I$ and $V$ can't both be constant while the resistance $R$—the ratio of $V$ to $I$—changes. In most cases we work with sources of constant voltage, and then the power dissipated is inversely proportional to the resistance.

---

**EXAMPLE 24.5** **Power and Resistance: A Lightbulb**

The voltage in typical household wiring is 120 V. How much current does a 100-W lightbulb draw? What's the bulb's resistance under these conditions?

**INTERPRET** Here we're given voltage and power and asked for current and resistance. So this problem involves both electric power and Ohm's law.

**DEVELOP** Equation 24.7, $P = IV$, relates power to current and voltage, so we can solve this equation for the current $I$. Then we can solve Ohm's law (Equation 24.5), $I = V/R$, for the resistance.

**EVALUATE** Solving Equation 24.7 for $I$ gives

$$I = \frac{P}{V} = \frac{100 \text{ W}}{120 \text{ V}} = 0.833 \text{ A}$$

The resistance then follows from Ohm's law:

$$R = \frac{V}{I} = \frac{120 \text{ V}}{0.833 \text{ A}} = 144 \text{ }\Omega$$

**ASSESS** Make sense? Sure: With $V = 120$ V, it should take a little less than 1 A for a power of 100 W. You can also see here why Equations 24.8a and 24.8b are equivalent. We could have bypassed the current calculation and obtained $R$ directly from Equation 24.8b: $R = V^2/P = (120 \text{ V})^2/100 \text{ W} = 144 \text{ }\Omega$. Or, given the current, we could have used Equation 24.8a: $R = P/I^2 = 100 \text{ W}/(0.833 \text{ A})^2 = 144 \text{ }\Omega$. Incidentally, the filament's temperature is about 3000 K when it's on, so our 144-$\Omega$ resistance is much higher than the resistance you'd measure with the bulb off (recall Fig. 24.7). ∎

---

Power loss depends on wire resistance $R_w$ and on current $I$.

Wire $R_w$

Power plant — Power plant imposes voltage $V$ across transmission line.

$V$ — $I$ — $V_L$ — Load

Voltage $V_L$ across load is lower because of power lost in wires.

**FIGURE 24.14** Power loss in wires is $I^2R$, so a transmission line is most efficient with a high voltage $V$ between the two wires and a correspondingly low current.

## Electric-Power Transmission

Equation 24.7 shows that it's possible to achieve a given electric power with high voltage and low current, or low voltage and high current. Safety considerations dictate relatively low voltages at the electrical "load," meaning end-use devices like lamps and appliances (although standard 120-V household power is still far from "safe"). But Equation 24.8a shows that, for a given resistance, power dissipation in the transmission line increases as the square of the current. So to minimize energy loss, it's best to transmit electric power with low current—and that means high voltage between the two current-carrying wires (see Fig 24.14 and Problem 56). That's the reason long-distance power-transmission lines operate at hundreds of kilovolts. At the load end, the voltage is stepped down to safer levels using transformers, whose operation we'll explore in Chapter 28.

## 24.5 Electrical Safety

Whether you're in a lab hooking up electronic equipment, or in a hospital connecting instrumentation to a patient, or on a job designing electric devices, or simply at home plugging in appliances and tools, you should be concerned with electrical safety.

Everyone knows enough to be wary of "high voltage." People with a little more sophistication say, "It isn't the voltage but the current that kills you." In fact, both points of view are partially correct. Current through the body is dangerous, but as with any resistor it takes voltage to drive that current.

Table 24.3 shows typical effects of electric currents entering the body through skin contact. A primary danger is disturbance of the electric signals that pace heartbeat; this is reflected in the lethal zone of 100–200 mA at which the heart goes into fibrillation—uncontrolled spasms of the cardiac muscle. With electric signals applied internally, much lower currents can be lethal. Doctors performing cardiac catheterization worry about currents at the microampere level.

**TABLE 24.3** Effects of Externally Applied Current on Humans

| Current Range | Effect |
|---|---|
| 0.5–2 mA | Threshold of sensation |
| 10–15 mA | Involuntary muscle contractions; can't let go |
| 15–100 mA | Severe shock; muscle control lost; breathing difficult |
| 100–200 mA | Fibrillation of heart; death within minutes |
| >200 mA | Cardiac arrest; breathing stops; severe burns |

Above 200 mA, complete cardiac arrest may occur, breathing may stop, and burns may occur. Sometimes high currents are useful: Emergency defibrillators briefly apply a high enough current to stop the heart, which often restarts normal beating. The figures in Table 24.3 are rough averages and vary from person to person as well as with duration of the shock and whether alternating or direct current is involved. Very young children and people with heart conditions are at higher risk.

Under dry conditions, the typical human has a resistance of about $10^5 \ \Omega$ on unbroken skin. What voltages are dangerous to such a person? At $10^5 \ \Omega$ it takes

$$V = IR = (0.1 \ \text{A})(10^5 \ \Omega) = 10{,}000 \ \text{V}$$

to drive the fatal 100 mA. But a person who's wet or sweaty has a much lower resistance and may be electrocuted by 120-V household electricity or even lower.

To be dangerous, an electric circuit must have high voltage *and* be capable of driving sufficient current. For example, a car battery can deliver 300 A, but it can't electrocute you because its 12 V won't drive much current through you. On the other hand the 20,000 V that runs your car's spark plugs won't electrocute you either, since the high-voltage circuit can't deliver more than a few mA.

Because potential difference is a property of two points, receiving an electric shock requires that two parts of the body be in contact with conductors at different potentials; this chapter's opening photo provides a dramatic example. In typical 120-V wiring used throughout North America, one of the two wires is connected physically to the ground. This ground connection prevents the wiring from reaching arbitrarily high potentials, as might otherwise happen in a thunderstorm or if a short circuit occurred in a power line. At the same time it means that an individual contacting the "hot" side of the circuit and any grounded conductor such as the ground, a water pipe, or a bathtub will receive a shock.

A potentially dangerous situation occurs when the operator of a power tool or appliance contacts a grounded conductor. Examples include working outdoors with an electric drill or in a kitchen with an electric mixer. Suppose you're using a power tool that's plugged in through a regular two-wire cord. Normally exposed metal parts of the tool aren't connected to either wire. But suppose something goes wrong and a "hot" wire short-circuits to the metal case; suddenly the tool is 120 V above ground. If you're standing on the ground, or in a damp basement, or leaning on the kitchen sink, you'll receive a potentially lethal shock (Fig. 24.15a).

To avoid this danger many devices use three-wire cords (Fig. 24.16). The third wire runs from exposed metal parts to a grounded wire in the outlet, and normally carries no current. If a short circuit occurs, the third wire provides a very low-resistance path to ground (Fig. 24.15b). A large current flows and blows the circuit breaker or fuse, shutting off the current. Meanwhile the ground wire holds the metal case at ground potential, protecting the operator from dangerous voltages.

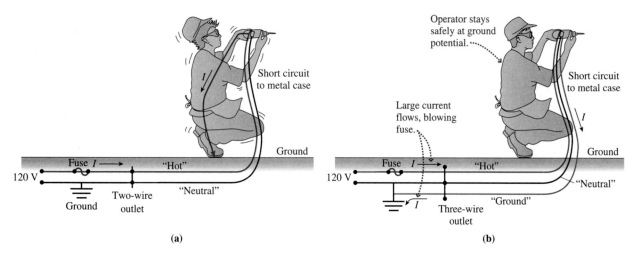

**FIGURE 24.15** (a) A short circuit in an ungrounded tool could result in a lethal shock. (b) With a grounded tool, the fuse blows and the operator is safe.

Some two-wire tools and other devices are "double-insulated" to provide an extra margin of safety. New appliances have "polarized plugs" that can be plugged in only one way (Fig. 24.16), ensuring that exposed metal parts are most likely to end up at ground potential. Finally, electronic devices called ground-fault interrupters (Fig. 24.17) are used in kitchen, bathroom, basement, and other hazardous circuits. These devices sense a slight imbalance in current flowing in the two wires of a circuit, and shut off the circuit in less than a millisecond on the assumption that the "missing" current is leaking to ground—perhaps through a person.

**FIGURE 24.16**  Plugs: (*left*) grounded; (*right*) polarized.

**FIGURE 24.17**  A ground-fault interrupter.

## Big Picture

The big idea here is **electric current**—the flow of electric charge—and its microscopic cousin, **current density**. With current we don't have electrostatic equilibrium, and there's usually an electric field in a current-carrying conductor. **Ohm's law** is an empirical statement—not a fundamental law of physics—that relates current and voltage, or current density and electric field.

## Key Concepts and Equations

Quantitatively, **current** is defined as the rate of charge flow:

$$I = \frac{\Delta Q}{\Delta t}$$

**Current density** is the current per unit area. Its magnitude is given by:

$$J = \frac{I}{A}$$

Charge $\Delta Q$ crosses this area in time $\Delta t$.

There are $n$ charge carriers per unit volume, with charge $q$ and drift velocity $v_d$.

Microscopically, current depends on the density of charge carriers, their charge, and the **drift velocity**:

$$I = nqAv_d \quad \text{and}$$
$$\vec{J} = nq\vec{v}_d.$$

The microscopic version of **Ohm's law** relates electric field, current density, and **conductivity** $\sigma$ (or its inverse, **resistivity** $\rho$):

$$\vec{J} = \sigma\vec{E}.$$

The macroscopic version relates voltage, current, and resistance:

$$I = V/R.$$

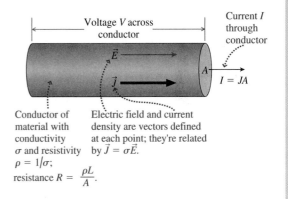

Voltage $V$ across conductor

Current $I$ through conductor

$I = JA$

Conductor of material with conductivity $\sigma$ and resistivity $\rho = 1/\sigma$; resistance $R = \frac{\rho L}{A}$.

Electric field and current density are vectors defined at each point; they're related by $\vec{J} = \sigma\vec{E}$.

Electric power is the product of voltage and current:

$$P = IV.$$

Using Ohm's law, this can also be written

$$P = I^2R$$

$$P = \frac{V^2}{R}$$

## Cases and Uses

Different types of conductors have different conduction mechanisms. In **metals**, free electrons carry the current; in **ionic solutions**, both positive and negative ions are involved; in **plasmas**, the charge carriers are free electrons and ions; and in **semiconductors**, both electrons and positive holes carry current, with semiconductor conduction properties readily adjustable. **Superconductors** are materials that exhibit zero resistance at sufficiently low temperatures.

Conduction in a metal

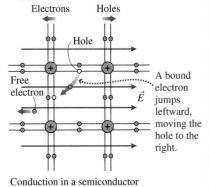

Electron and hole move oppositely in an electric field.

Electrons    Holes

Hole

Free electron

A bound electron jumps leftward, moving the hole to the right.

Conduction in a semiconductor

**Electrical safety** is a matter of avoiding currents high enough to cause biological harm, and that means avoiding voltages high enough to drive such currents.

# For Thought and Discussion

1. If you physically move an electric conductor, does this constitute a current?

2. What is the difference between current and current density?

3. A constant electric field generally produces a constant drift velocity. How is this consistent with Newton's assertion that force results in acceleration, not velocity?

4. When caught in the open in a lightning storm, a person should crouch low with feet close together rather than lie flat on the ground. Why?

5. Good conductors of electricity are often good conductors of heat. Why might this be?

6. Why can current persist forever in a superconductor with no applied voltage?

7. Does an electric stove burner draw more current when it's first turned on or when it's fully hot?

8. A person and a cow are standing in a field when lightning strikes the ground nearby. Why is the cow more likely to be electrocuted?

9. You put a 1.5-V battery across a piece of material and a 100-mA current flows through the material. With a 9-V battery the current increases to 400 mA. Is the material ohmic or not?

10. The resistance of a metal increases with increasing temperature, while the resistance of a semiconductor decreases. Why the difference?

11. A 50-W and a 100-W lightbulb are both designed to operate at 120 V. Which has the lower resistance?

12. Equation 24.8a suggests that no power can be dissipated in a superconductor because $R = 0$. But Equation 24.8b suggests the power should be infinite. Which is right, and why?

13. What's wrong with this news report: "A power-line worker was seriously injured when 4000 volts passed through his body"?

# Exercises and Problems

## Exercises

### Section 24.1 Electric Current

14. A wire carries 1.5 A. How many electrons pass through the wire in each second?

15. A 12-V car battery is rated at 80 ampere-hours, meaning it can supply 80 A of current for 1 hour before it becomes discharged. If you accidentally leave the headlights on until the battery discharges, how much charge moves through the lights?

16. Microbiologists measure the total current due to potassium ions $(K^+)$ moving through a cell membrane of a rock crab neuron cell as 30 nA. How many ions pass through the membrane each second?

17. The National Electrical Code specifies a maximum current of 10 A in 16-gauge (1.29-mm-diameter) copper wire. What is the corresponding current density?

### Section 24.2 Conduction Mechanisms

18. The electric field in an aluminum wire is 0.085 V/m. What is the current density in the wire?

19. What electric field is necessary to drive a 7.5-A current through a silver wire 0.95 mm in diameter?

20. A cylindrical tube of seawater carries a total electric current of 350 mA. If the electric field in the water is 21 V/m, what is the diameter of the tube?

21. A 1.0-cm-diameter rod carries a 50-A current when the electric field in the rod is 1.4 V/m. What is the resistivity of the rod material?

22. Use Table 24.1 to determine the conductivity of (a) copper and (b) seawater.

### Section 24.3 Resistance and Ohm's Law

23. What is the resistance of a heating coil that draws 4.8 A when the voltage across it is 120 V?

24. What voltage does it take to drive 300 mA through a 1.2-kΩ resistance?

25. What is the current in a 47-kΩ resistor with 110 V across it?

26. The "third rail" that carries the electric power to a subway train is a rectangular iron bar whose cross section measures 10 cm by 15 cm. What is the resistance of a 5.0-km length of this rail?

27. What current flows when a 45-V potential difference is imposed across a 1.8-kΩ resistor?

28. A uniform wire of resistance $R$ is stretched until its length doubles. Assuming its density and resistivity remain constant, what is its new resistance?

### Section 24.4 Electric Power

29. A car's starter motor draws 125 A with 11 V across its terminals. What is its power consumption?

30. A 4.5-W flashlight bulb draws 750 mA. (a) At what voltage does it operate? (b) What is its resistance?

31. A watch uses energy at the rate of 240 $\mu$W. How much current does it draw from its 1.5-V battery?

32. An electric stove burner with 35-Ω resistance consumes 1.5 kW of power. At what voltage does it operate?

33. What is the resistance of a standard 120-V, 60-W lightbulb?

### Section 24.5 Electrical Safety

34. Though rare, electrocution has been reported under wet conditions with voltages as low as 30 V. What resistance would be necessary to drive a fatal 100 mA at this voltage?

35. You touch a defective appliance while standing on the ground, and you feel the slight tingle of a 2.5-mA current. What's your resistance, assuming you're touching the "hot" side of the 120-V household wiring?

36. You have a typical resistance of 100 kΩ. (a) How much current could a 12-V car battery pass through you? (b) Would you feel this?

## Problems

37. An ion channel in a cell membrane carries 2.4 pA when it's open, but it's open only 20% of the time. (a) What is the average current in the channel? (b) If the channel opens for 1.0 ms, how many singly ionized ions pass through it in this time?

38. The filament in the lightbulb in Example 24.5 has a diameter of 0.050 mm. What is the current density in the filament? Compare it with the current density in a 12-gauge wire (diameter 2.1 mm) supplying current to the lightbulb.

39. A gold film in an integrated circuit measures 2.5 $\mu$m thick by 0.18 mm wide. It carries a current density of 0.68 MA/m². What is the total current?

40. A piece of copper wire joins a piece of aluminum wire whose diameter is twice that of the copper. The same current flows in both wires. The density of conduction electrons in copper is $1.1 \times 10^{29}$ m$^{-3}$; in aluminum it's $2.1 \times 10^{29}$ m$^{-3}$. Compare (a) the drift speeds and (b) the current densities in each.

41. In Fig. 24.18, a 100-mA current flows through a copper wire 0.10 mm in diameter, a 1.0-cm-diameter glass tube containing a salt solution, and a vacuum tube where the current is carried by an electron beam 1.0 mm in diameter. The density of conduction electrons in copper is $1.1 \times 10^{29}$ m$^{-3}$. The current in the solution is carried equally by positive and negative ions with charges $\pm 2e$; the density of each ion species is $6.1 \times 10^{23}$ m$^{-3}$. The electron density in the beam is $2.2 \times 10^{16}$ m$^{-3}$. Find the drift speed in each region.

**FIGURE 24.18** Problem 41

42. In a study of proteins mediating cell membrane transport, microbiologists measure current versus time through the cell membranes of oocytes (nearly mature egg cells) taken from the African clawed frog, *Xenopus*. The measured current versus time is given approximately by $I = 60t + 200t^2 + 4.0t^3$, with $t$ in seconds and $I$ in nA. Find the total charge that flows through the cell membrane in the interval from $t = 0$ to $t = 5.0$ s.

43. There is a potential difference of 2.5 V between opposite ends of a 6.0-m-long iron wire. (a) Assuming a uniform electric field in the wire, what is the current density? (b) If the wire diameter is 1.0 mm, what is the total current?

44. The maximum safe current in 12-gauge (2.1-mm-diameter) copper wire is 20 A. What are (a) the current density and (b) the electric field under these conditions?

45. Silver and iron wires of the same length and diameter carry the same current. How do the voltages across the two compare?

46. You have a cylindrical piece of material 2.4 cm long and 2.0 mm in diameter. When you attach a 9-V battery to the ends of the piece, a current of 2.6 mA results. Which material from Table 24.1 do you have?

47. How must the diameters of copper and aluminum wire be related if they are to have the same resistance per unit length?

48. Extension cords are often made from 18-gauge copper wire (diameter 1.0 mm). (a) Find the resistance per unit length of this wire. (b) An electric saw that draws 7.0 A is operated at the end of an 8.0-m-long extension cord. Find the potential difference between the wall outlet and the saw.

49. Engineers call for a power line with a resistance per unit length of 50 mΩ/km. What wire diameter is required if the line is made of (a) copper or (b) aluminum? (c) If the costs of copper and aluminum wire are \$4.65/kg and \$2.30/kg, respectively, which material is more economical? The densities of copper and aluminum are 8.9 g/cm$^3$ and 2.7 g/cm$^3$, respectively.

50. A solid, rectangular iron bar measures 0.50 cm by 1.0 cm by 20 cm. Find the resistance between each of the three pairs of opposing faces, assuming the faces in question are equipotentials.

51. Corrosion at battery terminals results in increased resistance and is a frequent cause of hard starting in cars. In an effort to diagnose hard starting, a mechanic measures the voltage between the battery terminal and the wire carrying current to the starter motor. While the motor is cranking, this voltage is 4.2 V. If the motor draws 125 A, what is the resistance at the battery terminal?

52. During a "brownout," the power-line voltage drops from 120 V to 105 V. By how much does the thermal power output of a 1500-W stove burner drop, assuming its resistance remains constant?

53. Two cylindrical resistors are made from the same material and have the same length. When connected across the same battery, one dissipates twice as much power as the other. How do their diameters compare?

54. A 2000-horsepower electric railroad locomotive gets its power from an overhead wire with 0.20 Ω/km. The potential difference between wire and track is 10 kV. Current returns through the track, whose resistance is negligible. (a) How much current does the locomotive draw? (b) How far from the power plant can the train go before 1% of the energy is lost in the wire?

55. A 100% efficient electric motor is lifting a 15-N weight at 25 cm/s. If the motor is connected to a 6.0-V battery, how much current does it draw?

56. A power plant produces 1000 MW to supply a city 40 km away. Current flows from the power plant on a single wire of resistance 0.050 Ω/km, through the city, and returns via the ground, assumed to have negligible resistance. At the power plant the voltage between the wire and ground is 115 kV. (a) What is the current in the wire? (b) What fraction of the power is lost in transmission?

57. A 240-V electric motor is 90% efficient, meaning that 90% of the energy supplied to it ends up as mechanical work. If the motor lifts a 200-N weight at 3.1 m/s, how much current does it draw?

58. A metal bar has a rectangular cross section 5.0 cm by 10 cm, as shown in Fig. 24.19. The bar has a nonuniform conductivity, ranging from zero at the bottom to a maximum at the top. As a result, the current density increases linearly from zero at the bottom to 0.10 A/cm$^2$ at the top. What is the total current in the bar?

**FIGURE 24.19** Problem 58

59. An immersion-type heating coil is connected to a 120-V outlet and immersed in a 250-ml cup of water initially at 10°C. The water comes to a boil in 85 s. What are (a) the power and (b) the resistance of the heater? Assume no heat loss, and neglect the mass of the heater.

60. The resistivity of copper as a function of temperature is given approximately by $\rho = \rho_0[1 + \alpha(T - T_0)]$, where $\rho_0$ is the value listed in Table 24.1 for 20°C, $T_0 = 20$°C, and $\alpha = 4.3 \times 10^{-3}$°C$^{-1}$. Find the temperature at which copper's resistivity is twice its room-temperature value.

61. Each atom in aluminum contributes about 3.5 conduction electrons. What is the drift speed in a 2.1-mm-diameter aluminum wire carrying 20 A?

62. A circular pan of radius $b$ has a plastic bottom and metallic side wall of height $h$. It is filled with a solution of resistivity $\rho$. A metal disk of radius $a$ and height $h$ is placed at the center of the pan, as shown in Fig. 24.20. The side and disk are essentially perfect conductors. Show that the resistance measured from side to disk is $\rho \ln(b/a)/2\pi h$.

**FIGURE 24.20** Problem 62

63. The radius of a wire, made from material with uniform resistivity $\rho$, increases linearly from $a$ to $b$ over its length $L$ (Fig. 24.21). Assuming that each circular cross section is an equipotential when the wire carries current, show that its resistance is the same as that of a wire made from the same material with a constant elliptical cross section, where $a$ and $b$ are the semi-axes.

**FIGURE 24.21** Problem 63

64. The current density in a particle beam with circular cross section of radius $a$ points along the beam axis with a magnitude that varies radially from a central value $J_0$ at $r = 0$ to half that value on the edge, $r = a$. What is the total current in the beam?

65. A cylindrical resistor is 5.0 mm in diameter and 1.5 cm long. It's made of a composite material whose resistivity varies from one end to the other according to the equation $\rho = \rho_0(1 + x/L)e^{x/L}$, for $0 \le x \le L$, where $\rho_0 = 2.41 \times 10^{-3} \; \Omega \cdot m$. What is its resistance?

66. Your roommate, also taking physics, calls from the college radio station where a small disaster has occurred. The transmitter has malfunctioned, and a capacitor is needed to make repairs. One cannot be obtained from the transmitter manufacturer for several days. Your roommate has fashioned a capacitor out of two sheets of metal, with area 10 cm², separated by a thin 1-mm sheet of glass.

A voltage of 100 V is placed across the homemade capacitor. Your roommate asks you to calculate the capacitance and amount of energy stored in the capacitor, and to verify that when the capacitor is disconnected, it will take at least 5 minutes to discharge. The resistivity of the glass is $\rho = 1.2 \times 10^{13} \; \Omega \cdot m$, and the dielectric constant is $\kappa = 5.6$. *Hint:* Remember that $I = dQ/dt$.

67. You're working in quality control for a major electronics company. The lead quality engineer is backlogged with work and asks you to check a calculation for a new integrated circuit. A circuit requires a small 470-$\Omega$ resistor. It will be 10 $\mu$m long, 1.4 $\mu$m wide, and 0.85 $\mu$m thick. The connections to the circuitry will be made at the ends of the 10-$\mu$m length. The engineer is suspicious of a prior calculation that calls for a material with resistivity $\rho = 8.45 \times 10^{-6} \; \Omega \cdot m$ and asks you to check the numbers. What should the correct resistivity be?

68. You're working for an architectural firm that is designing a new office building. The owner of the building has raised concerns about the amount of time it will take to travel from floor to floor and has put into the contract a stipulation that the time to travel 60 feet must not exceed 20 seconds. The elevator has a weight of 8.5 kN, is powered by a 480-V motor that draws 24 A, and is 85% efficient. Will the customer be happy with its performance?

69. You're the CEO of a courier company, and you decide to select an electric car for your fleet of vehicles. A particular car has a mass of 1500 kg and is powered by 26 12-V batteries connected in series, for a total of 312 V. Batteries are rated at 100 ampere-hours, which means they can deliver 100 A for 1 h, 1 A for 100 h, or any combination that equals 100 A·h. When the batteries are connected in series, the charge output of all the batteries is the same as one battery—in this case 100 A·h. The motor is 85% efficient in converting electrical energy to mechanical energy in the drive wheels. A test report says the car can climb a 10° slope at a speed of 45 km/h. Assuming the batteries are fully charged, how long can the car maintain this speed at that angle?

# Answers to Chapter Questions

## Answer to Chapter Opening Question

Collisions between electrons and the metal ions in the filament dissipate electric energy as heat. The energy results from the electrons' being accelerated by an electric field.

## Answers to GOT IT? Questions

24.1 (a) current, right to left; (b) current, up; (c) current, left; (d), (e) no current.

24.2 (a) $J_A < J_B$; (b) $E_A < E_B$; (c) $v_{dA} < v_{dB}$.

24.3 (a) (2); it's twice as long as (3) but with one-fourth the area and half the resistivity; (b) (1), because it has the lowest resistance.

■ Festive lights decorate a city. If one of them burns out, they all go out. Are they connected in series or in parallel?

An **electric circuit** is a collection of electrical components connected by conductors. Human-made circuits range from simple flashlights to computers. Electric circuits also exist in nature, including your own nervous system and Earth's atmospheric circuit in which thunderstorms are the batteries and the atmosphere a resistor. Understanding circuits helps you use effectively and safely the myriad electrical devices in your life, and can even help you design new devices or troubleshoot old ones.

## 25.1 Circuits, Symbols, and Electromotive Force

We diagram circuits using standard symbols for circuit components and lines to represent wires (Fig. 25.1). We usually approximate wires as perfect conductors; then all points connected by a wire are at the same potential and are electrically equivalent. Realizing this will help you understand circuit diagrams.

Resistor   Capacitor   Source of emf   Voltmeter   Switch   Ammeter   Variable resistor   Variable capacitor   Ground   Fuse

**FIGURE 25.1** Common circuit symbols.

Gravitational field $\vec{g}$ is analogous to electric field $\vec{E}$.

Mass is analogous to charge.

Lifting against gravity represents the energy–conversion process in an emf.

Charge "falls" through an external circuit, dissipating energy in collisions.

**FIGURE 25.2** Gravitational analog for emf.

It takes an electric field to drive current through a conductor with nonzero resistance. But unless we actively maintain the field, charge will quickly move to establish electrostatic equilibrium, with no field inside the conductor and no current. To prevent this we need a device that can maintain a fixed potential difference and therefore an electric field in a current-carrying conductor. Such a device is called a source of **electromotive force**, or **emf**. (The name "force" here is inaccurate and is used only for historical reasons.) Most sources of emf have two **terminals** for connection to other circuit components. An emf converts some other form of energy to electrical energy by separating charge to maintain a fixed potential difference between its terminals. The most familiar example is a battery, in which chemical reactions drive charge to the two terminals. Others include electric generators, which convert mechanical to electrical energy; photovoltaic cells, which use sunlight to separate charge; and cell membranes, which separate charge to control ion flow into and out of the cell.

When a source of emf is connected to an external circuit, current flows through the circuit from the emf's positive terminal to the negative terminal. Energy-conversion processes in the emf then "lift" charge against the emf's internal electric field, maintaining a fixed potential difference across its terminals. The charge then "falls" through the external circuit, dissipating its energy in the circuit resistance. The result is a steady current, driven by the constant voltage across the emf. Figure 25.2 shows a gravitational analogy for an emf connected across an external circuit.

Quantitatively, emf is the work per unit charge involved in "lifting" charge against the electric field. Its units are therefore volts. An **ideal emf** maintains the same voltage across its terminals under all conditions. Real emfs have internal energy losses, and the terminal voltage may not equal the rated emf.

In Fig. 25.3 an ideal battery of emf $\mathcal{E}$ drives current through resistor $R$. We're assuming the wires connecting the battery and the resistor are perfect conductors, so the voltage across the resistor is equal to the battery's emf. Ohm's law then gives the resistor current: $I = \mathcal{E}/R$. Energetically, this circuit is analogous to Fig. 25.2: Charge gains $\mathcal{E}$ joules per coulomb as it's "lifted" against the electric field inside the battery, then dissipates that energy in heating the resistor.

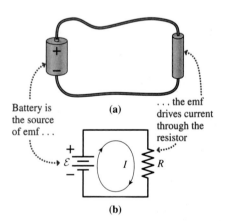

Battery is the source of emf . . .

**(a)**

. . . the emf drives current through the resistor

$\mathcal{E}$   $I$   $R$

**(b)**

**FIGURE 25.3** A circuit consisting of a battery and a resistor: (a) physical circuit; (b) schematic diagram.

---

✓ **TIP** Don't Get Hung Up on Wires

We approximate wires as perfect conductors, so it takes no potential difference to drive current through a wire. Thus all points on the wire are at the same potential (that is, the same voltage) and are electrically equivalent. That means there are many ways to draw the same circuit; as long as two points are connected by a wire, that's all that matters. Of course real wires have some resistance, but if it's negligible compared with other resistances in the circuit, then we can approximate the wires as being ideal.

---

**GOT IT? 25.1** The figure shows three circuits. Which two of them are electrically equivalent?

(a)          (b)          (c)

## 25.2 Series and Parallel Resistors

We considered series and parallel connections for capacitors in Section 23.3. Series and parallel are the two simplest ways to connect *any* electric components. Two components are in series if the current flowing through one component has nowhere to go but through the other component. Two components are in parallel if they're connected together at each end. Here we'll consider series and parallel resistors.

### Series Resistors

Figure 25.4 shows a circuit with two resistors in series. We'd like to know the current through and the voltage across each resistor. Neither is connected directly across the battery, so we can't argue that either resistor "sees" the battery emf. But the resistors are in series, and that means the only place for current to go after $R_1$ is through $R_2$. In a steady state, with no charge buildup in the circuit, that means the current through both resistors—and through the battery as well—must be the same. This is true whenever circuit components are in series:

> The current through circuit components in series is the same.

If $I$ is the current in Fig. 25.4, then by Ohm's law there must be a voltage $V_1 = IR_1$ across $R_1$ to drive the current through this resistor. Similarly, the voltage across $R_2$ is $V_2 = IR_2$. Thus, the voltage across the two resistors together is $V_1 + V_2 = IR_1 + IR_2$. But the battery is connected directly across this series combination, so we have $IR_1 + IR_2 = \mathcal{E}$, or

$$I = \frac{\mathcal{E}}{R_1 + R_2}$$

Comparison with Ohm's law in the form $I = V/R$ shows that the two resistors in series behave like an equivalent resistance equal to the sum of their resistances. In an obvious generalization to more resistors in series, we have

$$R_{\text{series}} = R_1 + R_2 + R_3 + \cdots \quad \text{(series resistors)} \quad (25.1)$$

In other words, resistors in series add.

Given the current, we can use Ohm's law in the form $V = IR$ to solve for the voltage across each resistor:

$$V_1 = \frac{R_1}{R_1 + R_2}\mathcal{E} \quad \text{and} \quad V_2 = \frac{R_2}{R_1 + R_2}\mathcal{E} \quad (25.2\text{a, b})$$

(a)

$R_1$ · Current from $R_1$ can't go anywhere but through $R_2$ ...

$R_2$

... so current through $R_1$ and $R_2$ is the same.

(b)

**FIGURE 25.4** A battery and two resistors in series: (a) physical circuit; (b) schematic diagram.

These expressions show that the battery voltage divides between the two resistors in proportion to their resistance. For this reason a series combination of resistors is called a **voltage divider**. Figure 25.5 depicts the voltages throughout the circuit.

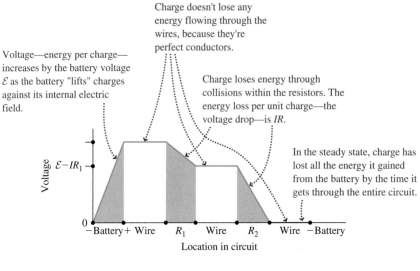

Charge doesn't lose any energy flowing through the wires, because they're perfect conductors.

Voltage—energy per charge—increases by the battery voltage $\mathcal{E}$ as the battery "lifts" charges against its internal electric field.

Charge loses energy through collisions within the resistors. The energy loss per unit charge—the voltage drop—is $IR$.

In the steady state, charge has lost all the energy it gained from the battery by the time it gets through the entire circuit.

**FIGURE 25.5** Voltages in the circuit of Fig. 25.4, with $V = 0$ at the negative battery terminal.

✓**TIP** How Does the Battery Know?

How does the battery in Fig. 25.4 "know" how much current to supply? For a brief instant when the circuit is first connected, it doesn't. But in a very short time an electric field is established throughout the wires and resistors, and the circuit settles into a steady state, with the same current everywhere. Later, with circuits including capacitors, we'll analyze the approach to the steady state; for now, assume that the circuit reaches that state essentially instantaneously.

**EXAMPLE 25.1** **Series Resistors: Designing a Voltage Divider**

A lightbulb with resistance 5.0 $\Omega$ is designed to operate at a current of 600 mA. To operate this lamp from a 12-V battery, what resistance should you put in series with it?

**INTERPRET** This problem is about a series circuit like Fig. 25.4, with a lightbulb and unknown resistance for the two resistors.

**DEVELOP** We've sketched the circuit in Fig. 25.6, taking $R_1$ as the unknown and $R_2$ as the 5-$\Omega$ lightbulb. The same current flows through

$R_1 = ?$

The lightbulb is a resistor.

$\mathcal{E} = 12\ V$

$R_2 = 5\ \Omega$

**FIGURE 25.6** Sketch for Example 25.1.

series resistors, so our plan is to find an expression for that current and then solve for the value of $R_1$ that will make the current 600 mA. Since resistors in series add, the current through both resistors follows from Ohm's law: $I = \mathcal{E}/(R_1 + R_2)$.

**EVALUATE** We solve for $R_1$ to get

$$R_1 = \frac{\mathcal{E} - IR_2}{I} = \frac{12\ V - (0.60\ A)(5.0\ \Omega)}{0.60\ A} = 15\ \Omega$$

**ASSESS** Make sense? The lightbulb's operating voltage is

$$V = IR_2 = (0.60\ A)(5.0\ \Omega) = 3.0\ V.$$

This is one-fourth of the battery voltage, so Equation 25.2b shows that the bulb's 5 $\Omega$ resistance should be one-fourth of the total. That makes the total 20 $\Omega$, leaving 15 $\Omega$ for $R_1$. Incidentally, this isn't a very efficient way to run the bulb, since a lot more energy gets dissipated in $R_1$ than goes into lighting the bulb. Better to use a 3-V battery and no resistor. ∎

**GOT IT? 25.2** Rank order the voltages across the identical resistors $R$ at the top of each circuit shown, and give the actual voltage for each. In (a) the second resistor has the same resistance $R$, and in (b) the gap is an open circuit (infinite resistance).

(a)  (b)  (c)

## Real Batteries

What's the difference between the two 1.5-V batteries in Fig. 25.7? If they were ideal, both would maintain 1.5 V across their terminals no matter how much current was flowing. But these are real batteries. Their chemical reaction rates limit the current; it's not surprising that the larger battery can deliver more current.

**FIGURE 25.7** These batteries are both rated at 1.5 V, but they have different internal resistances.

We model a real battery as an ideal emf in series with an **internal resistance** (Fig. 25.8). Of course there is no ideal emf! The internal resistance is intrinsic to the battery, and there's no way to circumvent it. Some of it is actual resistance of the battery's materials, but most represents the limited rate at which chemical reactions can separate charge. For a given battery voltage, lower internal resistance implies a more powerful battery—one that can deliver more current.

Figure 25.9 shows that the internal resistance $R_{int}$ is in series with the external load $R_L$ to which the battery supplies power; the resulting circuit is a voltage divider. If $R_{int}$ is small compared with $R_L$, Equation 25.2b shows that the voltage across the load will be very nearly the battery voltage. Then the battery is behaving nearly ideally because it has essentially $\mathcal{E}$ volts across its terminals. But if we lower $R_L$, more current flows and more voltage drops across $R_{int}$—and that leaves less voltage at the battery terminals and across the load. Even if we short-circuit the battery (not a good idea!), we won't get an infinite current; in fact, we'll get $I = \mathcal{E}/R_{int}$, the most possible current this battery can deliver.

**FIGURE 25.8** A real battery modeled as an ideal emf in series with an internal resistance.

Current $I$ means a voltage $IR_{int}$ across $R_{int}$ . . .

. . . that leaves a terminal voltage less than $\mathcal{E}$.

$V_{terminal}$

**FIGURE 25.9** A real battery connected to an external load. Some voltage drops across the internal resistance, making the terminal voltage less than the battery's rated voltage.

**EXAMPLE 25.2** **Internal Resistance: Starting a Car**

Your car has a 12-V battery with internal resistance 0.020 Ω. When the starter motor is cranking, it draws 125 A. What's the voltage across the battery terminals while starting?

**INTERPRET** This problem is about a real battery connected to a load, as in Fig. 25.9. We identify one resistor as the internal resistance and the load resistance as the starter motor.

**DEVELOP** Figure 25.10 is our sketch, showing the internal resistance in series with the load. The current is the same everywhere in a series circuit, so we can use Ohm's law to find the voltage across $R_{\text{int}}$. Subtracting that voltage from the battery's emf will then tell what's left across the load.

**EVALUATE** For the internal resistance, Ohm's law gives

$$V_{\text{int}} = IR_{\text{int}} = (125 \text{ A})(0.020 \text{ Ω}) = 2.5 \text{ V}.$$

That leaves 12 V − 2.5 V or 9.5 V across the battery terminals.

**ASSESS** Make sense? That 9.5 V is substantially less than the battery's 12-V rating, so we're hardly treating it ideally. But the starter motor

These points are the battery terminals.

**FIGURE 25.10** Sketch for Example 25.2.

runs only briefly; most of the time the load on the battery—headlights, ignition system, electronics, and so on—draws far less current and so the battery behaves essentially like an ideal 12-V emf. A battery voltage of 9–11 V is typical during starting; much less than 9 V indicates a weak battery, a defective starter, or very cold weather. ∎

## Parallel Resistors

Figure 25.11 shows two resistors in parallel, connected across an ideal battery. Since the two resistors are connected at top and bottom by ideal wires, the voltage across each must be the same. We made this point in Chapter 23 when we discussed parallel capacitors, and it's worth repeating here:

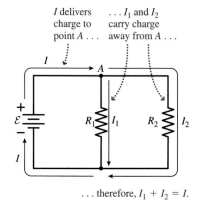

**FIGURE 25.11** Parallel resistors connected across a battery.

> The voltage across circuit elements in parallel is the same.

The parallel resistors are connected directly across the battery, so their common voltage is the battery emf $\mathcal{E}$. Applying Ohm's law then gives the current through each resistor:

$$I_1 = \frac{\mathcal{E}}{R_1} \quad \text{and} \quad I_2 = \frac{\mathcal{E}}{R_2}$$

At point $A$ in Fig. 25.11, a current $I$ brings in charge from the battery, while the currents $I_1$ and $I_2$ carry charge away. Charge can't accumulate at this point (see Problem 67), so the incoming and outgoing currents must be equal: $I = I_1 + I_2$. Using our expressions for the two resistor currents gives

$$I = \frac{\mathcal{E}}{R_1} + \frac{\mathcal{E}}{R_2} = \mathcal{E}\left(\frac{1}{R_1} + \frac{1}{R_2}\right)$$

Comparison with Ohm's law in the form $I = V/R$ shows that the equivalent resistance of the parallel combination is given by

$$\frac{1}{R_{\text{parallel}}} = \frac{1}{R_1} + \frac{1}{R_2}$$

This result readily generalizes to more parallel resistors:

$$\frac{1}{R_{\text{parallel}}} = \frac{1}{R_1} + \frac{1}{R_2} + \frac{1}{R_3} + \cdots \quad \text{(parallel resistors)} \tag{25.3a}$$

In other words, resistors in parallel add reciprocally. Equation 25.3a shows that the resistance of a parallel combination is always lower than that of the lowest resistance in the combination.

An analogy with highway traffic shows why this makes sense: adding a lane to a crowded highway eases congestion (i.e., lowers the overall resistance), allowing a greater traffic flow (e.g., greater current). Putting one resistor in parallel with another is like adding an extra traffic lane.

When there are only two parallel resistors, we can rewrite Equation 25.3a using a common denominator to obtain

$$R_{\text{parallel}} = \frac{R_1 R_2}{R_1 + R_2} \tag{25.3b}$$

Note that *parallel* resistors combine in the same way as *series* capacitors, and vice versa.

---

**GOT IT? 25.3**  The figure shows all four possible combinations of three identical resistors. Rank them from highest to lowest resistance.

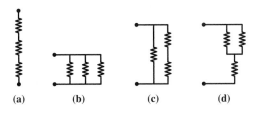

(a)        (b)        (c)        (d)

---

## Analyzing Circuits

Many circuits contain series and parallel combinations. We analyze these circuits using the tactics outlined next, following the approach we used with series and parallel capacitors in Example 23.3.

**TACTICS 25.1**  **Analyzing Circuits with Series and Parallel Components**

**1**  Identify series and parallel combinations. Remember that components are in parallel only if they're connected directly together at each end. Components are in series only if current through one component has no place to go but through the second. If you can't find at least one series or parallel combination, then you have to use the methods of Section 25.3.

**2**  Solve for the series and parallel equivalents using Equations 25.1 and 25.3 for resistors:

$$R_{\text{series}} = R_1 + R_2 + R_3 + \cdots \tag{25.1}$$

$$\frac{1}{R_{\text{parallel}}} = \frac{1}{R_1} + \frac{1}{R_2} + \frac{1}{R_3} + \cdots \tag{25.3a}$$

$$R_{\text{parallel}} = \frac{R_1 R_2}{R_1 + R_2} \tag{25.3b}$$

If you're dealing with capacitors, use Equations 23.6 and 23.5, respectively.

**3**  Redraw the circuit, replacing series and parallel combinations with their one-component equivalents.

**4**  Repeat Steps 1–3, each time identifying series and parallel combinations and then reducing each to a single equivalent. Continue until either you've found the quantity you're asked for or the circuit consists of just an emf and one other component. You can then solve for the current in this component.

**5**  Work backward, replacing series and parallel equivalents with combinations of individual components. At each point apply Ohm's law, $I = V/R$, to find the currents through and/or the voltages across the individual components. As you work backward, remember that series components carry the same current as their series equivalent, and parallel components have the same voltage as their parallel equivalent. Continue until you're able to evaluate the quantity you're asked for.

**EXAMPLE 25.3** **Analyzing a Circuit: Series and Parallel Components**

Find the current through the 2-$\Omega$ resistor in the circuit of Fig. 25.12a.

Following steps ❶ - ❸ of Tactics 25.1 gives the equivalent resistance of the parallel 2-$\Omega$ and 4-$\Omega$ resistors in (a).

Applying steps ❶ - ❸ again leads to the equivalent resistance of the series 1-$\Omega$, 1.33-$\Omega$, and 3-$\Omega$ resistors in (b).

**FIGURE 25.12** Analyzing a circuit.

**INTERPRET** This problem asks for the current in one resistor that's part of a more complex circuit. So it's about analyzing a circuit with series and parallel components.

**DEVELOP** We follow the steps in Tactics 25.1:
1. We identify the 2-$\Omega$ and 4-$\Omega$ resistors as being in parallel, and we find no other series or parallel resistor combinations. Note that the 1-$\Omega$ resistor, for example, is *not* in series with either the 2-$\Omega$ or 4-$\Omega$ resistor because current leaving the 1-$\Omega$ resistor can take either of two paths.
2. We apply Equation 25.3b, $R_{parallel} = R_1 R_2/(R_1 + R_2)$, to find the parallel combination: $(2\,\Omega)(4\,\Omega)/(2\,\Omega + 4\,\Omega) = 1.33\,\Omega$.
3. We redraw the circuit as Fig. 25.12b, replacing the two parallel resistors with their 1.33-$\Omega$ equivalent.
4. We repeat Steps 1–3 for the circuit in Fig. 25.12b, this time finding a series combination of three resistors. Applying Equation 25.1, $R_{series} = R_1 + R_2 + R_3$, gives 5.33 $\Omega$ for the equivalent resistance, and we redraw the circuit to get the simple circuit of Fig. 25.12c. Ohm's law, $I = V/R$, gives the current in the 5.33-$\Omega$ equivalent resistance: $I_{5.33\,\Omega} = (12\text{ V})/(5.33\,\Omega) = 2.25$ A.
5. Now we work backward, "unsimplifying" the circuit. That 5.33-$\Omega$ resistor is really the series combination in Fig. 25.12b; since the current through series components is the same, 2.25 A flows through each resistor—including the 1.33-$\Omega$ resistor that's really the parallel combination shown in Fig. 25.12a. We want the current in the 2-$\Omega$ member of that combination, and we could get that if we knew the voltage across it. But the voltage across parallel components is the same, and the same as the voltage across their parallel equivalent—in this case the 1.33-$\Omega$ resistance. We've found the current through that resistance, so Ohm's law gives the voltage: $V_{1.33\,\Omega} = I_{1.33\,\Omega} R_{1.33\,\Omega} = (2.25\text{ A})(1.33\,\Omega) = 3.0$ V.

**EVALUATE** Our last result is the voltage across each of the original parallel resistors, including the 2-$\Omega$ resistor whose current we want. So we're finally ready to compute our answer: $I_{2\,\Omega} = V_{2\,\Omega}/R_{2\,\Omega} = (3.0\text{ V})/(2.0\,\Omega) = 1.5$ A. Done!

**ASSESS** Make sense? A total of 2.25 A is flowing around the circuit; when it encounters the parallel combination, more should flow through the lower resistance, which is just what we found. Quantitatively, the current divides in inverse proportion to the parallel resistances, with 1.5 A through 2 $\Omega$, and half as much, 0.75 A, through 4 $\Omega$. Note how in solving this problem, we used Ohm's law to find, alternately, voltage and then current in different resistances. ∎

✓**TIP** Using Ohm's Law

Ohm's law relates the *voltage across a resistor* to the *current through that resistor*. It does *not* relate arbitrary voltages and currents anywhere in a circuit. Just because there's a 12-V battery in Fig. 25.12 doesn't mean there's 12 V across the 2-$\Omega$ resistor. And just because we found a total current of 2.25 A in Fig. 25.12c doesn't mean that's the current through the 2-$\Omega$ resistor.

**GOT IT? 25.4** The figure shows a circuit with three identical lightbulbs and a battery. (a) Which, if any, of the bulbs is brightest? (b) What happens to each of the other two bulbs if you remove bulb C?

## 25.3 Kirchhoff's Laws and Multiloop Circuits

Some circuits can't be simplified using series and parallel combinations. This happens when there's more than one emf, or when components are connected in complex ways. In Fig. 25.13, are $R_1$ and $R_2$ in parallel? No, because $R_3$ separates their lower ends. Are $R_1$ and $R_4$ in series? No, because current splits after leaving $R_1$. Analyzing such circuits requires a more general technique.

**FIGURE 25.13** This circuit can't be analyzed using series and parallel combinations.

### Kirchhoff's Laws

Charges moving through a circuit gain energy at emfs and lose energy in resistors or other circuit components, as we diagrammed in Fig. 25.5. If we go completely around a circuit, the *changes* in energy per unit charge—that is, increases and decreases in voltage—sum to zero. This is **Kirchhoff's loop law**, and it holds for any closed loop even if it's part of a more complex circuit: **The sum of voltage changes around a closed loop is zero.** The loop law is essentially a statement of energy conservation in circuits.

In analyzing parallel resistors, we saw that the current flowing into point A in Fig. 25.11 had to equal the current flowing out. That's because charge is conserved, and in a steady state charge can't be accumulating anywhere in a circuit. A junction of three or more wires, like point A in Fig. 25.11, is a **node**. If we count the currents flowing into a node as positive and those flowing out as negative, then we can state **Kirchhoff's node law: The sum of currents at any node is zero.**

### Multiloop Circuits

Kirchhoff's laws allow us to analyze even the most complex circuits; the following strategy details the approach.

---

**PROBLEM SOLVING STRATEGY 25.1** **Multiloop Circuits**

**INTERPRET**

- Identify circuit loops and nodes. A loop is any complete closed path; a node is any point where three or more wires meet.
- Label the currents at each node, assigning a direction to each. The directions are arbitrary, and the actual direction may not be obvious.

**DEVELOP**

- For all but one node, write equations expressing Kirchhoff's node law: The sum of the currents at each node is zero. Take a current flowing into the node as positive, a current flowing out as negative.
- For as many independent loops as necessary, write equations expressing Kirchhoff's loop law: The sum of the voltage changes around a closed loop is zero. You can go either clockwise or counterclockwise, following these rules:
  - ☐ The voltage change going through a battery from the negative to the positive terminal is $+\mathcal{E}$; the voltage change from $+$ to $-$ is $-\mathcal{E}$.
  - ☐ For resistors traversed in the direction you've assigned to the current, the voltage change is $-IR$; for the opposite direction, it's $+IR$.
  - ☐ For other circuit elements (which we haven't yet encountered) use the characteristics of that element to determine the voltage change.
- You don't need equations for all the nodes and loops because some are redundant, as you'll see in the next example.

**EVALUATE** Solve the equations to determine the unknown currents or other quantities.

**ASSESS** Assess your answer to see that it makes sense, paying particular attention to signs. A negative answer for a current means that the current actually flows opposite the direction you arbitrarily assigned.

---

**EXAMPLE 25.4** **Using Kirchhoff's Laws: A Multiloop Circuit**

Find the current in resistor $R_3$ of Fig. 25.14a, following the steps in Problem-Solving Strategy 25.1.

(a)

(b)

**FIGURE 25.14** Example 25.4.

**INTERPRET** In Fig. 25.14b we've redrawn the circuit with three loops and two nodes identified. We also labeled three distinct currents that flow into or out of node A. The directions shown are completely arbitrary and may or may not be the actual directions of current flow. Because currents in series elements are the same, we can put the current labels anywhere on the series paths leading to the node. Our drawing shows that the same currents flow at node B; that's why one of the node equations is redundant. Note also that loop 3 comprises parts of loops 1 and 2, so equations for any two of these loops contain all the information we need. So here one of the loop equations is redundant.

**DEVELOP** We need a current equation for one of the two nodes. Given the arbitrary current directions, the equation at node A is

$$-I_1 + I_2 + I_3 = 0 \quad \text{(node A)}$$

To get the equation for loop 1, let's go counterclockwise around the loop, as shown. Starting at node A, we first encounter a positive voltage change $+\mathcal{E}_1$, then $-I_1 R_1$, then $-I_3 R_3$. So the loop 1 equation is $\mathcal{E}_1 - I_1 R_1 - I_3 R_3 = 0$. Here it's simplest if we substitute the numerical values shown in Fig. 25.14a and temporarily drop the units to avoid clutter. Then we have

$$6 - 2I_1 - I_3 = 0 \quad \text{(loop 1)}$$

Loop 2 is similar except here we're going "backward" through $R_2$, so its term is positive:

$$9 + 4I_2 - I_3 = 0 \quad \text{(loop 2)}$$

**EVALUATE** We want $I_3$, so we eliminate the other two currents. The node equation gives $I_1 = I_2 + I_3$; substituting in the loop 1 equation gives $6 - 2I_2 - 3I_3 = 0$ or $I_2 = \frac{1}{2}(6 - 3I_3)$. We use this result in the loop 2 equation to get $9 + 2(6 - 3I_3) - I_3 = 0$, or $21 - 7I_3 = 0$. Solving gives our answer: $I_3 = 3$ A.

**ASSESS** We assigned $I_3$ an upward direction through $R_3$, so our positive answer means that this is indeed the direction of the current in $R_3$. This makes sense because both batteries have their negative terminals at node A. If either battery had been reversed, however, the situation wouldn't have been so clear and we would have had to rely on the algebraic sign to determine the current direction. Even with the circuit as shown in Fig. 25.14, the directions of $I_1$ and $I_2$ depend on the relative strengths of the two batteries. With the values we're given, $I_2$ comes out $-1.5$ A, showing that the current actually flows downward in $R_2$. But if we reduce $\mathcal{E}_2$ to 2 V, $I_2$ becomes zero; lower still, and it flows upward and "backward" through $\mathcal{E}_2$. ∎

---

**APPLICATION** **The Cell Membrane**

Many natural systems can be modeled as electric circuits. In 1952, Alan L. Hodgkin and Andrew F. Huxley developed a circuit model for the cell membrane; their work won them a share of the 1963 Nobel Prize for Physiology or Medicine. The figure shows a simplified version of the Hodgkin–Huxley model. The batteries $\mathcal{E}_K$, $\mathcal{E}_{Na}$, and $\mathcal{E}_L$ represent the electrochemical effects of potassium, sodium, and other ions, respectively; their emfs have values in the tens of millivolts. $R_K$, $R_{Na}$, and $R_L$ are the resistances the cell membrane offers to each ionic species. The currents $I_K$, $I_{Na}$, and $I_L$ represent ion flows across the membrane, and their values and signs follow from solving a multiloop-circuit problem similar to Example 25.4. The voltage $V_M$ is the membrane potential between the inside and outside of the cell.

The Hodgkin–Huxley model also contains a capacitor, which causes time-dependent behavior of the sort we'll see in Section 25.5.

**FIGURE 25.15** Correct (a) and incorrect (b) ways to measure the voltage across $R_2$.

## 25.4 Electrical Measurements

### Voltmeters

A **voltmeter** is a device that indicates the potential difference across its two terminals. The indicator is usually a digital readout, although older meters use a moving needle. Potential difference—voltage—is a property of two points, and therefore to measure the voltage between two points, we connect the two terminals of the voltmeter to those points. So to measure the voltage across resistor $R_2$ in Fig. 25.15a we connect the voltmeter *across* $R_2$, as shown. We do *not* break the circuit and insert the meter as in Fig. 25.15b, for then we wouldn't be measuring the voltage *across* the resistor; in fact, we would radically alter the circuit.

How good is our voltage measurement? That depends on two things: The first, obviously, is the accuracy of the meter in reading the voltage across it. The second is more subtle: Unless the voltmeter in Fig. 25.15a has infinite resistance, it alters the circuit and changes the voltage across $R_2$. Even if the meter is perfectly accurate, it won't read exactly what we want—namely, the voltage across $R_2$ *before* we connected the meter. An ideal voltmeter, therefore, should have infinite resistance so it doesn't draw any current and therefore doesn't alter the circuit it's measuring. Modern digital meters come close to this ideal, with typical meter resistances of 10 M$\Omega$ or higher.

---

**EXAMPLE 25.5** **Measuring Voltage: Two Voltmeters**

You want to measure the voltage across the 40-$\Omega$ resistor in Fig. 25.16. (a) What reading would an ideal voltmeter give? (b) What reading would a voltmeter with a resistance of 1000 $\Omega$ give?

**FIGURE 25.16** What's the voltage across the 40-$\Omega$ resistor?

**INTERPRET** Since the voltmeter behaves like a resistance, this is a problem about two circuits with different voltmeter resistances. We've learned that the right way to use a voltmeter is to connect it across the component whose voltage we want. Here that's the voltage across the 40-$\Omega$ resistor, so the voltmeter goes in parallel with the resistor.

**DEVELOP** Figure 25.17 is our drawing, showing the circuit with the voltmeter across the 40-$\Omega$ resistor. Equation 25.3b, $R_{parallel} = R_1 R_2/(R_1 + R_2)$, determines the equivalent resistance of the resistor and voltmeter in parallel. With a single equivalent resistance we'll have a voltage divider, for which Equation 25.2b, $V_2 = [R_2/(R_1 + R_2)]\mathcal{E}$, gives the voltage.

**FIGURE 25.17** A nonideal voltmeter ($R_m$) alters the circuit.

**EVALUATE** (a) An ideal voltmeter makes no change in the circuit, so here we have a voltage divider with just the 40-$\Omega$ resistor for $R_2$. Equation 25.2b then gives

$$V_{40\,\Omega} = \left(\frac{40\ \Omega}{80\ \Omega + 40\ \Omega}\right)(12\ \text{V}) = 4.00\ \text{V}$$

(b) With the 1000-$\Omega$ meter, we first use Equation 25.3b to find the parallel resistance:

$$R_{parallel} = \frac{(40\ \Omega)(1000\ \Omega)}{40\ \Omega + 1000\ \Omega} = 38.5\ \Omega$$

Now we have a voltage divider with $R_2 = 38.5\ \Omega$, giving a voltage across the parallel combination—and therefore across the 40 $\Omega$ resistor—of

$$V_{40\,\Omega} = \left(\frac{38.5\ \Omega}{80\ \Omega + 38.5\ \Omega}\right)(12\ \text{V}) = 3.95\ \text{V}$$

**ASSESS** The reading with the nonideal voltmeter is lower than with the ideal one. That makes sense because the parallel combination of resistor and meter has a lower resistance than the resistor alone, and thus draws more current. That means a larger voltage drop across the 80-$\Omega$ resistor, leaving less voltage across the 40-$\Omega$ resistor. The error here is a significant 2.5%. ∎

---

**GOT IT? 25.5** If an ideal voltmeter is connected between points $A$ and $B$ in the figure, what will it read? All the resistors have the same resistance $R$.

**FIGURE 25.18** Correct (a) and incorrect (b) ways to connect an ammeter.

## Ammeters

An **ammeter** measures the current flowing *through* itself. To measure the current through a circuit element, it's necessary to break the circuit and insert the ammeter in *series* with that element (Fig. 25.18a); only then will all the current also go through the meter. Connecting the ammeter across the resistor as in Fig. 25.18b is wrong because then the current through the resistor isn't going through the meter.

If the ammeter has any resistance, the total resistance of the circuit will increase with the meter connected in series. This in turn will decrease the current, giving an incorrect reading. So an ideal ammeter should have zero resistance. In practice, ammeter resistance should be much lower than typical resistances in the circuit being measured.

✓**TIP** Watch Your Language

A voltmeter measures potential difference *between* two points; hence, we connect it *across*—that is, in parallel with—the circuit element whose voltage we wish to measure. An ammeter measures the current *through* itself; hence, we connect it in *series* with the circuit element whose current we wish to measure. If you get used to voltages appearing *across* things and currents flowing *through* them, you'll have no trouble connecting meters. The ways to connect meters, and the words *across* for voltage and *through* for current, go back to the definitions of potential difference as a property of two points and of current as a flow.

## Ohmmeters and Multimeters

Often we want the resistance of a particular component. Connecting a known voltage in series with an ammeter and the unknown resistance gives both current and voltage, letting us calculate the unknown resistance. A meter used for this purpose can be calibrated directly in ohms even though it's really measuring current; it's then an **ohmmeter**. The functions of voltmeter, ammeter, and ohmmeter are often combined in a single instrument called a **multimeter**.

## 25.5 Capacitors in Circuits

So far we've considered only circuits with steady current. A flashlight is a good example: Turn it on and current starts almost instantaneously, and it then continues flowing steadily until you turn the flashlight off.

Capacitors alter this simple picture, causing circuit quantities to change more slowly. Recall that a capacitor is a pair of insulated conductors with charge and voltage related by $Q = CV$, where $Q$ is the magnitude of the charge on either conductor, $V$ is the potential difference between them, and $C$ is the capacitance. Because charge and voltage are proportional in a capacitor, a change in voltage requires a change in charge. Charge changes when current flows through the wires connecting the capacitor to the rest of a circuit, and the magnitude of the current gives the rate at which capacitor charge increases or decreases. Since the current in any real circuit is finite, the charge on the capacitor cannot change instantaneously. But capacitor voltage is proportional to charge, so:

The voltage across a capacitor cannot change instantaneously.

This statement is the key to understanding circuits with capacitors. It says that the voltage on a capacitor can't jump abruptly from one value to another; mathematically, capacitor voltage $V_C$ must be a continuous function of time, its derivative always finite. Just how rapidly the voltage can change depends on the capacitance and other circuit quantities, as we'll now see.

We consider an **RC circuit**, one that includes a resistor and capacitor. RC circuits are ubiquitous, appearing everywhere from microbiological structures to stereo amplifiers to giant energy-storage systems. We examine separately the two cases in which the capacitor is (1) charging and (2) discharging.

## The *RC* Circuit: Charging

In Fig. 25.19 the capacitor is initially uncharged, so the voltage across it is zero. Closing the switch connects the left side of the resistor to the battery, bringing its potential immediately to $\mathcal{E}$ volts (we're taking $V = 0$ at the negative battery terminal). The right end of the resistor remains at the same voltage as the upper capacitor plate—and that's still zero because the voltage across the capacitor can't change instantaneously. So there's $\mathcal{E}$ volts across the resistor, and therefore a current $I = \mathcal{E}/R$ through it. This current delivers positive charge to the upper capacitor plate and negative charge to the lower plate.

As charge accumulates on the plates, the capacitor voltage increases proportionately. But the capacitor and resistor voltages sum to the battery voltage $\mathcal{E}$, so as the capacitor voltage increases, the resistor voltage drops. By Ohm's law, the resistor current $I = V/R$ drops as well. This in turn decreases the *rate* at which charge accumulates in the capacitor. The capacitor voltage continues to increase as charge accumulates, but at an ever-slower rate.

Eventually the capacitor voltage approaches the battery voltage, and the resistor voltage and current tend to zero; so, therefore, does the rate at which charge accumulates on the capacitor. The whole system approaches a final state in which the capacitor is charged to the full battery voltage and the current in the circuit is zero. Figure 25.20 summarizes the interplay among current, charge, and voltage.

**FIGURE 25.19** An *RC* circuit. The switch is closed at time $t = 0$.

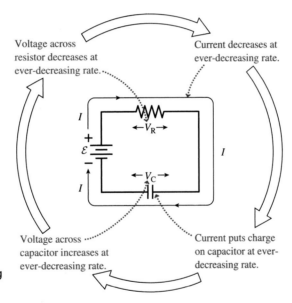

Voltage across resistor decreases at ever-decreasing rate.

Current decreases at ever-decreasing rate.

Voltage across capacitor increases at ever-decreasing rate.

Current puts charge on capacitor at ever-decreasing rate.

**FIGURE 25.20** Interrelationships among quantities in a charging *RC* circuit.

We can analyze the circuit of Fig. 25.19 quantitatively using the loop law. Going clockwise around the loop, we first encounter a voltage increase $\mathcal{E}$ across the battery, then a drop $IR$ across the resistor, then a drop $V_C$ from the upper to lower capacitor plate (Fig. 25.21). But $V_C = Q/C$, so the loop equation becomes

$$\mathcal{E} - IR - \frac{Q}{C} = 0$$

This equation contains two unknowns, $I$ and $Q$, but they're related because the current is the rate at which charge is accumulating on the capacitor: $I = dQ/dt$. To use this relation, we take the time derivative of the loop equation:

$$-R\frac{dI}{dt} - \frac{1}{C}\frac{dQ}{dt} = 0$$

The battery voltage $\mathcal{E}$ doesn't appear here because it's constant. Using $I = dQ/dt$ and rearranging the equation give

$$\frac{dI}{dt} = -\frac{I}{RC} \tag{25.4}$$

**FIGURE 25.21** Voltage changes in a charging *RC* circuit.

This equation shows that the rate of change of current is proportional to the current itself. Equations like this arise whenever a quantity changes at a rate proportional to the quantity itself. Population growth, the increase of money in a bank account, and the decay of a radioactive element are all described by similar equations.

Like the equation for simple harmonic motion in Chapter 13, Equation 25.4 is a *differential equation* because the unknown quantity *I* occurs in a derivative. The solution to a differential equation isn't a single number but a function expressing the relation between the unknown quantity—in this case current—and the independent variable—in this case time. We can solve this particular differential equation by multiplying both sides by $dt/I$ in order to collect all terms involving *I* on one side of the equation. This gives

$$\frac{dI}{I} = -\frac{dt}{RC}$$

We then integrate both sides, noting that *RC* is constant:

$$\int_{I_0}^{I} \frac{dI}{I} = -\frac{1}{RC} \int_{0}^{t} dt$$

where $I_0 = \mathcal{E}/R$ is the initial current at the time $t = 0$ just after the switch is closed and where the integration runs to an arbitrary time *t*. The integral on the left is the natural logarithm, and on the right it's just *t*. Then we have

$$\ln\left(\frac{I}{I_0}\right) = -\frac{t}{RC}$$

where we used $\ln I - \ln I_0 = \ln(I/I_0)$. To get an equation for *I* we exponentiate both sides, recalling that $e^{\ln x} = x$. This gives $I/I_0 = e^{-t/RC}$, or, since $I_0 = \mathcal{E}/R$,

$$I = \frac{\mathcal{E}}{R}e^{-t/RC} \tag{25.5}$$

Thus the current in the circuit decreases exponentially with time, in agreement with our qualitative analysis. The capacitor voltage is $V_C = \mathcal{E} - V_R$, or, since $V_R = IR = \mathcal{E}e^{-t/RC}$,

$$V_C = \mathcal{E}(1 - e^{-t/RC}) \quad \text{(RC circuit, charging)} \tag{25.6}$$

Equation 25.6 shows the capacitor voltage starting at zero and rising, with its rate of rise ever slowing as it gradually approaches the battery voltage $\mathcal{E}$—just as we reasoned in our qualitative analysis. Figure 25.22 plots capacitor voltage and current using the equations we've just derived.

When is the capacitor fully charged? Never, according to our equations! But the rate at which it approaches full charge is determined by the so-called **time constant**, *RC*—a characteristic time for changes to occur in a circuit containing a capacitor. Equation 25.6 shows that in one time constant, the voltage rises to $\mathcal{E}(1 - 1/e)$, or to about two-thirds of the battery voltage. A practical rule of thumb says that in five time constants $(t = 5RC)$ a capacitor is 99% charged (see Exercise 35). The *RC* time constant clarifies our statement that the voltage across a capacitor can't change instantaneously. We can now say that the voltage can't change appreciably in times small compared with the time constant. On the other hand, after many time constants, we'll find essentially no current flowing to the capacitor. We've shown quantitatively the role of the time constant *RC* by marking the time in units of *RC* on Fig. 25.22.

Resistors and capacitors are available in a wide range of values, so practical values for *RC* span many orders of magnitude. *RC* circuits with time constants from microseconds to hours are widely used in electronic devices to control the rates at which electric quantities vary. For example, circuits with *RC* many times the sixtieth-of-a-second period of standard AC power produce steady, direct-current power for audio and video equipment. Equalizers in audio systems are variable resistances in *RC* circuits; changing the resistance changes the time constant and therefore the way the circuit handles rapidly changing audio signals. Sometimes, though, the time constant can be a nuisance. Capacitance in audio systems can

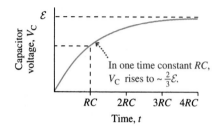

In one time constant *RC*, $V_C$ rises to ~ $\frac{2}{3}\mathcal{E}$.

In one time constant *RC*, *I* drops to about $\frac{1}{3}$ of its initial value $\mathcal{E}/R$.

**FIGURE 25.22** Time dependence of capacitor voltage and circuit current in a charging *RC* circuit. The approximate values $\frac{2}{3}$ and $\frac{1}{3}$ are actually $1 - 1/e$ and $1/e$, respectively.

limit high-frequency response, decreasing the quality of music reproduction. With computer speeds in the GHz range—meaning basic operations occur billions of times a second—even tiny $RC$ time constants associated with the resistance of wires and the capacitance of adjacent conductors can cause trouble.

## The $RC$ Circuit: Discharging

Suppose we connect a charged capacitor across a resistor, as shown in Fig. 25.23. If the capacitor voltage is initially $V_0$, then when the circuit is connected, this voltage will drive a current $I_0 = V_0/R$ through the resistor. This current transfers charge from the positive to the negative capacitor plate, lowering the charge on the capacitor. Since capacitor charge and voltage are proportional, the capacitor voltage drops, too. So, therefore, does the current and therefore the rate at which the capacitor discharges. We expect both the voltage and current in this circuit to decay toward zero. In terms of energy, that happens because the energy stored in the capacitor's electric field is gradually dissipated as heat in the resistor.

**FIGURE 25.23** A discharging $RC$ circuit.

The loop equation for Fig. 25.23 is particularly simple; going clockwise, we have $Q/C - IR = 0$, where the two terms are the voltage changes across the capacitor and resistor, respectively. Since we've indicated positive current in Fig. 25.23 in the direction that would *reduce* the capacitor charge $Q$, the rate of change $dQ/dt$ and the current have opposite signs: $I = -dQ/dt$. Differentiating our loop equation and substituting this expression for $I$ give $dI/dt = -I/RC$. This is Equation 25.4; the solution is therefore Equation 25.5, but with $I_0 = V_0/R$ instead of $\mathcal{E}/R$:

$$I = \frac{V_0}{R}e^{-t/RC} \tag{25.7}$$

In this circuit the capacitor and resistor voltages are the same, since the two are in parallel. Because the resistor voltage and the current are proportional, the voltage across the capacitor and resistor is

$$V = V_0 e^{-t/RC} \qquad (RC \text{ circuit, discharging}) \tag{25.8}$$

Equations 25.7 and 25.8 show that the capacitor discharges with the same characteristic time constant $RC$ that governs its charging.

---

**EXAMPLE 25.6** **Charging Capacitors: A Camera Flash**

A camera flash gets its energy from a 150-$\mu$F capacitor and requires 170 V to fire. If the capacitor is charged by a 200-V source through an 18-k$\Omega$ resistor, how long must the photographer wait between flashes? Assume the capacitor is fully discharged with each flash.

**INTERPRET** This is a problem about a charging capacitor, and we want to find the time to reach a given voltage.

**DEVELOP** Equation 25.6, $V_C = \mathcal{E}(1 - e^{-t/RC})$, gives the voltage across a charging capacitor, so our plan is to solve this equation for the time $t$.

**EVALUATE** First we solve for the exponential term that contains the time:

$$e^{-t/RC} = 1 - \frac{V_C}{\mathcal{E}}$$

Then we take the natural logarithm of both sides, recalling that $\ln e^x = x$, so

$$-\frac{t}{RC} = \ln\left(1 - \frac{V_C}{\mathcal{E}}\right)$$

Solving for $t$ and setting $V_C = 170$ V, $\mathcal{E} = 200$ V, $R = 18$ k$\Omega$, and $C = 150$ $\mu$F give

$$t = -RC \ln\left(1 - \frac{V_C}{\mathcal{E}}\right) = 5.1 \text{ s}$$

**ASSESS** The time constant here is $RC = 2.7$ s, and 170 V is well over two-thirds of the 200-V source. Therefore we expect a charging time longer than one time constant. Our 5.1-s answer is nearly $2RC$. Problem 69 explores energy and power in this circuit. ∎

---

## $RC$ Circuits: Long- and Short-Term Behavior

It's not always necessary to solve exponential equations in analyzing $RC$ circuits. If we're interested only in times much shorter than the time constant, then it's enough to remember that the voltage across a capacitor can't change instantaneously. And after many time constants, a capacitor has essentially reached its final voltage, and no current is flowing to it. These conditions are sufficient to analyze circuits on short and long time scales.

**EXAMPLE 25.7** **An *RC* Circuit: Long and Short Times**

The capacitor in Fig. 25.24a is initially uncharged. Find the current through $R_1$ (a) the instant the switch is closed and (b) a long time after the switch is closed.

(a)

(b)

(c)

**FIGURE 25.24** Original (a) and equivalent short-term (b) and long-term (c) circuits for Example 25.7.

**INTERPRET** We interpret "the instant the switch is closed" to mean a time much shorter than the time constant *RC*, and "a long time" to mean a time much longer than *RC*. Then this is a problem involving the long- and short-term behavior of a circuit containing a capacitor.

**DEVELOP** We follow Tactics 25.2 and first redraw the circuit with the capacitor replaced by a short circuit (Fig. 25.24b). Solving this circuit will give the current in $R_1$ right after the switch is closed. For the long-term behavior we redraw the circuit with the capacitor an open circuit (Fig. 25.24c).

**EVALUATE** There can't be any voltage across a short circuit—a perfect conductor—so there's no voltage across $R_2$ in Fig. 25.24b. Thus for part (a) the entire battery voltage appears across $R_1$, giving a current $I = \mathcal{E}/R_1$. In Fig. 25.24c we have two resistors in series and the current in both is $I = \mathcal{E}/(R_1 + R_2)$, our answer to part (b).

**ASSESS** The current through $R_1$ starts out at $\mathcal{E}/R_1$ and gradually drops to $\mathcal{E}/(R_1 + R_2)$. That makes sense because the uncharged capacitor initially "shorts out" $R_2$, making it irrelevant. But as the capacitor charges, current starts flowing through $R_2$ and its presence is "felt." Without solving more complicated equations, we can't describe the intermediate behavior of the circuit, but getting the short- and long-term behavior is straightforward. ∎

---

**GOT IT? 25.6** A capacitor is charged to 12 V and then connected between points *A* and *B* in the figure, with its positive plate at *A*. What is the current through the 2-kΩ resistor (a) immediately after the capacitor is connected and (b) a long time after it's connected?

## Big Picture

The big idea here is the **electric circuit**—an interconnection of electric components that usually includes one or more sources of electric energy, such as batteries.

## Key Concepts and Equations

A source of **emf** (or an emf, for short) is a battery or other device that imparts energy to electric charge flowing through it. The value of the emf, $\mathcal{E}$, is the energy imparted per unit charge, measured in volts. An ideal emf maintains a fixed potential difference (voltage) across its terminals.

$\mathcal{E}$ volts

Resistors in series add:

$$R_{\text{series}} = R_1 + R_2 + R_3 + \cdots$$

Resistors in parallel add reciprocally:

$$\frac{1}{R_{\text{parallel}}} = \frac{1}{R_1} + \frac{1}{R_2} + \frac{1}{R_3} + \cdots$$

Simple circuits are analyzed by evaluating series and parallel combinations.

To analyze more complicated circuits, use Kirchhoff's node and loop laws.

Current into node $A$ is 0:
$-I_1 + I_2 + I_3 = 0$
Voltage charges around loops sum to 0:
loop 1: $\mathcal{E}_1 - I_1 R_1 - I_3 R_3 = 0$
loop 2: $\mathcal{E}_2 + I_2 R_2 - I_3 R_3 = 0$

**Capacitors** result in time-changing behavior of circuit quantities.

The **time constant** $RC$ governs the time scales.

$V_C = \mathcal{E}(1 - e^{-t/RC})$

Charging

$V_C = V_0 e^{-t/RC}$

Discharging

## Cases and Uses

Real batteries and other electric-energy sources have **internal resistance**. When they supply current, their terminal voltage is therefore less than their rated voltage $\mathcal{E}$.

A **voltmeter** measures the voltage *across* its two terminals. It goes in parallel with the component whose voltage you want to measure.

Reads voltage across $R_2$

An **ammeter** measures the current *through* itself. It goes in series with the component whose current you want to measure.

Reads current through $R_2$

# For Thought and Discussion

1. Are the electrical outlets in a home connected in series or parallel? How do you know?
2. In which of the circuits of Fig. 25.25 does the battery supply the same current? All the resistors have the same resistance.

**FIGURE 25.25** For Thought and Discussion 2

3. Can the voltage across a battery's terminals differ from the rated voltage of the battery? Explain.
4. Can the voltage across a battery's terminals be higher than the rated voltage of the battery? Explain.
5. In some cities, streetlights are wired in such a way that when one light burns out, they all go out. Are the lights in series or parallel?

6. When the switch in Fig. 25.26 is open, what is the voltage across the resistor? Across the switch?

**FIGURE 25.26** For Thought and Discussion 6

7. Two identical resistors in series dissipate equal power. How can this be, when electric charge loses energy in flowing through the first resistor?
8. When a large electric load such as a washing machine, oven, or oil burner comes on, lights throughout a house often dim. Why is this? *Hint:* Think about real wires.
9. If the node law were not obeyed at some node in an electric circuit, what would happen to the voltage at that node?
10. How would you connect a pair of equal resistors across an ideal battery in order to get the greatest power dissipation?
11. You have a battery whose voltage and internal resistance are unknown. Using an ideal voltmeter and an ideal ammeter, how would you determine both these battery characteristics?
12. A student who is confused about voltage and current hooks a nearly ideal ammeter across a car battery. What happens?
13. A student who is confused about voltage and current tries to measure the voltage across a lighted lightbulb by inserting a voltmeter in series with the bulb. What happens to the bulb? Explain.

# Exercises and Problems

## Exercises

### Section 25.1 Circuits, Symbols, and Electromotive Force

14. Sketch a circuit diagram for a circuit that includes a resistor $R_1$ connected to the positive terminal of a battery, a pair of parallel resistors $R_2$ and $R_3$ connected to the lower-voltage end of $R_1$ and then returned to the battery's negative terminal, and a capacitor across $R_2$.
15. A circuit consists of two batteries, a resistor, and a capacitor, all in series. Sketch this circuit. Does the description allow any flexibility in how you draw the circuit?
16. Resistors $R_1$ and $R_2$ are connected in series, and this series combination is in parallel with $R_3$. This parallel combination is connected across a battery. Draw a diagram representing this circuit.
17. What is the emf of a battery that delivers 27 J of energy as it moves 3.0 C between its terminals?
18. A 1.5-V battery stores 4.5 kJ of energy. How long can it light a flashlight bulb that draws 0.60 A?
19. If you accidentally leave your car headlights (current drain 5 A) on for an hour, how much of the 12-V battery's chemical energy is used up?

### Section 25.2 Series and Parallel Circuits

20. A 47-kΩ resistor and a 39-kΩ resistor are in parallel, and the pair is in series with a 22-kΩ resistor. What is the resistance of the combination?
21. What resistance should be placed in parallel with a 56-kΩ resistor to make an equivalent resistance of 45 kΩ?
22. A defective starter motor in a car draws 300 A from the car's 12-V battery, dropping the battery terminal voltage to 6 V. A good starter motor should draw only 100 A. What will the battery terminal voltage be with a good starter?
23. What is the internal resistance of the battery in Exercise 22?
24. When a 9-V battery is temporarily short-circuited, a 200-mA current flows. What is the internal resistance of the battery?

25. What possible resistance combinations can you form using three resistors whose values are 1.0 Ω, 2.0 Ω, and 3.0 Ω? *Note:* Use all three resistors.

### Section 25.3 Kirchhoff's Laws and Multiloop Circuits

26. In the circuit of Fig. 25.27 it makes no difference whether the switch is open or closed. What is $\mathcal{E}_3$ in terms of the other quantities shown?

**FIGURE 25.27** Exercise 26

27. Find all three currents in the circuit of Fig. 25.14, but now with $\mathcal{E}_2 = 1.0$ V.
28. What is the current through the 3-Ω resistor in the circuit of Fig. 25.28? *Hint:* This is trivial. Can you see why?

**FIGURE 25.28** Exercise 28

29. Find $I_2$ in Example 25.4 for the case $\mathcal{E}_2 = 2.0$ V.

### Section 25.4 Electrical Measurements

30. A voltmeter with 200-kΩ resistance is used to measure the voltage across the 10-kΩ resistor in Fig. 25.29. By what percentage is the measurement in error because of the finite meter resistance?

**FIGURE 25.29** Exercises 30 and 31

31. An ammeter with 100-$\Omega$ resistance is inserted in the circuit of Fig. 25.29. By what percentage is the measured current in error because of the nonzero meter resistance?

32. A neophyte mechanic foolishly connects an ammeter with 0.1-$\Omega$ resistance directly across a 12-V car battery whose internal resistance is 0.01 $\Omega$. What is the power dissipation in the meter? No wonder it gets destroyed!

### Section 25.5 Capacitors in Circuits

33. Show that the quantity $RC$ has the units of time (seconds).

34. If capacitance is given in $\mu$F, what will be the units of the time constant $RC$ when resistance is given in (a) $\Omega$, (b) k$\Omega$, and (c) M$\Omega$? *Note:* Your answers eliminate the need for tedious power-of-10 conversions.

35. Show that a capacitor is charged to approximately 99% of the applied voltage in five time constants ($5RC$).

36. An uncharged 10-$\mu$F capacitor and a 470-k$\Omega$ resistor are connected in series, and 250 V is applied across the combination. How long does it take the capacitor voltage to reach 200 V?

37. Find an expression for the voltage across the capacitor in Example 25.7 when it's fully charged.

## Problems

38. In Fig. 25.30 all resistors have the same value, $R$. What will be the resistance measured (a) between $A$ and $B$ or (b) between $A$ and $C$?

39. In Fig. 25.30, take all resistors to be 1.0 $\Omega$. If a 6.0-V battery is connected between points $A$ and $B$, what will be the current in the vertical resistor?

**FIGURE 25.30** Problems 38 and 39

40. Three 1.5-V batteries, with internal resistances of 0.01 $\Omega$, 0.1 $\Omega$, and 1 $\Omega$, each have 1-$\Omega$ resistors connected across their terminals. To three significant figures, what is the voltage across each resistor?

41. A partially discharged car battery can be modeled as a 9-V emf in series with an internal resistance of 0.08 $\Omega$. Jumper cables are used to connect this battery to a fully charged battery, modeled as a 12-V emf in series with a 0.02-$\Omega$ internal resistance. How much current flows through the discharged battery?

42. You have several 50-$\Omega$ resistors, each capable of dissipating 0.50 W without overheating. How many resistors would you need, and how would you connect them, to make a 50-$\Omega$ combination that can be connected safely across a 12-V battery?

43. A 6.0-V battery has an internal resistance of 2.5 $\Omega$. If the battery is short-circuited, what is the rate of energy dissipation in its internal resistance?

44. How many 100-W, 120-V lightbulbs can be connected in parallel before they blow a 20-A circuit breaker?

45. Take $\mathcal{E} = 12$ V and $R_1 = 270$ $\Omega$ in the voltage divider of Fig. 25.4. (a) What should be the value of $R_2$ so that 4.5 V appears across $R_2$? (b) What will be the power dissipation in $R_2$?

46. A voltage divider consists of two 1.0-k$\Omega$ resistors connected in series across a 160-V emf. If a 10-k$\Omega$ resistor is connected across one of the 1.0-k$\Omega$ resistors, what will be the voltage across it?

47. In the circuit of Fig. 25.31, $R_1$ is a variable resistor and the other two resistors have equal resistances $R$. (a) Find an expression for the voltage across $R_1$, and (b) sketch a graph of this quantity as a

function of $R_1$ as $R_1$ varies from 0 to $10R$. (c) What is the limiting value as $R_1 \rightarrow \infty$?

**FIGURE 25.31** Problem 47

48. In the circuit of Fig. 25.32, find (a) the current supplied by the battery and (b) the current through the 6-$\Omega$ resistor.

**FIGURE 25.32** Problems 48 and 49

49. In the circuit of Fig. 25.32, how much power is dissipated in the 4-$\Omega$ resistor?

50. What is the current through the ammeter in Fig. 25.33?

**FIGURE 25.33** Problem 50

51. In Fig. 25.34, what is the equivalent resistance measured between points $A$ and $B$?

**FIGURE 25.34** Problem 51

52. Find all three currents in the circuit of Fig. 25.14 with the values given, but with battery $\mathcal{E}_2$ reversed.

53. In Fig. 25.35, take $\mathcal{E}_1 = 6.0$ V, $\mathcal{E}_2 = 1.5$ V, and $\mathcal{E}_3 = 4.5$ V. Take $R_1 = 270$ $\Omega$, $R_2 = 150$ $\Omega$, $R_3 = 560$ $\Omega$, and $R_4 = 820$ $\Omega$. Find the current in $R_3$, and give its direction.

**FIGURE 25.35** Problems 53 and 54

54. With all the values except $\mathcal{E}_2$ in Fig. 25.35 as given in Problem 53, find the condition on $\mathcal{E}_2$ that will make the current in $R_3$ flow upward.

55. The voltage across the 30-k$\Omega$ resistor in Fig. 25.36 is measured with (a) a 50-k$\Omega$ voltmeter, (b) a 250-k$\Omega$ voltmeter, and (c) a digital meter with 10-M$\Omega$ resistance. To two significant figures, what does each read?

**FIGURE 25.36** Problem 55

56. In Fig. 25.37, what are the meter readings when (a) an ideal voltmeter or (b) an ideal ammeter is connected between points $A$ and $B$?

**FIGURE 25.37** Problem 56

57. A resistor draws 1.00 A from an ideal 12.0-V battery. (a) If an ammeter with 0.10-$\Omega$ resistance is inserted in the circuit, what will it read? (b) If this current is used to calculate the resistance, how will the calculated value compare with the actual value?

58. The voltage across a charging capacitor in an $RC$ circuit rises to $1 - 1/e$ of the battery voltage in 5.0 ms. (a) How long will it take to reach $1 - 1/e^3$ of the battery voltage? (b) If the capacitor is charging through a 22-k$\Omega$ resistor, what is its capacitance?

59. A 1.0-$\mu$F capacitor is charged to 10.0 V. It is then connected across a 500-k$\Omega$ resistor. How long does it take (a) for the capacitor voltage to reach 5.0 V and (b) for the energy stored in the capacitor to decrease to half its initial value?

60. A capacitor used to provide steady voltages in the power supply of a stereo amplifier charges rapidly to 35 V every 1/60 second. It must then hold that voltage to within 1.0 V for the next 1/60 s while it discharges through the amplifier circuit. If the circuit draws 1.2 A from the 35-V supply, (a) what is its effective resistance, and (b) what value of capacitance is needed?

61. A capacitor is charged until it holds 5.0 J of energy. It is then connected across a 10-k$\Omega$ resistor. In 8.6 ms, the resistor dissipates 2.0 J. What is the capacitance?

62. A 2.0-$\mu$F capacitor is charged to 150 V. It is then connected to an uncharged 1.0-$\mu$F capacitor through a 2.2-k$\Omega$ resistor, by closing switch $S$ in Fig. 25.38. Find the total energy dissipated in the resistor as the circuit comes to equilibrium. *Hint:* Think about charge conservation.

**FIGURE 25.38** Problem 62

63. For the circuit of Example 25.7, take $\mathcal{E} = 100$ V, $R_1 = 4.0$ k$\Omega$, and $R_2 = 6.0$ k$\Omega$, and assume the capacitor is initially uncharged. What are the currents in both resistors and the voltage across the capacitor (a) just after the switch is closed, and (b) a long time after the switch is closed? Long after the switch is closed it's again opened. What are $I_1$, $I_2$, and $V_C$ (c) just after this switch opening, and (d) a long time later?

64. In the circuit of Fig. 25.39, the switch is initially open and both capacitors are initially uncharged. All resistors have the same value $R$. Find expressions for the current in $R_2$ (a) just after the switch is closed, and (b) a long time after the switch is closed. (c) Describe qualitatively how you expect the current in $R_3$ to behave after the switch is closed.

**FIGURE 25.39** Problems 64 and 79

65. A battery's voltage is measured with a voltmeter whose resistance is 1000 $\Omega$; the result is 4.36 V. When the measurement is repeated with a 1500-$\Omega$ meter, the result is 4.41 V. What are (a) the battery voltage and (b) its internal resistance?

66. Find the resistance needed in an $RC$ circuit to bring a 20-$\mu$F capacitor from zero charge to 45% of full charge in 140 ms.

67. Suppose the currents into and out of a circuit node differed by 1 $\mu$A. If the node consists of a small metal sphere with diameter 1 mm, how long would it take for the electric field around the node to reach the breakdown field in air (3 MV/m)?

68. A resistance $R$ is connected across a battery with internal resistance $R_{int}$. Show that the maximum power dissipation in $R$ occurs when $R = R_{int}$. *Note:* This is not the way to treat a battery! But it is the basis for matching loads in amplifiers and other devices; for example, a stereo amplifier designed to drive 8-$\Omega$ speakers has internal resistance close to 8 $\Omega$.

69. Of the total energy drawn from a battery in charging an $RC$ circuit, show that only half ends up as stored energy in the capacitor. *Hint:* What happens to the rest of it? You'll need to integrate.

70. What is the equivalent resistance between $A$ and $B$ in each of the circuits shown in Fig. 25.40? *Hint:* In (c), think about symmetry and the current that would flow through $R_2$.

**FIGURE 25.40** Problem 70

71. A 50-$\Omega$ resistor is connected across a battery, and a 26-mA current flows. When the 50-$\Omega$ resistor is replaced with a 22-$\Omega$ resistor, a 43-mA current flows. What are the battery's voltage and internal resistance?

72. Obtain an expression for the rate $(dV/dt)$ at which the voltage across a charging capacitor increases. Evaluate your result at time $t = 0$, and show that if the capacitor continued charging steadily at this rate, it would be fully charged in exactly one time constant.

73. The circuit in Fig. 25.41 extends forever to the right, and all the resistors have the same value $R$. Show that the equivalent resistance measured across the two terminals at left is $\frac{1}{2}R(1 + \sqrt{5})$. *Hint:* You don't need to sum an infinite series.

**FIGURE 25.41** Problem 73

74. Figure 25.42 shows the voltage across a capacitor that's charging through a 4700-$\Omega$ resistor in the circuit of Fig. 25.19. Use the graph to determine (a) the battery voltage, (b) the time constant, and (c) the capacitance.

**FIGURE 25.42** Problem 74

75. Figure 25.43 shows a portion of a circuit used to model the electrical behavior of long, cylindrical biological cells such as muscle cells or the axons of neurons. Find the current through the emf $\mathcal{E}_3$, given that all resistors have the same value $R = 1.5$ M$\Omega$ and $\mathcal{E}_1 = 75$ mV, $\mathcal{E}_2 = 45$ mV, and $\mathcal{E}_3 = 20$ mV. Be sure to specify the direction of the current.

**FIGURE 25.43** Problems 75 and 76

76. An electrochemical impulse traveling along the cell modeled in Fig. 25.43 changes the value of emf $\mathcal{E}_3$ so that now it supplies an upward current of 40 nA. Assuming the rest of the circuit remains as described in Problem 75, what is the new value of $\mathcal{E}_3$?

77. A parallel-plate capacitor has plates of 10 cm$^2$ area separated by a 0.10-mm layer of glass insulation with resistivity $\rho = 1.2 \times 10^{13}$ $\Omega$·m and dielectric constant $\kappa = 5.6$. Because of the finite resistivity, current can leak through the insulation. (a) How can such a leaky capacitor be represented in a circuit diagram? (b) Find the time constant for this capacitor to discharge through its insulation, and show that it depends only on the properties of the insulating material and not on its dimensions.

78. Write the node and loop laws for the circuit in Fig. 25.24a for Example 25.7, and find the time constant.

79. In Problem 64, take $C_1 = C_2 = C$, and find the current through $R_2$ as a function of time. *Hint:* Use the node and loop laws to get a differential equation for the current, and use the initial conditions for the current and its derivative to evaluate the constants of integration.

80. You're about to purchase a battery. Normally, batteries are rated in ampere-hours—the total amount of charge they can deliver. The 6-V battery you see at an online business is rated at 50 watt-hours, meaning the total amount of energy it can deliver, but the device you want to use it in requires a 5-A·h battery. Will this battery work?

81. You have a dilemma. You're building a small circuit. You need a 3.44-k$\Omega$ resistor. You have three resistors, 3.30 k$\Omega$, 4.70 k$\Omega$, and 1.50 k$\Omega$. How can you connect them to get the required 3.44 k$\Omega$?

82. You're reviewing the specifications for an electronics contract. A particular capacitor must be able to store energy for a specified time before the stored energy leaks out. The design calls for the energy stored in the capacitor to be at least half its initial amount after 25 hours. The dielectric constant of the capacitor is 2.6; the resistivity of the dielectric is $10^{16}$ $\Omega$·m. An engineering handbook says the time constant of this leaky capacitor is $\tau = \kappa\rho\epsilon_0$. Does the capacitor meet the requirements?

# Answers to Chapter Questions

## Answer to Chapter Opening Question
Series.

## Answers to GOT IT? Questions
25.1 (a) and (b).
25.2 (c) 6 V > (a) 3 V > (b) 0 V.
25.3 $R_a > R_d > R_c > R_b$.
25.4 (a) A is brightest because it carries more current; after A the current splits between B and C. (b) A and B become equally bright, with A dimming and B brightening relative to when C was in the circuit.
25.5 There's no current through the top right resistor, so the voltage is divided evenly across the other two resistors and the meter reads $\frac{1}{2}\mathcal{E}$.
25.6 (a) 6 mA; (b) 2 mA.

# Magnetism: Force and Field

■ This ultraviolet image shows delicate loops of million-kelvin ionized gas in the Sun's atmosphere. What force shapes the gas into such intricate structures, and why don't we see similar things in Earth's atmosphere?

People are fascinated with magnets and the mysterious, invisible force they produce. Magnetism plays essential roles in technology and throughout the natural universe. We use magnetism for everything from holding notes on refrigerators to storing computer data to propelling high-speed trains. Earth's magnetism protects us from dangerous solar radiation, which itself originates in violent magnetic storms on the Sun. Without magnetism we wouldn't even see, for light itself arises in an interaction between magnetism and electricity. In fact, magnetism and electricity are intimately related, and you'll soon come to see them as inseparable aspects of the same underlying phenomenon.

## 26.1 What Is Magnetism?

You know from experience that magnets exert forces on each other and on certain materials, like iron. As we did with the gravitational and electric forces, it's convenient to describe this interaction in terms of a **magnetic field** (symbol $\vec{B}$). One magnet produces a magnetic field, and another responds to the field in its vicinity. We use field lines to picture the field, and we can trace those lines using small iron filings that align with the field (Fig. 26.1). In our illustrations, we'll use color for magnetic field lines to distinguish them from electric fields.

But the magnetism you're most familiar with is only one manifestation of a much more fundamental and universal phenomenon that's intimately linked with electricity. Here we'll go straight to the essence of magnetism; later, we'll see how familiar magnets fit into the big picture.

In Chapter 20 we introduced electric charge, a fundamental property of matter, and we described its interactions using the concept of electric field. Magnetism, too, is based in *electric* charge. One crucial point both distinguishes and relates electricity and magnetism:

> The phenomena of magnetism involve *moving* electric charge.

In particular, *moving* electric charge is the source of magnetic fields, and *moving* electric charge is what responds to magnetic fields.

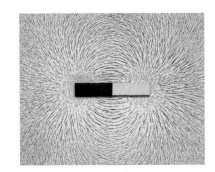

**FIGURE 26.1** Iron filings align with the magnetic field, tracing out the field of a bar magnet.

## 26.2 Magnetic Force and Field

In Chapter 20 we defined the electric field $\vec{E}$ with the equation $\vec{F}_E = q\vec{E}$, where $\vec{F}_E$ is the electric force on a charge $q$. Now consider a region where there's no electric field, but where there is a magnetic field. You could confirm the presence of the field and determine its direction with a compass, which is just a small magnet free to pivot into alignment with the field. Or, more fundamentally, you could explore the behavior of an electric point charge $q$ in this field. If the charge is at rest, nothing happens. But if it's moving, it experiences a **magnetic force** as shown in Fig. 26.2. Experiment shows that:

1. The magnetic force is always at right angles to both the velocity $\vec{v}$ of the charge and the magnetic field $\vec{B}$.
2. The strength of the force is proportional to the product of the charge $q$, its speed $v$, and the magnetic field strength $B$.
3. The force is greatest when the charge moves at right angles to the field and is zero for motion parallel to the field. In general, the force is proportional to $\sin\theta$, where $\theta$ is the angle between the velocity $\vec{v}$ and the field $\vec{B}$.

Putting these facts together lets us write the magnetic force compactly using the vector cross product introduced in Chapter 11:

$$\vec{F}_B = q\vec{v} \times \vec{B} \qquad \text{(magnetic force)} \qquad (26.1)$$

Recall that the cross product $\vec{v} \times \vec{B}$ is a vector of magnitude $vB \sin\theta$, so the magnitude of the magnetic force is

$$|\vec{F}_B| = |q|vB \sin\theta$$

The direction of $\vec{v} \times \vec{B}$ is given by the right-hand rule (Fig. 26.3), and Equation 26.1 shows that the magnetic force has the same direction for positive $q$ and the opposite direction for negative $q$.

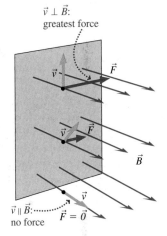

**FIGURE 26.2** The magnetic force on a charged particle is perpendicular to both the particle's velocity $\vec{v}$ and the magnetic field $\vec{B}$.

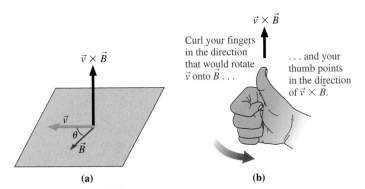

**FIGURE 26.3** Finding the direction of the cross product $\vec{v} \times \vec{B}$ with the right-hand rule.

Equation 26.1 shows that the units of magnetic field are N·s/(C·m), a unit given the name **tesla** (T) after the Serbian–American inventor Nikola Tesla (1856–1943). One tesla is a strong field, and a smaller unit called the gauss (G), equal to $10^{-4}$ T, is often used. Earth's magnetic field is a little less than 1 G, while the field of a refrigerator magnet is about 100 G. The fields used in magnetic resonance imaging (MRI) may be as strong as several tesla, while the incredibly dense, rapidly rotating collapsed stars called magnetars have fields up to $10^{11}$ T.

**GOT IT? 26.1**　The figure shows a proton in a magnetic field. For which of the three proton velocities shown will the magnetic force be greatest? What will be the direction of the force in all three cases?

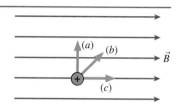

---

**EXAMPLE 26.1**　**Finding the Magnetic Force: Steering Protons**

Figure 26.4 shows three protons entering a 0.10-T magnetic field. All three are moving at 2.0 Mm/s. Find the magnetic force on each.

**FIGURE 26.4** What's the magnetic force on each proton?

**INTERPRET** This problem is about the magnetic force on moving charged particles with the same speed but in different directions of motion.

**DEVELOP** Equation 26.1, $\vec{F}_B = q\vec{v} \times \vec{B}$, gives the magnetic force, so we'll apply it to each of the particles.

**EVALUATE** Proton 2 is moving parallel to the field, so $\vec{v}_2 \times \vec{B} = 0$ and it experiences no magnetic force. Protons 1 and 3 are moving at right angles to the field, so $\sin\theta = 1$, and the magnitude of the force on each is $F_B = qvB\sin\theta = qvB$. Using the proton charge $q = e = 1.6 \times 10^{-19}$ C and the given values for $B$ and $v$ yields $F = 32$ fN. Since protons are positive, the direction of $\vec{F}_B$ is the same as that of $\vec{v} \times \vec{B}$; applying the right-hand rule shows that the direction is out of the page for proton 1 and into the page for proton 3.

**ASSESS** Our answer 32 fN $(32 \times 10^{-15}$ N$)$ is a tiny force, but that's not surprising given the proton's tiny charge. Note that the magnetic field alone doesn't determine the force; in this example identical particles experience different forces because they're moving in different directions relative to the field. ∎

---

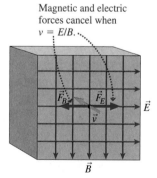

Magnetic and electric forces cancel when $v = E/B$.

**FIGURE 26.5** A velocity selector. The electric and magnetic fields cancel when $qE = qvB$, so only particles with speed $v = E/B$ pass through undeflected. The velocity $\vec{v}$ points into the page.

Although electricity and magnetism are related, the electric and magnetic forces are distinct. Both may be present simultaneously, in which case a charged particle experiences both an electric force $\vec{F}_E = q\vec{E}$ and a magnetic force $\vec{F}_B$ given by Equation 26.1. The result is an **electromagnetic force**:

$$\vec{F} = q\vec{E} + q\vec{v} \times \vec{B} \quad \text{(electromagnetic force)} \qquad (26.2)$$

Because the magnetic force depends on velocity but the electric force doesn't, it's possible to use perpendicular electric and magnetic fields to select particles of a particular velocity (Fig. 26.5). Such *velocity selectors* serve to prepare particle beams with uniform velocity as well as to analyze charged-particle populations in interplanetary space.

## 26.3 Charged Particles in Magnetic Fields

Like any force, the magnetic force deflects a particle from the straight-line path Newton says it would otherwise follow. Magnetic forces "steer" charged particles in a host of practical devices ranging from TV picture tubes to giant particle accelerators, and they shape particle trajectories throughout the astrophysical universe.

The magnetic force always acts at right angles to a particle's velocity. **Therefore it changes the direction of motion but not the speed, and it does no work.** In the special case of a particle moving at right angles to a uniform field, the magnetic force has a constant magnitude and the result, as Fig. 26.6 shows, is uniform circular motion. With $\vec{v}$ perpendicular to $\vec{B}$, the magnetic force of Equation 26.1 has magnitude $qvB$. This force provides the acceleration $v^2/r$ that describes circular motion with radius $r$. Then Newton's law, $F = ma$, reads $qvB = mv^2/r$. We can solve for the radius of the particle's circular path to get

$$r = \frac{mv}{qB} \qquad (26.3)$$

This result makes sense: The greater the particle's momentum $mv$, the harder it is for the magnetic force to bend its path and the larger the radius. On the other hand, a larger charge or field increases the force and makes a tighter orbit.

---

**GOT IT? 26.2** A uniform magnetic field points out of this page. Will an electron that's moving in the plane of the page circle (a) clockwise or (b) counterclockwise as viewed from above the page?

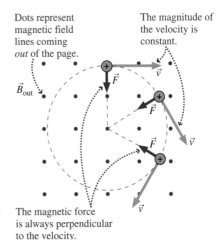

Dots represent magnetic field lines coming *out* of the page.

The magnitude of the velocity is constant.

The magnetic force is always perpendicular to the velocity.

**FIGURE 26.6** A charged particle moving at right angles to a uniform magnetic field describes circular motion.

---

**EXAMPLE 26.2** **Magnetic Deflection: A Mass Spectrometer**

A mass spectrometer separates ions according to their ratio of charge to mass. Such devices are widely used in science and engineering to analyze unknown mixtures and to separate isotopes of chemical elements. Figure 26.7 shows ions of charge $q$ and mass $m$ first being accelerated from rest through a potential difference $V$ and then entering a region of uniform magnetic field $B$ pointing out of the page. Only the magnetic force acts on the ions in this region, so they undergo circular motion and, after half an orbit, land on a detector. Find an expression for the horizontal distance $x$ from the entrance slit to the point where an ion lands on the detector.

**FIGURE 26.7** A mass spectrometer.

**INTERPRET** This problem is about charged particles undergoing circular motion in a uniform magnetic field. The distance we're asked for is the diameter of the particles' circular path.

**DEVELOP** Equation 26.3, $r = mv/qB$, shows that the path radius depends on the field and on the particle's mass, charge, and speed. We know everything but the speed, so this becomes a two-step problem in which we'll first find the speed. We're given the potential difference—energy per unit charge—so we can use energy conservation to find the kinetic energy and hence the ions' speed in the magnetic-field region. Then we'll use Equation 26.3 to find the radius of the ions' circular path.

**EVALUATE** A charge $q$ gains kinetic energy $qV$ in "falling" through a potential difference $V$, so an ion's kinetic energy once it enters the magnetic field is $\frac{1}{2}mv^2 = qV$. Solving for $v$ gives $v = \sqrt{2qV/m}$. Our answer, the path diameter $x$, is then twice the radius given in Equation 26.3:

$$x = 2r = \frac{2mv}{qB} = \frac{2m\sqrt{2qV/m}}{qB} = \frac{2}{B}\sqrt{\frac{2mV}{q}}$$

**ASSESS** Make sense? The greater the mass or speed—which increases with the accelerating voltage $V$—the harder it is to deflect the ion and the larger the diameter of its semicircular path. The larger the field or charge, the larger the force and the smaller the semicircle. Note that for a fixed voltage and magnetic field, this device sorts ions by their charge-to-mass ratio $q/m$. ∎

---

## The Cyclotron Frequency

What's the period of a particle's circular orbit in a uniform magnetic field? The orbit circumference is $2\pi r$, so the period is $T = 2\pi r/v$. Using Equation 26.3 for the radius gives

$$T = \frac{2\pi r}{v} = \frac{2\pi}{v}\frac{mv}{qB} = \frac{2\pi m}{qB}$$

Remarkably, the period is independent of the particle's speed and orbital radius. Equation 26.3 shows why: The higher the speed $v$, the larger the radius $r$ and hence the circumference. So a faster particle describes a larger circle and ends up taking the same amount of time to go around.

Equivalently, we can describe the particle's circular motion in terms of its frequency $f$, in revolutions per second, which is just the inverse of the period:

$$f = \frac{qB}{2\pi m} \quad \text{(cyclotron frequency)} \tag{26.4}$$

This quantity is the **cyclotron frequency**. Because it depends only on the field and the charge-to-mass ratio, cyclotron motion provides astrophysicists with a direct measure of magnetic fields in distant objects. Conversely, a fixed magnetic field guarantees a specific cyclotron frequency regardless of the particles' speeds. Microwave ovens exploit this fact, with their microwaves generated by electrons circling 2.4 billion times per second in a special tube called a magnetron.

## Particle Trajectories in Three Dimensions

When a charged particle moves in an arbitrary direction, we consider velocity components both perpendicular and parallel to the magnetic field. Our previous analysis applies to the perpendicular component, giving circular motion in a plane perpendicular to the field. And because there's no magnetic force with velocity parallel to the field, the parallel component is unaffected. The result, in a uniform field, is a spiral path about the field direction (Fig. 26.8).

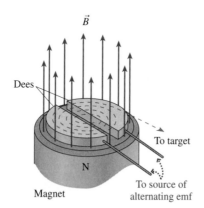

Motion parallel to the field is not affected by the magnetic force.

$\vec{B}$

**FIGURE 26.8** A particle in a uniform magnetic field describes a spiral path.

---

**APPLICATION** **The Cyclotron**

Physicists use high-energy particles to probe the structure of matter; engineers and physicians need high-energy particle beams in a variety of manufacturing, diagnostic, and therapeutic procedures. The easiest way to produce such particles is to accelerate ions through a potential difference, but the difficulties of handling high voltages make that impractical for all but the lowest energies. One of the earliest and most successful devices to circumvent this problem is the **cyclotron**, whose essential parts are shown in the figure. The device consists of an evacuated chamber between the poles of a magnet. Ions are produced at the center and undergo circular motion in the magnetic field.

$\vec{B}$

Dees

To target

N

Magnet

To source of alternating emf

Also in the chamber are two hollow conducting structures shaped like the letter D. A modest potential difference is applied across these "dees," and it alternates polarity at the cyclotron frequency. As ions circle around inside the cyclotron, they gain energy from the strong electric field associated with the potential difference at the gap. Inside the hollow conducting dee there's no electric field, so here the parti-

cles follow circular paths in the magnetic field. Halfway around they again encounter the dee gap. Because the potential is changing polarity in step with the particles' cyclotron motion, they again gain energy as they cross the gap. They move faster and in ever-larger circles, but always with the same orbital period. When they approach the edge of the machine, an electric field deflects the ions and they emerge as a high-energy ion beam.

Cyclotrons produce ions with energies of millions of electronvolts. This is high enough to cause nuclear reactions, and many medically useful radioactive isotopes are made using cyclotrons. In particular, the diagnostic procedure called PET (positron emission tomography) relies on cyclotron-produced radioisotopes; the photo shows a hospital-based cyclotron used for this purpose. At higher energies the theory of relativity alters our conclusion that the cyclotron frequency is independent of energy, and the cyclotron becomes useless. An alternative design is the **synchrotron**, in which both the magnetic field and the frequency vary to account for increasing particle energy.

The absence of magnetic force in the field direction means it's easy to move charged particles along the field. But try to push a charged particle at right angles to the field, and it goes into circular motion; push harder and the circle only gets bigger. As a result, charged particles are effectively "frozen" to the field lines and move along the field like beads strung on a wire (Fig. 26.9). Nonuniform fields and particle collisions make this "freezing" less than perfect, but in many cases particle density is low enough that the "frozen" assumption is an excellent approximation. The coronal loops in this chapter's opening photo are a beautiful example of charged particles "frozen" to the solar magnetic field. Similarly, high-energy particles from the Sun get trapped on Earth's magnetic field lines; where the field intersects the atmosphere, particles collide with atmospheric nitrogen and oxygen to produce the spectacular displays we call the aurora. Here on Earth, trapping of charged particles on magnetic field lines enables researchers to confine plasmas at temperatures of 100 MK in attempts to harness the energy of nuclear fusion.

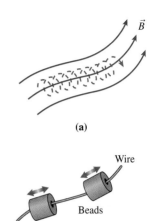

**(a)**

**(b)**

**FIGURE 26.9** (a) Charged particles undergoing spiral motion about the magnetic field are "frozen" to the field like (b) beads sliding along a wire.

## 26.4 The Magnetic Force on a Current

An electric current consists of charges in motion, so a current in a magnetic field should experience a magnetic force. Figure 26.10 shows a straight wire in a magnetic field $\vec{B}$. Charges in the wire are moving about with thermal motions, but because these are random, the magnetic force on all the charges averages to zero. But if there's a current $I$ in the wire, then the charges share a common drift velocity $\vec{v}_d$, and thus each experiences a magnetic force given by Equation 26.1: $\vec{F}_q = q\vec{v}_d \times \vec{B}$. If the wire has cross-sectional area $A$ and contains $n$ charges per unit volume, then the force on all the charge carriers in a length $L$ of wire is $\vec{F} = nALq\vec{v}_d \times \vec{B}$. But $nAqv_d$ is the current, $I$, as we found in Chapter 24. If we define a vector $\vec{L}$ whose magnitude is the wire length $L$ and whose direction is along the current, then we can write

$$\vec{F} = I\vec{L} \times \vec{B} \quad \text{(magnetic force on a current)} \quad (26.5)$$

This force is perpendicular to both the current and the magnetic field, or out of the page in Fig. 26.10. The direction of the magnetic force doesn't depend on the sign of the charge carriers; they could be negative electrons moving leftward, opposite the current direction in Fig. 26.10, or positive charges moving rightward. For a given current, changing the sign of the charge carriers reverses both the sign of $q$ and the direction of $\vec{v}_d$, leaving the force unchanged.

Equation 26.5 gives the net force on the charge carriers in the wire. In a physical wire, the magnetic force deflects charge carriers to one side of the wire, producing a charge separation and an electric field that results in a force on the rest of the wire (Fig. 26.11). Although its origin is not entirely magnetic, we loosely call the force in Equation 26.5 "the magnetic force on a wire." The magnetic force on a current-carrying wire is the basis for many practical devices, including loudspeakers and the electric motors that start cars and run refrigerators, disk drives, subway trains, pumps, food processors, power tools, and myriad other instruments of modern society.

Equation 26.5 holds for straight wires in uniform magnetic fields. In other cases we apply Equation 26.5 to very short segments of a wire that's either curved or in a nonuniform field, and we integrate to find the net force. Problem 58 explores this situation.

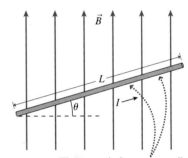

The magnetic force acts on all moving charges and points out of the page.

**FIGURE 26.10** A straight wire carrying current $I$ through a uniform magnetic field.

Electrons moving leftward are deflected upward by the magnetic force . . .

$\vec{B}_{\text{into page}}$

$\vec{F}_{\text{mag. on electrons}}$  $\vec{F}_{\text{elec. on ions}}$

. . . the resulting charge separation leads to an upward electric force on the rest of the wire.

**FIGURE 26.11** Origin of the magnetic force on a current-carrying wire.

---

**GOT IT? 26.3** The figure shows a flexible wire passing through a magnetic field that points out of the page. The wire is deflected upward, as shown. Is the current flowing (a) to the left or (b) to the right?

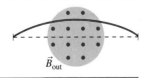

$\vec{B}_{\text{out}}$

---

**EXAMPLE 26.3** **Magnetic Force: A Power Line**

A power line runs along Earth's equator, where the magnetic field points horizontally from south to north and has strength $50\ \mu T$. The current in the line is 500 A, flowing from west to east. Find the magnitude and direction of the magnetic force on each kilometer of the power line.

**INTERPRET** This is a problem about the magnetic force on a straight wire in a uniform field.

**DEVELOP** Figure 26.12 shows the situation. We've taken eastward to be the $x$ direction, northward the $y$ direction, and upward the $z$ direction. Using the right-hand rule with current in the $\hat{\imath}$ direction and magnetic field in the $\hat{\jmath}$ direction, we see that the force points upward, in the $\hat{k}$ direction. To find the magnitude, we use Equation 26.5, $\vec{F} = I\vec{L} \times \vec{B}$, with $\vec{L} = 1.0\hat{\imath}$ km and $\vec{B} = 50\hat{\jmath}\ \mu T$.

**EVALUATE** Equation 26.5 gives

$$\vec{F} = I\vec{L} \times \vec{B} = (500\ A)(1.0\hat{\imath}\ km) \times (50\hat{\jmath}\ \mu T) = 25\hat{k}\ N$$

**FIGURE 26.12** Sketch for Example 26.3.

**ASSESS** Our answer shows that the force is upward. But there's no danger that the power line will break free and go soaring into the air: That 25-N force is far less than the line's weight, on the order of 10 kN.

## The Hall Effect

We noted earlier that the direction of the magnetic force depends on the direction of the current, not on the sign of the charge carriers. However, there's a subtle difference between two conductors that have the same current yet different charge carriers. In Fig. 26.13, moving charge carriers of either sign are deflected to the upper surface of the conductor. Again, that's because the signs of both charge and velocity are opposite in the two cases. In both cases the result is a small electric field and its associated potential difference *across* the wire. The *direction* of the electric field and the *sign* of the potential difference depend on the sign of the charge carriers.

The separation of charges across a current-carrying wire is the **Hall effect**, and the potential difference is the **Hall potential**. In a steady state, the magnetic force on the charge carriers is just balanced by the electric force associated with charge separation, giving $qE = qv_d B$, or simply $E = v_d B$. In the rectangular conductor of Fig. 26.13, the electric field is uniform and the Hall potential is then $V_H = Eh = v_d Bh$. Using $I = nqAv_d$ and solving for $v_d$, we can then write $V_H = IBh/nAq$. Since $A = ht$, with $t$ the conductor thickness in the field direction (see Fig. 26.13), this becomes

$$V_H = \frac{IB}{nqt} \quad \text{(Hall potential)} \tag{26.6}$$

The quantity $1/nq$ is the **Hall coefficient**. Measuring the Hall coefficient gives information on the nature and density of the charge carriers. Alternatively, measuring $V_H$ in a material of known coefficient carrying a known current gives a direct measure of the magnetic field strength.

## 26.5 Origin of the Magnetic Field

Electric charges respond to electric fields, and electric charges produce electric fields. So it is with magnetism. We've just explored how *moving* electric charges respond to magnetic fields; we'll now see how *moving* electric charges produce magnetic fields. The first inkling of a relation between electricity and magnetism came in 1820 when the Danish scientist Hans Christian Oersted discovered that a compass needle is deflected by an electric current. A month after Oersted's discovery became known in Paris, the French scientists Jean Baptiste Biot (rhymes with "Leo") and Félix Savart (rhymes with "bazaar") had experimentally determined the form of the force arising from a steady current.

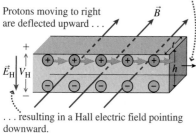

Electrons moving to left are deflected upward . . .

Magnetic field points into page.

$\vec{B}$

$\vec{E}_H \mid V_H$

. . . resulting in a Hall electric field pointing upward.

**(a)**

Protons moving to right are deflected upward . . .

$\vec{B}$

Current direction is the same.

$\vec{E}_H \mid V_H$

. . . resulting in a Hall electric field pointing downward.

**(b)**

**FIGURE 26.13** The Hall electric field $\vec{E}_H$ and Hall potential $V_H$ arise from the magnetic deflection of charge carriers. In both (a) and (b) the current is to the right, carried in (a) by negative charge moving to the left and in (b) by positive charge moving to the right.

## The Biot–Savart Law

**The Biot–Savart law** gives the contribution $d\vec{B}$ to the magnetic field at a point $P$ due to a small element of current, in much the way that Coulomb's law gives the electric field $d\vec{E}$ due to a charge element $dq$. Figure 26.14 shows a wire carrying a steady current $I$ and the contribution $d\vec{B}$ to the field at $P$ from a small length $dL$ of the wire. The current element $I\,d\vec{L}$ is the source of the field; it plays the same role as the charge $dq$ in Coulomb's law. That the product $I\,d\vec{L}$ has the units of charge times velocity is a reminder that magnetism is associated with *moving* electric charge. The magnetic field decreases with the inverse square of the distance, just as in Coulomb's law for the electric field.

There are important differences between the Coulomb and Biot–Savart laws. Charge—the source of electric field—is a scalar quantity. But *moving* charge—the source of magnetic field—has direction. The Biot–Savart law accounts for that direction by defining a vector $d\vec{L}$ along the current; then the field contribution $d\vec{B}$ from the source element $I\,d\vec{L}$ depends on the sine of the angle between $d\vec{L}$ and the unit vector $\hat{r}$ from the source toward the point where we're evaluating the field. Mathematically, all this is summarized in a compact vector equation:

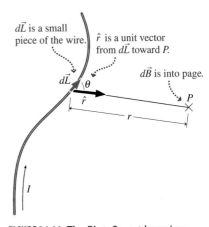

**FIGURE 26.14** The Biot–Savart law gives the magnetic field $d\vec{B}$ at the point $P$ arising from the current $I$ flowing along the infinitesimal vector $d\vec{L}$.

$$d\vec{B} = \frac{\mu_0}{4\pi}\frac{I\,d\vec{L} \times \hat{r}}{r^2} \quad \text{(Biot–Savart law)} \tag{26.7}$$

where $\mu_0$ is the **permeability constant**, whose exact value is $4\pi \times 10^{-7}\ \text{N/A}^2$ (equivalent and often-used units are T·m/A).

Besides the more complicated directionality evidenced by the cross product in the Biot–Savart law, there's another distinction between the Coulomb and Biot–Savart laws. Both describe fields of localized structures—namely, point charges and current elements. It makes sense to talk about an isolated point charge. But an isolated current element is impossible in the steady state; any steady current must flow in a complete circuit. So a Biot–Savart calculation necessarily involves the fields produced by current elements around an entire circuit. The magnetic field obeys the superposition principle, so the net field at any point is the vector sum, or integral, of the field contributions of all the individual current elements:

$$\vec{B} = \int d\vec{B} = \frac{\mu_0}{4\pi}\int \frac{I\,d\vec{L} \times \hat{r}}{r^2} \quad \text{(Biot–Savart law, integrated)} \tag{26.8}$$

The field given in Equation 26.8 depends on the details of the current distribution, but the directionality associated with the cross product means that, quite generally, magnetic field lines encircle the current that is their source (Fig. 26.15). The next two examples use the Biot–Savart law; later we'll find a simpler way to calculate magnetic fields for some current distributions.

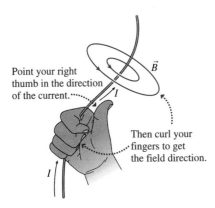

**FIGURE 26.15** Magnetic field lines generally encircle a current, with direction given by the right-hand rule.

---

**EXAMPLE 26.4** **Using the Biot–Savart Law:**
**The Field of a Current Loop**

Find the magnetic field at an arbitrary point $P$ on the axis of a circular loop of radius $a$ carrying current $I$.

**INTERPRET** This is a problem involving the magnetic field produced by a specified current distribution.

**DEVELOP** Figure 26.16a shows the current loop with the point $P$ a distance $x$ along the axis. The Biot–Savart law determines the field at $P$, and we've identified the vectors $d\vec{L}$ and $\hat{r}$ that appear in the law. As

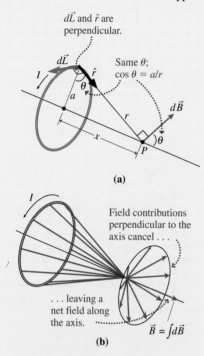

(a)

(b)

**FIGURE 26.16** Finding the magnetic field on the axis of a current loop.

Fig. 26.16b shows, the individual field components perpendicular to the axis cancel, giving a net field that's along the axis. So our plan is to find an expression for the $x$ components of the field contributions $d\vec{B}$, and then integrate to get the net field.

**EVALUATE** Figure 26.16a shows that the $x$ component of any $d\vec{B}$ is $dB_x = dB\cos\theta$, where $\cos\theta = a/r = a/\sqrt{x^2 + a^2}$. The figure also shows that $d\vec{L}$ and $\hat{r}$ are perpendicular; since $\hat{r}$ is a unit vector, the product $d\vec{L} \times \hat{r}$ has magnitude $dL$. Then the term $d\vec{L} \times \hat{r}/r^2$ in the Biot–Savart law has magnitude $dL/(x^2 + a^2)$, and we have

$$B = \int dB_x = \frac{\mu_0 I}{4\pi} \int_{\text{loop}} \frac{dL}{x^2 + a^2} \frac{a}{\sqrt{x^2 + a^2}}$$

$$= \frac{\mu_0 I a}{4\pi(x^2 + a^2)^{3/2}} \int_{\text{loop}} dL$$

where the integral reduces to a simple form because the distance $x$ is the same for all points on the loop. The remaining integral is the sum of infinitesimal lengths around the loop, or the loop circumference $2\pi a$. So we have

$$B = \frac{\mu_0 I a^2}{2(x^2 + a^2)^{3/2}} \qquad (26.9)$$

The direction of the field, as suggested in Fig. 26.16b, is along the axis.

**ASSESS** The field is strongest right at the loop center $(x = 0)$ because here we're closest to the loop and so the contributions from all segments of the loop are greatest. The field decreases as we move away from the loop. In general the field is a complicated function of distance, but for large distances $(x \gg a)$ it falls off as $1/x^3$. That should remind you of the field we found for an electric dipole in Chapter 20. We'll have a great deal more to say about this dipole-like behavior in Section 26.7. ∎

---

**EXAMPLE 26.5** **Using the Biot–Savart Law:**
**The Field of a Straight Wire**

Find the magnetic field produced by an infinitely long straight wire carrying steady current $I$.

**INTERPRET** This example, too, is about the field produced by a specified current distribution.

**DEVELOP** Figure 26.17 is our drawing of the wire on a coordinate system with the $x$ axis along the wire. Since the wire is infinite, the field magnitude must be the same at all points equidistant from the wire. We show one such point $P$, a distance $y$ from the wire. We also show an infinitesimal segment $d\vec{L}$ of the wire and the unit vector $\hat{r}$ toward the field point. Our plan is to calculate the field contributions $d\vec{B}$ from all such current elements, and then integrate to find the field $\vec{B}$.

**EVALUATE** Both $d\vec{L}$ and $\hat{r}$ lie in the plane of the page, so at $P$ the quantity $d\vec{L} \times \hat{r}$ in the Biot–Savart law is out of the page. This is true for any segment of the wire. Therefore we can sum the magnitudes of the contributions $d\vec{B}$ to find the magnitude of the net field, and we know its direction at $P$ will be out of the page. With $\hat{r}$

**FIGURE 26.17**
Calculating the magnetic field at $P$ due to an infinite straight wire carrying current $I$ along the $x$ axis.

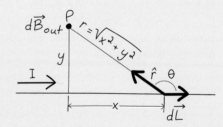

a unit vector, $|d\vec{L} \times \hat{r}| = dL\sin\theta$, where Fig. 26.17 shows that $\sin\theta = y/r = y/\sqrt{x^2 + y^2}$. Then the Biot–Savart law gives a field contribution of magnitude

$$dB = \frac{\mu_0 I}{4\pi} \frac{|d\vec{L} \times \hat{r}|}{r^2} = \frac{\mu_0 I}{4\pi} \frac{dL\sin\theta}{r^2} = \frac{\mu_0 I}{4\pi} \frac{y\,dL}{(x^2 + y^2)^{3/2}}$$

*cont'd.*

Since the segment $d\vec{L}$ lies along the $x$ axis, $dL = dx$. Also, $y$ is a constant here, so the net field becomes

$$B = \int dB = \frac{\mu_0 I y}{4\pi} \int_{-\infty}^{\infty} \frac{dx}{(x^2 + y^2)^{3/2}}$$

where we chose the limits to include the entire infinite wire. The integral is a standard one, given in the integral tables of Appendix A; the result is

$$B = \frac{\mu_0 I}{2\pi y} \qquad (26.10)$$

**ASSESS** This result for the *magnetic* field of a long current-carrying wire should remind you of our earlier finding for the *electric* field of a line charge; both decrease as the inverse of the distance from the line. But where the electric field of a line charge points radially outward, the magnetic field of a line current encircles the current, as shown in Fig. 26.18. Of course, an infinite line current is impossible, but our result is a good approximation to the fields of finite wires if we're close compared with the wire's length. ∎

**FIGURE 26.18** Magnetic field lines encircle a straight wire, with their direction given by the right-hand rule.

## The Magnetic Force Between Conductors

In Section 26.4 we found the force on a current-carrying wire in a magnetic field: $\vec{F} = I\vec{L} \times \vec{B}$. Now we find that a straight wire produces a magnetic field. That means two parallel current-carrying wires exert magnetic forces on each other. Figure 26.19 shows the situation for wires carrying currents in the same direction. The wires are a distance $d$ apart, so the field of wire 1 at the location of wire 2 follows from Equation 26.10: $B_1 = \mu_0 I_1 / 2\pi d$. The field is perpendicular to wire 2, so the force on a length $L$ of wire 2 is

$$F_2 = I_2 L \times B_1 = \frac{\mu_0 I_1 I_2 L}{2\pi d} \quad \text{(magnetic force between two wires)} \qquad (26.11)$$

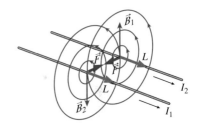

**FIGURE 26.19** The magnetic force between parallel currents in the same direction is attractive.

Figure 26.19 shows that the direction of this force is toward wire 1, so the parallel currents *attract*. Analyzing the force on wire 1 from wire 2 amounts to interchanging the subscripts 1 and 2, giving an attractive force of the same magnitude. Reversing one of the currents would change the signs of both forces, showing that antiparallel currents *repel*.

The force between nearby conductors can be quite large, so engineers who design high-strength electromagnets must provide enough physical support to withstand the magnetic force. The hum you often hear around electrical equipment comes from the mechanical vibration of nearby conductors in transformers and other devices, as they respond to the changing force associated with 60-Hz alternating current.

The magnetic force between conductors is the basis for the SI definition of the ampere and, consequently, the coulomb. One ampere is defined as the current in two long, parallel conductors 1 m apart when they experience a magnetic force of $2 \times 10^{-7}$ N per meter of length. Then 1 C is the amount of charge passing in 1 s through a wire carrying 1 A.

**GOT IT? 26.4**  A flexible wire is wound into a flat spiral as shown in the figure. If a current flows in the direction shown, will the coil (a) tighten or (b) become looser? Does your answer depend on the current direction?

**FIGURE 26.20** Finding the direction of a current loop's magnetic dipole moment.

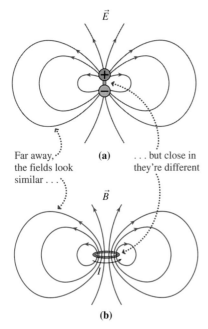

**FIGURE 26.21** (a) The electric field of an electric dipole and (b) the magnetic field of a current loop. Far from their sources, both have the shape and the $1/r^3$ dependence of the dipole field.

**FIGURE 26.22** Sunspots are regions of intense magnetic field. The large sunspot group below the solar equator was responsible for some of the most energetic solar outbursts ever recorded, which occurred in late 2003.

## 26.6 Magnetic Dipoles

The infinite line current of Example 26.5 is an impossibility. But the current loop of Example 26.4 is eminently realistic, and it shows the essential characteristic of all steady-state currents—namely, a closed loop in which current is everywhere the same. We found that the result of Example 26.4 reduces to a dipole-like inverse-cube dependence of the field on distance for points far from the current loop. Let's take a closer look at Equation 26.9: $B = \mu_0 I a^2/[2(x^2 + a^2)^{3/2}]$. For $x \gg a$, we can ignore $a^2$ compared with $x^2$ in the denominator to get $B \approx \mu_0 I a^2/2x^3$. Multiply both sides by $2\pi$, and the equation becomes $B \simeq 2 \mu_0 I A/4\pi x^3$, where $A$ is the loop area $\pi a^2$. Now compare this result with the field on the axis of an electric dipole, Equation 20.6b: $E = 2kp/x^3$. Both show the inverse-cube dependence characteristic of the dipole field. Both involve fundamental constants from the laws that relate fields and their sources: $k$ from Coulomb's law and $\mu_0/4\pi$ from the Biot–Savart law. Both have a factor of 2 in the numerator. Where the electric-field expression contains the electric dipole moment $p$, the product of charge and separation, the magnetic-field expression contains $IA$, the product of the loop current and loop area. This suggests that we identify $IA$ as the magnitude of the current loop's **magnetic dipole moment**. We designate the magnetic dipole moment by a vector $\vec{\mu}$, whose magnitude is $IA$ and whose direction follows from the right-hand rule shown in Fig. 26.20. If $\vec{A}$ is a vector whose magnitude is the area $A$ and whose direction is the same as that of $\vec{\mu}$ in Fig. 26.20, then we can write the magnetic dipole moment as $\vec{\mu} = I\vec{A}$. Practical current loops often have multiple turns wound from a single piece of wire. Since the individual turns are in series, each carries the same current. If a loop has $N$ turns, the fields of the individual turns superpose. That makes an effective current $NI$, and the magnetic dipole moment becomes

$$\vec{\mu} = NI\vec{A} \qquad \left(\begin{array}{c}\text{magnetic dipole moment,}\\ N\text{-turn current loop}\end{array}\right) \qquad (26.12)$$

Although we've found the magnetic field for a current loop only on the loop axis, a more elaborate calculation would show that the magnetic field anywhere far from the loop has exactly the same configuration as the electric field far from an electric dipole. And although we developed the magnetic dipole moment for a circular loop, Equation 26.12 in fact gives the dipole moment of *any* closed loop of current. We conclude that **any current loop constitutes a magnetic dipole**, and that far from any current loop, the magnetic field will be the characteristic field of a dipole. Electric and magnetic dipoles are analogous: Both have the same field configuration and mathematical form far from their sources (Fig. 26.21), and both sources are characterized by their respective dipole moments. But although the two fields have the same configuration, they aren't the same. One is an electric field, with its origin in static electric charge; the other is a magnetic field, with its origin in *moving* charge—specifically, charge moving in a closed loop. And the similarity in field configurations holds only at large distances; as Fig. 26.21 shows, the fields near electric and magnetic dipoles are very different, reflecting the different structures that give rise to each.

Current loops are ubiquitous, and so are dipole magnetic fields. Multiple turns of current-carrying wire produce the strong magnetic fields of electromagnets, and superconducting loops provide the fields in MRI scanners. At the atomic level, orbiting and spinning electrons become miniature magnetic dipoles. Currents flowing in Earth's liquid outer core produce the magnetic field that helps us navigate and that, for reasons we discussed in Section 26.3, protects us from high-energy solar particles. Reversal of Earth's magnetic field at irregular intervals on the order of a million years provides geologists with a magnetic record of seafloor spreading. Currents in the gaseous Sun and its atmosphere produce a solar magnetic field that varies much more rapidly and dramatically than Earth's. The overall solar field reverses roughly every 11 years, coinciding with the rise and fall in the number of sunspots—regions of strong magnetic field that often give rise to violent solar outbursts (Fig. 26.22). The origin and variation of magnetism in astrophysical objects are associated with electric currents in rotating conductive fluids, but the details are still under active investigation.

## Dipoles and Monopoles

*Electric* dipoles are important because many molecules exhibit dipole-like charge distributions. Other structures, from atoms to radio antennas, often have an essentially dipolar character. But every electric dipole ultimately consists of more fundamental entities—namely, point charges. It's the separation of two opposite point charges or charged regions that makes an electric dipole.

With magnetism the situation is different. No one has ever found an isolated magnetic north or south pole analogous to an electric point charge. Electromagnetic theory doesn't rule out the existence of such **magnetic monopoles**, and indeed some elementary-particle theories suggest that monopoles might have formed in the Big Bang. But they've never been found. All magnetic fields we've ever seen come from moving *electric* charge—that is, from electric current. As we'll see in Section 26.7, that includes the fields of permanent magnets. And because steady currents form closed loops, which produce dipole-like magnetic fields, that means the simplest magnetic entity is the dipole.

Electric field lines generally begin or end on electric charges. But there aren't any "magnetic charges"—magnetic monopoles. Magnetic field lines don't begin or end, but form closed loops encircling the moving electric charges that are their sources. In Chapter 21 we developed Gauss's law as a way of quantifying the statement that the number of electric field lines emerging from any closed surface depends only on the charge enclosed. Because there's no magnetic charge, the net number of magnetic field lines and therefore the **magnetic flux** $\oint \vec{B} \cdot d\vec{A}$ emerging from any closed surface is always zero. Thus **Gauss's law for magnetism** is

$$\oint \vec{B} \cdot d\vec{A} = 0 \qquad \text{(Gauss's law for magnetism)} \qquad (26.13)$$

Like Gauss's law for electricity, Equation 26.13 is one of the four fundamental laws that govern all electromagnetic phenomena in the universe. We'll meet the remaining two laws shortly. Although Gauss's law for magnetism has zero on its right side, it is not devoid of content; rather, it says that all magnetic fields must be configured so that their field lines have no beginnings or endings.

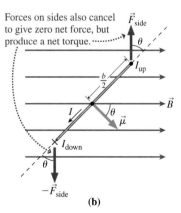

**FIGURE 26.23** (a) A rectangular current loop in a uniform magnetic field. (b) Top view of the loop, showing that magnetic forces on the vertical sides result in a net torque.

**GOT IT? 26.5** The figure shows two fields. Which could be a magnetic field?

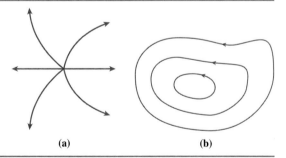

(a)                              (b)

## The Torque on a Magnetic Dipole

In Section 20.5 we found that an electric dipole of dipole moment $\vec{p}$ in a uniform electric field $\vec{E}$ experiences a torque given by $\vec{\tau} = \vec{p} \times \vec{E}$; in a nonuniform field there's a net force as well. The same is true for a magnetic dipole in a magnetic field, as we can see by considering the rectangular current loop in a uniform field shown in Fig. 26.23a. Current flowing along the top and bottom of the loop results in upward and downward forces of equal magnitude, and neither a net force nor a net torque is associated with these forces. Currents flowing along the vertical sides also result in equal but opposite forces. However, as Fig. 26.23b shows, these forces result in a net torque about a vertical axis through the center of the loop. The vertical sides have length $a$ and the currents are perpendicular to the horizontal magnetic field, so the force on each has magnitude $F_{side} = IaB$. The vertical sides are half the loop width $b$ from the axis, so the torque due to each is $\tau_{side} = \frac{1}{2}bF_{side}\sin\theta = \frac{1}{2}bIaB\sin\theta$. Torques on the two sides are in the same direction (out of the page in Fig. 26.23b), so the net torque is $\tau = IabB\sin\theta = IAB\sin\theta$, with $A$ the loop area. We've already identified $IA$ as the magnitude of the loop's magnetic dipole moment $\vec{\mu}$, and given the direction of $\vec{\mu}$ as shown in Figs. 26.20 and 26.23b, we can incorporate the directionality and the factor $\sin\theta$ into a cross product:

$$\vec{\tau} = \vec{\mu} \times \vec{B} \quad \text{(torque on a magnetic dipole)} \quad (26.14)$$

analogous to the torque on an electric dipole.

The magnetic torque of Equation 26.14 causes magnetic dipoles—current loops—to align with their dipole moment vectors along the magnetic field. It takes work to rotate a dipole out of alignment with the field, and in analogy with Equation 20.11 the associated potential energy is

$$U = -\vec{\mu} \cdot \vec{B} \quad (26.15)$$

If the magnetic field is not uniform, then a dipole experiences a net force as well. It's this force that causes the nonuniform field near the poles of a bar magnet to attract magnetic materials that, as we'll see in the next section, contain magnetic dipoles.

The torque on a magnetic dipole is the basis of many important technologies, including electric motors and MRI imaging. Some satellites use the torque produced by Earth's magnetic field to orient themselves in space; with electricity generated from solar panels used to power current loops, there's no fuel to run out.

---

**APPLICATION** **Electric Motors**

called the commutator. The current loop rotates to align with the field, but just as it does so, the brushes cross the gaps in the commutator and reverse the loop's current and therefore its dipole moment vector. Now the loop swings another 180° to its new "desired" position, but again the commutator reverses the current and so the loop rotates continuously. Real motors usually contain multiple coils at different angles to ensure smooth rotation and a nearly constant torque; you can see several coils in the photo of a disassembled motor. A rigid shaft spinning with the coils delivers mechanical energy to an external load. Thus the motor is a device that converts electrical energy to mechanical energy; the magnetic field is an intermediary in this energy conversion.

Electric motors are so much a part of our lives that we hardly think of them. Yet refrigerators, disk drives, subway trains, vacuum cleaners, power tools, food processors, fans, washing machines, water pumps, hybrid cars, and most industrial machinery would be impossible without electric motors.

At the heart of every electric motor is a current loop in a magnetic field. But instead of a steady current, the loop carries a current that reverses to keep the loop always spinning. In direct-current (DC) motors, this is achieved through the electrical contacts that provide current to the loop. The figure shows how current flows to the loop through a pair of stationary brushes that contact rotating conductors

**EXAMPLE 26.6**  **Torque on a Current Loop:**
**Designing a Hybrid-Car Motor**

Toyota's Prius gas–electric hybrid car uses a 50-kW electric motor that develops a maximum torque of 400 N·m. Suppose you want to produce this same torque in a motor like the one shown in the preceding Application, consisting of a 950-turn rectangular coil of wire measuring 30 cm by 20 cm, in a uniform field of 50 mT. How much current does the motor need?

**INTERPRET** This is a problem about an electric motor, which according to the Application is essentially a current loop in a magnetic field. We're given the torque and asked for the current.

**DEVELOP** Equation 26.14, $\vec{\tau} = \vec{\mu} \times \vec{B}$, determines the torque on a current loop. Figure 26.24 is a sketch of the loop at the point of maximum torque, $\tau_{max} = \mu B$, which occurs when $\sin\theta = 1$. To solve for the current, we need the magnetic dipole moment from Equation 26.12, $\mu = NIA$. Using this value for $\mu$ in our expression for the maximum torque gives $\tau_{max} = NIAB$.

**EVALUATE** Solving for $I$ using the maximum torque and the loop dimensions gives

$$I = \frac{\tau_{max}}{NAB} = \frac{400 \text{ N} \cdot \text{m}}{(950)(0.30 \text{ m})(0.20 \text{ m})(0.050 \text{ T})} = 140 \text{ A}$$

**FIGURE 26.24** Loop for the motor of Example 26.6, shown in the position of maximum torque.

**ASSESS** That's a large current, but propelling a car is a big job. The actual Prius motor operates at 500 V, so its 50-kW power output requires current $I = P/V = 100$ A, close to our answer.  ∎

## 26.7 Magnetic Matter

In this chapter on magnetism, we've said remarkably little about magnets. That's because, as we've emphasized, magnetism is fundamentally about moving electric charge. Magnets and magnetic matter are just a minor manifestation of this more universal phenomenon.

The magnetism of everyday magnets and of magnetic materials like iron results from atomic-scale current loops. An electron orbiting a nucleus constitutes a simple current loop and therefore has a magnetic dipole moment (Fig. 26.25). More importantly, an electron possesses an intrinsic magnetic dipole moment associated with a quantum-mechanical angular momentum called "spin." Interactions among these magnetic moments determine the magnetic properties of atoms and of bulk matter. The details necessarily involve quantum mechanics; here we give a qualitative overview of magnetism in matter, which manifests itself in three distinct ways: ferromagnetism, paramagnetism, and diamagnetism.

**FIGURE 26.25** In the classical model of the atom, the circling electron constitutes a miniature current loop. The current is opposite the motion because the electron is negative. Drawing is only suggestive; the electron's intrinsic magnetic dipole moment is usually more important than its orbital motion.

### Ferromagnetism

The magnetism you're familiar with is **ferromagnetism**, which is limited to a few substances, including iron, nickel, cobalt, and some alloys and compounds. Strong interactions among atomic magnetic moments result in **magnetic domains**, regions that contain $10^{17}$–$10^{21}$ atoms whose magnetic moments are all aligned in the same direction. Normally the magnetic moments of different domains point in random directions, so there's no net magnetic moment. But when an external magnetic field is applied, the domains all align and the material acquires a net magnetic moment. If the field is nonuniform, it then experiences a net force, which is why ferromagnetic materials are attracted to magnets.

So-called **hard** ferromagnetic materials retain their magnetism even after the applied field is removed; the result is a permanent magnet. A bar magnet, for example, has its internal magnetic moments aligned along its long dimension. You can think of its field as arising from currents circulating around the surface of the magnet (Fig. 26.26)—currents that ultimately result from the superposition of individual atomic current loops. Computer disks, credit card strips, and audio/video tapes use hard ferromagnetic materials that retain

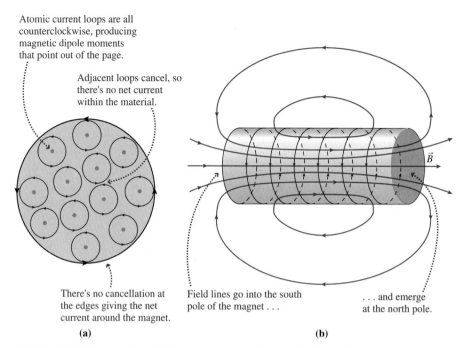

Atomic current loops are all counterclockwise, producing magnetic dipole moments that point out of the page.

Adjacent loops cancel, so there's no net current within the material.

There's no cancellation at the edges giving the net current around the magnet.

Field lines go into the south pole of the magnet . . .

. . . and emerge at the north pole.

$\vec{B}$

(a)

(b)

**FIGURE 26.26** (a) Cross section of a bar magnet, showing atomic current loops all aligned the same way and making a net current around the magnet. (b) Side view showing the field that results from this magnetization current.

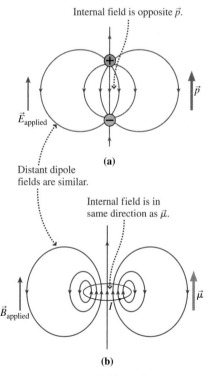

Internal field is opposite $\vec{p}$.

$\vec{E}_{applied}$

$\vec{p}$

(a)

Distant dipole fields are similar.

Internal field is in same direction as $\vec{\mu}$.

$\vec{B}_{applied}$

$I$

$\vec{\mu}$

(b)

**FIGURE 26.27** Internal fields of electric and magnetic dipoles have opposite directions. (a) Electric dipoles reduce an applied electric field; (b) magnetic dipoles increase an applied magnetic field.

information as patterns of permanent magnetization. **Soft** ferromagnetic materials, in contrast, don't hold magnetization. They're used where magnetization must be turned on and off rapidly, as in the "heads" that write information to computer disks and tapes. Ferromagnetism disappears at the so-called **Curie temperature**, as random thermal motions disrupt the organized alignment of magnetic dipoles; for iron this phase transition occurs at 1043 K.

## Paramagnetism

Many substances that aren't ferromagnetic nevertheless consist of atoms or molecules that have permanent magnetic dipole moments. There's no strong interaction among the individual dipoles, so these **paramagnetic** materials respond only weakly to external magnetic fields. Paramagnetic effects are generally significant only at very low temperatures.

## Diamagnetism

Materials without intrinsic magnetic moments can have moments induced by changes in an applied magnetic field. Whereas ferromagnetic and paramagnetic materials are attracted to magnets, these **diamagnetic** materials are repelled. We'll explore the origins of diamagnetism in Chapter 27.

## Magnetic Permeability and Susceptibility

We found in Chapter 20 that the alignment of molecular electric dipoles reduces the electric field in a material, and in Chapter 23 we introduced the dielectric constant as the factor describing this reduction. In paramagnetic and ferromagnetic materials, alignment of magnetic dipoles causes an *increase* in the field. Figure 26.27 shows that this difference occurs because the magnetic field within a current loop points in the *same* direction as the loop's magnetic dipole moment, whereas the internal field of an electric dipole is *opposite* the dipole moment. The magnetic response of ferromagnetic materials is further complicated because it depends on the past history, which is what makes permanent magnets possible. Coils for electromagnets and VCR and computer disk "heads" are usually wound on ferromagnetic cores to provide a much larger magnetic field than the coil current alone could produce.

## 26.8 Ampère's Law

Computing electric fields with Coulomb's law in Chapter 20 was cumbersome for all but the simplest charge distributions. In Chapter 21 we saw how Gauss's law greatly simplified electric-field calculations for symmetric charge distributions. Is there an analogous approach for magnetic fields? Gauss's law for magnetism, Equation 26.13, won't do because it doesn't relate a magnetic field to its source—namely, moving charge.

Figure 26.28 shows two of the circular magnetic field lines surrounding a long wire carrying a current $I$ out of the page. Imagine moving around the inner circle, and as you go a little way, take the product of the displacement $dr$ with the magnetic field in the direction you're going. Here you're moving in the direction of the field, so that product is $B\,dr$; more generally, it's the dot product $\vec{B}\cdot d\vec{r}$. Now add up all these products around the circle. Formally, the result is the line integral $\oint \vec{B}\cdot d\vec{r}$, where the circle indicates that we're integrating around a closed path. In this case the integral becomes just $\oint B\,dr$ because $\vec{B}$ and $d\vec{r}$ are in the same direction. But here the field magnitude is given by Equation 26.10: $B = \mu_0 I/2\pi r$, where we've replaced $y$ with the radius $r$. Since $r$ has the constant value $r_1$ on the inner circle in Fig. 26.28, the integral becomes $(\mu_0 I/2\pi r_1)\oint dr$. Now $\oint dr$ is the total length of the circular path, or its circumference $2\pi r_1$. So the value of $\oint \vec{B}\cdot d\vec{r}$ is $\mu_0 I$. If you try the same thing for the outer circle in Fig. 26.28, $r_2$ replaces $r_1$, but the result is the same: $\oint \vec{B}\cdot d\vec{r} = \mu_0 I$, independent of the radius.

We get the same result even if the path doesn't coincide with a field line, as Fig. 26.29 suggests. On the radial segments of the path shown, $\vec{B}\cdot d\vec{r} = 0$ and there's no contribution to the integral. On segment $AB$, the field is stronger than if we had stayed on $CD$, but the segment is proportionately shorter and the integral remains unchanged. We could approximate any arbitrary path as a sequence of radial segments and circular arcs, showing that the value of $\oint \vec{B}\cdot d\vec{r}$ is independent of path as long as the path surrounds the current $I$. The value of that integral is simply $\mu_0 I$. Magnetic fields obey the superposition principle, so this result must be true for any current distribution, not just a single line current. That is, the line integral $\oint \vec{B}\cdot d\vec{r}$ around *any* closed path is directly proportional to the current encircled by that path. This result is **Ampère's law**, a universal statement about current and magnetic field:

$$\oint \vec{B}\cdot d\vec{r} = \mu_0 I_{\text{encircled}} \quad \text{(Ampère's law, steady currents)} \qquad (26.16)$$

Ampère's law is another of the four fundamental laws of electromagnetism, although in the form of Equation 26.16 it's limited to steady currents; it also provides a decent approximation for slowly varying currents. Later we'll generalize Ampère's law to remove the restriction to steady currents.

Ampère's law relates the magnetic field to its source—namely, moving charge in the form of electric current—as does the Biot–Savart law. In fact, the laws of Ampère and Biot–Savart are related in the same way as Gauss's and Coulomb's laws. Coulomb and Biot–Savart show how fields arise from pointlike sources—charge elements $dq$ and current elements $I\,d\vec{L}$. Gauss and Ampère are global descriptions, telling how the field must behave over a geometric structure (a closed surface for Gauss, a closed loop for Ampère) in relation to the source (charge or current) enclosed or encircled by that structure. In both cases the field $\vec{E}$ or $\vec{B}$ that appears in the integral is the *net* field arising from *all* sources, not just the enclosed charge or encircled currents.

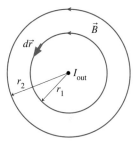

**FIGURE 26.28** Two magnetic field lines surrounding a wire carrying current out of the page.

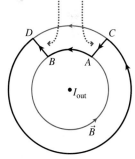

**FIGURE 26.29** A closed loop that does not coincide with a field line. The line integral $\oint \vec{B}\cdot d\vec{r}$ around this loop has the same value $\mu_0 I$ that it has around a circular loop.

Ampère's law, like Gauss's, is a truly universal statement. It holds in the electromagnetic devices we build, in atomic and molecular systems, and in distant astrophysical objects. Find a path around which $\oint \vec{B} \cdot d\vec{r}$ isn't zero, and you've found a place where electric current must be flowing (Fig. 26.30). Although it's difficult to show mathematically, the Biot–Savart law follows from Ampère's law just as Coulomb's law follows from Gauss's.

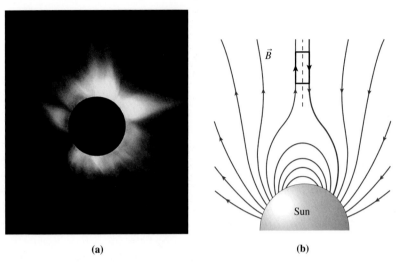

**(a)** **(b)**

**FIGURE 26.30** (a) Coronal streamers in the Sun's atmosphere contain oppositely directed magnetic fields. (b) A model calculation of the magnetic field in a single streamer. Since $\oint \vec{B} \cdot d\vec{r}$ is clearly nonzero around the loop shown, there must be an encircled current.

---

**EXAMPLE 26.7**  **Ampère's Law: Solar Currents**

The long dimension of the rectangular loop in Fig. 26.30b is 400 Mm, and the magnetic field strength near the loop has a constant magnitude of 2 mT. Estimate the total current encircled by the rectangle.

**INTERPRET** This is a problem about currents encircled by a loop, so it must involve Ampère's law.

**DEVELOP** Figure 26.30b provides our drawing. Ampère's law (Equation 26.16) equates $\oint \vec{B} \cdot d\vec{r}$ to $\mu_0 I_{\text{encircled}}$, so we want to evaluate the integral around the loop shown and then solve for $I_{\text{encircled}}$.

**EVALUATE** On the short segments of the path, $\vec{B} \perp d\vec{r}$, so there's no contribution to the integral. On both long segments, $\vec{B}$ is constant and lies in the direction we're traversing the path, so here $\int \vec{B} \cdot d\vec{r}$ becomes simply $BL$, where $L$ is the length of the path segments. Each side contributes this much, so $\oint \vec{B} \cdot d\vec{r} = 2BL$. Ampère's law equates this quantity to $\mu_0 I_{\text{encircled}}$, so we can solve to get

$$I_{\text{encircled}} = \frac{2BL}{\mu_0} = \frac{(2)(2 \text{ mT})(400 \text{ Mm})}{4\pi \times 10^{-7} \text{ N/A}^2} = 10^{12} \text{ A}$$

**ASSESS** This is a large current, but we're dealing with a region much larger than the Earth, so that shouldn't be too surprising. We can get

the direction of the current from the right-hand rule: Curl your fingers around the loop in the direction that gives positive $\oint \vec{B} \cdot d\vec{r}$, and your thumb points in the direction of the current. Here that's into the page. In three dimensions this current actually flows around the Sun in approximately the equatorial plane. Note that our result depends crucially on the field reversing across the equatorial plane. In a truly uniform field, one side of the loop would have contributed $BL$ to the line integral, the other $-BL$. That would make the integral zero and imply no encircled current. This shows that we can have a uniform field in a current-free region, but not a field that reverses sign—at least not abruptly as in this solar example.

✓ **TIP**  Amperian Loops

The loop used with Ampere's law is truly arbitrary. It doesn't need to coincide with a field line. In this example, the rectangular loop coincided with the straight field lines over its long sides but not along its shorter ends. The loop used with Ampere's law is called an **amperian loop**. Don't confuse amperian loops with field lines; they might coincide, but they don't have to.

∎

---

**GOT IT? 26.6** The figure shows three parallel wires carrying current of the same magnitude $I$, but in one of them the current direction is opposite that of the other two. If $\oint \vec{B} \cdot d\vec{r} \neq 0$ around loop 2, (a) what is $\oint \vec{B} \cdot d\vec{r}$ around loop 1, and (b) which current is the opposite one?

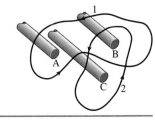

## Using Ampère's Law

For charge distributions with sufficient symmetry, we used Gauss's law to solve for the electric field in a simple and elegant way. We can do the same with Ampère's law for sufficiently symmetric current distributions. Here's the approach:

**PROBLEM SOLVING STRATEGY 26.1** **Ampère's Law**

**INTERPRET** Interpret the problem to be sure it's about magnetic field and current. Identify the symmetry.

**DEVELOP** Based on the symmetry, sketch some field lines. Then find an amperian loop over which you'll be able to evaluate $\oint \vec{B} \cdot d\vec{r}$. That means the field should be constant and parallel to the loop over all or part of the loop; where they're not parallel they should be perpendicular. Like a gaussian surface, the amperian loop is a purely mathematical construct; it need not correspond to anything physical. Draw your loop.

**EVALUATE**
- Evaluate $\oint \vec{B} \cdot d\vec{r}$ for your loop. This should be easy because you'll be able to take $B$ outside the integral over those segments where it's constant, and segments where $\vec{B}$ is perpendicular to the loop won't contribute to the integral.
- Evaluate the *encircled* current.
- Equate your result for $\oint \vec{B} \cdot d\vec{r}$ to $\mu_0 I_{\text{encircled}}$, and solve for $B$. Symmetry should give you the direction of $\vec{B}$.

**ASSESS** Does your answer make sense in terms of what you know about the fields of simple charge and current distributions?

---

**EXAMPLE 26.8** **Using Ampère's Law: Outside and Inside a Wire**

A long, straight wire of radius $R$ carries a current $I$ distributed uniformly over its cross section. Find the magnetic field (a) outside and (b) inside the wire.

**INTERPRET** We follow our strategy and identify this as a situation with line symmetry. Therefore the field depends only on the radial distance from the wire's central axis.

**DEVELOP** Magnetic field lines encircle their source, so the only field lines consistent with the symmetry are concentric circles. We sketched some in Fig. 26.31. The field is tangent to the field lines and, by symmetry, has the same magnitude $B$ over a field line. So the field lines themselves make good amperian loops.

**FIGURE 26.31** Cross section of a long cylindrical wire. Any field line can serve as an amperian loop.

**EVALUATE**
- The field is everywhere parallel to a circular amperian loop, and its magnitude is constant on the loop, so for a loop of radius $r$, $\oint \vec{B} \cdot d\vec{r}$ becomes just $2\pi r B$. This is true both outside and inside the wire.

We'll first answer (a):
- For any loop *outside* the wire, the encircled current is the total current $I$.
- Equating our expression for $\oint \vec{B} \cdot d\vec{r}$ to $\mu_0$ times the encircled current gives $2\pi r B = \mu_0 I$, so

$$B = \frac{\mu_0 I}{2\pi r} \quad \left( \begin{array}{l} \text{field outside any current} \\ \text{distribution with line symmetry} \end{array} \right) \quad (26.17)$$

Now on to (b):
- *Inside* the wire, a circular amperian loop encloses only some of the current. How much? With the current uniformly distributed over the wire's cross section, there's a uniform current density $J = I/A = I/\pi R^2$. The encircled current is the current density times the area $\pi r^2$ within our loop, so $I_{\text{encircled}} = I(r^2/R^2)$.
- Then we have $2\pi r B = \mu_0 I(r^2/R^2)$, which gives

$$B = \frac{\mu_0 I r}{2\pi R^2} \quad \left( \begin{array}{l} \text{field inside a uniform current} \\ \text{distribution with line symmetry} \end{array} \right) \quad (26.18)$$

In both cases application of the right-hand rule shows that the field circles counterclockwise, as shown in Fig. 26.31.

**ASSESS** Equation 26.17 is identical to our result for the line current of Example 26.5, and shows that the field outside *any* current distribution with line symmetry is the same as the field of a line current at the symmetry axis. We found the same thing for the electric fields outside cylindrical charge distributions, including the $1/r$ decrease with distance from the axis. Inside the wire, meanwhile, the field increases linearly with distance as we pick up more current in proportion to $r^2$, while the field decreases as $1/r$. You found a similar result for a uniformly charged cylinder if you worked Problem 54 in Chapter 21. Of course, the electric and magnetic fields of cylindrical distributions look very different—$\vec{E}$ is radial, while $\vec{B}$ forms circles—but the dependence on distance is the same in both cases.

✓**TIP** Symmetry Is Crucial

Our use of Ampère's law to derive the field of a long wire depends crucially on symmetry. We can't arbitrarily pull $B$ outside the integral unless we know—as we do here from symmetry—that it's constant in magnitude and in direction relative to our amperian loop.

**EXAMPLE 26.9** **Ampère's Law: A Current Sheet**

An infinite flat sheet carries current out of this page. The current is distributed uniformly along the sheet, with current per unit width given by $J_s$. Find the magnetic field of this sheet.

**INTERPRET** We follow our strategy, identifying the current distribution as having plane symmetry. Then the only thing the field might depend on is the distance from the current-carrying sheet.

**DEVELOP** The only field lines consistent with the symmetry are straight lines parallel to the plane; we've drawn the current and some field lines in Fig. 26.32. The situation is similar to Example 26.7, and a suitable amperian loop is a rectangle with sides along the field lines and perpendicular edges; we've sketched one such rectangle of width $L$.

**FIGURE 26.32** A current sheet extends infinitely to the left and right, as well as in and out of the page. Field lines and a rectangular amperian loop are shown.

We drew the field in opposite directions on either side of the sheet; it had better be that way if, as we discussed in Example 26.7, $\oint \vec{B} \cdot d\vec{r}$ is to be nonzero. And it has to be nonzero because we know that our amperian loop encircles current.

**EVALUATE**

- We evaluate $\oint \vec{B} \cdot d\vec{r}$ just as in Example 26.7, getting $2BL$.
- The sheet carries current $J_s$ per unit width, so our rectangle of width $L$ encircles a current $I_{\text{encircled}} = J_s L$.
- Equating our expression for $\oint \vec{B} \cdot d\vec{r}$ to $\mu_0 I_{\text{encircled}}$ gives $2BL = \mu_0 J_s L$, or

$$B = \tfrac{1}{2}\mu_0 J_s \quad \text{(field of an infinite current sheet)} \quad (26.19)$$

**ASSESS** Make sense? Like the electric field of an infinite plane charge, the magnetic field of an infinite current sheet doesn't depend on distance from the sheet. Of course, there's no truly infinite sheet, so our result is an approximation valid near a finite sheet but not close to its edges. As Fig. 26.33 shows, the lines of a finite sheet wrap around the ends to form closed loops, and far from the loop the field begins to resemble that of a wire. But close in, Equation 26.19 holds. ∎

Far out, the field lines become nearly circular.

$I_{\text{out}}$

Close in, the field resembles that of an infinite sheet.

**FIGURE 26.33** Field of a finite-width current sheet.

(a)

(b)

(c)

(d)

**FIGURE 26.34** As the coil gets longer, the interior field stays nearly constant but the exterior field weakens as the field lines spread ever farther apart.

## Fields of Simple Current Distributions

We've just used Ampère's law to calculate the fields of two symmetric current distributions, and we compared them with analogous results for electric fields. Table 26.1 summarizes these and other analogies. Although the magnetic and electric fields may look different, they exhibit the same general relationships between geometry and the way the fields decrease with distance. Real distributions are more complicated, but may often be approximated by these simple cases. Far from *any* current loop, for example, its field approximates that of a dipole. Very near *any* wire, its field is essentially that of a long straight wire. Very near *any* flat sheet of current, the field is essentially that of Example 26.9.

## Solenoids

We found in Chapter 23 that there's an essentially uniform electric field inside a parallel-plate capacitor. Here we explore a current configuration that produces an analogously uniform magnetic field.

Figure 26.34a shows a single current loop and its magnetic field. Figure 26.34b shows a few turns of wire forming a tight coil; the field is not much different from that of the single loop. But add more turns (Fig. 26.34c) and the region of strongest field is increasingly confined within the coil. With a very long coil (Fig. 26.34d), the field is strong and uniform deep within the coil and very weak outside. The limit of an infinitely long, tightly wound coil would produce a uniform field within and no field outside.

**TABLE 26.1** Fields of Some Simple Charge and Current Distributions

| Field Dependence on Distance[a] | Charge Distribution | Electric Field | Current Distribution | Magnetic Field |
|---|---|---|---|---|
| $\dfrac{1}{r^3}$ | Electric dipole | | Magnetic dipole | |
| $\dfrac{1}{r^2}$ | Point charge or spherically symmetric | | Impossible for steady current | |
| $\dfrac{1}{r}$ | Charge distribution with line symmetry | | Current distribution with line symmetry | |
| Uniform field; no variation | Infinite flat sheet of charge | | Current sheet | |

[a]For field *outside* distribution

A long, tightly wound coil is a **solenoid**. In the long-solenoid limit, we can use Ampère's law to find the magnetic field inside the solenoid. Figure 26.35 shows a cross section through a solenoid, with a rectangular amperian loop of width $L$. Since the field is zero outside, the only contribution to $\oint \vec{B} \cdot d\vec{r}$ is from the interior segment parallel to the field, and with a uniform field that gives $BL$. If the solenoid carries current $I$ and consists of $n$ turns of wire per unit length, then Fig. 26.35 shows that our amperian loop encircles a total current $nLI$. So Ampère's law reads $BL = \mu_0 nLI$, or

$$B = \mu_0 nI \quad \text{(solenoid field)} \tag{26.20}$$

Since the vertical dimension of the amperian rectangle never entered the calculation, the field has this same magnitude everywhere inside the solenoid. Although Fig. 26.34 depicts circular coils, neither Fig. 26.35 nor our calculation requires that, so Equation 26.20 holds for a solenoid of any cross section.

Solenoids are used to produce uniform magnetic fields in a variety of applications; among the most obvious is the long cylindrical "tunnel" of an MRI scanner. Because the field becomes nonuniform at the ends of a solenoid, ferromagnetic materials will be attracted into the coil. Small solenoids are used in this way to produce limited straight-line motion of an iron plunger. One application is the solenoid on a car starter, which engages

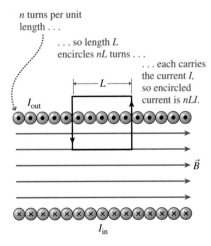

**FIGURE 26.35** Cross section of a long solenoid, showing a rectangular amperian loop straddling the region where solenoid coils emerge from the plane of the page.

the starter motor's gear with the gasoline engine. Solenoid-operated valves are widely used in controlling fluid flows; the valves that admit water to your washing machine and dishwasher are solenoid valves.

---

**EXAMPLE 26.10** **A Solenoid: The Current in an MRI Scanner**

The solenoid used in an MRI scanner is 2.4 m long and 95 cm in diameter. It's wound from superconducting wire 2.0 mm in diameter, with adjacent turns separated by an insulating layer of negligible thickness. Find the current that will produce a 1.5-T magnetic field inside the solenoid.

**INTERPRET** This is a problem about a solenoid, which involves relating current and field.

**DEVELOP** Equation 26.20, $B = \mu_0 nI$, provides the relation we need. But to use it we have to find $n$, the number of turns per unit length. A close-up sketch (Fig. 26.36) shows how we find $n$ from the wire diameter. Knowing $n$, we can use Equation 26.20 to find the current.

**EVALUATE** Figure 26.36 shows that $n = 500$ turns per meter. So now we can solve Equation 26.20 to get

$$I = \frac{B}{\mu_0 n} = \frac{1.5\ \text{T}}{(4\pi \times 10^{-7}\ \text{N/A}^2)(500\ \text{m}^{-1})} = 2.4\ \text{kA}$$

Wire diameter is 2 mm = $\frac{1}{500}$ m . . .

. . . therefore 500 wire diameters occupy 1 m, so $n = 500$ turns/meter.

**FIGURE 26.36** Finding $n$.

**ASSESS** That's a large current, but it's readily handled by the niobium–titanium superconductor used in the MRI scanner. Notice that more turns per unit length would reduce the current demand; that's because each turn carries the same current $I$, so cramming in more turns increases the encircled current and thus the field for a given total current. ∎

---

With its current flowing around an essentially cylindrical surface, the solenoid might remind you of the bar magnet in Fig. 26.26. There, atomic current loops produce a magnetization current flowing around the cylindrical magnet. Indeed, a solenoid and a bar magnet are very similar, and they produce similar magnetic fields (Fig. 26.37).

**FIGURE 26.37** Iron filings trace the magnetic field of a loosely wound solenoid. Compare with the field of a bar magnet shown in Fig. 26.1.

## Toroids

Wrap a solenoid around on itself and you've got a **toroid**—a donut-shaped coil (Fig. 26.38). Toroids are the favored magnetic geometry for devices that would extract the immense energy of nuclear fusion—making each gallon of seawater the energy equivalent of some 300 gallons of gasoline. More on fusion in Chapter 38.

Figure 26.39 shows that the field lines of a toroid must be circular, and symmetry therefore requires that the field magnitude $B$ is constant on an amperian loop of radius $r$ that coincides with a field line; for such a loop, $\oint \vec{B} \cdot d\vec{r}$ therefore becomes $2\pi r B$. Figure 26.39 also shows that our amperian loop surrounds all $N$ turns of the toroidal coil, for an encircled current of $NI$. Ampère's law equates $\oint \vec{B} \cdot d\vec{r}$ to $\mu_0$ times the encircled current, giving $2\pi r B = \mu_0 NI$. Therefore

$$B = \frac{\mu_0 NI}{2\pi r} \quad \text{(toroid field)} \tag{26.21}$$

This result holds within the toroidal coil itself. Draw an amperian loop inside the inner edge of the coils, and there's no encircled current and therefore no field. Beyond the outer edge, an amperian loop encircles equal but opposite currents, giving zero net current and again no field. So the toroid confines the field entirely to its interior. Unlike the solenoid field, however, the toroidal field isn't uniform; Equation 26.21 shows that it decreases as $1/r$.

**FIGURE 26.38** A toroidal coil.

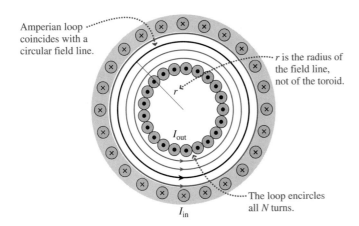

Amperian loop coincides with a circular field line.

$r$ is the radius of the field line, not of the toroid.

$I_{out}$

$I_{in}$

The loop encircles all $N$ turns.

**FIGURE 26.39** Applying Ampère's law to a toroid, shown here in cross section.

## Big Picture

The big new idea here is magnetism—an interaction that fundamentally involves *moving electric charge*. Moving charge produces magnetic fields, and moving charges respond to magnetic fields by experiencing a magnetic force.

## Key Concepts and Equations

The **magnetic force** on a charge $q$ moving with velocity $\vec{v}$ in a magnetic field $\vec{B}$ is

$$\vec{F} = q\vec{v} \times \vec{B} \quad \text{(magnetic force)}$$

The force acts at right angles to both $\vec{v}$ and $\vec{B}$, and therefore it does no work.

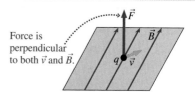

Force is perpendicular to both $\vec{v}$ and $\vec{B}$.

The **Biot–Savart law** gives the magnetic field $d\vec{B}$ arising from a small element of steady current:

$$d\vec{B} = \frac{\mu_0}{4\pi} \frac{I d\vec{L} \times \hat{r}}{r^2}$$

Here $\mu_0$ is the **permeability constant**, with value $4\pi \times 10^{-7}$ N/A$^2$.

$d\vec{L}$ is a small piece of the wire. $\hat{r}$ is a unit vector from $d\vec{L}$ toward $P$. $d\vec{B}$ is into page.

**Ampère's law** provides a more global description of how magnetic fields arise from currents, relating the line integral around any closed loop to the encircled current:

$$\oint \vec{B} \cdot d\vec{r} = \mu_0 I_{\text{encircled}}$$

Ampère's law in this form applies only to steady currents.

**Gauss's law for magnetism** expresses the fact that there are no magnetic monopoles—magnetic analogs of electric charge—and that magnetic field lines therefore do not begin or end:

$$\oint \vec{B} \cdot d\vec{A} = 0$$

Static electric fields, in contrast, always begin or end on electric charges.

Magnetic field      Electric field

## Cases and Uses

A charged particle moving perpendicular to a uniform magnetic field undergoes circular motion with the **cyclotron frequency** $f = qB/2\pi m$. More generally, charged particles in magnetic fields follow spiral paths, "trapped" on the field lines.

The magnetic force on a length $L$ of a straight wire carrying current $I$ in a uniform magnetic field is $\vec{F} = I\vec{L} \times \vec{B}$. Parallel wires a distance $d$ apart experience forces from each other's magnetic field: $F = \dfrac{\mu_0 I_1 I_2 L}{2\pi d}$.

The force is attractive for currents in the same direction, repulsive for currents in opposite directions.

$I_2$

$I_1$

A current loop gives rise to a magnetic field that, at distances large compared with the loop's size, is a dipole field. The loop's magnetic dipole moment has magnitude $\mu = IA$, with $A$ the loop area, and the loop responds to an external magnetic field by experiencing the torque typical of a dipole: $\vec{\tau} = \vec{\mu} \times \vec{B}$.

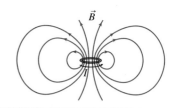

$\vec{B}$

Fields of simple current distributions:

Line current: $B = \dfrac{\mu_0 I}{2\pi r}$     Current sheet: $B = \frac{1}{2}\mu_0 J_s$     Solenoid: $B = \mu_0 n I$

$\vec{B}$

Magnetism in matter arises from the interactions of atomic-scale current loops. **Ferromagnetic** materials have strong interactions and exhibit the bulk magnetism associated with permanent magnets and with magnetic materials like iron. **Paramagnetism** and **diamagnetism** are weaker manifestations of magnetism in matter.

# For Thought and Discussion

1. A charged particle moves through a region containing only a magnetic field. Under what condition will the particle experience no force?

2. An electron moving with velocity $\vec{v}$ through a magnetic field $\vec{B}$ experiences a magnetic force $\vec{F}$. Which of the vectors $\vec{F}$, $\vec{v}$, and $\vec{B}$ must be at right angles?

3. A magnetic field points out of this page. Will a positively charged particle moving in the plane of the page circle clockwise or counterclockwise as viewed from above?

4. Do particles in a cyclotron gain energy from the electric field, the magnetic field, or both? Explain.

5. An electron and a proton moving at the same speed enter a region containing a uniform magnetic field. Which is deflected more from its original path?

6. Two identical particles carrying equal charge are moving in opposite directions along a magnetic field, when they collide elastically head-on. Describe their subsequent motion.

7. In what two senses does a current loop behave like a magnetic dipole?

8. The Biot–Savart law shows that the magnetic field of a current element decreases as $1/r^2$. Could you put together a complete circuit whose field exhibits this $1/r^2$ decrease? Why or why not?

9. Do currents in the same direction attract or repel? Explain.

10. If a current is passed through an unstretched spring, will the spring contract or expand? Explain.

11. The field of a long, straight wire consists of circular field lines. Does Ampère's law hold for a square loop surrounding the wire? For a circular loop not concentric with the wire? See Fig. 26.40.

**FIGURE 26.40** For Thought and Discussion 11

12. If the line integral around a closed loop is zero, does that mean the magnetic field on the loop must be zero?

13. Figure 26.41 shows some magnetic field lines associated with two parallel wires carrying equal currents perpendicular to the page. Are the currents in the same or opposite directions? How can you tell?

**FIGURE 26.41** For Thought and Discussion 13

14. Why is a piece of iron attracted into a solenoid?

15. Would there be a net magnetic force on a piece of iron deep inside a long solenoid? Explain.

16. An unmagnetized piece of iron has no net magnetic dipole moment, yet it's attracted to either pole of a bar magnet. Why?

# Exercises and Problems

## Exercises

### Section 26.2 Magnetic Force and Field

17. (a) What is the minimum magnetic field needed to exert a $5.4 \times 10^{-15}$ N force on an electron moving at $2.1 \times 10^7$ m/s? (b) What magnetic field strength would be required if the field were at 45° to the electron's velocity?

18. An electron moving at right angles to a 0.10-T magnetic field experiences an acceleration of $6.0 \times 10^{15}$ m/s². (a) What is the electron's speed? (b) By how much does its *speed* change in 1 ns $(10^{-9}$ s)?

19. What is the magnitude of the magnetic force on a proton moving at $2.5 \times 10^5$ m/s (a) at right angles; (b) at 30°; (c) parallel to a magnetic field of 0.50 T?

20. The magnitude of Earth's magnetic field is a little less than 1 G near Earth's surface. What is the maximum possible magnetic force on an electron with kinetic energy of 1 keV? Compare with the gravitational force on the same electron.

21. A velocity selector uses a 60-mT magnetic field perpendicular to a 24-kN/C electric field. At what speed will charged particles pass through the selector undeflected?

### Section 26.3 Charged Particles in Magnetic Fields

22. Find the radius of the circular path described by a proton moving at 15 km/s in a plane perpendicular to a 400-G magnetic field.

23. How long does it take an electron to complete a circular orbit at right angles to a 1.0-G magnetic field?

24. Radio astronomers detect electromagnetic radiation at a frequency of 42 MHz from an interstellar gas cloud. If this radiation is caused by electrons spiraling in a magnetic field, what is the field strength?

25. Electrons and protons with the same kinetic energy are moving at right angles to a uniform magnetic field. How do their orbital radii compare?

26. Microwaves in a microwave oven are produced by electrons circling in a magnetic field at a frequency of 2.4 GHz. (a) What is the magnetic field strength? (b) The electrons' motion takes place inside a special tube called a magnetron. If the magnetron can accommodate electron orbits with a maximum diameter of 2.5 mm, what is the maximum electron energy?

27. Two protons, moving in a plane perpendicular to a uniform magnetic field of 500 G, undergo an elastic head-on collision. How much time elapses before they collide again? *Hint:* Draw a picture.

### Section 26.4 The Magnetic Force on a Current

28. What is the magnitude of the force on a 50-cm-long wire carrying 15 A at right angles to a 500-G magnetic field?

29. A wire carrying 15 A makes a 25° angle with a uniform magnetic field. The magnetic force per unit length of wire is 0.31 N/m. (a) What is the magnetic field strength? (b) What is the maximum force per unit length that could be achieved by reorienting the wire in this field?

30. In a high-magnetic-field experiment, a conducting bar carrying 7.5 kA passes through a 30-cm-long region containing a 22-T magnetic field. If the bar makes a 60° angle with the field direction, what force is necessary to hold it in place?

31. A piece of wire with mass per unit length 75 g/m runs horizontally at right angles to a horizontal magnetic field. A 6.2-A current in the wire results in its being suspended against gravity. What is the magnetic field strength?

### Section 26.5 Origin of the Magnetic Field

32. A wire carries 15 A. You form the wire into a single-turn circular loop with magnetic field 80 $\mu$T at the loop center. What is the loop radius?

33. A single-turn wire loop is 2.0 cm in diameter and carries a 650-mA current. Find the magnetic field strength (a) at the loop center and (b) on the loop axis, 20 cm from the center.

34. A 2.2-m-long wire carrying 3.5 A is wound into a tight, loop-shaped coil 5.0 cm in diameter. What is the magnetic field at its center?

35. What is the current in a long wire if the magnetic field strength 1.2 cm from the wire's axis is 67 $\mu$T?

36. In standard household wiring, parallel wires about 1 cm apart carry currents of about 15 A. What is the magnitude of the force per unit length between such wires?

### Section 26.6 Magnetic Dipoles

37. Earth's magnetic dipole moment is $8.0 \times 10^{22}$ A·m². What is the magnetic field strength on Earth's surface at either pole?

38. A single-turn square wire loop 5.0 cm on a side carries a 450-mA current. (a) What is the magnetic dipole moment of the loop? (b) If the loop is in a uniform 1.4-T magnetic field with its dipole moment vector at 40° to the field direction, what is the magnitude of the torque it experiences?

39. An electric motor contains a 250-turn circular coil 6.2 cm in diameter. If it is to develop a maximum torque of 1.2 N·m at a current of 3.3 A, what should be the magnetic field strength?

### Section 26.8 Ampère's Law

40. The line integral of the magnetic field on a closed path surrounding a wire has the value 8.8 $\mu$T·m. What is the current in the wire?

41. The magnetic field shown in Fig. 26.42 has uniform magnitude 75 $\mu$T, but its direction reverses abruptly. How much current is encircled by the rectangular loop shown?

**FIGURE 26.42** Exercise 41

42. (a) What is the magnetic field strength 0.10 mm from the axis of a 1.0-mm-diameter wire carrying 5.0 A distributed uniformly over its cross section? (b) What is the field strength at the surface of the wire?

43. Show that Equations 26.17 and 26.18 give the same results when evaluated at the surface of the wire.

44. A superconducting solenoid has 3300 turns per meter and can carry a maximum current of 4.1 kA. Find the magnetic field strength in the solenoid.

## Problems

45. A particle carrying a 50-$\mu$C charge moves with velocity $\vec{v} = 5.0\hat{i} + 3.2\hat{k}$ m/s through a uniform magnetic field $\vec{B} = 9.4\hat{i} + 6.7\hat{j}$ T. (a) Find the magnetic force on the particle. (b) Form the dot products $\vec{F} \cdot \vec{v}$ and $\vec{F} \cdot \vec{B}$ to show explicitly that the force is perpendicular to both $\vec{v}$ and $\vec{B}$.

46. A proton moving with velocity $\vec{v}_1 = 3.6 \times 10^4 \hat{j}$ m/s experiences a magnetic force of $7.4 \times 10^{-16} \hat{i}$ N. A second proton moving on the $x$ axis experiences a magnetic force of $2.8 \times 10^{-16} \hat{j}$ N. Find the magnitude and direction of the magnetic field, and the velocity of the second proton.

47. A 1.4-$\mu$C charge moving at 185 m/s experiences a magnetic force $\vec{F}_B = 2.5\hat{i} + 7.0\hat{j}$ $\mu$N in a magnetic field $\vec{B} = 42\hat{i} - 15\hat{j}$ mT. Find the angle between the particle's velocity and the magnetic field.

48. A beam of electrons moving in the $x$ direction at $8.7 \times 10^6$ m/s enters a region where a uniform magnetic field of 180 G points in the $y$ direction. How far into the field region does the beam penetrate?

49. Show that the orbital radius of a charged particle moving at right angles to a magnetic field $B$ can be written as $r = \sqrt{2Km}/qB$, where $K$ is the kinetic energy in joules, $m$ the particle mass, and $q$ its charge.

50. A cyclotron is designed to accelerate deuterium nuclei. (Deuterium has one proton and one neutron in its nucleus.) (a) If the cyclotron uses a 2.0-T magnetic field, at what frequency should the dee voltage be alternated? (b) If the vacuum chamber has a diameter of 0.90 m, what is the maximum kinetic energy of the deuterons? (c) If the magnitude of the potential difference between the dees is 1500 V, how many orbits do the deuterons complete before achieving the energy of part (b)?

51. An electron is moving in a uniform magnetic field of 0.25 T; its velocity components parallel and perpendicular to the field are both equal to $3.1 \times 10^6$ m/s. (a) What is the radius of the electron's spiral path? (b) How far does it move along the field direction in the time it takes to complete a full orbit about the field direction?

52. A wire of negligible resistance is bent into a rectangle as in Fig. 26.43, and a battery and resistor are connected as shown. The right-hand side of the circuit extends into a region containing a uniform magnetic field of 38 mT pointing into the page. Find the magnitude and direction of the net force on the circuit.

**FIGURE 26.43** Problem 52

53. A 20-cm-long conducting rod with mass 18 g is suspended by wires of negligible mass, as in Fig. 26.44. The rod is in a region containing a uniform magnetic field of 0.15 T pointing horizontally into the page, as shown. An external circuit supplies current between the support points $A$ and $B$. (a) What is the minimum current necessary to move the bar to the upper position, so it's supported against gravity? (b) In which direction should the current flow?

**FIGURE 26.44** Problem 53

54. A rectangular copper strip measures 1.0 mm in the direction of a uniform 2.4-T magnetic field. When the strip carries a 6.8-A current at right angles to the field, the Hall potential across the strip is 1.2 $\mu$V. Find the number density of free electrons in the copper.

55. A single-turn wire loop 10 cm in diameter carries a 12-A current. It experiences a torque of 0.015 N·m when the normal to the loop plane makes a 25° angle with a uniform magnetic field. What is the magnetic field strength?

56. A simple electric motor consists of a 100-turn coil 3.0 cm in diameter, mounted between the poles of a magnet that produces a 0.12-T field. When a 5.0-A current flows in the coil, what are (a) its magnetic dipole moment and (b) the maximum torque developed by the motor?

57. Nuclear magnetic resonance (NMR) is a technique for analyzing chemical structures and also the basis of magnetic resonance imaging used for medical diagnosis. The NMR technique relies on sensitive measurements of the energy needed to flip atomic nuclei upside down in a given magnetic field. In an NMR apparatus with a 7.0-T magnetic field, how much energy is needed to flip a proton ($\mu = 1.41 \times 10^{-26}$ A·m$^2$) from parallel to antiparallel to the field?

58. A wire carrying 1.5 A passes through a region containing a 48-mT magnetic field. The wire is perpendicular to the field and makes a quarter-circle turn of radius 21 cm as it passes through the field region, as shown in Fig. 26.45. Find the magnitude and direction of the magnetic force on this section of wire.

**FIGURE 26.45** Problem 58

59. You have a spool of thin wire that can handle a maximum current of 0.50 A. If you wind the wire into a looplike coil 20 cm in diameter, how many turns should the coil have if the magnetic field at its center is to be 2.3 mT at this maximum current?

60. A single piece of wire is bent so that it includes a circular loop of radius $a$, as in Fig. 26.46. A current $I$ flows in the direction shown. Find an expression for the magnetic field at the center of the loop.

**FIGURE 26.46** Problem 60

61. A power line carries a 500-A current toward magnetic north and is suspended 10 m above the ground. The horizontal component of Earth's magnetic field at the power line's latitude is 0.24 G. If a magnetic compass is placed on the ground directly below the power line, in what direction will it point?

62. Part of a long wire is bent into a semicircle of radius $a$, as in Fig. 26.47. A current $I$ flows in the direction shown. Use the Biot–Savart law to find the magnetic field at the center of the semicircle (point $P$).

**FIGURE 26.47** Problem 62

63. Three parallel wires 4.6 m long each carry 20 A in the same direction. They're spaced at the vertices of an equilateral triangle 3.5 cm on a side. Find the magnitude of the force on each wire.

64. A long, straight wire carries 20 A. A 5.0-cm by 10-cm rectangular wire loop carrying 500 mA is located 2.0 cm from the wire, as shown in Fig. 26.48. Find the net magnetic force on the loop.

**FIGURE 26.48** Problem 64

65. A long conducting rod of radius $R$ carries a nonuniform current density given by $J = J_0 r/R$, where $J_0$ is a constant and $r$ is the radial distance from the rod's axis. Find expressions for the magnetic field strength (a) inside and (b) outside the rod.

66. A long, hollow conducting pipe of radius $R$ carries a uniform current $I$ along the pipe, as shown in Fig. 26.49. Use Ampère's law to find the magnetic field strength (a) inside and (b) outside the pipe.

**FIGURE 26.49** Problem 66

67. Two large, flat conducting plates lie parallel to the $x$-$y$ plane. They carry equal currents, one in the $+x$ and the other in the $-x$ direction. In each plate the current per meter of width in the $y$ direction is $J_s$. Find the magnetic field strength (a) between and (b) outside the plates.

68. The coaxial cable shown in Fig. 26.50 consists of a solid inner conductor of radius $a$ and a hollow outer conductor of inner radius $b$ and thickness $c$. The two carry equal but opposite currents $I$, uniformly distributed. Find expressions for the magnetic field strength as a function of radial position $r$ (a) within the inner conductor, (b) between the inner and outer conductors, and (c) beyond the outer conductor.

**FIGURE 26.50** Problems 68 and 73

69. A solenoid used in a plasma physics experiment is 10 cm in diameter and 1.0 m long and carries a 35-A current to produce a 100-mT magnetic field. (a) How many turns are in the solenoid? (b) If the solenoid resistance is 2.7 Ω, how much power does it dissipate?

70. You have 10 m of 0.50-mm-diameter copper wire and a battery capable of passing 15 A through the wire. What magnetic field strengths could you obtain (a) inside a 2.0-cm-diameter solenoid wound with the wire as closely spaced as possible and (b) at the center of a single circular loop made from the wire?

71. Derive Equation 26.20 for the field of a solenoid by considering the solenoid to be made of a large number of adjacent current loops. Use Equation 26.9 for the field of a current loop, and integrate over all loops.

72. The largest lightning strikes have peak currents of around 250 kA, flowing in essentially cylindrical channels of ionized air. How far from such a flash would the resulting magnetic field be equal to Earth's magnetic field strength, about 50 $\mu$T?

73. A coaxial cable like that shown in Fig. 26.50 consists of a 1.0-mm-diameter inner conductor and an outer conductor of inner diameter 1.0 cm and 0.20 mm thickness. A 100-mA current flows down the center conductor and back along the outer conductor. Find the magnetic field strength (a) 0.10 mm, (b) 5.0 mm, and (c) 2.0 cm from the cable axis.

74. A circular wire loop of radius 15 cm and negligible thickness carries a 2.0-A current. Use suitable approximations to find the magnetic field of this loop (a) in the loop plane, 1.0 mm outside the loop, and (b) on the loop axis, 3.0 m from the loop center.

75. A long, flat conducting bar of width $w$ carries a total current $I$ distributed uniformly, as shown in Fig. 26.51. Use suitable approximations to write expressions for the magnetic field strength (a) near the conductor surface $(r \ll w)$ and (b) very far from the conductor $(r \gg w)$.

**FIGURE 26.51** Problem 75

76. A long, hollow conducting pipe of radius $R$ and length $L$ carries a uniform current $I$ flowing around the pipe, as shown in Fig. 26.52. Find expressions for the magnetic field (a) inside and (b) outside the pipe. *Hint:* What configuration does this pipe resemble?

**FIGURE 26.52** Problem 76

77. A solid conducting wire of radius $R$ runs parallel to the $z$ axis and carries a current density given by $\vec{J} = J_0(1 - r/R)\hat{k}$, where $J_0$ is a constant and $r$ is the radial distance from the wire axis. Find expressions for (a) the total current in the wire, (b) the magnetic field strength for $r > R$, and (c) the magnetic field strength for $r < R$.

78. A disk of radius $a$ carries a uniform surface charge density $\sigma$ and is rotating with angular speed $\omega$ about the central axis perpendicular to the disk. Show that the magnetic field at the disk's center is $\frac{1}{2}\mu_0 \sigma \omega a$.

79. A wire of length $L$ carries a current $I$. (a) Find an expression for the magnetic dipole moment that results when the wire is wound into an $N$-turn circular coil. (b) For what integer value of $N$ is this moment a maximum?

80. In 2.0 $\mu$s, an electron moves 15 cm in the direction of a 0.10-T magnetic field. If the electron's velocity components perpendicular and parallel to the field are equal, (a) what is the length of its actual spiral trajectory, and (b) how many orbits about the field direction does it complete?

81. The structure shown in Fig. 26.53 is made from conducting rods. The upper horizontal rod is free to slide vertically on the uprights, while maintaining electrical contact with them. The upper rod has mass 22 g and length 95 cm. A battery connected across the insulating gap at the bottom of the left-hand upright drives a 66-A current through the structure. At what height $h$ will the upper wire be in equilibrium?

**FIGURE 26.53** Problem 81

82. A long, flat conducting ribbon of width $w$ is parallel to a long, straight wire, with its near edge a distance $a$ from the wire (Fig. 26.54). Wire and ribbon carry the same current $I$; this current is distributed uniformly over the ribbon. Use integration to show that the force per unit length between the two has magnitude $\frac{\mu_0 I^2}{2\pi w} \ln\left(\frac{a+w}{a}\right)$.

**FIGURE 26.54** Problem 82

83. Two coaxial circular loops (oriented like the one in Fig. 26.16a) of radius $a$, carrying parallel currents $I$, have centers a distance $b$ apart on the $x$ axis, at $x = \pm b/2$. Because the magnetic field on the $x$ axis is an even function of $x$ (about the midpoint $x = 0$), its first derivative vanishes at $x = 0$; that is, $(dB/dx)_0 = 0$. What value of $b$ also makes the second derivative vanish at $x = 0$; that is, $(d^2B/dx^2)_0 = 0$? This arrangement, called Helmholtz coils, produces a region of nearly uniform magnetic field.

84. Four long, straight wires parallel to the $z$ axis pass through the points $(\pm 7.5 \text{ cm}, \pm 7.5 \text{ cm})$ at the corners of a square in the $x$-$y$ plane. Each carries a current of 2.5 A, with the upper two currents in the $-z$ direction and the lower two in the $+z$ direction. Find the magnetic field at the center of the square.

85. For the wires in Problem 84, find the net force per unit length on the wire at the lower left corner of the square.

86. A straight segment of wire of length $L$, extending along the $x$ axis from $x_0$ to $x_0 + L$, carries a steady current $I$ in the $+x$ direction (fed by other wires not under consideration). Adapt the calculation of Example 26.5 to find the magnitude and direction of the magnetic field at the point $(0, y)$ in the $x$-$y$ plane due to the segment alone.

87. Use the result of Problem 86 to find the magnetic field strength at the center of a square loop of side $a$ carrying current $I$.

88. Repeat the calculation in Problem 71 for a solenoid of finite length $L$ and cross-sectional radius $a$ to find the magnetic field strength at the center of the solenoid's axis.

89. A magnetic dipole $\vec{\mu} = \mu \hat{i}$ is located on the axis of a circular current loop of radius $a$, oriented as shown in Fig. 26.16$a$, a distance $x$ from the center. Differentiate Equation 26.15 to find the force on the dipole, and evaluate its magnitude for $x = a$. Is the direction attractive or repulsive?

90. You're taking a class in space weather physics. Space weather deals with the dynamics of the far upper atmosphere and the magnetic regions surrounding the Earth. You're preparing a term paper on the van Allen belts, regions where high-energy particles are trapped in the Earth's magnetic field. Your textbook says the magnetic field strength at the belts is 0.1 G. To impress your professor, you calculate the radii of the spiral paths of 0.1-MeV, 1-MeV, and 10-MeV protons in the van Allen belts. What values do you get?

91. Your space weather project (see Problem 90) is going well, but you must add a section on the generation of the Earth's magnetic field. One highly simplified model suggests the Earth's field arises from a single loop of current at the outer edge of the planet's liquid core (radius 3000 km) concentric with the center. What current will give the measured field of 62 $\mu$T at the north magnetic pole?

92. You have a summer job working for a major manufacturer of audio components. The lead designer of loudspeakers asks you to make a calculation for a prototype speaker. The speaker coil consists of 100 turns of wire, 3.5 cm in diameter. The current in the speaker will be 2.1 A when the force on the speaker coil is 14.8 N. What strength magnetic field will give this force?

93. You have a job with an international consulting corporation that is assisting the European Space Agency with the design and launch of an interplanetary space probe. You must calculate (integrate!) the equation for the magnetic field of an infinite current sheet. You consult your physics book and decide to consider the sheet to be made up of infinitesimal line currents, as shown in Figure 26.55. Show that your result is consistent with that of Example 26.9.

**FIGURE 26.55** A current sheet modeled as a row of infinitesimal line currents (Problem 93).

# Answers to Chapter Questions

## Answer to Chapter Opening Question

Magnetic force shapes the structure of the solar atmosphere. Magnetism is fundamentally an interaction involving moving electric charge, and the hot, ionized gas of the solar atmosphere contains free charge that responds to magnetism. Earth's cooler atmosphere consists of neutral molecules that don't experience a magnetic force.

## Answers to GOT IT? Questions

26.1 Greatest for ($a$), 0 for ($c$); direction for ($a$) and ($b$) is into the page.

26.2 ($b$); it's a *negative* charge.

26.3 ($a$).

26.4 ($a$), because adjacent currents are in the same direction. Changing the current direction doesn't matter because the currents are still parallel.

26.5 ($b$), because the field lines form closed loops

26.6 ($a$) 0; ($b$) current A.

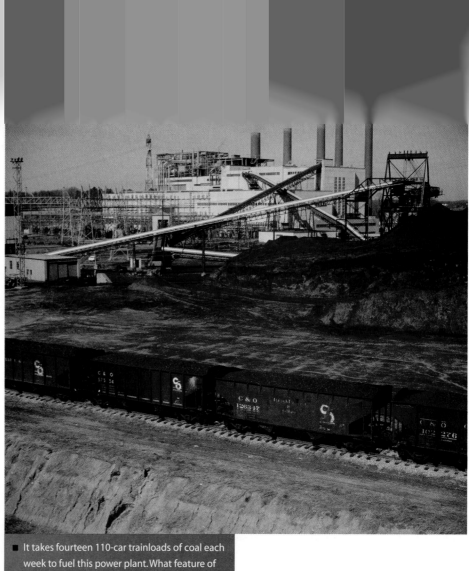

■ It takes fourteen 110-car trainloads of coal each week to fuel this power plant. What feature of the equation $\mathcal{E} = -d\Phi_B/dt$ demands this prodigious fuel consumption?

The electric and magnetic fields we've encountered so far originated in electric charge, either stationary or moving. We recognized a link between electricity and magnetism that lies in their common involvement with electric charge. In the remainder of our study, we'll explore a more intimate relation between electricity and magnetism, in which the fields themselves interact directly. This relation is the basis for new electromagnetic technologies, reveals the nature of light, and points toward the theory of relativity.

## 27.1 Induced Currents

In 1831, the English scientist Michael Faraday and the American Joseph Henry independently found that electric currents arose in circuits subjected to changing magnetic fields. Here are four experiments that illustrate this phenomenon:

1.  (Fig. 27.1) Move a bar magnet in the presence of a circuit consisting of a wire coil and an ammeter. There's no battery or other obvious source of emf. As long as you hold the magnet stationary, there's no current. But move the magnet, and the ammeter registers a current—which we call an **induced current**. Move the magnet faster, and the induced current increases. Reverse the direction of motion, and the induced current reverses.

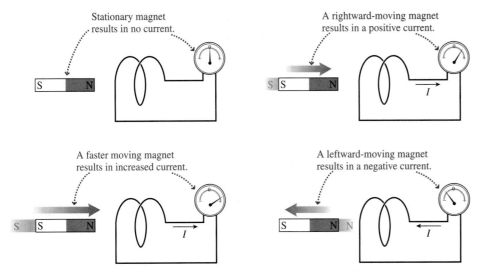

FIGURE 27.1 When a magnet moves near a closed circuit, current flows in the circuit.

2. (Fig. 27.2) Move a coil near a stationary magnet, and a similar induced current results. So the effect is the same whether it's the magnet that moves, or the coil. All that matters is the relative motion between magnet and coil.

3. (Fig. 27.3) Replace the bar magnet with a second coil, this one carrying a steady current from a battery. The new coil creates a magnetic field like that of a bar magnet and, not surprisingly given the results of experiments 1 and 2, an induced current arises in the original coil when the two coils move relative to one another.

FIGURE 27.2 Moving the coil instead of the magnet gives the same result.

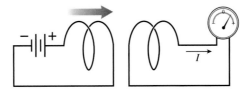

FIGURE 27.3 An induced current also results when a current-carrying circuit replaces the magnet.

4. (Fig. 27.4) Hold both coils stationary, and there's no induced current. But now open the switch connecting the battery to the left-hand coil. The current drops quickly to zero, and during that brief interval the ammeter registers a current in the right-hand coil. Then the induced current ceases as the current in the left-hand coil remains at zero. Now close the switch again; as current briefly rises in the left-hand coil, the ammeter registers an induced current in the right-hand coil—and its direction is opposite what it was when you opened the switch. Once the current in the left coil reaches a steady value, the induced current in the right coil again ceases.

FIGURE 27.4 A current is also induced when the current in an adjacent circuit changes.

The common feature in these experiments is a *changing magnetic field*. It doesn't matter whether the field changes because a magnet moves, or a circuit moves, or because the current giving rise to the field changes. In each case, an induced current appears in a circuit subjected to a changing magnetic field. Here's a new phenomenon—**electromagnetic induction**—whereby electrical effects arise from *changing* magnetic fields.

## 27.2 Faraday's Law

It takes a source of emf, such as a battery, to drive current in a circuit. With induced currents there's no battery, but there still must be an emf. This **induced emf** isn't usually localized, as with a battery, but may be spread through the conductors making up the circuit.

### Magnetic Flux

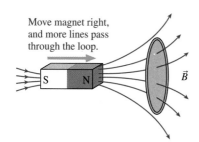

**FIGURE 27.5** A circular wire loop in the magnetic field of a bar magnet. As the magnet moves closer, the flux through the loop increases.

We found in Chapter 26 that the magnetic flux through any *closed* surface is zero. Here we're interested in the flux through *open* surfaces, which need not be zero (Fig. 27.5). Like the electric flux defined in Chapter 21, magnetic flux is the integral of the magnetic field over a surface:

$$\Phi_B = \int \vec{B} \cdot d\vec{A} \qquad \text{(magnetic flux)} \qquad (27.1a)$$

With electromagnetic induction, we're interested in the flux through a surface bounded by a circuit. For a loop like the one in Fig. 27.5, that surface can be the circular disk whose circumference is the loop. More generally, it can be *any* surface bounded by the loop. For a flat surface in a uniform magnetic field, Equation 27.1a reduces to

$$\Phi_B = \vec{B} \cdot \vec{A} = BA \cos\theta \qquad \text{(magnetic flux, uniform field and flat area)} \qquad (27.1b)$$

where $\theta$ is the angle between the field and the normal to the area. When the field and area are perpendicular, as in the next example, Equation 27.1b reduces further to $\Phi_B = BA$.

---

**EXAMPLE 27.1**  **Magnetic Flux: A Solenoid**

A solenoid of circular cross section has radius $R$, consists of $n$ turns per unit length, and carries current $I$. Find the magnetic flux through each turn of the solenoid.

**INTERPRET** The solenoid creates a uniform magnetic field, and we're asked for the flux of this field through an area bounded by one turn of the solenoid.

**DEVELOP** The solenoid field is perpendicular to the turns of wire that make up the solenoid, as we've drawn in Fig. 27.6. So we have a uniform field at right angles to a flat area, and the flux becomes $\Phi_B = BA$. Equation 26.20 gives the solenoid field, $B = \mu_0 nI$, and the area is that of a circle, $\pi R^2$.

**EVALUATE** $\Phi_B = BA = \mu_0 nI\pi R^2$

**ASSESS** The flux increases with any factor that increases either the field or the area, so this result makes sense. We're being a little loose in thinking of a single turn of the solenoid as a closed loop, but if the solenoid is tightly wound, then this is an excellent approximation. What we've found is the flux through one turn; if the solenoid has $N$ turns, then the flux through the entire solenoid is $N$ times our result. ∎

**FIGURE 27.6** Sketch for Example 27.1.

---

**EXAMPLE 27.2**  **Magnetic Flux: A Nonuniform Field**

A long, straight wire carries current $I$. A rectangular wire loop of dimensions $l$ by $w$ lies in a plane containing the wire, with its closest edge a distance $a$ from the wire and its dimension $l$ parallel to the wire. Find the magnetic flux through the loop.

**INTERPRET** The long, straight wire gives rise to a magnetic field, and we're asked for the flux of this field through an adjacent rectangular area.

**DEVELOP** Figure 27.7 shows the situation. Field lines encircle the long wire, and at the rectangular loop they're pointing into the page, perpendicular to the loop area. Thus $\vec{B} \cdot d\vec{A}$ in Equation 27.1a becomes

**FIGURE 27.7** A rectangular loop in the magnetic field of a long wire.

*cont'd.*

just $B\,dA$. Equation 26.17 gives the field strength: $B = \mu_0 I/2\pi r$. Since this varies with distance from the wire, we have to integrate. We divide the rectangle into thin strips of width $dr$ and area $dA = l\,dr$. Knowing $B$ and $dA$, we can integrate over all such strips.

**EVALUATE** We have

$$\Phi_B = \int B\,dA = \int_a^{a+w} \frac{\mu_0 I}{2\pi r} l\,dr = \frac{\mu_0 Il}{2\pi} \int_a^{a+w} \frac{dr}{r}$$

The integral is the natural logarithm, so

$$\Phi_B = \frac{\mu_0 Il}{2\pi} \ln r \Big|_a^{a+w} = \frac{\mu_0 Il}{2\pi} \ln\left(\frac{a+w}{a}\right)$$

**ASSESS** This result is directly proportional to the current, which determines the field strength, and to the loop length $l$. But it isn't proportional to the width $w$ because the field strength falls off, and increasing $w$ would expand the loop into regions of weaker field, contributing less to the overall flux. ∎

## Flux and Induced emf

Having developed the notion of magnetic flux, we can now state **Faraday's law of induction**, another of the four basic laws of electromagnetism:

> The induced emf in a circuit is proportional to the rate of change of magnetic flux through any surface bounded by that circuit.

This statement is a special case of Faraday's law that describes electromagnetic induction specifically in circuits; later we'll present a more general form that applies even when no circuit is present. The induced emf tends to oppose the change in flux—a crucial point to which we'll devote all of Section 27.3—and so in SI the proportionality between emf and rate of change of flux is $-1$. Thus Faraday's law is

$$\mathcal{E} = -\frac{d\Phi_B}{dt} \quad \text{(Faraday's law)} \tag{27.2}$$

where $\mathcal{E}$ is the induced emf in a circuit and $\Phi_B$ is the magnetic flux through any surface bounded by that circuit.

Faraday's law relates the induced emf to the *change* in flux. It isn't magnetic field or flux that causes an induced emf—it's the *change* in flux. The flux in a uniform field is given by Equation 27.1b, $\Phi_B = \vec{B}\cdot\vec{A} = BA\cos\theta$, which shows that we can change flux by changing the field strength $B$, the area $A$, or the angle $\theta$ describing the orientation between area and field.

---

**PROBLEM SOLVING STRATEGY 27.1**  **Faraday's Law and Induced emf**

**INTERPRET** Make sure the problem involves a circuit in which current flows because of a changing magnetic flux. Identify the circuit and the cause of the changing flux. Possibilities include

- A changing magnetic field, caused either by relative motion between the circuit and a magnet or by a changing current in an adjacent circuit. Alternatively, the problem may simply state that a magnetic field is changing at some specified rate, without giving the cause.
- A changing area, caused by the circuit expanding or contracting in the presence of a magnetic field.
- A changing orientation of the circuit relative to the field, causing a change in $\cos\theta$.

**DEVELOP** Find an appropriate expression for the magnetic flux through your circuit. If the field varies with position, you'll have to set up the integral in Equation 27.1a: $\Phi_B = \int \vec{B}\cdot d\vec{A}$; if not, you can use the simpler expression of Equation 27.1b: $\Phi_B = \vec{B}\cdot\vec{A} = BA\cos\theta$. Since the flux is changing, your expression for flux should either have an explicit time dependence or contain a quantity whose rate of change you're given.

**EVALUATE** Differentiate the flux with respect to time. Faraday's law, $\mathcal{E} = -d\Phi_B/dt$, then gives the induced emf. If you're asked about the circuit current, you can find it using Ohm's law: $I = \mathcal{E}/R$, with $R$ the circuit resistance.

**ASSESS** Does your answer make sense? Does the induced emf or current increase with an increased rate of whatever quantity is changing? Do the induced effects vanish if you set the rate of change to zero?

EXAMPLE 27.3 **Induced Current: A Changing Magnetic Field**

A wire loop of radius 10 cm has resistance 2.0 Ω. The plane of the loop is perpendicular to a uniform magnetic field $\vec{B}$ that's increasing at 0.10 T/s. Find the magnitude of the induced current in the loop.

**INTERPRET** We apply our problem-solving strategy, noting that this is a problem about induction in a circular loop, with the flux change caused by a changing magnetic field.

**DEVELOP** Figure 27.8 shows the loop with a field pointing into the page. With the field uniform and perpendicular to the loop area, we have $\Phi_B = BA = B\pi r^2$. Although there's no time dependence here, we're given the rate of change $dB/dt$, so we can evaluate the derivative $d\Phi_B/dt$.

**EVALUATE** The rate of change of flux is

$$\frac{d\Phi_B}{dt} = \frac{d}{dt}(B\pi r^2)$$

Since the radius isn't changing, this becomes

$$\frac{d\Phi_B}{dt} = \pi r^2 \frac{dB}{dt}$$

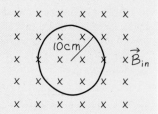

**FIGURE 27.8** A circular conducting loop in a plane perpendicular to a uniform magnetic field.

We're given $dB/dt = 0.10$ T/s and $r = 10$ cm. So with $\mathcal{E} = -d\Phi_B/dt$, the magnitude of the induced emf evaluates to 3.14 mV. Then Ohm's law gives the current: $I = \mathcal{E}/R = 3.14$ mV/2.0 Ω = 1.6 mA.

**ASSESS** Make sense? The induced emf and hence the current scale directly with the value $dB/dt$, confirming that the changing magnetic field is indeed the cause of the induced effects. Does it bother you that we took $\vec{B}$ as uniform even though it's changing, thus avoiding the integral of Equation 27.1a? The field is indeed changing, but that change is in *time*, not *space*, and the integral for the flux is over space. So at each instant the field is uniform, and we can dispense with the integral. ∎

EXAMPLE 27.4 **Induced Current: A Changing Area**

Two parallel conducting rails a distance $l$ apart are connected at one end by a resistance $R$. A conducting bar completes the circuit, joining the two rails electrically but free to slide along them. The whole circuit is perpendicular to a uniform magnetic field $\vec{B}$, as shown in Fig. 27.9. Find the current when the bar is pulled to the right with constant speed $v$.

**FIGURE 27.9** Pulling the bar to the right increases the circuit area, increasing the magnetic flux and inducing an emf that drives a current.

**INTERPRET** Here the circuit is formed by the rails, the resistance, and the conducting bar. The circuit area increases as the bar slides along the rails, so we've got a case of induction caused by a changing magnetic flux resulting from a changing area.

**DEVELOP** In this case of a uniform field perpendicular to the circuit, the flux is the product $\Phi_B = BA$. We can express this flux in terms of the changing position $x$ of the sliding bar; since we're given the bar's speed, we'll be able to evaluate the rate of change of flux. If we take $x = 0$ at the left end of the rails, then the circuit area is $A = lx$, so the flux is $\Phi_B = BA = Blx$.

**EVALUATE** Differentiating the flux with respect to time gives

$$\frac{d\Phi_B}{dt} = Bl\frac{dx}{dt} = Blv$$

since $dx/dt$ is the bar's velocity $v$. Faraday's law says that $Blv$ is the magnitude of the induced emf $\mathcal{E}$, so the current in the circuit becomes

$$I = \frac{\mathcal{E}}{R} = \frac{Blv}{R}.$$

**ASSESS** Make sense? Sure: The faster the bar moves, the greater the rate of change of flux, and so the greater the induced emf and current. ∎

✓**TIP** It's the Change That Matters

You may wonder why, in problems like the preceding two examples, you're not given values for the magnetic field itself or for the location of the sliding bar—quantities that determine the magnetic flux. But the flux itself doesn't matter, only its *rate of change*. And in both cases the rate followed from the given information: in one case the rate of change of the field and in the other the speed of the bar.

Examples 27.3 and 27.4 take care of two ways to change magnetic flux. The third—changing orientation—is at the heart of an important electromagnetic technology, and we'll do an example in the next section.

## 27.3 Induction and Energy

Move a bar magnet toward a wire loop, as in Fig. 27.10. An induced current flows, dissipating energy as it heats the loop. Where did that energy come from? It came from work you did in moving the magnet.

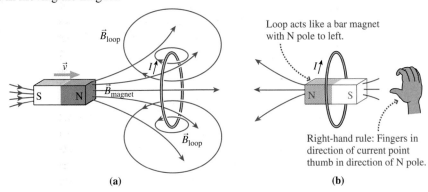

(a)                                                              (b)

**FIGURE 27.10** Conservation of energy determines the direction of the induced current. (a) Fields of bar magnet and loop. (b) The loop acts like a magnet with north to left, making it hard to move the bar magnet.

Normally it doesn't take work to move with constant speed. But the induced current makes the loop a magnetic dipole whose field, as Fig. 27.10 shows, *opposes* the field of the approaching magnet. You have to do positive work to overcome the resulting repulsive force. It had better be this way! Otherwise, you'd get something for nothing, heating the loop without any source of energy.

You can always find the direction of induced emfs and currents by asking: What direction of induced current will make it hard to move the magnet? The answer for Fig. 27.10 is a current that makes the loop a magnet with its north pole on the left, to repel the approaching bar magnet. By the right-hand rule, that gives the current direction shown: into the page at the top of the loop and out at the bottom. If, on the other hand, you move the magnet away from the loop, then the current flows in the opposite direction, putting the loop's south pole on the left and attracting the magnet, making it hard to pull the magnet away (Fig. 27.11).

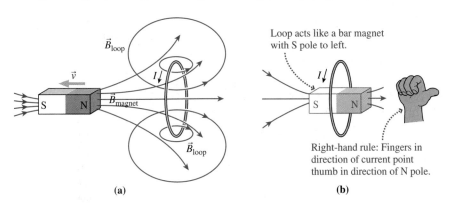

(a)                                                              (b)

**FIGURE 27.11** Now the direction of the induced current puts the loop's south pole to the left, making it hard to pull the magnet away.

This discussion is ultimately about energy conservation in the context of electromagnetic induction. **Lenz's law** summarizes what we've found:

> The direction of an induced emf or current is such that the magnetic field created by the induced current opposes the change in magnetic flux that created the current.

Mathematically, Lenz's law is contained in the minus sign that appears in Faraday's law, but it's usually easier to use Faraday's law to find the magnitude of the induced emf and then reason out the direction using energy conservation.

**GOT IT? 27.1**  You push a bar magnet toward a loop, with the north pole toward the loop as in Fig. 27.10. If you keep pushing the magnet straight through the loop, what will be the direction of the current as you pull it out the other side? Will you need to do work, or will work be done on you?

## Motional emf and Lenz's Law

When a conductor moves through a magnetic field, we can understand the origin of the induced emf in terms of the magnetic force on charge carriers. In this case of **motional emf**, we'll show explicitly that Lenz's law requires energy conservation.

Consider a square conducting loop of side $l$ and resistance $R$ pulled with constant speed $v$ out of a uniform magnetic field $\vec{B}$ (Fig. 27.12). The magnetic flux through the loop is changing, so there's an induced emf that drives a current. Energy is dissipated as heat and so, as we've just argued, the agent pulling the loop must do work. We'll now demonstrate quantitatively that energy is conserved by showing that the rate of heating in the loop is exactly equal to the rate at which the agent pulling the loop does work.

Pulling the loop to the right moves its free electrons through the magnetic field; the magnetic force $q\vec{v} \times \vec{B}$ on these electrons is downward in Fig. 27.12 (opposite $\vec{v} \times \vec{B}$ since electrons are negative). The resulting motion of the negative electrons in the left-hand side of the loop constitutes an upward current, which continues clockwise around the loop.

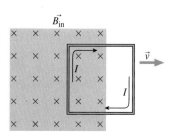

**FIGURE 27.12** A conducting loop being pulled out of a magnetic field.

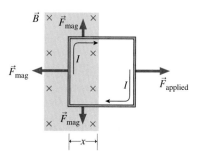

**FIGURE 27.13** Forces on the loop.

We found in Chapter 26 that the magnetic force on a current-carrying conductor of length $l$ is $\vec{F} = I\vec{l} \times \vec{B}$. Applying this expression to our conducting loop shows that there's no magnetic force on the right-hand side, where $\vec{B} = \vec{0}$, and that the forces on the top and bottom cancel (Fig. 27.13). So the magnetic force on the loop is that on the left side alone. The magnitude of this force is $IlB$, and the right-hand rule shows that it points to the left. This leftward magnetic force cancels the rightward applied force, giving the zero net force that Newton's law requires for the loop to move with constant velocity.

We could equally well determine the current direction from magnetic-flux considerations. As the loop leaves the field region, the flux decreases. The direction of the induced current is such as to oppose this decrease. Therefore the magnetic field of the induced current points *into* the page, as the induced current tries to maintain the flux. By the right-hand rule, a field within the loop and into the page requires that the induced current flow clockwise.

To calculate the current, we first find the induced emf. With the field perpendicular to the loop, and uniform in the region where it's nonzero, the magnetic flux is the product of the magnetic-field strength and the loop area that lies within the field: $\Phi_B = Blx$. Here $x$ is the distance between the left edge of the loop and the right edge of the magnetic-field region. The magnetic field remains constant, but as the loop moves, the distance $x$ decreases at the rate $dx/dt = -v$, where the minus indicates a decrease. Then the rate of change of flux is

$$\frac{d\Phi_B}{dt} = \frac{d(Blx)}{dt} = Bl\frac{dx}{dt} = -Blv$$

so Faraday's law gives

$$\mathcal{E} = -\frac{d\Phi_B}{dt} = Blv$$

This induced emf drives a current $I$ around the loop, where $I = \mathcal{E}/R = Blv/R$. The rate of energy dissipation in the loop is the product of the emf and the current (Equation 24.7):

$$P = I\mathcal{E} = \frac{Blv}{R}Blv = \frac{B^2l^2v^2}{R} \quad \text{(electrical power dissipated in loop)}$$

We've found that the magnetic force on the loop has magnitude $F = IlB$; since the loop is moving with constant velocity, this is also the magnitude of the applied force. Equation 6.19 gives $P = \vec{F} \cdot \vec{v}$ for the power supplied. Here, with $\vec{F}$ and $\vec{v}$ in the same direction, we have

$$P = Fv = IlBv = \frac{Blv}{R}lBv = \frac{B^2l^2v^2}{R} \quad \text{(mechanical power supplied to pull loop)}$$

the same as our expression for the power dissipated in the loop. Thus, all the work done by the agent pulling the loop ends up heating the resistor, showing explicitly that energy is indeed conserved.

---

**GOT IT? 27.2**    What will be the direction of the current when the loop in Fig. 27.12 first enters the field, coming in from the left side?

---

Electromagnetic induction is the principle behind many important technologies, from credit cards to electric-power generation. Induction also gives us the flexibility to transform voltage levels in electric-power systems, and to provide wireless charging systems for devices ranging from electric cars to toothbrushes.

---

**APPLICATION**    **Electric Generators**

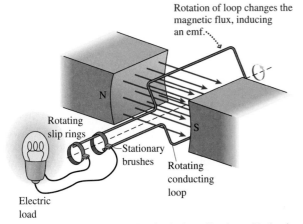

Rotation of loop changes the magnetic flux, inducing an emf.

Probably the most important technological application of induction is the electric generator. Humanity uses electrical energy at the phenomenal rate of about 2 TW $- 2 \times 10^{12}$ W and roughly equal to the power output of 20 billion human bodies—and virtually all this power comes from generators. A **generator** is just a system of conducting loops in a magnetic field, as shown in the figure. Mechanical energy rotates the conductors, resulting in a changing magnetic flux and therefore an induced emf. Current flows through the generator and on to whatever electrical loads are connected to it. Because the changing flux results from a change in the orientation of the loop relative to the field—that is, a change in $\theta$ in the expression $\Phi_B = BA\cos\theta$—a generator such as the one shown here produces an alternating emf that varies sinusoidally with time.

Any source of mechanical energy can power the generator, but the most common is steam from burning fossil fuels or from nuclear fission. Electrical energy is also generated from kinetic energy of water or wind. A small electric generator, driven by the car's engine, is used to recharge a car's battery.

Lenz's law, the conservation of energy in electromagnetic induction, is very much applicable to electric generators. Were it not for Lenz's law generators would turn on their own and happily supply electricity without coal, oil, or uranium! The voluminous quantities of fuel consumed by power plants are dramatic testimony to the minus sign on the right-hand side of Equation 27.2—as suggested in this chapter's opening photo.

Turn a hand-cranked or pedal-driven generator, and you can literally feel Lenz's law. Without any electrical load, turning the generator is easy. Switch on increasingly heavy loads, and the generator gets harder to turn. Most people find they can just sustain a 100-W lightbulb with a hand generator. Think about this next time you leave a light on!

If the diagram here reminds you of the motor in the Application in Chapter 26, that's no coincidence. Motors and generators are similar devices, just run in opposite ways. A motor converts electrical energy to mechanical energy; a generator converts mechanical energy to electrical energy. Often the same physical device serves both purposes. In a hybrid car, for example, an electric motor takes energy from a battery to provide propulsion. When the car brakes, the wheels turn the motor, which then acts as a generator and puts the car's energy back into the battery instead of dissipating it as heat. Such **regenerative braking** is one of the hybrid's several means of achieving greater energy efficiency.

---

**GOT IT? 27.3**    If you lower the electrical resistance connected across a generator while turning the generator at a constant speed, will the generator get easier or harder to turn?

**EXAMPLE 27.5** **Induction: Designing a Generator**

An electric generator consists of a 100-turn circular coil 50 cm in diameter. It's rotated at $f = 60$ rev/s to produce standard 60-Hz alternating current like that used throughout North America. Find the magnetic-field strength needed for a peak output voltage of 170 V (which is the actual peak in standard 120-V household wiring).

**INTERPRET** Here we have a conducting coil rotating in a fixed magnetic field, so this is an induction problem where changing flux results from a changing orientation.

**DEVELOP** We sketched the coil and magnetic field in Fig. 27.14. With a uniform field and flat, circular area, the flux through one turn of the coil is given by Equation 27.1b, $\Phi_{1\,\text{turn}} = \vec{B} \cdot \vec{A} = BA\cos\theta = B\pi r^2 \cos\theta$. The angle $\theta$ changes as the loop rotates, and with it the flux. We need to express the total flux as a function of time so we can evaluate its derivative and thus the emf. Because the loop rotates with constant angular speed $\omega = 2\pi f$, its angular position is $\theta = 2\pi ft$. Then the flux through each turn is $B\pi r^2 \cos(2\pi ft)$, and the total flux through all $N = 100$ turns is $NB\pi r^2 \cos(2\pi ft)$.

**EVALUATE** Faraday's law equates the induced emf with the rate of change of this flux:

$$\mathcal{E} = -\frac{d\Phi_B}{dt} = -NB\pi r^2 \frac{d}{dt}[\cos(2\pi ft)] = -NB\pi r^2[-2\pi f \sin(2\pi ft)]$$

The emf has its peak value when the sine is 1, so $\mathcal{E}_{\text{peak}} = 2\pi^2 r^2 NBf$. We want this value to be 170 V; using $r = 25$ cm, $N = 100$ turns, and $f = 60$ rev/s then gives $B = 23$ mT.

**FIGURE 27.14** Coil in the generator of Example 27.5; at this instant the normal to the coil makes an angle $\theta$ with the magnetic field.

**ASSESS** This value is about 200 G, typical of the field strength near the poles of a permanent magnet. Note that you don't need a value for time $t$ to find the peak emf; when a quantity varies sinusoidally, its peak occurs when the sine or cosine function is 1, so the peak value is the magnitude of whatever quantity multiplies the sine or cosine. ∎

---

Electromagnetic induction is also the basis of magnetic recording, once the dominant means of storing audio, video, and computer information but now more common in credit cards and similar applications. The magnetic strip on your credit card and the tapes in audio- and videocassettes consist of ferromagnetic material that stores information in regions of differing magnetization. Swiping your card or moving magnetic tape past a wire coil induces current in the coil, which extracts the stored information as an electrical signal (Fig. 27.15). Early computer disks worked on the same principle, although in today's disks the magnetic field of the spinning disk causes changes in electrical resistance in the "head" that reads the disk information.

**FIGURE 27.15** Swiping a credit card. Patterns of magnetization on the strip induce currents in the coil.

## Eddy Currents

Induced currents aren't limited to conducting loops and circuits. They also occur in solid conductors subject to changing magnetic flux. The resistance of a solid conductor is low, which can result in large induced currents and significant power dissipation. That can make it hard to move a conducting material into or out of a magnetic field, as it's subject to a changing flux. The result is a kind of magnetic friction that can sap energy. On the other hand, the effect can be useful in providing an alternative to friction brakes in moving machinery. Rapidly rotating saw blades or train wheels, for example, can be stopped quickly by turning on a nearby electromagnet; the resulting eddy currents dissipate the rotational kinetic energy. The mechanical resistance you feel in exercise machines like elliptical trainers or stationary cycles results from a magnet positioned near the machine's rotating parts. And eddy currents are guardians of our security, as the next Application shows.

**APPLICATION**    **Metal Detectors**

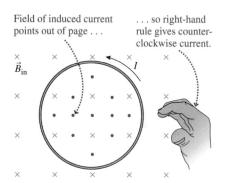

Nothing between coils — Alternating current, $I$, Transmitter coil

Induced current, $I$, Receiver coil — Current detector — Strong current: no conductor between coils

Metal between coils — $I$, $I$ — Weak current triggers alarm.

Eddy currents in conductor reduce changing flux at receiver.

Metal detectors used in airports and other security checkpoints rely on eddy currents. In one type of detector, shown in the figure, an alternating current in one coil—the transmitter—produces a changing magnetic field that induces a current in a second coil, the receiver. A detector, basically an electronic ammeter, monitors the receiver current. Eddy currents are induced in any conducting material that comes between the two coils, and the direction of the induced currents is, as always, such as to reduce the changing flux. The superposition of the transmitter's changing flux with the changing flux from the eddy currents therefore reduces the changing flux at the receiver, dropping the receiver current and triggering an alarm. Other detectors have a single coil, using a short pulse of current to induce eddy currents and then "listening" for the currents induced back in the coil. Either way, you can thank Faraday's law if you've ever been stopped while going through a metal detector!

**GOT IT? 27.4**    A copper penny falls on a path that takes it between the poles of a magnet. Does it hit the ground going (a) faster, (b) slower, or (c) at the same speed as if the magnet weren't present?

## Closed and Open Circuits

Figure 27.16 shows a closed, conducting loop in a magnetic field that points into the page. Suppose the field is increasing in strength; then in order to oppose this change, the induced current must be in such a direction as to oppose the increase. Here that means the field in the interior of the loop needs to come out of the page, and by the right-hand rule that means the induced current is counterclockwise. It's not that the induced field always opposes the inducing field—rather, it opposes the *change*. If the field in Fig. 27.16 had been decreasing, then the induced current would have "tried" to reinforce it by flowing clockwise to make additional field into the page.

What if we have an open circuit, like the conducting loop with a small gap shown in Fig. 27.17? Then there's no induced current whose effects can oppose a change in the field. But we can imagine what would happen if the circuit were completed; as in Fig. 27.16, current would flow counterclockwise. Open the gap, and that means positive charge accumulates at the upper end of the gap and negative charge at the bottom. Charge buildup continues until the potential difference at the gap opposes the induced emf's tendency to move charge. The result is a steady state in which the gap voltage equals the induced emf.

**GOT IT? 27.5**    A long wire carries a current $I$ as shown. What's the direction of the current in the circular conducting loop when $I$ is (a) increasing and (b) decreasing?

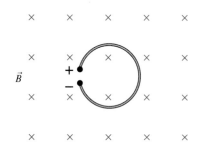

Field of induced current points out of page . . .     . . . so right-hand rule gives counterclockwise current.

$\vec{B}_{in}$     $I$

**FIGURE 27.16**  The field $\vec{B}$ is into the page and increasing; the induced current is counterclockwise, so its field opposes the increase.

$\vec{B}$     $+$  $-$

**FIGURE 27.17**  In a changing magnetic field, the induced emf results in charge buildup at the gap of an open circuit. The polarity shown results when the field is increasing.

## 27.4 Inductance

There are many ways to change magnetic flux and thus induce emfs and currents. We can move a magnet, or move or rotate the circuit. Or, as in Fig. 27.4 or Fig. 27.18, we can change the magnetic flux by changing the current in a circuit and therefore the magnetic field it produces. In that case we speak of the **inductance** of a circuit or circuits.

**FIGURE 27.18** Mutual inductance. A changing current in either coil induces an emf in the other coil.

### Mutual Inductance

Figure 27.18 shows two coils in proximity. If we send a changing current through the left-hand coil, the resulting magnetic field gives rise to a changing magnetic flux in the right-hand coil. There's then an induced emf in the right-hand coil and, if it's connected in a complete circuit, an induced current as well.

The two coils in Fig. 27.18 have **mutual inductance**, meaning a changing current in one coil produces a changing flux at the other coil, thereby inducing an emf. Just how strong this effect is depends on the construction and orientation of the coils; for maximum inductance they should be arranged so that most of the flux from each coil goes through the other. Often coils are wound on iron cores to concentrate the flux and increase the mutual inductance.

Mutual inductance is the basis of transformers, which change voltage levels in alternating-current circuits; more on that in Chapter 28. Your car's ignition coil uses mutual inductance to produce the tens of kilovolts needed to fire the spark plugs and ignite the gasoline–air mixture in the engine. Current from the car's 12-V battery flows through a primary coil with relatively few turns. It's surrounded by a coil with many more turns, through each of which passes the flux from the primary coil. Current in the primary coil is interrupted abruptly at the point when the gasoline is to ignite; the result is a rapid decrease in current and therefore a rapid change in magnetic flux. This induces emfs in each turn of the secondary coil, resulting in a large total emf and a high voltage across the spark plug. Dielectric breakdown occurs at the plug, and the resulting spark ignites the gasoline. A more mundane application occurs in electric toothbrushes, portable phones, and other devices whose batteries charge without electrical connection to a power source. Instead, a small coil in the device is placed in proximity with a coil in the charging base, and alternating current in the base coil transfers energy via mutual inductance to provide the charging current.

### Self-Inductance

Inductance isn't limited to two-coil systems. Magnetic flux from current in a single coil or circuit passes through that circuit itself (Fig. 27.19). If the current changes, so does the flux—and that induces an emf in the circuit. As always, the induced emf opposes the *change* that produces it. Suppose, for example, that the current in Fig. 27.19 is increasing. Then the induced emf will be in the direction that opposes the current increase—clockwise, or opposite the current in Fig. 27.19. The induced emf therefore makes it harder to increase the current. On the other hand, if the current in Fig. 27.19 is decreasing, then the induced emf will try to drive additional current to counter the decrease; the induced emf is therefore in the same direction as the current. Either way, the induced emf makes it hard to change the current in a circuit.

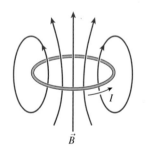

**FIGURE 27.19** Magnetic flux from a current loop passes through the loop itself; a change in the current induces an emf that opposes the change.

This property whereby a circuit's own magnetic field opposes changes in the circuit current is called **self-inductance**. All circuits have self-inductance, but it's most important in circuits that encircle a great deal of their own magnetic flux, or when currents change rapidly. A simple piece of wire has little impact on the 60-Hz alternating current used for electric power. But in TVs and computers, where currents change billions of times per second, even the slightest self-inductance can have deleterious effects.

An **inductor** is designed specifically to exhibit self-inductance. Inductors have many uses in electric circuits, including establishing the frequencies of radio transmitters and helping "steer" high- and low-frequency signals to the tweeters and woofers of loudspeaker systems. We'll explore some of these uses in the next chapter. A typical inductor consists of a wire coil, sometimes wound on an iron core to promote flux concentration. Ideally, the only electrical property of an inductor is its inductance, but real inductors have resistance as well.

As long as the current in an inductor is steady, the magnetic flux is constant, so there's no induced emf and the inductor acts like a wire. But when the current changes, the changing magnetic flux induces an emf that opposes the change in current. The more rapidly the current changes, the greater the rate of change of flux and so the greater the emf. The induced emf depends also on how much of its own magnetic flux the inductor encircles; consequently, we define self-inductance, $L$, as the ratio of magnetic flux through the inductor to current in the inductor:

$$L = \frac{\Phi_B}{I} \quad \text{(self-inductance)} \tag{27.3}$$

Equation 27.3 shows that the units of self-inductance are T·m²/A. This unit is given the name **henry** (H) in honor of the American scientist Joseph Henry (1797–1878). Inductances in common electronic circuits usually range from microhenrys ($\mu$H) up to several henrys.

Inductance is a constant determined by the physical design of an inductor. In principle we can calculate the inductance of any inductor, but in practice this is difficult unless the geometry is particularly simple.

---

**EXAMPLE 27.6**    **Calculating Inductance: A Solenoid**

A long solenoid of cross-sectional area $A$ and length $l$ has $n$ turns per unit length. Find its self-inductance.

**INTERPRET** We're asked for self-inductance, which Equation 27.3 shows is the ratio of magnetic flux through the solenoid to current in the solenoid.

**DEVELOP** We'll assume a current in the solenoid and find the resulting magnetic flux. Then we can take their ratio to get the self-inductance. We need the magnetic field of a solenoid, which follows from Equation 26.20: $B = \mu_0 nI$. The field is uniform and perpendicular to the solenoid coils, as we showed in our drawing for Example 27.1, so the flux through each turn follows from Equation 27.1b: $\Phi_{1\,\text{turn}} = BA$.

**EVALUATE** With $n$ turns per unit length, the solenoid contains a total of $nl$ turns, so the flux through all the turns is

$$\Phi_B = nlBA = (nl)(\mu_0 nI)A = \mu_0 n^2 AlI$$

Equation 27.3 gives the self-inductance as the ratio of flux to current, so

$$L = \frac{\Phi_B}{I} = \mu_0 n^2 Al \quad \text{(inductance of solenoid)} \tag{27.4}$$

**ASSESS** Make sense? As the area increases, so does the flux and therefore the inductance. As the length increases, so does the number of turns, so again the total flux increases. And as the number of turns per unit length increases, two things happen. First, the magnetic field of Equation 26.20 increases, increasing the flux $BA$ through each turn. Second, the total number of turns increases, again increasing the total flux. That's why the inductance, $L$, depends on $n$ *squared*. ∎

---

The induced emf in an inductor is determined by Faraday's law, which relates emf to the rate of change of magnetic flux: $\mathcal{E} = -d\Phi_B/dt$. Differentiating Equation 27.3, the definition of inductance, gives

$$\frac{d\Phi_B}{dt} = L\frac{dI}{dt}$$

Then Faraday's law becomes

$$\mathcal{E}_L = -L\frac{dI}{dt} \quad \text{(inductor emf)} \tag{27.5}$$

This equation gives the emf $\mathcal{E}$ induced in an inductor $L$ when the current in the inductor is changing at the rate $dI/dt$. The minus sign again tells us that the emf *opposes* the change in current. For this reason the inductor emf is often called a **back emf**; it works *against* changes brought about by an externally applied emf. Figure 27.20 shows how to interpret the signs in Equation 27.5.

When the current in an inductor is steady, $dI/dt = 0$ and there's no induced emf. In this case, the inductor acts like a piece of wire. But when the current changes, the inductor responds by producing a back emf that opposes the change. Now the inductor acts very much like a battery, with the magnitude of its emf dependent on how fast the current changes. If we try to start or stop current suddenly, $dI/dt$ is very large and a very large back emf appears. This isn't just mathematics! Rapid switching of inductive devices such as solenoids, solenoid valves, or motors can destroy delicate electronic devices. And people have been killed opening switches in circuits containing large inductors. In the next section we'll take a closer look at circuits that include inductors.

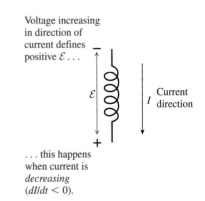

Voltage increasing in direction of current defines positive $\mathcal{E}$...

...this happens when current is *decreasing* ($dI/dt < 0$).

**FIGURE 27.20** The direction of induced emf in Equation 27.5 depends on whether the current is increasing or decreasing. The coil is the circuit symbol for an inductor.

**GOT IT? 27.6** Current flows from left to right through the inductor shown. A voltmeter connected across the inductor gives a constant reading, and shows that the left end of the inductor is positive. Is the current in the inductor (a) increasing, (b) decreasing, or (c) steady? Why?

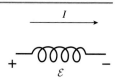

---

**EXAMPLE 27.7** **Back emf: A Dangerous Inductor**

A 5.0-A current is flowing in a 2.0-H inductor. The current is then reduced steadily to zero over 1.0 ms. Find the magnitude and direction of the inductor emf during this time.

**INTERPRET** There's an emf in the inductor because the current is changing; we want the magnitude and direction of that emf.

**DEVELOP** Figure 27.21 shows the situation, complete with an external circuit that's the source of the decreasing current. Equation 27.5, $\mathcal{E}_L = -L(dI/dt)$, gives the inductor emf in terms of the rate of change of current. Since the current changes steadily, the latter is just the change in current $(-5\,\text{A})$ divided by the time involved.

**EVALUATE**

$$\mathcal{E}_L = -L\frac{dI}{dt} = -(2.0\,\text{H})\left(\frac{-5.0\,\text{A}}{1.0\,\text{ms}}\right) = 10{,}000\,\text{V}$$

That this answer is positive tells us that the emf increases in the same direction as the current, as we indicated in Fig. 27.21.

**FIGURE 27.21** Sketch for Example 27.7.

**ASSESS** That's a potentially lethal voltage! Our answer is unrelated to the battery or whatever is supplying the inductor current. One could have a 6-V battery and still be electrocuted opening a circuit with a large inductance. Note that the direction we deduced is consistent with Lenz's law; here the inductor emf opposes the decrease in current, and that means it provides an emf in the direction that would keep current flowing in the external circuit—just the opposite of the situation in Got It? 27.6. ∎

## Inductors in Circuits

In Chapter 25 we found that the voltage across a capacitor can't change instantaneously. We can make an analogous statement for inductors. Because the inductor emf depends on the rate of change of current and because an infinite emf is impossible, **the current through an inductor can't change instantaneously.** Much of our understanding of capacitors applies to inductors if we interchange the words "voltage" and "current."

Figure 27.22 shows a circuit with a battery, switch, resistor, and inductor. With the switch open there's no current (Fig. 27.22a). Close the switch, and the current at that instant is still zero because the inductor current can't change instantaneously. With no current, there's no voltage across the resistor, so the inductor must be producing a back emf equal in magnitude to the battery emf (Fig. 27.22b). Although at this instant there's zero current in the inductor, the nonzero emf $\mathcal{E}_L = -L(dI/dt)$ shows that the *rate of change* of current, $dI/dt$, isn't zero.

**FIGURE 27.22** An *RL* circuit at three different times.

So the inductor current rises from zero, and with it the resistor current and therefore the resistor voltage $IR$. The battery emf $\mathcal{E}_0$ is constant, so as $IR$ increases, the magnitude of the inductor emf drops. Equation 27.5 shows that the rate of change of current drops as well. Eventually the whole circuit reaches a steady state in which $dI/dt$ and therefore the inductor emf are both zero (Fig. 27.22c). At this point the inductor acts like a wire, and the resistor determines the current: $I = \mathcal{E}_0/R$. Figure 27.23 summarizes this analysis of the *RL* circuit.

**FIGURE 27.23** Interrelationships among circuit quantities as current builds up in an *RL* circuit. Compare with Fig. 25.20 for a charging capacitor.

We can analyze the circuit quantitatively using the loop law. Going clockwise, we encounter a voltage increase $\mathcal{E}_0$ at the battery, a decrease $-IR$ at the resistor, and a change $\mathcal{E}_L$ at the inductor. This change is actually a decrease, but we'll let Equation 27.5 take care of the signs. Then the loop law reads $\mathcal{E}_0 - IR + \mathcal{E}_L = 0$. The battery emf is constant, so if we differentiate this equation, we get

$$\frac{d\mathcal{E}_L}{dt} = R\frac{dI}{dt}$$

But Equation 27.5 gives $dI/dt = -\mathcal{E}_L/L$, so

$$\frac{d\mathcal{E}_L}{dt} = -R\frac{\mathcal{E}_L}{L}$$

This looks like Equation 25.4 for the *RC* circuit, but with $\mathcal{E}_L$ in place of current $I$, $L$ in place of $C$, and $1/R$ in place of $R$. So the solution is that of Equation 25.4 with the appropriate substitutions:

$$\mathcal{E}_L = -\mathcal{E}_0 e^{-Rt/L} \tag{27.6}$$

This shows that the inductor emf decays exponentially from its initial value $-\mathcal{E}_0$ (negative because the inductor emf opposes the battery) to zero. Using the undifferentiated loop equation, we can now solve for the current:

$$I = \frac{\mathcal{E}_0 + \mathcal{E}_L}{R} = \frac{\mathcal{E}_0}{R}\left(1 - e^{-Rt/L}\right) \tag{27.7}$$

With a capacitor, we characterized time-changing quantities with the capacitive time constant $RC$. Here the **inductive time constant** is $L/R$. In contrast to the capacitor case, the inductive time constant depends *inversely* on resistance. That's because a lower resistance means a higher steady-state current, which therefore requires a longer time to build up. Significant changes in current can't occur on time scales much shorter than $L/R$. Wait many time constants, and the circuit approaches a steady state with $\mathcal{E}_L = 0$. Figure 27.24 summarizes the time-dependent behavior of circuit quantities in an *RL* circuit.

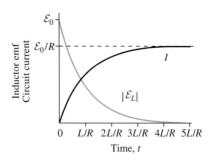

**FIGURE 27.24** Inductor current and emf as functions of time.

---

**EXAMPLE 27.8** The Inductive Time Constant:
Firing Up an Electromagnet

A large electromagnet used for lifting scrap iron has self-inductance $L = 56$ H. It's connected to a constant 440-V power source; the total resistance of the circuit is 2.8 Ω. Find the time it takes for the current to reach 75% of its final value.

**INTERPRET** This is a problem about the buildup of current in an $RL$ circuit.

**DEVELOP** Equation 27.7, $I = (\mathcal{E}_0/R)(1 - e^{-Rt/L})$, determines the current; here $\mathcal{E}_0/R$ is the final current, and we want to solve for the time $t$ when $I$ is 75% of this final value. That is, we want $0.75 = 1 - e^{-Rt/L}$.

**EVALUATE** Rearranging, we have $e^{-Rt/L} = 0.25$; then taking natural logs of both sides and using $\ln e^x = x$ gives $-Rt/L = \ln(0.25)$, or

$$t = -\frac{L}{R} \ln(0.25) = -\frac{56\ \text{H}}{2.8\ \Omega} \ln(0.25) = 28\ \text{s}$$

**ASSESS** This is a little more than 1 time constant $(L/R = 20\ \text{s})$—not surprising because we found with capacitors that we reach approximately two-thirds of the full charge in 1 time constant. Analogously, with inductors, we reach about two-thirds of the final current in 1 time constant. ∎

---

Figure 27.25 shows a circuit with a two-way switch. Throw the switch to $A$, and current builds up as we just described. Throw it to $B$, and current continues to flow through the inductor and resistor because the inductor current can't change instantaneously. We won't go through the math, but it's straightforward to show that the current decays exponentially with the same time constant $L/R$:

$$I = I_0 e^{-Rt/L} \tag{27.8}$$

This is analogous to our result for the discharging capacitor.

**FIGURE 27.25** Buildup and decay of current in an $RL$ circuit.

As with capacitors, it's not necessary to use exponential equations to analyze the short- and long-term behavior of circuits with inductors. All you need to remember is that for short times inductor current can't change instantaneously, and for long times inductors have no emfs and therefore act like wires. The next example explores this situation.

---

**EXAMPLE 27.9** Inductors: Short Times, Long Times

The switch in Fig. 27.26a is initially open. Find the current in $R_2$ immediately and a long time after the switch is closed. Long after the switch is closed, it's reopened. Find the current in $R_2$ immediately after this second switch opening.

**INTERPRET** This is a problem about the short- and long-term behavior of a circuit containing an inductor. The principles we need are that the inductor current can't change instantaneously and that after long times the inductor acts like a wire.

**DEVELOP** We've drawn the circuit in each of the situations we need to analyze. There's no current in the inductor before the switch closes, so there's no current right afterward; the inductor might as well be an open circuit. That's why we drew Fig. 27.26b with only the two resistors. After a long time the inductor becomes a wire, as we show in Fig. 27.26c. Finally, whatever current was flowing in the inductor continues to flow after we reopen the switch. Opening the switch takes out $R_1$ and the battery, so in Fig. 27.26d we show only the inductor and $R_2$ with current continuing to flow. Our job is to analyze each of these circuits.

**FIGURE 27.26** Example 27.9.

*cont'd.*

**EVALUATE** Figure 27.26b is a simple series circuit, with the current in both resistors given by $I = \mathcal{E}_0/(R_1 + R_2)$. In Fig. 27.26c there's no inductor emf, which is why we represent the inductor by a wire. So there's no voltage across $R_2$, and the current in $R_2$ is therefore zero. But there is current in the inductor, given by $I = \mathcal{E}_0/R_1$. In Fig. 27.26d that same current $\mathcal{E}_0/R_1$ continues to flow right after the switch opens, but now it goes *upward* through $R_2$. This current then decays exponentially to zero. Figure 27.27 summarizes the time dependence of the currents in $R_2$ and $L$.

**ASSESS** The last case may seem surprising, with the current in $R_2$ determined entirely by the battery voltage and the resistance $R_1$. Doesn't $R_2$ have any say in the matter? No: Right after the switch opens, the inductor current must be whatever it was before—and $R_1$ had determined that current. With current $I = \mathcal{E}_0/R_1$, the voltage across $R_2$ is $IR_2 = \mathcal{E}_0(R_2/R_1)$ and is equal in magnitude to the inductor emf. The inductor will produce whatever emf is needed to keep the current from changing instantaneously. If $R_2$ is large compared with $R_1$, that implies a voltage much higher than the battery voltage. If $R_2$ isn't in the

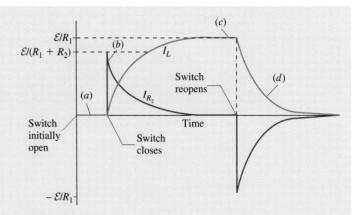

**FIGURE 27.27** Currents in $R_2$ and $L$ for Example 27.9.

circuit, then the voltage will rise dangerously high as the inductor tries at all cost to keep the current flowing. Resistors are often placed in parallel with large inductors to alleviate this danger. ∎

## 27.5 Magnetic Energy

In Figs. 27.25b and 27.26d, current flows in circuits containing only a resistor and an inductor. Energy is dissipated, heating the resistor. Where does this energy come from?

Because there's a current in the inductor, there's also a magnetic field. The change in that magnetic field is what produces the emf that drives the current. As the current decreases, so does the magnetic field. Eventually the circuit reaches a state where there's no current, no magnetic field—and a hot resistor. So where did the resistor's thermal energy come from? It came from the magnetic field.

Like the electric field, the magnetic field contains stored energy. Our decaying $RL$ circuit is analogous to a discharging $RC$ circuit, in which the electric field between the capacitor plates disappears as thermal energy appears in the resistor. As in the electric case, magnetic energy isn't limited to circuits: *Any* magnetic field contains energy. Release of magnetic energy drives a number of practical devices and also powers violent events throughout the universe (Fig. 27.28).

**FIGURE 27.28** This eruption of a huge prominence from the sun's surface releases energy stored in magnetic fields.

### Magnetic Energy in an Inductor

We can find the stored energy by reconsidering the buildup of current in the inductor. Earlier we wrote the loop law for the circuit of Fig. 27.22; if we multiply by the current $I$, we get $I\mathcal{E}_0 - I^2R + I\mathcal{E}_L = 0$ or, using Equation 27.5 for $\mathcal{E}_L$,

$$I\mathcal{E}_0 - I^2R - LI\frac{dI}{dt} = 0$$

The three terms here have the units of voltage times current, or power. The first shows that the battery supplies power *to* the circuit at the rate $I\mathcal{E}_0$. The second, $-I^2R$, is the rate of energy dissipation in the resistor; the negative sign means the resistor takes energy *from* the circuit. The current is increasing $(dI/dt > 0)$, so the third term is also negative; it describes energy the inductor takes from the circuit. But the inductor doesn't dissipate this energy; rather, it stores the energy in its magnetic field. The rate at which the inductor stores energy is thus

$$P = LI\frac{dI}{dt}$$

Suppose we increase the current in an inductor by some small amount $dI$ over a small time interval $dt$. Since the power is the rate of energy storage, the energy $dU$ stored during this time is $dU = P\, dt = LI\, dI$. We find the total energy stored in bringing the inductor current from zero to some final value $I$ by summing—that is, integrating—all the $dU$ values:

$$U = \int dU = \int P\, dt = \int_0^I LI\, dI = \tfrac{1}{2}LI^2 \Big|_0^I$$

Evaluating at the two limits then gives the stored energy:

$$U = \tfrac{1}{2}LI^2 \qquad \text{(energy stored in inductor)} \qquad (27.9)$$

This much energy is therefore released when the magnetic field decays.

---

**EXAMPLE 27.10** **Magnetic Energy: An MRI Disaster**

Superconducting electromagnets like the solenoids in MRI scanners store a lot of magnetic energy. Loss of coolant can be dangerous because the current is suddenly left without its zero-resistance path and quickly decays. The result is an explosive release of magnetic energy. A particular MRI solenoid carries 2.4 kA and has a 0.53-H inductance. When it loses superconductivity, its resistance goes abruptly to 31 m$\Omega$. Find (a) the stored magnetic energy and (b) the rate of energy release at the instant superconductivity is lost.

**INTERPRET** We're asked first for the total stored energy and then for the power dissipated in the resistance of the coils at the instant they cease to be superconducting.

**DEVELOP** Equation 27.9, $U = \tfrac{1}{2}LI^2$, determines the stored energy, while $P = I^2R$ determines the resistor power. Just before the coolant loss, the MRI solenoid carries 2.4 kA; since current can't change instantaneously in an inductor, this current remains momentarily unchanged. Therefore we have everything we need to find the power dissipation.

**EVALUATE** (a) Equation 27.8 gives
$$U = \tfrac{1}{2}LI^2 = \left(\tfrac{1}{2}\right)(0.53\text{ H})(2.4\text{ kA})^2 = 1.5\text{ MJ}$$
while for (b) we have
$$P = I^2R = (2.4\text{ kA})^2(31\text{ m}\Omega) = 0.18\text{ MW}$$

**ASSESS** This is substantial power, equivalent to 1800 100-W lightbulbs burning in the space of this roughly human-size device! You can show in Problem 58 that it takes some 20 s before 90% of the energy has dissipated. To prevent explosive energy release, superconducting wires generally incorporate copper or silver to carry the current in the event of coolant loss that quenches the superconductivity. ∎

---

## Magnetic-Energy Density

In Example 27.6 we found the inductance of a solenoid with length $l$ and cross-sectional area A: $L = \mu_0 n^2 Al$. Equation 27.9 then gives the magnetic energy stored in the solenoid:

$$U = \tfrac{1}{2}LI^2 = \tfrac{1}{2}\mu_0 n^2 AlI^2 = \frac{1}{2\mu_0}(\mu_0 nI)^2 Al = \frac{B^2}{2\mu_0}Al$$

where we recognized the quantity $\mu_0 nI$ as $B$, the magnetic field in the solenoid (Equation 26.20). The quantity $Al$ is the volume containing this field, so the energy per unit volume—the **magnetic-energy density**—is

$$u_B = \frac{B^2}{2\mu_0} \qquad \text{(magnetic-energy density)} \qquad (27.10)$$

Although we derived this expression for the field of a solenoid, it is, in fact, a universal expression for the local magnetic-energy density. Wherever there's a magnetic field, there's stored energy.

Equation 27.10 is similar to Equation 23.7 for the energy density in an electric field: $u_E = \tfrac{1}{2}\epsilon_0 E^2$. Each energy density is proportional to the *square* of the field strength, and each contains the appropriate constant, $\mu_0$ or $\epsilon_0$. That the constant appears in the numerator in one case and in the denominator in the other is merely a consequence of the way SI units are defined.

## 27.6 Induced Electric Fields

So far we've been talking about induction in terms of emfs and circuits. But what, really, is emf? In the case of a battery, it results from chemical reactions that separate charge. With motional emf (Section 27.3), magnetic forces on a moving conductor act to separate charge. But what causes the emf in a conducting loop subject to a changing magnetic field? There's no motion, yet there must be a force on the free charges in the conductor. The only force we know that acts on stationary charges is the electric force, which results from electric fields. Therefore there must be an electric field—an **induced electric field**—in the conducting loop. This field has the same effect on charges, exerting a force $q\vec{E}$, as did the electric fields we considered earlier. But the induced field originates not in electric charge but in **changing magnetic field**.

An induced electric field results whenever a magnetic field changes with time—whether or not an electric circuit is present. If there is a circuit, then the field drives induced currents. But the induced field, not the current, is fundamental. A single, stationary electron in a changing magnetic field experiences an *electric* force—clear evidence for the existence of the induced electric field.

We wrote Faraday's law as Equation 27.2, giving the relation between induced emf and changing magnetic flux. But the induced electric field is more fundamental, and emf simply means the work per unit charge gained as charge goes around a circuit—or for that matter any closed loop. Thus we can write $\mathcal{E} = \oint \vec{E} \cdot d\vec{r}$, and Faraday's law becomes

$$\oint \vec{E} \cdot d\vec{r} = -\frac{d\Phi_B}{dt} \quad \text{(Faraday's law)} \tag{27.11}$$

In this form, Faraday's law is a universal statement about electric fields and changing magnetic flux. The line integral on the left-hand side is over *any* closed loop, which need not coincide with a circuit or conductor. The flux on the right-hand side is the surface integral of the magnetic field over any open surface bounded by the loop on the left-hand side.

Faraday's law tells us that there's another source of electric fields besides electric charge—namely, changing magnetic field:

> A changing magnetic field creates an electric field.

This direct interaction between fields is the basis for many practical devices and, as we'll see in Chapter 29, is essential to the existence of light.

Faraday's law is similar to Ampère's law (Equation 26.16). On the left side, both involve the line integral of a field, $\vec{E}$ for Faraday and $\vec{B}$ for Ampère. On the right is a source of that field, changing magnetic field for $\vec{E}$ and moving electric charge—current—for $\vec{B}$. Both fields *encircle* their sources. That means the configuration of an induced electric field is very different from that of an electric field originating in charge. Field lines of an induced field have no beginnings or ends; they generally form closed loops encircling regions of changing magnetic field (Fig. 27.29).

When a changing magnetic field has sufficient symmetry, we can evaluate the induced electric field in the same way we did the magnetic field of a symmetric current distribution.

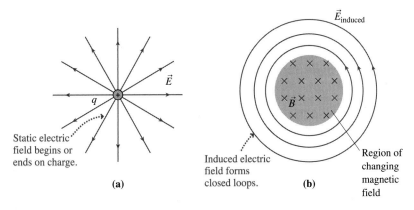

**FIGURE 27.29** (a) Static electric fields originate in charges and look very different from (b) induced fields that result from changing magnetic fields.

**EXAMPLE 27.11**  **Finding the Induced Electric Field: A Solenoid**

A long solenoid has circular cross section of radius $R$. The solenoid current is increasing, and as a result so is the magnetic field in the solenoid. The field strength is given by $B = bt$, where $b$ is a constant. Find the induced electric field outside the solenoid, a distance $r$ from the axis.

**INTERPRET** Here's a problem about a changing magnetic field producing an electric field—that is, about Faraday's law. We'll follow the steps in Chapter 26's strategy for Ampère's law, modifying as appropriate to Faraday's law. We begin by identifying the symmetry, which here is line symmetry.

**DEVELOP** Symmetry requires that encircling electric field lines be circular. We've drawn some field lines in Fig. 27.30 and marked one of them as a loop for the integration in Faraday's law. We chose a

loop coinciding with a field line because symmetry requires that the field strength $E$ be constant over a circle concentric with the symmetry axis.

**EVALUATE** The situation on the left-hand side of Faraday's law is just as in Example 26.8, except with $\vec{E}$ instead of $\vec{B}$. So $\oint \vec{E} \cdot d\vec{r}$ evaluates immediately to $2\pi r E$. Instead of current on the right-hand side, we have changing magnetic flux $-d\Phi_B/dt$. Here the loop is outside the solenoid, so it encircles the entire flux, which is $\Phi_B = BA = bt\pi R^2$. The right-hand side of Faraday's law has the rate of change of flux, which is $d\Phi_B/dt = \pi R^2 b$. Equating the left and right sides of Faraday's law then gives $2\pi r E = -\pi R^2 b$, or

$$E = -\frac{R^2 b}{2r}$$

**ASSESS** The $1/r$ dependence here shouldn't surprise you; we found the same thing for other fields resulting from distributions with line symmetry. The minus sign tells about the direction, but as usual it's easiest to invoke Lenz's law to reason out the direction. If a current were flowing as a result of the induced electric field, its direction would be such as to *oppose* the increase in the solenoid's magnetic field. So it would have to produce a field out of the page in Fig. 27.30, and that means a counterclockwise current. So the induced electric field goes counterclockwise. Calculating the field *inside* the solenoid would be similar, but a given field line would encircle only part of the magnetic flux; Exercise 34 covers this situation.  ∎

**FIGURE 27.30** Cross section of a solenoid whose magnetic field points into the page and is increasing. Field lines of the induced electric field are circles concentric with the solenoid axis.

Loop for Faraday's law

## Conservative and Nonconservative Electric Fields

Static electric fields—those beginning and ending on stationary charge distributions—are *conservative*, meaning that the work required to move a charge between two points is path independent. A consequence is that it takes no work to move around a closed path in an electrostatic field; mathematically, we express this by writing

$$\oint \vec{E} \cdot d\vec{r} = 0 \qquad \text{(electrostatic field)}$$

In contrast, induced electric fields generally form closed loops, and here Faraday's law shows that the line integral of the electric field around a closed path is decidedly not zero. That means the induced electric field does work on a charge moved around a *closed* path and that the work done in moving between two points cannot be independent of the path taken (Fig. 27.31). The induced electric field is therefore not conservative.

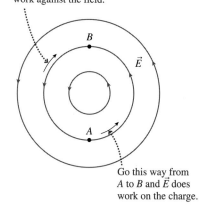

Move charge this way from $A$ to $B$, and you do work against the field.

$B$

$\vec{E}$

$A$

Go this way from $A$ to $B$ and $\vec{E}$ does work on the charge.

**FIGURE 27.31** The work done to move charge in an induced electric field isn't path independent, so the induced field isn't conservative.

**GOT IT? 27.7** The figure shows three resistors in series surrounding an infinitely long solenoid with a changing magnetic field; the resulting induced electric field drives a current counterclockwise, as shown. Two identical voltmeters are shown connected to the *same* points A and B. What does each read? Explain any apparent contradiction. *Hint:* This is a challenging question!

## Diamagnetism

We briefly described diamagnetism in Chapter 26 but couldn't explain it until now because diamagnetism results from induced electric fields. In a purely diamagnetic material, current loops associated with pairs of atomic electrons exactly cancel, leaving atoms with no intrinsic magnetic moments. Figure 27.32a shows a simplified model of such an atom. Don't take this picture too literally, for it uses classical physics to describe a phenomenon properly within the domain of quantum mechanics. Nevertheless, it shows qualitatively how diamagnetism involves electromagnetic induction.

The electron orbiting clockwise, on the right side of Fig. 27.32a, has its magnetic moment pointing out of the page. In the absence of an applied magnetic field, this moment is exactly canceled by the equal but oppositely directed moment of the other electron. But now apply a magnetic field—for example, by moving the north pole of a magnet down toward the plane of Fig. 27.32. The changing magnetic field results in an induced electric field that alters the speeds of the orbiting electrons. The direction of this effect must be such as to oppose the increase in field. Thus the electron whose magnetic moment points out of the page has its moment increase, while the moment pointing into the page decreases (Fig. 27.32b).

Once the field is applied and the orbital speeds of the two electrons are no longer identical, the atom has a net magnetic moment that points out of the page, opposing the incoming magnet and resulting in a repulsive force. This repulsion is the distinguishing characteristic of diamagnetism.

A superconductor is perfectly diamagnetic, meaning that the magnetic field resulting from induced currents completely cancels any applied field. Since these induced currents persist in the zero-resistance superconductor, the material completely excludes magnetic fields from its interior, a phenomenon known as the Meissner effect (Fig. 27.33). The repulsive force associated with the magnetic moments of a permanent magnet and a nearby superconductor results in the widely publicized phenomenon of magnetic levitation (Fig. 27.34).

No net magnetic moment

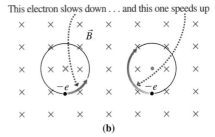

**(a)**
Increasing $\vec{B}$ induces $\vec{E}$ that changes electrons' motions.

This electron slows down . . . and this one speeds up

**(b)**
Result: nonzero net magnetic moment

**FIGURE 27.32** A simple model for diamagnetism.

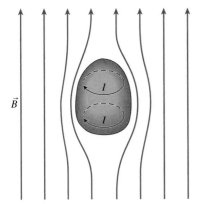

**FIGURE 27.33** Induced currents in a superconductor completely cancel an applied magnetic field.

**FIGURE 27.34** A small magnet levitates above a wafer of high-temperature superconductor in a bath of liquid nitrogen.

## Big Picture

The big idea here is **electromagnetic induction**, a phenomenon in which **a changing magnetic field produces an electric field**. Applied to circuits, induction results in induced emfs that drive induced currents.

Here a moving magnet produces the changing magnetic field.

Here a change in current produces the changing magnetic field.

## Key Concepts and Equations

**Faraday's law** describes induction quantitatively, relating the line integral of the induced electric field to changing magnetic flux; $\oint \vec{E} \cdot d\vec{r} = -d\Phi_B/dt$.

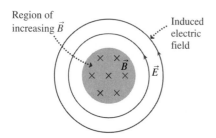

Region of increasing $\vec{B}$

Induced electric field

In conductors, Faraday's law gives the induced emf: $\mathcal{E} = -d\Phi_B/dt$.

**Lenz's law**: Electromagnetic induction is consistent with conservation of energy, which requires that induced effects act to oppose the changes that give rise to them.

Counterclockwise current makes loop magnetic moment out of page, *opposing* increase in $\vec{B}$.

Increasing $\vec{B}$ into page

Magnetic fields contain stored energy, as do electric fields. The **magnetic-energy density** is

$$u_B = \frac{B^2}{2\mu_0}.$$

## Cases and Uses

**Electric generators** convert mechanical to electrical energy by moving conductors in magnetic fields to induce emfs that drive currents.

Rotation of loop changes the magnetic flux, inducing an emf.

Rotating slip rings

Stationary brushes

Rotating conducting loop

Electric load

**Inductors** are wire coils that encircle their own magnetic flux, giving **self-inductance** $L = \Phi_B/I$. An inductor opposes changes in current, producing an emf given by $\mathcal{E} = -L(dI/dt)$. Circuit quantities in a simple *RL* circuit change with **inductive time constant** $L/R$.

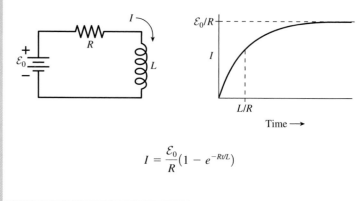

$$I = \frac{\mathcal{E}_0}{R}\left(1 - e^{-Rt/L}\right)$$

**Diamagnetism** occurs when electromagnetic induction results in atoms acquiring net magnetic moments; the result is a repulsive interaction.

# For Thought and Discussion

1. A bar magnet is moved toward a conducting ring, as shown in Fig. 27.35. What is the direction of the induced current in the ring?

**FIGURE 27.35** For Thought and Discussion 1

2. Figure 27.36 shows two concentric conducting loops, the outer connected to a battery and a switch. The switch is initially open. It's then closed, left closed for a while, and then reopened. Describe the currents in the inner loop during the entire procedure.

**FIGURE 27.36** For Thought and Discussion 2

3. A constant, uniform magnetic field points into the page. A flexible, circular conducting ring lies in the plane of the page. If the ring is stretched, maintaining its circular shape, what's the direction of the induced current?

4. Fluctuations in Earth's magnetic field due to changing solar activity can wreak havoc with communications, even those using underground cables. How is this possible?

5. In Chapter 26, we pointed out that a static magnetic field cannot change the energy of a charged particle. Is this true of a changing magnetic field? Discuss.

6. Can an induced electric field exist in the absence of a conductor?

7. A car battery has an emf of 12 V, yet energy from the battery provides the 30,000-V spark that ignites the gasoline. How is this possible?

8. You have a fixed length of wire to wind into an inductor. Will you get more inductance if you wind a short coil with large diameter, or a long coil with small diameter?

9. In wiring circuits that operate at high frequencies, like TV sets or computers, it's important to avoid extraneous loops in wires. Why?

10. In a popular demonstration of induced emf, a lightbulb is connected across a large inductor in an *RL* circuit, as shown in Fig. 27.37. When the switch is opened, the bulb flashes brightly and burns out. Why?

**FIGURE 27.37** For Thought and Discussion 10

11. List some similarities and differences between inductors and capacitors.

12. A 1-H inductor carries 10 A, and a 10-H inductor carries 1 A. Which inductor contains more stored energy?

13. The field of a magnetic dipole extends to infinity. Is an infinite amount of energy stored in the dipole field? Why or why not?

14. It takes work to push two bar magnets together with like poles facing each other. Where does this energy go?

# Exercises and Problems

## Exercises

### Sections 27.2 Faraday's Law and 27.3 Induction and Energy

15. Show that the volt is the correct SI unit for the rate of change of magnetic flux, making Faraday's law dimensionally correct.

16. Find the magnetic flux through a circular loop 5.0 cm in diameter oriented with the loop normal at 30° to a uniform 80-mT magnetic field.

17. A circular wire loop 40 cm in diameter has 100-$\Omega$ resistance and lies in a horizontal plane. A uniform magnetic field points vertically downward, and in 25 ms it increases linearly from 5.0 mT to 55 mT. Find the magnetic flux through the loop at (a) the beginning and (b) the end of the 25-ms period. (c) What is the loop current during this time? (d) Which way does this current flow?

18. A conducting loop of area 240 cm$^2$ and resistance 12 $\Omega$ lies at right angles to a spatially uniform magnetic field. The loop carries an induced current of 320 mA. At what rate is the magnetic field changing?

19. The magnetic field inside a 20-cm-diameter solenoid is increasing at the rate of 2.4 T/s. How many turns should a coil wrapped around the outside of the solenoid have so that the emf induced in the coil is 15 V?

### Section 27.4 Inductance

20. Find the self-inductance of a solenoid 50 cm long and 4.0 cm in diameter that contains 1000 turns of wire.

21. The current in an inductor is changing at the rate of 100 A/s, and the inductor emf is 40 V. What is its self-inductance?

22. A 2.0-A current is flowing in a 20-H inductor. A switch is opened, interrupting the current in 1.0 ms. Find the induced emf in the inductor.

23. A cardboard tube measures 15 cm long by 2.2 cm in diameter. How many turns of wire must be wound on the full length of the tube to make a 5.8-mH inductor?

24. What inductance should you put in series with a 100-$\Omega$ resistor to give a time constant of 2.2 ms?

25. The current in a series *RL* circuit increases to 20% of its final value in 3.1 $\mu$s. If $L = 1.8$ mH, what is the resistance $R$?

## Section 27.5 Magnetic Energy

26. How much energy is stored in a 5.0-H inductor carrying 35 A?

27. What's the current in a 10-mH inductor when the stored energy is 50 $\mu$J?

28. A 220-mH inductor carries 350 mA. How much energy must be supplied to the inductor in raising the current to 800 mA?

29. A 500-turn solenoid is 23 cm long and 1.5 cm in diameter and carries 65 mA. How much magnetic energy does it contain?

30. Show that the quantity $B^2/2\mu_0$ has the units of energy density $(J/m^3)$.

31. The Alcator fusion project at MIT has a 50-T magnetic field. What is the magnetic-energy density in Alcator?

32. What is the magnetic-field strength in a region where the magnetic-energy density is 7.8 $J/cm^3$?

## Section 27.6 Induced Electric Fields

33. The induced electric field 12 cm from the axis of a solenoid with a 10-cm radius is 45 V/m. Find the rate of change of the solenoid's magnetic field.

34. Find an expression for the electric-field strength *inside* the solenoid of Example 27.11, a distance $r$ from the axis.

## Problems

35. A conducting loop of area $A$ and resistance $R$ lies at right angles to a spatially uniform magnetic field. At time $t = 0$ the magnetic field and loop current are both zero. Subsequently, the current increases according to $I = bt^2$, where $b$ is a constant with the units $A/s^2$. Find an expression for the magnetic-field strength as a function of time.

36. A conducting loop with area 0.15 $m^2$ and resistance 6.0 $\Omega$ lies in the x-y plane. A spatially uniform magnetic field points in the $z$ direction. The field varies with time according to $B_z = at^2 - b$, where $a = 2.0$ $T/s^2$ and $b = 8.0$ T. Find the loop current when (a) $t = 3.0$ s and (b) $B_z = 0$.

37. A square wire loop of side $l$ and resistance $R$ is pulled with constant speed $v$ from a region of no magnetic field until it is fully inside a region of constant, uniform magnetic field $\vec{B}$ perpendicular to the loop plane. The boundary of the field region is parallel to one side of the loop. Find an expression for the total work done by the agent pulling the loop.

38. A 5-turn coil 1.0 cm in diameter is rotated at 10 rev/s about an axis perpendicular to a uniform magnetic field. A voltmeter connected to the coil through rotating contacts reads a peak value of 360 $\mu$V. What is the magnetic-field strength?

39. A magnetic field is given by $\vec{B} = B_0(x/x_0)^2 \hat{k}$, where $B_0$ and $x_0$ are constants. Find an expression for the magnetic flux through a square of side $2x_0$ that lies in the x-y plane with one corner at the origin and two sides coinciding with the positive $x$ and $y$ axes.

40. A square wire loop 3.0 m on a side is perpendicular to a uniform magnetic field of 2.0 T. A 6-V lightbulb is in series with the loop, as shown in Fig. 27.38. The magnetic field is reduced steadily to zero over a time $\Delta t$. (a) Find $\Delta t$ such that the light will shine at full brightness during this time. (b) Which way will the loop current flow?

**FIGURE 27.38** Problem 40

41. In Example 27.2 take $a = 1.0$ cm, $w = 3.5$ cm, and $l = 6.0$ cm. Suppose the rectangular loop is a conductor with resistance 50 m$\Omega$ and the current $I$ in the long wire is increasing at the rate of 25 A/s. Find the induced current in the loop. In what direction does it flow?

42. A 2.0-m-long solenoid is 15 cm in diameter and consists of 2000 turns of wire. The current in the solenoid is increasing at the rate of 1.0 kA/s. (a) Find the current in a wire loop with diameter 10 cm and resistance 5.0 $\Omega$, lying inside the solenoid in a plane perpendicular to the solenoid axis. (b) Repeat for a similarly oriented loop with diameter 25 cm, lying entirely outside the solenoid.

43. A magnetic field is described by $\vec{B} = B_0 \sin \omega t \, \hat{k}$, where $B_0 = 2.0$ T and $\omega = 10$ $s^{-1}$. A conducting loop with area 150 $cm^2$ and resistance 5.0 $\Omega$ lies in the x-y plane. Find the induced current in the loop (a) at $t = 0$ and (b) at $t = 0.10$ s.

44. A uniform magnetic field is given by $\vec{B} = bt \, \hat{k}$, where $b = 0.35$ T/s. Find the current in a conducting loop with area 240 $cm^2$ and resistance 0.20 $\Omega$ that lies in the x-y plane. In what direction is the current, as viewed from the positive $z$ axis?

45. A car alternator consists of a 250-turn coil 10 cm in diameter in a magnetic field of 0.10 T. If the alternator is turning at 1000 rpm, what is its peak output voltage?

46. A generator consists of a rectangular coil 75 cm by 1.3 m, spinning in a 0.14-T magnetic field. If it is to produce a 60-Hz alternating emf (i.e., $\mathcal{E} = \mathcal{E}_0 \sin 2\pi ft$, where $f = 60$ Hz) with peak value 6.7 kV, how many turns must it have?

47. Figure 27.39 shows a pair of parallel conducting rails a distance $l$ apart in a uniform magnetic field $\vec{B}$. A resistance $R$ is connected across the rails, and a conducting bar of negligible resistance is being pulled along the rails with velocity $\vec{v}$ to the right. (a) What direction is the current in the resistor? (b) At what rate does the agent pulling the bar do work?

**FIGURE 27.39** Problems 47-50, and 74

48. The resistor in Problem 47 is replaced by an ideal voltmeter. (a) To which rail should the positive meter terminal be connected if the meter is to indicate a positive voltage? (b) At what rate does the agent pulling the bar do work?

49. A battery of emf $\mathcal{E}$ is inserted in series with the resistor in Fig. 27.39, with its positive terminal toward the top rail. The bar is initially at rest, and now no agent pulls it. (a) Describe the bar's subsequent motion. (b) The bar eventually reaches a constant speed. Why? (c) What is that constant speed? Express it in terms of the magnetic field, the battery emf, and the rail spacing $l$. Does the resistance $R$ affect the final speed? If not, what role does it play?

50. In Fig. 27.39, $l = 10$ cm, $B = 0.50$ T, $R = 4.0$ $\Omega$, and $v = 2.0$ m/s. Find (a) the current in the resistor, (b) the magnetic force on the bar, (c) the power dissipation in the resistor, and (d) the mechanical power supplied by the agent pulling the bar. Compare your answers to parts (c) and (d).

51. The magnetic field inside a solenoid of circular cross section is given by $\vec{B} = bt \, \hat{k}$, where $b = 2.1$ T/ms. At time $t = 0.40$ $\mu$s, a proton is inside the solenoid at the point $x = 5.0$ cm, $y = 0$, $z = 0$ and is moving with velocity $\vec{v} = 4.8 \times 10^6 \, \hat{j}$ m/s. Find the net electromagnetic force on the proton.

52. An electron is inside a solenoid, 28 cm from the solenoid axis. It experiences an electric force of magnitude 1.3 fN. At what rate is the solenoid's magnetic field changing?

53. A circular wire loop of resistance $R$ and radius $a$ lies with its plane perpendicular to a uniform magnetic field. The field strength changes from an initial value $B_1$ to a final value $B_2$. Show, by integrating the loop current over time, that the total charge that moves around the ring is

$$ Q = \frac{\pi a^2}{R}(B_2 - B_1) $$

Note that this result is independent of how the field changes with time.

54. A *flip coil* is a small coil used to measure magnetic fields. The flip coil is placed in a magnetic field with its plane perpendicular to the field, and then rotated abruptly through 180° about an axis in the plane of the coil. The coil is connected to instrumentation to measure the total charge $Q$ that flows during this process. If the coil has $N$ turns of area $A$ and if its rotation axis is perpendicular to the magnetic field, show that the field strength is given by $B = QR/2NA$, where $R$ is the coil resistance.

55. The current in a series $RL$ circuit rises to half its final value in 7.6 s. What is the time constant?

56. In a series $RL$ circuit like Fig. 27.22a, $\mathcal{E}_0 = 45$ V, $R = 3.3$ Ω, and $L = 2.1$ H. If the current is 9.5 A, how long has the switch been closed?

57. In Fig. 27.22a, $R = 2.5$ kΩ and $\mathcal{E}_0 = 50$ V. When the switch is closed, the current through the inductor rises to 10 mA in 30 μs. (a) What is the inductance? (b) What will the current in the circuit be after many time constants?

58. How long does it take to dissipate 90% of the magnetic energy in Example 27.10?

59. A series $RL$ circuit like Fig. 27.22a has $\mathcal{E}_0 = 60$ V, $R = 22$ Ω, and $L = 1.5$ H. Find the rate of change of the current (a) immediately after the switch is closed and (b) 0.10 s later.

60. Resistor $R_2$ in Fig. 27.40 is to limit the emf that develops when the switch is opened. What should its value be so that the inductor emf does not exceed 100 V?

**FIGURE 27.40** Problem 60

61. In Fig. 27.25, $\mathcal{E}_0 = 12$ V, $R = 2.7$ Ω, and $L = 20$ H. Initially the switch is in position B and there's no current anywhere. At $t = 0$ the switch is thrown to position A, and at $t = 10$ s it is thrown back to position B. Find the inductor current at (a) $t = 5.0$ s and (b) $t = 15$ s.

62. In Fig. 27.41, $\mathcal{E}_0 = 12$ V, $R_1 = 4.0$ Ω, $R_2 = 8.0$ Ω, $R_3 = 2.0$ Ω, and $L = 2.0$ H. Find the current $I_2$ (a) immediately after the switch is first closed and (b) a long time after the switch is closed. (c) After a long time the switch is reopened. Now what's $I_2$?

**FIGURE 27.41** Problem 62

63. A battery, switch, resistor, and inductor are connected in series. When the switch is closed, the current rises to half its steady-state value in 1.0 ms. How long does it take for the magnetic energy in the inductor to rise to half its steady-state value?

64. When a nonideal 1.0-H inductor is short-circuited, its magnetic energy drops to one-fourth of its original value in 3.6 s. What is its resistance?

65. A superconducting solenoid with inductance $L = 3.5$ H carries 1.8 kA. Copper is embedded in the coils to carry the current in the event of a quench (see Example 27.10). (a) What is the magnetic energy in the solenoid? (b) What is the maximum resistance of the copper that will limit the power dissipation to 100 kW immediately after a loss of superconductivity? (c) With this resistance, how long will it take the power to drop to 50 kW?

66. The magnetic field of a neutron star is about $10^8$ T. How does the energy density in this field compare with the energy density stored in (a) gasoline and (b) pure uranium-235 (mass density $19 \times 10^3$ kg/m³)? Consult Appendix C.

67. A single-turn loop of radius $R$ carries current $I$. How does the magnetic-energy density at the loop center compare with that of a long solenoid of the same radius, carrying the same current, and consisting of $n$ turns per unit length?

68. A wire of radius $R$ carries current $I$ distributed uniformly over its cross section. Find an expression for the total magnetic energy per unit length *within* the wire.

69. (a) Use Equation 27.8 to write an expression for the power dissipation in the resistor as a function of time, and (b) integrate from $t = 0$ to $t = \infty$ to show that the total energy dissipated is equal to the energy initially stored in the inductor—namely, $\frac{1}{2}LI_0^2$.

70. An electric field and a magnetic field have the same energy density. Obtain an expression for the ratio $E/B$, and evaluate this ratio numerically. What are its units? Is your answer close to any of the fundamental constants listed inside the front cover?

71. A rectangular conducting loop of resistance $R$, mass $m$, and width $w$ falls into a uniform magnetic field $\vec{B}$, as shown in Fig. 27.42. If the loop is long enough and the field region has a great enough vertical extent, the loop will reach a terminal speed. (a) Why? (b) Find an expression for the terminal speed. (c) What will the direction of the loop current be as the loop enters the field?

**FIGURE 27.42** Problem 71.

72. A conducting disk with radius $a$, thickness $h$, and resistivity $\rho$ is inside a solenoid of circular cross section. The disk axis coincides with the solenoid axis. The magnetic field in the solenoid is given by $B = bt$, with $b$ a constant. Find expressions for (a) the current density in the disk as a function of the distance $r$ from the disk center and (b) the rate of power dissipation in the entire disk. *Hint:* Consider the disk to be made up of infinitesimal conducting loops.

73. A small bar magnet is moved steadily from left to right along the axis of a wire loop, as in Fig. 27.5. Sketch qualitatively the current and power dissipation in the loop as functions of time, indicating the time the magnet passes through the plane of the loop. Take positive current in the loop as counterclockwise when viewed from the right.

74. The bar in Problem 47 has mass $m$ and is initially at rest. A constant force $\vec{F}$ to the right, parallel to the rails, is applied to the bar. Formulate Newton's second law for the bar, and find its velocity as a function of time.

75. A coil, consisting of $N$ turns of tightly wound wire carrying current $I$, has the shape of a toroid with average radius $A$ and circular cross section of smaller radius $R$. Its magnetic field has line symmetry about the axis of the toroid and is zero outside the toroid. Find the self-inductance by calculating the total magnetic flux in the toroid and using Equation 27.3.

76. Show that in the limit $R/A \ll 1$, the self-inductance of the toroid in Problem 75 is approximately that of a long, thin solenoid of length $2\pi A$ and cross-sectional area $\pi R^2$.

77. Use the node and loop laws to determine the current in $R_2$ as a function of time after the switch is first closed in the circuit in Example 27.9 and Figure 27.26a.

78. A long, straight coaxial cable consists of two thin tubular conductors, the inner of radius $a$ and the outer of radius $b$. Current $I$ flows out along one conductor and back along the other. Find the cable's self-inductance per unit length. *Hint:* This problem can be done using either a rectangular area between the conductors and Equation 27.3, or the volume of a cylindrical shell between the conductors and Equations 27.8 and 27.9.

79. You're evaluating the performance specs of several electrical devices. A large toroid is used to store magnetic energy. The toroid has an inner radius of 1.5 m and a square cross section 22.8 cm on each side. It's wound with 2500 turns of wire and carries a current of 63 A. The manufacturer claims the toroid will store 100 J of energy. Is this correct? *Hint:* Use Equation 26.21 and integrate.

80. You and your roommate are headed to Cancun for spring break. Your roommate, who has had only high school physics, has read that an emf can be induced in the wings of an airplane and wonders whether this would give enough voltage to power a portable music player. What's your answer? You make these assumptions: The wingspan of your 747 is 60 m, the plane is flying at 600 mi/h, and you estimate Earth's magnetic field at 0.2 G.

81. People with crazy inventions seem to seek you out. The latest idea is a windmill for generating electricity without actually using a generator. The designer suggests that conducting blades of length $l$ rotating with angular speed $\omega$ about a horizontal axis in the Earth's magnetic field could generate an emf. The horizontal component of the Earth's field at the windmill is $B_x$. Find an equation for the induced emf and then estimate its value assuming 50-m blades rotating at 1 rev/s, with 0.1 G for the Earth's horizontal field component.

82. Your ninth-grade nephew has asked you to come to his physical science class to explain how a credit card magnetic-strip reader works. You explain that as the strip is pulled past the reader's head, the magnetic field generates an emf in the reader's 2-mm diameter, 5000-turn coil. The induced voltage is 7.07 mV. To impress the students with the small field involved, calculate the rate at which the field changes, in $\mu$T/ms.

# Answers to Chapter Questions

## Answer to Chapter Opening Question

The minus sign, which expresses conservation of energy in electromagnetic induction.

## Answers to GOT IT? Questions

27.1 Opposite the direction shown in Fig. 27.10, but you'll still have to do work.

27.2 Counterclockwise.

27.3 It gets harder to turn. Constant rate implies a fixed peak emf, so lowering the resistance increases the current and therefore the power.

27.4 (b), because eddy currents dissipate some of its kinetic energy.

27.5 Changing current in the long wire produces an increasing magnetic field that points into the page at the loop. (a) The loop current opposes the increase in this field by producing a field in its interior that's out of the page. Therefore the loop current is counterclockwise. (b) The loop current opposes the decrease in the into-the-page field, so it's clockwise.

27.6 (a), because the inductor emf is in such a direction as to oppose the current supplied by the external circuit.

27.7 Left-hand meter reads $2IR$, right-hand meter reads $IR$—even though they're electrically connected to the same points. There's no contradiction because the field isn't conservative, and electric potential therefore can't be defined unambiguously.

Why does alternating current facilitate the transmission and distribution of electric power?

So far we've considered electric circuits energized by steady sources like batteries. But many circuits—from household power to audio and video signals to the "clock" that orchestrates events inside your computer—involve time-varying electrical quantities. Here we consider such **alternating-current** (AC) circuits.

## 28.1 Alternating Current

We saw in Chapter 27 how rotational motion in electric generators naturally leads to voltage and current that vary sinusoidally with time. Audio, video, and computer signals have more complicated time dependence, but, as we showed in Fig. 14.17, those signals can be analyzed as sums of sinusoidal terms. Studying circuits with sinusoidally varying electrical quantities therefore provides insights into all AC circuits.

A sinusoidal AC voltage or current is characterized by its amplitude, frequency, and phase constant—the same quantities we developed in Chapter 13 to describe simple harmonic motion. Amplitude is specified by the peak value $(V_p, I_p)$ or the **root-mean-square** value $(V_{rms}, I_{rms})$. The rms is an average obtained by squaring the signal, taking the time average, and then taking the square root. For a sine wave, rms and peak values are related by

$$V_{rms} = \frac{V_p}{\sqrt{2}} \quad \text{and} \quad I_{rms} = \frac{I_p}{\sqrt{2}} \tag{28.1}$$

The 120 V of household wiring, for example, is the rms value (see Fig. 28.1).

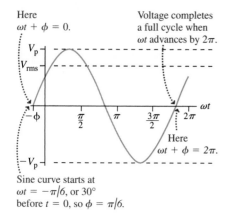

Here
$\omega t + \phi = 0$.

Voltage completes
a full cycle when
$\omega t$ advances by $2\pi$.

Here
$\omega t + \phi = 2\pi$.

Sine curve starts at
$\omega t = -\pi/6$, or 30°
before $t = 0$, so $\phi = \pi/6$.

**FIGURE 28.1** A sinusoidally varying AC voltage, showing peak and rms amplitudes and phase $\phi$.

In practical situations we usually describe frequency $f$ in cycles per second, or hertz (Hz). In mathematical analysis it's more convenient to use the angular frequency $\omega$ in radians per second or, equivalently, inverse seconds $(\text{s}^{-1})$. The relation between the two

$$\omega = 2\pi f \qquad (28.2)$$

is the same as for rotational and simple harmonic motion, and for the same reason: A full cycle contains $2\pi$ radians.

The phase constant $\phi$ of an AC signal tells when the sine curve crosses zero with positive slope (Fig. 28.1). A full mathematical description of an AC voltage or current then includes its amplitude $(V_p, I_p)$, frequency $(\omega)$, and phase constant $(\phi)$:

$$V = V_p \sin(\omega t + \phi_V) \quad \text{and} \quad I = I_p \sin(\omega t + \phi_I) \qquad (28.3)$$

Here we've labeled the phase constants with subscripts $V$ and $I$ to indicate that voltage and current—even in the same circuit element—do not need to have the same phase.

---

**EXAMPLE 28.1** AC: Characterizing Household Voltage

Standard household wiring in North America supplies 120 V rms at 60 Hz. Express this mathematically in the form of Equation 28.3, assuming the voltage is rising through zero at time $t = 0$.

**INTERPRET** We're given an AC voltage in "practical" units, and we're supposed to express it in the more mathematical form of Equation 28.3. We identify 120 V as the amplitude $V_{rms}$, 60 Hz as the frequency $f$, and the information about timing as describing the phase.

**DEVELOP** Equation 28.3, $V = V_p \sin(\omega t + \phi_V)$, contains the peak amplitude $V_p$ and angular frequency $\omega$. Equations 28.1, $V_{rms} = V_p/\sqrt{2}$, and 28.2, $\omega = 2\pi f$, determine these quantities from the values we're given.

**EVALUATE** Equation 28.1 gives $V_p = \sqrt{2}V_{rms} = (\sqrt{2})(120\text{ V}) = 170\text{ V}$, and Equation 28.2 gives $\omega = 2\pi f = (2\pi)(60\text{ Hz}) = 377\text{ s}^{-1}$. We don't have an equation for phase, but the fact that the sine curve rises through zero at $t = 0$ tells us that $\phi = 0$. So Equation 28.3's description of this AC voltage becomes $V = 170 \sin(377t)$ V.

**ASSESS** Make sense? Both the peak voltage and the angular frequency are numerically greater than their more familiar counterparts. That's because the rms voltage is a kind of average, lower than the peak, and because the angular frequency measures radians per second rather than full cycles. Incidentally, wires entering your house actually carry 240 V rms, which is split into separate 120-V circuits except for major appliances like stoves, dryers, and water heaters; these operate at the full 240 V rms. In Europe, standard household voltage is 230 V rms at 50 Hz, and much of the rest of the world uses 220-V, 50-Hz power. ∎

---

## 28.2 Circuit Elements in AC Circuits

Here we examine separately the AC behavior of resistors, capacitors, and inductors so we can subsequently understand what happens when we combine these elements in AC circuits.

### Resistors

An ideal resistor is a device whose current and voltage are always proportional: $I = V/R$. Figure 28.2 shows a resistor $R$ connected across an AC generator, making the voltage across the resistor equal to the generator voltage. The generator voltage is described by Equation 28.3, where we take $\phi_V = 0$. Then the current is

$$I = \frac{V}{R} = \frac{V_p \sin \omega t}{R} = \frac{V_p}{R} \sin \omega t$$

The current has the same frequency as the voltage, and, since its phase constant is also zero, the voltage and current are *in phase*; that is, they peak at the same time. The peak current is the peak voltage divided by the resistance: $I_p = V_p/R$. Because both voltage and current are sinusoidal, their rms values are in the same ratio as their peak values; thus $I_{rms} = V_{rms}/R$.

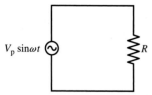

$V_p \sin\omega t$

$R$

**FIGURE 28.2** A resistor connected across an AC generator (symbol ⊚).

## Capacitors

Figure 28.3 shows a capacitor connected across an AC generator. In Chapter 23, we saw that voltage and charge are directly proportional in a capacitor: $q = CV$. Differentiating this relation gives

$$\frac{dq}{dt} = C\frac{dV}{dt}$$

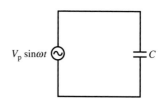

**FIGURE 28.3** A capacitor across an AC generator.

But $dq/dt$ is the current flowing to the capacitor plates (which we'll call the "capacitor current" even though charge doesn't actually flow through the space between the plates). So we have $I = C(dV/dt)$. The generator voltage $V_p \sin\omega t$ appears directly across the capacitor, so

$$I = C\frac{d}{dt}(V_p \sin\omega t)$$

$$= \omega CV_p \cos\omega t = \omega CV_p \sin\left(\omega t + \frac{\pi}{2}\right) \tag{28.4}$$

Because the cosine curve is just a sine curve shifted to the left by $\pi/2$ or 90°, Equation 28.4 tells us that **the current in a capacitor leads the voltage across the capacitor by 90°** (see Fig. 28.4).

The term $\omega CV_p$ multiplying the cosine in Equation 28.4 is the amplitude of the current, so we can write $I_p = \omega CV_p$ or, in a form resembling Ohm's law,

$$I_p = \frac{V_p}{1/\omega C} = \frac{V_p}{X_C} \tag{28.5}$$

**FIGURE 28.4** The current in a capacitor leads the voltage by one-fourth of a cycle, $\pi/2$ radians or 90°.

where we've defined $X_C = 1/\omega C$.

Equation 28.5 shows that the capacitor acts somewhat like a resistance $X_C = 1/\omega C$. But not quite! This "resistance" gives the relation between peak voltage and current, but it doesn't tell the whole story. The capacitor also introduces a phase difference between voltage and current. This phase difference reflects a fundamental physical difference between resistors and capacitors. A resistor dissipates electric energy as heat. A capacitor stores and releases electric energy. Over a complete cycle, the agent turning the generator in Fig. 28.3 does no net work, while the agent turning the generator with the resistive load of Fig. 28.2 continuously does work that gets dissipated as heat in the resistor. We give the quantity $X_C$ in Equation 28.5 the name **capacitive reactance**. Like resistance, reactance is measured in ohms $(\Omega)$.

Does it make sense that $X_C$ depends on frequency? Yes. As frequency goes to zero, $X_C$ goes to infinity. At zero frequency nothing is changing; there's no charge moving on or off the plates, and the capacitor might as well be an open circuit. As frequency increases, larger currents flow to move charge on and off the plates in ever-shorter times, so the capacitor looks increasingly like a short circuit. To summarize, a capacitor at low frequencies acts like an open circuit, while at high frequencies it acts like a short circuit.

Why does the capacitor current *lead* the voltage? Because the capacitor voltage is proportional to its charge, and it takes current to move charge onto the capacitor plates. Therefore current flows *before* the voltage changes significantly.

## Inductors

Figure 28.5 shows an inductor connected across an AC generator. The loop law for this circuit is $V_p \sin\omega t + \mathcal{E}_L = 0$. From Chapter 27 we know that the inductor emf is $\mathcal{E}_L = -L(dI/dt)$, so the loop law becomes

$$V_p \sin\omega t = L\frac{dI}{dt}$$

**FIGURE 28.5** An inductor across an AC generator.

To obtain a relation involving the current $I$ rather than its derivative, we integrate:

$$\int V_p \sin\omega t\, dt = \int L\frac{dI}{dt}dt = \int L\, dI$$

The integral of sine is the negative cosine, so

$$-\frac{V_p}{\omega}\cos\omega t = LI$$

Here we've set the integration constants to zero because nonzero values would represent a DC emf and current that aren't in this circuit. Solving for $I$ gives

$$I = -\frac{V_p}{\omega L}\cos\omega t = \frac{V_p}{\omega L}\sin\left(\omega t - \frac{\pi}{2}\right) \qquad (28.6)$$

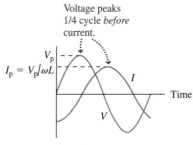

Voltage peaks 1/4 cycle *before* current.

$I_p = V_p/\omega L$

**FIGURE 28.6** The voltage across an inductor leads the current by $\pi/2$ or 90°.

where the last step follows because $\sin(\alpha - \pi/2) = -\cos\alpha$ for any $\alpha$.

Equation 28.6 shows that current in the inductor lags the voltage by $\pi/2$ or 90°. Equivalently, **the voltage across an inductor leads the inductor current by 90°** (Fig. 28.6). Equation 28.6 also shows that the peak current is

$$I_p = \frac{V_p}{\omega L} = \frac{V_p}{X_L} \qquad (28.7)$$

Again, this equation resembles Ohm's law, with **inductive reactance** $X_L = \omega L$. As with the capacitor, no power is dissipated; instead, energy is alternately stored and released as the inductor's magnetic field builds and decays.

Does it make sense that inductive reactance increases with increasing $\omega$ and $L$? Through its induced back emf, an inductor opposes changes in current. The greater the inductance, the greater the opposition. And the more rapidly the current is changing, the more vigorously the inductor opposes the change, so the inductive reactance increases at high frequencies. At very high frequencies, an inductor looks like an open circuit. But at very low frequencies, it looks more and more like a short circuit, until with direct current (zero frequency), an inductor exhibits zero reactance because there's no change in current.

Why does the inductor voltage *lead* the current? Because a changing current in an inductor induces an emf. *Before* the current can build up significantly, there must first, therefore, be voltage across the inductor.

Table 28.1 summarizes the amplitude and phase relations in resistors, capacitors, and inductors.

**TABLE 28.1** Amplitude and Phase Relations in Circuit Elements

| Circuit Element | Peak Current versus Voltage | Phase Relation |
|---|---|---|
| Resistor | $I_p = \dfrac{V_p}{R}$ | $V$ and $I$ in phase |
| Capacitor | $I_p = \dfrac{V_p}{X_C} = \dfrac{V_p}{1/\omega C}$ | $I$ leads $V$ by 90° |
| Inductor | $I_p = \dfrac{V_p}{X_L} = \dfrac{V_p}{\omega L}$ | $V$ leads $I$ by 90° |

**GOT IT? 28.1** A capacitor and an inductor are connected across separate but identical electric generators, and the same current flows in each. If the frequency of the generators is doubled, which component will carry more current?

**EXAMPLE 28.2** **Inductors and Capacitors: Equal Currents?**

A capacitor is connected across a 60-Hz, 120-V rms power line, and an rms current of 200 mA flows. (a) Find the capacitance. (b) What inductance, connected across the same power line, would result in the same current? (c) How would the phases of the inductor and capacitor currents compare?

**INTERPRET** We're being asked about the relation between AC voltage and current in capacitors and inductors. The idea here is that the voltage–current relation depends not only on the component values but also on frequency, and it involves phase as well as amplitude.

**DEVELOP** Equations 28.5, $I_{C_p} = V_{C_p}\omega C$, and 28.7, $I_{L_p} = V_{L_p}/\omega L$, relate the peak current and voltage in the two devices. Since rms and peak values are proportional, these equations also relate rms current and voltage. The equations and associated phase relations also appear in Table 28.1.

**EVALUATE** (a) For the capacitor, we know the voltage and current. Equation 28.5 then gives $C = I_{Crms}/\omega V_{Crms} = 4.42\ \mu F$, where we used $I_{Crms} = 0.20$ A and $\omega = 2\pi f = 377\ \text{s}^{-1}$ as we found in Example 28.1 for 60-Hz AC power. (b) For an inductor to pass the same current, it must have the same reactance; comparing Equations 28.5 and 28.7 shows that $\omega L = 1/\omega C$, or

$$L = \frac{1}{\omega^2 C} = \frac{1}{(377\ \text{s}^{-1})^2 (4.42\ \mu F)} = 1.6\ \text{H}$$

(c) Table 28.1 shows that the capacitor current *leads* the voltage by 90°, while the inductor current *lags* by 90°; therefore, the capacitor and inductor currents must be out of phase by 180° or $\pi$ radians.

**ASSESS** Our expression for $L$ shows that a larger capacitor would require a smaller inductor for the same current. That's because a larger capacitor has *lower* reactance and so passes more current at a given frequency. But an inductor is the opposite: A larger inductor has *higher* reactance. So at a fixed frequency, the inductance required for a given current scales inversely with the capacitance required for the same current. ∎

## Phasor Diagrams

Phase and amplitude relations in AC circuits may be summarized graphically in **phasor diagrams**. A **phasor** is an arrow whose fixed length represents the amplitude of an AC voltage or current. The phasor rotates counterclockwise about the origin with the angular frequency $\omega$ of the AC quantity. The component of the phasor on either axis can then represent the sinusoidally varying AC signal. We'll use the vertical axis to be consistent with Equation 28.3. Others, especially electrical engineers, may use the horizontal axis.

Figure 28.7a shows phasors for the current and voltage in a resistor. The lengths of the phasors are related by Ohm's law, $V_p = I_p R$. The current and voltage phasors point in the same direction, showing that current and voltage in the resistor are in phase. For the capacitor and inductor, the two phasors are at right angles, indicating the phase relations between current and voltage (Fig. 28.7b, c). Their lengths are related by the appropriate reactance, so $V_p = I_p X$. As phasors rotate, their vertical components would trace out current and voltage graphs like those of Figs. 28.4 and 28.6. You should convince yourself that all the relations of Table 28.1 are correctly described by the phasor diagrams of Fig. 28.7. Although phasor diagrams don't add much to our understanding of AC circuits that contain only one circuit element, they will greatly simplify the analysis of more complicated circuits.

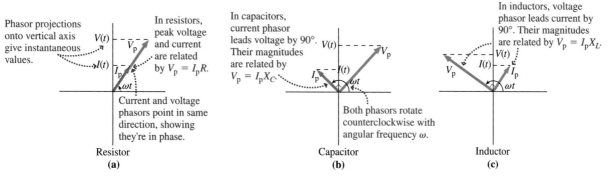

**FIGURE 28.7** Phasor diagrams showing voltage and current in (a) a resistor, (b) a capacitor, and (c) an inductor.

## Capacitors and Inductors: A Comparison

Capacitors and inductors are complementary devices. A capacitor opposes instantaneous changes in voltage, while an inductor opposes instantaneous changes in current. In an *RC* circuit, voltage builds up across the capacitor. In the analogous *RL* circuit, current builds up in the inductor. Similar curves describe capacitor voltage and inductor current over time. A capacitor stores electric energy given by $\frac{1}{2}CV^2$. An inductor stores magnetic energy given by $\frac{1}{2}LI^2$. A capacitor acts like an open circuit at low frequencies; an inductor like a short circuit at low frequencies. Each exhibits the opposite behavior at high frequencies. These comparisons reflect a deeper complementarity between electric and magnetic fields. Any verbal description of a capacitor applies to an inductor if we replace the words "capacitor" with "inductor," "electric" with "magnetic," and "voltage" with "current." Table 28.2 summarizes the complementary aspects of capacitors and inductors.

**TABLE 28.2** Capacitors and Inductors

|  | Capacitor | Inductor |
|---|---|---|
| Defining relation | $C = \dfrac{q}{V}$ | $L = \dfrac{\phi_B}{I}$ |
| Defining relation; differential form | $I = C\dfrac{dV}{dt}$ | $\mathcal{E} = -L\dfrac{dI}{dt}$ |
| Opposes changes in | Voltage | Current |
| Energy storage | In electric field | In magnetic field |
|  | $U = \frac{1}{2}CV^2$ | $U = \frac{1}{2}LI^2$ |
| Behavior in low-frequency limit | Open circuit | Short circuit |
| Behavior in high-frequency limit | Short circuit | Open circuit |
| Reactance | $X_C = 1/\omega C$ | $X_L = \omega L$ |
| Phase | Current leads by 90° | Voltage leads by 90° |

---

APPLICATION **Loudspeaker Systems**

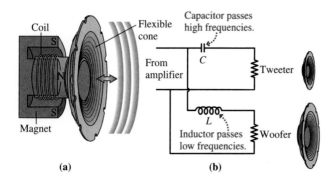

Capacitor passes high frequencies.

From amplifier

*C*

Tweeter

Inductor passes low frequencies.

*L*

Woofer

**(a)**  **(b)**

Loudspeakers convert electrical energy to sound, using the magnetic force on a coil that fits loosely around a permanent magnet. Part (a) of the figure shows that the coil is attached to a flexible cone. Both cone and coil move back and forth as AC current corresponding to the audio signal flows in the coil. The moving cone disturbs the air, producing sound waves. (In the figure, the size of the magnet and coil are exaggerated for clarity.)

Good loudspeaker systems include at least two separate units. A small *tweeter* produces high-frequency sound, while a larger, more massive *woofer* handles the low frequencies. A *crossover network* uses inductors and capacitors to "steer" the high- and low-frequency signals to the appropriate speakers. As the circuit diagram in part (b) shows, an inductor in series with the woofer blocks high frequencies but lets low frequencies pass unimpeded; a capacitor in series with the tweeter does the opposite. This circuit is an example of a *filter*, used in electronic systems to pass preferentially a range of frequencies. Bass and treble controls on audio systems provide another example, varying filter properties to boost low or high frequencies as desired.

## 28.3 *LC* Circuits

Suppose we charge a capacitor to some voltage $V_p$ and corresponding charge $q_p$, and then connect it across an inductor, as shown in Fig. 28.8. The capacitor contains stored electric energy, but initially there's no current in the inductor and so no stored magnetic energy (Fig. 28.9a). The capacitor begins to discharge through the inductor, but slowly at first because the inductor opposes changes in current. Gradually the current rises, and with it the magnetic energy in the inductor. The capacitor voltage, charge, and stored energy decrease. At some time the initial energy is divided equally between capacitor and inductor (Fig. 28.9b). But the capacitor keeps discharging, eventually reaching zero charge (Fig. 28.9c). Now all the energy that was originally in the electric field of the capacitor is in the magnetic field of the inductor.

Does everything stop at this point? No, because there's current in the inductor, and inductor current can't change instantaneously. So the current keeps flowing and begins piling positive charge on the bottom plate of the capacitor (Fig. 28.9d). Stored electric energy increases, and current and magnetic energy both decrease. Eventually the capacitor is fully charged but opposite its initial state (Fig. 28.9e). Again all the energy is in the capacitor. Now the capacitor begins to discharge, and the process repeats, with a counterclockwise current (Fig. 28.9f). All the energy is transferred to the inductor (Fig. 28.9g), and then back to the capacitor (Fig. 28.9a again). The circuit is now back to its initial state. Provided there's no energy loss, the oscillation repeats indefinitely.

**FIGURE 28.8** An *LC* circuit.

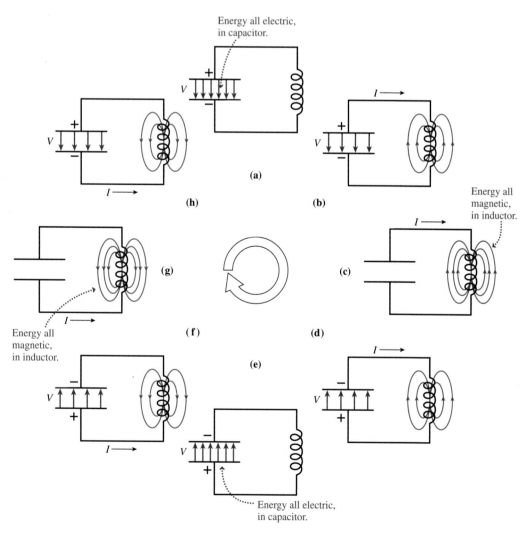

**FIGURE 28.9** *LC* oscillations transfer energy between electric and magnetic fields.

Capacitor $C$ corresponds to spring $k$.

Current $I$ corresponds to velocity $v$.

Inductor $L$ corresponds to mass $m$.

**FIGURE 28.10** An *LC* circuit is the electrical analog of a mass–spring system.

This *LC* oscillation should remind you of the mass–spring system of Chapter 13. There, energy went back and forth between kinetic energy of the mass and potential energy of the spring. Here, energy goes between magnetic energy of the inductor and electric energy of the capacitor. The mass–spring system oscillates with frequency determined by the mass $m$ and spring constant $k$. The *LC* circuit oscillates with frequency determined by the inductance $L$ and capacitance $C$, as we'll show next. Figure 28.10 illustrates this analogy between the mass–spring system and the *LC* circuit.

## Analyzing the *LC* Circuit

The total energy in the *LC* circuit is the sum of magnetic and electric energy:

$$U = U_B + U_E = \tfrac{1}{2}LI^2 + \tfrac{1}{2}CV^2$$

In an ideal *LC* circuit this quantity doesn't change, so its derivative is zero:

$$\frac{dU}{dt} = \frac{d}{dt}\left(\tfrac{1}{2}LI^2 + \tfrac{1}{2}CV^2\right) = 0$$

Carrying out the differentiation, we have

$$LI\frac{dI}{dt} + CV\frac{dV}{dt} = 0$$

We substitute $V = q/C$, $dV/dt = (1/C)\,dq/dt$, $I = dq/dt$, and $dI/dt = d^2q/dt^2$ and then divide by $I$ to get

$$L\frac{d^2q}{dt^2} + \frac{1}{C}q = 0 \tag{28.8}$$

This is a differential equation describing capacitor charge $q$ as a function of time. We encountered a similar equation in Chapter 13 for the mass–spring system: $m(d^2x/dt^2) + kx = 0$. The solution there was a sinusoidal oscillation with angular frequency $\omega = \sqrt{k/m}$. In Equation 28.8, $L$ replaces $m$ and $1/C$ replaces $k$; therefore the solution to Equation 28.8 is a sinusoidal oscillation:

$$q = q_\mathrm{p}\cos\omega t \tag{28.9}$$

with angular frequency

$$\omega = \frac{1}{\sqrt{LC}} \tag{28.10}$$

**FIGURE 28.11** Electric and magnetic energies in an *LC* circuit sum to a constant total energy.

Equation 28.9 readily provides other electrical quantities in the *LC* circuit. Using $q = CV$ gives the voltage, and differentiating gives $I = dq/dt$. From there you can get the electric and magnetic energies, $U_E = \tfrac{1}{2}CV^2$ and $U_B = \tfrac{1}{2}LI^2$. Sum them to verify that the total energy remains constant (Fig. 28.11; see Problem 62 for details).

---

**GOT IT? 28.2** You have an *LC* circuit that oscillates at a typical AM radio frequency of 1 MHz. You want to change the capacitor so it oscillates at a typical FM frequency, 100 MHz. Should you make the capacitor (a) larger or (b) smaller? By what factor?

---

**EXAMPLE 28.3** An *LC* Circuit: Tuning a Piano

You wish to make an *LC* circuit oscillate at 440 Hz (A above middle C) to use in tuning pianos. You have a 20-mH inductor. (a) What value of capacitance should you use? (b) If you charge the capacitor to 5.0 V, what will be the peak current in the circuit?

**INTERPRET** This problem is about designing an *LC* circuit for a given frequency; in (b) we want the peak current, that is, the current when all the energy is in the inductor.

**DEVELOP** Equation 28.10, $\omega = 1/\sqrt{LC}$, relates frequency, capacitance, and inductance; we're given $L$ and $f$. With $\omega = 2\pi f$, we can solve for $C$. We've recognized that the peak current comes when all the energy is the magnetic energy $\frac{1}{2}LI^2$ of the inductor. Given the initial capacitor voltage, we can equate this with the initial electric energy $\frac{1}{2}CV^2$ and solve for $I$.

**EVALUATE** Equation 28.10 gives $C = 1/\omega^2 L = 1/4\pi^2 f^2 L = 6.54\ \mu\text{F}$, where $f = 440$ Hz and $L = 20$ mH. Now that we know $C$, we equate the peak magnetic energy with the peak electric energy to get $\frac{1}{2}LI^2 = \frac{1}{2}CV^2$. Solving gives

$$I = \sqrt{\frac{C}{L}}\,V = \sqrt{\frac{6.54\ \mu\text{F}}{20\ \text{mH}}}\,(5.0\ \text{V}) = 90\ \text{mA}$$

**ASSESS** Our expression shows that the higher the initial voltage, the greater the current. Obviously, the higher the voltage, the greater the energy, so the greater the current needed when this energy becomes all magnetic. A larger capacitance also raises the electric energy $\frac{1}{2}CV^2$, while a larger inductance lowers the current needed to achieve the same magnetic energy $\frac{1}{2}LI^2$. ∎

## Resistance in *LC* Circuits—Damping

Real inductors, capacitors, and wires have resistance (Fig. 28.12). If the resistance is low enough that only a small fraction of the energy is lost in each cycle, then the analysis in the preceding discussion applies. The circuit oscillates at a frequency given very nearly by Equation 28.10, but the oscillation amplitude slowly declines as energy is dissipated in the resistance.

We can analyze an *RLC* circuit by evaluating $dU/dt$ as before, this time setting the result not to zero but to the rate of energy dissipation, $-I^2R$, where the minus sign indicates energy loss from the circuit:

$$\frac{dU}{dt} = \frac{d}{dt}\left(\tfrac{1}{2}LI^2 + \tfrac{1}{2}CV^2\right) = -I^2R$$

Making the same substitutions as before leads to

$$L\frac{d^2q}{dt^2} + R\frac{dq}{dt} + \frac{q}{C} = 0$$

This is mathematically identical to Equation 13.16 for damped harmonic motion, with $L$ again replacing $m$, $1/C$ replacing $k$, and now $R$ replacing the damping constant $b$. The solution follows by analogy with Equation 13.17, which is the solution to Equation 13.16:

$$q(t) = q_p e^{-Rt/2L}\cos\omega t \qquad (28.11)$$

Voltage and current behave similarly, with oscillation amplitude decaying exponentially with time constant $2L/R$ (Fig. 28.13).

As the resistance increases, oscillations decay more rapidly and the frequency of oscillation decreases. Eventually, when the time constant $2L/R$ equals the inverse of the frequency given in Equation 28.10, we have **critical damping**. Then all circuit quantities decay to zero without oscillation, just as we found for mechanical systems. In circuits designed to oscillate, like radio transmitters or TV tuners, engineers obviously want to minimize damping. But in situations where oscillations would be a nuisance, it's important that circuits have enough resistance to suppress oscillation.

## 28.4 Driven *RLC* Circuits and Resonance

Figure 28.14 shows an *RLC* circuit connected across an AC generator. Adding the generator is like adding the external driving force on the mechanical oscillator that we considered in Section 13.7. We'll call the generator frequency $\omega_d$, the driving frequency, just as we did in Chapter 13. Pursuing the mechanical analogy, we expect the driven *RLC* circuit to exhibit resonant behavior as we discussed in Section 13.7. Such electrical resonance is crucial to the operation of radio, TV, and other frequency-specific devices.

**FIGURE 28.12** An *RLC* circuit.

**FIGURE 28.13** An oscilloscope displays the capacitor voltage in an *RLC* circuit.

**FIGURE 28.14** A series *RLC* circuit driven by an AC generator.

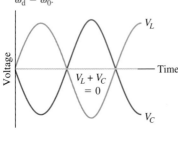

**FIGURE 28.15** Capacitor and inductor voltages are 180° out of phase, but their relative magnitudes vary with frequency.

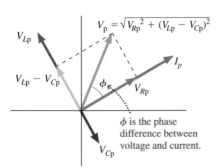

**FIGURE 28.16** Phasor diagram for the driven *RLC* circuit.

## Resonance in the *RLC* Circuit

Suppose we vary the generator frequency $\omega_d$ in Fig. 28.14 while keeping the generator's peak voltage constant. At low frequencies the capacitor acts almost like an open circuit (its reactance $X_C = 1/\omega C$ is large), so little current flows. At high frequencies the inductor acts almost like an open circuit (its reactance $X_L = \omega L$ is large), so little current flows. At some intermediate frequency the current must be a maximum. We now show that this **resonant frequency** is the undamped natural frequency $\omega_0 = 1/\sqrt{LC}$.

Figure 28.14 is a series circuit, so the *same* current flows through all components. The voltage in a capacitor lags the current by 90°, while the voltage in an inductor leads by 90°. Since the same current flows in both components, the inductor and capacitor voltages are therefore 180° out of phase and thus they tend to cancel (Fig. 28.15). But that cancellation is complete only when the two voltages have the same peak value. Since the current is the same in both components, comparison of Equations 28.5 for the capacitor, $I_p = V_p/X_C$, and 28.7 for the inductor, $I_p = V_p/X_L$, shows that the peak voltages are the same when the capacitive reactance $X_C = 1/\omega C$ equals the inductive reactance $X_L = \omega L$. Equating the reactances gives

$$\omega_0 = \frac{1}{\sqrt{LC}} \qquad \text{(resonant frequency)}$$

This is precisely the undamped natural frequency of Equation 28.10.

At resonance the capacitor and inductor voltages completely cancel. The voltage across the pair together is zero and—at the resonant frequency only—the pair might just as well be a wire. At resonance the resistance alone determines the circuit current. At any other frequency the effects of capacitance and inductance don't cancel, and the current is lower.

---

**GOT IT? 28.3** You measure the capacitor and inductor voltages in a driven *RLC* circuit, and find 10 V for the rms capacitor voltage and 15 V for the rms inductor voltage. Is the driving frequency (a) above or (b) below resonance?

---

## Frequency Response of the *RLC* Circuit

Here we'll use a phasor diagram (recall Section 28.2) to find the current in an *RLC* circuit as a function of the driving frequency (Fig. 28.16). Since the same current flows through all elements of this series circuit, a single phasor of length $I_p$ represents the current. The resistor voltage is in phase with the current, so its phasor, $V_{Rp}$, is in the same direction as $I_p$. But the inductor voltage leads the current and the capacitor voltage lags, each by 90°, so their phasors, $V_{Lp}$ and $V_{Cp}$, are perpendicular to the current. At each instant the three voltages sum to give the generator voltage; Fig. 28.16 shows that this sum has magnitude $V_p = \sqrt{V_{Rp}^2 + (V_{Lp} - V_{Cp})^2}$. Expressing this in terms of the common current $I_p$ and the resistance and reactances gives $V_p = \sqrt{I_p^2 R^2 + (I_p X_L - I_p X_C)^2}$. Solving for $I_p$, we then have

$$I_p = \frac{V_p}{\sqrt{R^2 + (X_L - X_C)^2}} = \frac{V_p}{Z} \qquad (28.12)$$

where we've defined the **impedance**, $Z$, as $Z = \sqrt{R^2 + (X_L - X_C)^2}$. Impedance is a generalization of resistance to include the frequency-dependent effects of capacitance and inductance. Equation 28.12 is the corresponding generalization of Ohm's law. Impedance is lowest when $X_L = X_C$ or $\omega = 1/\sqrt{LC}$ as we argued earlier; then it's equal to the resistance alone. But $Z$ becomes large at high frequencies, where $X_L = \omega L$ becomes large, and at low frequencies, where $X_C = 1/\omega C$ is large.

Figure 28.17 plots resonance curves from Equation 28.12, showing peak current versus frequency for three resistance values. At low resistance, the curve peaks sharply. Such a

**high-Q** (for high-quality) circuit does a good job distinguishing its resonance frequency from nearby frequencies. High-Q circuits are important in applications such as radio, TV, and cell phones, where many signals occupy nearby frequencies without interfering. With higher resistance, the resonance curve broadens and the circuit responds to a range of frequencies; such a circuit has low Q. Problem 71 gives a rigorous definition of Q.

Equation 28.12 relates peak current and voltage in the *RLC* circuit, but it doesn't tell the whole story. As Fig. 28.16 shows, current and voltage are out of phase by the angle $\phi$. Trigonometry gives $\tan \phi = (V_{Lp} - V_{Cp})/V_{Rp}$ or, since voltages are proportional to reactances and resistance,

$$\tan \phi = \frac{X_L - X_C}{R} = \frac{\omega L - 1/\omega C}{R} \qquad (28.13)$$

where $\phi = \phi_V - \phi_I$ is the phase difference between voltage and current. Positive $\phi$ means voltage leads current; negative $\phi$ means current leads voltage.

At resonance, $X_L = X_C$ and $\phi = 0$. Here capacitance and inductance cancel, and the circuit behaves like a pure resistance. At low frequencies, capacitive reactance dominates; here $\phi$ is negative and the current leads the voltage. This is just what we expect in a capacitive circuit. The opposite is true at high frequencies, where the inductor dominates. Figure 28.18 shows the phase difference as a function of frequency for three resistance values.

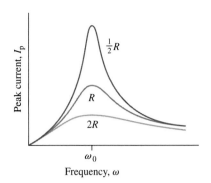

**FIGURE 28.17** Resonance curves for an *RLC* circuit with three different resistances.

✓ **TIP** Phase Matters

You can't analyze AC circuits by treating resistors, capacitors, and inductors all as "resistors" with resistances $R$, $X_C$, and $X_L$. That's because each component has a different phase relation between current and voltage. Phasor diagrams correctly account for these relations, which show up in the minus sign joining capacitive and inductive reactance, and in the Pythagorean addition of resistance and reactance in Fig. 28.16 and Equation 28.12.

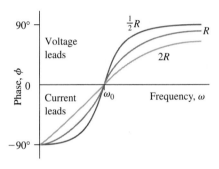

**FIGURE 28.18** Phase relations for the *RLC* circuits whose resonance curves are shown in Fig. 28.17.

---

**EXAMPLE 28.4** An *RLC* Circuit: Designing a Loudspeaker System

Current flows to the midrange speaker in a loudspeaker system through a 2.2-mH inductor in series with a capacitor. (a) What should the capacitance be so that a given voltage produces the greatest current at 1 kHz? (b) If the same voltage produces half this current at 618 Hz, what is the speaker's resistance? (c) If the peak output voltage of the amplifier is 20 V, what will the peak capacitor voltage be at 1 kHz?

**INTERPRET** This is a problem about the peak current and voltage in a series *RLC* circuit, where we identify the speaker as $R$ and the amplifier as the generator in Fig. 28.14.

**DEVELOP** The peak current is at the resonant frequency of Equation 28.10, $\omega = 1/\sqrt{LC}$, so in (a) we can solve this equation for $C$. Equation 28.12 relates peak voltage and current to the component values and the frequency, so in (b) we can solve for $R$. In (c) we'll need to find the current at 1 kHz, and then use Equation 28.5, $I_p = V_p/X_C$, which relates peak voltage and current in a capacitor, to find $V_{Cp}$.

**EVALUATE** (a) We solve Equation 28.10 for $C$, using $\omega = 2\pi f$: $C = 1/[(2\pi f)^2 L]$; with $f = 1000$ Hz and $L = 2.2$ mH, this gives $C = 11.5$ μF. (b) Equation 28.12 shows that we'll have half the peak current when $Z$ is twice its value $Z = R$ at resonance. So we want $Z = \sqrt{R^2 + (X_L - X_C)^2} = 2R$. Squaring and solving for $R$ give

$$R = \frac{1}{\sqrt{3}}\left|\omega L - \frac{1}{\omega C}\right| = 8.0 \ \Omega$$

where we used $X_L = \omega L$ and $X_C = 1/\omega C$ for the reactances. For (c), note that the impedance at the 1-kHz resonant frequency is just $R$, so the peak current is $I_p = V_p/R$. Then Equation 28.5 gives the peak capacitor voltage:

$$V_{Cp} = I_p X_C = \left(\frac{V_p}{R}\right)\left(\frac{1}{\omega C}\right) = 35 \ \text{V}$$

where $V_p$ is the 20-V peak voltage applied to the circuit and $\omega = 2\pi f$ with $f = 1$ kHz. (This answer is for the 1-kHz resonant frequency; Problem 76 shows that $V_{Cp}$ is actually somewhat higher than 35 V at frequencies just below resonance.)

**ASSESS** Our 35-V answer for (c) is greater than the 20-V peak output of the amplifier, so how can it be right? Remember that there's another source of emf in the circuit—the inductor, whose emf depends on the rate of change of current. Although the capacitor and inductor voltages cancel at resonance, individually both can be higher than the applied voltage. In this low-Q circuit, the peak capacitor voltage isn't too much higher than the applied voltage, but in high-Q circuits like radio transmitters, capacitors may have to withstand voltages hundreds of times the applied voltage. Incidentally, that 8-Ω answer in (b) is typical of loudspeakers. ∎

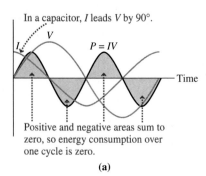

In a capacitor, $I$ leads $V$ by 90°.

$P = IV$

Time

Positive and negative areas sum to zero, so energy consumption over one cycle is zero.

**(a)**

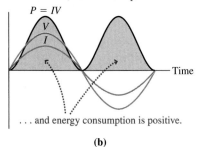

In a resistor, $I$ and $V$ are in phase . . .

$P = IV$

Time

. . . and energy consumption is positive.

**(b)**

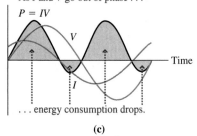

As $I$ and $V$ go out of phase . . .

$P = IV$

Time

. . . energy consumption drops.

**(c)**

**FIGURE 28.19** Energy consumption over one cycle is the area under the curve $P = IV$, with areas below the axis counted as negative.

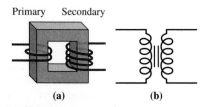

Primary    Secondary

**(a)**            **(b)**

**FIGURE 28.20** (a) A transformer consisting of two coils wound on an iron core. (b) Transformer circuit symbol.

## 28.5  Power in AC Circuits

Capacitors and inductors don't dissipate energy; rather, in an AC circuit they alternately store and release it. Therefore the average power consumption over one cycle is zero in a purely reactive circuit—one containing only capacitance and/or inductance. You can see this mathematically in Fig. 28.19a, which shows current, voltage, and power—the product $IV$—over one cycle for a capacitor. Although the capacitor absorbs energy during part of the cycle—when the power is positive—it returns the same amount later (negative power), giving zero net power over the cycle. That's because current and voltage are out of phase, so their product can be negative or positive at different times. In contrast, a resistor has $V$ and $I$ in phase (Fig. 28.19b), so the power is always positive and the resistor takes net energy from the circuit. In the more general case of a circuit containing resistance, capacitance, and inductance, the phase relation between current and voltage depends on the circuit details. Figure 28.19c shows a case where $I$ and $V$ are only slightly out of phase; the result is a net power consumption, but less than with a pure resistance.

We can develop a general expression for power in AC circuits by considering the time-average product of voltage and current with arbitrary phase difference $\phi$:

$$\langle P \rangle = \langle [I_p \sin(\omega t - \phi)][V_p \sin \omega t] \rangle$$

where $\langle \rangle$ indicates a time average over one cycle. Expanding the current term using a trig identity (see Appendix A) gives

$$\langle P \rangle = I_p V_p \langle (\sin^2 \omega t)(\cos \phi) - (\sin \omega t)(\cos \omega t)(\sin \phi) \rangle$$

The average of $(\sin \omega t)(\cos \omega t)$ is zero, as we've just shown for two signals 90° out of phase. The quantity $\sin^2 \omega t$ swings from 0 to 1 and is symmetric about $\frac{1}{2}$, so its average value is $\frac{1}{2}$. Then we have $\langle P \rangle = \frac{1}{2} I_p V_p \cos \phi$. Writing the peak values as $\sqrt{2}$ times the rms values gives

$$\langle P \rangle = \frac{1}{2} \sqrt{2} I_{rms} \sqrt{2} V_{rms} \cos \phi = I_{rms} V_{rms} \cos \phi \qquad (28.14)$$

This confirms our earlier graphical arguments. When the voltage and current are in phase, the average power is the product $I_{rms} V_{rms}$. But with current and voltage out of phase, the average power is lower; at 90° phase difference it's zero.

The factor $\cos \phi$ is called the **power factor**. A purely resistive circuit has a power factor of 1, while a circuit with only inductance and capacitance has a power factor of zero. In circuits containing resistance along with capacitance and/or inductance, the power factor generally depends on frequency; in the series $RLC$ circuit, for example, it's 1 at resonance but lower at other frequencies.

Equation 28.14 shows that a given average power at a fixed voltage requires higher current when the power factor is smaller. Because energy is dissipated in the $I^2 R$ losses of transmission lines, low power factors result in wasted energy. In extreme cases, high current associated with too much "reactive power" can bring down a power system. The widespread blackout of August 2003 in the United States and Canada left 50 million people without power and resulted in part from too low a power factor. Special devices included in the power grid are supposed to add inductance or capacitance as needed to keep the power factor close to 1, but such reactive power management was inadequate in the 2003 blackout and in several earlier power failures.

## 28.6  Transformers and Power Supplies

A **transformer** is a pair of wire coils, often wound on an iron core to concentrate magnetic flux (Fig. 28.20). A changing current in the **primary** coil results in a changing magnetic flux through the **secondary**, and this induces an emf in the secondary. The induced emf, in turn, drives current in any circuit connected across the secondary. Thus the device transfers electric power between two circuits without direct electrical contact.

**FIGURE 28.21** Transformers change voltage levels throughout the power distribution network.

The transformer in Fig. 28.20 is a **step-up transformer** because it has more turns in its secondary. Since each turn encircles the same changing magnetic flux, each gets the same induced emf and therefore the emf across the secondary is greater than across the primary. Interchanging primary and secondary in Fig. 28.20 would give a **step-down transformer**. In general, the ratio of the peak (or rms) secondary voltage $V_2$ to the peak (or rms) primary voltage $V_1$ is the same as the ratio of turns in the two coils:

$$V_2 = \frac{N_2}{N_1} V_1 \qquad (28.15)$$

Aren't we getting something for nothing with a step-up transformer? No. A step-up transformer increases voltage, but not power. An ideal transformer passes all the power supplied to the primary on to the secondary, so $I_1 V_1 = I_2 V_2$. If voltage goes up, current goes down, and vice versa. Real transformers have losses, but good engineering holds these to a few percent of the total power.

Transformers work only with AC because they use electromagnetic induction and therefore require *changing* current. One reason for the near-universal use of AC for electric power is the ease of changing voltage levels (Fig. 28.21). Relatively low voltages are safer and easier for the end user to handle. But since power $P = IV$, using a higher voltage between conductors in long-distance transmission means lower current. Power dissipated in the conductors themselves is $I^2 R$, so that in turn means less power lost in transmission. Transformers readily handle the voltage conversions in AC power systems. Changing the voltage from a DC source, in contrast, requires first interrupting the DC to produce a changing current; a car's ignition system is one example.

### Direct-Current Power Supplies

Lightbulbs and electric heaters work equally well on AC or DC, but electronic equipment requires DC. In Chapter 24 we found that a junction between *P*- and *N*-type semiconductors passes current in one direction but not the other. A **diode** is a *PN* junction designed specifically to serve as a "one-way valve" for electric current. An ideal diode acts like a short circuit to current flowing in the preferred direction, and like an open circuit in the opposite direction (Fig. 28.22).

**FIGURE 28.22** Circuit symbol for a diode, with preferred current direction indicated.

Figure 28.23a shows a diode and resistor connected to the secondary of a transformer. The resistor represents the load of the circuit needing DC power. The transformer brings the 120-V AC power line to a level appropriate to the load. The diode passes current in only one direction, resulting in the load current shown in Fig. 28.23b.

(a)                                    (b)

**FIGURE 28.23** (a) A simple DC power supply consists of a transformer and a diode, supplying current to the load resistance. (b) The load current flows in only one direction but still varies.

Although the load current in Fig. 28.23b flows in one direction, it still varies drastically with time. So we smooth, or **filter**, the output of the diode. The simplest filter is a capacitor, as shown in Fig. 28.24a. As the voltage on the left side of the diode rises, the capacitor voltage rises rapidly because of the short time constant associated with the low resistance of the diode in its preferred direction. But then the AC voltage begins to fall, and the diode "turns off." The capacitor can't discharge from right to left through the diode but only through the load resistance. If the time constant $RC$ is long enough, the capacitor voltage will drop only slightly before the AC voltage again rises and sends a new surge of current through the diode to bring the capacitor to its maximum charge (Fig. 28.24b). By making the capacitance large enough, we can make load current and voltage arbitrarily smooth. Large capacitors are expensive, so in practice additional filters—often using transistors or integrated circuits—are added to the simple capacitive filter in critical applications.

(a)                                    (b)

**FIGURE 28.24** (a) Adding a capacitor reduces variations in the load voltage and current. (b) Load voltage as a function of time.

## Big Picture

The big idea here is **alternating current** (AC), which in its simplest form exhibits sinusoidal variation in current and voltage. Resistors respond to AC as to DC, with current directly proportional to voltage. In capacitors and inductors, the current–voltage relation depends on frequency, and the current and voltage are out of phase.

## Key Concepts and Equations

An AC voltage or current is characterized by its amplitude (rms or peak value), its frequency, and phase:

$$V = V_p \sin(\omega t + \phi)$$

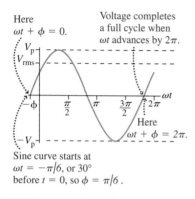

Here $\omega t + \phi = 0$.

Voltage completes a full cycle when $\omega t$ advances by $2\pi$.

Here $\omega t + \phi = 2\pi$.

Sine curve starts at $\omega t = -\pi/6$, or $30°$ before $t = 0$, so $\phi = \pi/6$.

**Reactance**, $X$, characterizes the relation between peak (or rms) current and voltage in capacitors and inductors:

$$I_p = \frac{V_p}{X}$$

Capacitor:

$X_C = 1/\omega C$

Inductor:

$X_L = \omega L$

where $\omega = 2\pi f$ is the angular frequency of the AC voltage and current.

**Phasors** are arrows used to describe time-varying AC quantities. They rotate with angular velocity equal to the angular frequency $\omega$.

Capacitor: Current leads voltage by $90°$.

Inductor: Current lags voltage by $90°$.

## Cases and Uses

In an **LC circuit**, energy oscillates between electric and magnetic forms with frequency

$$\omega_0 = \frac{1}{\sqrt{LC}}$$

In a series **RLC circuit**, capacitance and inductance cancel at the **resonant frequency**, $\omega_0$. Here the circuit exhibits the minimum **impedance**, $Z = \sqrt{R^2 + (X_L - X_C)^2}$, and passes the maximum current. The phase difference between voltage and current is $\tan \phi = (X_L - X_C)/R$.

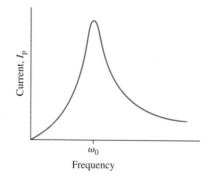

The average power in an AC circuit depends on the cosine of the phase difference, also called the **power factor**:

$$\langle P \rangle = I_{rms} V_{rms} \cos \phi$$

**Transformers** use electromagnetic induction to change voltage levels, transferring electric power between two circuits. **Diodes** and capacitive filters change AC to DC.

$N_1$ turns   $N_2$ turns

$V_1$   $V_2 = \dfrac{N_2}{N_1} V_1$

# For Thought and Discussion

1. Two AC signals have the same amplitude but different frequencies. Are their rms amplitudes the same?
2. What is meant by this statement: A capacitor acts like a DC open circuit?
3. How can current keep flowing in an AC circuit containing a capacitor? After all, a capacitor contains a gap between two conductors, and no charge can cross this gap.
4. Why does it make sense that inductive reactance increases with frequency?
5. The same AC voltage appears across a capacitor and a resistor, and the same rms current flows in each. Is the power dissipation the same in each?
6. When a particular inductor and capacitor are connected across the same AC voltage, the current in the inductor is higher than in the capacitor. Is this true at all frequencies?
7. An inductor and a capacitor are connected in series across an AC generator, and the rms voltage across the inductor is found to be higher than that across the capacitor. Is the generator frequency above or below resonance?
8. When the capacitor voltage in an undriven *LC* circuit reaches zero, why don't the oscillations stop?
9. Why is Equation 28.5 not a full description of the relation between voltage and current in a capacitor?
10. In a series *RLC* circuit, the applied voltage lags the current. Is the frequency above or below resonance?
11. The voltage across two circuit elements in series is zero. Is it possible that the voltages across the individual elements are nevertheless not zero? Give an example.
12. If you measure the rms voltages across the resistor, capacitor, and inductor in a series *RLC* circuit, will they add to the rms value of the generator voltage?
13. A step-up transformer increases voltage, or energy per unit charge. How is this possible without violating conservation of energy?

# Exercises and Problems

## Exercises

### Section 28.1 Alternating Current

14. Much of Europe uses AC power at 230 V rms and 50 Hz. Express this AC voltage in the form of Equation 28.3, taking $\phi_V = 0$.
15. An rms voltmeter connected across the filament of a TV picture tube reads 6.3 V. What is the peak voltage across the filament?
16. An industrial electric motor runs at 208 V rms and 400 Hz. What are (a) the peak voltage and (b) the angular frequency?
17. An AC current is given by $I = 495 \sin(9.43t)$, with $I$ in milliamperes and $t$ in milliseconds. Find (a) the rms current and (b) the frequency in Hz.
18. What are the phase constants for each of the signals shown in Fig. 28.25?

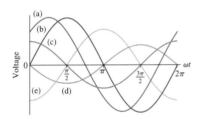

**FIGURE 28.25** Exercise 18

### Section 28.2 Circuit Elements in AC Circuits

19. Find the rms current in a 1.0-$\mu$F capacitor connected across 120-V rms, 60-Hz AC power.
20. A 470-$\Omega$ resistor, 10-$\mu$F capacitor, and 750-mH inductor are each connected across 6.3-V rms, 60-Hz AC power sources. Find the rms current in each.
21. Find the reactance of a 3.3-$\mu$F capacitor at (a) 60 Hz, (b) 1.0 kHz, and (c) 20 kHz.
22. A 15-$\mu$F capacitor carries an rms current of 1.4 A. What is the minimum safe voltage rating for the capacitor if the frequency is (a) 60 Hz and (b) 1.0 kHz?
23. A capacitor and a 1.8-k$\Omega$ resistor pass the same current when each is separately connected across a 60-Hz power line. What is the capacitance?

24. A 50-mH inductor is connected across a 10-V rms AC generator, and an rms current of 2.0 mA flows. What is the generator frequency?

### Section 28.3 *LC* Circuits

25. Find the resonant frequency of an *LC* circuit consisting of a 0.22-$\mu$F capacitor and a 1.7-mH inductor.
26. An *LC* circuit with $C = 18$ mF undergoes *LC* oscillations with period 2.4 s. Find the inductance.
27. You have a 2.0-mH inductor and wish to make an *LC* circuit whose resonant frequency spans the AM radio band (550–1600 kHz). What range of capacitance should your variable capacitor cover?
28. An *LC* circuit includes a 20-$\mu$F capacitor and has a period of 5.0 ms. The peak current is 25 mA. Find (a) the inductance and (b) the peak voltage.
29. An FM radio transmitter requires an *LC* circuit that oscillates at the 89.5-MHz transmitter frequency. What should the inductance be if the circuit capacitance is 47 pF?

### Section 28.4 Driven *RLC* Circuits and Resonance

30. A series *RLC* circuit has $R = 75\ \Omega$, $L = 20$ mH, and a resonant frequency of 4.0 kHz. (a) What is the capacitance? (b) What is the impedance of the circuit at resonance? (c) What is the impedance at 3.0 kHz?
31. Find the impedance of a circuit consisting of a 1.5-k$\Omega$ resistor, 5.0-$\mu$F capacitor, and 50-mH inductor in series, at a frequency of 10 kHz.
32. A series *RLC* circuit has $R = 18$ k$\Omega$, $C = 14\ \mu$F, and $L = 0.20$ H. (a) At what frequency is its impedance lowest? (b) What is the value of the impedance at this frequency?
33. If the peak voltage applied to produce the curves in Fig. 28.17 is 100 V and if $R = 10$ k$\Omega$, what are the values of the peak currents at resonance for the three curves shown?

### Sections 28.5 Power in AC Circuits and 28.6 Transformers and Power Supplies

34. An electric drill draws 4.6 A rms at 120 V rms. If the current lags the voltage by 25°, what is the drill's power consumption?
35. A 40-W fluorescent lamp has a power factor of 0.85 and operates from the 120-V rms AC power line. How much current does it draw?

36. An electric water heater draws 20 A rms at 240 V rms and is purely resistive. A large AC motor has the same current and voltage, but inductance causes the voltage to lead the current by 20°. Find the power consumption in each device.

37. A transformer is designed to step 230-V European power down to 120 V needed to operate North American devices. (a) If the primary has 1000 turns, how many should the secondary have? (b) If the primary current is 7.3 A, what is the secondary current?

## Problems

38. (a) A 2.2-H inductor is connected across the 120-V rms, 60-Hz power line. Find the rms inductor current. (b) Repeat if the same inductor is connected across the 230-V rms, 50-Hz power line commonly used in Europe.

39. A 2.0-$\mu$F capacitor has a capacitive reactance of 1.0 k$\Omega$. (a) What is the frequency of the applied voltage? (b) What inductance would give the same value for inductive reactance at this frequency? (c) How would the two reactances compare if the frequency were doubled?

40. Show that the unit of both capacitive and inductive reactance is the ohm.

41. A 1.2-$\mu$F capacitor is connected across a generator whose output is given by $V = V_p \sin 2\pi f t$, where $V_p = 22$ V, $f = 60$ Hz, and $t$ is in seconds. (a) What is the peak current? What are the magnitudes of (b) the voltage and (c) the current at $t = 6.5$ ms?

42. At 10 kHz an inductor has 10 times the reactance of a capacitor. At what frequency will their reactances be equal?

43. A 0.75-H inductor is in series with a fluorescent lamp, and the series combination is across the 120-V rms, 60-Hz power line. If the rms inductor voltage is 90 V, what is the rms lamp current?

44. A 2.2-nF capacitor and a capacitor of unknown capacitance are connected in parallel across a 10-V rms sine-wave generator. At 1.0 kHz, the generator supplies a total current of 3.4 mA rms. The generator frequency is then decreased until the rms current has dropped to 1.2 mA. Find (a) the unknown capacitance and (b) the lower frequency.

45. The FM radio band covers the frequency range 88–108 MHz. If the variable capacitor in an FM receiver ranges from 10.9 pF to 16.4 pF, what inductor should be used to make an $LC$ circuit whose resonant frequency spans the FM band?

46. An $LC$ circuit includes a 0.025-$\mu$F capacitor and a 340-$\mu$H inductor. (a) If the peak voltage on the capacitor is 190 V, what is the peak current in the inductor? (b) How long after the voltage peak does the current peak occur?

47. One-eighth of a cycle after the capacitor in an $LC$ circuit is fully charged, what are each of the following as fractions of their peak values: (a) capacitor charge, (b) energy in the capacitor, (c) inductor current, (d) energy in the inductor?

48. The 2000-$\mu$F capacitor in Fig. 28.26 is initially charged to 200 V. (a) Describe how you would manipulate switches $A$ and $B$ to transfer all the energy from the 2000-$\mu$F capacitor to the 500-$\mu$F capacitor. Include the times you would throw the switches. (b) What will be the voltage across the 500-$\mu$F capacitor once you've finished?

**FIGURE 28.26** Problem 48

49. A damped $LC$ circuit consists of a 0.15-$\mu$F capacitor and a 20-mH inductor with resistance 1.6 $\Omega$. How many cycles will the circuit oscillate before the peak voltage on the capacitor drops to half its initial value?

50. A damped $RLC$ circuit includes a 5.0-$\Omega$ resistor and a 100-mH inductor. If half the initial energy is lost after 15 cycles, what is the capacitance?

51. An $RLC$ circuit includes a 1.5-H inductor and a 250-$\mu$F capacitor rated at 400 V. The circuit is connected across a sine-wave generator whose peak voltage is 32 V. What minimum resistance must the circuit have to ensure that the capacitor voltage does not exceed its rated value when the generator is at the resonant frequency?

52. An $RLC$ circuit includes a 10-$\Omega$ resistor, 1.5-$\mu$F capacitor, and 50-mH inductor. The capacitor is rated at 1200 V. The circuit is driven by an AC source whose peak voltage is 100 V. (a) What would the peak capacitor voltage be at resonance? (b) Draw a graph of the peak capacitor voltage as a function of frequency, and from it determine the frequency range that should be avoided for the capacitor to stay within its voltage rating.

53. Figure 28.27 shows the phasor diagram for an $RLC$ circuit. (a) Is the driving frequency above or below resonance? (b) Complete the diagram by adding the applied voltage phasor, and from your diagram determine the phase difference between applied voltage and current.

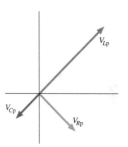

**FIGURE 28.27** Problem 53

54. An AC voltage of fixed amplitude is applied across a series $RLC$ circuit. The component values are such that the current at half the resonant frequency is half the current at resonance. Show that the current at twice the resonant frequency is also half the current at resonance.

55. A series $RLC$ circuit has resistance 100 $\Omega$ and impedance 300 $\Omega$. (a) What is the power factor? (b) If the rms current is 200 mA, what is the power dissipation?

56. A series $RLC$ circuit has power factor 0.80 and impedance 100 $\Omega$ at 60 Hz. (a) What is the circuit resistance? (b) If the inductance is 0.10 H, what is the resonant frequency?

57. A power plant produces 60-Hz power at 365 kV rms and 200 A rms. The plant is connected to a small city by a transmission line with total resistance 100 $\Omega$. What fraction of the power is lost in transmission if the city's power factor is (a) 1.0 and (b) 0.60? (c) Is it more economical for the power company if the load has a high power factor or a low one? Explain.

58. A car-battery charger runs off the 120-V rms AC power line and supplies 10-A DC at 14 V. (a) If the charger is 80% efficient in converting the line power to the DC power it supplies to the battery, how much current does it draw from the AC line? (b) If electricity costs 9.5¢/kWh, how much does it cost to run the charger for 10 hours? Assume the power factor is 1.

59. A power supply like that in Fig. 28.24 is supposed to deliver 22-V DC at a maximum current of 150 mA. The transformer's peak output voltage is appropriate to charge the capacitor to a full 22 V, and the primary is supplied with 60-Hz AC. What value of capacitance will ensure that the output voltage stays within 3% of the rated 22 V?

60. An *RLC* circuit includes a 3.3-$\mu$F capacitor and a 27-mH inductor. The capacitor is charged to 35 V, and the circuit begins oscillating. Ten full cycles later the capacitor voltage peaks at 28 V. What is the resistance?

61. A series *RLC* circuit with $R = 1.3\ \Omega$, $L = 27$ mH, and $C = 0.33\ \mu$F is connected across a sine-wave generator. If the capacitor's peak voltage rating is 600 V, what is the maximum safe value for the generator's peak output voltage when it's tuned to resonance?

62. Differentiate Equation 28.9 to find the current in the *LC* circuit, and use $q = CV$ to find the voltage. From these obtain the electric energy in the capacitor and the magnetic energy in the inductor, and sum to show that the total energy remains constant. *Hint:* You'll need to use Equation 28.10 and a familiar trig identity.

63. Find a second frequency where the speaker current in Example 28.4 has half its maximum value.

64. A sine-wave generator delivers a signal whose peak voltage is independent of frequency. Two identical capacitors are connected in parallel across the generator, and the generator supplies a peak current $I_p$ at frequency $f_1$. The capacitors are then connected in series across the generator. To what frequency should the generator be tuned to bring the current back to $I_p$?

65. Two capacitors are connected in parallel across a 10-V rms, 10-kHz sine-wave generator, and the generator supplies a total rms current of 30 mA. When the capacitors are rewired in series, the rms generator current drops to 5.5 mA. Find the values of the two capacitances.

66. An *LC* circuit starts at $t = 0$ with its 2000-$\mu$F capacitor at its peak voltage of 14 V. At $t = 35$ ms the voltage has dropped to 8.5 V. (a) What will be the peak current? (b) When will the peak current occur?

67. You wish to make a "black box" with two input connections and two output connections, as shown in Fig. 28.28. When you put a 12-V rms, 60-Hz sine wave across the input, a 6.0-V, 60-Hz signal should appear at the output, with the output voltage leading the input voltage by 45°. Design a circuit that could be used in the "black box."

**FIGURE 28.28** Problem 67

68. A series *RLC* circuit with $R = 47\ \Omega$, $L = 250$ mH, and $C = 4.0\ \mu$F is connected across a sine-wave generator whose peak output voltage is independent of frequency. Find the frequency range over which the peak current will exceed half its value at resonance. *Hint:* You can solve this problem graphically or, with appropriate algebraic manipulations, using quadratic equations.

69. A sine-wave generator with peak output voltage of 20 V is applied across a series *RLC* circuit. At the resonant frequency of 2.0 kHz, the peak current is 50 mA; at 1.0 kHz, it is 15 mA. Find $R$, $L$, and $C$.

70. Use phasor analysis to show that the parallel *RLC* circuit of Fig. 28.29 has impedance

$$Z = \left[\frac{1}{R^2} + \left(\frac{1}{X_L} - \frac{1}{X_C}\right)^2\right]^{-1/2}$$

**FIGURE 28.29** Problem 70

71. For *RLC* circuits in which the resistance is not too high, the $Q$ factor may be defined as the ratio of the resonant frequency to the difference between the two frequencies where the power dissipated in the circuit is half the power dissipated at resonance. Show, using suitable approximations, that this definition leads to the expression $Q = \omega_0 L/R$, with $\omega_0$ the resonant frequency.

72. Figure 28.30 shows a **low-pass filter**. When an alternating voltage is applied at the $V_{in}$ terminals, the output voltage $V_{out}$ depends on frequency. (a) Show that $V_{out} = V_{in}/\sqrt{1 + (RC\omega)^2}$, where the voltages are either peak or rms values (there is also a phase difference). (b) At what frequency is the output voltage down from the input voltage by a factor of $1/\sqrt{2}$? (This is called the *half-power point* since the power—proportional to $V^2$—is down by a factor of $\frac{1}{2}$.) *Hint:* You can repeat the phasor analysis of Section 28.4 but without the inductor. Or you can start from Equation 28.12, with $X_L$ set to zero.

**FIGURE 28.30** Problem 72

73. Find the rms voltage for the sawtooth signal shown in Fig. 28.31, in terms of the peak voltage $V_p$.

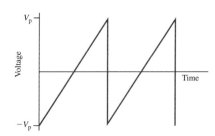

**FIGURE 28.31** Problem 73

74. Substitute the expression for $q(t)$ in Equation 28.11 into the differential equation for an *LC* circuit with resistance, and find an expression for the angular frequency of the damped oscillations in terms of $R$, $L$, and $C$.

75. An AC generator supplies an rms voltage of 120 V at 60 Hz, given by Equation 28.3 with $\phi_V = 0$, to a 50-$\Omega$ resistor in series with a 500-mH inductor and a 20-$\mu$F capacitor. At the instant when the generator voltage is decreasing through 45 V, what is the instantaneous voltage across each circuit element?

76. Although the maximum current flows in the speaker circuit of Example 28.4 at the resonant frequency of 1 kHz, the peak voltage across the capacitor is a maximum at a somewhat lower frequency. What is that frequency, and what is the corresponding peak voltage?

77. Your calculus professor, in whose class you're using graphing calculators, has given you an assignment on graphing trigonometric functions. Your assignment is to graph two trig functions on the same graph, and you decide to plot two 1.0-kHz AC voltage signals. One has an amplitude of 10 V rms and is 90° ahead of the other, which has an amplitude of 7.1 V. Sketch the signals, using a graphing calculator if you have one.

78. A friend working on a psychology experiment asks you for help with the design of a simple circuit. She needs an *LC* timing circuit in which the voltage on a fully charged capacitor drops to zero in 15 s. She has a 25-H inductor available, but doesn't know what value of capacitor to use. What do you recommend?

79. You're concerned about the functioning of a circuit that will be used in a remote communications installation. The series *RLC* circuit with $R = 5.5\ \Omega$, $L = 180$ mH, and $C = 0.12\ \mu$F is connected across a sine-wave generator. The inductor can safely handle 1.5 A of current. The peak generator output when it's tuned to resonance will be 8.0 V. Will the inductor current stay within its safe limit?

80. Your professor tells you about pre–digital computer days when engineers used electric circuits to model mechanical systems. They could modify parameters of the circuit elements and see the effect on the circuit's behavior. Suppose a 5.0-kg mass is connected to a spring with $k = 1.44$ kN/m. This is then modeled by an *LC* circuit with $L = 2.5$ H. What should *C* be in order for the *LC* circuit to have the same resonant frequency as the mass–spring system?

# Answers to Chapter Questions

## Answer to Chapter Opening Question

Electromagnetic induction lets us change the voltage levels in circuits, making for efficient power transmission but safe end-use voltages.

## Answers to GOT IT? Questions

28.1 The capacitor will carry twice as much current because its reactance goes down; the inductor will carry only half as much because its reactance doubles.

28.2 (b), by a factor of $10^{-4}$.

28.3 (a), because the inductor's reactance must be greater.

# 29 Maxwell's Equations and Electromagnetic Waves

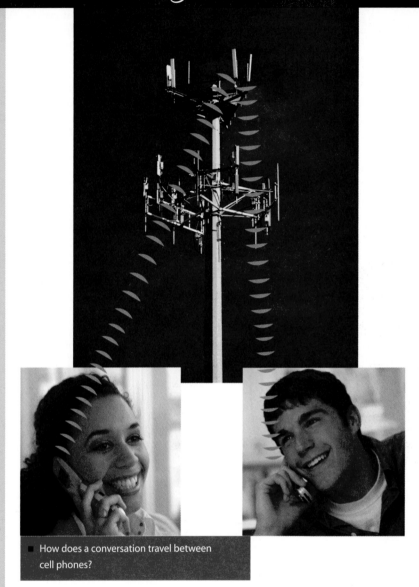

■ How does a conversation travel between cell phones?

▶ **To Learn**

By the end of this chapter you should be able to

■ State both mathematically and in words the four fundamental equations that govern all electromagnetic phenomena (29.1, 29.3).

■ Describe electromagnetic waves in terms of frequency, wavelength, amplitude, speed, and polarization (29.4, 29.5).

■ Recognize the wide range of frequencies and wavelengths that make up the electromagnetic spectrum (29.6).

■ Explain how electromagnetic waves are produced (29.7).

■ Calculate energy and momentum in electromagnetic waves (29.8).

◀ **To Know**

■ This chapter builds on the four fundamental laws of electromagnetism as we've introduced them throughout Part 4 of this text:
  ● Gauss's law for electricity (21.3)
  ● Gauss's law for magnetism (26.5)
  ● Ampère's law (26.8, so far for steady currents only)
  ● Faraday's law for the induced electric field (27.6)

■ We'll use the mathematical description of waves (14.2), and you may want to brush up on differentiating sines and cosines.

At this point we've introduced the four fundamental laws of electromagnetism—Gauss's law for electricity, Gauss's law for magnetism, Ampère's law, and Faraday's law—that govern the behavior of electric and magnetic fields throughout the universe. We've seen how these laws describe the electric and magnetic interactions that make matter act as it does, and we've explored practical electromagnetic devices. Here we extend the fundamental laws to their most general form and show how they predict the existence of electromagnetic waves. These include the light, radio, TV, microwaves, X rays, ultraviolet, and infrared with which we see, communicate, cook our food, diagnose our diseases, learn about the universe beyond Earth, and perform a myriad of other tasks from mundane to profound.

# 29.1 The Four Laws of Electromagnetism

Table 29.1 summarizes the four laws as we introduced them in earlier chapters. Look at these laws and you'll notice some similarities. On the left-hand sides of the equations, the two laws of Gauss are identical except for the interchanging of $\vec{E}$ and $\vec{B}$; the same is true for Ampère's and Faraday's laws.

The right-hand sides are more different. Gauss's law for electricity involves the charge enclosed by the surface of integration, while Gauss's law for magnetism has zero on the right-hand side. Actually, though, these laws are similar. Since we have no experimental evidence for the existence of isolated magnetic charge, the enclosed magnetic charge on the right-hand side of Gauss's law for magnetism is zero. If and when magnetic monopoles are discovered, then the right-hand side of Gauss's law for magnetism would be nonzero for any surface enclosing net magnetic charge.

The right-hand sides of Ampère's and Faraday's laws are distinctly different. Ampère's law has current—the flow of electric charge—as a source of magnetic field. We can understand the absence of a similar term in Faraday's law because we've never observed a flow of magnetic monopoles. If we had such a flow, then we would expect this magnetic current to produce an electric field.

Two of the differences among the laws of electromagnetism would be resolved if we knew for sure that magnetic monopoles exist. That current theories of elementary particles suggest the existence of monopoles is a tantalizing hint that there may be a fuller symmetry between electric and magnetic phenomena.

**TABLE 29.1**  Four Laws of Electromagnetism (still incomplete)

| Law | Mathematical Statement | What It Says |
|---|---|---|
| Gauss for $\vec{E}$ | $\oint \vec{E} \cdot d\vec{A} = \dfrac{q}{\epsilon_0}$ | How charges produce electric field; field lines begin and end on charges. |
| Gauss for $\vec{B}$ | $\oint \vec{B} \cdot d\vec{A} = 0$ | No magnetic charge; magnetic field lines don't begin or end. |
| Faraday | $\oint \vec{E} \cdot d\vec{r} = -\dfrac{d\Phi_B}{dt}$ | Changing magnetic flux produces electric field. |
| Ampère (steady currents only) | $\oint \vec{B} \cdot d\vec{r} = \mu_0 I$ | Electric current produces magnetic field. |

# 29.2 Ambiguity in Ampère's Law

There's one difference that magnetic monopoles won't resolve. On the right-hand side of Faraday's law is the term $d\Phi_B/dt$ that describes changing magnetic flux as a source of electric field. There's no comparable term in Ampère's law. Are we missing something? Is it possible that a changing electric flux produces a magnetic field? So far, we've given no experimental evidence for such a conjecture. It's suggested only by our sense that the near-symmetry between electricity and magnetism is not a coincidence. If a changing electric flux did produce a magnetic field, just as a changing magnetic flux produces an electric field, then we would expect a term $d\Phi_E/dt$ on the right-hand side of Ampère's law.

When we first stated Ampère's law in Chapter 26, we emphasized that it applied only to *steady* currents. Why that restriction? Figure 29.1 shows a situation in which current is *not* steady—namely, an *RC* circuit. Current in this circuit carries charge onto the capacitor plates. The current gradually decreases to zero as the capacitor charges. While it's flowing, the current should produce a magnetic field. Let's try to use Ampère's law to calculate that field.

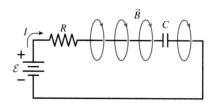

**FIGURE 29.1**  A charging *RC* circuit, showing some magnetic field lines surrounding current-carrying wire.

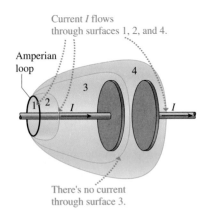

Current $I$ flows through surfaces 1, 2, and 4.

Amperian loop

There's no current through surface 3.

**FIGURE 29.2** Four surfaces bounded by the same circular amperian loop. Surface 1 is a flat, circular disk. The others are like soap bubbles in the process of being blown; they're open at the left end, so if current does pass through a surface, it does so at the right end only.

Ampère's law says that the line integral of the magnetic field around any closed loop is proportional to the encircled current:

$$\oint \vec{B} \cdot d\vec{r} = \mu_0 I.$$

The encircled current is the current through *any open surface* bounded by the loop. Figure 29.2 shows four such surfaces. The same current flows through surfaces 1, 2, and 4 because a current-carrying wire pierces each surface. But no current pierces surface 3 because it's in the gap between the capacitor plates. Charge flows onto the plates of the capacitor, but it doesn't flow through that gap. So for surfaces 1, 2, and 4 the right-hand side of Ampère's law is $\mu_0 I$, but for surface 3 it's zero. Thus Ampère's law is ambiguous in this case of a changing current.

This ambiguity doesn't arise with steady currents. In an *RC* circuit the steady-state current is zero, and thus the right-hand side of Ampère's law is zero for *any* surface. It's only when currents are changing with time that Ampère's law becomes ambiguous. That's why the form of Ampère's law we've used until now is strictly valid only for steady currents.

Can we extend Ampère's law to cover unsteady currents without affecting its validity in the steady case? Symmetry between Ampère's and Faraday's laws suggests that a changing electric flux might produce a magnetic field. Between the plates of a charging capacitor is an electric field whose magnitude is increasing. That means there's a changing electric flux through surface 3 of Fig. 29.2.

It was the Scottish physicist James Clerk Maxwell who, in about 1860, suggested that a changing electric flux should give rise to a magnetic field. Since that time many experiments, including direct measurement of the magnetic field inside a charging capacitor, have confirmed Maxwell's remarkable insight. Maxwell quantified his idea by introducing a new term into Ampère's law:

$$\oint \vec{B} \cdot d\vec{r} = \mu_0 I + \mu_0 \epsilon_0 \frac{d\Phi_E}{dt} \quad \text{(Ampère's law with Maxwell's modification)} \quad (29.1)$$

Now there's no ambiguity. The integral is taken around any loop, $I$ is the current through *any* surface bounded by the loop, and $\Phi_E$ is the electric flux through that surface. With our charging capacitor, Equation 29.1 gives the same magnetic field no matter which surface we choose. For surfaces 1, 2, and 4 of Fig. 29.2, the current $I$ makes all the contribution to the right-hand side of the equation. For surface 3, the right-hand side of Equation 29.1 comes entirely from the changing electric flux.

Although changing electric flux isn't the same thing as electric current, it has the same effect in producing a magnetic field. For this reason Maxwell called the term $\epsilon_0(d\Phi_E/dt)$ the **displacement current**. The word "displacement" has historical roots that don't provide much physical insight. But the word "current" is meaningful because the effect of displacement current is indistinguishable from that of real current in producing magnetic fields. Although we developed the idea of displacement current using the specific example of a charging capacitor, we emphasize that Ampère's law in its now complete form (Equation 29.1) is truly universal: *Any* changing electric flux results in a magnetic field. That fact will prove crucial in establishing the existence of electromagnetic waves.

---

**EXAMPLE 29.1** **Displacement Current: A Capacitor**

A parallel-plate capacitor with plate area $A$ and spacing $d$ is charging at the rate $dV/dt$. Show that the displacement current is equal to the current in the wires feeding the capacitor.

**INTERPRET** This is about a comparison between a familiar quantity—current—and a new quantity, namely, displacement current.

**DEVELOP** We're given the rate at which the capacitor voltage increases. Given that $q = CV$ for a capacitor, we can find the rate of charge

buildup—and that's equal to the current $I$ delivering charge to the capacitor. Equation 29.1 shows that the displacement current is $\epsilon_0(d\Phi_E/dt)$, so we'll need the rate of change of electric flux. A parallel-plate capacitor produces an essentially uniform field $E = V/d$. Since the field is uniform, the electric flux through a surface within the capacitor is simply the field strength times the plate area.

*cont'd.*

**EVALUATE** For the current, we differentiate the capacitor relation $q = CV$ to get $dq/dt = I = C\, dV/dt$. For the flux, we multiply the electric field by the plate area: $\Phi_E = EA = VA/d$. The rate of change of flux is then $d\Phi_E/dt = (A/d)(dV/dt)$, so the displacement current becomes

$$I_\mathrm{d} = \epsilon_0 \frac{d\Phi_E}{dt} = \frac{\epsilon_0 A}{d}\frac{dV}{dt}.$$

But $\epsilon_0 A/d$ is the capacitance of a parallel-plate capacitor (Equation 23.3), so the displacement current is $I_\mathrm{d} = C\, dV/dt$, the same as the actual current $I$.

**ASSESS** Make sense? It had better be this way, or Ampère's law would still be ambiguous. For any surface pierced by the wire in Fig. 29.2, the only contribution to the right-hand side of Ampère's law is from the current $I$. For any surface between the capacitor plates, the only contribution is from the displacement current $I_\mathrm{d} = \epsilon_0(d\Phi_E/dt)$. For Ampère's law to give the same magnetic field whichever surface we choose, $I$ and $I_\mathrm{d}$ had better be the same. ∎

## 29.3 Maxwell's Equations

It was Maxwell's genius to recognize that Ampère's law needed modifying to reflect the symmetry suggested by Faraday's law. To honor Maxwell, the four complete laws of electromagnetism are called **Maxwell's equations**. The complete set of equations, first published in 1864, governs the behavior of electric and magnetic fields everywhere. Table 29.2 summarizes Maxwell's equations.

**TABLE 29.2** Maxwell's Equations

| Law | Mathematical Statement | What It Says | Equation Number |
|---|---|---|---|
| Gauss for $\vec{E}$ | $\oint \vec{E}\cdot d\vec{A} = \dfrac{q}{\epsilon_0}$ | How charges produce electric field; field lines begin and end on charges. | (29.2) |
| Gauss for $\vec{B}$ | $\oint \vec{B}\cdot d\vec{A} = 0$ | No magnetic charge; magnetic field lines don't begin or end. | (29.3) |
| Faraday | $\oint \vec{E}\cdot d\vec{r} = -\dfrac{d\Phi_B}{dt}$ | Changing magnetic flux produces electric field. | (29.4) |
| Ampère | $\oint \vec{B}\cdot d\vec{r} = \mu_0 I + \mu_0\epsilon_0\dfrac{d\Phi_E}{dt}$ | Electric current and changing electric flux produce magnetic field. | (29.5) |

These four compact statements are all it takes to describe classical electromagnetic phenomena. Everything electric or magnetic that we've considered and will consider—from polar molecules to electric current; resistors, capacitors, inductors, and transistors; solar flares and cell membranes; electric generators and thunderstorms; computers and TVs; iPods and the northern lights—can be described using Maxwell's equations. And despite this wealth of phenomena, we have yet to discuss a most important manifestation of electromagnetism—namely, electromagnetic waves. We've put off waves until now because they depend crucially on Maxwell's extension of Ampère's law. It's easiest to understand electromagnetic waves when they propagate through empty space, so before beginning our study of waves, we'll simplify Maxwell's equations for the case of a vacuum.

### Maxwell's Equations in Vacuum

To express Maxwell's equations in vacuum, we simply remove all reference to matter—that is, to electric charge and current:

$$\oint \vec{E} \cdot d\vec{A} = 0 \quad (\text{Gauss, } \vec{E}) \quad (29.6) \qquad \oint \vec{B} \cdot d\vec{A} = 0 \quad (\text{Gauss, } \vec{B}) \quad (29.7)$$

$$\oint \vec{E} \cdot d\vec{r} = -\frac{d\Phi_B}{dt} \quad (\text{Faraday}) \quad (29.8) \qquad \oint \vec{B} \cdot d\vec{r} = \mu_0\epsilon_0 \frac{d\Phi_E}{dt} \quad (\text{Ampère}) \quad (29.9)$$

In vacuum the symmetry is complete, with electric and magnetic fields appearing on an equal footing. With charge and current absent in vacuum, the only source of either field is a change in the other field—as shown by the time derivatives on the right-hand sides of Faraday's and Ampère's laws.

## 29.4 Electromagnetic Waves

Faraday's law shows that a changing magnetic field induces an electric field. Ampère's law shows that a changing electric field induces a magnetic field. Together, the two suggest the possibility of **electromagnetic waves**, in which each type of field continuously induces the other, resulting in an electromagnetic disturbance that propagates through space as a wave. We'll now confirm this suggestion with a rigorous demonstration, directly from Maxwell's equations, that electromagnetic waves are indeed possible. In the process we'll discover the properties of electromagnetic waves and come to a deep understanding of the nature of light.

### A Plane Electromagnetic Wave

Here we describe the simplest type of electromagnetic wave—a plane wave in vacuum. A plane wave's properties don't vary in directions perpendicular to the wave propagation, so its wavefronts are planes. A plane wave is an approximation to the more realistic case of a spherical wave expanding from a localized source, and it's a good approximation at distances from the wave source that are large compared with the wavelength. Light waves from the Sun, for example, or radio waves miles from the transmitter are essentially plane waves.

In vacuum, it turns out that the electric and magnetic fields of an electromagnetic wave are perpendicular. They're both also perpendicular to the direction of wave propagation—making the electromagnetic wave a transverse wave, as defined in Chapter 14. To be concrete, we'll take the $x$ direction to be the direction of propagation, the $y$ direction that of the electric field, and the $z$ direction that of the magnetic field (Fig. 29.3). We won't prove that a configuration like this is the only one possible for an electromagnetic wave (although in vacuum it is; see Problem 46). What we will do is prove that this configuration satisfies Maxwell's equations—thus showing that such electromagnetic waves are indeed possible. But first we need a mathematical description of our plane electromagnetic wave.

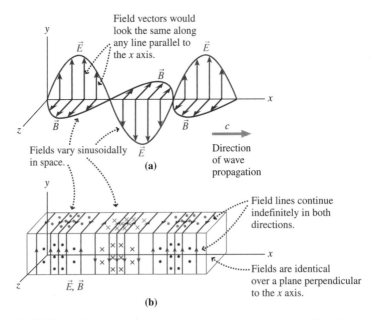

**FIGURE 29.3** Fields of a plane electromagnetic wave, shown at a fixed instant of time. (a) Field vectors for points on the $x$ axis show sinusoidal variation in the fields. (b) A partial representation of the field lines in a rectangular slab. Lines on the facing surfaces of the slab are shown as arrows; lines going through the slab appear as dots or crosses. Spacing of the field lines reflects the sinusoidal variation shown in part (a).

In Chapter 14 we described a sinusoidal wave propagating in the $x$ direction by a function of the form $A \sin(kx - \omega t)$, where $A$ is the wave amplitude, $k$ the wave number, and $\omega$ the angular frequency. For mechanical waves, $A \sin(kx - \omega t)$ described some physical quantity such as the height of a water wave or the pressure variation in a sound wave. In an electromagnetic wave, the corresponding physical quantities are the electric and magnetic fields. It turns out that these two wave fields, though perpendicular, are in phase—meaning that their peaks and troughs coincide, as shown in Fig. 29.3a. Having chosen the $y$ direction for the electric field and $z$ for the magnetic field, we can write the fields of our plane electromagnetic wave as

$$\vec{E}(x, t) = E_p \sin(kx - \omega t)\hat{j} \qquad (29.10)$$

and

$$\vec{B}(x, t) = B_p \sin(kx - \omega t)\hat{k} \qquad (29.11)$$

where the amplitudes $E_p$ and $B_p$ are constants and where $\hat{j}$ and $\hat{k}$ are unit vectors in the $y$ and $z$ directions. Figure 29.3a is a "snapshot" of some field vectors of this wave at points on the $x$ axis, shown at a fixed instant of time. That $\vec{E}$ and $\vec{B}$ are perpendicular is obvious from the figure, as is the fact that they're perpendicular to the propagation direction $x$. You can also see the sinusoidal variation, as the field vectors get alternately longer, then shorter, then reverse direction, and so on. And you can see that $\vec{E}$ and $\vec{B}$ are in phase because their peaks coincide. We emphasize that Fig. 29.3a shows field *vectors* for points on the $x$ axis only; the fields extend forever throughout space, and because this is a plane wave, a picture of field vectors along any line parallel to the $x$ axis would look the same.

We can also draw field *lines* for our wave, in contrast to the field vectors of Fig. 29.3a. We can't draw complete field lines because they extend forever in both directions. So in Fig. 29.3b we've shown the field lines only in a rectangular slab; that's enough to give a picture of what the fields look like everywhere. You should convince yourself that Fig. 29.3a and b show exactly the same thing—namely, a plane electromagnetic wave described by Equations 29.10 and 29.11. In one case we use field *vectors*, whose lengths are proportional to the field magnitudes, and in the other we use field *lines*, which extend forever and whose spacing indicates the field magnitudes.

We'll now show that the electric and magnetic fields pictured in Fig. 29.3 and described by Equations 29.10 and 29.11 satisfy Maxwell's equations. We've chosen a sinusoidal waveform for our wave fields because of its mathematical simplicity. But the superposition principle holds for electric and magnetic fields, and we know from Section 14.5 that we can represent *any* waveform by superposing sinusoids. So our proof that electromagnetic waves can exist holds for any wave shape. That means we can use electromagnetic waves to communicate the complex waveforms of music, TV images, and computer data.

## Gauss's Laws

In vacuum, Gauss's laws for electric and magnetic fields both have zero on the right-hand side, reflecting the absence of charge. That means the electric and magnetic flux through *any* closed surface must be zero, and therefore the field lines can't begin or end. With our plane wave, the field lines shown partially in Fig. 29.3b extend straight forever in both directions. So they don't begin or end, and therefore the fields satisfy Gauss's laws.

## Faraday's Law

To see that Faraday's law is satisfied, look directly toward the $x$-$y$ plane in Fig. 29.3b. You see electric field lines going up and down and magnetic field lines coming straight in and out, as shown in Fig. 29.4. Consider the small rectangular loop of height $h$ and infinitesimal width $dx$ shown in the figure. Evaluating the line integral of the electric field $\vec{E}$ around this loop, we get no contribution from the short ends at right angles to the field. Going around counterclockwise, we get a contribution $-Eh$ as we go down the left side against the field direction. Then we get a positive contribution going up the right side. Because the

The electric field is parallel to the plane of the page, and its strength and direction vary sinusoidally with position $x$.

The magnetic field is perpendicular to the page, and its strength and direction vary sinusoidally with position $x$.

**FIGURE 29.4** View of Fig. 29.3b in the $x$-$y$ plane, with a rectangular loop for evaluating the line integral in Faraday's law.

field varies with position, the field on the right side of the loop is different from that on the left. Let the change be $dE$, so the field on the right side is $E + dE$, giving a contribution $(E + dE)h$ to the line integral. Then the line integral of $\vec{E}$ around the loop is

$$\oint \vec{E} \cdot d\vec{r} = -Eh + (E + dE)h = h \, dE$$

This nonzero line integral implies an induced electric field. Induced by what? By a changing magnetic flux through the loop. The electric field of the wave arises because of the changing magnetic field of the wave. The area of the loop is $h \, dx$, and the magnetic field $\vec{B}$ is at right angles to this area, so the magnetic flux through the loop is $\Phi_B = Bh \, dx$. The rate of change of flux through the loop is then

$$\frac{d\Phi_B}{dt} = h \, dx \frac{dB}{(dt)}$$

Faraday's law relates the line integral of the electric field to the rate of change of flux:

$$\oint \vec{E} \cdot d\vec{r} = -\frac{d\Phi_B}{dt}$$

or, using our expressions for the line integral and the rate of change of flux, $h \, dE = -h \, dx(dB/dt)$. Dividing through by $h \, dx$ gives $dE/dx = -dB/dt$. In deriving this equation, we considered changes in $E$ with position at a fixed instant of time. Similarly, the change in $B$ with respect to time is taken at a fixed position. The derivatives are then *partial derivatives*—rates of change with respect to one variable while another is held fixed. If you've studied partial derivatives in calculus, you know that the symbol $\partial$ designates a partial derivative. In any event, our equation $dE/dx = -dB/dt$ should be written with partial derivatives:

$$\frac{\partial E}{\partial x} = -\frac{\partial B}{\partial t} \qquad (29.12)$$

This equation—which is Faraday's law applied to our electromagnetic wave—says that the rate at which the electric field changes with *position* depends on the rate at which the magnetic field changes with *time*.

## Ampère's Law

Now look at Fig. 29.3b from above. You see the magnetic field lines in the *x-z* plane and electric field lines emerging perpendicular to the *x-z* plane (Fig. 29.5). Apply Ampère's law (Equation 29.9) to the rectangle shown. In the line integral there's no contribution from the short sides because they're perpendicular to the field. Going down the left side gives $Bh$. Going up the right, against the field, gives $-(B + dB)h$, where $dB$ is the change in $B$ across the rectangle. So the line integral in Ampère's law is

$$\oint \vec{B} \cdot d\vec{r} = Bh - (B + dB)h = -h \, dB$$

The electric flux through the rectangle is $Eh \, dx$, so the rate of change of electric flux is

$$\frac{d\Phi_E}{dt} = h \, dx\left(\frac{dE}{dt}\right)$$

Ampère's law relates the line integral of the magnetic field to this time derivative of the electric flux, giving $-h \, dB = \epsilon_0 \mu_0 h \, dx(dE/dt)$. Dividing by $h \, dx$ and again using partial derivatives, we have

$$\frac{\partial B}{\partial x} = -\epsilon_0 \mu_0 \frac{\partial E}{\partial t} \qquad (29.13)$$

Equations 29.12 and 29.13—derived from Faraday's and Ampère's laws—express fully the requirements that Maxwell's universal laws of electromagnetism pose on the field

The magnetic field is parallel to the plane of the page, and its strength and direction vary sinusoidally with position *x*.

The electric field is perpendicular to the page, and its strength and direction vary sinusoidally with position *x*.

**FIGURE 29.5** View of Fig. 29.3b in the *x-z* plane, with a rectangular loop for evaluating the line integral in Ampère's law.

structure of Fig. 29.3. Each describes an induced field that arises from the changing of the other field. That other field, in turn, arises from the changing of the first field. Thus we have a self-perpetuating electromagnetic structure, whose fields exist and change without the need for charged matter. If Equations 29.10 and 29.11, which describe the fields in Fig. 29.3, can be made consistent with Equations 29.12 and 29.13, then we'll have shown that our electromagnetic wave satisfies Maxwell's equations and is thus a possible configuration of electric and magnetic fields.

## Conditions on Wave Fields

To see that Equation 29.12 is satisfied, we differentiate the electric field of Equation 29.10 with respect to $x$, and the magnetic field of Equation 29.11 with respect to $t$:

$$\frac{\partial E}{\partial x} = \frac{\partial}{\partial x}\left[E_p \sin(kx - \omega t)\right] = kE_p \cos(kx - \omega t)$$

and

$$\frac{\partial B}{\partial t} = \frac{\partial}{\partial t}\left[B_p \sin(kx - \omega t)\right] = -\omega B_p \cos(kx - \omega t)$$

Putting these expressions in for the derivatives in Equation 29.12 gives

$$kE_p \cos(kx - \omega t) = -\left[-\omega B_p \cos(kx - \omega t)\right]$$

The cosine cancels, showing that the equation holds if

$$kE_p = \omega B_p \tag{29.14}$$

To see that Equation 29.13 is also satisfied, we differentiate the magnetic field of Equation 29.11 with respect to $x$ and the electric field of Equation 29.10 with respect to $t$:

$$\frac{\partial B}{\partial x} = kB_p \cos(kx - \omega t) \quad \text{and} \quad \frac{\partial E}{\partial t} = -\omega E_p \cos(kx - \omega t)$$

Using these expressions in Equation 29.13 then gives

$$kB_p \cos(kx - \omega t) = -\epsilon_0 \mu_0\left[-\omega E_p \cos(kx - \omega t)\right]$$

Again, the cosine cancels, so this equation is satisfied if

$$kB_p = \epsilon_0 \mu_0 \omega E_p \tag{29.15}$$

Our analysis has shown that electromagnetic waves whose form is given by Fig. 29.3 and Equations 29.10 and 29.11 can exist, provided that the amplitudes $E_p$ and $B_p$, and the frequency $\omega$ and wave number $k$, are related by Equations 29.14 and 29.15. Physically, the existence of these waves is possible because a change in either field—electric or magnetic—induces the other field, giving rise to a self-perpetuating electromagnetic-field structure. Maxwell's theory thus leads to the prediction of an entirely new phenomenon— electromagnetic waves. We'll now explore some properties of these waves.

## 29.5 Properties of Electromagnetic Waves

### Wave Speed

In Chapter 14 we found that the speed of a sinusoidal wave is the ratio of the angular frequency to the wave number: speed $= \omega/k$. To determine the speed of our electromagnetic wave, we use Equation 29.14 to get $E_p = \omega B_p/k$, and then use this expression in Equation 29.15:

$$kB_p = \epsilon_0 \mu_0 \omega E_p = \frac{\epsilon_0 \mu_0 \omega^2 B_p}{k}$$

The amplitude $B_p$ cancels, and we solve for the wave speed $\omega/k$ to get

$$\text{wave speed} = \frac{\omega}{k} = \frac{1}{\sqrt{\epsilon_0 \mu_0}} \quad \text{(EM wave speed in vacuum)} \qquad (29.16a)$$

This result shows that the speed of an electromagnetic wave in vacuum depends only on the electric and magnetic constants $\epsilon_0$ and $\mu_0$. All electromagnetic waves in vacuum, regardless of frequency or amplitude, share this speed. Although we derived this result for sinusoidal waves, the superposition principle ensures that it holds for any wave shape.

We can easily evaluate the speed given in Equation 29.16a:

$$\frac{1}{\sqrt{\epsilon_0 \mu_0}} = \frac{1}{\sqrt{(8.85 \times 10^{-12} \, \text{C}^2/\text{N} \cdot \text{m}^2)(4\pi \times 10^{-7} \, \text{N/A}^2)}} = 3.00 \times 10^8 \, \text{m/s}$$

But this is the speed of light! During the two centuries before Maxwell, scientists had measured increasingly accurate values for light's speed. They had also recognized, thanks to Thomas Young's 1801 interference experiment, that light consists of waves. Then, in the 1860s, came Maxwell. Using a theory developed from laboratory experiments on electricity and magnetism, with no reference to optics or light, Maxwell showed how the interplay of electric and magnetic fields results in electromagnetic waves. The wave speed—calculated from the constants $\epsilon_0$ and $\mu_0$—was the known speed of light. Maxwell's conclusion was inescapable: Light is an electromagnetic wave.

Maxwell's identification of light as an electromagnetic phenomenon is a classic example of the unification of knowledge in science. With one simple calculation, Maxwell brought the entire science of optics under the umbrella of electromagnetism. Maxwell's work stands as a crowning intellectual triumph, one whose implications are still expanding our view of the universe.

Maxwell's discovery lets us recast Equation 29.16a in the form

$$\frac{\omega}{k} = c \quad \text{(EM wave speed in vacuum: the speed of light, } c!) \qquad (29.16b)$$

where $c = 1/\sqrt{\epsilon_0 \mu_0}$ is the speed of light. Equivalently, in terms of the more familiar wavelength and frequency,

$$f\lambda = c \quad \text{(frequency, wavelength, and the speed of light)} \qquad (29.16c)$$

As we saw in Chapter 1, the 1983 definition of the meter gives $c$ the exact value 299,792,458 m/s.

## Wave Amplitude

The amplitudes $E_p$ and $B_p$ dropped out of our analysis, showing that an electromagnetic wave's speed is independent of amplitude. But the field strengths $E$ and $B$ aren't independent. Using $\omega/k = c$, we can recast Equation 29.14 to show that

$$E = \frac{\omega}{k} B = cB \quad \text{(}E, B \text{ relation in vacuum EM wave)} \qquad (29.17)$$

Here we dropped the "peak" subscript because $E_p$ and $B_p$ multiply identical cosine terms in our wave description, so Equation 29.14 applies whether or not we're at the peak field.

## Phase, Orientation, and Waves in Matter

The wave of Fig. 29.3 and Equations 29.10 and 29.11 has $\vec{E}$ and $\vec{B}$ in phase in *time*—peaking at the same time—but perpendicular in *space* and also perpendicular to the propagation direction. Our derivation of the wave speed used these properties, so we've confirmed that electromagnetic waves in vacuum are transverse waves with $\vec{E}$ and $\vec{B}$ perpendicular and in phase. Specifically, the direction of propagation is that of the cross product $\vec{E} \times \vec{B}$. These geometrical properties also apply to electromagnetic waves in common materials like air and glass. The wave speed in these materials is lower than in vacuum, although for air the difference is minuscule. Electromagnetic waves in more complex materials can have very different properties and propagation speeds.

---

**GOT IT? 29.1**   At a particular point the electric field of an electromagnetic wave points in the $+y$ direction, while the magnetic field points in the $-z$ direction. Is the propagation direction (a) $x$; (b) $-x$; (c) either $+x$ or $-x$ but you can't tell which; (d) $-y$; (e) $+z$; or (f) not along any of the coordinate axes?

---

**EXAMPLE 29.2**   **Electromagnetic-Wave Properties: Laser Light**

A laser beam with wavelength 633 nm is propagating through air in the $+z$ direction. Its electric field is parallel to the $x$ axis and has amplitude 6.0 kV/m. (a) Find the wave frequency, (b) the amplitude of the magnetic field, and (c) the direction of the magnetic field.

**INTERPRET** Light is an electromagnetic wave, so the laser beam shares the wave properties we've just discussed. As we noted, its speed in air is nearly the same as in vacuum. Here we're given the wavelength and peak electric field, $E_p$.

**DEVELOP** Equation 29.16c, $f\lambda = c$, relates wavelength and frequency, so we'll use that for (a). Equation 29.17, $E = cB$, relates $E$ and $B$, so we can get (b) from the given value of $E_p$. For (c), we'll draw a reoriented version of Fig. 29.3a to help infer the direction of $\vec{B}$.

**EVALUATE** (a) Solving for the frequency gives

$$f = c/\lambda = (3.0 \times 10^8 \text{ m/s})/(633 \text{ nm}) = 4.7 \times 10^{14} \text{ Hz.}$$

(b) Solving for the magnetic-field amplitude gives $B_p = E_p/c = 20 \ \mu\text{T}$.
(c) Figure 29.6 shows that with propagation in the $z$ direction and $\vec{E}$ along the $x$ axis, $\vec{B}$ must be parallel to the $y$ axis.

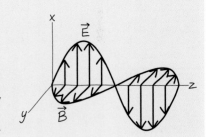

**FIGURE 29.6** Reoriented version of Fig. 29.3a for Example 29.2. It's important that we still have a right-handed coordinate system, with the $x$, $y$, and $z$ axes in the same relation as in Fig. 29.3. Equivalently, the direction of $\vec{E} \times \vec{B}$ is the propagation direction.

**ASSESS** That $10^{14}$-Hz frequency sounds huge, but light has such a short wavelength that its frequency is indeed high; more on this shortly. Notice both in Fig. 29.3 and in our reoriented wave of Fig. 29.6 that the vectors $\vec{E}$ and $\vec{B}$ and the propagation direction form a right-handed coordinate system, so any two of those vectors determine the direction of the third.

---

## Polarization

Although $\vec{E}$ and $\vec{B}$ are necessarily perpendicular, their orientation is still arbitrary within a plane perpendicular to the propagation direction. **Polarization** specifies the direction of the electric field and thus determines the perpendicular magnetic-field direction as well (Fig. 29.7).

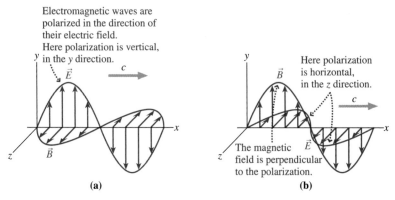

**FIGURE 29.7** The polarization direction is the direction of the wave's electric field.

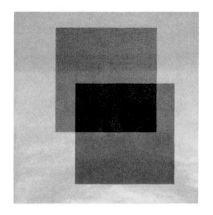

**FIGURE 29.8** Two pieces of polarizing material with their transmission axes at right angles. Where they overlap, no light gets through.

Electromagnetic waves used in radio and TV originate from antennas that give the waves a definite polarization. Most laser light is also polarized. In contrast, light from hot sources like the Sun or a lightbulb is unpolarized, consisting of a mix of waves with random field orientations. Unpolarized light becomes polarized when it reflects off surfaces or passes through substances whose structure has a preferred direction. Many crystals and synthetic materials such as the plastic Polaroid have this property. Light reflecting off a car's hood becomes partially polarized in the horizontal direction, and Polaroid sunglasses, with their transmission axis vertical, block the resulting glare.

A polarizing material passes unattenuated only the component of the wave field $\vec{E}$ that lies along the preferred direction—namely, $E\cos\theta$, where $\theta$ is the angle between the field and the preferred direction. We'll show shortly that the intensity of an electromagnetic wave is proportional to the square of the field strength. As a result, a wave of intensity $S_0$ emerges from a polarizer with intensity given by the **Law of Malus**:

$$S = S_0 \cos^2\theta \qquad (29.18)$$

Thus electromagnetic waves are blocked completely by a polarizer oriented perpendicular to the waves' polarization (Fig. 29.8).

Measuring polarization tells us about sources of electromagnetic waves and about materials through which they propagate. Many astrophysical processes produce polarized waves; their polarization gives clues to mechanisms operating in the cosmos. Geologists pass polarized light through thin sections of rock to reveal the rocks' composition, and engineers use polarization to locate stresses in mechanical structures. Polarization is essential in many technologies, including the ubiquitous liquid crystal displays (LCDs) in our cell phones, digital cameras, computers, and flat-panel TVs (Fig. 29.9).

---

**GOT IT? 29.2** Two polarizers are oriented at right angles, so no light gets through the combination. A third polarizer is inserted between the two, with its preferred direction at 45° to the others. Will this "sandwich" of polarizers still block a beam of initially unpolarized light?

---

Vertical polarizer passes only light with its electric field vertical.

Horizontal polarizer passes the now horizontally polarized light.

Electric field of incident light points in all directions perpendicular to the light's propagation.

Liquid crystal molecules align with striated plates and rotate the light's polarization.

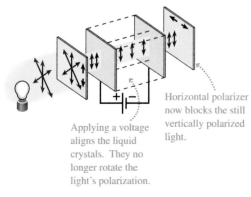

**FIGURE 29.9** Polarization plays a central role in the operation of a liquid crystal display. Multiple units like the one shown—millions in a TV or computer screen—produce the individual pixels on an LCD.

Applying a voltage aligns the liquid crystals. They no longer rotate the light's polarization.

Horizontal polarizer now blocks the still vertically polarized light.

## 29.6 The Electromagnetic Spectrum

Although Equations 29.16 relate an electromagnetic wave's frequency and wavelength, one of these quantities remains arbitrary. That means electromagnetic waves can have any frequency or, equivalently, any wavelength. Visible light occupies a wavelength range from about 400 nm to 700 nm, corresponding to frequencies of $7.5 \times 10^{14}$ Hz to $4.3 \times 10^{14}$ Hz. The different wavelengths or frequencies correspond to different colors, with red at the long-wavelength, low-frequency end of the visible region and blue at the short-wavelength, high-frequency end (see the enlargement in Fig. 29.10).

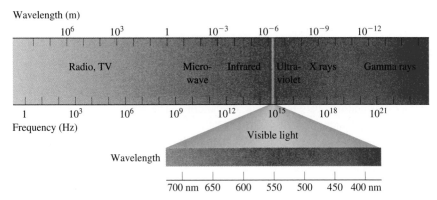

**FIGURE 29.10** The electromagnetic spectrum ranges from radio waves to gamma rays, with visible light occupying only a narrow range of wavelengths and frequencies on a logarithmic scale.

Figure 29.10 shows the entire **electromagnetic spectrum**, including frequencies and wavelengths that differ by many orders of magnitude from those of visible light. The invisible electromagnetic waves beyond the narrow visible range were unknown in Maxwell's time. A brilliant confirmation of Maxwell's theory came in 1888, when the German physicist Heinrich Hertz succeeded in generating and detecting electromagnetic waves of much lower frequency than visible light. Hertz intended his work only to verify Maxwell's modification of Ampère's law, but the practical consequences have proven enormous. In 1901, the Italian scientist Guglielmo Marconi transmitted electromagnetic waves across the Atlantic Ocean, creating a public sensation. From the pioneering work of Hertz and Marconi, spurred by the theoretical efforts of Maxwell, came the entire technology of radio, television, and microwaves that so dominates modern society. We now consider all electromagnetic waves in the frequency range from a few Hz to about $3 \times 10^{11}$ Hz as radio waves, with ordinary AM radio at about $10^6$ Hz, FM radio at $10^8$ Hz, and microwaves used for radar, cooking, cell phones, and satellite communications at $10^9$ Hz and above.

Between radio waves and visible light lies the infrared frequency range. Electromagnetic waves in this region are emitted by warm objects, even when they're not hot enough to glow visibly. For this reason, infrared cameras are used to determine subtle body-temperature differences in medical diagnosis, to examine buildings for heat loss, and to study the birth of stars in clouds of interstellar gas and dust.

Beyond the visible region are the ultraviolet rays responsible for sunburn, then the highly penetrating X rays, and finally the gamma rays whose primary terrestrial source is radioactive decay. All these phenomena, from radio to gamma rays, are fundamentally the same: All are electromagnetic waves, differing only in frequency and wavelength. All travel with speed $c$ in vacuum, and all consist of electric and magnetic fields produced from each other through the induction processes described by Faraday's and Ampère's laws. Naming the different types of electromagnetic waves is just a convenience; there are no gaps in the continuous range of frequencies and wavelengths. Practical differences arise because waves of different wavelengths interact differently with matter; in particular, shorter wavelengths tend to be generated and absorbed most efficiently by smaller systems.

Earth's atmosphere is transparent to visible light and to most radio frequencies. But it's opaque to most infrared, the higher-frequency ultraviolet, X rays, and all but the highest-frequency gamma rays. Earth's surface would be hazardous to life if ultraviolet weren't absorbed by ozone gas high in the atmosphere, and our planet would be a lot cooler if water vapor, carbon dioxide, and other gases didn't absorb outgoing infrared. But that same infrared absorption is at the heart of our worries about global climate change, because we humans are increasing the levels of infrared-trapping gases. Until the space age, our view of the universe beyond Earth was limited to visible light and some weak radio signals. Today, spacecraft observe the cosmos in all wavelengths from radio through gamma rays. Our picture of the universe is far richer than it would be if the only electromagnetic waves we could receive were those that made it through Earth's atmosphere.

## 29.7 Producing Electromagnetic Waves

All it takes to produce an electromagnetic wave is a changing electric or magnetic field. Once there's a change in one field, induction provides the second field, and together the changing fields continuously regenerate one another. The wave is on its way! Ultimately, changing fields of either type result when we alter the motion of electric charge. Therefore, **accelerated charge is the source of electromagnetic waves.**

In a radio transmitter, the accelerated charges are electrons moving back and forth in an antenna, driven by alternating voltage from an *LC* circuit (Fig. 29.11). In an X-ray tube, high-energy electrons decelerate rapidly as they slam into a target; their deceleration is the source of the electromagnetic waves, now in the X-ray region of the spectrum. In the magnetron tube of a microwave oven, electrons circle in a magnetic field; their centripetal acceleration is the source of the microwaves that cook your food. And the altered movement of electrons in atoms—although described accurately only by quantum mechanics—is the source of most visible light. If the motion of the accelerated charges is periodic, then the wave frequency is that of the motion; more generally, systems are most efficient at producing (and receiving) electromagnetic waves whose wavelength is comparable to the size of the system. That's why TV antennas are on the order of 1 m in size, while nuclei—some $10^{-15}$ m in diameter—produce gamma rays.

Calculation of electromagnetic waves from accelerated charges presents challenging but important problems for physicists and engineers. Figure 29.12 shows the field of an oscillating dipole—a configuration approximated by many systems from antennas to atoms and molecules. Note that the waves are strongest in the direction at right angles to the acceleration of the charge distribution and that there's no radiation in the direction of the acceleration. This accounts for, among other phenomena, the directionality of radio and TV antennas, which transmit and receive most effectively perpendicular to the long direction of the antenna.

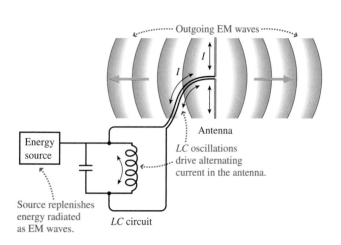

**FIGURE 29.11** Simplified diagram of a radio transmitter.

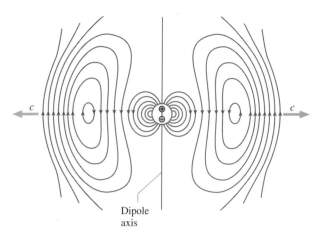

**FIGURE 29.12** "Snapshot" showing the electric field of an oscillating electric dipole at an instant in time.

The field shown in Fig. 29.12 seems to bear little resemblance to the plane-wave fields of Fig. 29.3 that we used to demonstrate the possibility of electromagnetic waves. We could produce true plane waves only with an infinite sheet of accelerated charge—an obvious impossibility. But far from the source, the curved field lines in Fig. 29.12 would appear straight, and the wave would approximate a plane wave. So our plane-wave analysis is a valid approximation at great distances—typically many wavelengths—from a localized wave source. Closer to the source more complicated expressions for the wave fields apply, but these, too, satisfy Maxwell's equations.

## 29.8 Energy and Momentum in Electromagnetic Waves

We showed in earlier chapters that electric and magnetic fields contain energy. Electromagnetic waves are combinations of electric and magnetic fields; as they propagate, they transport the energy contained in those fields.

### Wave Intensity

In Chapter 14 we defined wave intensity as the rate at which a wave transports energy across a unit area; its units are $W/m^2$. We can calculate the intensity $S$ of a plane electromagnetic wave by considering a rectangular box of thickness $dx$ and cross-sectional area $A$ with its face perpendicular to the wave propagation (Fig. 29.13). Within this box are wave fields $\vec{E}$ and $\vec{B}$ whose energy densities are given by Equations 23.7 and 26.9: $u_E = \frac{1}{2}\epsilon_0 E^2$ and $u_B = B^2/2\mu_0$. If $dx$ is sufficiently small, the total energy in the box is the sum of the electric and magnetic energy densities multiplied by the box volume $A\,dx$:

$$dU = (u_E + u_B)A\,dx = \frac{1}{2}\left(\epsilon_0 E^2 + \frac{B^2}{\mu_0}\right)A\,dx$$

This energy moves with speed $c$, so all the energy moves out of the box in a time $dt = dx/c$. The rate at which energy moves through the cross-sectional area $A$ is then

$$\frac{dU}{dt} = \frac{1}{2}\left(\epsilon_0 E^2 + \frac{B^2}{\mu_0}\right)\frac{A\,dx}{dx/c} = \frac{c}{2}\left(\epsilon_0 E^2 + \frac{B^2}{\mu_0}\right)A$$

So the intensity $S$, or rate of energy flow per unit area, is

$$S = \frac{c}{2}\left(\epsilon_0 E^2 + \frac{B^2}{\mu_0}\right)$$

We can recast this equation in simpler form using $E = cB$ and $B = E/c$ for an electromagnetic wave. Replacing one of the $E$'s in $E^2$ with $cB$ and one of the $B$'s in $B^2$ with $E/c$, we have

$$S = \frac{c}{2}\left(\epsilon_0 cEB + \frac{EB}{\mu_0 c}\right) = \frac{1}{2\mu_0}(\epsilon_0\mu_0 c^2 + 1)EB$$

But $c = 1/\sqrt{\epsilon_0\mu_0}$, so $\epsilon_0\mu_0 c^2 = 1$, giving

$$S = \frac{EB}{\mu_0} \qquad (29.19a)$$

Although we derived Equation 29.19a for an electromagnetic wave, it is in fact a special case of the more general result that nonparallel electric and magnetic fields entail a flow of electromagnetic energy. In general, the rate of energy flow per unit area is given by

$$\boxed{\vec{S} = \frac{\vec{E} \times \vec{B}}{\mu_0}} \quad \text{(Poynting vector)} \qquad (29.19b)$$

Here the vector $\vec{S}$ gives the direction of the energy flow as well as its magnitude. For an electromagnetic wave in vacuum, with $\vec{E}$ and $\vec{B}$ at right angles, Equation 29.19b reduces to Equation 29.19a, with the direction of energy flow the same as the direction of wave travel. The vector intensity $\vec{S}$ is called the **Poynting vector** after the English physicist J. H. Poynting, who suggested it in 1884. Problem 64 explores an important application of the Poynting vector to fields that don't constitute an electromagnetic wave.

**FIGURE 29.13** A box of length $dx$ and cross-sectional area $A$ at right angles to the propagation of an electromagnetic wave.

In an electromagnetic wave the fields oscillate, and so does the intensity. We're usually not interested in this rapid oscillation. For example, an engineer designing a solar collector doesn't care that sunlight intensity oscillates at about $10^{14}$ Hz. What she really wants is the *average* intensity, $\overline{S}$. Because the instantaneous intensity of Equation 29.19a contains a product of sinusoidally varying terms, which are in phase, the average intensity is just half the peak intensity:

$$\overline{S} = \frac{\overline{EB}}{\mu_0} = \frac{E_p B_p}{2\mu_0} \qquad \text{(average intensity)} \qquad (29.20a)$$

Typical values for $\overline{S}$ in visible light range from a few $W/m^2$ in the faint light of a candle to many $MW/m^2$ in the most intense laser beams.

We wrote Equation 29.20a in terms of both the electric and magnetic fields, but we can use the wave condition $E = cB$ to eliminate either field in terms of the other:

$$\overline{S} = \frac{E_p^2}{2\mu_0 c} \qquad \text{and} \qquad \overline{S} = \frac{cB_p^2}{2\mu_0} \qquad (29.20b, c)$$

---

**GOT IT? 29.3** Lasers 1 and 2 emit light of the same color, and the electric field in the beam from laser 1 is twice as strong as the field in laser 2's beam. How do their (a) magnetic fields, (b) intensities, and (c) wavelengths compare?

---

**EXAMPLE 29.3** **Fields and Power: Solar Energy**

The average intensity of noontime sunlight on a clear day is about 1 $kW/m^2$. (a) What are the peak electric and magnetic fields in sunlight? (b) At this intensity, what area of 40% efficient solar collectors would you need to replace a 4.8-kW water heater?

**INTERPRET** In (a) we're asked for the peak electric and magnetic fields, and we identify 1 $kW/m^2$ as the average intensity, $\overline{S}$. In (b) we want to use solar collectors to replace an electric heater whose power we're given.

**DEVELOP** For (a), Equations 29.20b ($\overline{S} = E_p^2/2\mu_0 c$) and 29.20c ($\overline{S} = cB_p^2/2\mu_0$), relate peak fields and average intensity. We could solve both of these or, more easily, solve one of them and then use Equation 29.17, $E = cB$, to get the other field. In (b) we'll need to use the 40% efficiency to get the effective power per square meter of solar collector, and then find the area needed to replace the 4.8-kW power of the electric heater.

**EVALUATE** (a) We solve Equation 29.20c for $B_p$: $B_p = \sqrt{2\mu_0 \overline{S}/c} = 2.9\,\mu T$. Then Equation 29.17 gives $E = cB = 0.87$ kV/m. (b) At 40% efficiency, each square meter of solar collector in 1-$kW/m^2$ sunlight produces 0.40 kW. So to get 4.8 kW, we need $(4.8\text{ kW})/(0.40\text{ kW/m}^2) = 12\text{ m}^2$ of collector area.

**ASSESS** The fields we've calculated are relatively modest, showing that even bright sunlight doesn't entail large electric and magnetic fields. That 12-$m^2$ collector area is also modest—much smaller than the 100 $m^2$ of a typical house roof—showing that water heating is a practical use of solar technology. Indeed, your author's house, in cloudy Vermont, gets 95% of its summertime hot water from just 9 $m^2$ of solar collectors. ∎

## Waves from Localized Sources

When a wave originates in a localized source such as an atom, a radio transmitting antenna, a lightbulb, or a star, its wavefronts aren't planes but expanding spheres (recall Fig. 14.13). As it expands, the wave's energy is spread over the area of an ever-larger sphere—whose area increases as the square of the distance from the source. Therefore the power per unit area—the intensity—decreases as the inverse square of the distance:

$$S = \frac{P}{4\pi r^2} \qquad (29.21)$$

Here $S$ and $P$ can be either peak or average intensity and power, and $r$ is the distance from the source. The intensity decreases not because electromagnetic waves "weaken" and lose energy but because their energy gets spread ever more thinly.

Because the intensity of an electromagnetic wave is proportional to the *square* of the field strengths (Equations 29.20), Equation 29.21 shows that the *fields* of a spherical wave decrease as $1/r$. Contrast that with the $1/r^2$ decrease in the electric field of a stationary point charge, and you can see why the wave fields associated with an accelerated charge dominate in all but the immediate vicinity of the charge.

**EXAMPLE 29.4** Electromagnetic-Wave Intensity:
Cell-Phone Reception

A cell phone's typical average power output is about 0.6 W. If the receiver at a cell tower can handle signals with peak electric fields as weak as 1.2 mV/m, what is the maximum allowable distance from cell phone to tower?

**INTERPRET** We're asked to find the distance from the 0.6-W cell phone to a cell tower on the condition that the electric field of the cell phone's electromagnetic wave is no weaker than 1.2 mV/m when measured at the tower.

**DEVELOP** Assuming the 0.6-W signal spreads in all directions, Equation 29.21, $\overline{S} = \overline{P}/4\pi r^2$, gives the average intensity at a distance $r$ from the phone. (Here the bars indicate we're dealing with average quantities.) Using this value for $\overline{S}$ in Equation 29.20b gives $\overline{P}/4\pi r^2 = E_p^2/2\mu_0 c$. Our plan is to solve for the distance $r$ that gives $E_p = 1.2$ mV/m.

**EVALUATE** Solving for $r$ gives $r = \sqrt{2\mu_0 c P/4\pi E_p^2} = 5$ km, using $P = 0.6$ W and $E_p = 1.2$ mV/m.

**ASSESS** This answer is about 3 miles, a bit more than the cell radius discussed in the Application below. That's enough to provide a margin of safety, ensuring reliable communications for all phones within the cell. ∎

---

**APPLICATION** Cell Phones

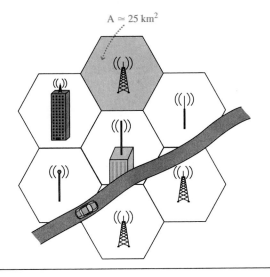

A ≈ 25 km²

Your cell phone contains a tiny, low-power radio transmitter whose signal intensity decreases as the inverse square of the distance from the phone. The cell-phone network consists of antennas and associated circuits that receive and transmit signals from and to individual phones. Because of the phones' low power, antennas need to be closely spaced so a phone is rarely out of range. The figure shows a typical urban cell-phone network consisting of multiple cells—hence the term "cell" phone—each with an antenna mounted on a tower or building. Cells are typically hexagonal regions about 25 km² in area; approximating them as circles gives a radius of about 2.8 km—roughly the maximum distance between a phone and an antenna. As you move through an urban area, the network automatically "hands off" your phone to the nearest cell tower. Cell phones transmit on one frequency and receive on another, allowing two-way communications with both parties able to talk at once. The system uses hundreds to thousands of frequency channels, and thus a single cell tower can handle many simultaneous calls. Cell towers are more widely spaced in rural regions, and phones automatically boost their power to compensate.

---

## Momentum and Radiation Pressure

Moving objects carry not only energy but also momentum. So do electromagnetic waves. Maxwell showed that wave energy $U$ and momentum $p$ are related by $p = U/c$. The wave intensity $\overline{S}$ is the average rate at which the wave carries energy per unit area, and therefore the wave carries momentum per unit area at the rate $\overline{S}/c$. An object that absorbs the wave energy (like a black object exposed to sunlight) absorbs this momentum as well. Newton's law in its general form $F = dp/dt$ shows that the object then experiences a force. Since $\overline{S}/c$ is the rate of momentum absorption per unit area, the result is a **radiation pressure**:

$$P_{rad} = \frac{\overline{S}}{c} \qquad \text{(radiation pressure)} \qquad (29.22)$$

**FIGURE 29.14** Cosmos 1 was to be the first solar-1 sailing spacecraft. Launched in 2005, it unfortunately failed to reach orbit.

Radiation pressure doubles if an object reflects electromagnetic waves, just as bouncing a basketball off the backboard changes the ball's momentum by $2mv$ and therefore delivers momentum $2mv$ to the backboard.

The pressure of ordinary light is tiny and difficult to measure, but high-power lasers can actually levitate small particles. Light pressure has been suggested as a means of driving interstellar "sailing ships" (Fig. 29.14). Finally, the idea that electromagnetic waves carry momentum played a crucial role in Einstein's development of his famous equation $E = mc^2$.

## Big Picture

The big idea here—and one of the biggest ideas in physics—is that electric and magnetic fields together form self-regenerating structures that propagate through space as **electromagnetic waves**. What makes these waves possible is that changing magnetic fields induce electric fields (Faraday's law), and changing electric fields induce magnetic fields (Ampère's law, with Maxwell's modification). Electromagnetic (EM) waves in vacuum consist of electric and magnetic fields perpendicular to each other and to the direction of wave propagation, and in phase.

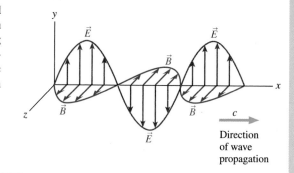

Direction of wave propagation

## Key Concepts and Equations

**Maxwell's equations** describe completely the behavior of electric and magnetic fields in classical physics:

| Law | Mathematical Statement | What It Says |
|---|---|---|
| Gauss for $\vec{E}$ | $\oint \vec{E} \cdot d\vec{A} = \dfrac{q}{\epsilon_0}$ | How charges produce electric field; field lines begin and end on charges. |
| Gauss for $\vec{B}$ | $\oint \vec{B} \cdot d\vec{A} = 0$ | No magnetic charge; magnetic field lines don't begin or end. |
| Faraday | $\oint \vec{E} \cdot d\vec{r} = -\dfrac{d\Phi_B}{dt}$ | Changing magnetic flux produces electric field. |
| Ampère | $\oint \vec{B} \cdot d\vec{r} = \mu_0 I + \mu_0 \epsilon_0 \dfrac{d\Phi_E}{dt}$ | Electric current and changing electric flux produce magnetic field. |

Maxwell's equations show that electromagnetic waves are possible and that their speed in vacuum, the speed of light $c$, is related to the electric and magnetic constants $\epsilon_0$ and $\mu_0$:

$$c = \frac{1}{\sqrt{\epsilon_0 \mu_0}}$$

The value of $c$ is very nearly $3.00 \times 10^8$ m/s. Its exact value, used in defining the meter, is 299,792,458 m/s.

In vacuum, the electric and magnetic fields of a wave are related by

$$E = cB$$

The wave's frequency and wave length are related by

$$f\lambda = c.$$

EM waves can have any wavelength; the whole range constitutes the **electromagnetic spectrum**.

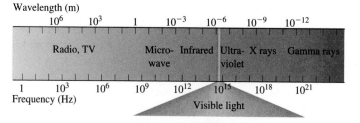

## Cases and Uses

**Polarization** describes the direction of an EM wave's electric field and is a property widely used in scientific research and in technological devices including the ubiquitous liquid crystal displays. When polarized light of intensity $S_0$ is incident on a polarizer with its preferred direction at angle $\theta$ to the polarization, the light emerges with intensity

$$S = S_0 \cos^2\theta$$

EM waves carry both energy and momentum. The **Poynting vector**

$$\vec{S} = \frac{\vec{E} \times \vec{B}}{\mu_0}$$

describes the rate of energy flow per unit area, while the momentum flow results in a **radiation pressure:**

$$P_{rad} = \frac{\bar{S}}{c}$$

# For Thought and Discussion

1. Why is Maxwell's modification of Ampère's law essential to the existence of electromagnetic waves?
2. The presence of magnetic monopoles would require modification of Gauss's law for magnetism. Which of the other Maxwell equations would also need modification?
3. Is there displacement current in an electromagnetic wave? Is there ordinary conduction current?
4. List some similarities and differences between electromagnetic waves and sound waves.
5. The speed of an electromagnetic wave is given by $c = \lambda f$. How does the speed depend on frequency? On wavelength?
6. When astronomers observe a supernova explosion in a distant galaxy, they see a sudden, simultaneous rise in visible light and other forms of electromagnetic radiation. How is this evidence that the speed of light is independent of frequency?
7. Turning a TV antenna so its rods point vertically may change the quality of your TV reception. Why? *Hint:* Think about polarization.
8. The Sun emits about half of its electromagnetic-wave energy in the visible region of the spectrum. Where in the spectrum is most of the remainder?
9. An *LC* circuit is made entirely from superconducting materials, yet its oscillations eventually damp out. Why?
10. If you double the field strength in an electromagnetic wave, what happens to the intensity?
11. The intensity of light falls off as the inverse square of the distance from the source. Does this mean that electromagnetic-wave energy is lost? Explain.
12. Electromagnetic waves don't readily penetrate metals. Why might this be?

# Exercises and Problems

## Exercises

### Section 29.2 Ambiguity in Ampère's Law

13. A uniform electric field is increasing at the rate of 1.5 V/m·µs. What is the displacement current through an area of 1.0 cm² at right angles to the field?
14. A parallel-plate capacitor has square plates 10 cm on a side and 0.50 cm apart. If the voltage across the plates is increasing at the rate of 220 V/ms, what is the displacement current in the capacitor?

### Section 29.4 Electromagnetic Waves

15. The fields of an electromagnetic wave are $\vec{E} = E_p \sin(kz + \omega t)\hat{j}$ and $\vec{B} = B_p \sin(kz + \omega t)\hat{i}$. Give a unit vector in the direction of propagation.
16. The electric field of a radio wave is given by $\vec{E} = E \sin(kz - \omega t)(\hat{i} + \hat{j})$. (a) Find the peak amplitude of the electric field. (b) Give a unit vector in the direction of the magnetic field at a place and time where $\sin(kz - \omega t)$ is positive.

### Section 29.5 Properties of Electromagnetic Waves

17. A light-minute is the distance light travels in 1 minute. Show that the Sun is about 8 light-minutes from Earth.
18. Your intercontinental telephone call is carried by electromagnetic waves routed via a satellite in geosynchronous orbit at an altitude of 36,000 km. Approximately how long does it take before your voice is heard at the other end?
19. An airplane's radar altimeter works by bouncing radio waves off the ground and measuring the round-trip travel time. If that time is 50 µs, what is the altitude?
20. Roughly how long does it take light to go 1 foot?
21. If you speak via radio from Earth to an astronaut on the Moon, how long is it before you can get a reply?
22. What are the wavelengths of (a) a 100-MHz FM radio wave, (b) a 3.0-GHz radar wave, (c) a $6.0 \times 10^{14}$-Hz light wave, and (d) a $1.0 \times 10^{18}$-Hz X ray?
23. A 60-Hz power line emits electromagnetic radiation. What is the wavelength?
24. A microwave oven operates at 2.4 GHz. What is the distance between wave crests in the oven?
25. An electromagnetic wave is propagating in the *z* direction. What is its polarization direction if its magnetic field is in the *y* direction?
26. Polarized light is incident on a sheet of polarizing material, and only 20% of the light gets through. Find the angle between the electric field and the polarization axis of the material.
27. Vertically polarized light passes through a polarizer whose polarization axis is oriented at 70° to the vertical. What fraction of the incident intensity emerges from the polarizer?

### Section 29.8 Energy and Momentum in Electromagnetic Waves

28. A typical laboratory electric field is 1000 V/m. What is the average intensity of an electromagnetic wave with this value for its peak field?
29. What would be the average intensity of a laser beam so strong that its electric field produced dielectric breakdown of air (which requires $E_p = 3 \times 10^6$ V/m)?
30. Estimate the peak electric field inside a 625-W microwave oven under the simplifying approximation that the microwaves propagate as a plane wave through the oven's 750-cm² cross-sectional area.
31. A radio receiver can pick up signals with peak electric fields as weak as 450 µV/m. What is the average intensity of such a signal?
32. A laser blackboard pointer delivers 0.10-mW average power in a beam 0.90 mm in diameter. Find (a) the average intensity, (b) the peak electric field, and (c) the peak magnetic field.
33. A 1.0-kW radio transmitter broadcasts uniformly in all directions. Find the intensity of its signal at a distance of 5.0 km from the transmitter.

## Problems

34. A parallel-plate capacitor has circular plates with radius 50 cm and spacing 1.0 mm. A uniform electric field between the plates is changing at the rate of 1.0 MV/m·s. Find the magnetic field between the plates (a) on the symmetry axis, (b) 15 cm from the axis, and (c) 150 cm from the axis.

35. An electric field points into the page and occupies a circular region of radius 1.0 m, as shown in Fig. 29.15. There are no electric charges in the region, but there is a magnetic field forming closed loops pointing clockwise, as shown. The magnetic-field strength 50 cm from the center of the region is 2.0 $\mu$T. (a) What is the rate of change of the electric field? (b) Is the electric field increasing or decreasing?

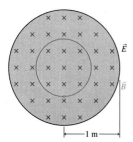

**FIGURE 29.15** Problem 35

36. Vertically oriented antennas for transmitting and receiving electromagnetic radiation usually have lengths on the order of a quarter-wavelength. Measure your cell phone's antenna when it's extended, and estimate the wavelength and frequency of the cell-phone signal.

37. What would be the electric field strength in an electromagnetic wave whose magnetic field equaled that of Earth, about 50 $\mu$T?

38. Dielectric breakdown in air occurs when the electric field is approximately $3 \times 10^6$ V/m. What would be the peak magnetic field in an electromagnetic wave with this value for its peak electric field?

39. A radio receiver can detect signals with electric fields as weak as 320 $\mu$V/m. What is the corresponding magnetic field?

40. A polarizer blocks 75% of a polarized light beam. What is the angle between the beam's polarization and the polarizer's axis?

41. An electro-optic modulator is a device that switches a laser beam rapidly from off to on by switching the polarization direction through 90° when a voltage is applied. But a power-supply failure results in only enough voltage for a 72° rotation. What fraction of the laser light is transmitted when it is supposed to be fully on?

42. Unpolarized light of intensity $S_0$ passes first through a polarizer with its polarization axis vertical and then through one with its axis at 35° to the vertical. Find the light intensity after the second polarizer.

43. Vertically polarized light passes through two polarizers, the first at 60° to the vertical and the second at 90° to the vertical. What fraction of the light gets through?

44. Unpolarized light of intensity $S_0$ is incident on a "sandwich" of three polarizers. The outer two have their transmission axes perpendicular, while the middle one has its axis at 45° to the others. Find the light intensity emerging from this "sandwich."

45. Show by differentiation and substitution that Equations 29.12 and 29.13 are satisfied by fields of the form $E(x, t) = E_p f(kx \pm \omega t)$ and $B(x, t) = B_p f(kx \pm \omega t)$, where $f$ is any function of its argument $kx \pm \omega t$.

46. Show that it's impossible for an electromagnetic wave in vacuum to have a time-varying component of its electric field in the direction of its magnetic field. *Hint:* Assume $\vec{E}$ does have such a component, and show that you can't satisfy both Gauss and Faraday.

47. The United States' safety standard for continuous exposure to microwave radiation is 10 mW/cm². The glass door of a microwave oven measures 40 cm by 17 cm and is covered with a metal screen that blocks microwaves. What fraction of the oven's 625-W microwave power can leak through the door window without exceeding the safe exposure to someone right outside the door? Assume the power leaks uniformly through the window area.

48. Use the fact that sunlight intensity at Earth's orbit is 1368 V/m to calculate the Sun's total power output.

49. A quasar 10 billion light-years from Earth appears the same brightness as a star 50,000 light-years away. How does the power output of the quasar compare with that of the star?

50. At 1.5 km from the transmitter, the peak electric field of a radio wave is 350 mV/m. (a) What is the transmitter's power output, assuming it broadcasts uniformly in all directions? (b) What is the peak electric field 10 km from the transmitter?

51. Find the peak electric and magnetic fields 1.5 m from a 60-W lightbulb that radiates equally in all directions.

52. A typical fluorescent lamp is a little more than 1 m long and a few cm in diameter. How do you expect the light intensity to vary with distance (a) near the lamp but not near either end and (b) far from the lamp?

53. A camera flash delivers 2.5 kW of light power for 1.0 ms. Find (a) the total energy and (b) the total momentum carried by the flash.

54. A laser produces an average power of 7.0 W in a light beam 1.0 mm in diameter. Find (a) the average intensity and (b) the peak electric field of the laser light.

55. What radiation pressure is exerted on a light-absorbing surface by a laser beam whose intensity is 180 W/cm²?

56. A laser beam shines vertically upward. What laser power is necessary for this beam to support a flat piece of aluminum foil with mass 30 $\mu$g and diameter equal to that of the beam? Assume the foil reflects all the light.

57. Serious proposals have been made to "sail" spacecraft to the outer solar system using the pressure of sunlight. What sail area must a 1000-kg spacecraft have if its acceleration at Earth's orbit is to be 1 m/s²? *Note:* Assume the sails are made from reflecting material. You can neglect the sun's gravity. (Why?)

58. A 65-kg astronaut is floating in empty space. If she shines a 1.0-W flashlight in a fixed direction, how long will it take her to accelerate to a speed of 10 m/s?

59. A "photon rocket" emits a beam of light instead of the hot gas of an ordinary rocket. How powerful a light source would be needed for a photon rocket with thrust equal to that of a space shuttle (35 MN)? Compare your answer with humanity's total electric power-generating capability, about $10^{12}$ W.

60. A white dwarf star is approximately the size of Earth but radiates about as much energy as the Sun. Estimate the radiation pressure on a light-absorbing object at the white dwarf's surface.

61. Use appropriate data from Appendix E to calculate the radiation pressure on a light-absorbing object at the Sun's surface.

62. A radar system produces pulses consisting of 100 full cycles of a sinusoidal 70-GHz electromagnetic wave. The average power while the transmitter is on is 45 MW, and the waves are confined to a beam 20 cm in diameter. Find (a) the peak electric field, (b) the wavelength, (c) the total energy in a pulse, and (d) the total momentum in a pulse. (e) If the transmitter produces 1000 pulses per second, what is its average power output?

63. The peak electric field measured at 8.0 cm from a light source is 150 W/m², while at 12 cm it measures 122 V/m. Describe the shape of the source.

64. A cylindrical resistor of length $L$, radius $a$, and resistance $R$ carries a current $I$. Calculate the electric and magnetic fields at the surface of the resistor, assuming the electric field is uniform throughout, including at the surface. Calculate the Poynting vector and show that it points into the resistor. Calculate the flux of the Poynting

vector (that is, $\int \vec{S} \cdot d\vec{A}$) over the surface of the resistor to get the rate of electromagnetic energy flow into the resistor, and show that the result is $I^2R$. Your result shows that the energy heating the resistor comes from the fields surrounding it. These fields are sustained by the source of electric energy that drives the current.

65. The "crest" of the wave in Equation 29.10 or 29.11 is a wavefront parallel to the $y$-$z$ plane at position $x$ and time $t$ with phase equal to $\phi(x, t) = kx - \omega t = \pi/2$. Show by direct differentiation that the crest moves in the direction of the wave at speed $\omega/k$.

66. "Ghosts" on a TV screen occur when part of the signal goes directly from transmitter to receiver, while part takes a longer route, reflecting off mountains or buildings. The picture on a 50-cm-wide TV tube is produced by scanning the electron beam from left to right across the screen in about $10^{-4}$ s. If a "ghost" image appears horizontally displaced about 1 cm from the main image, what is the difference in path length between the direct and indirect signals?

67. In a stack of polarizing sheets, each sheet has its polarization axis rotated 14° with respect to the preceding sheet. If the stack passes 37% of the incident, unpolarized light, how many sheets does it contain?

68. Maxwell's equations in a dielectric resemble those in vacuum (Equations 29.6–29.9) but with $\epsilon_0$ replaced by $\kappa\epsilon_0$, where $\kappa$ is the dielectric constant introduced in Chapter 23. Show that the speed of electromagnetic waves in such a dielectric is $c/\sqrt{\kappa}$.

69. A friend buys a used pickup truck that comes complete with a CB radio. However, the antenna is broken off, and your friend asks you to help make one out of a steel rod that he will affix to the rear bumper. You know that the frequency is 27.3 MHz and that the antenna must be a quarter-wavelength long. How long should you make the rod?

70. A classmate who is majoring in political science is preparing a term paper on global warming. He hasn't taken physics but wants to include some data on human energy use compared with the energy received from the Sun. He knows the human race consumes energy at a rate of about $10^{13}$ W. He read an editorial, in a not-so-reputable newspaper, that this is more power than we receive from the Sun. Your classmate is skeptical about this claim. You make a few assumptions: About two-thirds of the solar energy at Earth's orbit $(1368 \text{ W/m}^2)$ reaches the planet's surface. The energy is incident on the cross-sectional area of the Earth. Was the editorial correct?

71. You're taking astronomy. Your professor mentions the very low-power (21 W) radio transmitter on the Voyager 2 spacecraft that sent data back from Neptune, $4.5 \times 10^9$ km from Earth. The energy is received by the 70-m dish antenna of the Goldstone Deep Space Network. You wonder: How long would the antenna have to collect radio energy to have enough to light a 100-W lightbulb for 1 s? Assume the dish is circular.

72. Your friend who works at the college radio station must make electric-field measurements for a report to be filed with the station's application for license renewal. The measurement is made 4.6 km from the antenna, where your friend finds the electric field to be 380 V/m. The station is allowed to broadcast with 55-kW power. Is the station broadcasting legally? Assume the power spreads equally in all directions.

# Answers to Chapter Questions

## Answer to Chapter Opening Question

Electromagnetic waves, comprising changing electric and magnetic fields, carry not only cell-phone conversations but also TV shows, the energy of sunlight, and signals from physical processes in the farthest reaches of the universe.

## Answers to GOT IT? Questions

29.1 (b).

29.2 No; the middle polarizer lets through half the light that gets through the first one, and the final polarizer lets through half of that light.

29.3 (a) $B_1 = 2B_2$; (b) $S_1 = 4S_2$; (c) $\lambda_1 = \lambda_2$

# Electromagnetism

**Electromagnetism** is a fundamental force of nature. The strong attraction between positive and negative charge makes most bulk matter electrically neutral, and hides from us the essential role electricity and magnetism play in the structure of matter.

Electromagnetic interactions are best described in terms of **electric fields** and **magnetic fields**. Electric charges create electric fields, and electric charges respond to the fields of other charges.

**Moving** electric charges create magnetic fields, and moving electric charges respond to magnetic fields. Both electric and magnetic fields store energy.

A changing magnetic field creates an electric field, and vice versa. Together, changing fields combine to make **electromagnetic waves**—self-replicating structures propagating through empty space at the speed of light, c. Light itself is an electromagnetic wave.

---

**Maxwell's equations** are the four fundamental laws of electromagnetism:

| Law | Mathematical Statement | What It Says |
|---|---|---|
| Gauss for $\vec{E}$ | $\oint \vec{E} \cdot d\vec{A} = \dfrac{q}{\epsilon_0}$ | How charges produce electric field; field lines begin and end on charges. |
| Gauss for $\vec{B}$ | $\oint \vec{B} \cdot d\vec{A} = 0$ | No magnetic charge; magnetic field lines don't begin or end. |
| Faraday | $\oint \vec{E} \cdot d\vec{r} = -\dfrac{d\Phi_B}{dt}$ | Changing magnetic flux produces electric field. |
| Ampère | $\oint \vec{B} \cdot d\vec{r} = \mu_0 I + \mu_0 \epsilon_0 \dfrac{d\Phi_E}{dt}$ | Electric current and changing electric flux produce magnetic field. |

**Coulomb's law** and the **Biot–Savart law** provide alternatives to Gauss's and Ampère's laws for determining electric and magnetic fields of pointlike elements of charge and moving charge, respectively:

$$\vec{E} = \frac{kq}{r^2}\hat{r}$$

$$d\vec{B} = \frac{\mu_0}{4\pi}\frac{I\,d\vec{L} \times \hat{r}}{r^2}$$

$d\vec{L}$ is a small piece of the wire. $\hat{r}$ is a unit vector from $d\vec{L}$ toward P. $d\vec{B}$ is into page.

---

The **electromagnetic force** on a charged particle consists of the **electric force** and the **magnetic force**. Both are proportional to the charge and to the appropriate field; the magnetic force depends also on the particle's velocity $\vec{v}$:

$$\vec{F}_{\text{EM}} = \vec{F}_E + \vec{F}_B = q\vec{E} + q\vec{v} \times \vec{B}$$

The **electric potential difference** describes the work per unit charge needed to move charge between two points in an electric field; its units are N/C or **volts** (V):

$$V_{AB} = -\int_A^B \vec{E} \cdot d\vec{r}$$

---

**Electric current** is a flow of electric charge:

$$I = \frac{\Delta Q}{\Delta t} = nAqv_{\text{d}}$$

$\vec{v}_{\text{d}} \longrightarrow$
n charges/unit volume, each charge q

In ohmic materials, **Ohm's law** relates voltage, current, and resistance $I = V/R$.

**Electric circuits** are interconnections of electric components, including batteries, resistors, and others. They can often be analyzed by considering **series** and **parallel** combinations.

$$R_1 + R_2$$     Series

$$\frac{1}{R} = \frac{1}{R_1} + \frac{1}{R_2}$$     Parallel

---

**Electromagnetic induction**, described by Faraday's law, is the basis of electric generators and a host of other electromagnetic technologies and natural phenomena.

A rightward-moving magnet results in a positive current.

**Electromagnetic waves** result from changing electric and magnetic fields. EM waves include light, and all EM waves propagate in vacuum at the speed of light, $c = 1/\sqrt{\mu_0 \epsilon_0}$.

Direction of wave propagation

---

**Part Four challenge problem**

A wire of length $L$ and resistance $R$ forms a rectangular loop twice as long as it is wide. It's mounted on a nonconducting horizontal axle parallel to its longer dimension, as shown in the figure. A uniform magnetic field $\vec{B}$ points into the page. A long string of negligible mass is wrapped many times around a drum of radius $a$ attached to the axle, and a mass $m$ is attached to the string. When the mass is released, it falls and eventually reaches a speed that, averaged over one cycle of the loop's rotation, is constant. Find an expression for that average speed.

# Optics

Imagine a world without light. We see because light reflects off objects, and our eyes form images because light refracts in our corneas and lenses. When our built-in optical systems aren't perfect, we correct them with additional lenses or we use lasers to reshape the cornea. Microscopes and telescopes extend the range of our vision. The more subtle phenomenon of interference makes possible some of the most precise measurements and is behind the operation of everyday technologies like CDs and DVDs. Light signals carry e-mail, web pages, telephone conversations, and computer data through the optical fibers that form the world's communications networks. Although the behavior of light is ultimately grounded in Maxwell's equations of electromagnetism, we can learn much about light from the simpler perspective of optics. The next three chapters explore the behavior of light, images and optical instruments, and phenomena associated with the wave nature of light.

■ Drops of dew act as miniature optical systems, with light refracting through the drops to form myriad images of the background flowers—which themselves are out of focus in the photographer's camera.

■ Why does the bee's image appear at left, and what does this have to do with e-mail and the Internet?

▶ **To Learn**

By the end of this chapter you should be able to

■ Describe the reflection of light from plane surfaces (30.1).

■ Describe quantitatively the refraction of light at the interface between two transparent materials (30.2).

■ Recognize total internal reflection, and determine quantitatively the circumstances when it occurs (30.3).

■ Explain how the wavelength dependence of refraction leads to dispersion (30.4).

◀ **To Know**

■ This chapter is, in principle, based on the electromagnetic nature of light as described by Maxwell's equations.

■ Here we'll develop a simpler approach that applies when light interacts with objects much larger than its wavelength.

Maxwell's brilliant work shows that the phenomena of **optics**—the behavior of light—are manifestations of electromagnetism. Except in the atomic realm, where quantum physics reigns, all optical phenomena are understandable in terms of electromagnetic-wave fields described by Maxwell's equations. But when objects with which light or any other wave interacts are much larger than the wavelength, light generally travels in straight lines called **rays**. **Geometrical optics** describes the behavior of light when the ray approximation is valid. Here we'll use geometrical optics to explore the behavior of light at interfaces between different materials. In Chapter 31 we'll see how geometrical optics explains lenses, the human eye, and many optical instruments.

## 30.1 Reflection

Some materials, notably metals, **reflect** nearly all the light incident on them. It's no coincidence that these materials are also good electrical conductors. The oscillating electric field of a light wave drives a metal's free electrons into oscillatory motion, which, in turn, produces electromagnetic waves. The net effect is to reradiate the wave back into the original medium. Other materials reflect only part of the incident light. Either way, reflection satisfies the same geometrical conditions: The incident ray, the reflected ray, and the normal to

**FIGURE 30.1** (a) Angles of reflection and incidence are equal. (b) In specular reflection, a smooth surface reflects a light beam undistorted. (c) A rough surface results in diffuse reflection.

the interface between two materials all lie in the same plane. The **angle of reflection** $\theta_1'$ that the reflected ray makes with the normal is the same as the **angle of incidence** $\theta_1$ made by the incident ray (Fig. 30.1a):

$$\theta_1' = \theta_1 \tag{30.1}$$

where the subscript 1 denotes the first medium.

In **specular reflection**, parallel rays reflect off a smooth surface and the entire beam is reflected without distortion (Fig. 30.1b). In contrast, a rough surface reflects individual rays in different directions (Fig. 30.1c)—even though each ray still obeys Equation 30.1. This is **diffuse reflection**. White paper is a diffuse reflector, while the aluminum coating of a mirror is an excellent specular reflector.

**EXAMPLE 30.1** **Reflection: The Corner Reflector**

Two mirrors join at right angles. Show that any light ray incident in the plane of the page will return antiparallel to its incident direction.

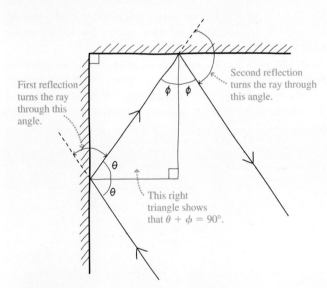

First reflection turns the ray through this angle.

Second reflection turns the ray through this angle.

This right triangle shows that $\theta + \phi = 90°$.

**FIGURE 30.2** A two-dimensional corner reflector.

**INTERPRET** We've sketched the situation in Fig. 30.2. We're asked to show that the lines representing incident and outgoing rays in the figure are parallel.

**DEVELOP** The outgoing ray will be antiparallel to the incident ray if the light turns through a total of 180°. Applying Equation 30.1 to each of the two reflections shows that the angles $\theta$ are the same, and so are the angles $\phi$. Figure 30.2 shows that the first reflection turns the incident ray through an angle $180° - 2\theta$. Similarly, the second turns it through $180° - 2\phi$.

**EVALUATE** The pair of mirrors thus turns the ray through a total angle of

$$(180° - 2\theta) + (180° - 2\phi) = 360° - 2(\theta + \phi).$$

But Fig. 30.2 shows that $\theta + \phi = 90°$, so the total angle is $360° - 180° = 180°$—which is what we set out to prove.

**ASSESS** Here we've explored a remarkable device, a pair of perpendicular mirrors that returns a light ray in exactly the direction from which it came—provided the light is in the plane perpendicular to both mirrors. Add a third mirror at right angles to the other two, and you have a **corner reflector**—a device that returns a light ray in the direction from which it came, period. Corner reflectors, often made with prisms rather than mirrors, are widely used in optics. A corner reflector left on the Moon allows laser-based measurements of the Moon's distance with an accuracy of a few centimeters. ∎

## Partial Reflection

Some light is reflected even at the interface with a transparent material. The detailed description of such partial reflection follows from Maxwell's equations, and is akin to the partial reflection of waves on strings that we described with Fig. 14.24; Fig. 14.25 showed partial reflection in light. In general, the least reflection occurs with normal incidence; for glass, about 4% of normally incident light is reflected. However, the reflection increases with larger angles of incidence. Camera lenses, solar photovoltaic cells, and other devices are often given special antireflection coatings to reduce light loss due to reflection.

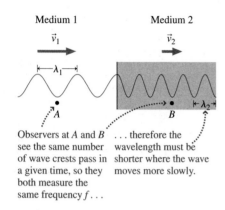

Medium 1      Medium 2

$\vec{v}_1$      $\vec{v}_2$

$\lambda_1$      $\lambda_2$

$A$      $B$

Observers at $A$ and $B$ see the same number of wave crests pass in a given time, so they both measure the same frequency $f$ . . .

. . . therefore the wavelength must be shorter where the wave moves more slowly.

**FIGURE 30.3** Wave frequency doesn't change as a wave goes from one medium to another, but wavelength does change.

**TABLE 30.1** Indices of Refraction*

| Substance | Index of Refraction, $n$ |
|---|---|
| Gases | |
|   Air | 1.000293 |
|   Carbon dioxide | 1.00045 |
| Liquids | |
|   Water | 1.333 |
|   Ethyl alcohol | 1.361 |
|   Glycerine | 1.473 |
|   Benzene | 1.501 |
|   Diiodomethane | 1.738 |
| Solids | |
|   Ice ($H_2O$) | 1.309 |
|   Polystyrene | 1.49 |
|   Glass | 1.5–1.9 |
|   Sodium chloride | |
|     (NaCl) | 1.544 |
|   Diamond (C) | 2.419 |
|   Rutile ($TiO_2$) | 2.62 |

*At 1 atm pressure and temperatures ranging from 0°C to 20°C, measured at a wavelength of 589 nm (the yellow line of sodium).

## 30.2 Refraction

We saw in Chapter 14 that wave speeds differ in different media. With light, the speed in transparent media is lower than in vacuum. We characterize a transparent medium by its **index of refraction**, defined as the ratio of the speed of light $c$ in vacuum to the speed of light $v$ in the medium:

$$n = \frac{c}{v} \quad \text{(index of refraction)} \tag{30.2}$$

Although the wave speed changes when light enters a new medium, Fig. 30.3 shows that its frequency $f$ can't change and therefore, since the wave speed is $v = f\lambda$, the wavelength must change. Equation 30.2 shows that the wavelength in a medium with refractive index $n$ is $\lambda = v/f = c/nf$; since $c$ and $f$ don't change, the wavelength is inversely proportional to $n$. Table 30.1 lists some refractive indices.

When light is incident at an angle on a transparent material, the light transmitted into the material undergoes **refraction**—a change in its propagation direction (Fig. 30.4). Figure 30.5 shows how refraction results from the change in wave speed and therefore wavelength. Here we assume the refractive index is higher in medium 2; our result $\lambda = c/nf$ then shows that the wavelength is shorter in medium 2. We've shaded two right triangles with a common hypotenuse and one side equal to the appropriate wavelength. The angles opposite these sides are the angles of incidence and refraction. In each case the hypotenuse is given by $\lambda/\sin\theta$. Equating expressions for this common hypotenuse gives $\lambda_1/\sin\theta_1 = \lambda_2/\sin\theta_2$. Since $\lambda = c/nf$ with $f$ the same in both media, we get **Snell's law**:

$$n_1 \sin\theta_1 = n_2 \sin\theta_2 \quad \text{(Snell's law)} \tag{30.3}$$

First developed geometrically in 1621 by van Roijen Snell of the Netherlands, and described analytically in the 1630s by René Descartes in France, Snell's law lets us predict what will happen at an interface given the refractive indices of the two media.

Snell's law applies whether light goes from a medium of lower to higher refractive index or vice versa. In going from a medium with a lower refractive index to a medium with a higher refractive index, the beam bends *toward* the normal. In the opposite case, where light travels from a higher refractive index to a lower refractive index, the beam bends *away* from the normal.

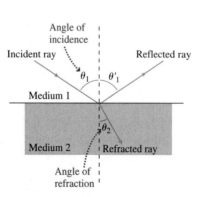

Angle of incidence

Incident ray      Reflected ray

$\theta_1$   $\theta_1'$

Medium 1

Medium 2    $\theta_2$

Refracted ray

Angle of refraction

**FIGURE 30.4** Refraction and reflection at an interface, here when medium 2 has the higher refractive index.

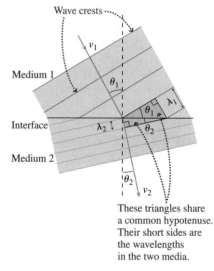

Wave crests

$v_1$

Medium 1

$\theta_1$

Interface    $\lambda_2$    $\theta_1$   $\lambda_1$

Medium 2    $\theta_2$

$\theta_2$   $v_2$

These triangles share a common hypotenuse. Their short sides are the wavelengths in the two media.

**FIGURE 30.5** Refraction occurs because wave speed and wavelength differ in the two media.

EXAMPLE 30.2    **Refraction: A Plane Slab**

A light ray propagating in air strikes a glass slab of thickness $d$ and refractive index $n$ at incidence angle $\theta_1$. Show that it emerges from the slab propagating parallel to its original direction.

**INTERPRET** This is a problem about refraction, which in this case occurs twice. We identify the two interfaces as first the air–glass interface and then the glass–air interface where the light exits the glass.

**DEVELOP** Figure 30.6 is a sketch showing the path of the light through the glass. There are two pairs of incidence and refractive angles, which we labeled $\theta_1$, $\theta_2$ and $\theta_3$, $\theta_4$. Our plan is to apply Snell's law at each interface and thus prove that $\theta_4 = \theta_1$. At the air–glass interface, we have $n_1 \sin \theta_1 = n_2 \sin \theta_2$, and at the glass–air interface, $n_3 \sin \theta_3 = n_4 \sin \theta_4$. Note that $\theta_1$ and $\theta_4$ are in the same medium (air), so we set $n_4 = n_1$. Similarly, $n_3 = n_2$ for $\theta_2$ and $\theta_3$ in glass. Then our equations become $n_1 \sin \theta_1 = n_2 \sin \theta_2$ and $n_2 \sin \theta_3 = n_1 \sin \theta_4$.

**EVALUATE** Taking $n_1 = 1$ for air and $n_2 = n$ for glass at the air–glass interface, we have $\sin \theta_2 = \sin \theta_1 / n$. But at the glass–air interface, $n_1 = n$ and $n_2 = 1$, so here $\sin \theta_4 = n \sin \theta_3$. But the slab faces are parallel, so $\theta_3 = \theta_2$. So $\sin \theta_4 = n \sin \theta_2$. Using our expression for $\theta_2$ from the first interface then gives

$$\sin \theta_4 = n \left( \frac{\sin \theta_1}{n} \right) = \sin \theta_1$$

showing that the incident and outgoing rays are indeed parallel.

**FIGURE 30.6** Light passing through a transparent slab.

**ASSESS** This result shows that light isn't deflected when it passes through a parallel-faced slab of transparent material. It is, however, displaced by the distance $x$ shown in Fig. 30.6. You can find that displacement in Problem 58. ∎

EXAMPLE 30.3    **Refraction: CD Music**

The laser beam that "reads" information from a compact disc is 0.737 mm wide when it strikes the disc, and it forms a cone with half-angle $\theta_1 = 27.0°$ as shown in Fig. 30.7. It then passes through a 1.20-mm-thick layer of plastic with refractive index 1.55 before reaching the reflective information layer near the disc's top surface. What is the beam diameter $d$ at the information layer?

**INTERPRET** Rays defining the outer edge of the beam refract toward the normal, making a smaller convergence angle within the disc. We're asked how much this converging beam narrows when it reaches the information layer.

**DEVELOP** Snell's law will give us the angles $\theta_2$ in the figure: $n_1 \sin \theta_1 = n_2 \sin \theta_2$. Given $\theta_2$, we can use trigonometry to find the distance $x$ marked in the drawing: $x = t \tan \theta_2$, with $t$ the 1.2-mm thickness. The beam diameter $d$ then follows from $d = D - 2x$, where $D$ is the 0.737-mm diameter when the beam strikes the disc surface.

**EVALUATE** With $n_1 = 1$ and $n_2 = 1.55$, Snell's law gives $\theta_2 = \sin^{-1}(\sin \theta_1 / n_2) = 17.03°$, so $d = D - 2x = D - 2t \tan \theta_2 = 1.80 \ \mu m$.

**ASSESS** This answer makes sense because $d$ is just a bit larger than the "pits" cut into the CD to store information. Narrowing of the laser

**FIGURE 30.7** Section through a compact disc, showing convergence of the laser beam to a narrow spot at the information layer.

beam plays a crucial role in keeping CDs noise free. The tiniest dust speck would blot out information at the $\mu$m-scale information layer, but at the point where the beam enters the disc, it would take mm-size dust to cause problems. We'll explore CD and DVD technology further in Chapter 32. ∎

**GOT IT? 30.1** The figure shows the path of a light ray through three different media. Rank the media according to their refractive indices.

## Multiple and Continuous Refraction

Light passes through multiple layers of transparent material by refracting at each interface, as suggested in Got It? 30.1. In some situations, including the human eye and Earth's atmosphere, the refractive index varies continuously. Then light refracts continuously, following a curved path. The shimmering mirages you see on a hot road result from this effect, as sky light refracts so it appears to be coming from the road surface (Fig. 30.8).

**FIGURE 30.8** Air's temperature-dependent refractive index results in the shimmering mirages you see on highways. What you're actually seeing is refracted sky light.

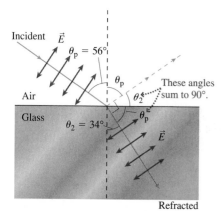

**FIGURE 30.9** The polarizing incidence angle $\theta_p$ occurs when the angles of incidence and refraction sum to 90°.

## Refraction, Reflection, and Polarization

Both reflection and refraction ultimately involve interactions between the incident light wave's electric field and charges in a material. With refraction, the oscillation of molecular dipoles in response to the field gives rise to both refracted and reflected light. It's not surprising, therefore, that details depend on the direction of the electric field—that is, on polarization. When the field lies in the plane defined by the incident and reflected rays, there's a special angle of incidence at which no reflection occurs. This is the **Brewster angle**, or **polarizing angle**, and it occurs when the reflected ray would be perpendicular to the transmitted ray (Fig. 30.9). Then the molecular dipoles are oscillating along the direction a reflected ray would take, and, as we saw in Section 29.7, there's no electromagnetic radiation from an oscillating dipole along the oscillation direction.

Figure 30.9 shows that the polarizing incidence angle $\theta_p$ occurs when $\theta_p$ and the angle $\theta_2$ of the reflected ray sum to 90°; equivalently, $\theta_2 = 90° - \theta_p$. Since $\sin\theta = \cos(90° - \theta)$, that means $\sin\theta_2 = \cos\theta_p$. Now, Snell's law gives $\sin\theta_2 = (n_1/n_2)\sin\theta_p$; substituting $\cos\theta_p$ for $\sin\theta_2$, this becomes $\cos\theta_p = (n_1/n_2)\sin\theta_p$. Multiplying both sides by $(n_2/n_1)$ and dividing by $\cos\theta_p$ then gives

$$\tan\theta_p = \frac{n_2}{n_1} \quad \text{(polarizing angle)} \tag{30.4}$$

For the air–glass interface shown in Fig. 30.9, $\theta_p$ is about 56°.

When unpolarized light is incident at the polarizing angle, only the component of the light's electric field that's perpendicular to the plane of Fig. 30.9 gets reflected, and the result is polarized light. Polarization using this effect is important in a number of technologies, including lasers. The window through which light emerges from a laser is usually cut at the polarizing angle, and as a result most laser light is intrinsically polarized. A similar polarizing phenomenon occurs for reflection from metals and other opaque surfaces, and it results in a glare that polarizing sunglasses can eliminate.

## 30.3 Total Internal Reflection

Light propagating from a medium with a higher refractive index to a medium with a lower one is bent *away* from the normal, as shown for a glass–air interface in Fig. 30.10. In other words, the angle of refraction in this case is larger than the angle of incidence. So at some incidence angle, the angle of refraction becomes 90°. Then what?

**FIGURE 30.10** Light propagating in glass is refracted away from the normal at the glass–air interface. Ray 3, incident at the critical angle $\theta_c$, just skims along the interface. At larger incidence angles (ray 4), the light undergoes total internal reflection. The rightmost incident beam in the photo undergoes two total internal reflections.

As Fig. 30.10 shows, light incident at this **critical angle** or larger cannot escape from the glass. Instead, **total internal reflection** occurs, returning the light to the medium with the larger refractive index. We can find the critical angle by setting $\theta_2 = 90°$ in Snell's law (Equation 30.3). The critical angle is then $\theta_1$, and we have

$$\sin\theta_c = \frac{n_2}{n_1} \quad \text{(critical angle)} \quad (30.5)$$

Total internal reflection makes uncoated glass an excellent reflector when it's oriented appropriately (Fig. 30.11). Binoculars owe their compact size to glass prisms that reflect light internally to provide a longer light path. For an underwater observer, the existence of the critical angle affects the view of the outside world, as the next example shows. Finally, total internal reflection is the basis of the optical fibers that carry signals over the global Internet, as the Application on the next page describes.

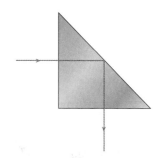

**FIGURE 30.11** Light undergoes total internal reflection in a glass prism.

---

**EXAMPLE 30.4** **Total Internal Reflection: Whale Watch!**

Planeloads of whale watchers fly over the ocean. Within what range of viewing angles can the whale see the planes?

**INTERPRET** The whale is underwater, so it's in a medium with a higher refractive index than air. Therefore some of the light it sees coming from the water surface is from above the water, and some is internally reflected from within the water. The critical angle marks the distinction. Figure 30.12 shows the situation.

**DEVELOP** The critical angle forms a cone about the whale, as shown. The whale needs to look within this cone to see objects above the water, so our plan is to calculate the critical angle.

**EVALUATE** Equation 30.5 gives $\theta_c = \sin^{-1}(1/1.333) = 48.6°$, where we found the refractive index of water in Table 30.1 and used $n = 1$ for air.

**ASSESS** If you've ever looked upward from underwater, you've shared the whale's sense that the entire outside world is compressed into a cone of half-angle $\theta_c$. Beyond this cone, the whale sees reflections of objects below the surface. ∎

**FIGURE 30.12** The whale sees the entire world above the surface in a cone of half-angle $\theta_c$; beyond that, it sees reflections of objects below the surface.

---

**GOT IT? 30.2** The glass prism in Fig. 30.11 has $n = 1.5$ and is surrounded by air $(n = 1)$. What would happen to the incident light ray shown if the prism were immersed in water $(n = 1.333)$?

---

**Optical Fibers**

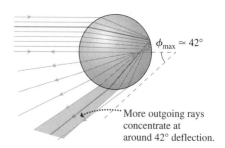

Refraction and total internal reflection are the basis for **optical fibers**, which today carry much of the world's communications. Optical fibers provide the physical connectivity of the global Internet and handle information ranging from telephone and television to light signals within medical, astronomical, and industrial instruments.

A typical fiber consists of a glass core only 8 $\mu$m in diameter, surrounded by a so-called cladding consisting of glass with a lower refractive index. Total internal reflection at the core–cladding interface guides light along the fiber, as shown in the figure. The glass used in optical fibers is so clear that a 1-km-thick slab would be as transparent as an ordinary window pane. Today's fibers carry light produced by semiconductor lasers at infrared wavelengths of 850, 1350, or 1550 nm.

The main advantage of optical fibers over copper wires or open-air electromagnetic-wave transmission is their huge rate of information flow, called **bandwidth**. Communicating information—audio, video, or digital data—requires a range of frequencies, and the greater the rate of information transfer, the wider that range must be. Telephone, for example, requires a few kHz, while a conventional TV signal occupies about 6 MHz. With its frequency of around $10^{14}$ Hz, light can accommodate a much wider swing in frequencies within a channel than can conventional microwave communication systems at $10^{10}$ Hz. A single optical fiber, for example, can carry tens of thousands of simultaneous telephone conversations. Fibers are also lighter and more rugged than copper cables, and they're less vulnerable than copper or open-air transmission to illicit tapping. And because they're made from insulators, optical fibers are less susceptible to electrical noise. The photo shows two cables that can carry information at the same rate. One consists of a few optical fibers while the other is a thick bundle of copper wires.

---

## 30.4 Dispersion

**FIGURE 30.13** Dispersion separates the colors in white light, with shorter-wavelength violet experiencing the greatest refraction.

Refraction ultimately involves the interaction of electromagnetic-wave fields with atomic electrons. It's not surprising that the electrons' behavior and therefore also the refractive index depend on frequency. That means different frequencies—different colors of visible light—refract through different angles. The classic example of this **dispersion** is Newton's demonstration that white light is a mixture of all colors in the visible spectrum (Fig. 30.13). The rainbow is a natural manifestation of dispersion (Fig. 30.14). The primary rainbow results from a single total internal reflection within raindrops (Fig. 30.15); the fainter secondary and tertiary rainbows involve multiple reflection. You can explore the rainbow further in Problems 59–61.

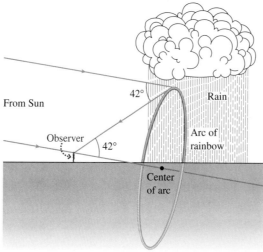

**FIGURE 30.14** The rainbow is a circular arc located at 42° from the line that connects the Sun, the observer, and the center of the arc.

**FIGURE 30.15** The primary rainbow results from total internal reflection in raindrops that concentrates light at approximately 42° deflection. Dispersion separates wavelengths slightly, resulting in the rainbow's colors.

Dispersion is the basis of **spectroscopy**, the analysis of light and other electromagnetic radiation in terms of its constituent wavelengths. Hot, dense objects emit a continuous range of wavelengths, while diffuse gases emit and absorb radiation at only a few specific wavelengths (Fig. 30.16). Such discrete spectra provide some of the strongest evidence for the nature of atoms, and today spectroscopy is a powerful tool throughout the sciences. Spectroscopy helps astronomers to determine the composition and motions of distant astrophysical objects, geologists to identify minerals, and chemists to study molecules. Although early spectroscopy used prisms, most modern instruments use instead diffraction gratings, which we'll describe in Chapter 32.

Dispersion can be a nuisance in optical systems. Glass lenses, for example, focus different colors at different points, resulting in a distortion known as *chromatic aberration*. Dispersion in optical fibers—based not on wavelength but on different paths taken by rays reflecting at different angles—can degrade the sharp pulses representing digital information. So-called single-mode fibers reduce this effect by passing only those rays that have a single specific reflection angle. On the other hand, dispersion of radio waves in Earth's atmosphere provides a crucial correction to the global positioning system (GPS). The ionization level in the upper atmosphere varies, and this introduces an uncertain but frequency-dependent variation in the travel time for radio waves from GPS satellites. It's this travel time that provides GPS location information. Sending waves at two different frequencies and comparing their travel times reveals the atmospheric conditions, and lets more sophisticated dual-frequency GPS receivers reduce a 5-m error to only a few centimeters.

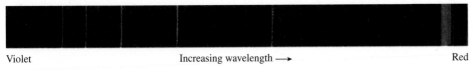

Violet           Increasing wavelength $\longrightarrow$           Red

**FIGURE 30.16** The spectrum of a diffuse gas—here hydrogen—consists of light at discrete wavelengths.

## Big Picture

The big idea here is that light can be considered to travel in straight **rays** when the objects with which it interacts are much larger than the wavelength. Under these conditions, light rays **reflect** and **refract** at interfaces between different materials.

## Key Concepts and Equations

The **angle of incidence** and **angle of reflection** are equal:

$$\theta_1' = \theta_1$$

Incident ray     Reflected ray

**Snell's law** relates the angle of incidence and angle of refraction:

$$n_1 \sin\theta_1 = n_2 \sin\theta_2$$

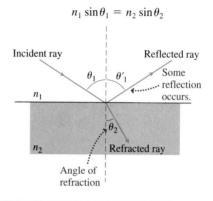

## Cases and Uses

**Total internal reflection** results when light is incident at greater than the **critical angle**, $\theta_c$, on an interface to a medium with lower refractive index:

$$\sin\theta_c = \frac{n_2}{n_1}$$

Light polarized in the plane of the incident and refracted rays undergoes no reflection at an interface; this special **polarizing angle**, $\theta_p$, is given by

$$\tan\theta_p = \frac{n_2}{n_1}$$

For an air–glass interface, $\theta_p \simeq 56°$.

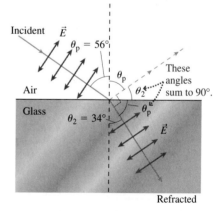

**Dispersion** results from the wavelength dependence of the refractive index and causes different colors to refract at different angles.

A combination of total internal reflection and dispersion in raindrops accounts for the rainbow.

More outgoing rays concentrate at around 42° deflection.

Total internal reflection guides signals in optical fibers.

# For Thought and Discussion

1. It's usually inappropriate to consider low-frequency sound waves as traveling in rays. Why? Why is it more appropriate for high-frequency sound and for light?

2. Describe why a spoon appears bent when it's in a glass of water.

3. Why do a diamond and an identically shaped piece of glass sparkle differently?

4. White light goes from air through a glass slab with parallel surfaces. Will its colors be dispersed when it emerges from the glass?

5. You send white light through two identical glass prisms, oriented as shown in Fig. 30.17. Describe the beam that emerges from the right-hand prism.

**FIGURE 30.17** For Thought and Discussion 5

6. What does a fish see as it looks around in directions above the horizontal? Explain.

7. In glass, which end of the visible spectrum has the smallest critical angle for total internal reflection?

8. Why can't you walk to the end of the rainbow?

9. What's wrong with the painting in Fig. 30.18, which shows rainbows over Niagara Falls? *Hint:* The rainbow subtends a half-angle of 42°.

**FIGURE 30.18** For Thought and Discussion 9. The painting is Harry Fenn's *Niagara*.

10. Why are polarizing sunglasses better than glasses that simply cut down on the total amount of light?

11. Under what conditions will the polarizing angle be smaller than 45°?

# Exercises and Problems

## Exercises

### Section 30.1 Reflection

12. Through what angle should you rotate a mirror so that a reflected ray rotates through 30°?

13. The mirrors in Fig. 30.19 make a 60° angle. A light ray enters parallel to the symmetry axis, as shown. (a) How many reflections does it make? (b) Where and in what direction does it exit the mirror system?

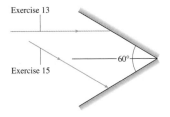

**FIGURE 30.19** Exercises 13 and 15 and Problem 29

14. To what angular accuracy must two ostensibly perpendicular mirrors be aligned so that an incident ray returns on a path within 1° of its incident direction?

15. If a light ray enters the mirror system of Fig. 30.19 propagating in the plane of the page and parallel to one mirror, through what angle will it be turned?

### Section 30.2 Refraction

16. In which substance in Table 30.1 does the speed of light have the value $2.292 \times 10^8$ m/s?

17. Information in a compact disc is stored in "pits" whose depth is essentially one-fourth the wavelength of the laser light used to "read" the information. That wavelength is 780 nm in air, but the wavelength on which the pit depth is based is measured in the $n = 1.55$ plastic that makes up most of the disc. Find the pit depth.

18. Light is incident on an air–glass interface, and the refracted light in the glass makes a 40° angle with the normal to the interface. The glass has refractive index 1.52. Find the incidence angle.

19. A light ray propagates in a transparent material at 15° to the normal to the surface. When it emerges into the surrounding air, it makes a 24° angle with the normal. Find the refractive index of the material.

20. Light propagating in the glass $(n = 1.52)$ wall of an aquarium tank strikes the interior edge of the wall with incidence angle 12.4°. What is the angle of refraction in the water?

21. Find the polarizing angle for diamond when light is incident from air.

22. Find the refractive index of a material for which the polarizing angle in air is 62°.

### Section 30.3 Total Internal Reflection

23. Find the critical angle for total internal reflection in (a) ice, (b) polystyrene, and (c) rutile. Assume the surrounding medium is air.

24. A drop of water is trapped in a block of ice. What is the critical angle for total internal reflection at the water–ice interface?

25. What is the critical angle for light propagating in glass with $n = 1.52$ when the glass is immersed in (a) water, (b) benzene, and (c) diiodomethane?

26. Total internal reflection occurs at an interface between plastic and air at incidence angles larger than 37°. What is the refractive index of the plastic?

### Section 30.4 Dispersion

27. Blue and red laser beams strike an air–glass interface with incidence angle 50°. If the glass has refractive indices of 1.680 and 1.621 for the blue and red light, respectively, what will be the angle between the two beams in the glass?

28. White light propagating in air is incident at 45° on the equilateral prism of Fig. 30.20. Find the angular dispersion $\gamma$ of the outgoing beam if the prism has refractive indices $n_{red} = 1.582$ and $n_{violet} = 1.633$.

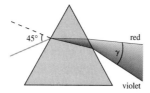

**FIGURE 30.20** Exercise 28 (angles of dispersed rays aren't accurate).

## Problems

29. Suppose the angle in Fig. 30.19 is changed to 75°. A ray enters the mirror system parallel to the axis. (a) How many reflections does it make? (b) Through what angle is it turned when it exits the system?

30. Two plane mirrors make an angle $\phi$. For what value of $\phi$ will a light ray reflecting once off each mirror be turned through the same angle $\phi$?

31. Two plane mirrors make an angle $\phi$. A light ray enters the system and is reflected once off each mirror. Show that the ray is turned through an angle $360° - 2\phi$.

32. A block of glass with $n = 1.52$ is submerged in one of the liquids listed in Table 30.1. For a ray striking the glass with incidence angle 31.5°, the angle of refraction is 27.9°. What is the liquid?

33. A meter stick lies on the bottom of the rectangular tank in Fig. 30.21, with its zero mark at the left edge of the tank. You look into the long dimension of the tank at a 45° angle, with your line of sight just grazing the top edge of the tank, as shown. What mark on the meter stick do you see when the tank is (a) empty, (b) half full of water, and (c) full of water?

**FIGURE 30.21** Problem 33

34. You look at the center of one face of a solid cube of glass, on a line of sight making a 55° angle with the normal to the cube face. What is the minimum refractive index of the glass for which you will see through the opposite face of the cube?

35. The cylindrical tank in a public aquarium is 10 m deep, 11 m in diameter, and full to the brim with water. If a flashlight shines on the tank from above, what is the smallest angle its beam can make with the horizontal if it is to illuminate part of the tank bottom?

36. You're standing 2.3 m horizontally from the edge of a 4.5-m-deep lake, with your eyes 1.7 m above the water surface. A diver holding a flashlight at the lake bottom shines the light so you can see it. If the light in the water makes a 42° angle with the vertical, at what horizontal distance is the diver from the edge of the lake?

37. You've dropped your car keys at night off the end of a dock into water 1.6 m deep. A flashlight held directly above the dock edge and 0.50 m above the water illuminates the keys when it's pointed at 40° to the vertical, as shown in Fig. 30.22. What is the horizontal distance $x$ from the edge of the dock to the keys?

**FIGURE 30.22** Problem 37

38. A light ray is propagating in a crystal where its wavelength is 540 nm. It strikes the interior surface of the crystal with an incidence angle of 34° and emerges into the surrounding air at 76° to the surface normal. Find (a) the light's frequency and (b) its wavelength in air.

39. The prism in Fig. 30.23 has $n = 1.52$ and $\alpha = 60°$ and is surrounded by air. A light beam is incident at $\theta_1 = 37°$. Find the angle $\delta$ through which the beam is deflected.

**FIGURE 30.23** Problems 39, 40, and 62

40. Repeat Problem 39 for the case $n = 1.75$, $\alpha = 40°$, and $\theta_1 = 25°$.

41. Find the minimum refractive index for the prism in Fig. 30.11 if total internal reflection is to occur as shown when the prism is surrounded by air.

42. Where and in what direction would the main beam emerge if the prism in Fig. 30.11 were made of ice?

43. What is the speed of light in a material for which the critical angle at an interface with air is 61°?

44. The prism of Fig. 30.11 has $n = 1.52$. When it's immersed in a liquid, a beam incident as shown in the figure ceases to undergo total reflection. What is the minimum value for the liquid's refractive index?

45. For the interface between air (refractive index 1) and a material with refractive index $n$, show that the critical angle and the polarizing angle are related by $\sin\theta_c = \cot\theta_p$.

46. A scuba diver sets off a camera flash a distance $h$ below the surface of water with refractive index $n$. Show that light emerges from the water surface through a circle of diameter $2h/\sqrt{n^2 - 1}$.

47. Suppose the red and blue beams of Exercise 27 are now propagating in the same direction *inside* the glass. For what range of incidence angles on the glass–air interface will one beam be totally reflected and the other not?

48. A compound lens is made from crown glass $(n = 1.52)$ bonded to flint glass $(n = 1.89)$. What is the critical angle for light incident on the flint–crown interface?

49. Find a simple expression for the speed of light in a material in terms of $c$ and the critical angle at an interface between the material and vacuum.

50. What is the polarizing angle for light incident from below on the surface of a pond?

51. Light propagating in air strikes a transparent crystal at incidence angle 35°. If the angle of refraction is 22°, what is the speed of light in the crystal?

52. A cylindrical tank 2.4 m deep is full to the brim with water. Sunlight first hits part of the tank bottom when the rising Sun makes a 22° angle with the horizon. Find the tank's diameter.

53. For what diameter tank in Problem 52 will sunlight strike some part of the tank bottom whenever the Sun is above the horizon?

54. Light is incident from air on the flat wall of a polystyrene water tank. If the incidence angle is 40°, what is the angle of refraction of the light ray in the water?

55. A parallel-sided slab with refractive index $n_2$ separates two media with indices $n_1$ and $n_3$. Show that a light ray incident on the $n_1$–$n_2$ interface then enters the third medium with the same angle of refraction it would have had if the slab had not been present. Assume total internal reflection doesn't occur at either interface.

56. The surfaces of a glass sheet are not quite parallel, but rather make an angle of 10°. The glass has refractive index $n = 1.52$. A light beam strikes one side of the sheet at incidence angle 35°,

coming in on the thicker side of the normal. Through what angle is the beam direction changed when it emerges on the opposite side of the glass sheet?

57. A cubical block is made from two equal-size slabs of materials with different refractive indices, as shown in Fig. 30.24. Find the index of the right-hand slab if a light ray is incident on the center of the left-hand slab and then describes the path shown.

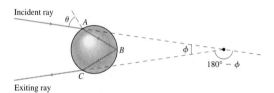

**FIGURE 30.24** Problem 57

58. Find an expression for the displacement $x$ in Fig. 30.6, in terms of $\theta_1$, $d$, and $n$.

59. Figure 30.25 shows a light ray passing through a spherical raindrop, undergoing refraction as it enters the drop, then total internal refraction, then another refraction as it leaves the drop. Show that the angle $\phi$ in the figure is given by $\phi = 4 \sin^{-1}(\sin\theta/n) - 2\theta$, where $\theta$ is the angle of incidence.

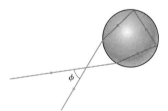

**FIGURE 30.25** Problem 59: A light ray passing through a spherical raindrop undergoes refraction at each interface and internal reflection at the back of the drop.

60. (a) Differentiate the result of Problem 59 to show that the maximum value of $\phi$ occurs when the incidence angle $\theta$ is given by $\cos^2\theta = \frac{1}{3}(n^2 - 1)$. (b) Use this result and that of Problem 59 to find the maximum value of $\phi$ in a raindrop $(n = 1.333)$. This is the angle $\phi_{\max}$ at which the rainbow appears, as shown in Figs. 30.14 and 30.15.

61. Figure 30.26 shows the approximate path of a light ray that undergoes internal reflection twice in a spherical water drop. Repeat Problems 59 and 60 for this case, thus finding the angle at which the secondary rainbow occurs.

**FIGURE 30.26** Problem 61

62. Light is incident with incidence angle $\theta_1$ on a prism with apex angle $\alpha$ and refractive index $n$, as shown in Fig. 30.23. Show that the angle $\delta$ through which the outgoing beam deviates from the incident beam is given by

$$\delta = \theta_1 - \alpha + \sin^{-1}\left\{n \sin\left[\alpha - \sin^{-1}\left(\frac{\sin\theta_1}{n}\right)\right]\right\}$$

Assume the surrounding medium has $n = 1$.

63. Show that a three-dimensional corner reflector (three mutually perpendicular mirrors, or a solid cube in which total internal reflection occurs) turns an incident light ray through 180°. *Hint:* Let $\vec{q} = q_x\hat{i} + q_y\hat{j} + q_z\hat{k}$ be a vector in the direction of propagation. How does this vector get changed on reflection by a mirror in a plane defined by two of the coordinate axes?

64. *Fermat's principle* states that the path of a light ray between two points is such that the time to traverse that path is an extremum (either a minimum or a maximum) when compared with the times for nearby paths. Consider two points $A$ and $B$ on the same side of a reflecting surface, and show that a light ray traveling from $A$ to $B$ via a point on the reflecting surface will take the least time if its path obeys the law of reflection. Thus, the law of reflection (Equation 30.1) follows from Fermat's principle.

65. Use Fermat's principle (see Problem 64) to show that a light ray going from point $A$ in one medium to point $B$ in a second medium will take the least time if its path obeys Snell's law. Thus, Snell's law follows from Fermat's principle.

66. Find the angle of incidence $\theta_1$ that minimizes the deviation of light passing through the prism in Problem 62, and determine the minimum deviation when $n = 1.5$ and $\alpha = 60°$.

67. You're working on a physics experiment and must determine the wavelength of a laser beam in a transparent material. The experiment design requires that the wavelength in the material be shorter than 400 nm, but longer than 350 nm. The incident beam has a wavelength of 633 nm and is incident on the material at an angle of 50°. You measure the angle of refraction to be 27°. Does the wavelength meet the requirements?

68. You're an automotive engineer charged with evaluating safety glass, which is made by bonding a layer of flexible plastic between two layers of more brittle glass, thus eliminating dangerous glass fragments during accidents. A new product uses glass with refractive index $n = 1.55$ and plastic with $n = 1.48$. You wonder whether total internal reflection at the glass–plastic interface could cause problems for the driver's visibility through the composite material. What do you conclude, and why?

69. A friend who is majoring in astronomy is making an optical device that uses a lens of titanium dioxide $(TiO_2; n = 2.62)$ bonded to glass with $n = 1.60$. In order for the device to pass all the light incident on it, the critical angle at the interface between the two materials must be larger than 60°. Will this lens work?

# Answers to Chapter Questions

## Answer to Chapter Opening Question

Light from the bee propagates along the solid, curved plastic "light pipe" via total internal reflection. This same process carries e-mail and other computer data on the fiber-optic cables that constitute the physical structure of the Internet.

## Answers to GOT IT? Questions

30.1 $n_3 > n_1 > n_2$.

30.2 Most would emerge into the water from the diagonal interface; some would still be reflected as shown.

# Images and Optical Instruments

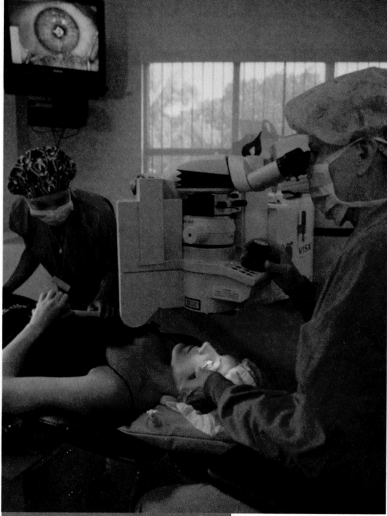

▶ **To Learn**

By the end of this chapter you should be able to

- Describe graphically how plane and curved mirrors form images (31.1).
- Explain both graphically and quantitatively how lenses form images (31.2).
- Describe the difference between virtual and real images (31.1, 31.2).
- Give the details of refraction at curved surfaces, and be able to calculate the focal lengths of thin lenses (31.3).
- Explain the operation of common optical instruments and the human eye, including vision correction (31.4).

◀ **To Know**

- This chapter is based on the optical phenomena of reflection and refraction introduced in Chapter 30 (30.1, 30.2).
- We'll be working extensively with geometrical constructions, including especially similar triangles.

Reflection and refraction alter the direction of light propagation, following laws developed in Chapter 30. Microscopes, telescopes, cameras, contact lenses, scanners, and your own eyes use reflection or refraction to form **images** that provide visual representations of reality. Here we study image formation using geometrical optics—a valid approximation provided the objects we're imaging are much larger than the wavelength of light.

When we view an object through an optical system, light reflects or refracts, so it doesn't propagate in straight paths from the object. As a result we see an image that may differ in size, orientation, or apparent position from the actual object. In some cases light actually comes from the image to our eyes; the image is then a **real image**. In other cases light only *appears* to come from the image location; then the image is a **virtual image**.

## 31.1 Images with Mirrors

### Plane Mirrors

In Fig. 31.1a we show three light rays that leave the tip of an arrowhead and reflect off a plane mirror to reach the observer's eye. The light looks to the observer like it's coming from a point behind the mirror, so that's the location of the arrowhead's image. The image is virtual because no light actually comes from behind the mirror—even though that's where the image is.

Since two lines define a point, we need only two rays to locate the arrowhead in Fig. 31.1a. We've repeated the image-location procedure in Fig. 31.1b, now using as one of the rays the ray that reflects normally. The same procedure locates the bottom of the arrow, and we could easily fill in to get the entire arrow; the resulting image is shown in Fig. 31.1b.

Because the angles of incidence and reflection are equal, angles $OQP$ and $O'QP$ in Fig. 31.1b are equal. The right triangles $OPQ$ and $O'PQ$ share a common side as well as this common angle, so they're congruent. That means the image is located behind the mirror as far as the object is in front of it. Furthermore, rays from the top and bottom of the arrow and normal to the mirror are parallel, so the image is the same height as the actual arrow.

Images in plane mirrors preserve an object's length and upright orientation, but they reverse the object. When you look in the mirror, you face the mirror. So does your image—meaning the reversal is front to back, not left to right as you might think. This front-to-back reversal makes the image of your right hand look like a left hand (Fig. 31.2). Mathematically, the mirror reverses the coordinate axis perpendicular to the mirror plane. This alters handedness, rotation, and all other phenomena connected with the right-handed coordinate systems we've been using.

**(a)**

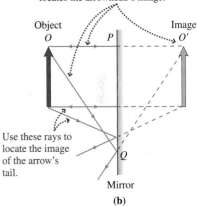

**(b)**

**FIGURE 31.1** Image formation in a plane mirror.

**FIGURE 31.2** The palm of a right hand faces the mirror. So does the image's palm. That makes the image look like a left hand, but it's still the image of a *right* hand.

**GOT IT? 31.1** You stand in front of a plane mirror whose top is at the same height as the top of your head. Approximately how far down must the mirror extend for you to see your full image?

### Curved Mirrors

In contrast to plane mirrors, curved mirrors form images that may be upright or inverted, virtual or real, large or small. The best curved mirrors are parabolic. That's because a parabola has the property that a line parallel to the parabola's axis makes the same angle to the normal of the parabola as does a second line drawn to a special point called the **focus** or **focal point** (Fig. 31.3). Because the angles of incidence and reflection are equal, this means a parabolic mirror reflects rays parallel to the mirror axis so they converge at the focus. This effect is used to concentrate light to high intensities or, conversely, to create a parallel beam from a point source of light at the focus.

Near the apex of the parabolic mirror in Fig. 31.3, you can't tell whether the shape is parabolic or spherical; a sphere closely approximates the parabola. Because a sphere is easier to form, many focusing mirrors are, in fact, spherical. The slight distortion this causes is called **spherical aberration**; a notable case is the Hubble Space Telescope mirror,

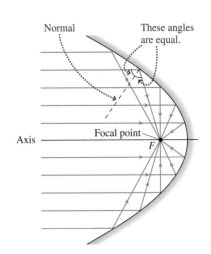

**FIGURE 31.3** A parabolic mirror reflects rays parallel to its axis to a common focus.

**FIGURE 31.4** Incorrect curvature gave the Hubble Space Telescope mirror substantial spherical aberration. Astronauts later installed corrective optics. Images are of the same galaxy before and after the repair.

which was ground to the wrong curve and had substantial aberration (Fig. 31.4). Normally spherical aberration is minimized by making the mirror only a tiny fraction of the entire sphere. In that case the **focal length**—the distance from mirror to focal point—is much larger than the mirror, and the only rays from distant objects that strike the mirror are those that are nearly parallel to the mirror axis. We assume such **paraxial rays** in what follows; it's only for them that the approximation of a parabola by a sphere results in accurate focusing. But for clarity our diagrams will often show mirrors with exaggerated curvature, and consequently not all rays will seem paraxial.

We can see how spherical mirrors form images by tracing two rays from each of several points on the object, as we did for plane mirrors. Some special rays simplify this process; their properties all follow from the law of reflection and the properties of a spherical mirror in the paraxial approximation.

**TACTICS 31.1** **Ray Tracing with Mirrors**

**1** Any ray parallel to the mirror axis reflects through the focal point.
**2** Conversely, any ray that passes through the focal point reflects parallel to the axis.
**3** Any ray that strikes the center of the mirror reflects symmetrically about the mirror axis.
**4** Any ray through the center of curvature strikes the mirror normal to the mirror surface, and thus returns on itself.

Figure 31.5 shows these special rays, any two of which suffice to locate an image.

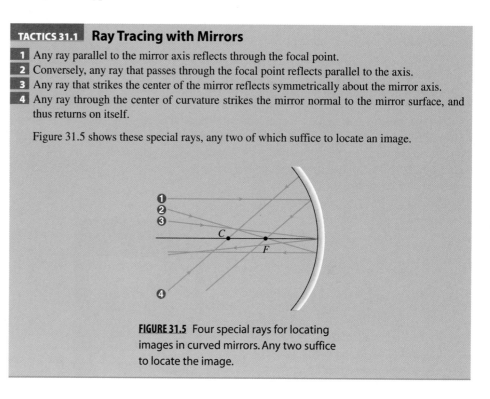

**FIGURE 31.5** Four special rays for locating images in curved mirrors. Any two suffice to locate the image.

Figure 31.6 shows ray tracings, using our special rays 1 and 2 that go through the focal point, to find the image location in three cases. In all three cases symmetry ensures that the bottom of the image arrow is on the axis, so we haven't bothered to trace it. In Fig. 31.6a

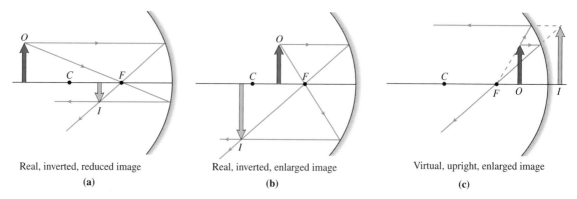

Real, inverted, reduced image
**(a)**

Real, inverted, enlarged image
**(b)**

Virtual, upright, enlarged image
**(c)**

**FIGURE 31.6** Image formation with a concave spherical mirror, using rays 1 and 2 described in Tactics 31.1. *O* denotes the object and *I* its image.

we see that an object beyond the mirror's center of curvature, *C*, forms a smaller, inverted image. Light actually emerges from this image, so it's a *real* image. If you looked from the left toward the mirror in Fig. 31.6a you would actually see the image in space in front of the mirror (Fig. 31.7).

As the object moves closer to the mirror, the real image grows; with the object between the center of curvature and the focus, the image is larger than the object and farther from the mirror (Fig. 31.6b). As the object moves toward the focus, the image grows larger and moves rapidly away from the mirror. With the object right at the focus, the rays emerge in a parallel beam and there's no image. Finally, rays from an object closer to the mirror than the focus diverge after reflection. To an observer they appear to come from a point behind the mirror. Thus there is a virtual image, in this case upright and enlarged (Fig. 31.6c).

**FIGURE 31.7** A bear meets its real image, formed by the concave mirror at the rear. Bear and image are both in *front* of the mirror.

**GOT IT? 31.2** Where would you place an object so that its real image is the same size as the object, as in Fig. 31.7?

## Convex Mirrors

A convex mirror reflects on the outside of its spherical curvature, causing light to diverge and therefore to form only virtual images (Fig. 31.8). Although the focus has less obvious physical significance in this case, its location still controls the geometry of reflected rays. As Fig. 31.8 shows, we can still draw a ray parallel to the axis and another ray that would go through the focus if the mirror weren't in its way. The reflected rays appear to diverge from a common point behind the mirror, showing a virtual image that's upright and re-duced in size. By considering different object positions, you can convince yourself that the image in a convex mirror always has these characteristics. Convex mirrors are widely used where an image of a broad region needs to be captured in a small space (Fig. 31.9).

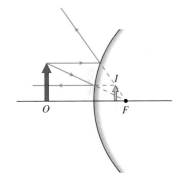

**FIGURE 31.8** Image formation with a convex mirror. The image is always virtual, upright, and reduced in size.

**FIGURE 31.9** A convex mirror gives a wide-angle view.

## The Mirror Equation

Drawing ray diagrams gives an intuitive feel for image formation. More precise image lo-cations and sizes follow from the **mirror equation**, which we now derive. This time we'll find the image using our special rays parallel to the axis and striking the mirror's center, as shown in Fig. 31.10a. The ray that strikes the center of the mirror reflects symmetrically about the axis; therefore, the two shaded triangles are similar. Then the **magnification** *M*—the ratio of image height *h'* to object height *h*—is the same as the ratio of image and object distances from the mirror. We'll consider the image height negative if the image is inverted; then from Fig. 31.10a we have

$$M = \frac{h'}{h} = -\frac{s'}{s} \quad \text{(magnification)} \tag{31.1}$$

Here object and image are both in front of the mirror, so we take object and image dis-tances *s* and *s'* as positive quantities; the negative sign in Equation 31.1 then shows that in this case the image is inverted. Also, for the object location of Fig. 31.10a, it's clear that $|M| < 1$, meaning the image is reduced rather than enlarged.

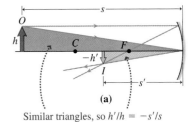

Similar triangles, so $h'/h = -s'/s$

Similar triangles, so $-h'/h = (s'-f)/f$

**FIGURE 31.10** Finding the image *I* using rays 1 and 3 of Tactics 31.1. For an inverted image the height *h'* is negative, so we've marked the arrow length—a positive quantity—as −*h'*.

Figure 31.10b is the same as Fig. 31.10a except that now we show only the ray that reflects through the focus. We've also labeled the focal length $f$ and shaded a different pair of similar triangles. From these you can see that $-h'/h = (s' - f)/f$. Here we put the minus sign on $h'$ because we've defined $h'$ as negative for the inverted image, but comparing similar triangles requires a ratio of positive quantities. Equation 31.1 shows that the ratio $h'/h$ is the magnification $M = -s'/s$. So we have $s'/s = (s' - f)/f$ or, dividing both sides by $s'$ and doing a little algebra,

$$\frac{1}{s} + \frac{1}{s'} = \frac{1}{f} \quad \text{(mirror equation)} \tag{31.2}$$

Although we derived the mirror equation using a real image, the equation applies to virtual images with the convention that a *negative* image distance $s'$ means the image is *behind* the mirror. And we can handle *convex* mirrors as well by taking the focal length to be a *negative* quantity. Table 31.1 summarizes image formation with mirrors, including these sign conventions.

**TABLE 31.1** Image Formation with Mirrors: Sign Conventions

| Focal Length, $f$ | Object Distance, $s$ | Image Distance, $s'$ | Type of Image | Ray Diagram |
|---|---|---|---|---|
| +<br>(concave) | +<br>(in front of mirror)<br>$s > 2f$ | +<br>(in front of mirror)<br>$s' < 2f$ | Real,<br>inverted,<br>reduced | |
| +<br>(concave) | +<br>(in front of mirror)<br>$2f > s > f$ | +<br>(in front of mirror)<br>$s' > 2f$ | Real,<br>inverted,<br>enlarged | |
| +<br>(concave) | +<br>(in front of mirror)<br>$s < f$ | −<br>(behind mirror) | Virtual,<br>upright,<br>enlarged | |
| −<br>(convex) | +<br>(in front of mirror) | −<br>(behind mirror) | Virtual,<br>upright,<br>reduced | |

A diagram similar to Fig. 31.10 but using a ray through the center of curvature gives another useful fact about curved mirrors: The magnitude of the focal length is half the radius:

$$|f| = \frac{R}{2} \tag{31.3}$$

You can prove this in Problem 76.

EXAMPLE 31.1 **A Concave Mirror: The Hubble Space Telescope**

During assembly of the Hubble Space Telescope, a technician stood 3.85 m in front of the telescope's concave mirror (Fig. 31.11). Given the telescope's 5.52-m focal length, find (a) the location and (b) the magnification of the technician's image.

**FIGURE 31.11** A technician standing in front of the Hubble Space Telescope mirror.

**INTERPRET** This problem is about image formation in a concave mirror. We identify the technician as the object, the 5.52-m focal length as $f$, and the 3.85-m distance as the object distance $s$.

**DEVELOP** We've sketched the situation in Fig. 31.12. With the object closer than the focal length, our sketch resembles Fig. 31.6c, so we anticipate an enlarged, virtual image, as shown. For (a), we'll solve the

**FIGURE 31.12** Sketch for Example 31.1, showing two rays that locate the virtual image of the technician's head.

mirror equation 31.2 for the image distance $s'$ to get the image location. Then for (b), we can find the magnification from Equation 31.1, $M = -s'/s$.

**EVALUATE** (a) Solving Equation 31.2 for $s'$ gives

$$s' = \frac{fs}{s - f} = \frac{(5.52 \text{ m})(3.85 \text{ m})}{3.85 \text{ m} - 5.52 \text{ m}} = -12.7 \text{ m}$$

(b) Using this result in Equation 31.1 gives the magnification:

$$M = -\frac{s'}{s} = -\frac{-12.7 \text{ m}}{3.85 \text{ m}} = 3.30$$

**ASSESS** The *negative* image distance confirms what our sketch anticipated: This is a virtual image, located behind the mirror. The negative distance cancels the minus sign in Equation 31.1, giving a positive threefold magnification. Thus the image is upright and enlarged, just as Fig. 31.11 shows. ∎

EXAMPLE 31.2 **A Convex Mirror: Jurassic Park**

In the film *Jurassic Park*, horrified passengers watch in a car's convex side-view mirror as a *Tyrannosaurus rex* pursues them. Printed on the mirror is the warning "OBJECTS IN MIRROR ARE CLOSER THAN THEY SEEM." If the mirror's curvature radius is 12 m and the *T. rex* is actually 9.0 m from the mirror, by what factor does the dinosaur appear reduced in size?

**INTERPRET** This is about a convex mirror, governed by the same basic equations as the concave mirror in Example 31.1. We identify the 12-m length as the curvature radius $R$ and the 9.0-m distance as the object distance $s$.

**DEVELOP** We've sketched the situation in Fig. 31.13. Since this resembles Fig. 31.8, we see that the image will indeed be reduced in size. Equation 31.1, $M = -s'/s$, gives the magnification we want, but to use it we need the image distance $s'$. In Example 31.1 we solved Equation 31.2 to get $s' = fs/(s - f)$. Although that mirror was concave and this one is convex, Equation 31.2 applies to both mirrors. So we can use $s' = fs/(s - f)$ in Equation 31.1 to get the magnification:

$$M = -\frac{s'}{s} = -\frac{fs/(s - f)}{s} = -\frac{f}{s - f}$$

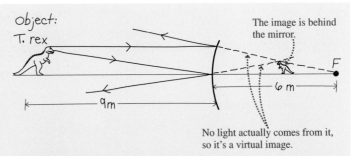

**FIGURE 31.13** Sketch for Example 31.2, showing two rays that locate the image of *T. rex*. Mirror is not to scale.

We aren't given the focal length, but Equation 31.3, $|f| = R/2$, shows that its magnitude is half the 12-m radius. Since this is a *convex* mirror, Table 31.1 shows that the focal length is *negative*, so $f = -6.0$ m.

**EVALUATE** Using our expression for $M$ gives

$$M = -\frac{f}{s - f} = -\frac{(-6.0 \text{ m})}{9.0 \text{ m} - (-6.0 \text{ m})} = 0.40$$

where the answer comes out positive because $f$ is negative.

**ASSESS** Our result shows that *T. rex* appears in the mirror at only 40% of its actual size; in other words, it looks farther away than it really is. ∎

## 31.2 Images with Lenses

A **lens** is a transparent piece of material that uses refraction to form images. Like mirrors, lenses can be either concave or convex. But light goes through a lens while it reflects off a mirror, so the roles of the two are reversed. A convex lens focuses parallel rays to a **focal point** and is therefore called a **converging lens** (Fig. 31.14). As we'll soon see, a convex lens can form real and virtual images, depending on the object location. A concave lens, in contrast, is a **diverging lens**; it refracts parallel rays so they appear to diverge from a common focus. Like a *convex* mirror, a *concave* lens forms only virtual images (Fig. 31.15).

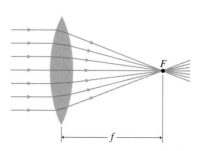

FIGURE 31.14 A convex lens brings parallel light rays to a focus at *F*.

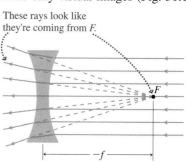

FIGURE 31.15 Parallel light passing through a concave lens diverges so it looks as though it's coming from a common focus.

We first explore a **thin lens**—one whose thickness is small compared with the curvature radii of its two surfaces. Although light refracts as it enters a lens and again as it exits, in the thin-lens approximation the two surfaces are so close that it suffices to consider that the light bends just once, as it crosses the center plane of the lens. Unlike a mirror, light can go either way through a lens; that means the lens has two focal points, one on either side. For a thin lens, the focal length proves to be the same in either direction, so it doesn't matter which way we orient the lens. We'll later justify the thin-lens approximation mathematically.

### Lens Images by Ray Tracing

As with mirrors, two rays serve to locate the image formed by a lens. For lenses, two special rays simplify ray tracing.

**TACTICS 31.2** **Ray Tracing with Lenses**

**1** Any ray parallel to the lens axis refracts through the focal point.
**2** Any ray through the center of the lens passes undeflected.

Figure 31.16 shows both these special rays.

FIGURE 31.16 Two special rays for locating images formed with lenses.

Figure 31.17 shows ray tracings for different object placements in relation to a converging lens. In Fig. 31.17a we see that an object farther out than two focal lengths produces a smaller, inverted, real image on the other side of the lens. Since light really emanates from this image, you could see it without looking through the lens. As the object moves toward the lens, the image moves away and grows. When the object is between one and two focal lengths from the lens, the image has moved beyond $2f$ and is enlarged (Fig. 31.17b). The image on a movie screen is formed in this way. Moving the object closer than the focal point produces an enlarged, virtual image that can be seen only by an observer looking *through* the lens (Fig. 31.17c).

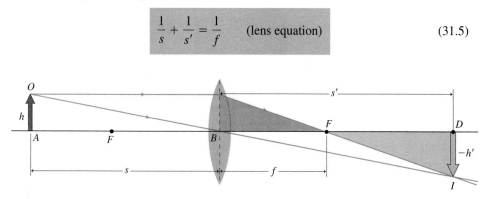

**FIGURE 31.17** Image formation with a converging lens, shown for three different object locations.

Figure 31.18 shows ray tracings for a diverging lens. Like a convex mirror, this lens produces only virtual images that are upright and reduced in size; they're visible only through the lens. The basic geometry of Fig. 31.18 doesn't change even if the object moves within the focal length.

**FIGURE 31.18** A diverging lens always forms a reduced, upright, virtual image, visible only through the lens.

## Getting Quantitative: The Lens Equation

Study Fig. 31.19 and you'll see that triangles *OAB* and *IDB* are similar. Therefore, as for mirrors, the image magnification is

$$M = \frac{h'}{h} = -\frac{s'}{s} \tag{31.4}$$

where again a negative height means an inverted image. The shaded triangles in Fig. 31.19 are also similar, so $-h'/(s'-f) = h/f$. Combining this result with Equation 31.4 and doing some algebra then give

$$\frac{1}{s} + \frac{1}{s'} = \frac{1}{f} \quad \text{(lens equation)} \tag{31.5}$$

**FIGURE 31.19** Ray diagram for deriving the lens equation. Triangles *OAB* and *IDB* are similar, as are the shaded triangles.

which is identical to the mirror equation 31.2. Although we derived Equation 31.5 for the case of a real image, it holds for virtual images if we consider the image distance negative; in that case the image is on the same side of the lens as is the object. And it holds for diverging lenses if we consider the focal length negative. Table 31.2 summarizes image formation with lenses, including these sign conventions. Figure 31.20 describes graphically the sizes and types of images formed at different object distances.

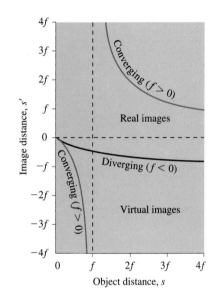

**FIGURE 31.20** Image distance versus object distance for lenses.

**TABLE 31.2** Image Formation with Lenses: Sign Conventions

| Focal Length, $f$ | Object Distance, $s$ | Image Distance, $s'$ | Type of Image | Ray Diagram |
|---|---|---|---|---|
| +<br>(convex) | +<br>$s > 2f$ | +<br>(opposite side of lens)<br>$2f > s' > f$ | Real,<br>inverted,<br>reduced | |
| +<br>(convex) | +<br>$2f > s > f$ | +<br>(opposite side of lens)<br>$s' > 2f$ | Real,<br>inverted,<br>enlarged | |
| +<br>(convex) | +<br>$s < f$ | −<br>(same side of lens) | Virtual,<br>upright,<br>enlarged | |
| −<br>(concave) | + | −<br>(same side of lens) | Virtual,<br>upright,<br>reduced | |

**GOT IT?  31.3**   You look through a lens at this page and see the words enlarged and right-side up. Is the image you observe real or virtual? Is the lens concave or convex?

**EXAMPLE 31.3** **Using the Lens Equation: Fine Print**

You're using a magnifying glass (a converging lens) with a 30-cm focal length to read a telephone book (Fig. 31.21). How far from the page should you hold the lens in order to see the print enlarged three times?

**INTERPRET** This is a problem involving image formation with a converging lens. The object is the phone book, so we identify the book-to-lens distance we're asked for as the object distance $s$. Since this is a converging (e.g., convex) lens, the focal length is positive: $f = +30$ cm. The factor-of-3 magnification we want is the quantity $M$.

**DEVELOP** The situation is like Fig. 31.17c, with an enlarged, upright, virtual image. We're given the focal length and magnification, but we don't know either the object distance $s$, which we're looking for, or the image distance $s'$. Equation 31.4, $M = -s'/s$, relates the two, so we can first use that equation to eliminate $s'$ in terms of $s$ and then solve the lens equation 31.5, $1/s + 1/s' = 1/f$, for $s$.

**EVALUATE** With $M = 3$, Equation 31.4 gives $s' = -3s$. Then Equation 31.5 becomes

$$\frac{1}{s} - \frac{1}{3s} = \frac{2}{3s} = \frac{1}{f} = \frac{1}{30 \text{ cm}}$$

so $s = (2)(30 \text{ cm})/3 = 20$ cm.

**ASSESS** Our answer is less than the focal length, as Fig. 31.17c shows is required for a virtual image. Figure 31.21 confirms that the image is enlarged, upright, and virtual, and it appears farther away than the object. It's also on the same side of the lens as the object, which explains the negative image distance $s' = -60$ cm. ∎

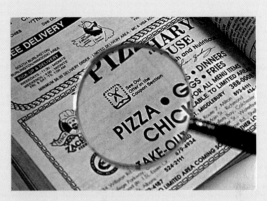

**FIGURE 31.21** Using a converging lens as a magnifying glass (Example 31.3).

## 31.3 Refraction in Lenses: The Details

So far we've treated lenses as being arbitrarily thin and neglected details of the refraction process. Here we develop a more general description of refraction in lenses, of which our thin-lens approximation is a special case.

### Refraction at a Curved Surface

Figure 31.22 shows a transparent material with refractive index $n_2$ and a curved surface of radius $R$. Outside the material is a medium with refractive index $n_1$. We'll now prove what Fig. 31.22a shows: that rays from a point object $O$ are refracted to a common image point $I$. Our proof is valid only in the paraxial approximation that all rays make small angles with the optic axis; as with mirrors, our drawings won't always show these angles as being small.

Figure 31.22b shows a single ray. With all the labeled angles small, we can approximate $\sin x \approx \tan x \approx x$, with $x$ in radians. Then Snell's law, $n_1 \sin\theta_1 = n_2 \sin\theta_2$, becomes $n_1\theta_1 = n_2\theta_2$. Triangles $BCI$ and $OBC$ give $\theta_2 = \beta - \gamma$ and $\theta_1 = \alpha + \beta$, so Snell's law becomes $n_1(\alpha + \beta) = n_2(\beta - \gamma)$. Furthermore, in the small-angle approximation the arc $BA$ is so close to a straight line that we can write $\alpha \approx \tan\alpha \approx BA/s$, with $s = OA$ the object's distance from the refracting surface. Similarly, $\beta \approx BA/R$ and $\gamma \approx BA/s'$. Thus our expression of Snell's law becomes

$$n_1\left(\frac{BA}{s} + \frac{BA}{R}\right) = n_2\left(\frac{BA}{R} - \frac{BA}{s'}\right)$$

(a)

(b)

**FIGURE 31.22** Refraction at an interface with a curved surface. All labeled angles are considered small, even though the drawing doesn't show them as such.

or, on canceling $BA$ and rearranging,

$$\frac{n_1}{s} + \frac{n_2}{s'} = \frac{n_2 - n_1}{R} \tag{31.6}$$

The angle $\alpha$ doesn't appear here, showing that this relation between object and image distances holds for *all* rays that satisfy the small-angle approximation. So Fig. 31.22a is correct: All such rays do indeed come to a common focus at $I$.

We derived Equation 31.6 for the case of a real image, but as usual it applies to virtual images if we take the image distance as negative. And it applies to concave surfaces if we take $R$ to be negative. It even works for flat surfaces, with $R = \infty$.

---

**EXAMPLE 31.4** **Refraction at a Curved Surface: A Cylindrical Aquarium**

An aquarium consists of a thin-walled plastic tube 70 cm in diameter. For a cat looking directly into the aquarium, what's the apparent distance to a fish 15 cm from the aquarium wall?

**INTERPRET** We interpret the cylindrical aquarium as a two-dimensional version of the spherical surface we analyzed with Fig. 31.22 and Equation 31.6. The plastic tube is thin, so we can neglect refraction within the plastic and consider that we have just a cylinder of water. The object is the fish, and since it's *inside* the water, the cylindrical edge of the aquarium is concave *toward* the object. Then the curvature radius is *negative*; we're given the diameter as 70 cm, so $R = -35$ cm. With the object in the water, $n_1 = 1.333$ from Table 30.1 and $n_2 = 1$ for air. The 15-cm distance from the edge to the fish is the object distance $s$.

**DEVELOP** Figure 31.23 shows the physical situation and a ray diagram viewed from above. Our plan is to solve Equation 31.6, $n_1/s + n_2/s' = (n_2 - n_1)/R$, for $s'$ and evaluate using $R = -35$ cm, $n_1 = 1.333$, $n_2 = 1$, and $s = 15$ cm.

**FIGURE 31.23** (a) A cylindrical aquarium. (b) Top view, showing the formation of a virtual image of a fish that's actually 15 cm from the edge.

**EVALUATE** Solving, we have

$$s' = n_2 \left( \frac{n_2 - n_1}{R} - \frac{n_1}{s} \right)^{-1} = -12.6 \text{ cm}$$

**ASSESS** Make sense? The fish is actually 15 cm from the edge, but refraction makes the image distance $s'$ shorter. The same effect occurs when you look down into a swimming pool or lake: Objects on the bottom look closer, and you can find out how much by applying Equation 31.6 with $R = \infty$ (see Exercise 30). ∎

---

## Lenses, Thick and Thin

Figure 31.24 shows a lens of thickness $t$ with refractive index $n$, surrounded by air with $n = 1$. Object $O_1$ lies a distance $s_1$ from the left-hand surface. This surface forms an image $I_1$, which we'll also call $O_2$ because it acts as an object for the right-hand surface.

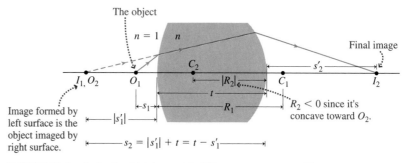

**FIGURE 31.24** Analysis of a thick lens with different curvature radii. $C_1$ and $C_2$ are the centers of curvature of left and right sides, respectively; $t$ is the lens thickness.

Refraction at that surface forms a second image $I_2$. We want to relate the original image $O_1$ and the final image $I_2$.

At the left-hand surface, the quantities in Equation 31.6 become

$$\frac{1}{s_1} + \frac{n}{s_1'} = \frac{n-1}{R_1} \qquad \text{(left-hand surface)}$$

We've placed $O_1$ so close that $I_1$ is a *virtual* image, so $s_1'$ is *negative*. Now we set up another instance of Equation 31.6, this time for the right-hand surface. Here $I_1$ is the object, whose distance $s_2$ is $t - s_1'$ because $s_1'$ is negative. Also at the right-hand surface, $s' = s_2'$, $n_1 = n$, and $R = R_2$, where, for the case shown, $R_2$ is negative because the right-hand surface is concave toward the object $(I_1, O_2)$ that's being imaged. So at the right-hand surface, Equation 31.6 reads

$$\frac{n}{t - s_1'} + \frac{1}{s_2'} = \frac{1-n}{R_2} \qquad \text{(right-hand surface)}$$

Now we'll let the lens become arbitrarily thin, so $t \to 0$. Then we add the two equations; the intermediate-image term $n/s_1'$ cancels, leaving only the first object distance and final image distance. So we drop the subscripts 1 and 2, and the result is

$$\frac{1}{s} + \frac{1}{s'} = (n-1)\left(\frac{1}{R_1} - \frac{1}{R_2}\right)$$

The left-hand side here is identical to the left-hand side of Equation 31.5, and Equation 31.5's right-hand side is $1/f$. Equating the two right-hand sides results in the **lensmaker's formula**, which gives the focal length:

$$\boxed{\frac{1}{f} = (n-1)\left(\frac{1}{R_1} - \frac{1}{R_2}\right)} \qquad \text{(lensmaker's formula)} \qquad (31.7)$$

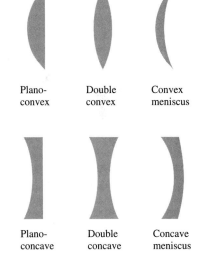

Plano-convex    Double convex    Convex meniscus

Plano-concave    Double concave    Concave meniscus

**FIGURE 31.25** Common lens types.

Again, the radii here can be positive or negative; in Fig. 31.24 $R_1$ is positive because the left-hand surface is convex toward the object, while $R_2$ is negative because the right-hand surface is concave toward its object, the intermediate image $I_1$. Although we derived Equation 31.7 for the case of a virtual intermediate image, the lensmaker's formula is a general result for the focal length of a thin lens.

Lenses come in a variety of shapes (Fig. 31.25). Those that are thicker in the center are converging lenses, for which Equation 31.7 gives a positive focal length. Those that are thinner in the center are diverging lenses, with negative $f$. These behaviors reverse if the medium surrounding the lens has a higher refractive index (see Problem 75).

---

**EXAMPLE 31.5** The Lensmaker's Formula: A Plano-Convex Lens

Find an expression for the focal length of the plano-convex lens in Fig. 31.25, given refractive index $n$ and radius $R$ for the one curved surface.

**INTERPRET** This is a thin lens, such as we just analyzed in deriving the lensmaker's formula. With an object at the left of the lens, we identify $R_1 = R$ and $R_2 = \infty$ for the flat surface.

**DEVELOP** The lensmaker's formula, Equation 31.7, relates the focal length, the lens surface radii, and the refractive index. Our plan is to solve for $f$.

**EVALUATE** With $R_1 = R$ and $R_2 = \infty$, Equation 31.7 gives

$$f = \left[(n-1)\left(\frac{1}{R} - \frac{1}{\infty}\right)\right]^{-1} = \frac{R}{n-1}$$

**ASSESS** Make sense? The smaller $R$, the more curved the lens and the more it bends light; the result is a shorter focal length. The higher $n$, the greater the refraction, and with $n$ in the denominator, the result is again a shorter focal length. We asserted earlier that a thin lens works the same either way. You can see that explicitly here by putting the object beyond the flat side; then $R_1 = \infty$ and $R_2 = -R$, but the result for $f$ is unchanged. ∎

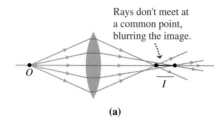

Rays don't meet at
a common point,
blurring the image.

**(a)**

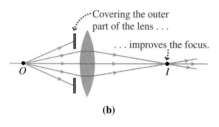

Covering the outer
part of the lens . . .

. . . improves the focus.

**(b)**

**FIGURE 31.26** (a) Spherical aberration. (b) Using only the central portion of the lens minimizes aberration, but at the expense of a dimmer image.

### Lens Aberrations

Lenses exhibit several optical defects. We described **spherical aberration** in mirrors; this same defect occurs with spherical lenses (Fig. 31.26a). Our lens analysis required that all rays make small angles with the lens axis; if not, then they don't share a common focus, causing spherical aberration. Small angles occur naturally with distant objects, but not with objects close to the lens. Using only the central portion of the lens can eliminate those rays with larger angles (Fig. 31.26b), leading to sharper focus. That's why a camera focuses over a wider range when it's "stopped down," with an opaque iris covering the outer part of the lens. The trade-off is that less light is available.

We mentioned **chromatic aberration** in Chapter 30; it occurs because the refractive index varies with wavelength, causing different colors to focus at different points. Quality optical systems minimize this effect by using composite lenses of materials with different refractive indices. **Astigmatism** occurs when a lens has different curvature radii in different directions. This is a common defect in the human eye, corrected with glasses or contact lenses that have compensating asymmetric curvature.

## 31.4 Optical Instruments

Numerous optical instruments use lenses, mirrors, or both to form images. All but the simplest have more than one optical element, but the principles we've developed here still apply. We analyze such instruments by tracing light through the sequence of optical elements, using the image formed by one element as the object for the next.

### The Eye

Our eyes are our primary optical instruments. Each eye is a complex optical system with several refracting surfaces and mechanisms to vary the focal length and amount of light admitted (Fig. 31.27). Light enters through the cornea and traverses a liquid called the aqueous humor before passing through the lens. It then traverses the vitreous humor, a liquid in the main body of the roughly 2.3-cm-diameter eyeball. Finally it strikes the retina, where special cells called rods and cones produce electrochemical signals that carry visual information to the brain.

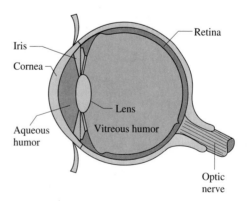

**FIGURE 31.27** The human eye.

A properly functioning eye produces well-focused real images on the retina, with the cornea providing most of the refractive focusing. Muscles adjust the lens, changing its focal length to compensate for different object distances. Other muscles adjust the iris, resizing the pupil opening to adjust for different light levels.

In nearsighted (myopic) people, the image forms in front of the retina, causing distant objects to appear blurred (Fig. 31.28a). Diverging corrective lenses produce closer intermediate images that the myopic eye can then focus (Fig. 31.28b). In farsighted (hyperopic) people, the image of nearby objects would form behind the retina, and converging corrective lenses are used (Fig. 31.29). Even normal eyes can't focus much closer than the so-called **near point** at about 25 cm, a distance that increases greatly with age.

**FIGURE 31.28** (a) A nearsighted eye focuses light from distant objects in front of the retina. (b) A diverging lens corrects the problem, creating a closer virtual image that the eye can focus.

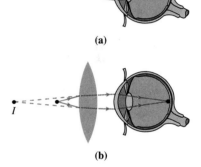

**FIGURE 31.29** (a) A farsighted eye can't focus light from nearby objects. (b) A converging lens produces a more distant image that the eye can focus.

Prescriptions for corrective lenses specify the corrective power, $P$, in **diopters**, which is the inverse of the focal length in meters. Thus a 1-diopter lens has $f = 1$ m, while a 2-diopter lens has $f = 0.5$ m and is more powerful in that it refracts light more sharply. Like $f$ itself, the sign of a lens's corrective power is positive or negative depending on whether the lens is converging or diverging.

It doesn't matter whether a corrective lens is several centimeters from the eye, as with glasses, or right on the cornea, as with contact lenses. Contact lenses can be thin because, as Equation 31.7 shows, it's the curvature radii and not the thickness that determine the focal length. A more radical approach to vision correction is laser surgery, described in the Application.

---

**APPLICATION** **Laser Vision Correction**

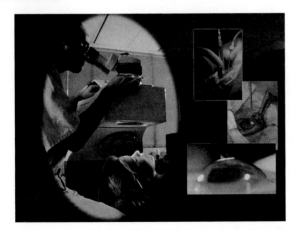

We noted that the cornea provides most of the eye's refractive power, with the adjustable lens compensating for different object distances. So a direct way to correct vision problems is to reshape the cornea. The figure depicts the popular LASIK procedure. Here the surgeon begins by mechanically cutting out a flap of the outermost corneal layer. Then the energy of a precision laser beam breaks the molecular bonds in the corneal tissue, vaporizing the material and reshaping the cornea

according to a prescription customized for the individual eye. With a nearsighted eye, the laser thins the central cornea, making it less sharply curved and thus reducing its refractive power. This has the same effect as the corrective lens in Fig. 31.28b, so this particular corneal reshaping corrects myopia (nearsightedness). It's harder, but still possible, to correct farsightedness with LASIK. This involves thinning a ring-shaped region around the central cornea, making the cornea more steeply curved and thus increasing its refractive power. The corrective lens in Fig. 31.29b accomplishes the same thing. The corneal reshaping doesn't have to be symmetric, so LASIK can also correct the asymmetric coronal shape that causes astigmatism. What it can't do is restore the ability to focus both near and far, since that's handled by the lens, which stiffens with age. So older LASIK patients still need reading glasses. It's possible to correct one eye for near vision and another for far vision, but then the patient loses some of the depth perception that comes with binocular vision.

The laser used in vision correction is no ordinary laser. It's a precisely controllable device called an excimer laser, which produces intense, short bursts of ultraviolet light. Each pulse removes only 0.25 $\mu$m of tissue—one four-thousandth of a millimeter. The laser is so precise that it can cut notches in a single human hair! The surgeon determines the necessary corneal adjustments and feeds the information to a computer that controls the laser itself. Thanks to the laser's precision, most patients achieve nearly complete vision correction.

**EXAMPLE 31.6** **The Power of Lenses: Lost Your Glasses!**

You're on vacation and have lost your reading glasses; without them, your eyes can't focus closer than 70 cm. Fortunately, you can buy nonprescription reading glasses at the pharmacy, where they come in 0.25-diopter increments. Which glasses should you buy so you can focus at the standard 25-cm near point?

**INTERPRET** Your eyes can't focus closer than 70 cm, so this problem is asking for the power of a lens that will make an object at 25 cm appear as if it's at 70 cm. In other words, when the object distance $s$ is 25 cm, the image distance $s'$ should be $-70$ cm. We put the minus sign here because, as Fig. 31.29b shows, the image is on the same side of the lens as the object, so this is a *virtual* image.

**DEVELOP** Equation 31.5, $1/s + 1/s' = 1/f$, relates the inverse of the focal length to the object and image distances. But with the focal length in meters, $1/f$ is just the power, $P$, in diopters. So we can get the required power directly from Equation 31.5.

**EVALUATE** Applying Equation 31.5 gives

$$P = \frac{1}{f} = \frac{1}{s} + \frac{1}{s'} = \frac{1}{0.25\ \text{m}} + \frac{1}{-0.70\ \text{m}} = 2.57\ \text{diopters}$$

**ASSESS** The closest available power is 2.5 diopters, so that's what you should buy. ∎

**GOT IT? 31.4** You and your roommate have gotten your boxes of disposable contact lenses mixed up. One box is marked "$-1.75$ diopter," the other "$+2.25$ diopter." You're farsighted and your roommate is nearsighted. Which lenses are yours?

## Cameras

A camera is much like the eye, except that an electronic detector or film replaces the light-sensitive retina. Where the eye changes the lens shape to accommodate different object distances, a camera moves its rigid lens to change the image distance. Simple "point and shoot" cameras use infrared beams to determine the object distance, and then automatically adjust the lens position for optimum focus. Similarly, the camera automatically adjusts the lens aperture and exposure time to the ambient light conditions. Zoom lenses have moveable elements that alter the focal length to provide views ranging from wide angle to telephoto.

## Magnifiers and Microscopes

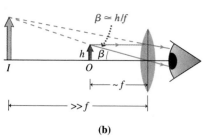

**FIGURE 31.30** Calculating the angular magnification $m = \beta/\alpha$.

Examining very small objects requires bringing them closer than the 25-cm near point below which the human eye can't focus. We therefore use lenses to put enlarged images at greater distances, where we can focus. What matters is not the actual image size, but how much bigger an object looks to us—and that depends on how much of our field of view it occupies. **Angular magnification**, $m$, is the ratio of the angle an object subtends when seen through a lens to the angle subtended when it's at the 25-cm near point and viewed with the naked eye. Figure 31.30a shows that the latter angle, measured in radians, is approximately $\alpha = h/25$ cm, where $h$ is the object height. We get the most comfortable viewing with the eye close to the lens and the object just inside the focal length, forming a large and distant virtual image. With this geometry, Fig. 31.30b shows that the image angle is very nearly $h/f$. Then the angular magnification is

$$m = \frac{\beta}{\alpha} = \frac{h/f}{h/25\ \text{cm}} = \frac{25\ \text{cm}}{f} \qquad \text{(simple magnifier)} \qquad (31.8)$$

Single lenses produce angular magnifications up to about 4 before aberrations compromise image quality. Greater magnification requires more than one lens. In a **compound microscope**, an **objective lens** of short focal length forms a magnified real image. This image is viewed through a second lens, the **eyepiece**, used as a simple magnifier (Fig. 31.31). The object being viewed is positioned just beyond the focus of the objective lens, and its image falls just inside the focal length of the eyepiece. If both focal lengths are small compared with the distance between the lenses, then the object distance for the objective lens is approximately the objective focal length $f_o$, and the resulting image distance is approximately the lens spacing $L$. The real image formed by the objective lens is larger than the object by the ratio of the image and object distances, or $M_o = -L/f_o$. The eyepiece

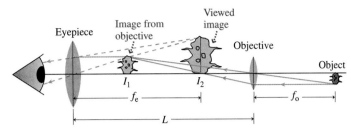

**FIGURE 31.31** Image formation in a compound microscope. Figure is not to scale; $L$ should be much greater than either focal length, and image $I_1$ should be very near the eyepiece's focus, resulting in greater magnification.

makes the real image look larger still, by a factor of its angular magnification $m = 25$ cm$/f_e$. So the overall magnification of the microscope is

$$M = M_o m_e = -\frac{L}{f_o}\left(\frac{25 \text{ cm}}{f_e}\right) \qquad \text{(compound microscope)} \qquad (31.9)$$

where, as usual, the minus sign signifies an inverted image.

Optical microscopes work well as long as the approximation of geometrical optics holds—that is, when the object is much larger than the wavelength of light. Viewing smaller objects requires shorter wavelengths than those of visible light. In the electron microscope, those "waves" are electrons, whose wavelike nature we'll examine in Chapter 34.

## Telescopes

A telescope collects light from distant objects, either forming an image or supplying light to instruments for analysis. Modern astronomical instruments are invariably **reflectors**, whose main light-gathering element is a mirror. Small handheld telescopes, binoculars, and telephoto lenses are **refractors**, which use lenses to gather light.

A simple refractor consists of an objective lens that images distant objects at essentially its focal point, followed by an eyepiece to view this image (Fig. 31.32). The focal points of objective and eyepiece are nearly coincident, so the real image at the objective's focus is then seen through the eyepiece as a greatly enlarged, virtual image. The angular magnification is the ratio of the angle $\beta$ subtended by the final image to the angle $\alpha$ subtended by the actual object; Fig. 31.32 shows that this ratio is

$$m = \frac{\beta}{\alpha} = \frac{f_o}{f_e} \qquad \text{(refracting telescope)} \qquad (31.10)$$

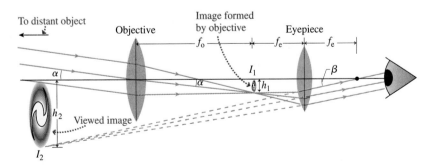

**FIGURE 31.32** Image formation in a refracting telescope. A distant object is imaged first at the focus of the objective lens (image $I_1$). An eyepiece with its focus at nearly the same point then gives an enlarged virtual image ($I_2$). The angles $\alpha$ and $\beta$ are given by $\alpha \simeq h_1/f_o$ and $\beta \simeq h_1/f_e$, leading to Equation 31.10.

Primary mirror

**(a)**

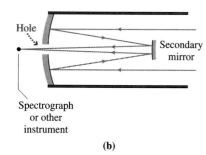

Hole

Secondary mirror

Spectrograph or other instrument

**(b)**

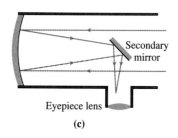

Secondary mirror

Eyepiece lens

**(c)**

**FIGURE 31.33** Reflecting telescopes. (a) A detector at the prime focus gives the best image quality. (b) The Cassegrain design is widely used in large telescopes. (c) The Newtonian design is used primarily in small telescopes.

Since a real image is inverted and a virtual image is upright, a two-lens refracting telescope gives an inverted image. This is fine for astronomical work, but telescopes designed for terrestrial use have an extra lens, a diverging eyepiece, or a set of reflecting prisms to produce an upright image.

Reflecting telescopes offer many advantages over refractors. Mirrors have reflective coatings on their front surfaces, eliminating chromatic aberration because light doesn't pass through glass. Reflectors can be much larger since mirrors are supported across their entire back surfaces—unlike lenses, which must be supported at the edges. Whereas the largest refracting telescope ever built has a 1-m-diameter lens, large reflectors boast diameters of 10 m or greater. These designs incorporate segmented and/or flexible mirrors whose shape can be adjusted under computer control for optimum focusing. With so-called adaptive optics, such systems may adjust rapidly enough to compensate for the atmospheric turbulence that has traditionally limited the resolution of ground-based telescopes.

The simplest reflecting telescope is a curved mirror with a detector at its focus. Superb image quality results, in principle limited only by wave effects we'll discuss in the next chapter. More often the telescope is used as a "light bucket," collecting light from distant sources too small to image even with today's large optical telescopes. Then a secondary mirror sends light to a focus at a point that's convenient for telescope-mounted instrumentation. Optical fibers may also be used to bring light collected by the primary mirror to fixed instruments. Figure 31.33 shows three common designs for reflecting telescopes.

Magnification is not a particularly important quantity in astronomical telescopes, which are used more for spectral and other analysis than for direct imaging. More important is the light-gathering power of the instrument, which is determined simply by the area of its objective lens or primary mirror. The 10-m Keck Telescope, for instance, has 100 times the light-gathering power of the 1-m Yerkes refractor and more than 17 times the power of the 2.4-m Hubble Space Telescope. The Giant Magellan Telescope, scheduled for operation in 2016, boasts seven mirrors equivalent to a single 21-m mirror, with more than four times the light-gathering power of the 10-m Keck instruments (Fig. 31.34).

**FIGURE 31.34** A painting of the Giant Magellan Telescope as it will look when it's completed in 2016.

## Big Picture

The big idea here is the use of reflection and refraction to form **images**. Images are **real** or **virtual** depending on whether or not light actually comes from the image location.

## Key Concepts and Equations

A curved mirror or lens has a **focal point**, $F$, at which parallel light rays converge:

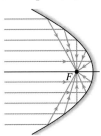

The mirror and lens equations are

$$\frac{1}{s} + \frac{1}{s'} = \frac{1}{f}$$

The table summarizes the sign conventions for each term.

| Value | Symbol | Condition | Sign |
|---|---|---|---|
| Object distance | $s$ | Object on same side as *incoming* light rays | $+$ |
| | | Object on opposite side from *incoming* light rays | $-$ |
| Image distance | $s'$ | Image on same side as *outgoing* light rays | $+$ |
| | | Image on opposite side from *outgoing* light rays | $-$ |
| Focal length | $f$ | Focus on same side as *outgoing* light rays | $+$ |
| | | Focus on opposite side from *outgoing* light rays | $-$ |

Image formation with mirrors and lenses:

The **lensmaker's formula** is

$$\frac{1}{f} = (n-1)\left(\frac{1}{R_1} - \frac{1}{R_2}\right)$$

Curvature radius $R_1$ (+ or −)   $n$   Curvature radius $R_2$ (+ or −)

## Cases and Uses

### Compound microscope

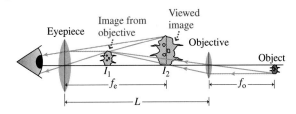

**Magnification:** $M = -\dfrac{L}{f_o}\left(\dfrac{25\ \text{cm}}{f_e}\right)$

### Refracting telescope

**Angular magnification:** $m = \dfrac{f_o}{f_e}$

# For Thought and Discussion

1. How can you see a virtual image, when it really "isn't there"?
2. You lay a magnifying glass (which is just a converging lens) on a printed page. Looking toward the lens, you move it away from the page. Explain the changes in what you see, especially as you move the lens beyond its focal length.
3. Under what circumstances will the image in a concave mirror be the same size as the object?
4. If you're handed a converging lens, what can you do to estimate quickly its focal length?
5. What is the meaning of a negative object distance? A negative focal length?
6. A diverging lens always makes a reduced image. Could you use such a lens to start a fire by focusing sunlight? Explain.
7. Is there any limit to the temperature you can achieve by focusing sunlight? *Hint:* Think about the second law of thermodynamics.
8. Can a concave mirror make a reduced real image? A reduced virtual image? An enlarged real image? An enlarged virtual image? Give conditions for each image that's possible.

9. If you placed a screen at the location of a virtual image, would you see the image on the screen? Why or why not?
10. Where should you place a flashlight bulb in relation to the focus of its reflector?
11. Is the image on a movie screen real or virtual? How do you know?
12. Does a fish in a spherical bowl appear larger or smaller than it actually is?
13. A block of ice contains a hollow, air-filled space in the shape of a double-convex lens. Describe the optical behavior of this space.
14. The refractive index of the human cornea is about 1.38. If you can see clearly in air, why can't you see clearly underwater? Why do goggles help?
15. Do you want a long or short focal length for the objective lens of a telescope? Of a microscope?
16. Give several reasons reflecting telescopes are superior to refractors.

# Exercises and Problems

## Exercises

### Section 31.1 Images with Mirrors

17. A shoe store uses small floor-level mirrors to let customers view prospective purchases. At what angle should such a mirror be inclined so that a person standing 50 cm from the mirror with eyes 140 cm off the floor can see her feet?
18. A candle is 36 cm from a concave mirror with focal length 15 cm and on the mirror axis. (a) Where is its image? (b) How do the image and object sizes compare? (c) Is the image real or virtual?
19. An object is five focal lengths from a concave mirror. (a) How do the object and image heights compare? (b) Is the image upright or inverted?
20. A virtual image is located 40 cm behind a concave mirror with focal length 18 cm. (a) Where is the object? (b) By how much is the image magnified?
21. (a) Where on the axis of a concave mirror would you place an object in order to produce a full-size image? (b) Will the image be real or virtual?

### Section 31.2 Images with Lenses

22. A lightbulb is 56 cm from a convex lens, and its image appears on a screen located 31 cm on the other side of the lens. (a) What is the focal length of the lens? (b) By how much is the image enlarged or reduced?
23. By what factor is the image magnified when an object is placed 1.5 focal lengths from a converging lens? Is the image upright or inverted?
24. A lens with 50-cm focal length produces a real image the same size as the object. How far from the lens are image and object?
25. By holding a magnifying glass 25 cm from my desk lamp, I can focus an image of the lamp's bulb on a wall 1.6 m from the lamp. What is the focal length of my magnifying glass?
26. A real image is four times as far from a lens as is the object. What is the object distance, measured in focal lengths?
27. A magnifying glass enlarges print by 50% when it's held 9.0 cm from a page. What is its focal length?

### Section 31.3 Refraction in Lenses: The Details

28. My magnifying glass is a double-convex lens with equal curvature radii of 32 cm. If the lens glass has $n = 1.52$, what is its focal length?
29. You're standing in a wading pool and your feet appear to be 30 cm below the surface. How deep is the pool?
30. The bottom of a swimming pool looks to be 1.5 m below the surface. Find the pool's actual depth.
31. A tiny insect is trapped 1.0 mm from the center of a spherical dew drop 4.0 mm in diameter. As you look straight into the drop at the insect, what is its apparent distance from the edge of the drop?
32. You're underwater, looking through a spherical air bubble (Fig. 31.35). What's its actual diameter if it appears, along your line of sight, to be 1.5 cm in diameter?

**FIGURE 31.35**
Exercise 32

### Section 31.4 Optical Instruments

33. You find that you have to hold a book 55 cm from your eyes for the print to be in sharp focus. What power lens is needed to correct your farsightedness?
34. What focal length is needed to make a simple magnifier with angular magnification 3?
35. A nearsighted person cannot see clearly beyond 80 cm. Prescribe a lens power that will image the most distant objects at 80 cm, giving clear vision at all distances.
36. A particular eye has a focal length of 2.0 cm instead of the 2.2 cm that would be required for a sharply focused image on the retina. (a) Is this eye nearsighted or farsighted? (b) What power of corrective lens is needed?

37. A compound microscope has objective and eyepiece focal lengths of 6.1 mm and 1.7 cm, respectively. If the lenses are 8.3 cm apart, what is the magnification of the instrument?

## Problems

38. (a) Find the focal length of a concave mirror if an object placed 50 cm in front of the mirror has a real image 75 cm from the mirror. (b) Where and what type will the image be if the object is moved to a point 20 cm from the mirror?

39. A 12-mm-high object is 10 cm from a concave mirror with focal length 17 cm. (a) Where, (b) how high, and (c) what type is its image?

40. Repeat Problem 39 for a convex mirror, assuming all numbers stay the same.

41. An object's image in a 27-cm-focal-length concave mirror is upright and magnified by a factor of 3. Where is the object?

42. What is the curvature radius of a concave mirror that produces a 9.5-cm-high virtual image of a 5.7-cm-high object, when the object is located 22 cm from the mirror?

43. When viewed from Earth, the Moon subtends an angle of 0.5° in the sky. How large an image of the Moon will be formed by the 3.6-m-diameter mirror of the Canada–France–Hawaii Telescope, which has a focal length of 8.5 m?

44. At what two distances could you place an object from a 45-cm-focal-length concave mirror in order to get an image 1.5 times the object's size?

45. (a) How far from a 1.2-m-focal-length concave mirror should you place an object in order to get an inverted image 1.5 times the size of the object? (b) Where will the image be?

46. Find the focal length of a concave mirror if an object 15 cm from the mirror has a virtual image 2.5 times the object's actual size.

47. How far from a page should you hold a lens with 32-cm focal length in order to see the print magnified 1.6 times?

48. A converging lens has a focal length of 4.0 cm. A 1.0-cm-high arrow is located 7.0 cm from the lens with its lowest point 5.0 mm above the lens axis. Make a full-scale drawing of the situation, and use ray tracing to locate both ends of the image. Confirm using the lens equation.

49. A lens has focal length $f = 35$ cm. Find the type and height of the image produced when a 2.2-cm-high object is placed at distances (a) $f + 10$ cm and (b) $f - 10$ cm.

50. How far apart are the object and image produced by a converging lens with 35-cm focal length when the object is (a) 40 cm and (b) 30 cm from the lens?

51. A candle and a screen are 70 cm apart. Find two points between candle and screen where you could put a convex lens with 17-cm focal length to give a sharp image of the candle on the screen.

52. An object is placed two focal lengths from a diverging lens. (a) What type of image forms, (b) what is the magnification, and (c) where is the image?

53. How far from a 25-cm-focal-length lens should you place an object to get an upright image magnified by a factor of 1.8?

54. An object and its lens-produced real image are 2.4 m apart. If the lens has 55-cm focal length, what are the possible values for the object distance and magnification?

55. An object is 68 cm from a plano-convex lens whose curved side has curvature radius 26 cm. The refractive index of the lens is 1.62. Where and what type is the image?

56. Use Equation 31.6 to show that an object at the center of a glass sphere will appear to be its actual distance—one radius—from the edge. Draw a ray diagram showing why this result makes sense.

57. Rework Example 31.4 for a fish 15 cm from the *far* wall of the tank.

58. Consider the inverse of Example 31.4: You're inside a 70-cm-diameter hollow tube containing air, and the tip of your nose is 15 cm from the wall of the tube. The tube is immersed in water, and a fish looks in. To the fish, what is the apparent distance from your nose to the tube wall?

59. Two specks of dirt are trapped in a crystal ball, one at the center and the other halfway to the surface. If you peer into the ball on a line joining the two specks, the outer one appears to be only one-third of the way to the other. What is the refractive index of the ball?

60. A contact lens is in the shape of a convex meniscus (see Fig. 31.25); the inner surface is curved to fit the eye, with a curvature radius of 7.80 mm. The lens is made from plastic with refractive index $n = 1.56$. If it's to have a focal length of 44.4 cm, what should be the curvature radius of its outer surface?

61. For what refractive index would the focal length of a plano-convex lens be equal to the curvature radius of its one curved surface?

62. An object is 28 cm from a double-convex lens with $n = 1.5$ and curvature radii 35 cm and 55 cm. Where and what type is the image?

63. A double-convex lens has equal curvature radii of 35 cm. An object placed 30 cm from the lens forms a real image at 128 cm. What is the refractive index of the lens?

64. A double-convex lens with equal 38-cm curvature radii is made from glass with refractive indices $n_{red} = 1.51$ and $n_{violet} = 1.54$ at the edges of the visible spectrum. If a point source of white light is placed on the lens axis at 95 cm from the lens, over what range will its visible image be smeared?

65. An object placed 15 cm from a plano-convex lens made of crown glass focuses to a virtual image twice the size of the object. If the lens is replaced with an identically shaped one made from diamond, what type of image will appear and what will be its magnification? See Table 30.1.

66. A camera's zoom lens covers the focal-length range 38–110 mm. You point the camera at a distant object and photograph it first at 38 mm and then with the camera zoomed out to 110 mm. Compare the sizes of its images on the two photos.

67. A camera can normally focus as close as 60 cm, but it has provisions for mounting additional lenses at the outer end of the main lens to provide closeup capability. What type and power of auxiliary lens will allow the camera to focus as close as 20 cm? The distance between the two lenses is negligible.

68. A 300-power compound microscope has a 4.5-mm-focal-length objective lens. If the distance from objective to eyepiece is 10 cm, what should be the focal length of the eyepiece?

69. To the unaided eye, the planet Jupiter has an angular diameter of 50 arc seconds. What will its angular size be when viewed through a 1-m-focal-length refracting telescope with an eyepiece whose focal length is 40 mm?

70. A Cassegrain telescope like that shown in Fig. 31.33b has 1.0-m focal length, and the convex secondary mirror is located 0.85 m from the primary. What should be the focal length of the secondary in order to put the final image 0.12 m behind the front surface of the primary mirror?

71. An object is held 6.0 cm from the surface of a reflecting ball, and its image appears three-quarters full size. What is the ball's diameter?

72. My contact lens prescription calls for +2.25-diopter lenses with an inner curvature radius of 8.6 mm to fit my cornea. (a) If the lenses are made from plastic with $n = 1.56$, what should be the outer curvature radius? (b) Wearing these lenses, I hold a newspaper 30 cm from my eyes. Where is its image as viewed through the lenses?

73. Show that the powers of closely spaced lenses add; that is, placing a 1-diopter lens in front of a 2-diopter lens gives the equivalent of a single 3-diopter lens.

74. Show that identical objects placed equal distances on either side of the focal point of a concave mirror or converging lens produce images of equal size. Are the images of the same type?

75. Generalize the derivation of the lensmaker's formula (Equation 31.7) to show that a lens of refractive index $n_{lens}$ in an external medium with index $n_{ext}$ has focal length given by

$$\frac{1}{f} = \left(\frac{n_{lens}}{n_{ext}} - 1\right)\left(\frac{1}{R_1} - \frac{1}{R_2}\right)$$

76. Draw a diagram similar to Fig. 31.10, but showing a ray from the arrowhead through the center of curvature. Using the fact that this ray reflects back on itself, draw similar triangles with object and image as their vertical sides, and show that the center of curvature is twice as far from the mirror as the focal point—that is, $R = 2f$, with $R$ the curvature radius.

77. Before Equation 31.7 are two equations describing refraction at the two surfaces of a lens with thickness $t$. Combine these equations to show that the object distance $s$ and image distance $s'$ for such a lens are related by

$$\frac{1}{s} + \frac{1}{s'} - \frac{[(n-1)s - R_1]^2 t}{sR_1[t(s+R_1) + ns(R_1 - t)]} = (n-1)\left(\frac{1}{R_1} - \frac{1}{R_2}\right)$$

78. Use the result of Problem 77 to find an expression for the focal length of a transparent sphere with radius $R$ and refractive index $n$, as measured from the exit surface.

79. Galileo's first telescope used the arrangement shown in Fig. 31.36, with a double-concave eyepiece placed slightly before the focus of the objective lens. Use ray tracing to show that this design gives an upright image, which makes this telescope useful in terrestrial observing.

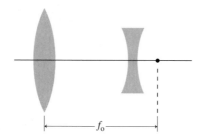

**FIGURE 31.36** Problem 79

80. A two-dimensional corner reflector is formed from two plane mirrors, as shown in Figure 30.2. Take the origin of a Cartesian coordinate system at the intersection of the mirrors, with the $x$ axis down along the left mirror and the $y$ axis to the right along the top mirror, with each mirror extending 4 m along an axis. Find the locations of all images of an object at $x = 2$ m, $y = 1$ m.

81. The maximum magnification of a simple magnifier occurs with the image at the 25-cm near point. Show that the angular magnification is then given by $m = 1 + (25 \text{ cm}/f)$, where $f$ is the focal length.

82. Chromatic aberration results from the variation of the refractive index with wavelength. Start with the lensmaker's formula, and find an expression for the fractional change $df/f$ in the focal length of a thin lens in terms of the change $dn$ in refractive index.

83. For wavelengths in the visible range, the refractive index of a thin glass lens is $n = n_0 - b\lambda$, where $n_0 = 1.546$ and $b = 4.47 \times 10^{-5}$ nm$^{-1}$. If its focal length is 30 cm at 550 nm, how much does the focal length vary over a spread in wavelength of 10 nm centered on 550 nm? *Hint:* See Problem 82.

84. You're at a very boring art show. You look into a reflecting sphere and see an image of your face that is one-third its normal size. You are 40 cm from the sphere. You wonder, Can I figure out the diameter of the sphere? Looking like you are intently concentrating on the art show, you calculate an answer. What do you get?

85. A friend who's an education major is working on a project for a science-teaching class. She wants her students to make a simple camera with just one lens. She will use a single converging lens. The lens-to-focal-plane distance will be 4.77 cm and the lens-to-object distance is 80 cm. What focal length lens should she use?

86. You're preparing a special exhibit on photographic equipment for an art class. To show how a slide projector works, you're making a simple model. It uses a double-convex lens with $n = 1.58$ and equal curvature radii. You find the specification for the curvature is 121 mm, but you don't know the focal length. After calculating the focal length, you need to know how far from the focal point to place the slide. What should that distance be in order to form the image on a screen 3.1 m from the lens?

# Answers to Chapter Questions

## Answer to Chapter Opening Question

High-intensity laser light reshapes the cornea, so that refracted light converges to produce sharp images.

## Answers to GOT IT? Questions

31.1 About half your height.

31.2 At the mirror's center of curvature, which is at twice the focal length.

31.3 Virtual image; convex lens.

31.4 $+2.25$ because they're converging lenses.

# 32 Interference and Diffraction

- How do optical principles ultimately govern the amount of information that can be stored on a DVD?

The preceding chapters described the behavior of light using geometrical optics—an approximation that's valid when we're dealing with length scales much larger than the wavelength of light, so we can ignore light's wave nature. We now turn to **physical optics**, which treats optical phenomena for which the wave nature of light plays an essential role. Two related phenomena, interference and diffraction, are central in physical optics.

## 32.1 Coherence and Interference

In Chapter 14 we showed how wave displacements add to produce **constructive interference** or **destructive interference**. Electromagnetic waves, including light, are no exception: Since electric and magnetic fields obey the superposition principle, the net fields at any point are the vector sums of individual wave fields. That sum may increase (constructive interference) or reduce (destructive interference) the net field.

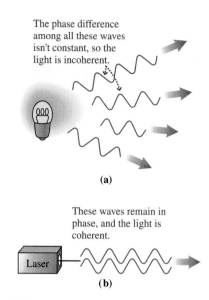

The phase difference among all these waves isn't constant, so the light is incoherent.

**(a)**

These waves remain in phase, and the light is coherent.

Laser

**(b)**

**FIGURE 32.1** (a) Lightbulbs emit incoherent light consisting of short wavetrains with random phases. (b) Lasers produce coherent light, which facilitates stable interference.

## Coherence

Steady interference patterns occur only when waves are **coherent**, meaning they maintain a constant phase relation. Ordinary light sources like lightbulbs or the Sun produce short wavetrains with random phases, so their light doesn't stay coherent for long (Fig. 32.1a). Lasers, in contrast, produce long wavetrains that maintain coherence over many wavelengths (Fig. 32.1b). The typical length of a wavetrain is the **coherence length** for a given light source.

It takes at least two waves to interfere, and even with lasers it's virtually impossible for two different sources to maintain coherence. Therefore interference usually occurs when light from a single source splits, travels different paths, and then recombines. Even light from an ordinary lightbulb will produce interference provided light in the recombining beams originated at very nearly the same place and time. Coherent laser light relaxes this restriction considerably. Coherence also requires that interfering light beams have exactly the same frequency and therefore wavelength or color. Again, lasers make things easier: Their light is very nearly **monochromatic**, consisting of a narrow band of wavelengths.

## Destructive and Constructive Interference

Consider light waves that originate together at a single source, travel two different paths, and then rejoin. Suppose one wave's path is exactly half a wavelength longer than the other. Then, when the waves recombine, they'll be out of phase by half a wavelength (Fig. 32.2a) and thus their superposition has smaller amplitude (zero, if the two interfering waves have exactly the same amplitude). If, on the other hand, the path lengths don't differ, or they differ by a full wavelength, then the two waves recombine in phase (Fig. 32.2b) and their superposition has larger amplitude. These two cases correspond, respectively, to destructive interference and constructive interference. It doesn't matter whether path lengths for the waves in Fig. 32.2a differ by half a wavelength, or $1\frac{1}{2}$ wavelengths, or $2\frac{1}{2}$ wavelengths; as long as the difference is an odd multiple of a half-wavelength, the waves recombine out of phase and destructive interference results. Thus **destructive interference results when light paths differ by an odd-integer multiple of a half-wavelength**. Similarly, it doesn't matter whether the path lengths in Fig. 32.2b are the same, or differ by 1, 2, 3, or any other integer number of wavelengths. Thus **constructive interference results when light paths differ by an integer multiple of the wavelength**.

There's one caveat to our statements: The path difference can't be greater than the coherence length; otherwise, the interfering waves won't be coherent. Once again, laser light has the advantage here because of its greater coherence length.

Of course, light paths don't have to differ by half or full multiples of the wavelength. In intermediate cases interfering waves superpose to make a composite wave whose amplitude may be enhanced or diminished, depending on the relative phase.

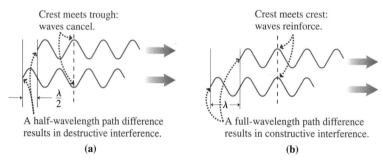

Crest meets trough: waves cancel.

$\frac{\lambda}{2}$

A half-wavelength path difference results in destructive interference.

**(a)**

Crest meets crest: waves reinforce.

$\lambda$

A full-wavelength path difference results in constructive interference.

**(b)**

**FIGURE 32.2** Two waves that start out in phase but travel different paths before rejoining.

**More CD Music**

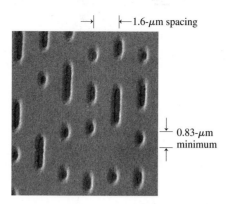

Example 30.3, "CD Music," showed how refraction helps focus the laser beam that reads information from a compact disc. Interference, too, plays a crucial role in reading a CD.

Information on a CD or a DVD is stored digitally in a sequence of pits stamped into a reflective metallic information layer, as shown in the photo. The pits' depth is very nearly one-quarter wavelength of the laser light used to read the disc. From the transparent underside of the disc, each pit appears as an elevated bump. Since the bumps stick down one-quarter wavelength, light reflecting off a bump follows a round-trip path that's shorter by half a wavelength compared with the path of light reflecting off the undisturbed information layer (see the figure). The laser beam is wider than the pit, so the reflected beam includes light both from the undisturbed disc and from the bump. The two interfere destructively, making the reflected beam less intense when a bump is present. As the disc spins, the result is a pattern of fluctuating light intensity conveying the information associated with the pattern of pits. A photodetector then converts that pattern to electrical signals that ultimately drive loudspeakers, headphones, or a video display.

## 32.2 Double-Slit Interference

In Chapter 14 we looked briefly at interference patterns produced by a pair of coherent sources. Such a pair can be made by passing light through two narrow slits. In 1801 Thomas Young used this approach in a historic experiment that confirmed the wave nature of light. Young first admitted sunlight to his laboratory through a hole small enough to ensure coherence of the incoming light. The light then passed through a pair of narrow, closely spaced slits, after which it illuminated a screen. Each slit acts as a source of cylindrical wavefronts that interfere in the region between slits and screen (Fig. 32.3a). Constructive and destructive interference produce **interference fringes**—alternating bright and dark bands (Fig. 32.3b).

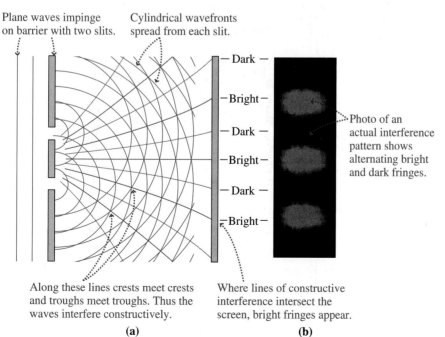

**FIGURE 32.3** Double-slit interference results when light from a single source passes through closely spaced slits.

**FIGURE 32.4** Geometry for finding locations of the interference fringes. In the blowup you can see that for $L \gg d$, the paths to $P$ are nearly parallel and differ by $d \sin \theta$.

Bright fringes represent constructive interference, and therefore they occur where the difference in the path length for light traveling from the two slits is a multiple of the wavelength. When the distance $L$ from slits to screen is much greater than the slit spacing $d$, Fig. 32.4 shows that the path difference to a point on the screen is $d \sin \theta$, where $\theta$ is the angular position of a point on the screen measured from an axis perpendicular to slits and screen. So our criterion for constructive interference—that this difference be an integer number of wavelengths—becomes

$$d \sin \theta = m\lambda \qquad (\text{bright fringes, } m = 0, 1, 2, \ldots) \qquad (32.1a)$$

The integer $m$ is the **order** of the fringe, with the central bright fringe being the zeroth-order fringe and with higher-order fringes on either side.

Waves interfere destructively when their path lengths differ by an odd-integer multiple of a half-wavelength:

$$d \sin \theta = \left(m + \tfrac{1}{2}\right)\lambda \qquad (\text{dark fringes, } m = 0, 1, 2, \ldots) \qquad (32.1b)$$

where $m$ is any integer.

In a typical double-slit experiment, $L$ may be on the order of 1 m, $d$ a fraction of 1 mm, and $\lambda$ the sub-$\mu$m wavelength of visible light. Then we have the additional condition that $\lambda \ll d$. This makes the fringes very closely spaced on the screen, so the angle $\theta$ in Fig. 32.4 is small even for large orders $m$. Then $\sin \theta \simeq \tan \theta = y/L$, and a fringe's position $y$ on the screen, measured from the central maximum, becomes

$$y_{\text{bright}} = m\frac{\lambda L}{d} \quad \text{and} \quad y_{\text{dark}} = \left(m + \tfrac{1}{2}\right)\frac{\lambda L}{d} \quad \left(\begin{array}{c}\text{fringe position,}\\ \lambda \ll d\end{array}\right) \quad (32.2a, b)$$

**GOT IT? 32.1** If you increase the slit separation in a two-slit system, do the interference fringes become closer together or farther apart?

---

**EXAMPLE 32.1** **Measuring Wavelength: Laser Light**

Two slits 0.075 mm apart are located 1.5 m from a screen. Laser light shining through the slits produces an interference pattern whose third-order bright fringe is 3.8 cm from the screen center. Find the light's wavelength.

**INTERPRET** The concept behind this problem is two-slit interference. The phrase "third order" tells us we're dealing with the $m = 3$ bright fringe located at $y_{\text{bright}} = 3.8$ cm.

**DEVELOP** Our plan is to use Equation 32.2a, $y_{\text{bright}} = m(\lambda L/d)$, and solve for $\lambda$. Since that equation requires $\lambda \ll d$, we'll then check to see whether our answer is consistent with this condition.

**EVALUATE** Solving, we have

$$\lambda = \frac{y_{\text{bright}} d}{mL} = \frac{(0.038 \text{ m})(0.075 \times 10^{-3} \text{ m})}{(3)(1.5 \text{ m})} = 633 \text{ nm}$$

**ASSESS** This is indeed much less than the slit spacing of 0.075 mm or 75,000 nm. Our 633-nm result is in fact the wavelength of the red light from widely used helium–neon lasers. ∎

---

### Intensity in the Interference Pattern

We found the maxima and minima in two-slit interference using geometrical arguments alone. To find the actual intensity we need to superpose the interfering waves, which means adding the wave fields. So consider again a point $P$ in the interference pattern (Fig. 32.5). In the approximation $d \ll L$, the difference in path lengths is so small that we can neglect any difference in the amplitudes of the two waves. However, the difference in phase is crucial; it's what causes the interference. So we consider waves whose electric fields at $P$ vary sinusoidally in time, with equal amplitude $E_p$ but an explicit phase difference $\phi$:

$$E_1 = E_p \sin \omega t \qquad \text{and} \qquad E_2 = E_p \sin(\omega t + \phi)$$

We aren't bothering with vectors because the two waves are polarized in the same direction and therefore their fields add algebraically. Then the net electric field at $P$ is

$$E = E_1 + E_2 = E_p[\sin\omega t + \sin(\omega t + \phi)]$$

Appendix A gives the trig identity $\sin\alpha + \sin\beta = 2\sin[(\alpha+\beta)/2]\cos[(\alpha-\beta)/2]$, which, with $\alpha = \omega t$ and $\beta = \omega t + \phi$, gives

$$E = 2E_p \sin\left(\omega t + \frac{\phi}{2}\right)\cos\left(\frac{\phi}{2}\right)$$

where we also used $\cos(-x) = \cos x$. Thus, the electric field at $P$ oscillates with the wave frequency $\omega$, and its amplitude is $2E_p\cos(\phi/2)$. Since the phase difference $\phi$ depends on the difference in path lengths from the two slits, this amplitude varies across the screen, giving the interference pattern.

We've seen that the path-length difference is $d\sin\theta$, with $d$ the slit spacing and $\theta$ the angle to $P$; under our approximation $d \ll L$, $\theta$ is small and $\sin\theta \simeq \tan\theta = y/L$, where $y$ is the position of $P$ as shown in Fig. 32.5. Then the path difference becomes $yd/L$. Now, that all-important phase difference $\phi$ is whatever fraction of a full cycle ($2\pi$ radians) this path difference is of the wavelength $\lambda$; that is, $\phi = 2\pi(yd/\lambda L)$. Then the amplitude $2E_p\cos(\phi/2)$ becomes $2E_p\cos(\pi yd/\lambda L)$. The average intensity follows from Equation 29.20b:

$$\overline{S} = \frac{[2E_p\cos(\pi yd/\lambda L)]^2}{2\mu_0 c} = 4\overline{S}_0\cos^2\left(\frac{\pi d}{\lambda L}y\right) \tag{32.3}$$

where $\overline{S}_0 = E_p^2/2\mu_0 c$ is the average intensity of either wave alone. Now, $\cos^2$ has its maximum value, 1, when its argument is a multiple of $\pi$. Thus Equation 32.3 gives maximum intensity when $yd/\lambda L$ is an integer $m$, or when $y = m\lambda L/d$. This is just the condition 32.2a, showing that our intensity calculation is fully consistent with the simpler geometrical analysis. But the intensity calculation tells more: It gives not only the fringe positions, but also the intensity variation in between.

## 32.3 Multiple-Slit Interference and Diffraction Gratings

Systems with multiple slits play a crucial role in optical instrumentation and in the analysis of materials. As we'll see, gratings with several thousand slits per centimeter make possible high-resolution spectroscopic analysis. At a much smaller scale the regularly spaced rows of atoms in a crystal act much like a multiple-slit system for X rays, and the resulting X-ray patterns reveal the crystal structure.

Figure 32.6 shows waves from three evenly spaced slits interfering at a screen. Maximum intensity requires that all three waves be in phase or, equivalently, travel paths differing by an integer number of wavelengths. Our criterion for the maximum in a two-slit pattern, $d\sin\theta = m\lambda$, ensures that waves from two adjacent slits will add constructively. Since the slits are evenly spaced with distance $d$ between each pair, waves coming through a third slit will be in phase with the other two if this criterion is met. So the criterion for a maximum in an $N$-slit system is still Equation 32.1a:

$$d\sin\theta = m\lambda \quad (\text{maxima in multiple-slit interference, } m = 0, 1, 2, \dots) \tag{32.1a}$$

With more than two waves, however, the criterion for destructive interference is more complicated. Somehow all the waves need to sum to zero. Figure 32.7 shows that this happens for three waves when each is out of phase with the others by one-third of a cycle. Thus, the path-length difference $d\sin\theta$ must be either $\left(m+\frac{1}{3}\right)\lambda$ or $\left(m+\frac{2}{3}\right)\lambda$, where $m$ is an

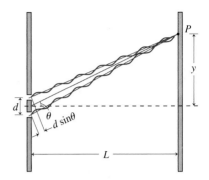

**FIGURE 32.5** Waves from the slits arrive at $P$ displaced by the path-length difference $d\sin\theta$. For $L \gg d$, $\sin\theta \simeq \tan\theta = y/L$.

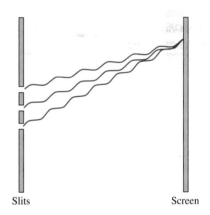

**FIGURE 32.6** Waves from three evenly spaced slits interfere constructively when they reach the screen in phase.

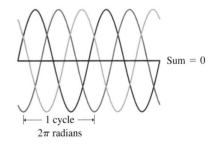

**FIGURE 32.7** Waves from three slits must be out of phase by one-third of a cycle in order to interfere destructively.

2 slit

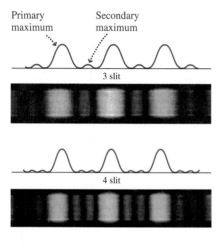

3 slit

4 slit

5 slit

**FIGURE 32.8** Interference patterns for multiple-slit systems with the same slit spacing. The bright fringes stay in the same place but become narrower and brighter as the number of slits increases. Intensity plots don't have the same vertical scale; peak intensity scales as the *square* of the number of slits.

integer. The case $\left(m + \frac{3}{3}\right)\lambda$ is excluded because then the path lengths differ by a full wavelength, giving constructive interference and thus a maximum in the interference pattern. More generally we can write

$$d \sin\theta = \frac{m}{N}\lambda \qquad (32.4)$$

for destructive interference in an $N$-slit system, where $m$ is an integer *but not an integer multiple of N*.

Figure 32.8 shows interference patterns and intensity plots from some multiple-slit systems. Note that the bright, or *primary*, maxima are separated by several minima and fainter, or *secondary*, maxima. Why this complex pattern? Our analysis of the three-slit system shows two minima between every pair of primary maxima; for example, we considered the minima at $d \sin\theta$ equal to $\left(m + \frac{1}{3}\right)\lambda$ or $\left(m + \frac{2}{3}\right)\lambda$, which lie between the maxima at $d \sin\theta$ equal to $m\lambda$ and $(m + 1)\lambda$. More generally, Equation 32.4 shows that there are $N - 1$ minima between each pair of primary maxima given by Equation 32.1a. The secondary maxima that lie between these minima result from interference that is neither fully destructive nor fully constructive. The figure shows that the primary maxima become brighter and narrower as the number of slits increases, while the secondary maxima become relatively less bright. With a large number $N$ of slits, then, we should expect a pattern of bright but narrow primary maxima, with broad, essentially dark regions in between.

## Diffraction Gratings

A set of many closely spaced slits is called a **diffraction grating** and proves very useful in the spectroscopic analysis of light. Diffraction gratings are commonly several centimeters across and have several thousand slits—usually called lines—per cm. Gratings are made by photoreducing images of parallel lines or by ruling with a diamond stylus on aluminum-plated glass. Gratings like the slit systems we've been discussing are **transmission gratings**, since light passes through the slits. **Reflection gratings** produce similar interference effects by reflecting incident light.

We've seen that the maxima of the multiple-slit interference pattern are given by the same criterion, $d \sin\theta = m\lambda$, that applies to a two-slit system. For $m = 0$ this equation implies that all wavelengths peak together at the central maximum, but for larger values of $m$ the angular position of the maximum depends on wavelength. Thus, a diffraction grating can be used in place of a prism to disperse light into its component wavelengths, and the integer $m$ is therefore called the **order** of the dispersion. Figure 32.9 shows a spectrometer that works on this principle. Because the maxima in $N$-slit interference are very sharp for large $N$ (recall Fig. 32.8), a grating with many slits diffracts individual wavelengths to very precise locations.

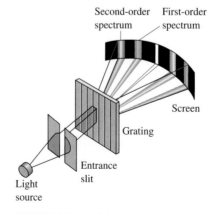

**FIGURE 32.9** Essential elements of a grating spectrometer. An electronic detector is normally used in place of the screen.

**EXAMPLE 32.2    Finding the Separation: A Grating Spectrometer**

Light from glowing hydrogen contains discrete spectral lines called H$\alpha$ (hydrogen-alpha) and H$\beta$ (hydrogen-beta), with wavelengths of 656.3 nm and 486.1 nm, respectively. Find the first-order angular separation between these wavelengths in a spectrometer that uses a grating with 6000 slits per cm.

**INTERPRET** The concept behind the grating spectrometer is multiple-slit interference, so our job is to find the angles to which the grating sends the given wavelengths. "First-order" means we have $m = 1$.

**DEVELOP** Equation 32.1a, $d \sin\theta = m\lambda$, gives the location of the interference maxima as a function of wavelength $\lambda$, order $m$, and slit spacing $d$. We're given $m$ and two values for $\lambda$, but we don't know $d$. However, we're told that there are 6000 slits per cm, so we can find the spacing. Our plan is first to calculate $d$, then use Equation 32.1a to find the angular positions for the two wavelengths, and finally take their difference to get the angular separation.

**EVALUATE** With 6000 slits/cm, the spacing is $d = 1/6000\,\text{cm} = 1.667\,\mu\text{m}$. Applying Equation 32.1a with $m = 1$ gives

$$\theta_\alpha = \sin^{-1}\left(\frac{\lambda}{d}\right) = \sin^{-1}\left(\frac{0.6563\,\mu\text{m}}{1.667\,\mu\text{m}}\right) = 23.2°$$

A similar calculation gives $\theta_\alpha = 17.0°$. Thus the angular separation is 6.2°.

**ASSESS** Our 6.2° result is certainly adequate to distinguish clearly these two wavelengths. For greater angular separation, or to separate closer wavelengths, we could look at the higher-order dispersion (see Exercise 20). ∎

## Resolving Power

The detailed shapes and wavelengths of spectral lines contain a wealth of information about the systems in which light originates. Studying these details requires a high dispersion in order to separate nearby spectral lines or to analyze the intensity-versus-wavelength profile of a single line. Suppose we pass light containing two spectral lines of nearly equal wavelengths $\lambda$ and $\lambda'$ through a grating. Figure 32.10 shows that we'll just be able to distinguish two spectral lines if the peak of one line corresponds to the first minimum of the other; any closer and the lines blur together. Suppose wavelength $\lambda$ has its $m$th-order maximum at angular position $\theta$. The criterion for this maximum is, as usual, $d \sin\theta_{\max} = m\lambda$. We can write this as $d \sin\theta_{\max} = (mN/N)\lambda$, with $N$ the number of slits in the grating. Equation 32.4 then shows that we get an adjacent minimum if we add 1 to the numerator $mN$. Thus, there is an adjacent minimum whose position satisfies

$$d \sin\theta_{\min} = \frac{mN + 1}{N}\lambda$$

Our criterion that the two wavelengths $\lambda$ and $\lambda'$ be just distinguishable is that the maximum for $\lambda'$ fall at the location of this minimum for $\lambda$. But the maximum for $\lambda'$ satisfies $d \sin\theta'_{\max} = m\lambda' = (mN/N)\lambda'$, so for $\theta'_{\max} = \theta_{\min}$ we must have $(mN + 1)\lambda = mN\lambda'$. It's convenient to express this result in terms of the wavelength difference $\Delta\lambda = \lambda' - \lambda$. Solving our equation relating $\lambda$ and $\lambda'$ for $\lambda'$ and using the result to write $\Delta\lambda$ in terms of $\lambda$ alone then gives

$$\frac{\lambda}{\Delta\lambda} = mN \qquad \text{(resolving power)} \qquad (32.5)$$

The quantity $\lambda/\Delta\lambda$ is the **resolving power** of the grating, a measure of its ability to distinguish closely spaced wavelengths. The higher the resolving power, the smaller the wavelength difference $\Delta\lambda$ that we can distinguish. Equation 32.5 shows that the resolving power increases with the number of lines, $N$, on the grating and also with the order, $m$, of the spectrum we observe.

**FIGURE 32.10** Intensity versus angular position for spectral lines with slightly different wavelengths, as dispersed with a grating.

**EXAMPLE 32.3** **Resolving Power: "Seeing" a Double Star**

A certain double-star system consists of a massive star essentially at rest, with a smaller companion in circular orbit. The system is far too distant for the pair to appear as anything but a single point even to the largest telescopes. Yet astronomers can "see" the companion star through the Doppler shift in wavelengths of its spectral lines. The Hα spectral line from the stationary massive star is at $\lambda = 656.272$ nm; for the companion when it's moving away from Earth, the Hα line Doppler-shifts to 656.329 nm (this corresponds to an orbital speed of about 26 km/s). If a spectrometer has 5000 lines, what order spectrum will resolve the Hα lines from the two stars?

**INTERPRET** The concept here is resolution of distinct spectral lines using a grating spectrometer.

**DEVELOP** Equation 32.5, $\lambda/\Delta\lambda = mN$, determines the resolving power. We can solve for $m$ to get the order: $m = \lambda/(N \, \Delta\lambda)$. We're given the two wavelengths, so we can readily find $\Delta\lambda$.

**EVALUATE** We have $\Delta\lambda = 656.329$ nm $- 656.272$ nm $= 0.057$ nm. Then

$$m = \frac{\lambda}{N \, \Delta\lambda} = \frac{656.272 \text{ nm}}{(5000)(0.057 \text{ nm})} = 2.3$$

**ASSESS** Since $m$ must be an integer, we'll have to use the third-order spectrum. ∎

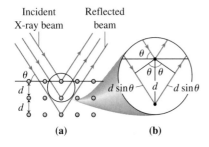

**FIGURE 32.11** (a) X rays reflecting off the planes of atoms in a crystal. (b) Constructive interference enhances the outgoing beam when the extra distance $2d \sin\theta$ is an integer multiple of the X-ray wavelength.

## X-Ray Diffraction

The wavelengths of X rays, on the order of 0.1 nm, are far too short for diffraction with gratings produced mechanically or photographically. Instead, **X-ray diffraction** occurs when X rays interact with the regularly spaced atoms in a crystal. At the microscopic level, reflection of an electromagnetic wave occurs when the wave's electric field sets electrons oscillating. The electrons re-radiate, producing the reflected beam. With X rays reflecting from a crystal, the regular atomic spacing results in interference that enhances the reflected radiation at certain angles. Figure 32.11a shows an X-ray beam interacting with the atoms in a crystal. In Fig. 32.11b we see that waves reflecting at one layer of atoms travel a distance $2d \sin\theta$ farther than those reflecting at the layer above, where $\theta$ is the angle between the incident beam and the atomic planes. Constructive interference occurs when this difference is an integer number of wavelengths:

$$2d \sin\theta = m\lambda \quad \text{(Bragg condition, } m = 1, 2, 3 \dots \text{)} \quad (32.6)$$

This **Bragg condition** lets us use a crystal with known spacing as a diffraction grating for X rays. More important is the converse: Much of what we know about crystal structure comes from probing crystals with X rays and using the resulting patterns to deduce positions of their atoms. X-ray diffraction measurements by British scientist Rosalind Franklin in 1952 were crucial in establishing the structure of DNA.

## Other Gratings

Anything with regularly spaced structures can act as a diffraction grating for waves of suitable wavelength. The rainbow of colors you see on the underside of a CD or DVD (see this chapter's opening photo) results because adjacent pits of CD tracks (shown in the Application earlier in this chapter) act as a diffraction grating. Sound waves in a solid set up regular patterns that become diffraction gratings; changing the wavelength of the sound changes the "slit" spacing and therefore the diffraction angle. **Acousto-optic modulators** (AOMs) based on this principle are widely used to "steer" light beams in light-wave communication and other opto-electronic technologies. Laser printers and digital copiers, for example, use an AOM to control the laser beam that "paints" the image of the printed page on a light-sensitive surface.

---

**GOT IT? 32.2** If you increase the number of slits in a grating while keeping the spacing the same, what happens to (a) the positions of the intensity maxima in the interference pattern for a given wavelength; (b) the intensity at the maxima; and (c) the width of the maxima?

---

## 32.4 Interferometry

Passing light through multiple slits isn't the only way to produce interference. So will any process that separates light into several beams, sends them on different paths, and then rejoins them. Such processes are the basis of **interferometry**, an exquisitely sensitive technique for measuring small displacements, time intervals, and other quantities.

## Thin Films

Light passing through thin, transparent films is partially reflected at both the front and back surfaces, and the resulting beams recombine to produce interference. In Section 14.6 we saw how waves on a string reflect where one string joins another with different properties; in particular, the reflected wave is inverted if the second string has greater mass per unit length. Light behaves analogously: It's reflected with a 180° phase change at the interface between a material with lower refractive index and one with higher refractive index. But it reflects without a phase change at an interface from higher to lower refractive index. For a thin film with refractive index $n$ higher than its surroundings, Fig. 32.12 shows that there's a 180° phase change at the first interface and no change at the second.

If the film in Fig. 32.12 has thickness $d$, there's also a phase change due to the additional path length for beam 2; for the case of normal incidence, the extra length is $2d$. Because reflected beam 1 forms with a 180° phase change but beam 2 has no phase change, it takes another 180° phase shift—half a wavelength path difference—to put beams 1 and 2 back in phase to give constructive interference. That occurs if beam 2's extra path length $2d$ is half a wavelength, or $1\frac{1}{2}$ wavelengths, or any odd-integer multiple of one-half wavelength: $2d = \left(m + \frac{1}{2}\right)\lambda_n$, where $m = 0, 1, 2, 3, \ldots$, and where the subscript $n$ indicates that this is the wavelength as measured in the material. In Chapter 30 we found that the wavelength in a material with refractive index $n$ is reduced by a factor $1/n$ from its value in air or vacuum; thus $\lambda_n = \lambda/n$, and our condition for constructive interference becomes

$$2nd = \left(m + \tfrac{1}{2}\right)\lambda \qquad \text{(constructive interference, thin film)} \qquad (32.7)$$

Thin-film interference accounts for the bands of color you see in a soap film or oil slick. Equation 32.7 shows that constructive interference occurs for different wavelengths at different thicknesses; a film of varying thickness therefore produces colored bands when illuminated with white light (Fig. 32.13). If part of the film is too thin to satisfy Equation 32.7 for any visible wavelength, then that part appears dark. You can see this effect at the top of the soap film in Fig. 32.13.

Interference in thin layers is the basis of some very sensitive optical techniques. The shape of a lens, for example, can be measured to within a fraction of a wavelength of light using interference in a thin "film" consisting of the air between the lens and a flat glass plate (Fig. 32.14).

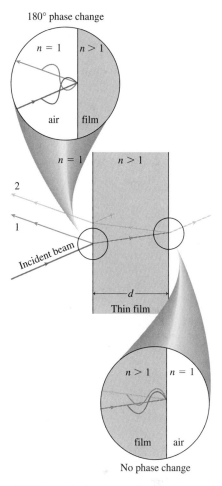

**FIGURE 32.12** Reflection and refraction at a thin film of transparent material, showing a 180° phase change at the first interface and none at the second.

**FIGURE 32.13** Interference in a soap film illuminated with white light. The film's thickness increases toward the bottom, resulting in constructive interference for a given wavelength occurring at several different vertical positions.

(a)

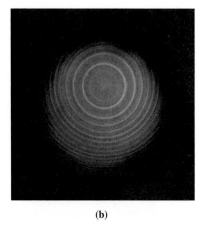

(b)

**FIGURE 32.14** (a) A portion of a lens sitting on a flat glass plate. (b) Newton's rings arise from the difference in path lengths between rays like 1 and 2, and provide precise information about the lens shape.

**EXAMPLE 32.4** **Interference: A Soap Film**

A wire loop 20 cm high is dipped in soap solution and then held vertically to produce a film whose thickness varies from essentially zero at the top to 1.0 $\mu$m at the bottom. The film is illuminated with 650-nm light. How many bright bands appear? The film's refractive index is that of water, $n = 1.333$.

**INTERPRET** The concept here is interference in thin films; we'll get bright bands where that interference is constructive.

**DEVELOP** Figure 32.15 is our sketch. Equation 32.7, $2nd = \left(m + \frac{1}{2}\right)\lambda$, gives the conditions for constructive interference. We've got a fixed $n$ and $\lambda$, but the film thickness $d$ varies. Furthermore, Equation 32.7 holds for any integer $m$. The larger $m$, the more half-wavelengths fit across the film. Therefore we'll get constructive interference with the highest $m$ value near the thickest point in the film—namely, the bottom where it's 1.0 $\mu$m thick. Constructive interference will also occur above that, for lower integer values of $m$. So our plan is to solve for $m$ with the greatest film thickness. The number of integers below that value will then be the number of bright bands.

**EVALUATE** Solving Equation 37.2 for $m$ gives

$$m = \frac{2nd}{\lambda} - \frac{1}{2} = \frac{(2)(1.333)(1.0\ \mu m)}{0.65\ \mu m} - \frac{1}{2} = 3.6$$

**FIGURE 32.15** Sketch for Example 32.4.

**ASSESS** $m$ must be an integer, so this tells us there's no bright band right at the bottom of the film. The lowest band is farther up, and it's the $m = 3$ band. The $m = 0, 1, 2$ bands are higher still, for a total of four bright bands. ∎

In analyzing thin films, we've considered only the first reflection at each interface. Actually, multiple reflections occur within the film, producing ever-weaker rays. A fuller treatment involving Maxwell's equations shows that when a film of refractive index $n_2$ is sandwiched between materials with indices $n_1$ and $n_3$, complete cancellation of reflected rays in the incident medium occurs if the thickness is right and if $n_2 = \sqrt{n_1 n_3}$. This is the basis of **antireflection coatings**, used to ensure maximum light transfer in camera lenses, solar photovoltaic cells, and other applications.

## The Michelson Interferometer

Several optical instruments use interference for precise measurement. Among the simplest and most important is the **Michelson interferometer**, invented by the American physicist Albert Michelson and used in the 1880s in a famous experiment that paved the way for the theory of relativity. We discuss this experiment in the next chapter; here we describe the interferometer, which is still used for precision measurements.

Figure 32.16 shows the basic Michelson interferometer. The key idea is that light from a monochromatic source is split into two beams by a half-silvered mirror called a **beam splitter**. The beam splitter is set at a 45° angle, so the reflected and transmitted beams travel perpendicular paths. Each then reflects off a flat mirror and returns to the beam splitter. The beam splitter again transmits and reflects half the light incident on it, with the result that some light from the originally separated beams is recombined. The recombined beams interfere, and the interference pattern is observed with a viewing lens at the bottom of Fig. 32.16.

If the path lengths for the two beams were exactly the same, they would recombine in phase and interfere constructively. In reality, the path lengths are never exactly the same, the mirrors are never exactly perpendicular, and the beams aren't perfectly parallel. But that's no problem: What happens is that different parts of the beams recombine with different phase differences, and the result is a pattern of light and dark interference fringes, as shown in Fig. 32.16. The distance between successive fringes corresponds to a path-length difference of one full wavelength.

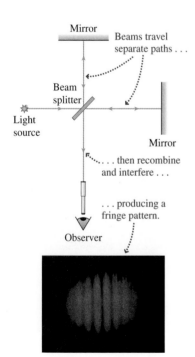

**FIGURE 32.16** Schematic diagram of a Michelson interferometer, with a photo of the interference fringes.

Now suppose one mirror moves slightly. The path-length differences change and therefore the interference pattern shifts. A mere quarter-wavelength mirror movement adds an extra half-wavelength to the round-trip path. That results in a 180° phase shift, moving dark fringes to where light ones were. Shifts a fraction of this amount are readily detected, allowing the measurement of mirror displacements to within a small fraction of a wavelength. A similar shift occurs if a transparent material is placed in one path, retarding the beam because of its refractive index. This provides accurate measures of the refractive indices of gases, which are so close to 1 that less-sensitive techniques don't work.

The largest Michelson interferometers ever built are twin instruments with 4-km arms (Fig. 32.17). These are part of LIGO, the Laser Interferometer Gravitational Wave Observatory. These instruments are designed to detect mirror displacements on the order of $10^{-18}$ m resulting from "ripples" in the structure of space and time caused by distant cosmic events. LIGO began observations in 2002. It will eventually be dwarfed by a space-based interferometer with arms 5 million km long!

**FIGURE 32.17** The LIGO instrument at Hanford, Washington, is an interferometer with 4-km arms. The light undergoes multiple reflections, giving an effective arm length of 300 km.

---

**EXAMPLE 32.5**  **An Interferometric Measurement: Sandstorm!**

A sandstorm has pitted the aluminum mirrors of a desert solar-energy installation, and engineers want to know the depths of the pits. They construct a Michelson interferometer with a sample from one of the pitted mirrors in place of one flat mirror. With 633-nm laser light, the interference pattern in Fig. 32.18 results. What is the approximate depth of the pit?

This distance corresponds to a path-length difference λ . . .

**FIGURE 32.18** Fringe pattern resulting from a pitted mirror.

. . . so this distance corresponds to ~0.2λ.

**INTERPRET** The concept here is interferometry—inferring distance from observations of light interference, in this case with the Michelson configuration of Fig. 32.16. The distortions of the interference fringes shown in Fig. 32.18 result because some of the light travels a little farther—namely, into the bottom of a pit and back out.

**DEVELOP** The full fringe spacing corresponds to a path difference of one wavelength, and Fig. 32.18 shows that the fringe distortion due to the pit gives a shift about one-fifth of the distance between fringes. We need to use this information to find the extra distance traveled by light reflecting from the bottom of the pit, and from that the pit depth. We're given the wavelength, so we can estimate the round trip as approximately $0.2\lambda$. The pit depth will be half this quantity.

**EVALUATE** The extra path length for the light reflecting off the pit is $0.2\lambda$, so the pit depth is about $0.1\lambda$. With $\lambda = 633$ nm, the pit depth is about 63 nm.

**ASSESS** Try measuring that with a meter stick! Interferometry provides an exquisitely sensitive measurement of small distances.  ∎

---

## 32.5 Huygens' Principle and Diffraction

The interference we've been studying in this chapter isn't the only optical phenomenon where the wave nature of light is important. There's also **diffraction**—the bending of light or other waves as they pass by objects. Interference and diffraction are closely related, and the double-and multiple-slit interference we've studied actually involves diffraction as well—hence the term "diffraction grating."

Diffraction, like other optical phenomena, is ultimately governed by Maxwell's equations. But we can understand diffraction more readily using **Huygens' principle**, articulated in 1678 by the Dutch scientist Christian Huygens, who was the first to suggest that light might be a wave. Huygens' principle states:

> All points on a wavefront act as point sources of spherically propagating "wavelets" that travel at the speed of light appropriate to the medium. At a short time $\Delta t$ later, the new wavefront is the unique surface tangent to all the forward-propagating wavelets.

Figure 32.19 shows how Huygens' principle accounts for the propagation of plane and spherical waves.

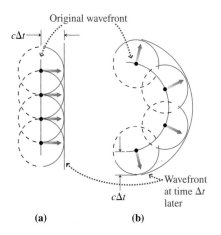

**FIGURE 32.19** Application of Huygens' principle to (a) plane and (b) spherical waves. In each case the wavefront acts like a set of point sources emitting circular waves that expand to produce a new wavefront.

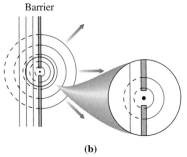

**FIGURE 32.20** Plane waves incident on an opaque barrier with a hole. (a) Diffraction is negligible for a hole large compared with the wavelength, but (b) pronounced for a small hole.

## Diffraction

Figure 32.20 shows plane waves incident on an opaque barrier containing a hole. Since the waves are blocked by the barrier, Huygens' wavelets produced near each barrier edge cause the wavefronts to bend at the barrier. When the width of the hole is much greater than the wavelength, as in Fig. 32.20a, this diffraction is of little consequence, and the waves effectively propagate straight through the hole in a beam defined by the hole size. But when the hole size and wavelength are comparable, wavefronts emerging from the hole spread in a broad pattern (Fig. 32.20b). Thus diffraction, although it always occurs, is significant only on length scales comparable to or smaller than the wavelength. That's why we could ignore diffraction and assume that light always travels in straight lines when we considered optical systems with dimensions much larger than the wavelength of light.

Diffraction ultimately limits our ability to image small objects and to focus light with arbitrary precision. Next, we'll see why this is so by examining the behavior of light as it passes through a single slit. The result will help us understand optical challenges ranging from telescopic imaging of distant astrophysical objects to the development of Blu-ray discs that hold 13 hours of movies on a disc the same size as a standard audio CD.

## Single-Slit Diffraction

In treating double-slit and multiple-slit interference, we assumed that plane waves passing through a slit emerged with circular wavefronts. According to Fig. 32.20b, that's true only if the slit width is small compared with the wavelength, so the slit can be treated as a single, localized source of new waves. When the slit width isn't small, Huygens' principle implies that we have to consider each point in the slit as a separate source—and then we can expect interference from waves originating at different points in the same slit. Thus a single wide slit is really like a multiple-slit system with infinitely many slits!

Figure 32.21a shows light incident on a slit of width $a$. Each point in the slit acts as a source of spherical wavelets propagating in all directions to the right of the slit. We focus on a particular direction described by the angle $\theta$, and we'll look at interference of light from the five points shown. Figure 32.21b concentrates on the points from which rays 1, 2, and 3 originate and shows that the path lengths for rays 1 and 3 differ by $\frac{1}{2}a\sin\theta$. These two beams will interfere destructively if this distance is half the wavelength—that is, if $\frac{1}{2}a\sin\theta = \frac{1}{2}\lambda$ or $a\sin\theta = \lambda$. But if rays 1 and 3 interfere destructively, so do rays 3 and 5, which have the same geometry, and so do rays 2 and 4, for the same reason. In fact, a ray leaving *any* point in the lower half of the slit will interfere destructively with the point located a distance $a/2$ above it. Therefore, an observer viewing the slit system at the angle $\theta$ satisfying $a\sin\theta = \lambda$ will see no light.

Similarly, the sources for rays 1 and 2 are $a/4$ apart and will therefore interfere destructively if $\frac{1}{4}a\sin\theta = \frac{1}{2}\lambda$, or $a\sin\theta = 2\lambda$. But then so will rays 2 and 3, and rays 3 and 4; in fact, any ray from a point in the lower three-quarters of the slit will interfere destructively with a ray from the point $a/4$ above it, and therefore, an observer looking at an angle $\theta$ satisfying $a\sin\theta = 2\lambda$ will see no light.

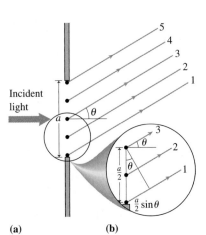

**FIGURE 32.21** Each point in a slit acts as a source of Huygens' wavelets, which interfere in the region to the right of the slit.

We could equally well have divided the slit into six sections with seven evenly spaced points; we would then have found destructive interference if $\frac{1}{6}a \sin\theta = \frac{1}{2}\lambda$, or $a \sin\theta = 3\lambda$. We could obviously continue this process for any number of points in the slit, and therefore, we conclude that destructive interference occurs for all angles $\theta$ satisfying

$$a \sin\theta = m\lambda \qquad \text{(destructive interference, single-slit diffraction)} \qquad (32.8)$$

with $m$ any nonzero integer and $a$ the slit width. Note that the case $m = 0$ is excluded; it produces not destructive interference but a central maximum in which all waves are in phase.

---

✓ **TIP** Interference and Diffraction

Equation 32.8 for the *minima* of a single-slit diffraction pattern looks just like Equation 32.1a for the *maxima* of a multiple-slit interference pattern, except that the slit width $a$ replaces the slit spacing $d$. Why does the same equation give the minima in one case and the maxima in another? Because we're dealing with two distinct but related phenomena. In the multiple-slit case, we considered each slit to be so narrow that it could be considered a single source, neglecting the interference of waves originating within the same slit. In the single-slit case, the diffraction pattern occurs precisely because of the interference of waves from different points within the same slit.

---

## Intensity in Single-Slit Diffraction

Geometry gave us the *positions* of the minima in single-slit diffraction, just as it did for multiple-slit interference. But to get the *intensity* across the diffraction pattern, we'd have to superpose electric fields of the interfering waves, as we did in deriving Equation 32.3 for two-slit interference. But now we've got infinitely many fields to sum, corresponding to waves from every part of the slit. It's possible to do this using a calculus-like graphical technique involving the phasor concept we introduced in Chapter 28. We won't go through the whole derivation; however, we'll motivate the result by pointing out that the diffraction pattern at any point occurs because of phase differences among waves originating at different parts of the slit. It's not surprising, therefore, that a key factor is the phase difference between waves originating from opposite ends of the slit. Applying the analysis of Fig. 32.21 to the entire slit width $a$ gives a path-length difference $a \sin\theta$ for these waves. As usual, this path difference is to the wavelength as the phase difference $\phi$ is to a full cycle, $2\pi$ radians. Thus

$$\phi = \frac{2\pi}{\lambda}a \sin\theta \qquad (32.9)$$

is the phase difference between rays from the ends of the slit to a point at angular position $\theta$. The phasor-summing process relates the amplitude of the net electric field to this phase difference $\phi$ and shows that the field is proportional to $\sin(\phi/2)/(\phi/2)$. Used in Equation 29.20 to get the intensity from the electric field, this result gives the intensity as a function of angle in single-slit diffraction:

$$\overline{S} = \overline{S}_0\left[\frac{\sin(\phi/2)}{\phi/2}\right]^2 \qquad (32.10)$$

Here $\overline{S}_0$ is the average intensity at the central maximum of the pattern $(\theta = \phi = 0)$, and $\phi$ is given by Equation 32.9. At $\theta = \phi = 0$, Equation 32.10 appears to be indeterminate, but using the limit $\sin x/x \to 1$ as $x \to 0$ shows that the result is indeed $\overline{S}_0$. Problem 74 explores another approach to Equation 32.10.

Figure 32.22 plots Equation 32.10 for three values of the slit width $a$ in relation to the wavelength $\lambda$. For wide slits—$a$ large compared with $\lambda$—the central peak is narrow and the secondary peaks are much lower and also half as wide as the central peak. Here diffraction is negligible, and the beam essentially propagates through the slit in the ray approximation of geometrical optics. But as the slit narrows, the diffracted beam spreads until, with $a = \lambda$, it covers an angular width of some 120°.

(a)

(b)

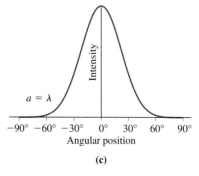

(c)

**FIGURE 32.22** Intensity in single-slit diffraction, as a function of the angle $\theta$ from the centerline, for three values of slit width $a$.

The intensity given by Equation 32.10 will be zero when the numerator on the right-hand side is zero—that is, when the argument of the sine function is an integer multiple of $\pi$. That occurs when $\phi/2 = (\pi a/\lambda)\sin\theta = m\pi$, or when $a\sin\theta = m\lambda$. Thus, we recover our result of Equation 32.8 for the angular positions where destructive interference gives zero intensity.

## Multiple Slits and Other Diffracting Systems

In treating multiple-slit systems in Section 32.2, we assumed the slits were so narrow compared with the wavelength that the central diffraction peak spread into the entire space beyond the slit system. When the slit width is not negligible, each slit produces a single-slit diffraction pattern. The result is a pattern that combines single-slit diffraction with multiple-slit interference (Fig. 32.23).

Maxima of single-slit diffraction

**FIGURE 32.23** When the slit width is not negligible, a double-slit system produces the regular variations of double-slit interference within a single-slit diffraction pattern.

Diffraction occurs any time light passes a sharp, opaque edge like the edges of the slits we've been considering. Close examination of the shadow produced by a sharp edge shows parallel fringes resulting from interference of the diffracting wavefronts (Fig. 32.24a). More complex diffraction patterns result from objects of different shape (Fig. 32.24b). Such diffraction limits our ability to form sharp optical images, as we show in the next section.

**FIGURE 32.24** Diffraction patterns from light passing sharp edges. (a) Straight edge of an opaque barrier. (b) Circular aperture with crosshairs.

## 32.6 The Diffraction Limit

Diffraction imposes a fundamental limit on the ability of optical systems to distinguish closely spaced objects. Consider two point sources of light illuminating a slit. The sources are so far from the slit that waves reaching the slit are essentially plane waves, but the different source positions mean the waves reach the slit at different angles. We assume the sources are incoherent, so they don't produce a regular interference pattern. Then light diffracting at the slit produces two single-slit diffraction patterns, one for each source. Because the sources are at different angular positions, the central maxima of these patterns don't coincide, as shown in Fig. 32.25.

If the angular separation between the sources is great enough, then the central maxima of the two diffraction patterns will be entirely distinct. In that case we can clearly distinguish the two sources (Fig. 32.25a). But as the sources get closer, the central maxima begin to overlap (Fig. 32.25b). They remain distinguishable as long as the total intensity pattern shows two peaks. Since the sources are incoherent, the total intensity is just the sum of the individual intensities. Figure 32.26 shows how that sum loses its two-peak structure as the diffraction patterns merge. In general, two peaks are barely distinguishable if the central maximum of one coincides with the first minimum of the other. This condition is called the **Rayleigh criterion**, and when it's met the two sources are just barely **resolved**.

Optical systems are analogous to the single slit we've just considered. Every system has an aperture of finite size through which light enters. That aperture may be an actual slit or hole, like the diaphragm that "stops down" a camera lens, or it may be the full size of a lens or a telescope mirror. So all optical systems ultimately suffer loss of resolution if two sources—or two parts of the same object—have too small an angular separation. Thus, diffraction fundamentally limits our ability to probe the structure of objects that are either very small or very distant. Figure 32.27 shows the loss of resolution as diffraction patterns overlap.

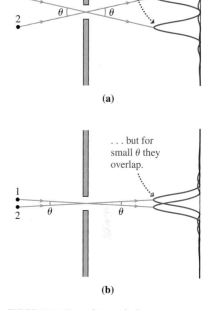

**(a)**

**(b)**

**FIGURE 32.25** Two distant light sources at different angular positions produce diffraction patterns whose central peaks have the same angular separation $\theta$ as the sources.

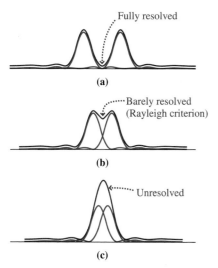

**(a)**

**(b)**

**(c)**

**FIGURE 32.26** Since the two sources are incoherent, the total intensity is just the sum (gray curve) of the intensities of the two diffraction patterns.

**FIGURE 32.27** Diffraction patterns produced by a pair of point sources. The angular separation of the sources decreases until they can't be resolved.

Figure 32.25 shows that the angular separation between the diffraction peaks is equal to the angular separation between the sources themselves. Then the Rayleigh criterion is just met if the angular separation between the two sources is equal to the angular separation between a central peak and the first minimum. We found earlier that the first minimum in single-slit diffraction occurs at the angular position given by $\sin\theta = \lambda/a$, with $a$ the slit width

**FIGURE 32.28** 3-D plot of intensity versus position in circular diffraction. Figure 32.27 shows the corresponding diffraction patterns.

and with $\theta$ measured from the central peak. In most optical systems the wavelength is much less than the size of any apertures, so we can use the small-angle approximation $\sin\theta \approx \theta$. Then the Rayleigh criterion—the condition that two sources be just resolvable—for single-slit diffraction becomes

$$\theta_{min} = \frac{\lambda}{a} \qquad \text{(Rayleigh criterion, slit)} \qquad (32.11a)$$

Most optical systems have circular apertures rather than slits. The diffraction pattern from such an aperture is a series of concentric rings (Fig. 32.28). Mathematical analysis shows that the angular position of the first ring and therefore the minimum resolvable source separation for a circular aperture is

$$\theta_{min} = \frac{1.22\lambda}{D} \qquad \text{(Rayleigh criterion, circular aperture)} \qquad (32.11b)$$

with $D$ the aperture diameter.

Equations 32.11 show that increasing the aperture size allows smaller angular differences to be resolved. In optical instrument design, that means larger mirrors or lenses. An alternative is to decrease the wavelength, which may or may not be an option depending on the source. In high-quality optical systems, diffraction is often the limiting factor preventing perfectly sharp image formation; such systems are said to be **diffraction limited**. For example, the diffraction limit sets a minimum size for objects resolvable with optical microscopes; that's why electron microscopes—with shorter effective wavelength—are used to image the smallest biological structures. Large ground-based telescopes are an exception to the diffraction limit; their image quality is limited by atmospheric turbulence, although this can be reduced with adaptive optics. From its vantage point above the atmosphere, the Hubble Space Telescope is the first large diffraction-limited astronomical telescope.

Astronomers can ease the diffraction limit by combining data from several telescopes to produce, in effect, a single instrument with aperture equal to the telescope separation. Such a setup, called an interferometer, requires combining the data with its proper phase relation. For that reason it's easier to accomplish with the relatively low frequency of radio waves, and today radio astronomers achieve exquisite resolution by combining telescopes distributed over the entire Earth. Optical and infrared interferometry is harder, with one ambitious project creating an 85-m aperture using the two Keck Telescopes on Hawaii's Mauna Kea.

**GOT IT? 32.3** You're a biologist trying to resolve details of structures within a cell, but they look fuzzy even at the highest power of your microscope. Which of the following might help: (a) substituting an eyepiece with shorter focal length, as suggested by Equation 31.10; (b) putting a red filter over the white light source used to illuminate the microscope slide; or (c) putting a blue filter over the white light source?

---

**EXAMPLE 32.6** **The Diffraction Limit: Asteroid Alert**

An asteroid $20 \times 10^6$ km away appears on a collision course with Earth. What's the minimum size for the asteroid that could be resolved with the 2.4-m-diameter diffraction-limited Hubble Space Telescope, using 550-nm reflected sunlight?

**INTERPRET** This is a problem about the diffraction limit with a circular aperture. We're after the minimum physical size for the asteroid at a given distance. We identify $D = 2.4$ m as the aperture size and $\lambda = 550$ nm as the wavelength of the light.

**DEVELOP** Equation 32.11b, $\theta_{min} = 1.22\lambda/D$, determines the diffraction limit, expressed as the minimum angular size that can be resolved. So our plan is to express the unknown physical size $l$ in terms of angular size $\theta$ and then apply Equation 32.11b.

**EVALUATE** If we consider the opposite ends of the asteroid to be like the two peaks in Fig. 32.25, then the angular size for small $\theta$ is just $l/L$, where $L$ is the distance to the asteroid. Then Equation 32.11b becomes

$$\frac{l}{L} = \frac{1.22\lambda}{D}$$

or $l = 1.22\lambda L/D = 5.6$ km using the numbers given.

**ASSESS** An object this size poses a grave danger, being comparable to the asteroid whose impact caused the extinction of the dinosaurs. If astronomers see only a fuzzy blur, then they'll have to wait until the asteroid is closer to resolve its physical size and assess the danger. ∎

**APPLICATION** **Movies on Disc: CD to DVD to Blu-ray**

The Application earlier in this chapter described how a CD encodes information in pits 1.6 $\mu$m apart and as short as 0.83 $\mu$m. CDs are read with 780-nm infrared laser light. The pit size and spacing are chosen so diffraction effects at that wavelength don't cause the CD player's optical system to confuse adjacent pits. The result is a maximum capacity of about 650 MB (megabytes; 1 byte is 8 binary bits, with a bit the fundamental piece of binary information represented by a digital 1 or 0). This translates into 74 minutes of audio.

CDs were developed in the 1980s, when inexpensive semiconductor lasers were available only in the infrared. By the 1990s inexpensive visible-light lasers became available, and that enabled the development of DVDs (for "digital video disc" or "digital versatile disc"). Read with 635-nm or 650-nm red light, DVDs can use smaller pit size

and spacing because of the lower diffraction limit. That, coupled with a two-layer structure and more sophisticated data-compression schemes, gives standard DVDs a capacity of about 4.7 GB—enough for 2 hours or more of video, depending on quality.

Despite their large capacity, DVDs aren't adequate for today's high-definition TV (HDTV) broadcasts. But improvements in laser technology give us a 405-nm violet laser that enables high-definition video discs. Again, the shorter wavelength and hence lower diffraction limit, along with other improvements, allow much more information to fit on a disc. Two competing technologies, Blu-ray and HD-DVD, store 25 GB and 15 GB, respectively, on a single-sided disc. For Blu-ray, that corresponds to 4.5 hours of high-definition video, or 12 hours of standard video. The figure compares CDs, DVDs, and Blu-ray discs.

## Big Picture

The big idea here is that—despite our use of the geometrical-optics approximation in the preceding chapters—light is indeed a wave and therefore exhibits the two related phenomena of **interference** and **diffraction**. These wave effects are important whenever light or any other wave interacts with objects whose size is comparable to or smaller than the wavelength.

## Key Concepts and Equations

**Constructive interference** occurs when two waves combine in phase:

**Destructive interference** occurs when two waves combine 180° out of phase:

When light of wavelength $\lambda$ passes through two or more narrow slits, the resulting interference shows maxima when

$$d \sin\theta = m\lambda,$$

where $m$ is an integer called the order. With multiple slits the maxima become stronger and narrow, but their position doesn't change.

2 slit

5 slit

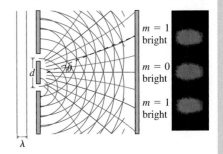

$m = 1$ bright

$m = 0$ bright

$m = 1$ bright

$\lambda$

**Diffraction** occurs because, according to **Huygens' principle**, each point on a wavefront acts as a source of spherical waves, causing light to bend as it encounters sharp edges. Waves from different parts of a wavefront interfere to produce diffraction patterns.

The **diffraction limit** is a fundamental restriction on our ability to image small or distant objects. For a circular aperture of diameter $D$, the **Rayleigh criterion** gives the minimum angular separation that can be resolved with light of wavelength $\lambda$:

$$\theta_{min} = \frac{1.22\lambda}{D}$$

Resolved    Barely resolved    Not resolved

## Cases and Uses

A **diffraction grating** consists of multiple slits or lines that result in constructive interference at different positions for different wavelengths. Diffraction gratings are used in spectrometers to disperse individual wavelengths. A grating's **resolving power**, the ratio of wavelength to the minimum resolvable difference in wavelengths, is given by

$$\frac{\lambda}{\Delta\lambda} = mN$$

where $N$ is the number of lines in the grating and $m$ is the order of the dispersion.

**X-ray diffraction** uses regularly positioned atoms as a grating, and is a powerful technique for analyzing crystal and molecular structure. Maximum intensity occurs when

$$2d \sin\theta = m\lambda.$$

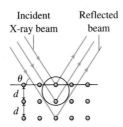

Incident X-ray beam    Reflected beam

$\theta$

$d$

$d$

The **Michelson interferometer** splits light into two beams that travel perpendicular paths. They recombine, and the resulting interference allows precision measurements.

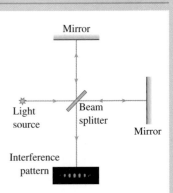

Mirror

Light source

Beam splitter

Mirror

Interference pattern

# For Thought and Discussion

1. A prism bends blue light more than red. Is the same true of a diffraction grating?
2. Why does an oil slick show colored bands?
3. Why does a soap bubble turn colorless just before it dries up and breaks?
4. Why don't you see interference effects between the front and back of your eyeglasses?
5. You can hear around corners, but you can't see around corners. Why the difference?
6. In deriving the intensity in double-slit interference, we couldn't simply add the intensities from the two slits. Why not?
7. The primary maxima in multiple-slit interference are in the same angular positions as those in double-slit interference. Why, then, do diffraction gratings have thousands of slits instead of just two?
8. When the Moon passes in front of a star, the intensity of the starlight fluctuates instead of dropping abruptly. Explain.
9. Sketch roughly the diffraction pattern you would expect for light passing through a square hole a few wavelengths wide.

# Exercises and Problems

## Exercises

### Section 32.2 Double-Slit Interference

10. A double-slit system is used to measure the wavelength of light. The system has slit spacing $d = 15$ $\mu$m and slit-to-screen distance $L = 2.2$ m. If the $m = 1$ maximum in the interference pattern occurs 7.1 cm from screen center, what is the wavelength?
11. A double-slit experiment with $d = 0.025$ mm and $L = 75$ cm uses 550-nm light. Find the spacing between adjacent bright fringes.
12. A double-slit experiment has slit spacing 0.12 mm. (a) What should be the slit-to-screen distance $L$ if the bright fringes are to be 5.0 mm apart when the slits are illuminated with 633-nm laser light? (b) What will be the fringe spacing with 480-nm light?
13. The interference pattern from two slits separated by 0.37 mm has bright fringes with angular spacing 0.065°. What is the wavelength of the light illuminating the slits?
14. The green line of gaseous mercury at 546 nm falls on a double-slit apparatus. If the fifth dark fringe is at 0.113° from the center-line, what is the slit separation?

### Section 32.3 Multiple-Slit Interference and Diffraction Gratings

15. In a five-slit system, how many minima lie between the zeroth-order and first-order maxima?
16. In a three-slit system, the first minimum occurs at an angular position of 5°. Where is the first maximum?
17. A five-slit system with 7.5-$\mu$m slit spacing is illuminated with 633-nm light. Find the angular positions of (a) the first two maxima and (b) the third and sixth minima.
18. Green light at 520 nm is diffracted by a grating with 3000 lines per cm. Through what angle is the light diffracted in the (a) first and (b) fifth order?
19. Light is incident normally on a grating with 10,000 lines per cm. What is the maximum order in which (a) 450-nm and (b) 650-nm light will be visible?
20. Find the second-order angular separation of the two wavelengths in Example 32.2.

### Section 32.4 Interferometry

21. Find the minimum thickness of a soap film $(n = 1.333)$ in which 550-nm light will undergo constructive interference.
22. Light of unknown wavelength shines on a precisely machined wedge of glass with refractive index 1.52. The closest point to the apex of the wedge where reflection is enhanced occurs where the wedge is 98 nm thick. Find the wavelength.
23. Monochromatic light shines on a glass wedge with refractive index 1.65, and enhanced reflection occurs where the wedge is 450 nm thick. Find all possible values for the wavelength in the visible range.

24. White light shines on a 100-nm-thick sliver of fluorite $(n = 1.43)$. What wavelength is most strongly reflected?
25. Suppose the soap-film thickness in Example 32.4 increases linearly from 0 at the top to 1.0 $\mu$m at the bottom. If the film is illuminated with white light, what portion of it will appear dark?

### Section 32.5 Huygens' Principle and Diffraction

26. For what ratio of slit width to wavelength will the first minima of a single-slit diffraction pattern occur at $\pm 90°$?
27. Light with wavelength 633 nm is incident on a 2.5-$\mu$m-wide slit. Find the angular width of the central peak in the diffraction pattern, taken as the angular separation between the first minima.
28. A beam of parallel rays from a 29-MHz citizen's band radio transmitter passes between two electrically conducting (hence opaque to radio waves) buildings located 45 m apart. What is the angular width of the beam when it emerges from between the buildings?
29. Find the intensity as a fraction of the central peak intensity for the second secondary maximum in single-slit diffraction, assuming the peak lies midway between the second and third minima.

### Section 32.6 The Diffraction Limit

30. Find the minimum angular separation resolvable with 633-nm laser light passing through a circular aperture of diameter 2.1 cm.
31. What is the minimum telescope aperture needed to resolve an object whose angular diameter is 0.35 arc second, observed at 500-nm wavelength? *Note:* 1 arc sec $= (1/3600)°$.
32. What is the longest wavelength of light you could use to resolve a structure with angular diameter 0.44 mrad, using a microscope with aperture 1.2 mm in diameter?
33. The movie *Patriot Games* has a scene in which CIA agents use spy satellites to identify individuals in a terrorist camp. Suppose that a minimum resolution for distinguishing human features is about 5 cm. If the spy satellite's optical system is diffraction limited, what diameter mirror or lens is needed to achieve this resolution from an altitude of 100 km? Assume a wavelength of 550 nm.

## Problems

34. What is the angular position $\theta$ of the second-order bright fringe in a double-slit system with 1.5-$\mu$m slit spacing if the light has wavelength (a) 400 nm and (b) 700 nm?
35. A double-slit experiment has slit spacing 0.035 mm, slit-to-screen distance 1.5 m, and wavelength 500 nm. What is the phase difference between two waves arriving at a point 0.56 cm from the center line?
36. For a double-slit experiment with slit spacing 0.25 mm and wavelength 600 nm, at what angular position is the path difference equal to one-fourth the wavelength?

37. A screen 1.0 m wide is located 2.0 m from a pair of slits illuminated by 633-nm laser light, with its center on the centerline of the slits. Find the highest-order bright fringe that will appear on the screen if the slit spacing is (a) 0.10 mm and (b) 10 $\mu$m.

38. Laser light at 633 nm falls on a double-slit apparatus with slit separation 6.5 $\mu$m. Find the separation between (a) the first and second and (b) the third and fourth bright fringes, as seen on a screen 1.7 m from the slits.

39. A tube of glowing gas emits light at 550 nm and 400 nm. In a double-slit apparatus, what is the lowest-order 550-nm bright fringe that will fall on a 400-nm dark fringe, and what are the corresponding orders?

40. On the screen of a multiple-slit system, the interference pattern shows bright maxima separated by 0.86° and seven minima between each bright maximum. (a) How many slits are there? (b) What is the slit separation if the incident light has wavelength 656.3 nm?

41. Find the angular separation between the red hydrogen-$\alpha$ spectral line at 656 nm and the yellow sodium line at 589 nm if the two are observed in third order with a 3500-lines/cm grating spectrometer.

42. Visible light has wavelengths between 400 nm and 700 nm. What is the lowest pair of consecutive orders for which there will be some overlap between the visible spectra as dispersed by a grating?

43. (a) What portions of the fourth- and fifth-order visible spectra overlap in a 3000-lines/cm grating spectrometer? (b) How would your answer change for a 1000-lines/cm grating? (c) For a 10,000-lines/cm grating?

44. Find the total number of lines in a 2.5-cm-wide diffraction grating whose third-order spectrum has the 656-nm hydrogen-$\alpha$ spectral line at an angular position of 37°.

45. A 400-lines/mm diffraction grating is 3.5 cm wide. Two spectral lines whose wavelengths average to 560 nm are just barely resolved in the fourth-order spectrum of this grating. What is the difference between their wavelengths?

46. What order is necessary to resolve wavelengths of 647.98 nm and 648.07 nm using a 4500-line grating?

47. A thin film of toluene $(n = 1.49)$ floats on water. What is the minimum film thickness if the most strongly reflected light has wavelength 460 nm?

48. What diameter optical telescope would be needed to resolve a Sun-sized star 10 light-years from Earth? Take $\lambda = 550$ nm. *Note:* Your answer shows why stars appear as point sources in optical astronomy.

49. Echelle spectroscopy uses relatively coarse gratings in high order. Compare the resolving power of an 80-lines/mm echelle grating used in 12th order with a 600-lines/mm grating used in first order, assuming the two have the same width.

50. You wish to resolve the calcium-H line at 396.85 nm from the hydrogen-$\epsilon$ line at 397.05 nm in a first-order spectrum. To the nearest hundred, how many lines should your grating have?

51. X-ray diffraction in potassium chloride (KCl) results in a first-order maximum when the X rays graze the crystal plane at 8.5°. If the X-ray wavelength is 97 pm, what is the spacing between crystal planes?

52. As a soap bubble $(n = 1.333)$ evaporates and thins, the reflected colors gradually disappear. (a) What is its thickness just as the last vestige of color vanishes? (b) What is the last color seen?

53. An oil film $(n = 1.25)$ floats on water, and a soap film $(n = 1.333)$ is suspended in air. Find the minimum thickness for each that will result in constructive interference with 500-nm green light.

54. An oil film with refractive index 1.25 floats on water. The film thickness varies from 0.80 $\mu$m to 2.1 $\mu$m. If 630-nm light is incident normally on the film, at how many locations will it undergo enhanced reflection?

55. Microwave ovens operate at a frequency of 2.4 GHz. Find the minimum thickness for a plastic tray with refractive index 1.45 that will cause enhanced reflection of microwaves incident normal to the plate.

56. Two perfectly flat glass plates are separated at one end by a piece of paper 0.065 mm thick. A source of 550-nm light illuminates the plates from above, as shown in Fig. 32.29. How many bright bands appear to an observer looking down on the plates?

**FIGURE 32.29** An air wedge
(Problems 56, 57, 68, and 78)

57. An air wedge like that shown in Fig. 32.29 shows N bright bands when illuminated from above. Find an expression for the number of bands that will appear if the air is replaced by a liquid of refractive index $n$ different from that of the glass.

58. A Michelson interferometer uses light from glowing hydrogen at 486.1 nm. As you move one mirror, 530 bright fringes pass a fixed point in the viewer. How far did the mirror move?

59. Find the wavelength of light used in a Michelson interferometer if 550 bright fringes go by a fixed point when the mirror moves 0.150 mm.

60. One arm of a Michelson interferometer is 42.5 cm long and is enclosed in a box that can be evacuated. The box initially contains air, which is gradually pumped out to create a vacuum. In the process, 388 bright fringes pass a fixed point in the viewer. If the interferometer uses light with wavelength 641.6 nm, what is the refractive index of the air?

61. Your personal stereo is in a dead spot caused by direct reception from an FM radio station at 89.5 MHz interfering with the signal reflecting off a wall behind you. How much farther from the wall should you move so that the interference is fully constructive?

62. A proposed "star wars" antimissile laser is to focus 2.8-$\mu$m-wavelength infrared light to a 50-cm-diameter spot on a missile 2500 km distant. Find the minimum diameter for a concave mirror that can achieve this spot size, given the diffraction limit. *Note:* Your answer suggests one of the many technical difficulties faced by antimissile defense systems.

63. Suppose one of the 10-m-diameter Keck Telescopes in Hawaii could be trained on San Francisco, 3400 km away. Would it be possible to read (a) newspaper headlines or (b) a billboard sign at this distance? Justify your answers by giving the minimum separation resolvable with Keck at 3400 km, assuming 550-nm light. (c) Repeat for the case of the Keck optical interferometer, formed from the two 10-m Keck Telescopes and several smaller ones, with an effective aperture of 50 m.

64. A camera has an $f/1.4$ lens, meaning that the ratio of focal length to lens diameter is 1.4. Find the smallest spot diameter (defined as the diameter of the first diffraction minimum) to which this lens can focus parallel light with 580-nm wavelength.

65. While driving at night, your eyes' irises have dilated to 3.1-mm diameter. If your vision were diffraction limited, what would be the greatest distance at which you could see as distinct the two headlights of an oncoming car, which are spaced 1.5 m apart? Take $\lambda = 550$ nm.

66. Under the best conditions, atmospheric turbulence limits the resolution of ground-based telescopes to about 1 arc second (1/3600 of a degree). For what aperture sizes is this limitation more severe than that of diffraction at 550 nm? *Note:* Your answer shows why large ground-based telescopes don't produce better images than small ones, although they do gather more light.

67. White light shines on a 250-nm-thick layer of diamond $(n = 2.42)$. What wavelength *of visible light* is most strongly reflected?

68. An air wedge like that of Fig. 32.29 displays 10,003 bright bands when illuminated from above. If the region between the plates is then evacuated, the number of bands drops to 10,000. Calculate the refractive index of the air.

69. A thin-walled glass tube of length $L$ containing a gas of unknown refractive index is placed in one arm of a Michelson interferometer using light of wavelength $\lambda$. The tube is then evacuated, and during the process $m$ bright fringes pass a fixed point as seen in the viewer. Find an expression for the refractive index of the gas.

70. The signal from a 103.9-MHz FM radio station reflects off a building 400 m away, effectively producing two sources of the same signal. You're driving at 60 km/h along a road parallel to a line between the station's antenna and the building, and located a perpendicular distance of 6.5 km from them. How often does the signal appear to fade when you're driving roughly opposite the transmitter and building?

71. Light is incident on a diffraction grating at an angle $\alpha$ to the normal. Show that the condition for maximum light intensity becomes $d(\sin\theta \pm \sin\alpha) = m\lambda$.

72. An arrangement known as Lloyd's mirror (Fig. 32.30) allows interference between direct and reflected beams from the same source. Find an expression for the separation of bright fringes on the screen, given the distances $d$ and $D$ in Fig. 32.30 and the wavelength $\lambda$ of the light. *Hint:* Think of two sources, one the virtual image of the other.

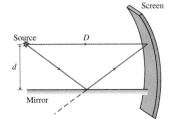

**FIGURE 32.30** Lloyd's mirror (Problem 72)

73. Microwaves shine on a pair of narrow slits whose spacing is three times the wavelength. Find the distances to the first- and second-order bright fringes on a screen 50 cm from the slits, measured from the central maximum. *Hint:* Do Equations 32.2 apply?

74. The intensity of the single-slit diffraction pattern can be calculated by summing the amplitudes of infinitely many field amplitudes corresponding to waves from every infinitesimal part of the slit. (a) Referring to Fig. 32.21, show that the field from an element of slit width $dy$, a distance $y$ from the bottom edge of the slit, is $dE = (E_p \, dy/a) \sin(\omega t + \phi(y))$, where $\phi(y) = (2\pi y/\lambda) \sin\theta$. (b) Integrate $dE$ over the entire slit from $y = 0$ to $y = a$, and use trig

identities from Appendix A, to find the total amplitude and from there show that the average intensity is given by Equation 32.10.

75. Figure 32.23 shows that a double-slit system actually produces an interference pattern combined with a diffraction pattern. Use the method of Problem 74 to find the intensity pattern for two slits of width $a$, separated by a center-to-center distance $d > a$. You'll need to integrate over both slits, from $y = 0$ to $a$ and from $y = d$ to $d + a$.

76. You're on an international panel charged with allocating "real estate" for communications satellites in geosynchronous orbit. The panel needs to know how many satellites could fit in geosynchronous orbit without receivers on the ground picking up multiple signals. Assume that all satellites broadcast at 12 GHz, and that receiver dishes are 45 cm in diameter. Begin by calculating the angular size of the beam associated with such a receiver dish, defined as the full width of its central diffraction peak. Use your result to find the number of satellites allowed in geosynchronous orbit if each receiver dish is to "see" just one satellite. *Hint:* Consult Example 8.3.

77. You have a summer job with an aerospace firm that has a contract to provide instrumentation for a deep space probe to study the Jovian planets. The probe will carry an ultraviolet spectrometer with a 2.0-cm-wide grating ruled at 102 lines/mm. The spectrometer must be able to resolve spectral features only 1 pm apart in twelfth order, when observing at 155 nm. Does it meet these specifications?

78. You're helping a member of the physics faculty who is giving a course for middle school teachers on teaching physics. She's showing how to demonstrate interference using a pair of glass plates like those shown in Fig. 32.29. She illuminates the plates from above with 500-nm light and then places her finger where a dark band appears. She depresses gently and the band becomes light, then dark, then light again. She asks you to show the teachers how to calculate the deflection of the plate. What's your answer?

79. You're working for the state environmental protection agency and are investigating an oil spill. You have a sample of an oil slick on a glass surface. You use the principle of thin-film interference to measure the thickness of the slick. For this type of oil, you know that $n = 1.38$. You know that $m = 1$ gives interference in the visible range and with a spectrometer, you determine that the wavelength of the most strongly reflected light is 580 nm. What do you report for the thickness of the oil slick?

80. You and some friends watch the film *Contact* starring Jodie Foster and based on the book by the late Carl Sagan. One of your friends, who is not a physics student, questions why such a big radio telescope dish is needed. You say, "To see fine details." "How fine?" your friend asks. So you explain that the 305-m-diameter Arecibo radio telescope is used to observe clouds of interstellar hydrogen gas at a wavelength of 21 cm. What do you tell your friend is the theoretically finest detail the radio telescope can resolve at this wavelength? Give your result as an angular resolution.

# Answers to Chapter Questions

## Answer to Chapter Opening Question

The individual bits of information need to be stored in physical structures no smaller than the wavelength of light. The progression to greater information storage from CD to DVD to Blu-ray disc is largely a result of shorter-wavelength lasers being used to read the discs.

## Answers to GOT IT? Questions

32.1 Closer together.

32.2 (a) No change; (b) intensity increases; (c) peaks become narrower.

32.3 (c) A blue filter means you're using a shorter wavelength, decreasing $\lambda$ and therefore $\theta$ in Equations 32.11, and thus increasing the resolution.

# Optics

**Optics** is the study of light and its behavior. **Geometrical optics** is an approximation that holds when the objects with which light interacts are much larger than its wavelength. Then the light generally travels in straight lines called **rays**. **Physical optics**, in contrast, treats light explicitly as a wave. Physical optics explains a host of phenomena that ultimately involve the interference of light waves.

When light rays are incident on an interface between two materials, they generally undergo **reflection** and, for transparent materials, **refraction**. The angles of incidence and reflection are equal. **Snell's law** relates the angles of incidence and refraction:

Reflection: $\theta'_1 = \theta_1$
Refraction (Snell's law):
$n_1 \sin \theta_1 = n_2 \sin \theta_2$

The **index of refraction** $n$ relates light's speed in a medium to its speed $c$ in vacuum:

$$n = \frac{c}{v}$$

**Lenses** and curved mirrors use refraction and reflection, respectively, to form images.

In both cases the object distance $s$, image distance $s'$, and **focal length** $f$ are related by $\frac{1}{s} + \frac{1}{s'} = \frac{1}{f}$.

With **real images**, shown in both figures above, light actually comes from the image. With **virtual images**, light only appears to come from the image:

The wave nature of light becomes important when light interacts with objects comparable in size to the wavelength, or when light travels different paths and recombines to produce interference.

**Destructive interference** occurs when waves are out of step by an odd-integer multiple of a half-wavelength.

**Constructive interference** occurs when waves are out of step by an integer multiple of the wavelength.

A system consisting of two narrow slits produces a pattern of **interference fringes** resulting from alternating regions of constructive and destructive interference:

Where lines of constructive interference intersect the screen, bright fringes appear.

Photo of an actual interference pattern shows alternating bright and dark fringes.

$$d \sin \theta = m\lambda \quad \text{(bright fringes)}$$
$$d \sin \theta = \left(m + \tfrac{1}{2}\right)\lambda \quad \text{(dark fringes)}$$

With multiple slits the bright fringes become narrower and brighter:

Two slits          Five slits

A multiple-slit system constitutes a **diffraction grating** and is used to separate different wavelengths in **spectroscopy**.

**Huygens' principle** explains the propagation of waves by stating that each part of a wavefront acts as a source of circular waves that spread out and interfere to propagate the wave. When light passes through small apertures or by sharp edges, Huygens' principle shows that the light **diffracts**, bending and producing interference fringes as waves from different points interfere.

Diffraction fundamentally limits our ability to resolve small objects or to see closely spaced but distant objects as separate.

Resolved    Barely resolved    Not resolved

For a circular aperture of diameter $d$ (such as a telescope with $d$ its mirror diameter), the **diffraction limit** gives the smallest angular separation that can be resolved at a given wavelength $\lambda$:

$$\theta_{min} = \frac{1.22\lambda}{d}$$

**Part Five challenge problem**

A double-slit system consists of two slits each of width $a$, with separation $d$ between the slit centers $(d > a)$. Light of intensity $S_0$ and wavelength $\lambda$ is incident on the system, perpendicular to the plane containing the slits. Find an expression for the outgoing intensity as a function of angular position $\theta$, taking into account both the slit width and the separation. Plot your result for the case $d = 4a$, and compare with Fig. 32.23.

# Modern Physics

W hat are the fundamental particles of matter? What holds them together to make protons, neutrons, nuclei, atoms, molecules, and solids? Is nature fundamentally predictable, or does uncertainty rule in the microscopic world? At the other extreme, how big is the universe? How did it begin, and how will it end? All these are questions for relativity and quantum physics—collectively called "modern physics" because they were developed after the turn of the 20th century. In Part 6 we give a brief account of Einstein's theory of relativity, followed by a glimpse at quantum physics and its applications. We end with an overview of the latest developments in fundamental physics, from the nature of elementary particles to surprising new understandings of the origin and composition of the universe.

■ The world's smallest electrical wire, a carbon nanotube, is only 10 atoms across. In this image made with an atomic force microscope, the nanotube wire runs across a backdrop of platinum electrodes. Our understanding of physics at the atomic and molecular level lets us construct an increasing variety of practical nanoscale devices.

# 33

# Relativity

▶ **To Learn**

By the end of this chapter you should be able to

- Recount the historical progression from 19th-century physics to relativity (33.1, 33.2).
- State the principle of special relativity (33.3).
- Work quantitative problems involving time dilation and length contraction (33.4, 33.5).
- Explain why events simultaneous in one frame of reference need not be simultaneous in another (33.5).
- Use the Lorentz transformations to find the space and time coordinates of events in different reference frames (33.6).
- Discredit the common misconception that Einstein's theories imply that "everything is relative" (33.6).
- Describe quantitatively the relations among mass, energy, and momentum in relativity (33.7).
- Explain qualitatively how relativity connects electricity and magnetism (33.8).
- Explain qualitatively why a generalization of relativity leads to Einstein's theory of gravity (33.9).

◀ **To Know**

- This chapter builds on your entire prior knowledge of physics, especially electromagnetism and electromagnetic waves (Chapter 29).
- We'll elaborate on the concept of reference frames, especially inertial reference frames (3.3, 4.2).
- Be prepared for really new ideas: Relativity alters the meanings of and relations among physics concepts like mass, momentum, and energy, as well as the underlying concepts of space and time.

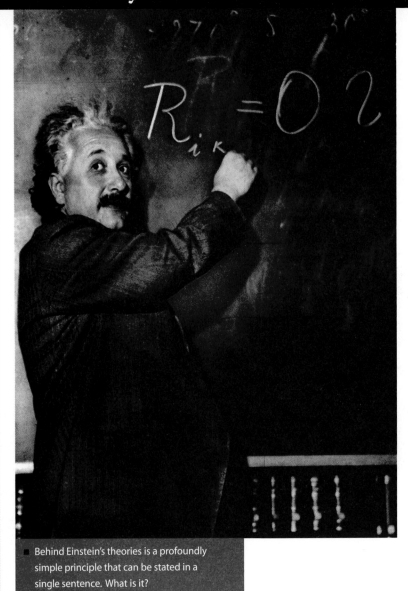

Behind Einstein's theories is a profoundly simple principle that can be stated in a single sentence. What is it?

Maxwell's electromagnetic theory was a crowning achievement of 19th-century physics, providing an understanding of the nature of light and enabling a host of practical technologies. At the same time, Maxwell's electromagnetism led to baffling questions and contradictions that shook the roots of physical understanding and even of common sense.

The resolution of these contradictions was relativity, a theory that radically altered our fundamental understanding of the physical world, and whose influence spilled over into all areas of human thought. Relativity stands as a monument to human intellect and imagination, and it reveals a universe far richer than earlier physicists could conceive of. We'll approach relativity historically, building on our understanding of electromagnetism. That way you'll get a sense of the baffling questions that electromagnetism posed to 19th-century physicists and of how Einstein's resolution of these questions was at once profoundly bold and sweepingly simple.

## 33.1  Speed *c* Relative to What?

Maxwell's equations show that electromagnetic waves in vacuum propagate with speed $c$. Speed $c$ relative to what? In Chapter 14 we found the speed of waves on a string; clearly the speed in that case was relative to the string. Similarly, the 340-m/s speed of sound in air is clearly relative to the air. If you move through the air, the speed of sound *relative to you* won't be the same. In these and other cases of mechanical waves, it's clear what the wave speed means: It's the speed relative to the medium in which the wave is a disturbance.

### The Ether Concept

What about light? Nineteenth-century physicists, their worldview built on the highly successful mechanical paradigm of Newtonian physics, supposed that light waves were like mechanical waves in requiring a medium. They postulated a tenuous substance called the **ether** that permeated the entire universe, allowing light from distant stars to reach us. Electric and magnetic fields were stresses in the ether, and electromagnetic waves were propagating disturbances moving through the ether at speed $c$.

The ether had to have some unusual properties. It must offer no resistance to material bodies, or the planets would lose energy and spiral into the Sun. It must be very stiff, to account for the high speed of light. And it had to be more jelly-like than fluid, to support transverse electromagnetic waves. These properties make ether a rather improbable substance, but to 19th-century physicists the ether was essential in understanding electromagnetic waves.

The speed of light $c$ follows from Maxwell's equations. But in the 19th-century view, light has speed $c$ only for an observer at rest with respect to the ether. Therefore Maxwell's equations could be correct only in the ether frame of reference. This put electromagnetism in a rather different position from mechanics. In mechanics, the concept of absolute motion is meaningless. You can eat your dinner, toss a ball, or do any mechanical experiment as well on an airplane moving steadily at 1000 km/h as you can when the plane is at rest on the ground. This is the principle of **Galilean relativity**, which states that the laws of mechanics are valid in all inertial reference frames—that is, frames of reference in uniform motion (Section 4.2). But the laws of electromagnetism seemed valid only in the ether's reference frame because only in this frame was the prediction of electromagnetic waves moving at speed $c$ correct.

So for 19th-century physicists, the laws for one branch of physics (mechanics) seemed to work in all inertial frames, while those of another branch (electromagnetism) could not. Despite this dichotomy, physicists had great faith in mechanical models and in the ether concept, for without the ether the question "speed $c$ relative to what?" seemed impossible to answer. Thus the late 19th century saw a flurry of experiments to detect the ether. Ultimately they failed, paving the way for the new worldview of relativity that, in Einstein's own words, "arose from necessity, from serious and deep contradictions in the old theory from which there seemed no escape."

## 33.2  Matter, Motion, and the Ether

It was natural for 19th-century physicists to ask about Earth's motion relative to the ether. If Earth is moving through the ether, then the speed of light should be different in different directions. On the other hand, Earth might be at rest relative to the ether. Because other planets, stars, and galaxies move with respect to Earth, it's hard to imagine that the ether is everywhere fixed with respect to Earth alone: This violates the Copernican view that Earth doesn't occupy a privileged spot in the universe. But maybe Earth drags with it the ether in its immediate vicinity. If this "ether drag" occurs, then the speed of light must be independent of direction, but if there's no ether drag, then the speed of light measured on Earth must depend on direction. Through observation and experiment, 19th-century physicists sought to resolve the question of Earth's motion through the ether.

Vertical
rain;
umbrella
overhead

**(a)**

Running;
tilt umbrella
to keep dry.

$\vec{v}$

**(b)**

Situation of
(b) from runner's
reference frame

**(c)**

With "air drag"
rain always
hits umbrella
vertically.

"Dragged" air

**(d)**

**FIGURE 33.1** A rain/umbrella analogy for aberration of starlight.

## Aberration of Starlight

Imagine standing in a rainstorm with rain falling vertically. To keep dry, you hold your umbrella with its shaft straight up, as shown in Fig. 33.1a. But if you run, as in Fig. 33.1b, you'll keep driest if you tilt your umbrella forward. Why? Because the direction of rainfall *relative to you* is at an angle, as shown in Fig. 33.1c. This assumes you don't drag with you a large volume of air. If such an "air drag" occurred, raindrops entering the region around you would be accelerated quickly in the horizontal direction by the air moving with you, so they would now fall vertically relative to you, as in Fig. 33.1d. No matter which way you ran, as long as you dragged air with you, you would point your umbrella vertically upward to stay dry.

This umbrella example is exactly analogous to the observation of light from stars, with the rain being starlight and the umbrella a telescope. If Earth doesn't drag ether, then the direction from which starlight comes will depend on the motion of Earth relative to the ether. But if "ether drag" occurs in analogy with Fig. 33.1d, then light from a particular star will always come from the same direction.

In fact we do observe a tiny change in the direction of starlight. As Earth swings around in its orbit, we must first point a telescope one way to see a particular star. Then, six months later, Earth's orbital motion is in exactly the opposite direction, and we must point the telescope in a slightly different direction. This phenomenon is called **aberration of starlight** and shows that *Earth does not drag the ether*.

## The Michelson–Morley Experiment

If we reject the pre-Copernican notion that Earth alone is at rest relative to the ether, then aberration of starlight forces us to conclude that Earth moves through the ether. Furthermore, the relative velocity of the motion must change throughout the year as Earth orbits the Sun.

In 1881–1887, American scientists Albert A. Michelson and Edward W. Morley attempted to determine Earth's velocity relative to the ether. They used Michelson's interferometer (Fig. 33.2), whose operation we described in Chapter 32. Recall that the interferometer produces a pattern of interference fringes that shifts if the round-trip travel time for light on one of its two perpendicular arms changes. The interference pattern reflects, among other things, possible differences in travel times that arise from differences in the speed of light in different directions—differences that should result from Earth's motion through the ether. Rotating the apparatus through 90° would interchange the directions of the arms and should therefore shift the interference pattern.

Now suppose Earth moves at speed $v$ relative to the ether. Then to an observer on Earth, there's an "ether wind" blowing past Earth. Suppose the Michelson–Morley apparatus is oriented with one light path parallel to the wind and the other perpendicular. Consider a light beam moving the distance $L$ at right angles to the wind. The beam must be aimed slightly upwind so that it will actually move perpendicular to the wind. The light moves in this direction at speed $c$ relative to the ether, but the ether wind sweeps it back so its path in

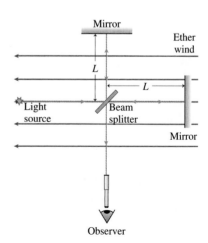

Mirror

Ether
wind

$L$

$L$

Light
source

Beam
splitter

Mirror

Observer

**FIGURE 33.2** Simplified diagram of the Michelson–Morley experiment. Ether wind should result in a longer travel time for light on the horizontal arm.

the Michelson–Morley apparatus is at right angles to the wind. From Fig. 33.3, we see that its speed relative to the apparatus is $u = \sqrt{c^2 - v^2}$, so the round-trip travel time is

$$t_{\text{perpendicular}} = \frac{2L}{u} = \frac{2L}{\sqrt{c^2 - v^2}} \tag{33.1}$$

Light sent a distance $L$ "upstream"—against the ether wind—travels at speed $c$ relative to the ether but at speed $c - v$ relative to Earth. It therefore takes time $t_{\text{upstream}} = L/(c - v)$. Returning, the light moves at $c + v$ relative to Earth, so $t_{\text{downstream}} = L/(c + v)$. The round-trip time parallel to the ether wind is then

$$t_{\text{parallel}} = \frac{L}{c - v} + \frac{L}{c + v} = \frac{2cL}{c^2 - v^2} \tag{33.2}$$

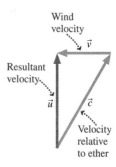

**FIGURE 33.3** Vector diagram for light moving at right angles to an ether wind.

The two round-trip travel times differ, with the trip parallel to the ether wind always taking longer (see Exercises 13, 14, and 27). Light on the parallel trip slows when it moves against the ether wind, then speeds up when it moves with the wind. But slowing always dominates because the light spends more time moving against the wind than with it.

The Michelson–Morley experiment of 1887 was sensitive enough to detect differences in the speed of light an order of magnitude smaller than Earth's orbital speed. The experiment was repeated with the apparatus oriented in different directions, and at different times throughout the year, and the same simple but striking result always emerged: There was never any difference in the travel times for the two light beams. In terms of the ether concept, the Michelson–Morley experiment showed that *Earth does not move relative to the ether.*

## A Contradiction in Physics

Aberration of starlight shows that Earth doesn't drag ether with it. Earth must therefore move relative to the ether. But the Michelson–Morley experiment shows that it doesn't. This contradiction is a deep one, rooted in the fundamental laws of electromagnetism and in the analogy between mechanical waves and electromagnetic waves. The contradiction arises directly in trying to answer the simple question: With respect to what does light move at speed $c$?

Physicists at the end of the 19th century made ingenious attempts to resolve the dilemma of light and the ether, but their explanations either were inconsistent with experiment or lacked sound conceptual bases.

## 33.3 Special Relativity

In 1905, at the age of 26, Albert Einstein (Fig. 33.4) presented his **special theory of relativity**, which resolved the dilemma but altered the very foundation of physical thought. Einstein declared simply that the ether is a fiction. But then with respect to what does light move at speed $c$? With respect, Einstein declared, to anyone who cares to observe it. This statement is at once simple, radical, and conservative. Simple, because its meaning is clear and obvious. Anyone who measures the speed of light will get the value $c = 3.0 \times 10^8$ m/s. Radical, because it alters our commonsense notions of space and time. Conservative, because it asserts for electromagnetism what had long been true in mechanics: that the laws of physics don't depend on the motion of the observer. Einstein summarized his new ideas in the **principle of relativity**, which is expressed in this simple sentence:

> The laws of physics are the same in all inertial reference frames.

Recall that inertial frames are unaccelerated—that is, frames in which the laws of *mechanics* were already valid. Einstein's statement encompasses *all* laws of physics, including mechanics and electromagnetism. The prediction that electromagnetic waves move at speed $c$ must, then, be a universal prediction that holds in *all* inertial reference frames. The *special* theory of relativity is special because it's valid only for the special case of inertial frames. Later we'll discuss the *general* theory of relativity, which removes this restriction.

**FIGURE 33.4** In 1905, when he formulated special relativity, Einstein was a 26-year-old father.

Einstein's relativity explains the result of the Michelson–Morley experiment: No matter what the speed of Earth relative to anything, an observer on Earth should measure the same speed for light in all directions. But at the same time, relativity flagrantly violates our commonsense notions of space and time. We'll see just how in Sections 33.4–33.6.

## 33.4 Space and Time in Relativity

A pedestrian stands by the roadside as a car drives by (Fig. 33.5). Driver and pedestrian each measure the speed of light from a blinking traffic light. The theory of relativity says they'll both get the same value, $c = 3.0 \times 10^8$ m/s, even though the car is moving toward the light source. How is that possible? Consider how each observer might make the measurement. Each has a meter stick and an accurate stopwatch. Suppose a light flash passes the front end of each meter stick just as they coincide. Each observer measures the time for the flash to traverse the 1-m stick and calculates *speed* = *distance/time*. They get the same answer—even though common sense suggests that the light should pass the far end of the "moving" meter stick sooner.

How can this be? Maybe the car's motion affects the driver's stopwatch, making it inaccurate. But no; this suggestion violates relativity's assertion that *all* uniformly moving reference frames are equally good vantage points for doing physics. There can't be anything special about the "moving" reference frame; in fact, it's meaningless to talk about the car as "moving" and the pedestrian "at rest." This is the point of relativity: The concept of absolute motion is meaningless.

The only way out, consistent with relativity, is to let go of absolute space and time. Our two observers' instruments are measuring different quantities that depend on their reference frame—namely, the distance and travel time for the light flash. Those quantities differ in just the right way to make the speed of light come out the same for both observers. This is certainly not what common sense tells us about space and time. But in relativity it's the laws of physics, not measures of space and time, that must be the same for all. Keep in mind the principle of relativity, and you'll see how the rest of special relativity's remarkable consequences follow.

### Time Dilation

Figure 33.6a shows a "light box," consisting of a box of length $L$ with a light source at one end and a mirror at the other. A light flash leaves the source, reflects off the mirror, and returns to the source. We want the time between two events: the emission of the flash and its return to the source. An **event** is an occurrence specified by giving its position and its time.

For concreteness, we'll imagine that the light box is in a spaceship moving past Earth at a uniform velocity. But don't think there's something special about space or spaceships. The whole point of relativity is that all inertial frames are equivalent places for doing physics, and our spaceship is just one inertial reference frame. We'll call that frame $S'$. There's also an accurate clock, $C$, in the spaceship, and $C$ reads zero just as the light flash is emitted.

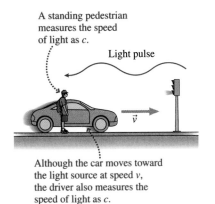

A standing pedestrian measures the speed of light as $c$.

Light pulse

Although the car moves toward the light source at speed $v$, the driver also measures the speed of light as $c$.

**FIGURE 33.5** Both driver and pedestrian measure the same speed $c$ for light, even though they're in relative motion.

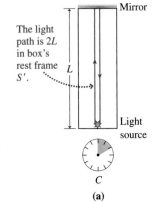

The light path is $2L$ in box's rest frame $S'$.

Mirror

$L$

Light source

$C$

**(a)**

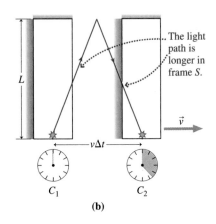

The light path is longer in frame $S$.

$L$

$\vec{v}$

$v\Delta t$

$C_1$ $C_2$

**(b)**

**FIGURE 33.6** A "light box" to explore time dilation shown (a) in a reference frame $S'$ at rest with respect to the box and (b) in a frame $S$ where the box is moving to the right.

In Fig. 33.6a we consider the light-box experiment viewed in the spaceship's reference frame. Since the light box is at rest in this frame, the light travels a round-trip distance $2L$ from source to mirror and back, giving a round-trip travel time of $\Delta t' = 2L/c$. This is the time read on the spaceship's clock $C$.

Now consider the situation as viewed in Earth's reference frame, which we'll call $S$. In this frame, spaceship and light clock are moving to the right with speed $v$, as shown in Fig. 33.6b. Suppose there are two clocks in Earth's frame, positioned so the light box passes clock $C_1$ just as the flash is emitted, and passes $C_2$ just as the flash returns to the source. The clocks are synchronized, and $C_1$ reads zero just as the light flash is emitted. We want to know $C_2$'s reading at the instant the light box passes it and the flash returns to its source; that will be the time, $\Delta t$, between the flash emission and return as measured in Earth's frame $S$.

Figure 33.6b shows that the box moves to the right a distance $v\,\Delta t$ in the time between emission and return of the light flash. Meanwhile the light takes a diagonal path up to the mirror of the moving box and then back down. The path length is twice the diagonal from source to mirror or, by the Pythagorean theorem, $2\sqrt{L^2 + (v\,\Delta t/2)^2}$. The time for light to go this distance is the distance divided by the speed of light, or $\Delta t = 2\sqrt{L^2 + (v\,\Delta t/2)^2}/c$. We explicitly used relativity here, assuming the speed of light remained $c$ in Earth's frame. If we didn't believe relativity, we would have vectorially added light's velocity $\vec{c}$ and the box's velocity $\vec{v}$. But that would make the spaceship's frame the only one in which the speed of light was $c$—in violation of the relativity principle.

The unknown $\Delta t$ appears on both sides of our expression; multiplying through by $c$ and squaring give

$$c^2(\Delta t)^2 = 4L^2 + v^2(\Delta t)^2.$$

We then solve for $(\Delta t)^2$ to get

$$(\Delta t)^2 = \frac{4L^2}{c^2 - v^2} = \frac{4L^2}{c^2}\left(\frac{1}{1 - v^2/c^2}\right)$$

Taking the square root of both sides, and noting that $2L/c$ is just the time $\Delta t'$ measured in the frame $S'$ at rest with respect to the box, we have $\Delta t = \Delta t'/\sqrt{1 - v^2/c^2}$ or

$$\Delta t' = \Delta t\sqrt{1 - v^2/c^2} \qquad \text{(time dilation)} \qquad (33.3)$$

Equation 33.3 describes the phenomenon of **time dilation**, in which the time between two events is always shortest in a frame of reference in which the two events occur at the same place. In our example, the events are the emission and return of the light flash, and they occur at the same place in the spaceship frame $S'$—namely, at the bottom of the box. They don't occur at the same place in the Earth frame $S$ because the box moves relative to this frame, so the bottom of the box isn't in the same place when the light flash returns as it was when the flash was emitted. Thus $\Delta t'$—the time interval measured in the spaceship frame $S'$, where the events occur at the same place—is shorter than $\Delta t$, as you can see from Equation 33.3 and as illustrated in Fig. 33.7.

The shorter time measured by a single clock present at two events is called a **proper time**. Here the word "proper" doesn't mean "correct" or "right"—that would violate relativity, since a time measurement in any inertial frame has equal claim to being valid. Rather, "proper" is used in the sense of "proprietary" in that proper time is the time that *belongs* to this one particular clock. In Earth frame $S$ there's no single clock present at both the emission and return of the light flash, so the time measurement in this frame doesn't "belong" to any one clock.

Time dilation is sometimes characterized by saying "moving clocks run slow," but this statement violates the spirit of relativity because it suggests that some frames are "really" moving and others aren't. The whole point of relativity is that all inertial frames are equally good for describing physical reality, so none can claim to be "at rest" while others are "moving." What the statement "moving clocks run slow" is trying to convey is what we've just seen: The time interval between two events is shortest in a reference frame where the two occur at the same place. There's no significance whatever to our putting the light box in a "moving" spaceship and comparing ship time with Earth time. We could equally well have

**FIGURE 33.7** Clock $C$ moves between clocks $C_1$ and $C_2$, which are at rest relative to each other and synchronized in their rest frame. $C$ measures a shorter elapsed time.

put the light box and its clock on Earth, and two separate clocks in the ship. Then Earth with its light box would be moving past the ship at speed $v$—so the Earth clock would measure the time $\Delta t'$ in Equation 33.3, and the two clocks in the ship frame would measure the longer time $\Delta t$. (That may sound like a contradiction, but it can't be because *there's nothing special about any inertial frame, including Earth's*. We'll return to this point shortly.)

We used a light box to illustrate time dilation. But time dilation isn't something that happens only when we use light to determine time intervals. It's something that happens to *time itself*. Take away the light box in Fig. 33.6, and the clocks will show the same discrepancy. Don't look for a physical mechanism that slows things down. All manifestations of time—the oscillations of the quartz crystal in a digital watch, the swing of a pendulum clock, the period of vibration of atoms in an atomic clock, biological rhythms, and human lifetimes—are affected in the same way.

---

**EXAMPLE 33.1**    **Calculating Time Dilation: Star Trek**

A spaceship leaves Earth on a one-way trip that earthbound observers judge will take 25 years. If the ship travels at $0.95c$ relative to Earth, how long will the trip to a distant star take as judged by observers in the ship?

**INTERPRET** This is a problem about time dilation. Since the events of departure from Earth and arrival at the star occur at the same place in the ship's reference frame, we identify the ship time as the proper time $\Delta t'$ and the Earth time as $\Delta t$ in our discussion of time dilation.

**DEVELOP** Equation 33.3, $\Delta t' = \Delta t\sqrt{1 - v^2/c^2}$, relates the two times. We're given the Earth time $\Delta t$, so we can use this equation to find $\Delta t'$.

**EVALUATE** With $v = 0.95c$, $v/c = 0.95$ and the quantity $v^2/c^2$ in Equation 33.3 becomes $0.95^2$. Then

$$\Delta t' = \Delta t\sqrt{1 - v^2/c^2} = (25\ \text{y})\sqrt{1 - 0.95^2} = 7.8\ \text{y}.$$

**ASSESS** This time is considerably shorter than 25 years, confirming our statement that the time between events is shortest as measured in a reference frame where the events occur at the same place. We'll soon explore what happens if the ship turns around and returns to Earth. ∎

---

We don't notice time dilation in our everyday lives because the factor $v^2/c^2$ is so small for even our fastest motion relative to Earth. Even in a jet airplane, the time difference amounts to a few milliseconds per century. This illustrates the important point that any results from relativity should agree with commonsense Newtonian physics when relative velocities are small compared with the speed of light. Since our intuition and common sense are built on experience at low relative velocities, it's not surprising that effects at high relative velocities seem counter to our common sense.

---

**APPLICATION**    **Mountains and Muons**

Lines represent paths of muons.

$v = 0.994\ c$

2000 m

Sea level

The lines stop where muons decay.

Time dilation is obvious in experiments with subatomic particles moving, relative to us, at speeds near $c$. In a classic experiment, the "clocks" are the lifetimes of particles called muons, which are created by the interaction of cosmic rays with Earth's upper atmosphere and subsequently decay. The experiment consists in counting the number of muons incident each hour on the top of Mt. Washington in New Hampshire, about 2000 m above sea level. The measurement is then repeated at sea level. The figure shows the situation in Earth's reference frame.

Using a detector that records only those muons moving at about $0.994c$ at the mountaintop altitude, the experiment shows that an average of about 560 muons with this speed are incident on the mountaintop each hour. If the mountain weren't there, the muons would travel from the mountaintop altitude to sea level in a time given by

$$\Delta t = \frac{2000\ \text{m}}{(0.994)(3.0 \times 10^8\ \text{m/s})} = 6.7\ \mu s$$

The muon's decay rate is such that one should expect only about 25 of the original 560 muons to remain after a 6.7-$\mu$s interval, so that's approximately the number we might expect to detect each hour at sea level. However, that 6.7-$\mu$s interval is measured in Earth's reference frame—not the muons'. In the muons' frame, time dilation should reduce that interval to

$$\Delta t' = (6.7\ \mu s)\sqrt{1 - 0.994^2} = 0.73\ \mu s$$

The muons' decay is determined by *their* measure of time, and their decay rate is such that we should expect 414 muons to survive for 0.73 $\mu$s.

So what happens? Observers count just over 400 muons per hour at sea level. This is no subtle effect. The difference between 25 and 414 is dramatic. At $0.994c$, the nonrelativistic description is hopelessly inadequate, and time dilation is obvious.

## The Twin Paradox

Time dilation lets us travel into the future! The famous "twin paradox" shows how. One twin boards a spaceship for a journey to a distant star; the other stays on Earth. There are clocks at both Earth and star, like clocks $C_1$ and $C_2$ in Fig. 33.7. There's a clock on the spaceship, like $C$ in Fig. 33.7. Ship clock and Earth clock read the same time as the ship departs (Fig. 33.8a), but when it arrives at the star, time dilation means the ship clock reads less time than the star clock (Fig. 33.8b). Now the ship turns around and returns home. Again, the situation is just like Fig. 33.7 or our light box of Fig. 33.6, so less time elapses on the ship (Fig. 33.8c). The traveling twin arrives home younger than her earthbound brother! Depending on how far and how fast she goes, that age difference can be arbitrarily large. But this is a one-way trip to the future. If the traveling twin doesn't like what she finds in the future, there's no going back.

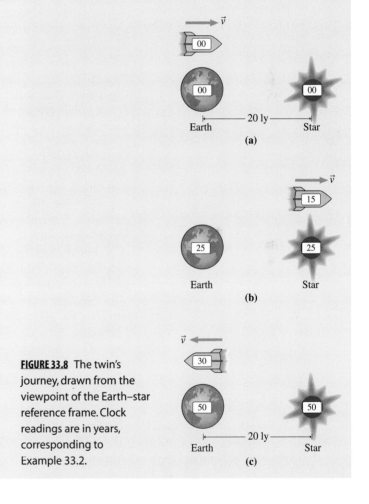

FIGURE 33.8 The twin's journey, drawn from the viewpoint of the Earth–star reference frame. Clock readings are in years, corresponding to Example 33.2.

---

**EXAMPLE 33.2  Time Dilation: The Twin Paradox**

Earth and a star are 20 light years (ly) apart, measured in a frame at rest with respect to Earth and star. Twin A boards a spaceship, travels at $0.80c$ to the star, and then returns immediately to Earth at $0.80c$. Determine the round-trip travel times in Earth and ship reference frames.

**INTERPRET** This problem involves time dilation, here applied to the two separate legs of the round-trip journey. We identify the ship clock as $C$ in Fig. 33.7, and the ship time as $\Delta t'$.

**DEVELOP** Earth–star time for the one-way journey follows from *distance* = *speed* × *time*, so that's how we'll find $\Delta t$. Then we can apply Equation 33.3, $\Delta t' = \Delta t \sqrt{1 - v^2/c^2}$, to get the ship time $\Delta t'$. We'll double both to get the round-trip times.

**EVALUATE** At $0.80c$, the time to go 20 ly is $\Delta t = (20\,\text{ly})/(0.80\,\text{ly/y}) = 25\,\text{y}$. Equation 33.3 then gives

$$\Delta t' = (25\ \text{y})\sqrt{1 - 0.80^2} = 15\ \text{y}.$$

Doubling these values gives round-trip times of 50 years and 30 years for the Earth and ship, respectively.

**ASSESS** The traveling twin returns younger, by 20 years! We've marked the various times on the clocks in Fig. 33.8.

✓**TIP**  Years, Light Years, and the Speed of Light

A light year (ly) is the distance light travels in one year. By definition, therefore, the speed of light is 1 ly/y. It's often easiest in relativity to work in units where the speed of light is 1, whether those units be light years and years, light seconds and seconds, or whatever.

■

---

Here's the paradox: From the ship's viewpoint, it looks like Earth recedes, turns around, and returns. So why isn't the Earth twin younger? The answer lies in what's *special* about special relativity—namely, it applies only to reference frames in uniform motion. The traveling twin accelerates when turning around, so briefly she's in a noninertial reference frame. Although relativity precludes us from saying that one twin is moving and the other is not, we can say that one twin's motion *changes* and the other's doesn't. This is obvious to the traveling twin, who experiences forces associated with the turnaround. The earthbound twin, of course, doesn't feel anything unusual when the ship turns around. During the journey the ship occupies two different inertial frames, separated by the turnaround acceleration, while Earth remains in a single inertial frame. It's that asymmetry that resolves the paradox. The traveling twin really is younger!

What if the traveling twin didn't turn around? Then the situation would be symmetric, and each could argue that the other's clocks "run slow." But unless they get together again at the same place, there's no unambiguous way to compare their clock readings or ages. And they can't get together without accelerating. As we'll soon see, clocks that are synchronized in one reference frame aren't synchronized in another—and that takes the seeming contradiction out of two observers each finding that the other's clocks "run slow."

**GOT IT? 33.1** Triplets A and B board spaceships and head away from Earth in opposite directions, each traveling the same distance at the same speed before returning to Earth. When they get back, how do their ages compare? How do they compare with triplet C, who remained on Earth?

## Length Contraction

In Example 33.2 the 20-ly distance, 25-y time, and $0.8c$ speed are related through the expression $\Delta x = v \, \Delta t$, where $\Delta x$ and $\Delta t$ are measured in the Earth frame. But relativity tells us that this relation must hold in *all* inertial reference frames, so it's also valid in the spaceship frame except during the turnaround. From the ship's viewpoint, Earth and star are moving at $v = 0.80c$, and it takes $\Delta t' = 15$ y for the Earth–star system to pass the ship. Then $\Delta x' = v \, \Delta t' = (0.80 \text{ ly/y})(15 \text{ y}) = 12$ ly is the Earth–star distance as measured in the ship frame. Thus measures of space as well as time depend on one's reference frame. Equation 33.3 gives $\Delta t' = \Delta t \sqrt{1 - v^2/c^2}$, so we can write generally that

$$\Delta x' = v \, \Delta t' = v \, \Delta t \sqrt{1 - v^2/c^2} = \Delta x \sqrt{1 - v^2/c^2} \qquad (33.4)$$

for the distance between two objects measured in a reference frame in which they move with speed $v$. Here $\Delta x$ is the distance in a reference frame at rest with respect to the two objects. Since $\sqrt{1 - v^2/c^2}$ is less than 1 for $v > 0$, Equation 33.4 shows that $\Delta x > \Delta x'$. Therefore the distance is greatest in this so-called rest frame. The two points in question could be the ends of a single object, in which case $\Delta x$ is the object's length. This phenomenon of **length contraction** is also called **Lorentz–Fitzgerald contraction**, after Dutch physicist H. A. Lorentz and Irish physicist George F. Fitzgerald, who independently proposed it as an ad hoc way of explaining the Michelson–Morley experiment. Only with Einstein's work did the contraction acquire a solid conceptual basis.

Length contraction shows that an object is longest in its own rest frame and is shorter to observers for whom it's moving. As with time dilation, don't go looking for a physical mechanism that squashes "moving" objects. That presupposes an absolute space with respect to which contraction occurs. Rather, it's space itself that's different for different observers. Accepting relativity means giving up notions of absolute space and time; length contraction and time dilation are necessary consequences.

---

**EXAMPLE 33.3** **Length Contraction and Time Dilation: SLAC**

At the Stanford Linear Accelerator Center (SLAC), subatomic particles are accelerated to high energies over a straight path whose length, in Earth's reference frame, is 3.2 km. For an electron traveling at $0.9999995c$, how long does the trip take as measured (a) in Earth's frame and (b) in the electrons' frame? (c) What's the length of the linear accelerator in the electrons' frame?

**INTERPRET** We're being asked about time dilation and length contraction. The Earth frame here is like the Earth frame of Example 33.2, with the ends of the accelerator replacing Earth and star, and $\Delta x = 3.2$ km in place of the 20-ly Earth–star separation. In (a) we're therefore being asked for $\Delta t$, in (b) for $\Delta t'$, and in (c) for $\Delta x'$.

**DEVELOP** As always $\Delta x = v \, \Delta t$ relates distance, time, and speed in a single reference frame. We're given $\Delta x$ and $v$, so we'll first solve for $\Delta t$. Then we can use Equation 33.3, $\Delta t' = \Delta t \sqrt{1 - v^2/c^2}$, for $\Delta t'$, and Equation 33.4, $\Delta x' = \Delta x \sqrt{1 - v^2/c^2}$, for $\Delta x'$.

**EVALUATE** The electrons' speed is so close to $c$ that it suffices to calculate the travel time with $v = c$. (a) We have $\Delta t = \Delta x/c = (3.2 \text{ km})/(3.0 \times 10^8 \text{ m/s}) = 11 \, \mu\text{s}$. (b) Equation 33.3 gives $\Delta t' = \Delta t \sqrt{1 - v^2/c^2}$; with $v/c = 0.9999995$, the square root works out to be $10^{-3}$, so $\Delta t' = 11$ ns. (c) Equation 33.4 shows that the length shrinks by the same factor, to 3.2 m.

**ASSESS** In this case of extremely relativistic speed, the relativistic factor $\sqrt{1 - v^2/c^2}$ is tiny, and the effects of time dilation and length contraction are dramatic. Note that we could approximate $v$ as $c$ in finding $\Delta t$, but not in working with the relativistic factor, where even the slightest difference from $c$ is crucial. As a check on our answer, note that $\Delta x' = v \, \Delta t'$, as required by the principle of relativity. ∎

Equations 33.3 and 33.4 show that relativistic effects are significant only at high relative velocities, with $v^2/c^2$ comparable to 1. We've no experience of such velocities in our everyday lives, so relativity seems counterintuitive. Had we grown up moving relative to our surroundings at speeds approaching $c$, the relativity of space and time would be as obvious as our commonsense notions seem now. For physicists working with high-energy particles or studying distant, rapidly moving galaxies, relativistic effects *are* obvious features of physical reality.

## 33.5 Simultaneity Is Relative

One remarkable consequence of relativity is that simultaneity of events and sometimes even their time order depend on one's reference frame. Here we explore how this comes about.

Figure 33.9a shows two identical rods approaching each other in a reference frame $S$ where they have the same speed $v$. Figure 33.9b shows that the right end of rod A passes the right end of rod B at the same instant that the left ends pass. Call the passing of the two right ends event $E_1$ and the passing of the left ends $E_2$. Clearly $E_1$ and $E_2$ are **simultaneous**, meaning they occur at the same time.

Now consider a reference frame $S'$ in which rod A is at rest and is therefore longer than it was in frame $S$. Rod B, meanwhile, is moving toward rod A with a greater speed relative to $S'$ than it had relative to $S$. Therefore it's shorter than it was in $S$. Figure 33.10 shows that, as a result of their different lengths, the right ends of the two rods coincide before the left ends; in other words, event $E_1$ precedes $E_2$. Now look at the situation from a reference frame in which rod B is at rest, and you'll see that $E_2$ precedes $E_1$ (Fig. 33.11). So events that are simultaneous in one reference frame aren't simultaneous in another frame.

**FIGURE 33.9** (a) In frame $S$, rods A and B have the same speed $v$ and both are contracted by the same amount. (b) Their ends coincide at the same time.

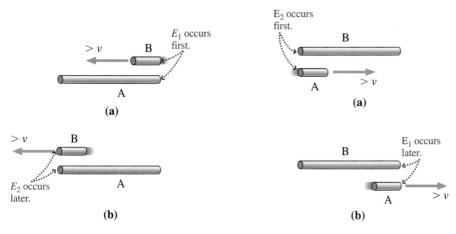

**FIGURE 33.10** The passing rods viewed in a reference frame $S'$ at rest with respect to rod A.

**FIGURE 33.11** The passing rods viewed in a reference frame at rest with respect to rod B.

Isn't this just an illusion resulting from apparent length differences due to the rods' motion? Isn't the picture in frame $S$ (Fig. 33.9) the "real" one? No! Relativity assures us that all inertial reference frames are equally valid for describing physical reality. The length differences and changes in the ordering of events aren't "apparent" and they aren't "illusions." They result from valid descriptions in different reference frames, and each has equal claim to reality. If you insist that one frame—say $S$—has more validity, then you're clinging to the 19th-century notion that there's one favored reference frame in which alone the laws of physics are valid.

But how can observers disagree about the *order* of events? After all, if one event causes another, we expect cause always to precede effect. As we'll soon show, the only events whose time order is different for different observers are those that are so far apart in space, and so close in time, that not even a light signal from one event could reach the location of the other event before it happened. There's no way for such events to influence each other, so they can't be causally related.

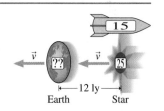

**GOT IT? 33.2** The figure shown is the same as Fig. 33.8b, showing the spaceship at the star but now as observed in the ship's frame. (a) Describe the differences between the two figures, and explain each. (b) Remember that events simultaneous in one frame need not be simultaneous in another. In this case that means the Earth and star clocks don't read the same time simultaneously; in other words, they aren't synchronized. We've drawn the star clock showing what we know it reads when the ship reaches it—namely, 25 years. What, then, should the Earth clock read in this figure? A correct answer will convince you that it's not a contradiction for observers in each frame to observe that clocks in the other frame "run slow."

## 33.6 The Lorentz Transformations

Events are determined by where (three spatial coordinates) and when (time) they occur. Our work with time dilation, length contraction, and the ordering of events suggests that these coordinates depend on one's frame of reference. Here we develop general expressions, called the **Lorentz transformations**, that relate the time and space coordinates of events in different reference frames.

At $t = 2$ s, the $x'$ axis has moved 2 m to the right, so $x'$ is 2 m less than $x$.

**FIGURE 33.12** Nonrelativistic picture of two coordinate axes in relative motion, shown at time $t = 2$ s. In general, $x' = x - vt$.

Consider coordinate axes in a reference frame $S$ and in another frame $S'$ moving in the $x$ direction with speed $v$ relative to $S$. The origins of the two systems coincide at time $t = t' = 0$. Given an event with coordinates $x, y, z, t$ in $S$, what are its coordinates $x', y', z', t'$ in $S'$? Were it not for relativity, we'd expect $y, z$, and $t$ to remain unchanged, while the relative motion along $x$ means a given $x$ in $S$ would correspond to $x' = x - vt$ in $S'$ (Fig. 33.12).

Relativity should alter the transformations for both time and the spatial coordinate, $x$, along the direction of relative motion. However, our results must reduce to the nonrelativistic results $x' = x - vt$ and $t' = t$ in the limit $v \ll c$. A simple form with this property is $x' = \gamma(x - vt)$, where $\gamma$ is a factor, still to be determined, that reduces to 1 as $v \to 0$. We could also transform the other way; the only difference is that the $x$ axis is moving in the negative direction relative to $x'$. Therefore we should have a similar form, with the sign of $v$ reversed: $x = \gamma(x' + vt')$. Now suppose a light flash goes off just as the origins of the two coordinate systems coincide. The coordinates of this event, $E_1$, are $x = 0, t = 0$ in $S$, and $x' = 0, t' = 0$ in $S'$. At some later time $t$, an observer at position $x$ in $S$ observes the light flash; call this event $E_2$. Since light travels with speed $c$, $x = ct$. In frame $S'$, $E_2$ has coordinates $x', t'$. But relativity requires that light travel with speed $c$ in *all* inertial reference frames, so we must have $x' = ct'$. Putting these expressions for $x$ and $x'$ in our proposed transformation equations gives $ct' = \gamma t(c - v)$ and $ct = \gamma t'(c + v)$. Multiplying these two equations yields $c^2 = \gamma^2(c - v)(c + v) = \gamma^2(c^2 - v^2)$. Therefore $\gamma = 1/\sqrt{1 - v^2/c^2}$. Taking $v \to 0$ in this expression shows that $\gamma \to 1$ in the nonrelativistic limit, as required. So we have our transformation equations for $x$.

What about $y, z$, and $t$? The $y$ and $z$ axes are perpendicular to the direction of motion, so there's no length contraction and therefore $y' = y$ and $z' = z$. The fact of time dilation makes clear that measures of time differ in different reference frames, so it's not surprising that $t' \neq t$. You can derive the transformation equations for $t$ from those for $x$ (see Problem 42). The results, along with the equations we found for $x, y$, and $z$, are summarized in Table 33.1.

**TABLE 33.1** The Lorentz Transformations

| S to S' | S' to S | |
|---|---|---|
| $y' = y$ | $y = y'$ | |
| $z' = z$ | $z = z'$ | where $\gamma = \dfrac{1}{\sqrt{1 - v^2/c^2}}$ |
| $x' = \gamma(x - vt)$ | $x = \gamma(x' + vt')$ | |
| $t' = \gamma(t - vx/c^2)$ | $t = \gamma(t' + vx'/c^2)$ | |

**EXAMPLE 33.4**    **Galactic Fireworks: Using the Lorentz Transformations**

Our Milky Way and the Andromeda galaxy are approximately at rest with respect to each other and are 2 million light years (Mly) apart. Supernova explosions occur simultaneously in both galaxies, as judged in the galaxies' reference frame. A spacecraft is traveling at $0.8c$ from the Milky Way toward Andromeda. Find the time between the supernova events as measured in the spacecraft's reference frame.

**INTERPRET** We're given two distant events—both supernova explosions—that are simultaneous in a particular reference frame. We're asked to find the time between them in a different reference frame. So this problem is about using the Lorentz transformations for time coordinates.

**DEVELOP** Following our strategy, we establish coordinate system $S$, the galaxy reference frame. We take the origin at the Milky Way, with the $x$ axis extending through Andromeda (here we're treating the galaxies as points). We take $t = 0$ to be the time of the supernova explosions, which are simultaneous in $S$. Then the coordinates of the two supernovas in $S$ are $x_{MW} = 0$, $t_{MW} = 0$ and $x_A = 2$ Mly, $t_A = 0$. Here the subscripts designate not just the galaxies but the specific events of the supernova explosions. We show the situation at time $t = 0$ in Fig. 33.13. The other reference frame is that of the spaceship, which we designate $S'$; it's moving at speed $v = 0.8c$ along the $x$ axis relative to frame $S$. Taking the two coordinate systems to coincide at time $t = 0$ gives $t'_{MW} = 0$, and we're in the situation we used to derive the Lorentz transformations of Table 33.1. So our plan is to apply the transformation equation that gives $t'$, namely $t' = \gamma(t - vx/c^2)$.

**FIGURE 33.13** Our sketch for Example 33.4, drawn in frame $S$ at time $t = 0$, when observers in $S$ judge both supernova explosions to occur. This picture does not apply in frame $S'$!

**EVALUATE** We first evaluate the factor $\gamma$, finding that $\gamma = 1/\sqrt{1 - 0.8^2} = 5/3$. Then we apply the transformation equation for $t'$ to get the time of the Andromeda supernova in the spacecraft's reference frame:

$$t'_A = \gamma\left(t_A - \frac{vx_A}{c^2}\right) = \left(\frac{5}{3}\right)\left(0 - \frac{(0.8 \text{ ly/y})(2 \text{ Mly})}{(1 \text{ ly/y})^2}\right) = -2.7 \text{ My}$$

**ASSESS** Since the Milky Way supernova goes off at $t'_{MW} = 0$, our *negative* answer means the Andromeda event occurs 2.7 million years *earlier* in the spacecraft's reference frame. Again, there's no problem with causality; since the distant events are simultaneous in *some* frame (the galaxy frame), they can't possibly be cause and effect. You can easily show that a spacecraft observer going the other way would judge the Andromeda supernova to occur 2.7 My later. Problems 39–41 explore the case when the supernovas occur far enough apart in time that they *could* be causally related. ∎

## Relativistic Velocity Addition

If you're in an airplane going 1000 km/h relative to the ground and you walk toward the front of the plane at 5 km/h, common sense suggests that you move at 1005 km/h relative to the ground. But measures of time and distance vary among frames of reference in relative motion. For this reason the velocity of an object with respect to one frame doesn't simply add to the relative velocity between frames to give the object's velocity with respect to another frame. In the airplane your speed with respect to the ground is actually a little less than 1005 km/h as you stroll down the aisle, although the difference is insignificant at such a low speed.

The correct expression for **relativistic velocity addition** follows from the Lorentz transformations. Consider a reference frame $S$ and another frame $S'$ moving in the positive $x$ direction with speed $v$ relative to $S$. Let their origins coincide at time $t = t' = 0$, so the Lorentz transformations of Table 33.1 apply.

Suppose an object moves with velocity $u'$ along the $x'$ axis in $S'$. We seek the velocity $u$ of the object relative to the frame $S$. (We're using $u$ and $u'$ for the object because we've already used $v$ for the relative velocity of the reference frames.)

In either frame, velocity is the ratio of change in position to change in time, or $u = \Delta x/\Delta t$. Designating the beginning of the interval $\Delta t$ by the subscript 1 and the end by 2, we can use the Lorentz transformations to write

$$\Delta x = x_2 - x_1 = \gamma\left[(x_2' - x_1') + v(t_2' - t_1')\right] = \gamma(\Delta x' + v\,\Delta t')$$

and

$$\Delta t = t_2 - t_1 = \gamma\left[(t_2' - t_1') + v(x_2' - x_1')/c^2\right] = \gamma(\Delta t' + v\,\Delta x'/c^2)$$

Forming the ratio of these quantities, we have

$$\frac{\Delta x}{\Delta t} = \frac{\Delta x' + v\,\Delta t'}{\Delta t' + v\,\Delta x'/c^2} = \frac{(\Delta x'/\Delta t') + v}{1 + v(\Delta x'/\Delta t')/c^2}$$

But $\Delta x'/\Delta t'$ is the velocity $u'$ of the object in frame $S'$, and $\Delta x/\Delta t$ is the velocity $u$, so

$$u = \frac{u' + v}{1 + u'v/c^2} \quad \text{(relativistic velocity addition)} \qquad (33.5a)$$

The numerator of this expression is just what we would expect from common sense. But this simple sum of two velocities is altered by the second term in the denominator, which is significant only when both the object's velocity $u'$ and the relative velocity $v$ are comparable with $c$. Solving Equation 33.5a for $u'$ in terms of $u$, $v$, and $c$ gives the inverse transformation:

$$u' = \frac{u - v}{1 - uv/c^2} \quad \text{(relativistic velocity addition)} \qquad (33.5b)$$

---

**EXAMPLE 33.5** **Relativistic Velocity Addition: Collision Course**

Two spacecraft approach Earth from opposite directions, each moving at $0.80c$ relative to Earth, as shown in Fig. 33.14a. How fast do the spacecraft move relative to each other?

**INTERPRET** The naïve answer, $1.6c$, isn't consistent with relativity. Instead, this is a problem involving relativistic velocity addition. We identify the Earth frame of reference as $S'$ and ship B's frame as $S$. Ship A is moving at $u' = 0.8c$ relative to the Earth frame $S'$, and we want to know its speed $u$ relative to ship B's frame $S$. Ship B is also moving toward Earth at $0.8c$, or, equivalently, Earth is moving toward ship B at this speed, so we identify $v = 0.8c$ as the relative velocity between frames.

**DEVELOP** With all terms identified, our plan is to apply Equation 33.5a to find the velocity $u$ of ship A relative to ship B.

**EVALUATE** $u = \dfrac{u' + v}{1 + u'v/c^2} = \dfrac{0.80c + 0.80c}{1 + (0.80c)(0.80c)/c^2} = \dfrac{1.6c}{1.64} = 0.98c.$

**ASSESS** The relative speed is less than the $1.6c$ we get from a naïve addition and also less than the speed of light. This result is quite general: Equations 33.5 imply that as long as an object moves with speed $u < c$ relative to some frame, then its speed relative to any other frame is less than $c$. And if you set $u = c$ to describe a light beam, you'll find that Equations 33.5 give $u' = c$ as well—reaffirming the relativistic point that the speed of light is the same in all inertial reference frames. ∎

**FIGURE 33.14** Sketch for Example 33.5 (a) in Earth's frame and (b) in spacecraft B's frame.

**GOT IT? 33.3** You're driving down the highway, and your speedometer reads exactly 30 km/h. A car passes you, going in the same direction at exactly 20 km/h relative to you. Does its speedometer—which measures the car's speed relative to the road—read more or less than 50 km/h?

### Is Everything Relative?

You already know the answer: The laws of physics aren't relative—that's the fundamental principle of relativity. Neither is the speed of light, whose existence and value follow from laws of physics—specifically, Maxwell's equations. And there are a host of other **relativistic invariants**, independent of reference frame. One such invariant is the **spacetime interval**, a kind of four-dimensional "distance" between two events in space and time. The spacetime interval is given by an expression that looks like a modified Pythagorean theorem:

$$(\Delta s)^2 = c^2(\Delta t)^2 - \left[ (\Delta x)^2 + (\Delta y)^2 + (\Delta z)^2 \right] \tag{33.6}$$

where the $\Delta$ quantities are the differences between the time and space coordinates of the events. The invariance of $\Delta s$ follows directly from the Lorentz transformations, as you can show in Problem 61.

The spacetime interval describes a relation between two events that's independent of reference frame. The invariance of the spacetime interval suggests that something absolute underlies the shifting sands of relativistic space and time. That something is **spacetime**—a four-dimensional framework linking space and time into a single continuum. The spacetime interval is the magnitude of a four-dimensional vector—a **4-vector**—whose components involve the three spatial distances $\Delta x$, $\Delta y$, $\Delta z$ and the time $\Delta t$ between two events. The individual space and time components differ in different reference frames, but they always conspire to give the same invariant interval. This is analogous to the vectors of ordinary two- and three-dimensional space, where the vector components depend on your choice of coordinate system. But the actual vector quantity—for example, a force—has a reality independent of your coordinate choices, and its magnitude doesn't depend on the coordinate system (Fig. 33.15). This analogy isn't perfect because of the negative sign in Equation 33.6's expression for the spacetime interval. That sign reflects the fact that the underlying geometry of spacetime isn't the Euclidean geometry you learned in high school.

Other 4-vectors play a role in more advanced treatments of relativity. These include a four-dimensional electric-current density, whose components involve charge density and the three components of ordinary current density; a four-dimensional "wave vector" that links frequency and wavelength and whose invariant magnitude yields the Doppler effect in its correct relativistic form; and a 4-potential that yields both the electric and magnetic fields. A particularly important example is the energy–momentum 4-vector. Its invariant magnitude is famously related to mass, as we'll see in the next section.

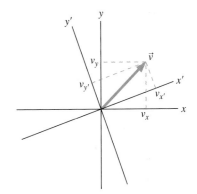

**FIGURE 33.15** Although the $x$ and $y$ components of an ordinary vector depend on the choice of coordinate system, the magnitude of the vector does not.

## 33.7 Energy and Momentum in Relativity

Conservation of momentum and conservation of energy are cornerstones of Newtonian mechanics, where they hold in any inertial reference frame. But momentum and energy are functions of velocity, and we've just seen that relativity alters the Newtonian picture of how velocities transform from one reference frame to another. How, then, can momentum and energy be conserved in all reference frames?

### Momentum

In Newtonian mechanics the momentum of a particle with mass $m$ and velocity $\vec{u}$ is $m\vec{u}$. (Here we use $\vec{u}$ for particle velocities, reserving $\vec{v}$ for the relative velocity between reference frames.) But if a system's momentum—the sum of its individual particles' momenta $m\vec{u}$—is conserved in one frame of reference, then relativistic velocity addition suggests that it won't be conserved in another. The problem here lies not with momentum conservation

**FIGURE 33.16** The ratio of relativistic momentum to Newtonian momentum *mu*. The curve follows Equation 33.7; crosses and circles mark experimental data.

but with the Newtonian expression for momentum. The expression $m\vec{u}$ is actually an approximation valid only for speeds $u$ much less than $c$. The measure of momentum valid at any speed is

$$\vec{p} = \frac{m\vec{u}}{\sqrt{1 - u^2/c^2}} = \gamma m\vec{u} \quad \text{(relativistic momentum)} \tag{33.7}$$

where $\gamma$ is the familiar relativistic factor. The momentum in Equation 33.7 is conserved in all reference frames, and at low velocities it reduces to the Newtonian expression $\vec{p} = m\vec{u}$.

As $u \rightarrow c$, the factor $\gamma$ grows arbitrarily large, and so does the relativistic momentum (Fig. 33.16). Since force is the rate of change of momentum, that means a very large force is required to produce even the slightest change in the velocity of a rapidly moving particle. This helps answer a common question about relativity: Why is it impossible to accelerate an object to the speed of light? The answer is that the object's momentum would approach infinity, and no matter how close to $c$ it was moving, it would still require infinite force to give it the last bit of speed.

## Energy and Mass

The most widely known result of relativity is the famous equation $E = mc^2$. Here we develop a general expression for relativistic energy and show just what $E = mc^2$ means. In the process we'll gain new insights into energy, momentum, and mass.

In Chapter 6 we derived the work–energy theorem, and thus developed the concept of kinetic energy, by calculating the work needed to accelerate a mass $m$ from rest to some final speed. We did that by integrating force over distance, using Newton's law $F = dp/dt$ to write force in terms of momentum. We'll do the same thing here, now using our relativistic expression for momentum. So we have

$$\frac{dp}{dt} = \frac{d}{dt}\left[\frac{mu}{\sqrt{1 - u^2/c^2}}\right] = \frac{m(du/dt)}{\left(1 - u^2/c^2\right)^{3/2}}$$

Then the kinetic energy gained as a particle accelerates from rest in our reference frame to a final speed $u$ is

$$K = \int \frac{dp}{dt}dx = \int \frac{dp}{dt}u\,dt = \int \frac{m(du/dt)}{\left(1 - u^2/c^2\right)^{3/2}}u\,dt = \int_0^u \frac{mu}{\left(1 - u^2/c^2\right)^{3/2}}du$$

where we used $u = dx/dt$ to replace $dx$ with $u\,dt$. The integral is readily evaluated using the fact that $u\,du = \frac{1}{2}d(u^2)$, giving

$$K = \frac{mc^2}{\sqrt{1 - u^2/c^2}} - mc^2 = \gamma mc^2 - mc^2 \tag{33.8}$$

for a particle's kinetic energy in a reference frame where the particle has speed $u$. Once again, $\gamma$ here is the relativistic factor $1/\sqrt{1 - u^2/c^2}$.

For speeds low compared with $c$, Equation 33.8 reduces to the Newtonian $K = \frac{1}{2}mu^2$, as you can show in Problem 59. But Equation 33.8 suggests, more generally, that kinetic energy is the difference of two energies—the velocity-dependent quantity $\gamma mc^2$ and the term $mc^2$ that depends only on mass, not velocity. Pursuing this interpretation, we identify $\gamma mc^2$ as the particle's **total energy** and $mc^2$ as its **rest energy**. Rearranging Equation 33.8 lets us write the total energy as the sum of the kinetic energy and the rest energy: $E = K + mc^2$ or, more simply

$$E = \gamma mc^2 = \frac{mc^2}{\sqrt{1 - u^2/c^2}} \quad \text{(total energy)} \tag{33.9}$$

What does all this mean? Put $u = 0$ in Equation 33.9 and $E = mc^2$—showing that the total energy of a stationary particle isn't zero but is directly proportional to its mass. Thus that particle has energy simply by virtue of having mass or, as Einstein first recognized, *mass and energy are equivalent*. The proportionality between mass and energy is a whopping big $c^2$—about $9 \times 10^{16}$ J/kg in SI units.

Although we developed Equation 33.9 by considering kinetic energy, the mass–energy equivalence $E = mc^2$ is universal. Energy, like mass, exhibits inertia. A hot object is slightly harder to accelerate than a cold one because of the inertia of its thermal energy. A stretched spring is more massive than an unstretched one because of its extra potential energy. When a system loses energy, it loses mass as well.

To the general public, $E = mc^2$ is synonymous with nuclear energy. This equation does describe mass changes that occur in nuclear reactions, but it applies equally well to chemical reactions and all other energy conversions. Weigh a nuclear power plant and then weigh it again a month later, and you'll find it weighs slightly less. Weigh a coal-burning power plant and all the coal and oxygen that go into it for a month, and then weigh all the carbon dioxide and other combustion products that come out, and you'll find a difference. If both plants produce the same amount of energy, the mass difference will be the same for both. The distinction lies in the amount of mass released as energy in each individual reaction. The fissioning of a single uranium nucleus involves about 50 million times as much energy, and therefore mass, as the reaction of a single carbon atom with oxygen to make carbon dioxide. That's why a coal-burning power plant consumes many hundred-car trainloads of coal each week, while a nuclear plant needs only a few truckloads of uranium every year or so. Incidentally, neither process converts very much of the fuel mass to energy; if we could convert *all* the mass in a given object to energy, ordinary matter would be an almost limitless source of energy. Such conversion is in fact possible, but only in the annihilation of matter and antimatter.

---

**EXAMPLE 33.6** **Mass–Energy Equivalence: Annihilation**

A positron is an antimatter particle with the same mass as the electron but the opposite electric charge. When an electron and positron meet, they annihilate and produce a pair of identical gamma rays (bundles of electromagnetic energy). Find the energy of each gamma ray.

**INTERPRET** This is a problem about mass–energy equivalence—in this case all the mass of the electron–positron pair changing to gamma-ray energy. We'll assume the pair has negligible kinetic energy $K$, so the total energy available is just their rest energy.

**DEVELOP** With $K = 0$, Equation 33.9 reduces to $E = mc^2$. Since there are two particles, each of mass $m$, and two identical gamma rays, each gets energy $mc^2$. So our plan is to evaluate this quantity. We'll need the electron mass $m$, which is given inside the cover of this book.

**EVALUATE** We have $E = mc^2 = (9.1 \times 10^{-31}\,\mathrm{kg})(3.0 \times 10^8\,\mathrm{m/s})^2 = 82\,\mathrm{fJ}$.

**ASSESS** High-energy physicists usually work in electronvolts, and our 82-fJ answer is equivalent to 511 keV. Detection of 511-keV gamma rays from laboratory or astrophysical sources is a sure sign of electron–positron annihilation. The medical technique PET (positron emission tomography) uses these annihilation gamma rays to image processes occurring inside the body. ∎

---

Given the fame of $E = mc^2$, it's easy to overlook the fact that the rest energy $mc^2$ is generally only part of a particle's total energy. For a particle moving at velocity $u$ that's small compared with $c$, the total energy is only slightly greater than $mc^2$; the extra is very nearly the Newtonian kinetic energy, $\frac{1}{2}mu^2$. (Here "at rest" and "moving" are, of course, relative to some inertial reference frame.) But when a particle moves with nearly the speed of light, the relativistic factor $\gamma = 1/\sqrt{1 - u^2/c^2}$ becomes much greater than 1, and the total energy $\gamma mc^2$ is many times the rest energy. Such a particle is termed **relativistic**.

**EXAMPLE 33.7** **Total Energy: A Relativistic Electron**

An electron has total energy 2.50 MeV. Find (a) its kinetic energy and (b) its speed.

**INTERPRET** Here we're given the electron's total energy and asked how much of that is the kinetic energy associated with its speed, which we're also asked for.

**DEVELOP** Equation 33.9 gives total energy, which is the sum of kinetic energy $K$ and rest energy $mc^2$. So our plan is to subtract the rest energy to find the kinetic energy. Equation 33.9 also expresses the total energy as $\gamma mc^2$, so we can find $\gamma = 1/\sqrt{1 - u^2/c^2}$ and then solve for the speed

$u$. We don't need to calculate $mc^2$ because we found it in Example 33.6: For the electron, $mc^2 = 511$ keV or 0.511 MeV.

**EVALUATE** (a) We have

$$K = E - mc^2 = 2.50 \text{ MeV} - 0.511 \text{ MeV} = 1.99 \text{ MeV}.$$

(b) Since $E = \gamma mc^2$, $\gamma = E/mc^2 = 2.50 \text{ MeV}/0.511 \text{ MeV} = 4.89$. But $\gamma = 1/\sqrt{1 - u^2/c^2}$, which we solve to get

$$u = c\sqrt{1 - 1/\gamma^2} = 2.94 \times 10^8 \text{ m/s}.$$

**ASSESS** Our answer for kinetic energy is considerably greater than the electron's rest energy, and our speed $u$ is close to $c$, both confirming that this is a relativistic electron. ∎

**GOT IT? 33.4** The rest energy of a proton is 938 MeV. Without doing any calculations, quickly estimate the speed of a proton with total energy 1 TeV $(10^{12}$ eV$)$.

## The Energy–Momentum Relation

In Newtonian physics the equations $p = mu$ and $K = \frac{1}{2}mu^2$ yield $p^2 = 2K/m$. Similarly, Problem 60 shows that in relativity we can combine the equations $p = \gamma mu$ and $E = \gamma mc^2$ to get

$$E^2 = p^2c^2 + (mc^2)^2 \quad \text{(energy–momentum relation)} \tag{33.10}$$

which involves $E$ rather than $K$ because in relativity the energy includes both kinetic and rest energies. For a particle at rest, $p = 0$ and Equation 33.10 shows that the total energy is just the rest energy. For highly relativistic particles, the rest energy is negligible and the total energy becomes very nearly $E = pc$. Some "particles"—like the photons that, in quantum physics, are "bundles" of electromagnetic energy—have no mass. These particles exist only in motion at the speed of light, and for them Equation 33.10 gives the exact relation $E = pc$.

Rearranging Equation 33.10 gives $(mc^2)^2 = E^2 - p^2c^2$. This should remind you of Equation 33.6 for the spacetime interval, whose square is the difference between the squares of the time component $c\,\Delta t$ and the spatial separation $\sqrt{(\Delta x)^2 + (\Delta y)^2 + (\Delta z)^2}$. Similarly, our rearranged Equation 33.10 gives the square of the rest energy $mc^2$ as the difference between the squares of the total energy and the magnitude of the momentum. We can therefore think of energy and momentum as the time and space components of a 4-vector. Your frame of reference determines how this energy–momentum 4-vector breaks out into time and space components—that is, into energy and momentum. In a particle's rest frame, for example, $p = 0$ and the vector has only a time component equal to the rest energy. But no matter what frame you're in, the magnitude of the 4-vector is the same, and it's equal to the rest energy $mc^2$. Therefore mass—the rest energy divided by the constant $c^2$—is a relativistic invariant.

## 33.8 Electromagnetism and Relativity

Historically, relativity arose from deep questions about the propagation of electromagnetic waves. We've seen that relativity alters concepts, such as space, time, energy, and momentum, that are fundamental to Newtonian physics. For that reason, Newtonian physics becomes an approximation valid at low speeds. But relativity is built on the premise that Maxwell's equations of electromagnetism are correct in all reference frames—including the prediction of electromagnetic waves propagating at speed $c$. Even the title of Einstein's famous 1905 paper—"On the Electrodynamics of Moving Bodies"—shows how intimately related are electromagnetism and relativity. Maxwell's equations are relativistically correct and require no modification.

Although electric and magnetic fields in any frame of reference obey the same Maxwell equations, the fields themselves aren't invariant. Sit in the rest frame of a point charge, and you see a spherically symmetric point-charge field. Move relative to the charge, and you see a magnetic field as well, associated with the moving charge. So electric and magnetic fields aren't absolutes; what one observer sees as a purely electric field another may see as a mix of electric and magnetic fields, and vice versa. You can think of the electric and magnetic fields as components of a more fundamental electromagnetic field; how that field breaks out into electric and magnetic fields depends on your frame of reference. Although the fields are different in different reference frames, there is one important electromagnetic quantity that's invariant—namely, electric charge.

We'll illustrate the deep relationship that relativity imposes on electricity and magnetism by considering the force on a positively charged particle moving relative to a current-carrying wire. For simplicity, assume the wire contains equal line-charge densities of positive and negative charge, both moving in opposite directions with the same speed $v$ relative to the wire (Fig. 33.17a). The resulting current produces a magnetic field that encircles the wire, and the charged particle moving to the right with velocity $\vec{u}$ as shown in Fig. 33.17a experiences a *magnetic* force $\vec{F} = q\vec{u} \times \vec{B}$ toward the wire. Because the positive and negative line-charge densities are equal, the wire is neutral, so there's no *electric* force.

Now look at the situation in the particle's reference frame. The positive charges in the wire have a lower speed relative to the particle than the negative charges, so in the particle's frame the distances between negative charges are contracted *more* than the distances between positive charges, as shown in Fig. 33.17b. But charge is invariant, so that means the charge per unit length is *greater* for the negative charges. In the frame of the charged particle, *the wire carries a net negative charge!* That results in an *electric* field pointing toward the wire, and therefore the charged particle experiences an *electric* force toward the wire. Of course there's still a magnetic field as well, but since the particle is at rest in its own frame, the magnetic force $q\vec{u} \times \vec{B}$ is zero.

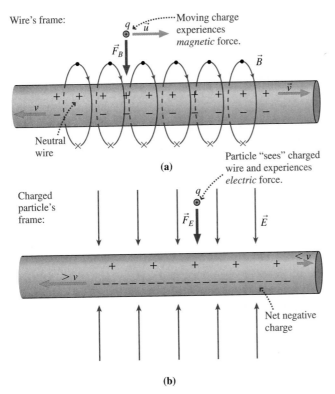

**FIGURE 33.17** The force on a charged particle is magnetic or electric, depending on the reference frame.

We've given two quite different descriptions of the force on the charged particle. In the wire's reference frame, the particle experiences a purely magnetic force that we could determine knowing the magnetic force $q\vec{u} \times \vec{B}$ and how magnetic fields arise from currents. In contrast, describing the force in the particle's reference frame requires no knowledge of magnetism whatever. We need to know only the electric force $q\vec{E}$ and how electric fields arise from charges. This illustrates a profound point: Electricity and magnetism aren't independent phenomena that happen to be related. Rather, they're two aspects of a single phenomenon—electromagnetism. Given the relativity principle, it's not even logically possible to have electricity without magnetism, or vice versa. Relativity provides the complete unification of electromagnetism that we've hinted at throughout our study of these phenomena.

## 33.9 General Relativity

The *special* theory of relativity is restricted to reference frames in uniform motion. Following special relativity, Einstein attempted to formulate a theory that would encompass observers in accelerated motion. But he recognized that it's impossible to distinguish the effects of uniform acceleration from those of a uniform gravitational field (Fig. 33.18). Consequently, Einstein's 1916 **general theory of relativity** became a theory of gravity. General relativity describes gravity as the geometrical curvature of four-dimensional spacetime. In this description, matter and energy curve spacetime in their vicinity, and objects moving through the curved spacetime follow the straightest possible paths—which aren't the straight lines of Euclidean geometry. Figure 33.19 shows a two-dimensional analogy for particles in curved spacetime.

Rocket at rest on Earth; ball accelerates downward.

Rocket accelerates in interstellar space; no gravity, but ball accelerates relative to rocket.

9.8 m/s²     9.8 m/s²

Earth

**FIGURE 33.18** It's impossible to distinguish the effects of uniform gravitation from acceleration, which is why general relativity is about gravity.

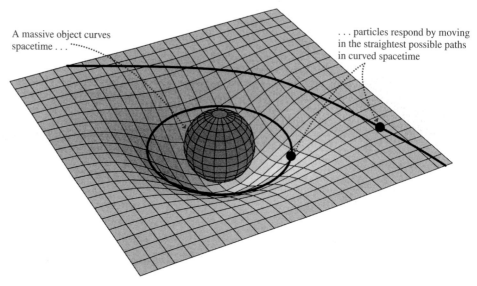

A massive object curves spacetime . . .

. . . particles respond by moving in the straightest possible paths in curved spacetime

**FIGURE 33.19** A two-dimensional analog of curved spacetime in general relativity.

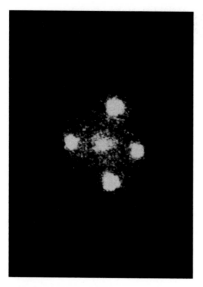

**FIGURE 33.20** The "Einstein Cross" comprises four images of the same quasar, formed as light follows different paths in the curved spacetime surrounding a massive galaxy (visible at center).

General relativity's predictions differ substantially from those of Newton's theory of gravity only in regions of very strong gravitational fields—far stronger than those found anywhere in our solar system. For that reason general relativity has become a cornerstone of modern astrophysics, describing the physics of such bizarre objects as neutron stars, black holes, and so-called gravitational lenses that can produce multiple images of astrophysical objects (Fig. 33.20). General relativity also addresses cosmological questions of the origin and ultimate fate of the universe. But we're not without terrestrial uses for general relativity; the Global Positioning System (GPS) would be off by several kilometers if the effects of general relativity weren't included in GPS calculations.

## Big Picture

The big idea here is a simple one: The laws of physics are the same for all observers and don't depend on one's state of motion. That's the **principle of relativity. Special relativity** is restricted to inertial reference frames; **general relativity** removes that restriction and in so doing becomes a theory of gravity. Both relativity theories radically alter our commonsense notions of space and time. In relativity, measures of space and time—but not the laws of physics—become dependent on one's reference frame.

## Key Concepts and Equations

Among the laws of physics are Maxwell's equations, with their prediction of electromagnetic waves (e.g., light) propagating at the speed of light $c$. Therefore the speed of light is the same in all inertial reference frames.

Invariance of $c$ leads to **time dilation** and **length contraction**:

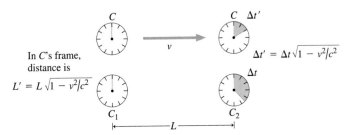

In $C$'s frame, distance is
$$L' = L\sqrt{1 - v^2/c^2}$$

$$\Delta t' = \Delta t\sqrt{1 - v^2/c^2}$$

Additionally, events simultaneous in one reference frame may not be simultaneous in another frame.

Time dilation and length contraction are specific instances of the **Lorentz transformations** of space and time coordinates of events observed in different reference frames. The transformations here apply to relative motion in the $x$ direction.

| $S$ to $S'$ | $S'$ to $S$ | |
|---|---|---|
| $y' = y$ | $y = y'$ | |
| $z' = z$ | $z = z'$ | where $\gamma = \dfrac{1}{\sqrt{1 - v^2/c^2}}$ |
| $x' = \gamma(x - vt)$ | $x = \gamma(x' + vt')$ | |
| $t' = \gamma(t - vx/c^2)$ | $t = \gamma(t' + vx'/c^2)$ | |

Underlying the changing measures of space and time is four-dimensional **spacetime**, in which exist **4-vectors** whose magnitude is independent of reference frame.

Energy, momentum, and mass are closely related in relativity:

$$\text{Momentum: } \vec{p} = \frac{m\vec{u}}{\sqrt{1 - u^2/c^2}} = \gamma m\vec{u}$$

$$\text{Energy: } E = \frac{mc^2}{\sqrt{1 - u^2/c^2}} = \gamma mc^2 = K + mc^2 = \sqrt{(pc)^2 + (mc^2)^2}$$

Kinetic energy     Rest energy

## Cases and Uses

The **Michelson–Morley experiment** of 1887 failed to detect any motion of Earth relative to the **ether**, the medium in which 19th-century physicists believed light propagated with speed $c$. This result helped pave the way for relativity, with no "light medium" and with $c$ the same in all inertial reference frames.

Relativity shows that velocities don't simply add; rather,

$$u = \frac{u' + v}{1 + u'v/c^2}$$

where $u'$ is an object's velocity relative to a reference frame $S'$, $u$ its velocity relative to frame $S$, and $v$ the relative velocity between $S$ and $S'$.

Einstein's famous $E = mc^2$ describes a universal interchangeability between matter and energy; contrary to common opinion, it isn't just about nuclear energy.

# For Thought and Discussion

1. Why was the Michelson–Morley experiment a more sensitive test of motion through the ether than independent measurements of the speed of light in two perpendicular directions?

2. Why was it necessary to repeat the Michelson–Morley experiment at different times throughout the year?

3. What is *special* about the *special* theory of relativity?

4. Does relativity require that the speed of sound be the same for all observers? Why or why not?

5. Time dilation is sometimes described by saying that "moving clocks run slow." In what sense is this true? In what sense does the statement violate the spirit of relativity?

6. If you're in a spaceship moving at 0.95c relative to Earth, do you perceive time to be passing more slowly than it would on Earth? Think! Is your answer consistent with the principle of relativity?

7. The Andromeda galaxy is 2 million light years from our Milky Way. Although nothing can go faster than light, it would still be possible to travel to Andromeda in much less than 2 million years. How is this possible?

8. Is matter converted to energy in a nuclear reactor? In a burning candle? In your body?

9. If you took your pulse while traveling in a high-speed spacecraft, would it be faster, slower, or the same as on Earth?

10. The rest energy of an electron is 511 keV. What is the approximate speed of an electron whose total energy is 1 GeV $(10^9\,\text{eV})$? *Note:* You need not do any calculations!

11. An atom in an excited state emits a burst of light. What happens to the mass of the atom?

12. The quantity $\vec{E} \cdot \vec{B}$ is invariant. What does this say about how different observers will measure the angle between $\vec{E}$ and $\vec{B}$ in a light wave?

# Exercises and Problems

## Exercises

### Section 33.2 Matter, Motion, and the Ether

13. Consider an airplane flying at 800 km/h airspeed between two points 1800 km apart. What is the round-trip travel time for the plane (a) if there's no wind (b) if a wind is blowing at 130 km/h perpendicular to a line joining the two points; and (c) if a wind is blowing at 130 km/h along a line joining the two points?

14. What would be the difference in light travel times on the two legs of the Michelson–Morley experiment if the ether existed and if Earth moved relative to it at (a) its orbital speed relative to the Sun (Appendix E); (b) $10^{-2}c$; (c) 0.5c; and (d) 0.99c? Assume each light path is exactly 11 m long and that the paths are oriented parallel and perpendicular to the ether wind.

### Section 33.4 Space and Time in Relativity

15. Two stars are 50 ly apart, measured in their common rest frame. How far apart are they to a spaceship moving between them at 0.75c?

16. How long would it take a spacecraft traveling at 65% of the speed of light to make the $5.8 \times 10^9$ km journey from Earth to Pluto according to clocks (a) on Earth and (b) on the spacecraft?

17. A spaceship goes by at half the speed of light, and you determine that it is 35 m long. What is its length as measured in its rest frame?

18. An extraterrestrial spacecraft whizzes through the solar system at 0.80c. How long does it take to go the 8.3-light-minute distance from Earth to the Sun (a) according to an observer on Earth, and (b) according to an alien aboard the ship?

19. How fast would you have to move relative to a meter stick for its length to measure 99 cm in your frame of reference?

20. Electrons in a TV tube move through the tube at 30% of the speed of light. By what factor is the tube foreshortened in the electrons' frame of reference, relative to its length at rest?

### Section 33.7 Energy and Momentum in Relativity

21. By what factor does the momentum of an object change if you double its speed when its original speed is (a) 25 m/s and (b) $1.0 \times 10^8$ m/s?

22. At what speed will the momentum of a proton (mass 1 u) equal that of an alpha particle (mass 4 u) moving at 0.5c?

23. At what speed will the Newtonian expression for momentum be in error by 1%?

24. A particle is moving at 0.90c. If its speed increases by 10%, by what factor does its momentum increase?

25. Find (a) the total energy and (b) the kinetic energy of an electron moving at 0.97c.

26. At what speed will the relativistic and Newtonian expressions for kinetic energy differ by 10%?

## Problems

27. Show that the time of Equation 33.2 is longer than that of Equation 33.1 when $0 < v < c$.

28. Suppose the speed of light differed by 100 m/s in two perpendicular directions. How long should the arms be in a Michelson interferometer if this difference is to cause the interference pattern to shift one-half cycle (i.e., a light fringe shifts to where a dark one was) relative to the pattern if there were no speed difference? Assume 550-nm light.

29. Earth and Sun are 8.3 light minutes apart, as measured in their rest frame. (a) What is the speed of a spacecraft that makes the trip in 5.0 min according to its on-board clocks? (b) What is the trip time as measured by clocks in the Earth–Sun reference frame?

30. The Andromeda galaxy is 2 million light years from Earth, measured in the common rest frame of Earth and Andromeda. Suppose you took a fast spaceship to Andromeda, so it got you there in 50 years as measured on the ship. If you sent a radio message home as soon as you reached Andromeda, how long after you left Earth would it arrive, according to timekeepers on Earth?

31. You wish to travel to a star $N$ light years from Earth. How fast must you go if the one-way journey is to occupy $N$ years of your life?

32. The nearest star beyond our solar system is about 4 light years away. If a spaceship can get to the star in 5 years, as measured on Earth, (a) how long would the ship's pilot judge the journey to take? (b) How far from Earth would the pilot find the star to be?

33. Twins A and B live on Earth. On their 20th birthday, twin B climbs into a spaceship and makes a round-trip journey at $0.95c$ to a star 30 light years distant, as measured in the Earth–star frame of reference. What are their ages when the twins are reunited?

34. Radioactive oxygen-15 decays at such a rate that half the atoms in a given sample decay every 2 min. If a tube containing 1000 $O^{15}$ atoms is moved at $0.80c$ relative to Earth for 6.67 min according to clocks at rest with respect to Earth, how many atoms will be left at the end of that time?

35. Two distant galaxies are receding from Earth at $0.75c$, in opposite directions. How fast does an observer in one galaxy measure the other to be moving?

36. Two spaceships are having a race. The "slower" one moves past Earth at $0.70c$, and the "faster" one moves at $0.40c$ relative to the slower one. How fast does the faster ship move relative to Earth?

37. Use relativistic velocity addition to show that if an object moves at speed $v < c$ relative to some uniformly moving frame of reference, then its speed relative to any other uniformly moving frame must also be less than $c$.

38. Earth and Sun are 8.33 light minutes apart. Event A occurs on Earth at time $t = 0$ and event B on the Sun at time $t = 2.45$ min, as measured in the Earth–Sun frame. Find the time order and time difference between A and B for observers (a) moving on a line from Earth to Sun at $0.750c$, (b) moving on a line from Sun to Earth at $0.750c$, and (c) moving on a line from Earth to Sun at $0.294c$.

39. Two civilizations are evolving on opposite sides of a galaxy whose diameter is $10^5$ light years. At time $t = 0$ in the galaxy frame of reference, civilization A launches its first interstellar spacecraft. Civilization B launches its first spacecraft 50,000 years later. A being from a more advanced civilization is traveling through the galaxy at $0.99c$, on a line from A to B. Which civilization does this being judge to have first achieved interstellar travel, and how much in advance of the other?

40. Repeat Problem 39, now assuming that civilization B lags A by 1 million years in the galaxy frame of reference.

41. Could there be observers who would judge the events in Problems 39 and 40 to be simultaneous? If so, how must each be moving relative to the galaxy?

42. Derive the Lorentz transformations for time from the transformations for space.

43. In the light box of Fig. 33.6, let event A be the emission of the light flash and event B its return to the source. Assign suitable space and time coordinates to these events in the frame in which the box moves with speed $v$. Apply the Lorentz transformations to show that the time $\Delta t'$ between the two events in the box frame is given by Equation 33.3.

44. Two spaceships are each 25 m long, as measured in their rest frames (Fig. 33.21). Ship A is approaching Earth at $0.65c$. Ship B is approaching Earth from the opposite direction at $0.50c$. Find the length of ship B as measured (a) in Earth's frame of reference and (b) in ship A's frame of reference.

**FIGURE 33.21** Problem 44; the drawing is in Earth's frame of reference.

45. How fast would you have to go to reach a star 200 light years distant in a 75-year human lifetime?

46. An advanced civilization has developed a spaceship that goes, with respect to the galaxy, only 50 km/s slower than light. (a) How long, according to the ship's crew, does it take to cross the 100,000-ly diameter of the galaxy? (b) What is the galactic diameter as measured in the ship's frame of reference?

47. A spaceship travels at $0.80c$ from Earth to a star 10 light years distant, as measured in the Earth–star reference frame. Let event A be the ship's departure from Earth and event B its arrival at the star. (a) Find the distance and time between the two events in the Earth–star reference frame. (b) Find the distance and time between the two events in the ship frame. *Hint:* The distance in the ship frame is the distance an observer has to move *with respect to that frame* to be at both events—not the same as the Lorentz-contracted distance between Earth and star. (c) Compute the square of the spacetime interval in both frames to show explicitly that it's invariant.

48. Use Equation 33.6 to calculate the square of the spacetime interval between the events (a) of Problem 39 and (b) of Problem 40. Comment on the signs of your answers in relation to the possibility of a causal relationship between the events.

49. A light beam is emitted at event A and arrives at event B. Show that the spacetime interval between the two events is zero.

50. Compare the momentum changes needed to boost a spacecraft (a) from $0.1c$ to $0.2c$ and (b) from $0.8c$ to $0.9c$.

51. Event A occurs at $x = 0$ and $t = 0$ in a frame of reference $S$. Event B occurs at $x = 3.8$ light years and $t = 1.6$ years in $S$. Find (a) the distance and (b) the time between A and B in a frame $S'$ moving at $0.80c$ along the $x$ axis of $S$.

52. When a particle's speed doubles, its momentum increases by a factor of 3. What was the original speed?

53. Find (a) the speed and (b) the momentum of a proton whose kinetic energy is 500 MeV.

54. Among the most energetic cosmic rays ever detected are protons with energies of about $10^{20}$ eV. Find the momentum of such a proton, and compare with that of a 25-mg insect crawling at 2 mm/s.

55. A large city consumes electrical energy at the rate of $10^9$ W. If you converted all the rest mass in a 1-g raisin to electrical energy, for how long could it power the city?

56. In a nuclear-fusion reaction, two deuterium nuclei ($^2$H) combine to give a helium nucleus ($^3$He) plus a neutron. The energy released in the process is 3.3 MeV. By how much do the combined masses of the helium nucleus and the neutron differ from the combined masses of the original deuterium nuclei?

57. Find the kinetic energy of an electron moving at (a) $0.0010c$, (b) $0.60c$, and (c) $0.99c$. Use suitable approximations where possible.

58. Find the speed of an electron with kinetic energy (a) 100 eV, (b) 100 keV, (c) 1 MeV, and (d) 1 GeV. Use suitable approximations where possible.

59. Use the binomial approximation (Appendix A) to show that Equation 33.8 reduces to the Newtonian expression for kinetic energy in the limit $u \ll c$.

60. Show that Equation 33.10 follows from the expressions for relativistic momentum and total energy.

61. Show from the Lorentz transformations that the spacetime interval of Equation 33.6 has the same value in all frames of reference.

62. How fast would you have to travel to reach the Crab Nebula, 6500 light years from Earth, in 20 years? Give your answer to seven significant figures.

63. At what speed are a particle's kinetic and rest energies equal?

64. A cosmic-ray proton with energy 20 TeV is heading toward Earth. What is Earth's diameter measured in the proton's frame of reference?

65. When the speed of an object increases by 5%, its momentum goes up by a factor of 5. What was the original speed?

66. Use the Lorentz transformations to show that if two events are separated in space and time so that a light signal leaving one event cannot reach the other, then there is an observer for whom the two events are simultaneous. Show that the converse is true as well: If a light signal can get from one event to the other, then no observer will find them simultaneous.

67. A source emitting light with frequency $f$ moves toward you at speed $u$. By considering both time dilation and the effect of wavefronts "piling up" as shown in Fig. 14.34, show that you measure a Doppler-shifted frequency given by

$$f' = f\sqrt{\frac{c + u}{c - u}}$$

Use the binomial approximation to show that this result can be written in the form of Equation 14.13 for $u \ll c$.

68. Equation 33.5a gives the transformation for the velocity $u$ of an object moving in the $x$ direction—the same direction as the relative velocity $v$ of the two reference frames. Now suppose the object's velocity also has a component $u_y$, perpendicular to the relative velocity $v$. Find the transformation from $u_y'$ to $u_y$.

69. Consider a relativistic particle of mass $m$ moving along a straight line. Use Equation 33.7 to find an expression for the force on the particle, defined as $F = dp/dt$, in terms of its acceleration $a = du/dt$.

70. A particle of mass $m$ is in uniform circular motion at a constant relativistic speed $u$. Use Equation 33.7 to find an expression for the centripetal force, $\vec{F} = d\vec{p}/dt$, in terms of the centripetal acceleration, $\vec{a} = d\vec{u}/dt$.

71. Find the speed of a particle whose relativistic kinetic energy is 50% greater than the Newtonian value calculated for the same speed.

72. A friend is working on a science fiction script that he hopes to sell to a major motion picture studio, and he asks for your help. An alien spaceship passes Earth moving at 0.274$c$. The spaceship then passes the Moon. Your friend needs to know how much time, according to the ship's clock, elapses between the time an alien looking out the ship's window sees Earth pass by and when the Moon passes by. What do you answer?

73. Your friend's science fiction screenplay has been tentatively accepted by a major movie studio. However, the studio wants more details in the technical notes so that their special effects designers can be as realistic as possible. Two spaceships are each 25 m long as measured in their rest frames. Captain Gorgonzoli's ship is approaching Earth at 0.65$c$. Sub Commander Krankenfelder's star cruiser is approaching Earth from the opposite direction at 0.50$c$. The studio wants the length of Krankenfelder's ship as measured (a) in Earth's frame of reference, and (b) in Gorgonzoli's frame of reference. What do you tell them?

74. It's the 24th century and you are working at the Starfleet Museum of Ancient Technology. Archaeologists have unearthed a device that was called a television, an archaic image receiver and display system. Your job is to figure out how it worked. You have one reference that says electrons were accelerated to 0.328$c$ and then bombarded a screen. In order to make the device work, you need to find the accelerating potential difference. In the 24th century, the rest energy of the electron is still 511 keV.

# Answers to Chapter Questions

## Answer to Chapter Opening Question

The laws of physics are the same for all observers, regardless of their state of motion.

## Answers to GOT IT? Questions

33.1 A and B are the same age, both younger than C.

33.2 Velocity arrows are on Earth and star rather than ship because the figure is drawn in the ship's rest frame. Earth, star, and the distance between them are reduced in the direction of relative motion (e.g., horizontally) to 60% of their dimensions in the Earth–star frame (the relativistic factor $\sqrt{1 - v^2/c^2}$ is 3/5). The ship is longer because it's depicted in its rest frame. Its length in Fig. 33.8 is 60% of its rest length, so here its length is greater by a factor of 5/3. (b) From the ship's frame, clocks in the Earth–star frame undergo time dilation, "running slow" by the relativistic factor 0.6. Earth and ship clocks agreed (both read 0) at the ship's departure from Earth, and now the ship clock reads 15 years. So Earth's clock must have advanced only 60% of 15 years, to read 9 years. From the ship's perspective, Earth and star clocks really do "run slow," but they also aren't synchronized; in fact, the star clock is 16 years ahead. Two very different explanations of the same events, but no contradiction!

33.3 Less, although insignificantly so at such low relative speeds.

33.4 Very nearly $c$.

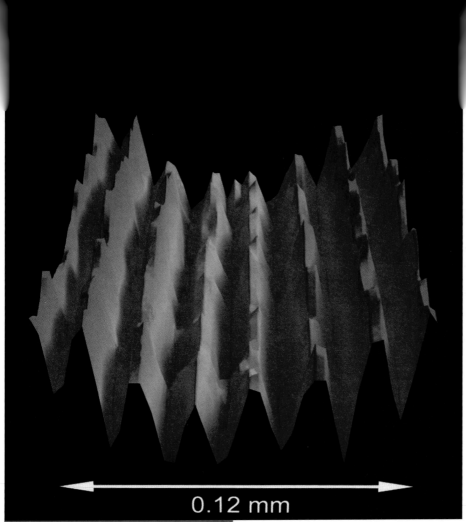

0.12 mm

■ Ridges and valleys represent bright and dark fringes in the interference pattern produced by two beams of ultracold sodium atoms. What does this picture tell us about the nature of matter?

Newtonian mechanics and Maxwell's electromagnetism constitute the core of **classical physics**, providing a deep understanding of physical reality. Although these theories were firmly established by the middle of the 19th century, they remain central to the work of many contemporary scientists and engineers.

Nevertheless, at the end of the 19th century a few seemingly minor phenomena defied classical explanation. Most physicists felt that it was only a matter of time before these, too, came under the classical umbrella. But that was not to be. We've seen how questions about light led to a radical restructuring of our concepts of space and time. Other questions, especially those concerning matter at the atomic scale, were to bring an even more radical transformation of physical thought.

This chapter explores some phenomena that led to the ideas of quantum physics and recounts early attempts to explain them. The next chapter gives a fuller account of quantum theory, and subsequent chapters explore its application to atoms, molecules, nuclei, and quantum-based technologies.

## 34.1 Toward Quantum Theory

Are matter and energy continuously divisible? The essential difference between classical and quantum physics is that the former answers this question "in general, yes," while the latter says definitively "no." Most physical quantities are **quantized**, coming in only certain discrete values.

The idea that physical quantities might come in discrete "chunks" is not new. Some 2400 years ago the Greek philosopher Democritus proposed that all matter consists of indivisible atoms. By the start of the 20th century a more scientifically grounded atomic theory was widely accepted. J. J. Thomson's discovery of the electron in 1897 showed that atoms might be divisible after all, but at the same time it revealed a finer division of matter into discrete "chunks." Robert A. Millikan's 1909 oil-drop experiment showed that electric charge is similarly quantized. Discovery of the proton and later the neutron further solidified the notion that matter comprises fundamental building blocks with definite values for their physical properties.

Quantization of matter into particles with discrete properties is not incompatible with classical physics as long as those particles behave according to classical laws—in particular, that they move continuously through space and can have *any* amount of energy. Add electromagnetism to the picture and the classical viewpoint requires that the fields be continuous, exerting forces on charged particles and sometimes changing, in a gradual and continuous way, the particles' energies.

The startling fact of quantum physics is that this classical behavior does not occur at the atomic scale; instead, energy itself is often quantized. Reconciling the implications of that fact with our commonsense notions of matter and motion has proved impossible; instead, the quantum world speaks a different language, one in which deeply ingrained ideas about causality and the solid reality of matter seem no longer to apply. Here we look at three distinct phenomena that force us to accept the idea that energy can be quantized.

## 34.2 Blackbody Radiation

Heat an object hot enough and it glows, emitting electromagnetic radiation in the form of light. As we saw in Section 16.3, the total power radiated is proportional to the fourth power of the temperature. There's also a change in wavelength with increasing temperature: The first visible glow is a dull red, changing with higher temperatures to orange and then yellow colors corresponding to ever-shorter wavelengths.

A perfect absorber of electromagnetic radiation is called a **blackbody** because it absorbs all light and thus appears black. When a blackbody is heated, it emits **blackbody radiation** at a range of wavelengths. Many objects—such as the Sun or an electric-stove burner—behave approximately like blackbodies. An excellent approximation to a blackbody is a hollow piece of *any* material with a small hole. As Fig. 34.1 shows, any radiation entering the hole undergoes multiple reflections and is eventually absorbed. The hole, therefore, is a nearly perfect absorber, so when the material is heated, the radiation emerging is blackbody radiation.

Experimental study of blackbody radiation shows three characteristic features:

1. The radiation covers a continuous range of wavelengths, with the total power radiated at all wavelengths combined given by the Stefan–Boltzmann law that we introduced in Chapter 16:

$$P_{\text{blackbody}} = \sigma A T^4 \tag{34.1}$$

where $A$ is the area of the radiating surface, $T$ is its temperature, and $\sigma = 5.67 \times 10^{-8} \, \text{W/(m}^2 \cdot \text{K}^4)$ is the Stefan–Boltzmann constant.

2. The radiation peaks at a wavelength that's inversely proportional to the temperature; this is known as **Wien's law**.

3. The distribution of wavelengths depends only on temperature, not on the material of which the blackbody is made.

A blackbody's **radiance** measures the radiated power as a function of wavelength. Because the blackbody emits a continuous spectrum, we have to express radiance as power in small spectral intervals. If we choose fixed intervals in wavelength, then the relation implied in feature 2 above gives a peak radiance at wavelength $\lambda_{\text{peak}}$ such that

$$\lambda_{\text{peak}} T = 2.898 \, \text{mm} \cdot \text{K} \quad \text{(Wien's law)} \tag{34.2a}$$

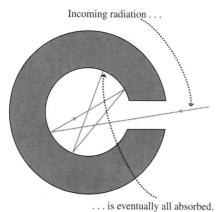

Incoming radiation . . .

. . . is eventually all absorbed.

**FIGURE 34.1** A cavity with a small hole absorbs nearly all incident radiation and hence is a near-perfect blackbody.

We emphasize, however, that the choice of fixed wavelength intervals is arbitrary. If we had chosen fixed frequency intervals, then the constant in Equation 34.2 would be different, and a plot of radiance versus wavelength would peak at a different wavelength (see Problem 74). So $\lambda_{peak}$ in Equation 34.2 is not some absolute measure of the wavelength at which the blackbody emits the "most" radiation, but rather the wavelength of the maximum radiation if you choose to keep track of power in intervals of fixed wavelength. A more physically based quantity is the median wavelength, below and above which half the power is radiated; it's given by

$$\lambda_{median}T = 4.11 \text{ mm} \cdot \text{K}. \tag{34.2b}$$

Whatever measure one chooses, though, the important point is that the peak wavelength is inversely proportional to temperature. In our subsequent discussion we'll adopt a definition of radiance as the power emitted per unit area per unit wavelength interval; then Equation 34.2a describes the peak wavelength. Figure 34.2 plots this measure of blackbody radiance at three temperatures.

Microscopically, blackbody radiation is associated with the thermal motions of atoms and molecules, so it's not surprising that the radiation increases with temperature. In the late 1800s physicists tried to apply the laws of electromagnetism and statistical mechanics to explain the experimental observations of blackbody radiation. They met with some success in describing such aspects as the $T^4$ dependence of the total energy radiated, and the shifting of the radiation distribution toward shorter wavelengths with increasing temperature, but they could not reproduce the actual observed distribution at all wavelengths.

In 1900 the German physicist Max Planck formulated an equation that fit the observed radiance curves for blackbody radiation at all wavelengths:

$$R(\lambda, T) = \frac{2\pi hc^2}{\lambda^5 \left(e^{hc/\lambda kT} - 1\right)} \tag{34.3}$$

Two familiar quantities here are Boltzmann's constant $k = 1.38 \times 10^{-23}$ J/K, introduced in Chapter 17, and the speed of light $c$. A new quantity is the constant $h$, whose value Planck chose to make the equation fit experimental data.

Planck first presented his law as a purely empirical equation describing blackbody experiments. Later he showed that his equation had a remarkable physical interpretation:

> The energy of a vibrating molecule is quantized, meaning it can have only certain discrete values. Specifically, if $f$ is the vibration frequency, then the energy must be an integer multiple of the quantity $hf$:
>
> $$E = nhf, \qquad n = 0, 1, 2, 3, \ldots \tag{34.4}$$

where $h$ is the constant Planck introduced in Equation 34.3. Today we know $h$ as one of the fundamental constants of nature and call it **Planck's constant**. Its value is approximately $6.63 \times 10^{-34}$ J·s, and it's because $h$ is so small that quantum phenomena are usually obvious only in the atomic and molecular realm. Planck's quantization of the energy of vibrating molecules implies further that a molecule can absorb or emit energy only in discrete "bundles" of size $hf$, and that in doing so it jumps abruptly from one of its allowed energy levels to another (Fig. 34.3). (Later developments showed that Planck was correct about the size of the energy jumps but that the factor $n$ in Equation 34.4 should actually be $n + \frac{1}{2}$.)

Planck himself was very conservative and reluctant to accept or elaborate on his theory's evident disagreement with classical physics; nevertheless, his revolutionary work won Planck the 1918 Nobel Prize. Other physicists subsequently emphasized the contrast between Planck's work and the classical treatment of blackbody radiation. That earlier treatment, based on the assumption that energy is shared equally among all possible vibrational modes, had led to the **Rayleigh–Jeans law** for the radiance of a blackbody:

$$R(\lambda, T) = \frac{2\pi ckT}{\lambda^4} \tag{34.5}$$

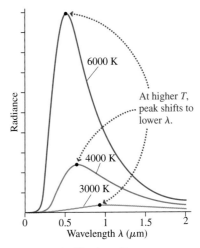

**FIGURE 34.2** Blackbody radiance—energy per unit wavelength interval—as a function of wavelength.

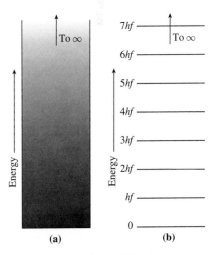

**FIGURE 34.3** (a) In classical physics, a vibrating molecule can have any energy. (b) Allowed energies in Planck's theory are integer multiples of $hf$. Energy-level diagrams like this are used frequently in quantum physics, and usually the horizontal axis has no physical significance.

Not only did the Rayleigh–Jeans law contradict experimental measurements, but it led to the absurd conclusion that every glowing object should emit electromagnetic energy at an infinite rate, with that energy concentrated at the shortest wavelengths (Fig. 34.4). Since the shortest wavelength known at the time was ultraviolet, this phenomenon was called the **ultraviolet catastrophe**. In Planck's equation the exponential term in the denominator grows rapidly with decreasing wavelength, diminishing the radiance and averting the ultraviolet catastrophe. Problems 68, 72, and 75 show that Planck's law reduces to the Rayleigh–Jeans law for longer wavelengths, and that it also leads to Wien's displacement law and the Stefan–Boltzmann law.

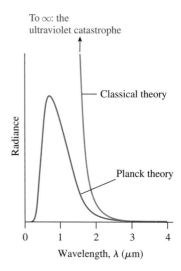

**FIGURE 34.4** Radiance versus wavelength for blackbody radiation at 6000 K, showing also the incorrect classical prediction.

---

**EXAMPLE 34.1** **Blackbody Radiation: Lightbulb Efficiency**

A standard incandescent lightbulb's filament temperature is about 3000 K. (a) Find the wavelength of peak radiance, and (b) compare the radiance at 550 nm—the approximate center of the visible spectrum—with the peak radiance.

**INTERPRET** This problem involves radiation from an object at a known temperature, and we identify the filament as a blackbody. We're asked for both the peak wavelength and a comparison of radiances at two different wavelengths. We're implicitly adopting our definition of radiance as power emitted per unit area per unit wavelength, and our answers will reflect this choice.

**DEVELOP** Equation 34.2a gives the peak wavelength, and Equation 34.3 gives the radiance as a function of wavelength. So our plan for (a) is to solve Equation 34.2a, $\lambda_{peak}T = 2.898$ mm·K, for $T = 3000$ K. For (b) we'll form a ratio of radiances from Equation 34.3, using the result of (a) and the given 550-nm visible wavelength.

**EVALUATE** (a) Equation 34.2a gives $\lambda = 2.898$ mm·K/3000 K $= 966$ nm. (b) To compare radiances, we form a ratio of the right-hand sides of Equation 34.3 evaluated at the wavelengths $\lambda_2 = 550$ nm and $\lambda_1 = 966$ nm. The numerators cancel, giving

$$\frac{R(\lambda_2, T)}{R(\lambda_1, T)} = \frac{\lambda_1^5\left(e^{hc/\lambda_1 kT} - 1\right)}{\lambda_2^5\left(e^{hc/\lambda_2 kT} - 1\right)} = 0.38$$

**ASSESS** Our 966-nm answer for (a) lies in the infrared, suggesting that incandescent lightbulbs aren't very efficient because their radiance spectrum peaks in the infrared. Our answer to (b) confirms this point, giving the bulb's radiance at the center of the visible spectrum as less than half its peak infrared radiance. And remember that we've defined radiance as power per unit area per unit wavelength interval; if we adopt the more physical median wavelength given by $\lambda_{median}T = 4.11$ mm·K, we find $\lambda_{median} = 1.4$ $\mu$m, well into the infrared. Since half the radiation occurs at wavelengths longer than this median, you can see that the bulb must emit far more than half its energy as invisible infrared. ∎

---

**GOT IT? 34.1** Two identical blackbodies are heated until A's temperature is twice B's. Compare their total radiated power and their wavelengths of peak radiance.

## 34.3 Photons

Planck showed that vibrating molecules could exchange energy with electromagnetic radiation only in quantized bundles of size $hf$. Is the radiation's energy similarly quantized?

### The Photoelectric Effect

In 1887 Heinrich Hertz observed that metals emit electrons when struck by light. Observations of this **photoelectric effect** continued with experiments involving metal electrodes in evacuated glass containers (Fig. 34.5). Illuminating one electrode causes it to emit electrons. Making the second electrode positive attracts the electrons, and the resulting current measures the number of electrons ejected. Make the second electrode sufficiently negative, on the other hand, and the electron energy isn't great enough to overcome the repulsive potential; then the current ceases. This so-called *stopping potential* provides a measure of the ejected electrons' maximum kinetic energy.

Classical physics suggests that the photoelectric effect should occur because an electron experiences a force in the oscillating electric field of a light wave. As the electron absorbs energy from the wave, the amplitude of its motion should grow until eventually it has enough energy to escape from the metal. Because the energy in a wave is spread throughout the entire wave, it might take a while for a single tiny electron to absorb enough energy. Increasing the light intensity should increase the electric field, resulting in the electron being ejected sooner and with more energy. Changing the wave frequency should have little effect.

The photoelectric effect does occur, but not in the way classical physics suggests. Figure 34.6 shows results from a photoelectric experiment, in the form of current versus voltage as read by the meters in Fig. 34.5. These results, along with observations made by varying the frequency of the incident light, show three major disagreements with the classical prediction:

1. Current begins immediately, showing that electrons are ejected immediately, even in dim light.
2. The maximum electron energy, as measured by the stopping potential $V_s$, is independent of the light intensity.
3. Below a certain cutoff frequency *no* electrons are emitted, no matter how intense the light. Above the cutoff frequency electrons are emitted with a maximum energy that increases in proportion to the light-wave frequency.

In 1905, the same year he formulated the special theory of relativity, Albert Einstein proposed an explanation for the photoelectric effect. Einstein suggested that an electromagnetic wave's energy is concentrated in "bundles" called **quanta** or **photons**. Einstein applied to these photons the same energy-quantization condition that Planck had already proposed for molecular vibrations: that photons in light with frequency $f$ have energy $hf$, where again $h$ is Planck's constant:

$$E = hf \quad \text{(photon energy)} \tag{34.6}$$

The more intense the light, the more photons—but the energy of each photon is unrelated to the light intensity.

Einstein's idea explains the nonclassical aspects of the photoelectric effect. Each material has a minimum energy—called the *work function*, $\phi$, required to eject an electron. (Table 34.1 lists work functions for selected elements.) Since the energy in a photon of light with frequency $f$ is $hf$, the photons in low-frequency light have less energy than the work function and are therefore unable to eject electrons—no matter how many photons there are. At the cutoff frequency, the photon energy equals the work function, and the photons have just enough energy to eject electrons. As the frequency increases still further, the electrons emerge with maximum kinetic energy $K$ equal to the difference between the photon energy and the work function:

$$K_{\text{max}} = hf - \phi \tag{34.7}$$

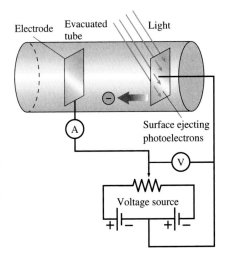

**FIGURE 34.5** Apparatus for studying the photoelectric effect.

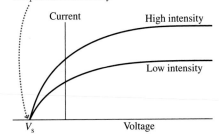

The stopping potential $V_s$ indicates the maximum electron energy, which is independent of intensity.

**FIGURE 34.6** Current versus voltage for the photoelectric experiment of Fig. 34.5, shown for two light intensities at the same frequency.

**TABLE 34.1** Work Functions

| Name (Symbol) | $\phi$ (eV) |
|---|---|
| Silver (Ag) | 4.26 |
| Aluminum (Al) | 4.28 |
| Cesium (Cs) | 2.14 |
| Copper (Cu) | 4.65 |
| Potassium (K) | 2.30 |
| Sodium (Na) | 2.75 |
| Nickel (Ni) | 5.15 |
| Silicon (Si) | 4.85 |

**FIGURE 34.7** Results of a photoelectric experiment, showing stopping potential as a function of light frequency and wavelength. The stopping potential in volts is a direct measure of the electron energy in electronvolts.

Thus, the electrons' maximum kinetic energy depends only on the photon energy—that is, on the light frequency but not on its intensity (Fig. 34.7). Finally, the immediate ejection of electrons follows because an individual photon delivers its entire bundle of energy to an electron all at once. Einstein received the 1921 Nobel Prize primarily for his explanation of the photoelectric effect rather than for his more controversial relativity theories. In 1914 Millikan, who had earlier demonstrated the quantization of electric charge, carried out meticulous photoelectric experiments that confirmed Einstein's hypothesis and helped earn Millikan the 1923 Nobel Prize.

---

**GOT IT? 34.2** If you replot Fig. 34.7 for a substance with a different work function, (a) will the slope of the line change? (b) Will the point at which it intersects the horizontal axis change?

---

Today, the photoelectric effect is used in extremely sensitive light detectors called **photomultipliers**. In these devices, electrons dislodge additional electrons in a series of electrodes called dynodes, resulting in a cascade of as many as a billion electrons for each incident photon (Fig. 34.8).

**FIGURE 34.8** A photomultiplier produces a large pulse of electrons from a single incident photon.

---

**EXAMPLE 34.2** **The Photoelectric Effect: Designing a Photomultiplier**

(a) Choose a suitable material from Table 34.1 for the light-sensitive surface in a photomultiplier that will respond to visible light at wavelengths of 575 nm and shorter. (b) Find the maximum kinetic energy of electrons ejected with the shortest-wavelength visible light, around 400 nm.

**INTERPRET** This problem is about the photoelectric effect. In (a) we're asked for a material in which 575-nm light can eject electrons. That means the work function can be no greater than the energy of a 575-nm photon. In (b) we need to find the excess electron energy, over the work function, for 400-nm light.

**DEVELOP** Equation 34.6, $E = hf$, relates the quantized energy of a photon to its frequency. Since $f\lambda = c$, we can rewrite Equation 34.6 as $E = hc/\lambda$. We'll use this to find the photon energy, and then we'll consult Table 34.1 for an appropriate material. Finally, we can use $f = c/\lambda$ in Equation 34.7, $K_{max} = hf - \phi$, to get the maximum kinetic energy of electrons when $\lambda = 400$ nm.

**EVALUATE** (a) At 575 nm, $E = hc/\lambda = 3.46 \times 10^{-19}$ J, or 2.16 eV, where $1$ eV $= 1.6 \times 10^{-19}$ J. An electron gaining this energy must overcome the work function to escape the metal; the only material in Table 34.1 for

which this is possible is cesium, with $\phi = 2.14$ eV. (b) At $\lambda = 400$ nm and $f = c/\lambda$, Equation 34.7 gives $K_{max} = hc/\lambda - \phi = 0.96$ eV. Here we converted the SI value of $hc/\lambda$ to electronvolts before subtracting $\phi$.

**ASSESS** Make sense? The work function we've chosen is just under the 2.16-eV photon energy at 575 nm, so electrons ejected with photons of this wavelength have negligible kinetic energy. The 400-nm minimum visible wavelength corresponds to roughly 50% higher frequency and therefore energy, or roughly 3 eV. It takes about 2 eV to overcome the work function, leaving about 1 eV of electron kinetic energy.

**✓TIP** Working with Electronvolts

Recall that 1 electronvolt (eV) is the *energy* gained by an electron across a 1-volt potential. So electronvolts are a unit of energy, but *not* the standard SI unit, which is the joule. We computed an energy $E = hc/\lambda$ in SI units and then converted to electronvolts using the factor 1 eV $= 1.6 \times 10^{-19}$ J. In general it's safest to work in SI units and then convert to eV as needed.

---

## Waves or Particles?

In positing the existence of photons, Einstein gave the first inklings of the **wave–particle duality**—the seemingly dual nature of light, which acts in some situations like a wave and in others, as in the photoelectric effect, more like a localized particle. We now turn to another phenomenon that demonstrates light's particle-like aspect. Later we'll see how the wave–particle duality encompasses not only light but matter as well.

## The Compton Effect

In 1923 the American physicist Arthur Holly Compton, at Washington University in St. Louis, did an experiment that dramatically confirmed the particle-like aspect of electromagnetic radiation. Although Compton's work came much later in the history of quantum theory than Einstein's, we include it here because it so strongly corroborates Einstein's photon hypothesis.

Compton was studying the interaction of X rays with electrons. Classically, an electron subject to an electromagnetic wave should undergo oscillatory motion, driven by the wave's oscillating electric field. Since accelerated charge is the source of electromagnetic waves, the electron should itself produce electromagnetic waves *of the same frequency as the incident waves* (Fig. 34.9a). As we saw in Section 29.7, the electron should radiate in all directions, with maximum radiation perpendicular to its oscillatory motion.

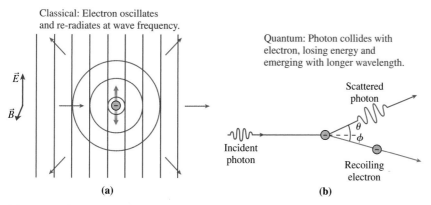

**FIGURE 34.9** Classical and quantum descriptions of the interaction between electromagnetic waves and free electrons.

Compton and his coworkers measured the intensity of scattered X rays as a function of wavelength for different scattering angles. Remarkably, they found the greatest concentration of scattered X rays at a wavelength *longer* than that of the incident radiation (Fig. 34.10). They interpreted their results as implying that particle-like photons had collided with electrons, losing energy to the electrons and therefore, since $E = hf$, emerging with lower frequency and correspondingly longer wavelength (Fig. 34.9b).

We can understand this **Compton effect** by treating the interaction as an elastic collision between the incident photon and a stationary electron. The photon moves at $c$, so it's necessary to use relativistic expressions for energy and momentum. You can work out the details in Problem 73; the result gives the **Compton shift**: $\Delta\lambda = \lambda - \lambda_0$—that is, the change from the photon's original wavelength $\lambda_0$:

$$\Delta\lambda = \frac{h}{mc}(1 - \cos\theta) \qquad \text{(Compton shift)} \qquad (34.8)$$

Figure 34.10 shows that this equation is in excellent agreement with experimental data.

The term $h/mc$ in Equation 34.8 is the **Compton wavelength** of the electron and gives the wavelength shift for a photon scattering at $\theta = 90°$. Its value is $\lambda_C = h/mc = 0.00243$ nm. Equation 34.8 shows that the largest wavelength shift will be $2\lambda_C$, occurring at $\theta = 180°$. For the shift to be noticeable it should be a significant fraction of the incident wavelength, which therefore can't be too many times the Compton wavelength. For X rays, $\lambda$ is in the range from approximately 0.01 nm to 10 nm, and therefore, detection of the Compton shift in X rays is already difficult. It would be totally impossible with visible light.

Today, Compton scattering with gamma rays is a widely used technique for studying the structure of matter. For example, abnormalities in human bone can be detected through Compton scattering of gamma rays emitted by a radioactive source embedded in bone. And the inverse Compton effect—the scattering of a rapidly moving electron off a photon—is a common process in high-energy astrophysical systems and is used in the laboratory to produce beams of gamma radiation.

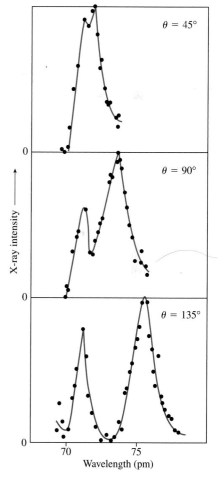

**FIGURE 34.10** Compton's results for scattering of X rays with $\lambda = 71$ pm. Right-hand peak shows the wavelength shift of the Compton effect. The unshifted left-hand peak is from photons scattering off tightly bound atomic electrons, which don't absorb significant energy. Solid curves are theoretical predictions.

The wavelength shift in Compton scattering admits no classical explanation. Coming after a decade of experimental and theoretical work that pointed increasingly to quantization as the essence of the atomic world, Compton's experimental results were for many physicists the convincing evidence for the reality of quanta.

**GOT IT? 34.3** Will the Compton wavelength shift be greater or less for photons of a given wavelength scattering off protons rather than electrons?

## 34.4 Atomic Spectra and the Bohr Atom

In Chapter 29 we found that accelerated charges are the source of electromagnetic radiation. By 1900 it was known that atoms contain negative electrons as well as regions of positive charge; by 1911 experiments by Ernest Rutherford and his colleague Hans Geiger and student Ernest Marsden had localized the positive charge in the tiny but massive nucleus. According to classical physics, electrons should orbit the nucleus under the influence of the electric force, radiating electromagnetic wave energy as they accelerate in their orbits. In fact, a classical calculation shows that atomic electrons will quickly radiate away all their energy and spiral into the nucleus. Thus, the very existence of atoms is at odds with classical physics.

### The Hydrogen Spectrum

A more subtle problem involving atoms dates to 1804, when William Wollaston noticed lines between some of the colors dispersed by a prism. Ten years later the German optician Josef von Fraunhofer dispersed the solar spectrum sufficiently that he could see hundreds of narrow, dark lines against the otherwise continuous spectrum. Studies of light emitted by diffuse gases excited by electric discharges show similar **spectral lines**, these bright against an otherwise dark background (recall Fig. 30.16). Such **emission spectra** result when atoms emit light of discrete frequencies. **Absorption spectra**, in contrast, arise when atoms in a diffuse gas absorb discrete frequencies of light from a continuous source. We emphasize the word "diffuse": Discrete spectra generally arise only when a gas is sufficiently diffuse that light from one atom stands a strong chance of escaping the gas before it interacts with other atoms. In dense gases multiple interactions result in the continuous spectrum of blackbody radiation.

Every element produces its own unique spectral lines, so analysis of spectra, even from the remote reaches of the cosmos, allows us to identify and characterize the material emitting the light. Spectral analysis led to the discovery of helium in the Sun's atmosphere before that element had been identified on Earth—hence the name, from the Greek word *helios* for Sun. Measuring the Doppler shift of spectral lines lets us "see" stars orbiting black holes in distant galaxies, and also gives direct evidence for the expansion of the universe. Back on Earth, the technique of atomic absorption spectroscopy uses spectral lines to determine the elemental composition of substances, helping identify pollutants or trace the flow of elements in biological samples.

In 1884, a Swiss schoolteacher named Johann Balmer realized that the wavelengths of the first four lines in the visible spectrum of hydrogen (Fig. 30.16) were related by the equation

$$\frac{1}{\lambda} = R_{\text{H}}\left(\frac{1}{2^2} - \frac{1}{n^2}\right)$$

where $n = 3, 4, 5, 6 \ldots$ and $R_{\text{H}}$ is the **Rydberg constant** for hydrogen, with a value of approximately $1.097 \times 10^7 \text{ m}^{-1}$. Other series of lines in the hydrogen spectrum were soon found, and Balmer's equation was generalized to

$$\frac{1}{\lambda} = R_{\text{H}}\left(\frac{1}{n_2^2} - \frac{1}{n_1^2}\right) \tag{34.9}$$

where $n_1 = n_2 + 1, n_2 + 2, \ldots$ . The Balmer series of lines has $n_2 = 2$; the Lyman series, in the ultraviolet, has $n_2 = 1$; and the infrared Paschen series has $n_2 = 3$. There are in fact infinitely many such series, corresponding to $n_2 = 1, 2, 3, \ldots$ .

But why should atoms emit discrete spectral lines? And why should the hydrogen lines form patterns with the simple regularity of Equation 34.9?

## The Bohr Atom

In 1913 the great Danish physicist Niels Bohr proposed an atomic theory that accounted for the spectral lines of hydrogen. In the **Bohr atom** the electron moves in a circular orbit about the nucleus, held by the electric force. Classically, any orbital radius and correspondingly any energy and angular momentum should be possible. But Bohr quantized the atom, stating that the only possible orbits were those with angular momentum an integer multiple of Planck's constant divided by $2\pi$. Angular momentum quantization implies energy quantization, which, as we'll show, leads to Equation 34.9 for the hydrogen spectral lines.

Bohr asserted that an electron in an allowed orbit does not radiate energy, in contradiction to the predictions of classical electromagnetism. But an electron can jump from one orbit to another, emitting or absorbing a photon whose energy is equal to the energy difference between the two orbital levels. We can therefore find the expected photon energies—and the corresponding wavelengths—if we know the allowed energy levels.

To find the quantized atomic energy levels in Bohr's model, consider a hydrogen atom consisting of a fixed proton and an electron in circular orbit. Treating the proton as fixed is a good approximation because its mass is nearly 2000 times the electron's. We consider only electron speeds much less than that of light, which is an excellent approximation in hydrogen.

In Example 11.1 we found that the angular momentum of a particle with mass $m$ and speed $v$, moving in a circular path of radius $r$, is $mvr$. Thus, Bohr's quantization condition reads

$$mvr = n\hbar \qquad \text{(quantization, Bohr atom)} \qquad (34.10)$$

where $n = 1, 2, 3, \ldots$ and where we define $\hbar \equiv h/2\pi$ (read "h bar"). We need to relate the electron's angular momentum to its energy so we can find out what Equation 34.10 implies about energy quantization.

We studied circular orbits for the inverse-square force of gravity in Chapter 8, where we found that kinetic and potential energies in a circular orbit are related by $K = -\frac{1}{2}U$, with the zero of potential energy at infinity. The total energy $K + U$ is therefore $\frac{1}{2}U$. These results hold for any $1/r^2$ force, including the electric force. In the electric case the potential energy $U$ is the point-charge potential of the proton, $ke/r$, multiplied by the electron charge, $-e$. Then the total energy is $E = \frac{1}{2}U = -ke^2/2r$. The minus sign means the electron is *bound* to the proton, in that it would take energy to separate them. Solving this equation for $r$ then gives

$$r = -\frac{ke^2}{2E} \qquad (34.11)$$

Since the kinetic energy is $K = -\frac{1}{2}U = -E$, we also have $\frac{1}{2}mv^2 = -E$ or $v = \sqrt{-2E/m}$. Using our expressions for $r$ and $v$ in the quantization condition 34.10 gives $m\sqrt{-2E/m}(-ke^2/2E) = n\hbar$. Solving for the energy $E$, we find

$$E = -\frac{k^2 e^4 m}{2\hbar^2 n^2}$$

It's convenient to define the **Bohr radius**, $a_0$, as

$$a_0 = \frac{\hbar^2}{mke^2} = 0.0529 \text{ nm}$$

With this definition the energy becomes

$$E = -\frac{ke^2}{2a_0}\left(\frac{1}{n^2}\right) \qquad \text{(energy levels, Bohr atom)} \qquad (34.12a)$$

Equation 34.12a gives us the allowed energy levels under Bohr's quantization condition. Evaluating this expression for the case $n = 1$ gives $E_1 = -2.18 \times 10^{-19}$ J $= -13.6$ eV; it's then convenient to write Equation 39.12a numerically in the form

$$E = -\frac{13.6 \text{ eV}}{n^2} \tag{34.12b}$$

where in both forms $n = 1, 2, 3, \ldots$. The lowest energy state, $n = 1$, is called the **ground state**; the others are **excited states**.

Now we have the allowed energy levels. What about spectra? When an electron jumps between energy levels, it emits or absorbs a photon whose energy $hf$ is equal to the energy difference between the levels. So imagine an electron going from a higher level $n_1$ to a lower level $n_2$. The energy difference, according to Equation 34.12a, is

$$\Delta E = -\frac{ke^2}{2a_0}\left(\frac{1}{n_1^2} - \frac{1}{n_2^2}\right) = \frac{ke^2}{2a_0}\left(\frac{1}{n_2^2} - \frac{1}{n_1^2}\right)$$

and this is equal to the energy of the emitted photon. But the photon energy is $\Delta E = hf = hc/\lambda$, and therefore $1/\lambda = \Delta E/hc$ or, using our expression for $\Delta E$,

$$\frac{1}{\lambda} = \frac{ke^2}{2a_0 hc}\left(\frac{1}{n_2^2} - \frac{1}{n_1^2}\right)$$

This looks just like Equation 34.9 for the hydrogen spectral lines, except that $ke^2/2a_0 hc$ replaces the Rydberg constant $R_H$. Evaluating this quantity gives $ke^2/2a_0 hc = 1.09 \times 10^7$ m$^{-1}$, which is very close to the experimentally observed Rydberg constant. The small discrepancy results from our approximation that the proton is stationary.

Bohr's theory of quantized angular momentum thus accounts brilliantly for the observed spectrum of hydrogen. We can understand the origin of the various spectral line series using Fig. 34.11, an **energy-level diagram** for the Bohr model of hydrogen. Allowed energy levels are shown as horizontal lines, and various possible transitions among levels as vertical arrows. Transitions with a common final state are grouped, and each group represents a different series of spectral lines.

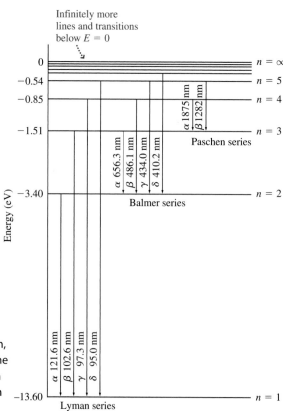

**FIGURE 34.11** Energy-level diagram for the Bohr model of the hydrogen atom, showing transitions responsible for the first three series of spectral lines. Each series arises from jumps to a common final state.

Knowing the energy levels of Equation 34.12, we can also find the radii of the allowed electron orbits, as given by Equation 34.11:

$$r = -\frac{ke^2}{2E} = \left(\frac{ke^2}{2}\right)\left(\frac{2a_0 n^2}{ke^2}\right) = n^2 a_0 \qquad (34.13)$$

Thus, the lowest energy orbit has a radius of one Bohr radius, with higher energy orbits growing rapidly with increasing $n$. A hydrogen atom in its ground state—$n = 1$—has a diameter of two Bohr radii, or about 0.1 nm. As we'll see in Chapter 35, the Bohr model's precise electron orbits aren't compatible with the fully developed theory of quantum mechanics; nevertheless, Equation 34.13 does give the approximate size of atoms.

---

**EXAMPLE 34.3** **The Bohr Model: Big Atoms**

Hydrogen atoms are normally in their ground state, with diameter approximately 0.1 nm. But in the diffuse gas of interstellar space, atoms exist in highly excited states with sizes approaching a fraction of a millimeter. Such **Rydberg atoms** can also be produced temporarily in the lab. Transitions among Rydberg states result in photons at radio wavelengths. One of the longest wavelengths observed corresponds to a transition from $n = 273$ to $n = 272$. (a) What is the diameter of a hydrogen atom in the $n = 273$ state? (b) At what wavelength should a radio telescope be set to observe this transition?

**INTERPRET** This is a problem about electron transitions in hydrogen atoms, albeit of unusual size. The Bohr model applies.

**DEVELOP** We'll use Equation 34.13, $r = n^2 a_0$, to find the atomic diameter, with $n = 273$. Equation 34.9, $1/\lambda = R_H(1/n_2^2 - 1/n_1^2)$, will give the transition wavelength with $n_1 = 273$ and $n_2 = 272$.

**EVALUATE** (a) The diameter is twice the radius, so Equation 34.13 gives $d = (2)(273^2)a_0 = 7.9 \ \mu m$. (b) Inverting Equation 34.9 to get the wavelength gives

$$\lambda = \left[R_H\left(\frac{1}{272^2} - \frac{1}{271^2}\right)\right]^{-1} = 92 \text{ cm}$$

with $R_H = 1.097 \times 10^7 \text{ m}^{-1}$.

**ASSESS** Our atomic diameter is some 75,000 times that of ground-state hydrogen and about the size of a red blood cell! A wavelength of 92 cm corresponds to a frequency $f = c/\lambda$ of about 300 MHz, which happens to lie in a gap between VHF TV channel 13 and UHF channel 14. ∎

---

Equation 34.12 shows, and Fig. 34.11 suggests, that there are infinitely many electron energy levels between the ground state at $-13.6$ eV and zero energy. It's possible to give an atomic electron enough energy to bring it above the $E = 0$ level, but then it's no longer bound to the proton. Removing an electron is **ionization**, and Equation 34.12b and Fig. 34.11 show that it takes 13.6 eV to ionize a hydrogen atom in its ground state. This quantity is the **ionization energy**.

## Limitations of the Bohr Model

Bohr's theory proved astoundingly successful in explaining the hydrogen spectrum. It also explains the spectra of hydrogen-like ions—atoms with all but one of their electrons removed—with the appropriate change in the value of the nuclear charge. And it has some success in predicting the spectra of atoms such as lithium and sodium that have a single valence electron beyond a group of more tightly bound electrons. But it can't account for the spectra of more complicated atoms, even two-electron helium. And with hydrogen, there are subtle spectral details that the Bohr model doesn't address. Furthermore, like Planck's original quantum hypothesis, Bohr's quantization of atomic energy levels lacked a convincing theoretical basis. We'll see in the next chapter how the much more comprehensive theory of quantum mechanics overcomes these limitations.

## 34.5 Matter Waves

In classical physics, light is purely a wave phenomenon. Einstein's photons gave light a particle-like quality as well. In 1923, ten years after Bohr's atomic theory, a French prince named Louis de Broglie (pronounced "de Broy") set forth a remarkable hypothesis in his doctoral thesis. If light has both wave-like and particle-like properties, he reasoned, why shouldn't matter also exhibit both properties?

We saw in Chapter 29 that light with energy $E$ also carries momentum $p = E/c$. Combined with Equation 34.6, that means a photon of light with frequency $f$ has momentum $p = hf/c$. Since $f\lambda = c$, the photon's momentum and wavelength are therefore related by

$$\lambda = \frac{h}{p} \quad \text{(de Broglie wavelength)} \tag{34.14}$$

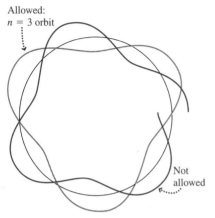

Allowed:
$n = 3$ orbit

Not allowed

**FIGURE 34.12** The allowed electron orbits in the Bohr atom are those that can fit an integral number of de Broglie wavelengths around the circular Bohr orbit.

De Broglie proposed that this same relation should hold for particles of matter; at nonrelativistic speed, for example, an electron should have associated with it a **de Broglie wavelength** given by $h/mv$.

De Broglie used his matter-wave hypothesis to explain why atomic electron orbits are quantized. He proposed that Bohr's allowed orbits were those in which standing waves could exist (Fig. 34.12), in much the same way that a violin string can support only certain frequencies of standing waves. Suppose that $n$ full wavelengths of a de Broglie electron wave fit around the circumference of the electron's circular orbit. Then we must have $n\lambda = n(h/p) = n(h/mv) = 2\pi r$, with $r$ the orbit radius. Multiplying both sides by $mv/2\pi$ then gives $mvr = nh/2\pi = n\hbar$, which is Bohr's quantization condition. Thus, de Broglie's hypothesis provides a natural explanation for the quantization of atomic energy levels.

If matter has wave-like properties, why don't we notice them? As in optics, wave effects become important only in systems whose size is comparable to or smaller than a wavelength. For macroscopic objects, and even for electrons moving at substantial speed in, say, a TV tube, the electron wavelength is so short that wave properties aren't noticeable. But when an electron is confined to small spaces—like the lower-energy orbits of the Bohr atom—its wave properties are dominant.

---

**EXAMPLE 34.4** **The de Broglie Wavelength: Electrons and Baseballs**

Find the de Broglie wavelength of (a) a 150-g baseball pitched at 45 m/s and (b) an electron moving at 1 Mm/s. Compare your results with the sizes of home plate and an atom, respectively.

**INTERPRET** This is a straightforward question about the de Broglie wavelengths of two particles.

**DEVELOP** Equation 34.14, $\lambda = h/p$, gives the de Broglie wavelength. We'll use $p = mv$ for the momentum; then Equation 34.14 becomes $\lambda = h/mv$.

**EVALUATE** (a) For the baseball with $m = 0.15$ kg and $v = 45$ m/s, we get $\lambda_{\text{baseball}} = h/mv = 9.8 \times 10^{-35}$ m. (b) For the electron, $\lambda = h/mv$ gives $\lambda_{\text{electron}} = 0.73$ nm. Recall that $h = 6.63 \times 10^{-34}$ J·s.

**ASSESS** The baseball's wavelength is about 34 orders of magnitude smaller than home plate, showing that wave effects involving the baseball are totally negligible. But the electron's 0.73-nm wavelength is several times the size of a ground-state hydrogen atom, so we expect wave effects to dominate this electron's interactions with atoms. ∎

---

**APPLICATION** **The Electron Microscope**

In Chapter 32 we found that light can't sharply image objects whose size is on the order of the wavelength or smaller—a factor that limits the resolving power of conventional microscopes. But Equation 34.14 shows that we can control the wavelength of electrons by adjusting their speed—and therefore we can achieve electron wavelengths much shorter than that of light. The **electron microscope** exploits this effect, providing resolutions down to about 1 nm and magnifications of $10^6$.

Electron microscopes accelerate electron beams to energies of 50–100 keV, with corresponding wavelengths of about 0.005 nm. Magnetic fields act as focusing lenses in the microscope's "optical" system, forming an image of whatever object is placed in the beam path. An electronic detector reads the image, which is then displayed on a screen.

Electron microscopes are indispensable tools in biology, chemistry, and materials science. A related device, the scanning electron microscope, produces dramatic three-dimensional images at magnifications of $10$–$10^5$, as shown in the photo of an ant carrying a microelectronic chip.

## Electron Diffraction and Matter-Wave Interference

In 1927 the American physicists Clinton Davisson and Lester Germer gave a convincing verification of de Broglie's matter–wave hypothesis. Davisson and Germer were studying the interaction of an electron beam with a nickel crystal, and they noticed regular intensity peaks reminiscent of X-ray diffraction. Shortly afterward, the Scottish physicist George Thomson observed electron diffraction directly, further evidence of the electron's wave nature (Fig. 34.13). Thomson was the son of J. J. Thomson, who had discovered the electron as a particle in 1897. Together their work captured the electron's wave–particle duality. Today, experiments with entire atoms and even larger clusters of matter exhibit wave interference—as shown in this chapter's opening photo.

## 34.6 The Uncertainty Principle

In classical physics it's possible, in principle, to know the exact position and velocity of a particle and therefore to predict with certainty its future behavior. But not so in quantum physics! In 1927 the German physicist Werner Heisenberg presented his famous **uncertainty principle**, which states that some pairs of quantities cannot be measured simultaneously with arbitrary precision. Position and momentum constitute one such pair; if we measure a particle's position to within an uncertainty $\Delta x$, then we can't simultaneously determine its momentum to an accuracy better than $\Delta p$, where

$$\Delta x\, \Delta p \geq \hbar \qquad \text{(uncertainty principle)} \qquad (34.15)$$

**FIGURE 34.13** Diffraction produced by passing an electron beam through a circular aperture shows that electrons have a wave-like character.

Why this limitation? The fundamental reason is quantization. To measure some property of a system requires interacting with the system—for example, shining light on it. Interaction involves energy, and the interaction energy disturbs the system slightly. As a result, values inferred from the measurement are no longer quite right. In classical physics the energy can be arbitrarily small, resulting in a negligible disturbance. But in quantum theory the minimum energy is a single quantum, like a photon of light, and thus the disturbance can't be arbitrarily small.

So why not use lower-frequency light, whose photon energy $hf$ is lower? Because lower frequency means longer wavelength and, as we found in Chapter 32, diffraction effects limit resolution at longer wavelengths. Heisenberg summarized this dilemma with the "thought experiment" illustrated in Fig. 34.14, which uses a single photon to observe an electron. A short-wavelength photon allows precise localization of the electron (Fig. 34.14a). But short wavelength means high frequency and thus high photon energy. The high-energy photon imparts considerable momentum to the electron, and thus the very act that fixes the electron's position degrades our knowledge of its momentum. We can decrease this disturbance with a lower-energy, longer-wavelength photon (Fig. 34.14b). But now diffraction precludes precisely determining the electron's position. So we can measure the electron's position accurately, at the expense of knowing its momentum. Or we can measure its momentum accurately, but then we can't know its position. With a photon of intermediate wavelength we could measure both quantities, but neither precisely. The uncertainty principle, Equation 34.15, describes this trade-off.

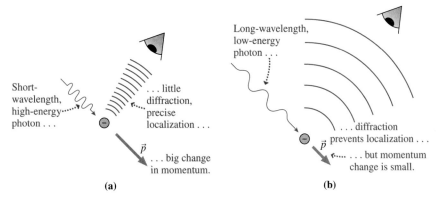

**FIGURE 34.14** Heisenberg's "quantum microscope" thought experiment.

The uncertainty principle is intimately connected with de Broglie's wave hypothesis. Suppose we pass an electron beam through a slit, as shown in Fig. 34.15. Then we know the electrons' vertical position to within the slit width. If the slit is much wider than the electrons' de Broglie wavelength, there's minimal diffraction. The electrons follow straight lines and we're quite sure of their vertical momentum, in this case zero (Fig. 34.15a). But with a wide slit we don't know much about the electrons' vertical position. Making the slit smaller gives a more precise position, but then diffraction spreads the beam, increasing the uncertainty in the electrons' vertical momentum (Fig. 34.15b). So the wave nature of matter ultimately imposes a trade-off: The more we know of a particle's position, the less we know of its momentum, and vice versa.

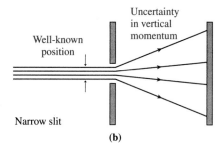

**FIGURE 34.15** The wave nature of matter is intimately related to the uncertainty principle, as shown here for beams of electrons passing through wide and narrow slits. In (b), diffraction introduces the uncertainty in vertical momentum.

---

**EXAMPLE 34.5** **The Uncertainty Principle: Microelectronics**

A beam of aluminum atoms is used to dope a semiconductor chip to set its electrical properties. If the atoms' velocity is known to within 0.2 m/s, how accurately can they be positioned?

**INTERPRET** This is a question about simultaneously knowing the atoms' position and velocity—paired quantities governed by the uncertainty principle.

**DEVELOP** We're given the velocity uncertainty $\Delta v$, from which we'll find the momentum uncertainty $\Delta p = m \Delta v$. Then we can use the uncertainty principle, Equation 34.15, $\Delta x \Delta p \geq \hbar$, to find the uncertainty $\Delta x$ in position. We'll need aluminum's atomic weight from Appendix D to find the mass.

**EVALUATE** We have

$$\Delta p = m \Delta v = (26.98\,\text{u})(1.66 \times 10^{-27}\text{kg/u})(0.20\,\text{m/s}) = 9 \times 10^{-27}\text{kg·m/s}.$$

Then Equation 34.15 gives the position uncertainty:

$$\Delta x = \hbar/\Delta p = 12\,\text{nm},$$

where, again, $\hbar = h/2\pi$.

**ASSESS** Our 12-nm answer is about 100 atomic diameters and shows that the uncertainty principle constrains our ability to fabricate very small microelectronic structures. ∎

---

It sounds like the uncertainty principle only limits our knowledge. But in fact it proves useful in estimating the size and energies of atomic-scale systems, as the next example shows.

**EXAMPLE 34.6** The Uncertainty Principle:
Estimating Atomic and Nuclear Energies

Use the uncertainty principle to estimate the minimum energy possible for (a) an electron confined to a region of atomic dimensions, about 0.1 nm, and (b) a proton confined to a region of nuclear dimensions, about 1 fm.

**INTERPRET** We're given the uncertainty in position; that's the width of the region in which the particles are confined. The particles can't be at rest, or we'd know that their momentum was exactly zero—in violation of the uncertainty principle. So they must have a minimum momentum and therefore energy. The problem is asking us to find that energy.

**DEVELOP** We need to find the minimum momentum consistent with the uncertainty principle and then the corresponding energy. Suppose a particle has momentum of magnitude $p$, but we don't know its direction. It could be going one way, with momentum $p$, or the other way, with momentum $-p$. Then the momentum itself is uncertain by

$\Delta p = p - (-p) = 2p$. The uncertainty principle says $\Delta p \geq \hbar/\Delta x$, so there's a minimum magnitude for the momentum given by $p \geq \hbar/2 \, \Delta x$. Using $p = mv$ and $K = \frac{1}{2}mv^2$ gives $K = p^2/2m$, and therefore the uncertainty principle requires

$$K \geq \frac{1}{2m}\left(\frac{\hbar}{2 \, \Delta x}\right)^2$$

**EVALUATE** Evaluating this constraint for an electron with $\Delta x = 0.1$ nm and for a proton with $\Delta x = 1$ fm gives minimum energies of about 1 eV and 5 MeV, respectively.

**ASSESS** Energies in electronvolts are typical of atomic-scale systems, as we saw in Fig. 34.11. Our result shows that nuclear energies are some 5 million times greater—indicating the dramatic difference between chemical and nuclear energy sources. ∎

## Energy–Time Uncertainty

A second pair of variables that defy simultaneous measurement are the energy of a system and the time it remains at that energy. The uncertainty in the energy $\Delta E$ is related to the time $\Delta t$ through the inequality

$$\Delta E \, \Delta t \geq \hbar \tag{34.16}$$

One effect of energy–time uncertainty is to render atomic and nuclear energy levels inexact and therefore to broaden spectral lines. If an atom were forever in a fixed energy state, we could take infinitely long to measure its energy and therefore make $\Delta E$ arbitrarily small. But excited states of atoms have characteristic lifetimes (typically $\sim 10^{-8}$ s), which limit the measurement time and therefore set a minimum uncertainty in the energy level. Problems 66 and 79 explore energy–time uncertainty.

## Observers, Uncertainty, and Causality

The uncertainty principle moves the observer from a passive onlooker to an active participant in physical events. To observe is necessarily to disturb, and quantum theory is therefore concerned with the role of the observer and the process of measurement. The uncertainty principle is fundamentally a statement about what can and cannot be learned through measurement.

Position and momentum cannot be measured simultaneously with perfect accuracy. Surely, though, a particle has well-defined values of both, even though we can't know them? The answer seems to be no. The standard interpretation of quantum mechanics suggests that it makes no sense to talk about what can't be measured, and recent experiments have ruled out "hidden variables" that might be active at a lower level to guide the particle in a deterministic path. Its wave aspect makes a particle a "fuzzy" thing, and it really makes no sense to think of it as a tiny ball with definite momentum and position. For that reason it also makes no sense to think of the particle's future as being fully determined in the sense that Newton's laws determine the future path of, say, a baseball. We're left with uncertainty—or indeterminacy, as Heisenberg's word also translates—as a fundamental fact of our universe.

## 34.7 Complementarity

One of the most disturbing aspects of quantum theory is the wave–particle duality—the seeming contradiction that matter and light have both wave-like and particle-like properties. If this bothers you, you're in good company: Heisenberg himself expressed frustration in trying to understand the quantum world:

> I remember discussions with Bohr which went through many hours till very late at night and ended almost in despair; and when at the end of the discussion I went alone for a walk in the neighboring park I repeated to myself again and again the question: Can nature possibly be as absurd as it seems to us in these atomic experiments?*

Bohr dealt with the wave–particle duality through his **principle of complementarity**. The wave and particle pictures, he said, are complementary aspects of the same reality. If we do an experiment to measure a wave-like property—for example, the diffraction of electrons—then we find wave properties but not particle properties. If we do an experiment to measure a particle-like property—for example, localizing an electron—then we won't find wave properties. The two measurements require different experiments, and we can't perform both simultaneously on the same entity. So we'll never catch wave and particle in an outright contradiction, and the answer to the question "Which is it, wave or particle?" has to be that it's both, and which you find depends on what experiment you choose to perform.

Bohr articulated a second principle that helps reconcile the seeming contradiction between classical and quantum physics. His **correspondence principle** states that the predictions of classical and quantum physics should agree in situations where the size of individual quanta is negligible. Taking $h \rightarrow 0$ in Planck's law, for example, gives the classical Rayleigh–Jeans law (see Problem 68). Or, for large $n$, the energies of adjacent atomic states in the Bohr model become so close that the levels appear essentially as a continuum of allowed energies—as expected in classical physics. Or consider a 1000-W radio beam; the photon energy $hf$ is so low that the beam contains an enormous number of photons, and we can consider the energy distributed essentially continuously over the beam. But in a 1000-W X-ray beam, the photon energy is much higher and the number of photons correspondingly fewer; it's therefore difficult to avoid the fundamental fact of energy quantization. Visible light lies somewhere in between; we can often treat its energy as being continuously distributed, except when it interacts with systems as small as individual atoms.

*Werner Heisenberg, *Physics and Philosophy: The Revolution in Modern Science* (New York: Harper & Brothers, 1962).

## Big Picture

The big ideas here are at the heart of quantum physics—a radically different view of reality at the atomic scale. **Quantization** means that some physical quantities—often including energy—come only in discrete values. Another fundamental aspect of quantum reality is the **wave–particle duality**, wherein light and matter exhibit both wave-like and particle-like aspects. Bohr's **complementarity principle** precludes these ever being in direct conflict. Finally, quantization and wave–particle duality lead to the **uncertainty principle**, which makes it impossible to measure simultaneously and with arbitrary precision a particle's position and momentum.

## Key Concepts and Equations

**Planck's constant,**
$$h = 6.63 \times 10^{-34} \text{ J} \cdot \text{s}$$
sets the fundamental scale of quantization.

It's also expressed
$$\hbar = h/2\pi$$

The energy of electromagnetic radiation with frequency $f$ is quantized in **photons** with energy
$$E = hf$$

Electron energies in the **Bohr model** of hydrogen are quantized according to
$$E = -\frac{ke^2}{2a_0}\left(\frac{1}{n^2}\right) \simeq -\frac{13.6 \text{ eV}}{n^2}$$
where $n$ is an integer and $a_0 = 0.0529$ nm is the **Bohr radius**.

The **de Broglie wavelength** of a particle with momentum $p$ is
$$\lambda = \frac{h}{p}$$

The **uncertainty principle** relates uncertainties in position and momentum:
$$\Delta x \, \Delta p \geq \hbar$$

## Cases and Uses

A correct description of **blackbody radiation** requires Planck's quantization hypothesis. The peak radiance—energy radiated per unit wavelength interval—from a blackbody at temperature $T$ occurs at a wavelength given by $\lambda T = 2.898$ mm·K.

The **photoelectric effect** involves the ejection of electrons from a metal surface illuminated with electromagnetic waves. Explanation of the effect led Einstein to propose photons as the quanta of electromagnetic-wave energy.

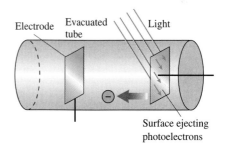

The **Compton effect** shows that photons interact with free electrons exactly like colliding particles, losing energy and emerging with longer wavelength.

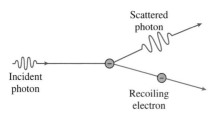

Quantization of atomic energy levels leads directly to **atomic spectra**. In the Bohr model of hydrogen, the spectral line produced in a transition from the $n_1$ to the $n_2$ energy level is given by
$$\frac{1}{\lambda} = R_H\left(\frac{1}{n_2^2} - \frac{1}{n_1^2}\right)$$

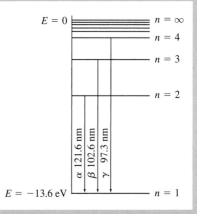

**625**

# For Thought and Discussion

1. Why does classical physics predict that atoms should collapse?
2. Looking at the night sky, you see one star that appears red, another yellow, and another blue. Compare their temperatures.
3. Imagine an atom that, unlike hydrogen, had only three energy levels. If these levels were evenly spaced, how many spectral lines would result? How would their wavelengths compare?
4. What colors of visible light have the highest energy photons?
5. Why is the immediate ejection of electrons in the photoelectric effect surprising from a classical viewpoint?
6. Suppose the Compton effect were significant at radio wavelengths. What problems might this present for radio and TV broadcasting?
7. How are wave–particle duality and the uncertainty principle related?
8. How many spectral lines are in the entire Balmer series?
9. Why are the lines of the Lyman series in the ultraviolet while some Balmer lines are visible?
10. Why does the photoelectric effect suggest that light has particle-like properties?
11. Energy–time uncertainty limits the precision with which we can know the mass of unstable particles (those that decay after a finite time). Why?
12. If you measure a particle's position with perfect accuracy, what do you know about its momentum?
13. How might our everyday experience be different if Planck's constant had the value 1 J·s?
14. Why is the energy given by Equations 34.12 negative?

# Exercises and Problems

## Exercises

### Section 34.2 Blackbody Radiation

15. If you increase the temperature of a blackbody by a factor of 2, by what factor does its radiated power increase?
16. The surface temperature of the star Rigel is $10^4$ K. (a) Find the power radiated per square meter of Rigel's surface. Find (b) $\lambda_{peak}$ and (c) $\lambda_{median}$ for Rigel.
17. Find $\lambda_{peak}$ and $\lambda_{median}$ for Earth treated as a 288-K blackbody.
18. Spacecraft instruments measure the radiation from an asteroid, measuring power in bins of fixed wavelength. They find the peak power at 40 $\mu$m. Assuming the asteroid is a blackbody, find its surface temperature.
19. The Sun approximates a blackbody at 5800 K. (a) Find the wavelength of peak radiance on the per-unit-wavelength basis implicit in Equation 34.2a. (b) Find the median wavelength, $\lambda_{median}$, below which half the radiation is emitted (Equation 34.2b). Identify the spectral region of each.

### Section 34.3 Photons

20. Find the energy in electronvolts of (a) a 1.0-MHz radio photon, (b) a $5.0 \times 10^{14}$-Hz optical photon, and (c) a $3.0 \times 10^{18}$-Hz X-ray photon.
21. What is the wavelength of a 6.5-eV photon? In what spectral region is this?
22. A microwave oven uses electromagnetic radiation at 2.4 GHz. (a) What is the energy of each microwave photon? (b) At what rate does a 900-W oven produce photons?
23. A red laser at 650 nm and a blue laser at 450 nm emit photons at the same rate. How do their total power outputs compare?
24. What is the maximum work function for a surface to emit electrons when illuminated with 900-nm infrared light?

### Section 34.4 Atomic Spectra and the Bohr Atom

25. Calculate the wavelengths of the first three lines in the Lyman series for hydrogen.
26. Which spectral line of the hydrogen Paschen series $(n_2 = 3)$ has wavelength 1282 nm?
27. What is the maximum wavelength of light that can ionize hydrogen in its ground state? In what spectral region is this?
28. At what energy level does the Bohr hydrogen atom have diameter 5.18 nm?

### Section 34.5 Matter Waves

29. Find the de Broglie wavelength of (a) Earth, in its 30-km/s orbital motion, and (b) an electron moving at 10 km/s.
30. How slowly must an electron be moving for its de Broglie wavelength to equal 1 mm?
31. A proton and an electron have the same de Broglie wavelength. How do their speeds compare, assuming both are much less than that of light?
32. Find the de Broglie wavelength of electrons with kinetic energies of (a) 10 eV, (b) 1.0 keV, and (c) 10 keV.

### Section 34.6 The Uncertainty Principle

33. A proton is confined to a space 1 fm wide (about the size of the atomic nucleus). What is the minimum uncertainty in its velocity?
34. Is it possible to measure an electron's velocity to an accuracy of $\pm 1.0$ m/s while simultaneously finding its position to an accuracy of $\pm 1.0$ $\mu$m? What about a proton?
35. A proton has velocity $v = (1500 \pm 0.25)$ m/s. What is the uncertainty in its position?
36. An electron is moving in the $+x$ direction with speed measured at $5.0 \times 10^7$ m/s, to an accuracy of $\pm 10\%$. What is the minimum uncertainty in its position?
37. Find the minimum energy for a neutron in a uranium nucleus whose diameter is 15 fm.

## Problems

38. Find the power per unit area emitted by a 3000-K incandescent lamp filament in the wavelength interval from 500 nm to 502 nm.
39. Treating the Sun as a 5.8-kK blackbody, how does its ultraviolet radiance at 200 nm compare with its visible radiance at its peak wavelength of 500 nm?
40. For a 2.0-kK blackbody, by what percentage is the Rayleigh–Jeans law in error at wavelengths of (a) 1.0 mm, (b) 10 $\mu$m, and (c) 1.0 $\mu$m?
41. The radiance of a blackbody peaks at 660 nm. (a) What is its temperature? (b) How does the radiance at 400 nm compare with that at 700 nm?
42. (a) Find the Compton wavelength for the proton. (b) Find the energy in electronvolts of a gamma ray whose wavelength equals the proton's Compton wavelength.

43. Find the rate of photon production by (a) a radio antenna broadcasting 1.0 kW at 89.5 MHz, (b) a laser producing 1.0 mW of 633-nm light, and (c) an X-ray machine producing 0.10-nm X rays with a total power of 2.5 kW.

44. Electrons in a photoelectric experiment emerge from an aluminum surface with maximum kinetic energy of 1.3 eV. What is the wavelength of the illuminating radiation?

45. (a) Find the cutoff frequency for the photoelectric effect in copper. (b) Find the maximum energy of the ejected electrons if the copper is illuminated with light of frequency $1.8 \times 10^{15}$ Hz.

46. The stopping potential in a photoelectric experiment is 1.8 V when the illuminating radiation has wavelength 365 nm. (a) What is the work function of the emitting surface? (b) What would be the stopping potential for 280-nm radiation?

47. Which materials in Table 39.1 exhibit the photoelectric effect *only* for wavelengths shorter than 275 nm?

48. Find the initial wavelength of a photon that loses half its energy when it Compton-scatters from an electron and emerges at 90° to its initial direction of motion.

49. When light shines on a potassium surface, the maximum speed with which electrons leave the surface is $4.2 \times 10^5$ m/s. Find the wavelength of the light.

50. The maximum electron energy in a photoelectric experiment is 2.8 eV. When the wavelength of the illuminating radiation is increased by 50%, the maximum electron energy drops to 1.1 eV. Find (a) the work function of the emitting surface and (b) the original wavelength.

51. A 150-pm X-ray photon undergoes Compton scattering with an electron, and emerges at 135° to its original direction of motion. Find (a) the wavelength of the scattered photon and (b) the kinetic energy of the electron.

52. An electron is initially at rest. What will its kinetic energy be after a 0.10-nm X-ray photon scatters from it at 90° to its original direction of motion?

53. A photocathode ejects electrons with maximum energy 0.85 eV when illuminated with 430-nm blue light. Will it eject electrons when illuminated with 633-nm red light? If so, what will be the maximum electron energy?

54. (a) Find the energy of the highest energy photon that can be emitted as the electron jumps between two adjacent allowed energy levels in the Bohr hydrogen atom. (b) Which energy levels are involved?

55. Find (a) the wavelength and (b) the energy in electronvolts of the photon emitted as a Rydberg hydrogen atom drops from the $n = 180$ level to the $n = 179$ level.

56. The wavelengths of a spectral line series tend to a limit as $n_1 \to \infty$. Evaluate the series limit for (a) the Lyman series and (b) the Balmer series in hydrogen.

57. A Rydberg hydrogen atom makes a downward transition to the $n = 225$ state. If the photon emitted has energy 9.32 $\mu$eV, what was the original state?

58. A hydrogen atom is in its ground state when its electron absorbs 48 eV in an interaction with a photon. What is the energy of the resulting free electron?

59. How much energy does it take to ionize a hydrogen atom in its first excited state?

60. Ultraviolet light with wavelength 75 nm shines on a gas of hydrogen atoms in their ground states, ionizing some of the atoms. What is the energy of the electrons freed in this process?

61. Helium with one of its two electrons removed acts very much like hydrogen, and the Bohr model successfully describes it. Find (a) the radius of the ground-state electron orbit and (b) the photon

energy emitted in a transition from the $n = 2$ to the $n = 1$ state in this singly ionized helium. *Hint:* The nuclear charge is $2e$.

62. Through what potential difference should you accelerate an electron from rest so its de Broglie wavelength will be the size of a hydrogen atom, about 0.1 nm?

63. What is the minimum electron speed that would make an electron microscope superior to an optical microscope using 450-nm light?

64. Give the ratio of de Broglie wavelength to diameter for (a) a 100-g ball, 10 cm in diameter, moving at 5 m/s; (b) a 4-u alpha particle, 3.4 fm in diameter, moving at 20 Mm/s; and (c) a 238-u uranium nucleus, 13 fm in diameter, moving at 1.4 km/s. (d) For which will wave properties be negligible, and for which will they be completely dominant?

65. An electron is trapped in a "quantum well" 20 nm wide. What is the minimum speed it could have?

66. Typically, an atom remains in an excited state for about $10^{-8}$ s before it drops to a lower energy state, emitting a photon in the process. What is the uncertainty in the energy of this transition?

67. An electron is moving at $10^6$ m/s, and you wish to measure its energy to an accuracy of $\pm 0.01\%$. What is the minimum time necessary for the measurement?

68. Use the series expansion for $e^x$ (Appendix A) to show that Planck's law (Equation 34.3) reduces to the Rayleigh–Jeans law (Equation 34.5) when $\lambda \gg hc/kT$.

69. A photon's wavelength is equal to the Compton wavelength of a particle with mass $m$. Show that the photon's energy is equal to the particle's rest energy.

70. Show that the frequency range occupied by the hydrogen spectral line series involving transitions ending at the $n$th level is given by $\Delta f = cR_H/(n + 1)^2$.

71. A photon undergoes a 90° Compton scattering off a stationary electron, and the electron emerges with *total* energy $\gamma m_e c^2$, where $\gamma$ is the relativistic factor introduced in Chapter 33. Find an expression for the initial photon energy.

72. Show that Wien's law (Equation 34.2a) follows from Planck's law (Equation 34.3). *Hint:* Differentiate Planck's law with respect to wavelength.

73. Consider an elastic collision between a photon initially with wavelength $\lambda_0$ moving in the $x$ direction and a stationary electron, as depicted in Figure 34.9b. Use relativistic expressions for energy and momentum from Chapter 33 to show that conservation of energy and momentum result in the equations $hc/\lambda_0 + mc^2 = hc/\lambda + \gamma mc^2$, $h/\lambda_0 = (h/\lambda)\cos\theta + \gamma mu\cos\phi$, and $0 = (h/\lambda)\sin\theta - \gamma mu\sin\phi$, where $\lambda$ is the post-collision photon wavelength and the angles $\theta$ and $\phi$ are as shown in Fig. 34.9b. Solve these equations to find the Compton shift (Equation 34.8).

74. What would the constant in Equation 34.2a be if blackbody radiance were defined for fixed intervals of frequency rather than wavelength? (*Hint:* First use $\lambda = c/f$ to express the radiance as $R(f, T)$, then differentiate to find the maximum, and then solve the resulting relation numerically. Express your answer in an equation similar to Equations 34.2.

75. Integrate Equation 34.3 over all wavelengths to get the total power radiated per unit area. Show that your result is equivalent to Equation 34.1, with the Stefan-Boltzmann constant given by $\sigma = 2\pi^5 k^4/15c^2 h^3$. *Hint:* Use $hc/\lambda kT$ as the integration variable and consult tables or software to evaluate the integral.

76. Perform a numerical integration of Equation 34.3 to the wavelength given by Equation 34.2b. Divide by the result of Problem 75, and thus verify that Equation 34.2b gives the wavelength above and below which a blackbody radiates half its energy.

77. Use the momentum conservation equations in Problem 73 and Equation 34.8 for the Compton shift to show that the electron's recoil angle in Fig. 34.9b is given by $\tan\phi = \lambda_0 \sin\theta/(\lambda - \lambda_0 \cos\theta)$.

78. Show that in the Bohr model, the frequency of a photon emitted in a transition between levels $n+1$ and $n$, in the limit of large $n$, is equal to the orbital frequency of the electron. (This is an example of Bohr's correspondence principle.)

79. You're reading about early observations of cosmic rays and are astonished to learn that cosmic-ray particles reacted for a mere 12 fs with the detector. Having just learned about the uncertainty principle, you realize this interaction time establishes a limit to the precision of a measurement of cosmic-ray energy. What is the minimum uncertainty in that energy?

80. It's the 24th century and curators at the Starfleet Museum of Ancient Technology are puzzling over the device once known as a TV tube. They ask you to determine the precision with which electrons in this device would have hit its screen; this will tell whether the device might have been used to produce focused images. The curators have determined that an electron beam was accelerated at the back of the tube and headed toward the center of the screen with horizontal velocity $2.2 \times 10^7$ m/s. As the electrons left the acceleration region, their vertical position was accurate to $\pm 45$ nm. Although the Heisenberg Compensators work well on the Starship Enterprise, the uncertainty principle was alive and well back in the TV era. What was the minimum angular spread in the TV tube's electron beam?

# Answers to Chapter Questions

## Answer to Chapter Opening Question

That matter, like light, behaves sometimes as waves.

## Answers to GOT IT? Questions

34.1 A emits 16 times as much radiation, with a peak wavelength half that of B.

34.2 (a) No, the slope remains $h/e$; (b) Yes; the horizontal intercept is the cutoff frequency, which depends on the work function.

34.3 Much less because $\lambda_C$ is inversely proportional to mass.

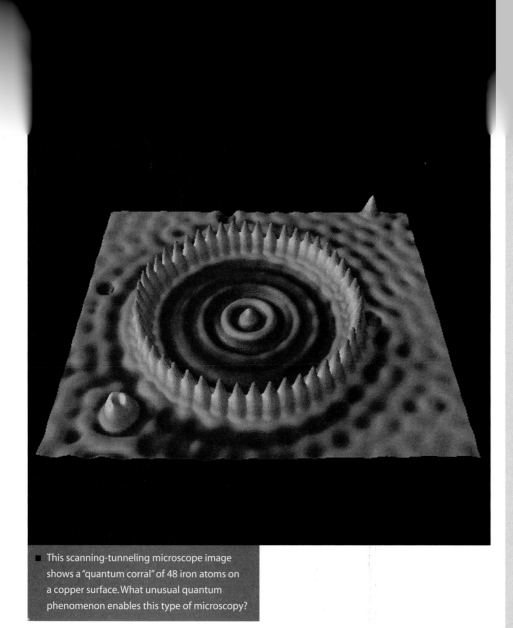

■ This scanning-tunneling microscope image shows a "quantum corral" of 48 iron atoms on a copper surface. What unusual quantum phenomenon enables this type of microscopy?

The ideas developed in the preceding chapter are at the core of the **old quantum theory**. The old quantum theory introduced the basic concepts of quantum physics and was successful in explaining a number of quantum phenomena—for example, blackbody radiation, the photoelectric effect, and the hydrogen spectrum. On the other hand, it couldn't treat even the simplest multielectron atoms, and it left some subtle spectral features unexplained. Furthermore, the old quantum theory was a hodgepodge of separate but loosely related propositions, each developed to explain a particular phenomenon; it lacked coherence and clear guiding principles.

Is there a more coherent theory that predicts the behavior of systems at the atomic and subatomic scales, and that offers a satisfying description of how such systems really work? The answer is at once an emphatic yes and a disappointing no. Yes, because **quantum mechanics**, developed in the 1920s, predicts with remarkable precision the observed properties of atomic systems, including their energies, the wavelengths of spectral lines, and the lifetimes of excited atoms. No, because quantum mechanics doesn't give a satisfying visual *picture* of the atomic and subatomic world. The uncertainty principle and wave–particle duality are essential aspects of quantum mechanics. Any picture we formulate of electrons or photons whizzing around like miniature balls with precise positions and momenta is inappropriate at the atomic scale. But quantum mechanics does provide a self-consistent description that lets us explore and predict the behavior of atoms, the organization of chemical elements, the physics of semiconductors and superconductors, the extraordinary behavior of matter at low temperature, the formation of white dwarf stars, the operation of lasers, and a host of other phenomena for which classical physics is at best inaccurate and at worst totally inappropriate. In this chapter we explore the mathematical structure and physical interpretation of quantum mechanics. In Chapters 36 and 37 we'll apply quantum mechanics first to the atom and then to more complex systems that involve the quantum-mechanical interaction of many atoms.

## 35.1 Particles, Waves, and Probability

### Photons and Light Waves

In Maxwell's electromagnetic theory we had a seemingly complete description of light as an electromagnetic wave. Now we find, through the photoelectric and Compton effects, that light sometimes manifests itself as particles. What's the connection between the wave and particle descriptions?

In a photoelectric experiment, the number of ejected electrons depends on the light's intensity. Since an electron is ejected when it absorbs a photon, we conclude that the number of photons in the incident light is proportional to light intensity. Now, the intensity of an electromagnetic wave depends on the square of the electric or magnetic field (Equations 29.20b, c). The fields, in turn, obey Maxwell's equations, so one aspect of a photoelectric experiment—namely, the number of electrons ejected—relates to Maxwell's description of light as an electromagnetic wave.

We can quantify the relation between wave intensity and number of photons, but only in a statistical sense. The ejection of individual electrons in a photoelectric experiment is quite random. The uncertainty principle prevents us from following a photon trajectory and predicting when and where an electron will be ejected. All we can say is that electrons are more likely to be ejected where the wave intensity is greater. Specifically, the probability that an electron will be ejected is directly proportional to the intensity of the incident electromagnetic waves—that is, to the square of the wave fields. More generally, the probability of finding a photon in a beam of electromagnetic waves is directly proportional to the wave intensity (Fig. 35.1).

In this quantum-mechanical description, the fields still evolve according to Maxwell's equations. For example, the fields of an electromagnetic wave undergoing double-slit interference develop regions of maximum and minimum wave intensity—the bright and dark bands of the interference pattern. But the wave fields determine only the *probability* that individual photons will be detected in the interference pattern. That's why a very short exposure or a low-intensity beam results not in a weak version of the interference pattern but in a seemingly arbitrary pattern. Only with large numbers of photons does the statistical pattern emerge (Fig. 35.2).

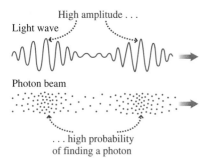

**FIGURE 35.1** The probability of finding a photon is directly proportional to the intensity of the electromagnetic wave. The figure is only suggestive because we can't depict photons as localized particles.

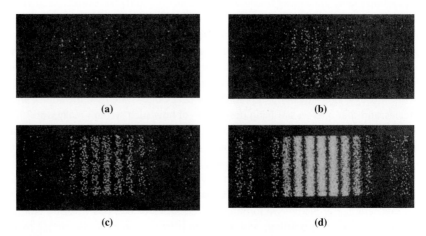

**FIGURE 35.2** Development of a two-slit interference pattern from random photon events: (a) 50 photons, (b) 250 photons, (c) 1000 photons, (d) 10,000 photons.

In quantum mechanics, then, the relation between the wave and particle aspects of light is this: As long as we don't try to detect the light, it propagates as a wave governed by Maxwell's equations. But when we detect the light, we do so through interactions involving individual photons. Those interactions are random events whose probability depends on the wave intensity—that is, on the square of the wave fields.

### Electrons and Matter Waves

In Chapter 34 we introduced de Broglie's remarkable hypothesis that matter, as well as light, exhibits both wave and particle properties. The wave–particle duality puts matter and light on essentially the same footing, and the statistical interpretation is the same for each. Figure 35.3 shows a beam of particles and its associated de Broglie matter wave. Just as the probability of finding a photon is proportional to the wave intensity—that is, the *square* of the electromagnetic-field amplitude—so the probability of finding a particle is directly proportional to the square of the matter-wave amplitude. And as with light, the particle nature of matter manifests itself only when we try to detect a particle; leave it alone, and the particle's behavior is governed by its wave nature.

Maxwell's equations determine the behavior of light waves, but what equation describes matter waves? In 1926 the Austrian physicist Erwin Schrödinger answered this question with his **Schrödinger wave equation**. In the same year Schrödinger showed that his wave theory was equivalent to a matrix-based theory that Heisenberg, Max Born, and Pascual Jordan had formulated in 1925. Heisenberg received the 1932 Nobel Prize in physics, and Schrödinger shared the 1933 Nobel Prize with Paul Dirac for their contributions to quantum theory.

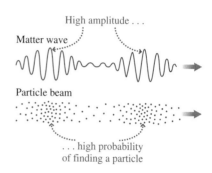

**FIGURE 35.3** A beam of particles and its associated matter wave.

## 35.2 The Schrödinger Equation

The Schrödinger equation describes matter waves in terms of a **wave function**, $\psi$ (Greek psi), which depends on both space and time. The solution of differential equations in two variables is beyond the mathematical level of this text, so here we'll consider only spatial variations, and for now we'll further restrict ourselves to one dimension.

We can understand the Schrödinger equation by considering a sinusoidal wave of the form $\psi(x) = A \sin kx$, where, as usual in describing waves, $k = 2\pi/\lambda$, with $\lambda$ the wavelength. Differentiating this expression twice gives

$$\frac{d^2\psi(x)}{dx^2} = -Ak^2 \sin kx = -k^2\psi(x)$$

But $k = 2\pi/\lambda$ and, for matter waves, de Broglie's hypothesis gives $\lambda = h/p$, with $h$ Planck's constant and $p$ the particle's momentum. Thus we can write $k$ in terms of momentum as $k = 2\pi p/h = p/\hbar$. Now in classical physics a particle of mass $m$ has kinetic energy $K$ and momentum $p$ related by $K = p^2/2m$. Furthermore, kinetic energy is the difference between the total energy $E$ and the potential energy $U$; thus $E - U = p^2/2m$. Putting this all together, we can write the quantity $k^2$ in our differentiated wave expression as

$$k^2 = \frac{p^2}{\hbar^2} = \frac{2m(E - U)}{\hbar^2}$$

Making this substitution and doing a little algebra then give

$$-\frac{\hbar^2}{2m}\frac{d^2\psi(x)}{dx^2} + U(x)\psi(x) = E\psi(x) \qquad \left(\begin{array}{c}\text{time-independent} \\ \text{Schrödinger equation}\end{array}\right) \qquad (35.1)$$

This is the **time-independent Schrödinger equation**, giving the spatial variation of matter waves in one dimension. A solution of the full time-dependent equation consists of a solution to Equation 35.1 multiplied by a sinusoidal oscillation with frequency $f = E/h$, where $E$ is the particle energy. We developed the time-independent Schrödinger equation by merging de Broglie's matter-wave hypothesis $\lambda = h/p$ with the Newtonian relation $K = p^2/2m$; for that reason, we expect the equation to hold only for nonrelativistic particles.

The Schrödinger equation provides a description of physical reality in remarkable agreement with experiments. As we'll see, Schrödinger's equation goes a long way toward explaining the structure of atoms, their chemical properties, and indeed the entire science of chemistry. Furthermore, the Schrödinger description agrees with Newtonian mechanics for macroscopic systems where quantum effects are small.

## The Meaning of $\psi$

What's the meaning of the wave function $\psi$? That's a deep question that physicists and philosophers continue to debate. In the standard interpretation, $\psi$ is not an observable quantity. It manifests itself only in the statistical distributions of particle detections. More specifically, the probability per unit volume—also called the **probability density**—that we'll find a particle is given by $\psi^2$. For a particle confined to one dimension, the probability density becomes the probability per unit length, and we interpret $\psi$ to mean that the probability $P(x)$ of finding the particle in a small interval $dx$ at position $x$ is

$$P(x) = \psi^2(x)\, dx \qquad \text{(probability and the wave function)} \qquad (35.2)$$

We can interpret Equation 35.2 in two ways. At face value, it gives the probability that a single experiment, with a detector at position $x$ set up to find particles in an interval of width $dx$, will detect the particle (Fig. 35.4). Or, if we do many such experiments, the equation gives the fraction of the experiments in which we'll find a particle in our detector.

But *what is $\psi$*? How can it be unobservable yet govern the behavior of matter? There can't be a direct causal link between the wave function and individual particles, since $\psi$ determines only the *probability* that a particle will behave in a certain way. Think about this! In quantum mechanics the outcome of an experiment isn't fully determined. The Schrödinger equation describes only the probability of a given outcome. The quantum world is so different, according to the standard interpretation, that our macroscopic

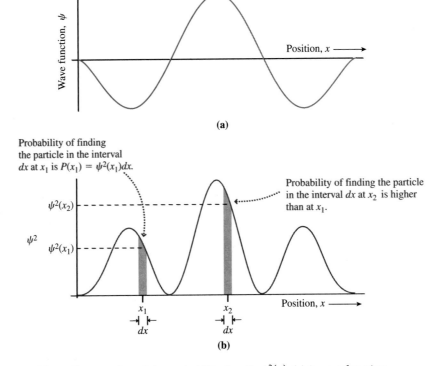

**FIGURE 35.4** The meaning of the probability density $\psi^2(x)$. (a) A wave function and (b) its square, which gives the probability density.

language, concepts, and pictorial models are simply inadequate. In particular, macroscopic causality gives way to microscopic indeterminancy in which quantum events are truly random; physical laws govern only the statistical pattern of events.

The philosophical implications of quantum mechanics have been a matter of debate since the theory was formulated. A central theme in this debate is the possibility of "hidden variables," physical quantities that might be hidden from us by the uncertainty principle but that might nevertheless govern the microscopic world in a fully deterministic way. Experiments of the early 1980s placed severe restrictions on such hidden-variable theories, but fascinating discussions on the interpretation of quantum mechanics continue. Here, however, we turn to the Schrödinger equation to see how it's used in analyzing quantum-mechanical systems.

## Normalization and Other Constraints on the Wave Function

In one dimension, the quantity $\psi^2\,dx$ represents the probability of finding the particle in the interval $dx$. But the particle *must* be *somewhere*. Therefore, if we sum the probabilities of finding the particle in all such intervals $dx$, the result must be 1; there must be a 100% chance that we'll find the particle somewhere. Since the probability density may vary with position, that sum becomes an integral:

$$\int_{-\infty}^{+\infty} \psi^2\,dx = 1 \qquad (\text{normalization condition}) \qquad (35.3)$$

Once we have a solution $\psi(x)$ to the Schrödinger equation 35.1, this **normalization condition** sets the overall amplitude of the function $\psi$.

The Schrödinger equation contains the second derivative of $\psi$. In order that this term be well defined, both $\psi$ itself and its first derivative must be continuous. (An exception to the continuity condition on $d\psi/dx$—possible only in unrealistic example situations—occurs if the potential energy $U$ becomes infinite.)

# 35.3 Particles and Potentials

## The Infinite Square Well

We first solve the Schrödinger equation for a particularly simple system—a particle trapped in one dimension between two perfectly rigid walls. Although unrealistic in some respects, this system nevertheless is a surprisingly good approximation to some real quantum systems, including some electronic devices and simple nuclei. More important, its analysis illustrates the general procedure for applying the Schrödinger equation and shows how energy quantization emerges from Schrödinger's theory.

In classical physics, a particle trapped between rigid walls moves back and forth with constant speed. In the absence of friction or other losses, the particle's energy remains constant at its initial value. And in classical physics, that value can be anything.

We can describe the particle's situation using its potential-energy curve. Since the particle experiences no forces while it's between the walls, its potential energy $U$ is constant in this region, and we can fix the arbitrary zero of potential energy by setting $U = 0$. If the walls are perfectly rigid, then the particle can't penetrate them, no matter what its energy. This means that the potential energy becomes abruptly infinite at the walls. Then the potential-energy curve for our particle looks like Fig. 35.5; you can see from the figure why this curve is called an **infinite square well**. In this case the well extends from $x = 0$ to $x = L$.

We now consider the quantum-mechanical description of a particle in the infinite square well. The particle has a wave function whose time-independent part is given by the Schrödinger equation (Equation 35.1):

$$-\frac{\hbar^2}{2m}\frac{d^2\psi}{dx^2} + U(x)\psi = E\psi$$

where the potential energy $U(x)$ is that of the square well in Fig. 35.5:

$$U = 0 \text{ for } 0 < x < L$$
$$U = \infty \text{ for } x < 0 \text{ or } x > L$$

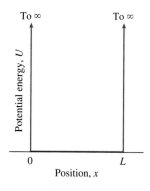

**FIGURE 35.5** Infinite square well potential-energy curve describes a particle constrained to move in one dimension between rigid walls.

Since there's no chance that the particle can penetrate the rigid walls, the function $\psi$ must be exactly zero in the region where $U = \infty$. All we need to calculate, then, is $\psi$ inside the well, where $0 \le x \le L$. To ensure that the particle is confined to the well, our solution must satisfy so-called **boundary conditions** $\psi = 0$ at $x = 0$ and at $x = L$.

Within the well, $U = 0$ and the Schrödinger equation becomes

$$-\frac{\hbar^2}{2m}\frac{d^2\psi}{dx^2} = E\psi \tag{35.4}$$

To find solutions, recall de Broglie's hypothesis that the allowed orbits in the Bohr atom are those for which standing waves just "fit" around the orbit. We have a similar situation with the infinite square well, in which the allowed solutions should be standing waves with nodes at the ends of the well—exactly analogous to standing waves on a string with both ends clamped that we discussed in Chapter 14. So we want a sinusoidal wave for $\psi(x)$, subject to the boundary conditions $\psi(0) = 0$ and $\psi(L) = 0$. The first condition is satisfied if we take a wave of the form $\psi = A\sin kx$, with $A$ and $k$ both constants. The second condition requires that $k = n\pi/L$, where $n$ is any integer—a condition equivalent to saying that an integer number of half-wavelengths fit in the well. So we propose a solution of the form

$$\psi(x) = A\sin\left(\frac{n\pi x}{L}\right)$$

with the constant $A$ still undetermined. This equation represents standing waves with nodes at the ends of the square well, but does it satisfy the Schrödinger equation? We can find out by substituting into Equation 35.4. We need not only $\psi$ but also its second derivative; twice differentiating our proposed solution gives

$$\frac{d^2\psi}{dx^2} = -A\frac{n^2\pi^2}{L^2}\sin\left(\frac{n\pi x}{L}\right)$$

Substituting $\psi$ and $d^2\psi/dx^2$ into Equation 35.4 gives

$$\left(-\frac{\hbar^2}{2m}\right)\left[-A\frac{n^2\pi^2}{L^2}\sin\left(\frac{n\pi x}{L}\right)\right] = EA\sin\left(\frac{n\pi x}{L}\right)$$

which reduces to

$$E = \frac{n^2\pi^2\hbar^2}{2mL^2} = \frac{n^2h^2}{8mL^2} \quad \text{(energy levels for an infinite square well potential)} \tag{35.5}$$

Equation 35.5 says that our proposed solution can indeed satisfy the Schrödinger equation—provided the particle energy $E$ has a value given by Equation 35.5, with $n$ an integer.

Our standing-wave solutions show how the quantization of energy arises naturally from the Schrödinger equation. Physically, the reason for quantization remains as de Broglie had postulated: Matter waves in a confined system must be standing waves with an integer number of half-wavelengths. Although de Broglie's hypothesis and the Schrödinger equation lead to exactly the same conclusion for the infinite square well, we'll see that with more complicated potential-energy functions only the Schrödinger equation can give us the full story.

The integer $n$ that appears in Equation 35.5 is the **quantum number** for the particle in the square well. The physical state of a quantum-mechanical system is its **quantum state**. Here one quantum number suffices to specify the quantum state, which then tells us everything quantum mechanics has to say about the situation. As far as the Schrödinger equation is concerned, it looks like all integer values of $n$ are allowed. The choice of negative or positive $n$ has no physical significance, since $\psi^2$ has the same value with either sign of $\psi$; for this reason, negative $n$'s are redundant. But $n = 0$ implies $\psi = 0$ everywhere, giving no chance of finding the particle anywhere. So we're left with positive integer values of $n$.

With only nonzero $n$'s allowed, Equation 35.5 shows that the particle's energy is always positive; zero energy isn't allowed. The lowest possible energy is $h^2/8mL^2$, obtained with $n = 1$. This is the **ground-state energy**; the corresponding wave function is the **ground-state wave function**. A nonzero ground-state energy is a common feature of quantum systems and one with no classical counterpart. Figure 35.6 is an energy-level diagram for the infinite square well.

**FIGURE 35.6** Energy-level diagram for a particle in an infinite square well. Energy is proportional to $n^2$, so the levels aren't evenly spaced.

Why isn't it possible to have a zero-energy particle in the square well? The answer lies in the uncertainty principle. We know for sure that the particle is confined within the well, giving it a position uncertainty $\Delta x = L$. If the energy is exactly zero, then we know that the momentum is exactly zero too. So there's no uncertainty $\Delta p$ in momentum, and the uncertainty principle $\Delta x \, \Delta p \geq \hbar$ implies infinite $\Delta x$—inconsistent with the particle's confinement in the well. We applied similar considerations in Example 34.6 to show that there's generally a minimum energy consistent with the uncertainty principle.

---

**EXAMPLE 35.1** **Transitions in a Square Well: Finding Photon Energy**

An electron is in the $n=2$ state of a 0.75-nm-wide infinite square well. It drops to the ground state, emitting a photon. Find the photon's energy.

**INTERPRET** This question is about an electron undergoing a transition from a higher to a lower energy level in an infinite square well. The photon carries off the extra energy, so the photon energy is the energy difference between the two states.

**DEVELOP** Equation 35.5, $E = n^2 h^2 / 8mL^2$, gives the energies of the different quantum states. Our plan is to subtract the energies of the two states, given by Equation 35.5 with $n=2$ for the initial state and $n=1$ for the ground state. The resulting difference is the photon energy.

**EVALUATE** Subtracting the two energies gives

$$\Delta E = E_2 - E_1 = \frac{h^2}{8mL^2}\left(2^2 - 1^2\right) = \frac{3h^2}{8mL^2}$$

which evaluates to $3.2 \times 10^{-19}$ J or 2.0 eV.

**ASSESS** Our system is several times the size of a hydrogen atom, and we're used to eV-range energies for atomic-scale systems, so our result seems reasonable. The wavelength corresponding to our 2-eV photon is $\lambda = c/f = hc/E = 620$ nm. This is red light, again consistent with our sense that atomic-size systems tend to emit visible or near-visible light. ∎

---

**GOT IT? 35.1** Electron A is confined to a square well 1 nm wide; electron B to a similar well only 1 pm wide. How do their ground-state energies compare?

---

## Normalization, Probabilities, and the Correspondence Principle

In our solution for the infinite square well, we still don't know the value of the constant $A$ in the wave function. We find this using the normalization condition 35.3: $\int_{-\infty}^{\infty} \psi^2 \, dx = 1$. Inside the well, $\psi = A \sin(n\pi x/L)$; outside, $\psi = 0$. So we can write the normalization condition as an integral over $0 < x < L$:

$$\int_0^L A^2 \sin^2\left(\frac{n\pi x}{L}\right) dx = 1$$

If we divided $\int_0^L \sin^2(n\pi x/L) \, dx$ by the well width $L$, we would have the average of sine squared over an integer number of half-cycles—or just $\frac{1}{2}$. So the integral of $\sin^2(n\pi x/L)$ from 0 to $L$ is $L$ times $\frac{1}{2}$, and therefore $A^2(L/2) = 1$, or $A = \sqrt{2/L}$. The complete wave function is then

$$\psi_n = \sqrt{\frac{2}{L}} \sin\left(\frac{n\pi x}{L}\right) \tag{35.6}$$

where the subscript $n$ refers to the function associated with the $n$th quantum state. Figure 35.7 shows wave functions for the ground state and three excited states.

Where are we likely to find the particle? Classically, it would move back and forth at constant speed and therefore would be equally likely to be anywhere in the well. Quantum-mechanically, the probability of finding it at some position $x$ is proportional to the probability density $\psi^2$ at that point. Figure 35.8 shows the probability densities given by squaring the wave functions of Fig. 35.7. For $n=1$ we're clearly more likely to find the particle near the middle of the well—in marked contrast to the classical prediction of equal probability everywhere. For other low-$n$ states there are obvious regions of high and low probability. But as the quantum number increases, the maxima and minima of the probability density get closer together. Any instrument we use to detect the electron has a finite

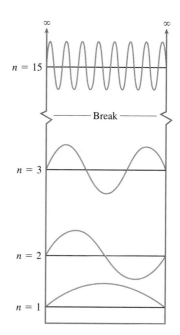

**FIGURE 35.7** Wave functions for a particle in an infinite square well, each centered on the corresponding energy level.

resolution, and once the periodicity of the wave function drops below that resolution, we measure an average probability, which is essentially constant over the interval (Fig. 35.8).

This is a manifestation of Bohr's correspondence principle: For large quantum numbers $n$, the interval between adjacent energy levels becomes small compared with the energy itself, and a measurement of the electron's position gives results in agreement with classical physics. But classical physics is totally inadequate at low $n$, where the nonclassical zero-point energy and quantization are most evident.

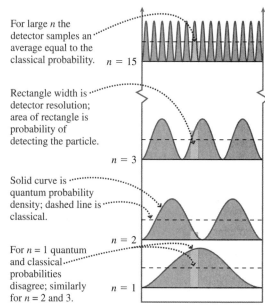

For large $n$ the detector samples an average equal to the classical probability.   $n = 15$

Rectangle width is detector resolution; area of rectangle is probability of detecting the particle.   $n = 3$

Solid curve is quantum probability density; dashed line is classical.   $n = 2$

For $n = 1$ quantum and classical probabilities disagree; similarly for $n = 2$ and 3.   $n = 1$

**FIGURE 35.8** Classical (dashed) and quantum (solid) probability densities for a particle in an infinite square well. The shaded area under each curve is 1, indicating that the particle must be somewhere in the well. Width of the colored rectangle is the resolution of an instrument used to detect the particle.

---

**EXAMPLE 35.2** **Quantum Probability:**
**The Square-Well Ground State**

A particle is in the ground state of an infinite square well. Find the probability that it will be found in the left-hand quarter of the well.

**INTERPRET** This is a question about probability, and we know that the probability density is the square of the wave function. So our solution is going to involve $\psi^2$.

**DEVELOP** The ground-state wave function from Equation 35.6 is $\psi_1 = \sqrt{2/L}\sin(\pi x/L)$. We normalized the wave function so the integral of $\psi_1^2$ over the entire well is 1, showing that the particle is *somewhere* in the well. That is, the area under the entire plot of $\psi_1^2$ is 1. We sketched $\psi_1^2$ in Fig. 35.9, showing that the probability of finding the particle in a specific region is the area under the curve in that region. So to find the probability that the particle is in the left-hand quarter of the well, we need to evaluate $\int \psi^2\,dx$ over the region from 0 to $\frac{1}{4}L$.

**EVALUATE** The probability becomes

$$P = \frac{2}{L}\int_0^{L/4} \sin^2\left(\frac{\pi x}{L}\right) dx$$

We can integrate using the table at the end of Appendix A; the result is

$$P = \frac{2}{L}\left(\frac{x}{2} - \frac{\sin(2\pi x/L)}{4\pi/L}\right)\Bigg|_0^{L/4} = \frac{2}{L}\left(\frac{L}{8} - \frac{L}{4\pi}\right) = 0.091$$

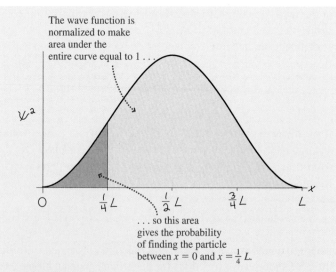

The wave function is normalized to make area under the entire curve equal to 1 . . .

. . . so this area gives the probability of finding the particle between $x = 0$ and $x = \frac{1}{4}L$.

**FIGURE 35.9** Sketch for Example 35.2.

**ASSESS** This is considerably lower than the probability $P = 0.25$ we would expect classically for finding the particle in any quarter of the well, and reflects the lower value of $\psi^2$ nearer the well ends. Problem 56 repeats the calculation of this example for arbitrary quantum numbers, showing that classical and quantum probabilities agree at large $n$.

**GOT IT? 35.2** Which of the following would be a reasonable answer if Example 35.2 had asked for the probability that the particle would be found in the central quarter of the well: (a) 0.091, (b) 0.25, (c) 0.475, (d) 0.90?

The infinite square well gives insights into important quantum phenomena shared by more realistic systems such as atoms. These include quantized energy levels, nonzero ground-state energy, nonclassical probabilities, and agreement with classical physics at large quantum numbers. In Chapter 36 we'll apply the Schrödinger equation to atoms, where we'll find many of the same phenomena. First, though, we look at some other simple systems that exhibit additional quantum behaviors.

## The Harmonic Oscillator

In Chapter 13 we studied simple harmonic motion, which occurs when a particle is subject to a restoring force that's directly proportional to the displacement from equilibrium. Such a *linear* restoring force implies a *quadratic* potential-energy function, and conversely, as we pointed out in Section 13.5, any system with a quadratic potential-energy function is a harmonic oscillator. That includes many systems at the atomic and molecular scale. Understanding the quantum-mechanical harmonic oscillator is therefore crucial in describing the behavior of matter on small scales.

A mass–spring system has potential energy $U = \frac{1}{2}kx^2$ and oscillates with angular frequency given by Equation 13.7: $\omega = \sqrt{k/m}$. Combining these equations gives $U = \frac{1}{2}m\omega^2 x^2$. Now, a quantum oscillator isn't a real spring but is likely to be an electron or an atom vibrating at the end of a molecular bond; therefore our new form for $U$, which makes no mention of springs, is more appropriate. Using this expression for $U(x)$ in the Schrödinger equation 35.1 gives

$$-\frac{\hbar^2}{2m}\frac{d^2\psi}{dx^2} + \frac{1}{2}m\omega^2 x^2 \psi = E\psi$$

Solving this equation requires more advanced math techniques, yielding a set of functions that *mathematically* satisfy our Schrödinger equation. However, most of these functions become infinite for large $x$, and therefore they can't be normalized and so don't represent valid *physical* solutions. The normalizable solutions—those that represent particle wave functions—occur for only discrete values of the energy $E$, given by

$$E_n = \left(n + \tfrac{1}{2}\right)\hbar\omega \tag{35.7}$$

where now $n = 0$ is the ground state. Figure 35.10 shows an energy-level diagram for the harmonic oscillator; note the even spacing implied by Equation 35.7. The additive factor $\frac{1}{2}$ in Equation 35.7 shows that Planck was wrong in his hypothesis that the allowed harmonic-oscillator energies should be multiples of $hf (=\hbar\omega)$. Planck's spectral distribution (Equation 34.3) is nevertheless correct, but he did not foresee the existence of nonzero ground-state energy.

The even spacing between the energy levels of the harmonic oscillator is in marked contrast to the situation in atoms (Fig. 34.11) or in the infinite square well (Fig. 35.6). A quantum harmonic oscillator emits or absorbs photons as it makes transitions among adjacent levels, and the even spacing implies that all transitions between adjacent levels of a pure harmonic oscillator involve photons of the same energy.

A classical harmonic oscillator moves slowest near its turning points, and therefore it's most likely to be found at the extremes of its motion. It's least likely to be at its equilibrium position, where it's moving fastest. As with the square well, the harmonic oscillator in low-$n$ states exhibits unclassical behavior; in the ground state it's *most* likely to be found at its equilibrium position! Figure 35.11 shows classical and quantum probability densities for the harmonic oscillator; note that for larger $n$ the two begin to agree, once again showing Bohr's correspondence principle at work.

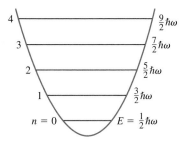

**FIGURE 35.10** Energy-level diagram for a quantum-mechanical harmonic oscillator, superposed on the parabolic potential-energy curve.

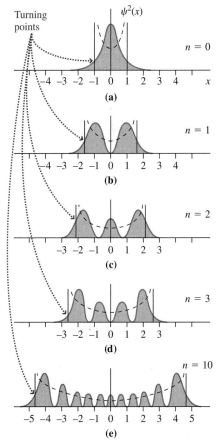

**FIGURE 35.11** Probability densities $\psi^2(x)$ for some states of the harmonic oscillator. Dashed curves are classical predictions. Increasing spread in the turning points reflects the higher energy of the higher-$n$ states.

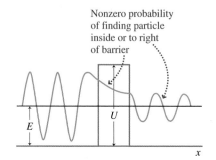

Nonzero probability of finding particle inside or to right of barrier

**FIGURE 35.12** A potential barrier of height $U$, showing the wave function for a particle incident from the left with energy $E$ lower than the barrier energy $U$.

**FIGURE 35.13** For a massive particle, the wave function drops rapidly in the barrier, giving negligible probability of penetration.

## Quantum Tunneling

One remarkable feature of Fig. 35.11 is the nonzero probability of finding a quantum-harmonic oscillator beyond its classical turning points—the points at which its kinetic energy has been converted entirely to potential energy. This unusual situation, which seems to violate energy conservation, has no counterpart in the classical description of matter.

Another example of penetration into a classically forbidden region is a particle encountering a potential barrier (Fig. 35.12). Examples of such barriers include electric potential differences associated with atomic nuclei, gaps between solid materials, and insulating layers in some semiconductor devices. Classically, a particle whose total energy is lower than the barrier energy is confined to one side of the barrier. If we solve the Schrödinger equation for this potential-energy curve, however, we find oscillatory solutions on either side of the barrier, joined according to the continuity conditions on $\psi$ and $d\psi/dx$ by exponential functions within the barrier (see Problems 57–59). Such a solution is shown superimposed on the barrier in Fig. 35.12. The probability density $\psi^2$ associated with this solution remains nonzero through the barrier and continues to give a nonzero probability of finding the particle on the far side—implying that a particle initially on one side of the barrier may later be found on the other side.

How likely is this phenomenon, called **quantum tunneling**? That depends on the relation of the particle energy to the barrier energy, and also on the width of the barrier. As you can show in Problem 50, the $\psi$ function inside the barrier involves exponential functions of the form $e^{\pm\sqrt{2m(U_0-E)}x/\hbar}$, where $U_0$ is the barrier height. In general, these exponentials drop very rapidly across the barrier width unless the particle energy $E$ is close to the barrier energy or the particle mass $m$ is small. The probability that a particle will be found on the far side of the barrier is therefore very low when the mass $m$ is large, so quantum tunneling is a microscopic phenomenon (Fig. 35.13).

It looks as if tunneling violates conservation of energy. But we're saved by the uncertainty principle. Suppose we try to catch a particle in the act of tunneling. If we know the particle is within the barrier, the uncertainty in its position is no greater than the barrier width. We know from Example 34.6 that this implies a minimum energy. A quantitative analysis shows that minimum to be such that we can no longer be sure the particle energy is lower than the barrier energy. But quantum tunneling really does occur, as we confirm experimentally by measuring particle energies on both sides of the barrier. They can both be lower than the barrier height, and yet somehow particles have gone from one side to the other.

If we don't try to detect a particle within the barrier, its penetration is a purely wave phenomenon to which our particulate energy considerations don't apply. Again we see the wave–particle duality at work: If we don't observe the particle, its behavior is governed by the associated waves and may result in most unparticle-like phenomena such as tunneling. If we do try to catch it in the act of such behavior, it ceases to be wave-like and the surprising phenomena cease.

Tunneling is important in a number of quantum-mechanical phenomena and technological devices. That the Sun shines—and therefore that we're alive—is a consequence of quantum tunneling of nuclei in the Sun's core. Classically, those nuclei don't have sufficient energy to get close enough to overcome their mutual electric repulsion. But they can tunnel through this "Coulomb barrier" and then fuse together to release the enormous energy that powers the Sun. An opposite nuclear process, alpha decay, occurs as alpha particles tunnel through a nuclear potential barrier that traps them inside large nuclei like uranium. Measurement of the alpha particles' energy shows it to be lower than the barrier energy, confirming that tunneling occurs. Semiconductor devices involving quantum tunneling hold the promise of a new generation of much faster electronic circuits. Finally, tunneling is the basis of the scanning-tunneling microscope (STM), a remarkable device that lets us image individual atoms.

---

**GOT IT? 35.3** A proton and an electron approach a barrier. Both have the same energy $E$, which is lower than the barrier potential $U$. Which is more likely to get through?

---

**The Scanning-Tunneling Microscope**

Developed in the 1980s by Heinrich Rohrer and Gerd Binnig of IBM Zurich Research Laboratory, the STM has become a vital tool for semiconductor engineers, biologists, chemists, and nanotechnologists. The STM works by quantum tunneling between an extraordinarily fine conducting tip and the surface under study. The photo shows a scanning electron microscope image of an STM tip, which may be only one atom wide. As in the barrier of Fig. 35.12, the electron wave function tapers off exponentially in the space outside the surface. Place a conducting tip near but not touching the surface, and there's a nonzero probability that electrons will tunnel through the gap to reach the tip,

resulting in an electric current. The exponential falloff of the wave function means this **tunneling current** is extremely sensitive to the tip-to-surface gap, and therefore changes significantly with surface irregularities.

A practical STM scans the tip over the surface, and feedback devices move the tip to keep the tunneling current constant despite surface irregularities, as shown in the figure. Therefore the tip traces out surface irregularities, and this information is used to construct an image of the surface (see this chapter's opening photo).

## Finite Potential Wells

Both the infinite square well and the harmonic oscillator have potential wells of infinite depth. A particle is necessarily bound in such a well; it can't escape to large distances. Its quantized energy states are therefore **bound states**. Provided they aren't too shallow, wells of finite depth also exhibit quantized bound states whose wave functions resemble those of the infinite square well (Fig. 35.14), although they show a small probability of tunneling into the classically forbidden region outside the well. Problems 57–59 explore the mathematics of the finite well.

Quantized bound states represent particles with energy lower than the well height. Particles with higher energy are free to move anywhere, and their wave functions are everywhere oscillatory. Furthermore, particles in these **unbound states** can have any energy whatsoever as long as it exceeds the well height; unbound energies aren't quantized. Rather, there's a **continuum** of allowed energies above the well top, in contrast to the discrete, quantized levels below (Fig. 35.15). We'll find both bound and unbound states again in the next chapter when we study the atom.

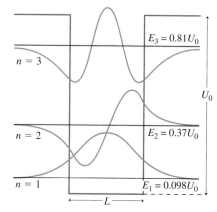

**FIGURE 35.14** Bound-state wave functions for a finite square well, superposed on the associated energy levels. For this particular well depth and width there are only three bound states.

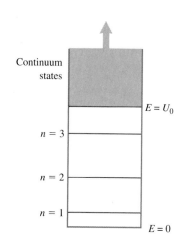

**FIGURE 35.15** Energy-level diagram for a finite square well shows discrete bound states and a continuum of unbound states.

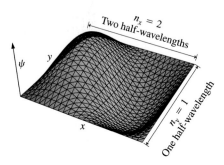

**FIGURE 35.16** The wave function for a particle confined to a square region in two dimensions. The function is $\sin(n_x \pi x/L) \sin(n_y \pi y/L)$, with $n_x = 2$ and $n_y = 1$.

## 35.4 Quantum Mechanics in Three Dimensions

One-dimensional quantum systems show important features of the quantum world, like energy quantization and tunneling. But atoms and most other quantum systems are three-dimensional. The wave function then depends on all three spatial variables, and the Schrödinger equation reflects this complexity. You can explore the three-dimensional Schrödinger equation in Problem 52; here we just point out some new features of three-dimensional quantum systems.

A single quantum number $n$ characterizes quantum states in one dimension. For an infinite square well, that reflects the requirement that an integer number of half-wavelengths fit in the well. Solving the Schrödinger equation gives a unique energy level associated with each value of $n$. Similarly, the potential-energy function of the harmonic oscillator leads to energy levels distinguished by a single quantum number.

In two or three dimensions, analogous constraints lead to independent quantum numbers for each dimension (Fig. 35.16). For each set of quantum numbers there's an associated energy level. For a particle of mass $m$ confined to a cubical box of side $L$, for example, a generalization of the one-dimensional square well leads to the energy levels

$$E = \frac{h^2}{8mL^2}\left(n_x^2 + n_y^2 + n_z^2\right) \tag{35.8}$$

where the $n$'s are the quantum numbers associated with each spatial dimension. As in one dimension, the allowed values for the $n$'s are positive integers. Thus, the ground state has $n_x = n_y = n_z = 1$. But what's the first excited state? It could be $n_x = 2$, $n_y = n_z = 1$. But it could equally well be $n_x = n_y = 1$, $n_z = 2$, or $n_x = n_z = 1$, $n_y = 2$, since all three of these combinations give the same energy.

When two or more quantum states have the same energy, those states are said to be **degenerate**. The first excited state of a particle confined to a cubical box, for example, is threefold degenerate, meaning there are three distinct states with the same energy. Degeneracy is often associated with symmetry of the quantum-mechanical system. In the cubical box, for example, the fact that the sides have equal lengths allows different combinations of quantum numbers to give the same energy. Making all three sides different would remove the degeneracy, splitting what was a single energy level into three (Fig. 35.17). The same thing happens in more realistic quantum systems. For example, imposing a magnetic field on an otherwise spherically symmetric atom breaks the symmetry and may split energy levels that were previously degenerate (Fig. 35.18). Detection of this splitting in optical spectra allows measurement of magnetic fields on the Sun and in other remote objects.

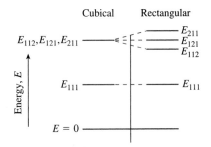

**FIGURE 35.17** Energy-level diagrams showing the ground state and first excited state for a particle in a three-dimensional box. Making the sides different lengths removes the degeneracy.

(a)

(b)

**FIGURE 35.18** (a) Spectral line at 404.66 nm produced by mercury atoms undergoing transitions from $n = 7$ to $n = 6$. The upper level is actually threefold degenerate. (b) A magnetic field breaks the symmetry and removes the degeneracy, splitting the spectral line.

## 35.5 Relativistic Quantum Mechanics

Like Newtonian physics, quantum mechanics based on the Schrödinger equation is not consistent with special relativity's requirement that the laws of physics be the same in all inertial reference frames. It's therefore an approximation valid for particle speeds $v$ much lower than $c$. For most applications in atomic, molecular, and condensed-matter physics, $v \ll c$ so the Schrödinger equation applies. But when particle speeds are a significant fraction of $c$, the Schrödinger equation becomes inadequate and must be replaced with a relativistic wave equation. And even for slowly moving particles, the requirement of relativistic invariance leads to some surprising new phenomena.

### The Dirac Equation and Antiparticles

In 1928 the English physicist Paul Dirac formulated a relativistic wave equation for electrons. In the process he encountered several unexpected mathematical requirements with deep physical significance.

Dirac replaced the Newtonian energy–momentum relation $K = p^2/2m$ with the relativistic expression $E^2 = (mc^2)^2 + p^2c^2$ that we saw in Chapter 33. But this expression implies two values for $E$, depending on which sign one chooses in taking the square root. Dirac argued that both roots are meaningful and that the negative root implies the existence of a particle identical in mass to the electron but carrying positive charge. The 1932 discovery of this **positron** vindicated Dirac's brilliant idea. Today we know that every particle has a corresponding **antiparticle**, identical in mass but opposite in electric, magnetic, and other properties.

Einstein's energy–mass equivalence implies that **pair creation** of a particle–antiparticle pair is possible, given energy $2mc^2$ equivalent to the mass of the pair. The opposite process, annihilation, occurs as particle and antiparticle meet and disappear to form a pair of photons. Although pair creation is rare today, it was commonplace in the hot, early universe, where thermal energy alone was high enough to create particle–antiparticle pairs. In those early times Einstein's mass–energy equivalence would have been obvious, and the number of particles in a closed volume wouldn't remain constant.

### Electron Spin

Another unexpected mathematical result of Dirac's work was that the wave function had to involve matrices. This, Dirac showed, implied physically that the electron must possess an intrinsic angular momentum—something physicists had already inferred from experiments, but without any theoretical grounding. This angular momentum, called **spin**, has enormous significance in quantum mechanics and particularly in atomic structure, as we'll see in the next chapter.

## Big Picture

The big idea here is the description of particles in the quantum realm using **wave functions**, whose square relates to the probability of finding a particle. Thus the link between the most thorough description physics can provide—the wave function—and the behavior of an individual particle is only statistical. The **Schrödinger equation** gives the wave function for nonrelativistic particles and leads to the energy quantization for confined particles.

## Key Concepts and Equations

The **time-independent Schrödinger equation** gives the wave function $\psi$ for a particle of mass $m$ with total energy $E$ and potential energy $U$:

$$-\frac{\hbar^2}{2m}\frac{d^2\psi}{dx^2} + U(x)\psi = E\psi$$

The square of the wave function is the **probability density**, and in one dimension the probability of finding the particle in some small interval $dx$ at position $x$ is

$$P(x) = \psi^2(x)\,dx.$$

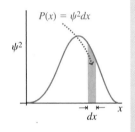

**Normalization**: A particle must be *somewhere*, so

$$\int_{-\infty}^{+\infty} \psi^2(x)\,dx = 1$$

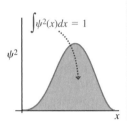

## Cases and Uses

### Infinite square well

Wave functions: $\psi_n = \sqrt{\dfrac{2}{L}}\sin\left(\dfrac{n\pi x}{L}\right)$

Energy levels: $E_n = \dfrac{n^2 h^2}{8mL^2}$

3-D well: $E = \dfrac{h^2}{8mL^2}\left(n_x^2 + n_y^2 + n_z^2\right)$

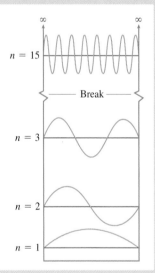

### Harmonic oscillator

Energy levels: $E_n = \left(n+\frac{1}{2}\right)\hbar\omega$

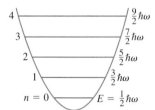

### Finite well

Discrete bound states; continuum of unbound states

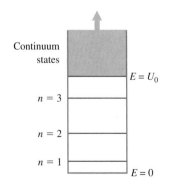

### Quantum tunneling

Nonzero probability of finding quantum particle in region forbidden by classical energy conservation

# For Thought and Discussion

1. Explain qualitatively why a particle confined to a finite region cannot have zero energy.
2. Does quantum tunneling violate energy conservation? Explain.
3. Bohr's correspondence principle states that quantum and classical mechanics must agree in a certain limit. What limit? Give an example.
4. The ground-state wave function for a quantum harmonic oscillator has a single central peak. Why is this at odds with classical physics?
5. What's the essential difference between the energy-level structures of infinite and finite square wells?
6. In terms of de Broglie's matter–wave hypothesis, how does making the sides of a box different lengths remove some of the degeneracy associated with a particle confined to that box?

7. A particle is confined to a two-dimensional box whose sides are in the ratio 1:2. Are any of its energy levels degenerate? If so, give an example. If not, why not?
8. What did Einstein mean by his remark, loosely paraphrased, that "God does not play dice"?
9. Some philosophers argue that the strict determinism of classical physics is inconsistent with human free will, but that the indeterminacy of quantum mechanics does leave room for free will. Others claim that physics has no bearing on the question of free will. What do you think?

# Exercises and Problems

## Exercises

### Section 35.2 The Schrödinger Equation

10. What are the units of the wave function $\psi(x)$ in a one-dimensional situation?
11. The wave function for a particle is given by $\psi = Ae^{-x^2/a^2}$, where $A$ and $a$ are constants. (a) Where is the particle most likely to be found? (b) Where is the probability per unit length half its maximum value?
12. Figure 35.19 shows the wave function $\psi(x)$ for a particle confined between $x = 0$ and $x = L$. Find an expression for $A$ in order that the wave function be normalized.

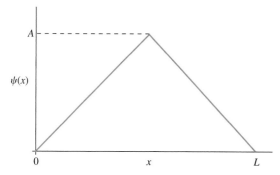

**FIGURE 35.19** Exercises 12 and 13

13. What is the probability of finding the particle of Fig. 35.19 in the region $0 < x < L/4$?
14. The solution to the Schrödinger equation for a particular potential is $\psi = 0$ for $|x| > a$ and $\psi = A\sin(\pi x/a)$ for $-a \le x \le a$, where $A$ and $a$ are constants. In terms of $a$, what value of $A$ is required to normalize $\psi$?

### Section 35.3 Particles and Potentials

15. What's the quantum number for a particle in an infinite square well if the particle's energy is 25 times the ground-state energy?
16. A particle in an infinite square well makes a transition from a higher to a lower energy state; the corresponding energy decrease is 33 times the ground-state energy. Find the quantum numbers of the initial and final states.
17. Determine the ground-state energy for an electron in an infinite square well of width 10.0 nm.
18. What is the width of an infinite square well in which a proton cannot have an energy lower than 100 eV?

19. One reason we don't notice the effects of quantum mechanics in everyday life is that Planck's constant $h$ is too small. Treating yourself as a particle in a room-sized one-dimensional infinite square well, how big would $h$ have to be if your minimum possible energy corresponded to a speed of 1.0 m/s? Take the width of the room to be 2.6 m and your mass to be 60 kg.
20. A particle is confined to an infinite square well of width 1.0 nm. If the energy difference between the ground state and the first excited state is 1.13 eV, is the particle an electron or a proton?
21. A snail of mass 3 g crawls between two rocks 15 cm apart at a speed of 0.5 mm/s. Treating this system like a particle in an infinite square well, determine the approximate quantum number. Does the correspondence principle permit the use of the classical approximation in this case?
22. An alpha particle (mass 4 u) is trapped in a uranium nucleus with diameter 15 fm. Treating the system as a one-dimensional square well, what would be the minimum energy for the alpha particle?
23. A particle in a harmonic oscillator potential has a ground-state energy of 0.14 eV. What would be the classical frequency $f$ of the oscillator?
24. What is the ground-state energy for a particle in a harmonic oscillator potential whose classical angular frequency is $\omega = 1.0 \times 10^{17} \text{ s}^{-1}$?
25. A harmonic oscillator emits a 1.1-eV photon as it undergoes a transition between adjacent states. What is its classical oscillation frequency?
26. The ground-state energy of a harmonic oscillator is 4.0 eV. What is the energy separation between adjacent quantum states?
27. Find the separation between the energy levels of a harmonic oscillator consisting of a 1-g mass on a spring with spring constant 80 N/m. Your answer shows why quantization isn't noticeable in macroscopic systems.

### Section 35.4 Quantum Mechanics in Three Dimensions

28. If all sides of a cubical box are doubled, what happens to the ground-state energy of a particle in that box?
29. A very crude model for an atomic nucleus is a cubical box about 1 fm on a side. What would be the energy of a gamma ray emitted if a proton in such a nucleus made a transition from its first excited state to the ground state?
30. An electron is confined to a cubical box. For what box width will a transition from the first excited state to the ground state result in emission of a 950-nm infrared photon?

## Problems

31. Find an expression for the normalization constant $A$ for the wave function given by $\psi = 0$ for the case $|x| > b$ and by $\psi = A(b^2 - x^2)$ for $-b \le x \le b$.

32. Use a table of definite integrals or symbolic math software to help evaluate the normalization constant $A$ in the wave function of Exercise 11.

33. Suppose $\psi_1$ and $\psi_2$ are both solutions of the Schrödinger equation for the same energy $E$. Show that $a\psi_1 + b\psi_2$ is also a solution, where $a$ and $b$ are arbitrary constants.

34. An electron is trapped in an infinite square well 25 nm wide. Find the wavelengths of the photons emitted in these transitions: (a) $n = 2$ to $n = 1$; (b) $n = 20$ to $n = 19$; (c) $n = 100$ to $n = 1$.

35. An electron drops from the $n = 7$ to the $n = 6$ level of an infinite square well 1.5 nm wide. Find (a) the energy and (b) the wavelength of the photon emitted.

36. Show explicitly that the difference between adjacent energy levels in an infinite square well becomes arbitrarily small compared with the energy of the upper level, in the limit of large quantum number $n$.

37. An electron is in a narrow molecule 4.4 nm long, a situation that approximates a one-dimensional infinite square well. If the electron is in its ground state, what is the maximum wavelength of electromagnetic radiation that can cause a transition to an excited state?

38. The ground-state energy for an electron in an infinite square well A is equal to the energy of the first excited state for an electron in well B. How do the widths of the two wells compare?

39. Electrons in an ensemble of many square-well systems are initially in the $n = 4$ state. The wells all have the same width, 0.10 nm. Find the wavelengths of all spectral lines emitted as the electrons cascade downward to the ground state through all possible transitions.

40. Sketch the probability density for the $n = 2$ state of an infinite square well extending from $x = 0$ to $x = L$, and determine where the particle is most likely to be found.

41. An infinite square well extends from $-L/2$ to $L/2$. (a) Find expressions for the normalized wave functions for a particle of mass $m$ in this well. You will need separate expressions for even and odd quantum numbers. (b) Find the corresponding energy levels.

42. A particle is in the ground state of an infinite square well. Find the probability of finding the particle in the left-hand third of the well.

43. A laser emits 1.96-eV photons. If this emission is due to electron transitions from the $n = 2$ to $n = 1$ states of an infinite square well, what's the well width?

44. What is the probability of finding a particle in the central 80% of an infinite square well, assuming it's in the ground state?

45. A 10-g mass is attached to an ideal spring with spring constant $k = 150$ N/m and set oscillating with amplitude 12 cm. To what quantum number does this oscillation correspond?

46. A hydrogen chloride molecule may be modeled as a hydrogen atom (mass $1.67 \times 10^{-27}$ kg) on a spring; the other end of the spring is attached to a rigid wall (the massive chlorine atom). If the minimum photon energy that will promote this molecule to its first excited state is 0.358 eV, find the "spring constant."

47. For what quantum numbers is the spacing between adjacent energy levels in a harmonic oscillator less than 1% of the actual energy?

48. A particle detector has a resolution 15% of the width of an infinite square well. What is the chance that the detector will find a particle in the ground state of the square well if the detector is centered on (a) the midpoint of the well and (b) a point one-fourth of the way across the well?

49. Find the probability that a particle in an infinite square well will be found in the central one-fourth of the well for the quantum states (a) $n = 1$, (b) $n = 2$, (c) $n = 5$, and (d) $n = 20$. (e) What is the classical probability in this situation?

50. A particle of mass $m$ is in a region where its total energy $E$ is less than its potential energy $U$. Although this situation is forbidden classically, show that the Schrödinger equation has nonzero solutions of the form $Ae^{\pm\sqrt{2m(U-E)}x/\hbar}$. Such solutions describe the wave function in quantum tunneling, beyond the turning points in a quantum harmonic oscillator, or beyond the well edges in a finite potential well.

51. (a) Use Equation 35.8 to draw an energy-level diagram for the first six energy levels of a particle in a cubical box, in terms of $\hbar^2/8mL^2$, and (b) give the degeneracy of each.

52. The generalization of the Schrödinger equation to three dimensions is

$$-\frac{\hbar^2}{2m}\left(\frac{\partial^2\psi}{\partial x^2} + \frac{\partial^2\psi}{\partial y^2} + \frac{\partial^2\psi}{\partial z^2}\right) + U(x, y, z)\psi = E\psi$$

(a) For a particle confined to the cubical region $0 \le x \le L$, $0 \le y \le L$, $0 \le z \le L$, show by direct substitution that the equation is satisfied by wave functions of the form $\psi(x, y, z) = A\sin(n_x\pi x/L)\sin(n_y\pi y/L)\sin(n_z\pi z/L)$, where the $n$'s are integers and $A$ is a constant. (b) In the process of working part (a), verify that the energies $E$ are given by Equation 35.8.

53. For what quantum state is the probability of finding a particle in the left-hand quarter of an infinite square well equal to 0.303?

54. A 9-W laser beam shines on an ensemble of $10^{24}$ electrons, each in the ground state of a one-dimensional infinite square well 0.72 nm wide. The photon energy is just high enough to raise an electron to its first excited state. How many electrons can be so excited if the beam shines for 10 ms?

55. A large number of electrons are all confined to infinite square wells 1.2 nm wide. They're undergoing transitions among all possible states. How many (a) visible (400 nm to 700 nm) and (b) infrared lines will there be in the spectrum emitted by this ensemble of square-well systems?

56. A particle is in the $n$th quantum state of an infinite square well. (a) Show that the probability of finding it in the left-hand quarter of the well is

$$P = \frac{1}{4} - \frac{\sin(n\pi/2)}{2n\pi}$$

(b) Show that for even $n$ the probability is exactly $\frac{1}{4}$. (c) Show that for odd $n$ the probability approaches the classical value $\frac{1}{4}$ as $n \to \infty$.

57. The next three problems illustrate the matching procedure necessary in solving the Schrödinger equation for potential wells of finite depth. Consider a particle of mass $m$ in the semi-infinite well shown in Fig. 35.20. It will be convenient to work with dimensionless forms of the particle energy $E$ and well depth $U_0$, in units of $\hbar^2/2mL^2$: $\mathcal{E} \equiv 2mL^2E/\hbar^2$ and $\mu \equiv 2mL^2U_0/\hbar^2$. Assume that $E < U_0$ or, equivalently, $\mathcal{E} < \mu$. Show by substitution that the following wave functions satisfy the Schrödinger equation in the regions indicated:

$$\psi_1 = A\sin(\sqrt{\mathcal{E}}x/L) \text{ for } 0 \le x \le L$$

$$\psi_2 = Be^{-\sqrt{\mu - \mathcal{E}}x/L} \text{ for } x \ge L$$

where $A$ and $B$ are arbitrary constants.

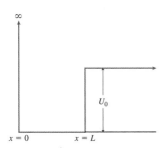

**FIGURE 35.20** Problems 57–59

58. The wave functions of Problem 57 satisfy the Schrödinger equation, but they don't necessarily meet the conditions that $\psi$ and its derivative be continuous; in particular, both quantities could change abruptly at the well edge $x = L$. To achieve continuity of $\psi$, evaluate $\psi_1$ and $\psi_2$ at $x = L$ and set them equal. To achieve continuity of $d\psi/dx$, first differentiate $\psi_1$ and $\psi_2$, then evaluate the derivatives at $x = L$, and then set them equal. The two continuity conditions should result in these equations:

$$A\sin(\sqrt{\mathcal{E}}) = Be^{-\sqrt{\mu - \mathcal{E}}}$$

$$\sqrt{\mathcal{E}}A\cos(\sqrt{\mathcal{E}}) = -\sqrt{\mu - \mathcal{E}}Be^{-\sqrt{\mu - \mathcal{E}}}$$

Use the first of these equations to eliminate the quantity $Be^{-\sqrt{\mu - \mathcal{E}}}$ from the second equation, and show that the result is

$$\tan(\sqrt{\mathcal{E}}) = -\sqrt{\frac{\mathcal{E}}{\mu - \mathcal{E}}}$$

This equation determines the values of $\mathcal{E}$ that give the quantized energy levels.

59. The equation derived in Problem 58 cannot be solved algebraically because the unknown $\mathcal{E}$ appears in a trig function and under the square root. It can be solved graphically, by plotting both sides on the same graph and determining where they intersect. It can also be solved by trial and error on a calculator, by using a calculator with a root-finding routine, or by computer. Use one of these methods to find all possible values of $\mathcal{E}$ for (a) $\mu = 2$, (b) $\mu = 20$, and (c) $\mu = 50$. *Note:* The number of solutions varies with $\mu$; there may be no solutions, meaning no bound states are possible, or there may be one or more bound states.

60. (a) Show by direct substitution in the Schrödinger equation for the one-dimensional harmonic oscillator that the wave function $\psi_0(x) = A_0 e^{-\alpha^2 x^2/2}$, where $\alpha^2 = m\omega/\hbar$, is a solution with energy corresponding to $n = 0$ in Equation 35.7. (b) Find the normalization constant $A_0$. (c) Find the classical turning points (where the classical probability is greatest) and show, in contrast, that the quantum probability density has a maximum at $x = 0$, as shown in Figure 35.11.

61. (a) Show by direct substitution in the Schrödinger equation for the one-dimensional harmonic oscillator that the wave function $\psi_1(x) = A_1 xe^{-\alpha^2 x^2/2}$, where $\alpha^2 = m\omega/\hbar$, is a solution with energy corresponding to $n = 1$ in Equation 35.7. (b) Find the normalization constant $A_1$. (c) Show that the probability density has a minimum at $x = 0$ and maxima at $x = \pm 1/\alpha$, corresponding to the classical turning points for the ground state $n = 0$, as shown in Figure 35.11.

62. Show that the wave function $\psi(x) = Ae^{ikx}$ is a solution of Equation 35.1 for a particle of mass $m$ and energy $E$, in a region where the potential energy is a constant $U_0 < E$. Here $i$ is the imaginary unit, $i = \sqrt{-1}$. Find an expression for $k$ in terms of $m$ and the energies $E$ and $U_0$, and also in terms of the particle's momentum and its de Broglie wavelength.

63. Wave functions like the one in Problem 62 can represent free particles moving with velocity $v = p/m$ in the $x$ direction. Consider a beam of such particles incident on a potential energy step $U(x) = 0$, for $x < 0$, and $U(x) = U_0 < E$, for $x > 0$. The wave function for $x < 0$ is $\psi_1(x) = Ae^{ik_1 x} + Be^{-ik_2 x}$, representing both incident and reflected particles. For $x > 0$ the wave function is $\psi_2(x) = Ce^{ik_2 x}$, representing transmitted particles. Use the conditions that both $\psi$ and its first derivative must be continuous at $x = 0$ to find the constants $B$ and $C$ in terms of $k_1$, $k_2$ and $A$.

64. You're researching a term paper for a history of science course. You read in an article on nuclear physics that the ground-state energy of a proton in a nucleus is about 200 MeV. You realize this energy would allow you to calculate the width of a square-well potential that would approximate the behavior of the proton. What is that width?

65. The author of the article you just read on nuclear physics also wrote about how the energy of photons could be used to find the width of an infinite square well in which an electron is localized. Your physics professor said something about the width of such a well being a few nanometers. Assume an electron drops from the $n = 5$ level to the $n = 2$ level of an infinite square well, emitting a 1.4-eV photon. Is your professor's estimate correct?

66. You're in a course for students considering medical school. You're learning about various imaging technologies including positron emission tomography (PET). In PET scans, a positron emitter is implanted in an organ to be imaged. The positrons are immediately annihilated when they come in contact with the more numerous electrons, with each annihilation producing a pair of oppositely directed gamma-ray photons. You're asked to find the photon energy and frequency. Why, you wonder, are two photons necessary, and why are they emitted in opposite directions?

67. You just watched the most fascinating program on TV about the early universe and the creation of matter from energy. The program narrator mentioned that shortly after the Big Bang the temperature was so high that thermal energy caused frequent pair production. However, the program failed to mention the temperature. What temperatures would be needed for proton–antiproton pair creation?

# Answers to Chapter Questions

## Answer to Chapter Opening Question

Quantum tunneling, the ability of particles to penetrate a barrier that classical physics says they don't have sufficient energy to overcome.

## Answers to GOT IT? Questions

35.1 B's ground-state energy is $10^6$ times higher than A's.

35.2 (c), which is almost twice the classical prediction.

35.3 The electron.

# 36

# Atomic Physics

■ How do the principles of quantum physics lead to the different chemical elements?

In Chapter 35 we applied the Schrödinger equation to simplified quantum systems. Here we turn to the more realistic case of the atom, and show how quantum mechanics explains atomic structure and the periodic table of the elements. We'll deal most thoroughly with the simplest atom, hydrogen, and we'll be more qualitative in describing multielectron atoms.

## 36.1 The Hydrogen Atom

Like a particle in a three-dimensional box, the electron in hydrogen is confined to a three-dimensional potential well. For the electron, the well results from the electrostatic attraction of the proton. From Chapter 22 we know that the electric potential of the proton, treated as a point charge $e$, is $V(r) = ke/r$, with $r$ the distance to the proton and the zero of potential at infinity. Electric potential is energy per unit charge, so we multiply by the electron charge $-e$ to get the potential energy of an electron in the presence of a proton:

$$U(r) = -\frac{ke^2}{r} \qquad (36.1)$$

We treat the massive proton as being at rest at the origin, so Equation 36.1 gives the electron's potential energy as a function of radial position $r$ (Fig. 36.1). We can therefore use Equation 36.1 as the potential energy in the Schrödinger equation for the hydrogen atom.

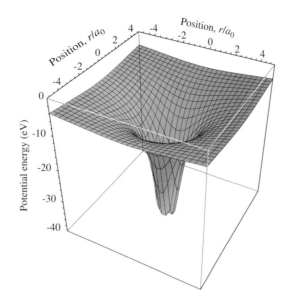

**FIGURE 36.1** Potential energy of the electron in a hydrogen atom, as a function of position in a plane containing the proton. Position is in units of the Bohr radius $a_0$ introduced in Chapter 34.

## The Schrödinger Equation in Spherical Coordinates

Because the electron's potential energy depends on radial distance $r$, it's best to work in spherical coordinates, where the position of a point is given by its distance $r$ from the origin along with two angles $\theta$ and $\phi$ that specify its orientation (Fig. 36.2). Converting to spherical coordinates is straightforward but tedious; the result is

$$-\frac{\hbar^2}{2mr^2}\left[\frac{\partial}{\partial r}\left(r^2\frac{\partial\psi}{\partial r}\right) + \frac{1}{\sin\theta}\frac{\partial}{\partial\theta}\left(\sin\theta\frac{\partial\psi}{\partial\theta}\right) + \frac{1}{\sin^2\theta}\frac{\partial^2\psi}{\partial\phi^2}\right] - \frac{ke^2}{r}\psi = E\psi \quad \left(\begin{array}{l}\text{Schrödinger equation,}\\ \text{spherical coordinates}\end{array}\right) \quad (36.2)$$

where we've used Equation 36.1 for the potential-energy function.

Although Equation 36.2 looks forbidding, it can be solved readily using advanced techniques. For total energy $E$ less than zero, corresponding to bound states in the potential well of Fig. 36.1, most solutions become infinite at large $r$ and therefore aren't normalizable. As a result, only certain values of the energy $E$ give acceptable bound-state solutions. For total energy greater than zero, the electron is unbound and any energy proves possible, as with the finite square well.

## The Hydrogen Ground State

In general, solutions to Equation 36.2 depend on all three variables $r$, $\theta$, $\phi$. But some solutions, including the ground state, are spherically symmetric—they depend only on $r$. Here we show that the ground state has the form of an exponential, and in the process derive the ground-state energy. Consider the function

$$\psi = Ae^{-r/a_0} \quad (36.3)$$

where $A$ and $a_0$ are as yet undetermined constants, the latter with the units of length. For this spherically symmetric function, nothing depends on the angular variables $\theta$ and $\phi$, so derivatives with respect to those variables are strictly zero. We're then dealing with a function of only one variable, so we can write total instead of partial derivatives. Equation 36.2 then becomes

$$-\frac{\hbar^2}{2mr^2}\frac{d}{dr}\left(r^2\frac{d\psi}{dr}\right) - \frac{ke^2}{r}\psi = E\psi \quad (36.4)$$

Substituting the proposed solution 36.3 for $\psi$ gives

$$-\frac{\hbar^2}{2ma_0^2} + \frac{\hbar^2}{mra_0} - \frac{ke^2}{r} = E_1$$

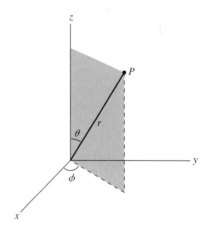

**FIGURE 36.2** Spherical coordinates $r$, $\theta$, $\phi$ provide an alternative to rectangular coordinates $x, y, z$.

(see Problem 42), where $E_1$ is the ground-state energy. This equation must be satisfied for all values of $r$, so the two $r$-dependent terms must cancel:

$$\frac{\hbar^2}{mra_0} = \frac{ke^2}{r}$$

or

$$a_0 = \frac{\hbar^2}{mke^2} = 5.29 \times 10^{-11}\ \text{m} = 0.0529\ \text{nm}$$

This is precisely the **Bohr radius** that we introduced in Chapter 34.

With the $r$ terms gone, our expression for the ground-state energy becomes $E_1 = -\hbar^2/2ma_0^2 = -13.6$ eV, where the minus sign shows that the atom is a bound system. Thus Equation 36.3 is indeed a solution to the Schrödinger equation for hydrogen, with energy $E_1 = -13.6$ eV.

In deriving expressions for $a_0$ and $E_1$, we've shown how Schrödinger's theory gives two fundamental parameters of atomic physics: the Bohr radius and the hydrogen ground-state energy. Both agree with the values we found in Chapter 34 using the simpler Bohr model. But Bohr's theory still clings to the notion of classical orbits, with $a_0$ the ground-state orbital radius. Schrödinger's theory is truly quantum mechanical, representing the electron with its wave function $\psi$ and associated probability distribution. The Bohr radius is no longer an actual orbital radius but instead determines atomic size only in a statistical sense.

## The Radial Probability Distribution

Because the ground-state wave function falls off exponentially as $e^{-r/a_0}$, we're unlikely to find the electron at distances far greater than the Bohr radius. But where are we most likely to find it? Although $\psi$ is greatest at $r = 0$, that's not the answer. In three dimensions the probability density $\psi^2$ is the probability *per unit volume* of finding the electron. In asking where we're most likely to find the electron, we want the probability *per unit radial distance*. Figure 36.3 shows a thin spherical shell with radius $r$ and therefore area $4\pi r^2$. It has thickness $dr$, so its volume is $dV = 4\pi r^2\, dr$. Then the probability of finding the electron in this shell is $\psi^2\, dV = 4\pi r^2\psi^2\, dr$. The **radial probability density**, $P(r)$, is the probability per unit radius, or

$$P(r) = 4\pi r^2\psi^2 \qquad \text{(radial probability density)} \qquad (36.5)$$

**FIGURE 36.3** A thin shell has volume $dV = 4\pi r^2\, dr$; thus the probability per unit radial distance is $4\pi r^2$ times the probability per unit volume.

For the hydrogen ground state, we use Equation 36.3 for $\psi$ to get $P_1 = 4\pi r^2 A^2 e^{-2r/a_0}$. Figure 36.4 is a plot of this probability density, which peaks at $r = a_0$. Thus the single most likely place to find the electron in ground-state hydrogen is one Bohr radius from the proton.

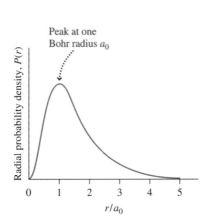

**FIGURE 36.4** Radial probability density for the hydrogen ground state.

**EXAMPLE 36.1**   **The Hydrogen Atom: Normalization and the Probability Distribution**

(a) Determine the normalization constant $A$ in Equation 36.3. (b) Use the resulting wave function to find the probability that the electron in the hydrogen ground state will be found beyond the Bohr radius.

**INTERPRET** The wave function 36.3 contains an undetermined constant $A$. This problem is asking us to apply the normalization condition to find $A$ and then use the concept of radial probability density to determine the likelihood of finding the electron beyond $r = a_0$.

**DEVELOP** The electron must be *somewhere* in the range $r = 0$ to $r = \infty$. Since $P(r)\,dr$ is the probability of finding the electron in a region of width $dr$, the normalization condition becomes $\int_0^\infty P(r)\,dr = 1$. So our plan is to evaluate this integral using the ground-state probability density $P_1 = 4\pi r^2 A^2 e^{-2r/a_0}$. We'll then solve for the unknown $A$. Then we can integrate again, this time from $r = a_0$ to $r = \infty$, to get the probability of finding the electron beyond $a_0$.

**EVALUATE** Using the probability density $P_1$, the normalization condition becomes

$$\int_{r=0}^{r=\infty} 4\pi r^2 A^2 e^{-2r/a_0}\,dr = 1$$

We could evaluate using integration by parts; however, the result is in the integral table at the end of Appendix A. Replacing $x$ by $r$ and $a$ by $-2/a_0$ in the table's expression for $\int x^2 e^{ax}\,dx$, we have

$$\int_0^\infty 4\pi A^2 r^2 e^{-2r/a_0}\,dr$$

$$= 4\pi A^2 \left\{ \frac{r^2 e^{-2r/a_0}}{(-2/a_0)} - \frac{2}{(-2/a_0)}\left[ \frac{e^{-2r/a_0}}{(-2/a_0)^2}\left( -\frac{2}{a_0}r - 1 \right) \right] \right\}\Bigg|_0^\infty = 1$$

The expression in curly brackets vanishes at $r = \infty$, and at $r = 0$ the exponentials are just 1, so we have $4\pi A^2\left[ 0 - \left(-\frac{1}{4}a_0^3\right) \right] = 1$, or $A = 1/\sqrt{\pi a_0^3}$. (b) Now we change the lower limit on the integral from 0 to $a_0$. The result is

$$P(r > a_0) = \int_a^\infty 4\pi r^2 A^2 e^{-2r/a_0}\,dr = 4\pi A^2 a_0^3\left(\tfrac{1}{2}e^{-2} + \tfrac{3}{4}e^{-2}\right) = 5\pi A^2 a_0^3 e^{-2}$$

With $A^2 = 1/\pi a_0^3$, this becomes $P(r > a_0) = 5e^{-2} \approx 0.677$.

**ASSESS** Our result shows that about two-thirds of the time, the electron will be found beyond the Bohr radius. So although it's reasonable to say that the atom's radius is roughly the Bohr radius, both Fig. 36.4 and our result here show that there's no sharp cutoff that marks the "size" of the atom. ∎

## Excited States of Hydrogen

So far we've examined only the ground state of hydrogen. But Equation 36.2 admits many more normalizable solutions, corresponding to the excited states of hydrogen.

In general, each energy level is associated with one spherically symmetric wave function and a number of nonsymmetric ones. For historical reasons, the spherically symmetric states are called *s* **states**. The distinct energy levels are labeled by the quantum number $n$, called the **principal quantum number**. The ground state, for example, is the 1*s* state. The energy of the $n$th level, derivable from the Schrödinger equation, turns out to agree exactly with the earlier Bohr theory:

$$E_n = -\frac{1}{n^2}\frac{\hbar^2}{2ma_0^2} = \frac{E_1}{n^2} = \frac{-13.6\text{ eV}}{n^2} \quad \text{(hydrogen energy levels)} \quad (36.6)$$

The spherically symmetric state with energy $E_2$—that is, the 2*s* state—has wave function given by

$$\psi_{2s} = \frac{1}{4\sqrt{2\pi a_0^3}}\left( 2 - \frac{r}{a_0} \right)e^{-r/2a_0} \quad (36.7)$$

By substituting this function into Equation 36.4, you can verify that the energy $E_2$ is given by Equation 36.5 (see Problem 63). The radial probability densities for the first three spherically symmetric states are plotted in Fig. 36.5; notice that the excited states correspond to larger, more "smeared-out" atoms.

Although we're discussing hydrogen, our results generalize to any single-electron atom—that is, to an atom of atomic number $Z$ ionized $Z - 1$ times. For such an atom the

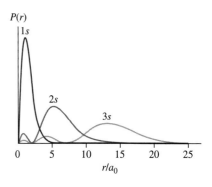

**FIGURE 36.5** Radial probability densities for the spherically symmetric 1*s*, 2*s*, and 3*s* states of hydrogen.

potential-energy function becomes $-kZe^2/r$, and our calculations go through as before except that the factor $e^2$ is replaced by $Ze^2$. Then the energy levels become

$$E_n = -\frac{Z^2}{n^2}\frac{\hbar^2}{2ma_0^2} = \frac{Z^2 E_1}{n^2} = -\frac{(13.6\ \text{eV})Z^2}{n^2} \tag{36.8}$$

reflecting the tighter binding of the more highly charged nucleus (see Problem 64).

---

**GOT IT? 36.1** Which is the most appropriate estimate for the radial "size" of a hydrogen atom in its $2s$ state: (a) $a_0$, (b) $2a_0$, (c) $5a_0$, (d) $15a_0$?

---

## Orbital Quantum Numbers and Angular Momentum

In the spherically symmetric $s$ states, it turns out that the **orbital angular momentum** associated with the electron's motion is zero. This is at odds with Bohr's prediction that angular momentum should be a multiple of $\hbar$. And it makes clear that we can't be talking here about classical orbits, since motion in an elliptical or circular path entails angular momentum. But there are other solutions to the Schrödinger equation for hydrogen, solutions that aren't spherically symmetric and that have nonzero angular momentum.

For a given principal quantum number $n$, there are in fact $n$ distinct solutions with different angular momenta. The **orbital quantum number** $l$ distinguishes these states and ranges from 0 to $n-1$. Thus the ground state $(n=1)$ corresponds to the single value $l=0$. Higher energy levels, however, are degenerate, meaning there's more than one $l$ value for each $n>1$. The orbital quantum number determines the magnitude $L$ of the electron's orbital angular momentum:

$$L = \sqrt{l(l+1)}\hbar \qquad \text{(quantization of orbital angular momentum)} \tag{36.9}$$

---

**EXAMPLE 36.2** **Orbital Angular Momentum: An Excited State**

Find the possible values for the orbital angular momentum of an electron in the $n=3$ state of hydrogen.

**INTERPRET** We're asked about the orbital angular momentum $L$, whose value follows from the orbital quantum number $l$. Thus we'll need the possible $l$ values for $n=3$.

**DEVELOP** There are $n$ distinct $l$ values, from 0 to $n-1$. For $n=3$, that means $l=0$, 1, or 2. So our plan is to evaluate $L$ using Equation 36.9, $L = \sqrt{l(l+1)}\hbar$, for these three $l$ values.

**EVALUATE** With $l=0$, Equation 36.9 gives $L=0$; for $l=1$, $L=\sqrt{2}\hbar$; and for $l=2$, $L=\sqrt{6}\hbar$.

**ASSESS** $l=0$ is the spherically symmetric $3s$ state, which we've seen has zero angular momentum. The higher-$l$ states have increasing angular momentum. ∎

---

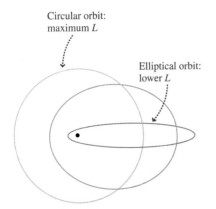

Circular orbit: maximum $L$

Elliptical orbit: lower $L$

**FIGURE 36.6** Classical electron orbits with the same energy but different angular momenta.

States with $l$ values 0, 1, 2, 3, 4, 5, ... are given the letter labels $s$, $p$, $d$, $f$, $g$, $h$, .... These combine with the principal quantum number $n$ to specify both the energy and angular momentum of a state. Thus, the ground state is $1s$, and the $n=2$ state with $l=1$ is the $2p$ state. (The lowercase letters $s$, $p$, $d$, ... are used in discussing individual electrons, while the corresponding capital letters denote orbital angular momentum states of an entire atom. For one-electron hydrogen, the two are the same.)

Quantization of orbital angular momentum is another nonclassical aspect of quantum mechanics. In classical physics, an electron of a given energy can have any angular momentum, up to the maximum of a circular orbit (Fig. 36.6). At high $n$, the number of $l$ values is so large that we don't notice quantization—another manifestation of Bohr's correspondence principle. But for low $n$, the quantum-mechanical discreteness of both energy and angular momentum is clearly evident.

## Space Quantization

Angular momentum is a vector, and the angular momentum vector is quantized not only in magnitude but also in direction—a phenomenon called **space quantization**. Space quantization of orbital angular momentum gives rise to a third quantum number, $m_l$. Space quantization becomes evident when an atom is in a magnetic field that establishes a preferred axis along which the angular momentum component can be measured; for this reason $m_l$ is called the **orbital magnetic quantum number**.

Space quantization requires that the component $L_z$ of orbital angular momentum along any chosen axis have only values given by

$$L_z = m_l \hbar \qquad \text{(space quantization)} \qquad (36.10)$$

where $m_l$ takes integer values from $-l$ to $l$. Thus an $l = 1$ state can have one of three possible $m_l$ values: $-1$, $0$, or $+1$, corresponding to angular momentum components $-\hbar$, $0$, or $+\hbar$ along some axis. Since the magnitude of the angular momentum in an $l = 1$ state is $\sqrt{2}\hbar$ (see Example 36.2), none of these values corresponds to full alignment with the axis. Instead, we can think geometrically of the angular momentum vectors as being constrained to lie at angles $\cos^{-1}(L_z/L)$ to the axis; for $l = 1$ these angles are $\pm 45°$ and $90°$ (Fig. 36.7). Although the angle is useful for diagramming the angular momentum vector, we emphasize that the quantum numbers $l$ and $m_l$ tell everything there is to know about quantized orbital angular momentum. Quantum physicists, therefore, aren't usually concerned with the orientation of angular momentum vectors.

**FIGURE 36.7** The three possible orientations for the angular momentum vector in the $l = 1$ state, where $L = \sqrt{2}\hbar$. Only the $z$ component is fixed; the $x$ and $y$ components are uncertain.

## 36.2 Electron Spin

Detailed observation of the hydrogen spectrum shows that spectral lines exhibit a fine splitting; where a lower-resolution spectrum shows one spectral line, at higher resolution there appears a closely spaced pair of lines. This splitting could not be explained using the three quantum numbers $n$, $l$, and $m_l$. In 1925 the Austrian physicist Wolfgang Pauli suggested that a fourth quantum number, capable of taking only two values, might be needed. Soon Samuel Goudsmit and George Uhlenbeck realized that the spectral splitting could be explained if this fourth quantum number were associated with an intrinsic angular momentum, or **spin**, carried by the electron. Later, as we indicated in Chapter 35, Paul Dirac showed that electron spin follows from the requirement of relativistic invariance. Spin is an inherently quantum-mechanical property with no classical analog. Although it can be visualized crudely by imagining the electron to be a small sphere spinning about an axis, this classical picture is really inappropriate.

Spin angular momentum is quantized similarly to orbital angular momentum. But unlike the orbital quantum number $l$ that takes a range of integer values, the electron spin quantum number $s$ has only the single value $s = \frac{1}{2}$. The electron is therefore a **spin-$\frac{1}{2}$** particle. The magnitude of the spin angular momentum is related to the spin quantum number in the same way that the magnitude of orbital angular momentum is related to the orbital quantum number $l$:

$$S = \sqrt{s(s + 1)}\hbar \qquad \text{(quantization of spin angular momentum)} \qquad (36.11)$$

Since $s$ takes only the value $\frac{1}{2}$, the magnitude of the electron spin angular momentum is $S = \frac{\sqrt{3}}{2}\hbar$.

Spin angular momentum also exhibits space quantization. That is, the component of spin along a chosen axis takes only the values

$$S_z = m_s \hbar \qquad (36.12)$$

where the quantum number $m_s$ has the two possible values $-\frac{1}{2}$ and $+\frac{1}{2}$. Figure 36.8 shows space quantization of electron spin.

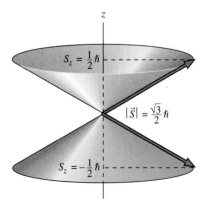

**FIGURE 36.8** Space quantization of electron spin.

## Magnetic Moment of the Electron

Together, the electron's spin and electric charge mean the electron behaves like a miniature current loop, with an intrinsic magnetic dipole moment. The dipole moment vector $\vec{M}$ associated with the spin angular momentum vector $\vec{S}$ is given by

$$\vec{M} = -\frac{e}{m}\,\vec{S} \tag{36.13}$$

with $e/m$ the electron's charge-to-mass ratio. Since the component of $\vec{S}$ on any axis can take only the values $\pm\frac{1}{2}\hbar$, the components of the magnetic moment can be only

$$M_z = \pm\frac{e\hbar}{2m} \tag{36.14}$$

The quantity $e\hbar/2m$ is a fundamental unit for measuring magnetic moments called the **Bohr magneton**; its value is $9.27 \times 10^{-24}$ A·m$^2$.

The ratio of magnetic moment to spin angular momentum is twice what we would expect classically for a charged particle in circular motion. Like spin itself, the factor of 2 is a relativistic effect first explained by Dirac. Actually, the factor is not quite 2 but approximately 2.00232, a result that follows from the theory of quantum electrodynamics.

## The Stern–Gerlach Experiment

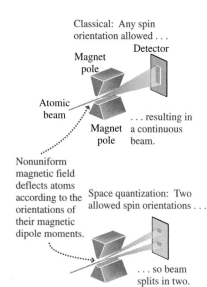

Classical: Any spin orientation allowed . . .

Magnet pole

Detector

Atomic beam

Magnet pole

. . . resulting in a continuous beam.

Nonuniform magnetic field deflects atoms according to the orientations of their magnetic dipole moments.

Space quantization: Two allowed spin orientations . . .

. . . so beam splits in two.

**FIGURE 36.9** The Stern–Gerlach experiment.

In 1922, Otto Stern and Walther Gerlach at the University of Hamburg demonstrated the quantization of atomic angular momentum vectors. The **Stern–Gerlach experiment** used a nonuniform magnetic field to separate a beam of silver atoms according to the component of their angular momentum along the field direction. T. E. Phipps and J. B. Taylor repeated the experiment in 1927, giving unambiguous verification of quantized electron spin. They used hydrogen atoms in the ground state; as we've seen, this state has zero orbital angular momentum, so the only angular momentum effects are due to electron spin. Classically a beam of hydrogen should be spread into a continuous band corresponding to angular momentum components from $\frac{-\sqrt{3}}{2}\hbar$ to $\frac{+\sqrt{3}}{2}\hbar$. But in fact the beam always splits in two, corresponding to the two angular momentum components $\pm\frac{1}{2}\hbar$. Figure 36.9 shows the experiment.

**GOT IT? 36.2** The nucleus of oxygen-17 has spin $\frac{5}{2}$. How many possible orientations are there for its spin angular momentum vector?

## Total Angular Momentum and Spin-Orbit Coupling

Orbital and spin angular momenta combine through the process of **spin-orbit coupling** to give an atom's total angular momentum, $\vec{J}$:

$$\vec{J} = \vec{L} + \vec{S} \tag{36.15}$$

The magnitude $J$ is quantized similarly to orbital and spin angular momenta:

$$J = \sqrt{j(j+1)}\,\hbar \qquad \text{(quantization of total angular momentum)} \tag{36.16}$$

For an atom with a single electron, the quantum number $j$ takes the values

$$j = l \pm \tfrac{1}{2} \qquad \text{for } l \neq 0 \tag{36.17a}$$

$$j = \tfrac{1}{2} \qquad \text{for } l = 0 \tag{36.17b}$$

The state of an atom with total angular momentum $J$ is specified by the principal quantum number, the capital letter designating the orbital angular momentum $(S, P, D, F, G, \ldots)$, and, as a subscript, the $j$ value. Thus a hydrogen atom with $n = 3$, $l = 2$, and $j = \frac{3}{2}$ is designated $3D_{3/2}$.

Total angular momentum also exhibits space quantization, with the component of $\vec{J}$ on some axis given by

$$J_z = m_j \hbar \qquad (36.18)$$

Here the quantum number $m_j$ takes the values $(-j, -j+1, \ldots, j-1, j)$.

Derivation of these so-called **angular momentum coupling rules** is not easy, but we can understand them in terms of simple vector diagrams like those shown in Fig. 36.10.

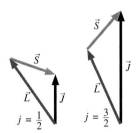

**FIGURE 36.10** Spin-orbit coupling with $l = 1$, for which Equation 36.17a gives $j = \frac{1}{2}$ or $j = \frac{3}{2}$.

---

**EXAMPLE 36.3** **Spin-Orbit Coupling: Finding the Angular Momentum**

(a) Find the possible magnitudes for the total angular momentum of hydrogen in the $l = 2$ state. (b) For each possible $J$, how many values are there for the component of $\vec{J}$ on a given axis?

**INTERPRET** We're asked about total angular momentum, which results from spin and orbital angular momentum contributions as shown in Fig. 36.10, and about the space quantization of $\vec{J}$.

**DEVELOP** (a) Equations 36.16, $J = \sqrt{j(j+1)}\hbar$, and 36.17a, $j = l \pm \frac{1}{2}$, determine $J$. With $l = 2$, our plan is to use Equation 36.17a to find the possible $j$ values. Then we can apply Equation 36.16 to get each corresponding $J$. (b) Equation 36.18, $J_z = m_j \hbar$, determines the $J_z$ values in terms of $m_j$, which ranges from $-j$ to $j$. So once we have the $j$'s, we can determine the number of $J_z$ values.

**EVALUATE** (a) With $l = 2$, Equation 36.17a gives $j = \frac{3}{2}$ or $j = \frac{5}{2}$. For $j = \frac{3}{2}$, Equation 36.16 yields $J = \sqrt{\frac{3}{2}\left(\frac{3}{2}+1\right)}\hbar = \frac{\sqrt{15}}{2}\hbar$; similarly, $j = \frac{5}{2}$ gives $J = \frac{\sqrt{35}}{2}\hbar$. (b) For $j = \frac{3}{2}$ there are four possible $m_j$ values from $-j$ to $j$: $-\frac{3}{2}, -\frac{1}{2}, \frac{1}{2}, \frac{3}{2}$, and correspondingly four $J_z$ values, given by Equation 36.18: $-\frac{3}{2}\hbar, -\frac{1}{2}\hbar, \frac{1}{2}\hbar, \frac{3}{2}\hbar$. Similar counting gives six values for $j = \frac{5}{2}$.

**ASSESS** Figure 36.11 shows how the spin and orbit angular momenta combine to give the two possible $j$ values in this example. ∎

**FIGURE 36.11** Vector diagrams for spin-orbit coupling with $l = 2$ (Example 36.3).

---

The two $j$ values for a given $l$ correspond to two distinct quantum states, and these states have slightly different energies. This energy difference is associated with the orientation of the electron's magnetic moment in a magnetic field that, in the electron's reference frame, results from the apparent motion of the positively charged nucleus around the electron—a field whose direction is that of the electron's orbital angular momentum $\vec{L}$. As Equation 36.13 shows, the electron's negative charge means its spin $\vec{S}$ and magnetic moment $\vec{M}$ have opposite directions. Because a magnetic dipole has the highest energy when it's oriented opposite the magnetic field, this means that the more-nearly-parallel alignment of $\vec{S}$ and $\vec{L}$—corresponding to $j = l + \frac{1}{2}$—has the higher energy.

In hydrogen, the magnitude of the energy difference between the $j = \frac{1}{2}$ and $j = \frac{3}{2}$ states of the first excited level is only $5 \times 10^{-5}$ eV, far smaller than the separation between this level and the ground state. Because the $n = 2$, $l = 1$ state is actually two states of slightly different energy, hydrogen atoms undergoing transitions from these states to the ground state emit two spectral lines slightly separated in wavelength. The term **fine structure** describes this and related spectral-line splittings. In the present example, the split spectral line is called a **doublet**. Figure 36.12 is an energy-level diagram showing the effect of spin-orbit splitting in hydrogen. Other fine-structure effects also alter energy levels; in hydrogen, for example, both levels in Fig. 36.12 shown actually drop below the Schrödinger prediction, and the $2S_{1/2}$ level is degenerate with $2P_{1/2}$.

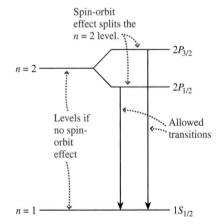

**FIGURE 36.12** Energy-level diagram showing spin-orbit splitting of the 2P levels in hydrogen. Other fine-structure effects lower the energies of both 2P states and of the ground state, and make the 2S state degenerate with $2P_{1/2}$.

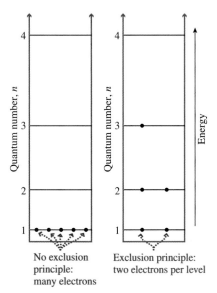

**FIGURE 36.13** Particles in a square well, showing the exclusion principle's effect on electron distribution.

The spin-orbit effect results from a magnetic field internal to the atom itself. But splitting of energy levels also occurs in an external magnetic field and is called the **Zeeman effect**. We showed an example of Zeeman splitting in Fig. 35.18.

Since it has zero orbital angular momentum, the ground state does not exhibit spin-orbit splitting. But interaction of the electron spin with the magnetic dipole moment of the nucleus results in an even finer splitting known as **hyperfine structure**. The transition between the two hyperfine levels of the hydrogen ground state—corresponding physically to a change in the orientation of the electron spin vector—involves a photon of 21-cm wavelength. Radio astronomers use the 21-cm hydrogen radiation to map interstellar hydrogen in the cosmos.

## 36.3 The Exclusion Principle

In trying to understand why atomic electrons distributed themselves as they did, Pauli in 1924 developed his **exclusion principle**, which, loosely, states that **two electrons cannot be in the same quantum state**. Since an electron's quantum state includes its spin orientation specified by $m_s$, the exclusion principle means that at most two electrons can occupy a state whose other quantum numbers $n$, $l$, and $m_l$ are the same.

The Pauli exclusion principle has profound implications for multielectron systems, requiring that most electrons remain in high-energy states (Fig. 36.13). If the exclusion principle didn't hold, atomic electrons would collapse to the ground state and there would be no such thing as chemistry or life! The exclusion principle even manifests itself at the cosmic scale, as the Application shows.

---

**APPLICATION** **White Dwarfs and Neutron Stars**

When a star exhausts its nuclear fuel, it collapses because there's no longer pressure to counter gravity. For a star with more than several times the Sun's mass, there's no force strong enough to halt the collapse, and the star becomes a black hole from which nothing can escape. But in less massive stars the collapse eventually halts because of a quantum-mechanical pressure associated with the exclusion principle.

When the Sun collapses some 5 billion years from now, its electrons will drop into the lowest available energy states. But as with the square well in Fig. 36.13, the exclusion principle requires that most of the Sun's $10^{57}$ electrons will end up in high-energy states. The associated **degenerate electron pressure**—independent of temperature, unlike the pressure of an ordinary gas—will stabilize the Sun as a **white dwarf**, about the size of Earth. For stars more massive than about 1.4 Sun masses, collapse proceeds until the protons and electrons merge to form neutrons. The neutrons, too, develop a degenerate pressure that stabilizes the resulting **neutron star** with a mass exceeding the Sun's crammed into a 20-km sphere! The figure compares the sizes of these stellar endpoints.

---

**FIGURE 36.14** Velocity distribution of atoms in a Bose–Einstein condensate shows a large peak at the near-zero velocity of atoms all in their common ground state. The three peaks show the evolution from a normal gas to the condensate.

The Pauli exclusion principle quickly became fundamental to the developing quantum mechanics of the late 1920s. But physicists remained dissatisfied invoking this seemingly ad hoc rule with no theoretical basis. Late in the 1930s, following detailed analysis of relativistic quantum theories, Pauli finally showed that the exclusion principle, like spin, is ultimately grounded in the requirement of relativistic invariance. Pauli found that particles whose spin quantum number $s$ is a half-integer (collectively called **fermions**) must necessarily obey the exclusion principle. On the other hand, particles with integer spin (called **bosons**) *do not* obey the exclusion principle. Photons, for example, are spin-1 particles and therefore an arbitrarily large number of them can occupy exactly the same quantum state. The laser, with its intense, coherent beam of light, is possible because the many photons that make up the beam are essentially all in the same state. In 1995 physicists at the University of Colorado first succeeded in producing an assemblage of bosonic matter all in the same quantum state (Fig. 36.14). This so-called **Bose–Einstein condensate** had been a goal of physicists since 1924, when the Indian physicist Satyendra Nath Bose first suggested the possibility. The Bose–Einstein condensate represents a truly new state of matter, in which thousands of atoms join quantum mechanically to behave as a single entity. Today physics labs around the world are experimenting with Bose–Einstein condensates, probing the fundamentals of quantum physics and developing applications including atom-beam analogs of the optical laser.

## 36.4 Multielectron Atoms and the Periodic Table

Our modern understanding of the chemical elements developed in the late 18th century, when chemists first distinguished compounds, such as water, from elements, defined as substances that couldn't be decomposed by chemical means. From the formulas for various compounds, chemists could determine the relative atomic masses of the elements. The first attempts to organize the elements systematically used atomic mass, but a breakthrough occurred in 1869, when the Russian chemist Dmitri Mendeleev set up a table with the approximately 60 elements then known. He left blanks where necessary to maintain the periodic occurrence of similar chemical properties. Elements filling the blanks were soon discovered, validating Mendeleev's periodic table and suggesting an underlying order in the composition of atoms. Then, early in the 20th century, studies of X-ray spectra led to a table organized by **atomic number** $Z$, the number of protons in the nucleus. When this was done, a number of elements missing from earlier periodic tables were identified. The modern periodic table is shown in Fig. 36.15 and is printed with atomic weights inside the back cover.

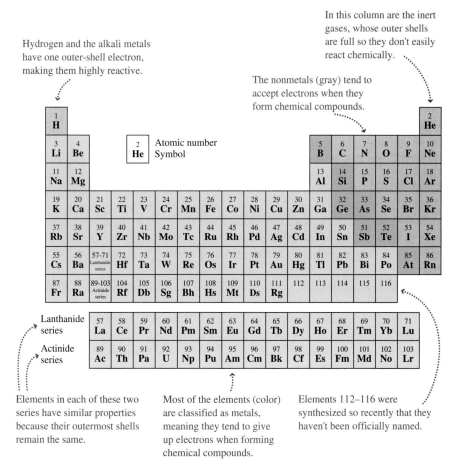

**FIGURE 36.15** The periodic table. A larger version, with atomic weights, is inside the back cover; the names of the elements are in Appendix D.

## Explaining the Periodic Table

The orderly arrangement of elements in the periodic table enhances our understanding of chemistry and our ability to formulate new and useful compounds. But why does nature exhibit this regularity? The answer lies in the Schrödinger equation and the exclusion principle.

Solution of the Schrödinger equation for multielectron atoms is complicated by the interactions among the electrons; analytic solutions like those for hydrogen aren't generally available. But qualitatively, we still find energy levels characterized by the principal quantum number $n$. Each such level is called a **shell**; for historical reasons, the shells $n = 1, 2, 3, \ldots$ are also labeled with the letters $K, L, M, \ldots$. As with hydrogen, an electron at the $n$th energy level can have any of the $n$ values $l = 0, 1, 2, \ldots, n - 1$ for the orbital quantum number. The different angular momentum states within a shell are termed **subshells**; subshells with the values $l = 0, 1, 2, 3, \ldots$ are labeled with the letters $s, p, d, f, \ldots$. Finally, for each subshell there are $2l + 1$ possible values of the magnetic orbital quantum number $m_l$, ranging from $-l$ to $l$. A state characterized by all three quantum numbers $n$, $l$, and $m_l$ is called an **orbital**. Table 36.1 summarizes shell-structure notation; for completeness, the table also lists the spin quantum number $m_s$.

**TABLE 36.1** Atomic Shell Structure

| Quantum Number | Shell Notation | Allowed Values | Letter Labels | Number of States |
|---|---|---|---|---|
| $n$ | Shell | $1, 2, 3, \ldots$ | $K, L, M, \ldots$ | Infinite |
| $l$ | Subshell | $0, 1, 2, \ldots, n-1$ | $s, p, d, f, \ldots$ | $n$ |
| $m_l$ | Orbital | $-l, -l+1, \ldots, l-1, l$ | — | $2l + 1$ |
| $m_s$ | — | $-\frac{1}{2}, +\frac{1}{2}$ | | 2 |

The structure of a multielectron atom is determined by the quantum states of its constituent electrons—their distribution among the shells, subshells, and orbitals. According to the exclusion principle, no two electrons can be in exactly the same quantum state; that is, they can't have the same values for all four quantum numbers $n$, $l$, $m_l$, and $m_s$. Since an atomic orbital is characterized by the three quantum numbers $n$, $l$, and $m_l$, the exclusion principle implies that at most two electrons can occupy a single orbital.

We're now ready to understand the ground-state electronic structure of multielectron atoms. The simplest is helium (He), with two electrons. The $K$ shell $(n = 1)$ is the lowest possible energy level. As Table 36.1 shows, only the zero-angular-momentum $s$ subshell is permitted within the $K$ shell, and within that subshell there's only the single orbital corresponding to $m_l = 0$. But that orbital can accommodate two electrons. So in the ground state of helium, both electrons occupy the $s$ subshell of the $K$ shell. We describe this with the notation $1s^2$, where 1 stands for the principal quantum number $n$, $s$ for the subshell, and the superscript 2 for the number of electrons in that subshell. The corresponding notation for hydrogen is $1s^1$.

After helium comes lithium (Li), with three electrons. From our analysis of helium, we know that the $K$ shell is full with two electrons. So the third electron goes into the $L$ shell, or $n = 2$ energy level. Of the subshells in the $L$ shell, the $s$ subshell turns out to have slightly lower energy than the others, so the third electron occupies the $s$ subshell. Then the electronic configuration of lithium is $1s^2 2s^1$—that is, a helium-like core with a single outer electron in the $s$ subshell of the $n = 2$ level.

Beryllium (Be), with four electrons, fills the $1s$ and $2s$ subshells; its designation is $1s^2 2s^2$. The fifth electron of boron (B) then goes into the $2p$ subshell, giving the structure $1s^2 2s^2 2p^1$. Table 36.1 shows that a $p$ subshell $(l = 1)$ allows three $m_l$ values—that is, three orbitals, capable of holding a total of six electrons. As we advance in atomic number, electrons continue to fill the $p$ subshell. Finally, at neon $(Z = 10)$, the $2p$ subshell is full. Only with the next element, sodium $(Z = 11)$, does the $n = 3$ shell begin to fill. Table 36.2 lists electronic configurations for the elements hydrogen $(Z = 1)$ through argon $(Z = 18)$, along with their ionization energies (the energy required to remove the outermost electron).

**TABLE 36.2** Electronic Configurations and Ionization Energies of Elements 1–18

| Atomic Number, $z$ | Element | Electronic Configuration | Ionization Energy (eV) |
|---|---|---|---|
| 1 | H | $1s^1$ | 13.60 |
| 2 | He | $1s^2$ | 24.60 |
| 3 | Li | $1s^2 2s^1$ | 5.390 |
| 4 | Be | $1s^2 2s^2$ | 9.320 |
| 5 | B | $1s^2 2s^2 2p^1$ | 8.296 |
| 6 | C | $1s^2 2s^2 2p^2$ | 11.26 |
| 7 | N | $1s^2 2s^2 2p^3$ | 14.55 |
| 8 | O | $1s^2 2s^2 2p^4$ | 13.61 |
| 9 | F | $1s^2 2s^2 2p^5$ | 17.42 |
| 10 | Ne | $1s^2 2s^2 2p^6$ | 21.56 |
| 11 | Na | $1s^2 2s^2 2p^6 3s^1$ | 5.138 |
| 12 | Mg | $1s^2 2s^2 2p^6 3s^2$ | 7.644 |
| 13 | Al | $1s^2 2s^2 2p^6 3s^2 3p^1$ | 5.984 |
| 14 | Si | $1s^2 2s^2 2p^6 3s^2 3p^2$ | 8.149 |
| 15 | P | $1s^2 2s^2 2p^6 3s^2 3p^3$ | 10.48 |
| 16 | S | $1s^2 2s^2 2p^6 3s^2 3p^4$ | 10.36 |
| 17 | Cl | $1s^2 2s^2 2p^6 3s^2 3p^5$ | 13.01 |
| 18 | Ar | $1s^2 2s^2 2p^6 3s^2 3p^6$ | 15.76 |

Gaps mark ends of periodic-table rows.

The chemical behavior of an atom is determined primarily by its outermost electrons, both because these electrons interact most directly with nearby atoms and because they're most weakly bound to their nuclei. Table 36.2 shows that the outer-electron configurations for lithium through neon (Ne) are the same as the corresponding configurations for sodium (Na) through argon (Ar). The chemical properties of the corresponding atoms are therefore similar. Both lithium and sodium, for example, have a single electron in their outermost shell. As the relatively low ionization energy suggests, this electron is loosely bound, so it readily interacts with other atoms, giving these elements their extreme reactivity. Neon and argon, in contrast, both have completely filled outermost shells. All the outer-shell electrons have essentially the same energy; the corresponding ionization energy is high; and there's little tendency for these electrons to interact with other atoms. As a result, argon and neon are **inert**; they don't readily form chemical compounds, and at normal temperatures they're gases. Other element pairs also share similar properties. Fluorine and chlorine, for instance, each need only one more electron to achieve the energetically favorable inert-gas configuration. Consequently these elements readily accept electrons. Materials such as common salt, NaCl, owe their high melting points to the strong bond that results when reactive sodium gives up its outer electron to electron-accepting chlorine, and the resulting positive and negative ions bind strongly by the electrostatic force. We'll consider molecular bonding in the next chapter.

Beyond argon $(Z = 18)$, shielding effects of the inner electrons result in the $4s$ states having lower energy than the $3d$ states. Potassium $(Z = 19)$ thus has the electronic configuration $1s^2 2s^2 2p^6 3s^2 3p^6 4s^1$ rather than $1s^2 2s^2 2p^6 3s^2 3p^6 3d^1$. After potassium comes calcium, with two electrons in its single $4s$ orbital. But the $4p$ orbitals do have higher energy than the $3d$, so elements beyond calcium begin filling the $3d$ orbitals. The next ten elements, scandium through zinc, have chemical properties that vary only slightly because their two outermost electrons remain $4s$ electrons; collectively, they're **transition elements**. (Chromium, $Z = 24$, and copper, $Z = 29$, are minor exceptions; in these an extra electron goes into the $3d$ orbitals, leaving only one $4s$ electron). Finally, elements 31 (gallium) through 36 (krypton) repeat the pattern of aluminum through argon shown in Table 36.2, as their $4p$ orbitals fill with electrons. Krypton, with its outer $p$ subshell full, is again an inert gas.

**EXAMPLE 36.4** **Atomic Structure: The Electronic Configuration of Iron**

Given that the 4s states fill before the 3d states in iron (Fe), determine the electronic configuration of iron.

**INTERPRET** We're asked for the symbolic configuration like those in Table 36.2.

**DEVELOP** Generally electrons fill in order of increasing shell number $n$, and within each shell with increasing $l$ ($s, p, d, f, \ldots$)—except we're told that in iron, like other transition elements, the 4s states fill before 3d. So our plan is to start listing the shells and orbitals in the appropriate order until we've accounted for all of iron's electrons. We don't really need to start from scratch because we can read the configuration of the innermost 18 electrons from Table 36.2's entry for argon.

**EVALUATE** The periodic table shows that $Z = 26$ for iron, so we need to accommodate 8 more electrons outside iron's 18-electron argon-like core. Argon's outermost electrons are the 3p electrons; normally 3d would come next, but we're told 4s fills first. An s shell can hold only 2 electrons (see Table 36.1), so the remaining 6 go into 3d. So we get the structure of iron by tacking $3d^6 4s^2$ onto argon, giving $1s^2 2s^2 2p^6 3s^2 3p^6 3d^6 4s^2$.

**ASSESS** As a check, note that the superscripts add to give a total of 26 electrons. ∎

We can now understand the organization of the periodic table. The vertical columns contain elements with the same outer-electron configurations, resulting in similar chemical properties. Each row starts with an element whose outermost shell contains a single s electron; these are hydrogen and the highly reactive alkali metals. Each row ends with an inert gas, its outermost p subshell full. Traversing the first row involves filling the 1s orbital only; since this orbital holds at most two electrons, there are only two elements in the first row. The second row has eight elements, associated with the filling of the 2s and 2p orbitals as shown in Table 36.2. The third row is like the second, but with the 3s and 3p orbitals filling. Because the 4s orbitals have lower energy than the 3d orbitals, the third row ends with an inert gas whose 3p orbitals are full, and the fourth row begins as the 4s orbital begins to fill. But then come the elements $Z = 21$ through $Z = 30$, in which the 3d orbitals are filling; these make for ten additional elements in the fourth row. The fifth row is a repeat of the fourth, as first the 5s orbitals fill, then the higher-energy 4d orbitals, then the remaining 5p orbitals. The sixth row begins with element 55, cesium, as the 6s orbitals fill before either the 5d or 4f orbitals. Two elements later, at lanthanum, the 4f orbitals begin to fill. Since their outer electrons remain 6s electrons, these elements have very similar properties. Together, they constitute the **lanthanide series**. Because its elements are so similar chemically, and to avoid stretching the sixth row of the periodic table, the lanthanide series is usually printed separately below the main table. Row seven repeats row six, with the **actinide series** again printed below. Elements beyond 92 (uranium) are radioactive with half-lives short compared with the age of the Earth. They're not found naturally but are produced in particle accelerators, fission reactors, and nuclear weapons. Elements 104 through 116 follow the actinide series and partially complete the seventh row of the periodic table. This group, some of which were first synthesized very recently, are generally short lived.

Note the crucial role the exclusion principle plays in our discussion of the chemical elements. Without that principle, every atom in its ground state would have all its electrons in the 1s orbital. There would be no qualitative distinction among the elements, and the science of chemistry would not exist. Nor would there be any chemists or physicists; life itself would be impossible without the rich diversity of chemical compounds formed from the different elements.

## 36.5 Transitions and Atomic Spectra

Emission and absorption of photons with specific energies provide the most direct manifestation of quantized atomic-energy levels, and give rise to the spectral lines that permit precise analysis of atomic systems from the laboratory to distant astrophysical objects. Even simple hydrogen exhibits a myriad of quantum states. In multielectron atoms, the possibilities for electronic excitation are even more numerous. The spectra of atoms reflect this rich array of available quantum states.

## Selection Rules

Not all energy-level transitions are equally likely. So-called **selection rules** determine which are **allowed transitions**—those most likely to occur. One rule limits allowed transitions to those for which the orbital quantum number $l$ changes by $\Delta l = \pm 1$; this and other selection rules are related to conservation of angular momentum. Quantum mechanics also provides a way of calculating transition probabilities or, equivalently, the mean lifetimes of excited states. For outer electrons, excited states that de-excite by allowed transitions have typical lifetimes on the order of $10^{-9}$ s.

Transitions that are not allowed by selection rules are called **forbidden transitions**; most are not strictly impossible but just extremely unlikely. States that can lose energy only by forbidden transitions are **metastable states**; their lifetimes are many orders of magnitude longer than the nanosecond timescale for allowed transitions. "Glow in the dark" phosphorescent materials emit light through the slow de-excitation of metastable states. Forbidden spectral lines are valuable probes of low-density astrophysical gases in which collisions are rare, and atoms can therefore remain in metastable states.

## Optical Spectra

Spectral lines in or near the visible involve transitions among the incompletely filled outer atomic shells. The alkali metals, with a single outer $s$ electron, therefore produce spectra qualitatively similar to that of hydrogen. However, the more complicated structure of a multielectron atom shifts some energy levels (Fig. 36.16). Many of the transitions in Fig. 36.16 are actually doublets or triplets resulting from spin-orbit splitting (Fig. 36.17). The energy-level structure is even more complicated for atoms with more than one outer-shell electron.

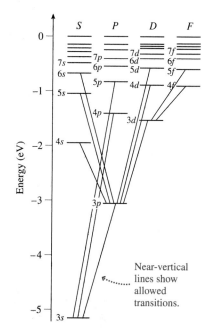

**FIGURE 36.16** Energy-level diagram for sodium, neglecting spin-orbit splitting. Note the widely separated 4s and 4p levels, with 3d between them; this explains why the 4s orbital fills before 3d.

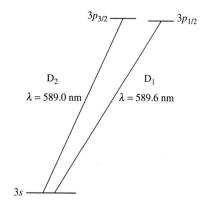

**FIGURE 36.17** Magnified portion of sodium's energy-level diagram, showing spin-orbit splitting of the 3p level. Transitions from the two states result in the slightly different wavelengths of the sodium D doublet.

---

**GOT IT? 36.3** Which of the transitions shown in Fig. 36.16 results in a photon of the shortest wavelength?

---

**EXAMPLE 36.5** **Atomic Spectra: The Sodium Doublet**

Use Fig. 36.17 to determine the energy difference between the 3p states of sodium.

**INTERPRET** We're asked about the energy difference between two atomic states ($3p_{1/2}$ and $3p_{3/2}$), and we're given the wavelengths of photons emitted in transitions from those states to a common end state ($3s$). We know that those photons carry off energy equal to the difference between the energies of the starting and ending states.

**DEVELOP** The quantization condition $E = hf$ relates photon energy and frequency; since $f\lambda = c$, we also have $E = hc/\lambda$. Our plan is to use this expression for the energies for the two transitions shown in Fig. 36.17, and then subtract to get the energy difference between the 3p levels.

**EVALUATE** We have $\Delta E_{3p} = \dfrac{hc}{589.0 \text{ nm}} - \dfrac{hc}{589.6 \text{ nm}} = 3.44 \times 10^{-22}$ J.

**ASSESS** Our answer is about 2 meV, much lower than the eV-range energies associated with optical transitions themselves. That's expected, given the small separation between the 3p states evident in Fig. 36.17. In sodium, states below 3s are all full, so 3s is the lowest end state for optical transitions. ∎

## Spontaneous and Stimulated Transitions

What makes an electron jump between energy levels? In an upward transition, the electron must absorb the appropriate amount of energy. Generally, that energy is supplied by a photon whose energy is equal to the energy difference between the two levels; the process is called **stimulated absorption** (Fig. 36.18a). (Upward transitions can result from other processes as well, as, for example, in an energetic collision between two atoms or the interaction of a free electron with atomic electrons.)

For most downward transitions, however, there's no specific cause. An electron spontaneously jumps from a higher to a lower energy level and a photon is emitted; this is **spontaneous emission** (Fig. 36.18b). Although an individual spontaneous emission is a random event, quantum mechanics gives the probability per unit time for that event to occur; the inverse of that probability is the mean lifetime of the excited state.

In 1917 Einstein recognized a third possibility: Excited atoms can be stimulated to drop into lower energy states by a photon, again of energy appropriate to the transition. A second photon is emitted in the process, with the same energy and phase as the stimulating photon, and in the same direction. This process, **stimulated emission**, is the reverse of stimulated absorption (Fig. 36.18c).

Spontaneous emission, stimulated absorption, and stimulated emission play major roles in the transfer of radiation through gases. And stimulated emission is responsible for an important technological development: the laser.

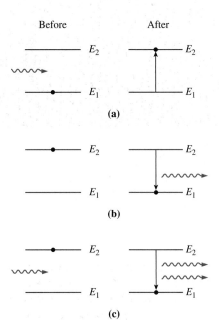

**FIGURE 36.18** Interaction of photons with atomic electrons. Horizontal lines denote two atomic energy levels, and the wave is a photon with energy equal to the difference between the two levels. (a) Stimulated absorption; (b) spontaneous emission; (c) stimulated emission.

**APPLICATION**   **The Laser**

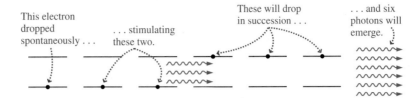

This electron dropped spontaneously . . .   . . . stimulating these two.   These will drop in succession . . .   . . . and six photons will emerge.

As Fig. 36.18c suggests, stimulated emission is a way to multiply photons with identical wavelength and phase. The **laser**, whose name derives from **l**ight **a**mplification by **s**timulated **e**mission of **r**adiation, exploits this effect to produce an intense beam of coherent light. The key to laser action is a **population inversion**, an unusual situation with many atoms in an excited state. The excited state is usually metastable, to prevent spontaneous emission from de-exciting the atoms. Atoms are first excited to a higher state from which they quickly drop by spontaneous emission to the metastable state, where they're "stuck" by the lack of allowed transitions downward. The excitation process is called **pumping**, and the excitation energy source is the **pump**. Laser pumps include flash lamps, sunlight, other lasers, electric currents, chemical reactions, and even nuclear explosions.

With a large number of excited atoms, it isn't long before one randomly de-excites even from the metastable state. It emits a photon that passes by other excited atoms, causing stimulated emission. That makes more photons and still more stimulated emission, as shown in the figure. The process snowballs, resulting in an intense beam of photons with the same wavelength and phase. In a laser, the radiating medium sits in a cavity with mirrors at the ends; as the photons reflect off the mirrors and traverse the medium, more and more stimulated emission results, building up the beam intensity. One mirror is only partially reflective to allow the laser beam to emerge. Some lasers produce a short burst of radiation before being pumped to prepare for another burst. Others are pumped continuously, resulting in a continuous beam.

The first laser, built in 1960, used a ruby rod as the lasing medium, surrounded by a coiled flash lamp for the pump. Since then a myriad of laser types have been developed. Almost anything can be used as the lasing medium, provided a population inversion is possible. Laser media include gases, solids, liquids, semiconductors, and ionized plasmas. Natural laser action occurs even in interstellar gas clouds. Some lasers, especially those using chemical dyes or temperature-sensitive semiconductors, are tunable over a range of wavelengths.

Laser light is monochromatic, since all photons have essentially the same energy. It's coherent because the photons all have the same phase. Coherence allows the beam to travel long distances with minimal spreading and enables very precise focusing. Finally, laser light can be made extremely intense, since stimulated emission extracts energy from many atoms simultaneously. Since photons are spin-1 particles that don't obey the exclusion principle, there's no limit to the number of photons in a laser beam. Small lasers like those used in laser pointers have power outputs in the sub-milliwatt range, while large continuous lasers in excess of 1 MW are available, and pulsed outputs of 100 MW and higher have been achieved. Multiple laser systems, like those used in nuclear fusion experiments, reach peak powers of $10^{14}$ W—100 times the output of all the world's electric generating plants.

Today lasers are ubiquitous. They're used in commonplace applications like bar-code scanners and CD/DVD drives. Medical lasers correct vision, whiten teeth, and perform bloodless surgery. Biologists use laser beams as "optical tweezers" to manipulate microscopic structures within cells. Lasers have replaced older technologies in surveying, leveling, and measuring instruments used in construction. Industrial lasers cut metal, shape gears, and harden surfaces. Semiconductor lasers drive the optical fibers that carry communications signals and Internet traffic. Military lasers lock on targets for precise weapons guidance. Ultrafast lasers probe chemical reactions that occur on femtosecond timescales. Lasers halt atoms' thermal motion, cooling materials to nanokelvin levels and enabling Bose–Einstein condensates. Laser beams reflected from the Moon measure its distance to within a few centimeters, testing Einstein's general relativity. Holograms capture interfering laser wavefronts, creating three-dimensional images. Future laser applications may include laser-driven spaceflight and the use of lasers to clear space debris.

## Big Picture

The big idea here is that atomic electrons are quantum particles trapped in the potential well associated with the electric force. Solving the Schrödinger equation then leads to quantized energy levels. Considerations of electron spin and angular momentum introduce subtle details into the energy-level structure. The **exclusion principle** permits only one atomic electron per quantum state, and this leads to the shell structure of atoms and the periodic table of the elements.

## Key Concepts and Equations

The **principal quantum number** $n$ determines the energy levels in hydrogen:

$$E_n = -\frac{1}{n^2}\frac{\hbar^2}{2ma_0^2} = \frac{E_1}{n^2} = \frac{-13.6\text{ eV}}{n^2}$$

For $n = 1$ the electron is most likely to be found one Bohr radius $a_0$ from the nucleus; in higher-energy states it's likely to be farther away.

Energy-level diagram

Radial probability density

---

The **orbital quantum number** $l$ determines the angular momentum:

$$L = \sqrt{l(l+1)}\,\hbar$$

where $l$ ranges from 0 to $n-1$.

The **orbital magnetic quantum number** $m_l$ determines the component of the angular momentum along any given axis:

$$L_z = m_l \hbar$$

This is **space quantization,** where $m_l$ ranges from $-l$ to $l$.

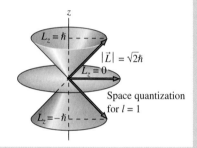

Space quantization for $l = 1$

---

Electrons are **spin-$\frac{1}{2}$** particles or **fermions;** the component of their spin angular momentum on a given axis is $\pm\frac{1}{2}\hbar$.

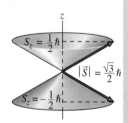

Spin-orbit coupling results in **fine-structure** splitting of atomic-energy levels.

Spin angular momentum $\vec{S}$ and total angular momentum $\vec{J}$ obey quantization rules similar to those of orbital angular momentum.

## Cases and Uses

**Bosons** are particles with integer spin. They don't obey the exclusion principle, allowing many particles to be in the same state, as happens in a **Bose–Einstein condensate** or a laser beam.

Forming a Bose-Einstein condensate

In stimulated absorption, an electron absorbs a photon and jumps to a higher energy level.

Electrons in excited states can drop to lower energy states by either **spontaneous** or **stimulated emission.**

# For Thought and Discussion

1. The electron in a hydrogen atom is very crudely like a particle confined to a three-dimensional box. In the atom, what plays the role of the confining box?

2. A friend who hasn't studied physics asks you the size of a hydrogen atom. How do you answer?

3. How many quantum numbers are required to specify fully the state of a hydrogen atom? *Note:* Don't forget about spin.

4. Both the Bohr and Schrödinger theories of the hydrogen atom predict the same ground-state energy. Do they agree about the angular momentum in the ground state? Explain.

5. Is it possible for a hydrogen atom to be in the $2d$ state? Explain.

6. Can the component of a quantized angular momentum measured on a given axis ever equal the magnitude of the angular momentum vector? Explain.

7. The electron is a spin-$\frac{1}{2}$ particle. Does this mean the intrinsic angular momentum of the electron is $\frac{1}{2}\hbar$? Explain.

8. How does the Stern–Gerlach experiment provide convincing evidence for space quantization?

9. Why is there no spin-orbit splitting in the ground state of hydrogen?

10. In what way is the exclusion principle responsible for the diversity of chemical elements?

11. Helium and lithium exhibit very different chemical behavior, yet they differ by only one unit of nuclear charge. Explain how the exclusion principle brings about the dramatic difference in chemical behavior.

12. Why is stimulated emission essential for laser action?

13. In what sense do the atoms in a Bose–Einstein condensate behave as one atom?

# Exercises and Problems

## Exercises

### Section 36.1 The Hydrogen Atom

14. Using physical constants accurate to four significant figures (see the inside front cover), verify the numerical values of the Bohr radius $a_0$ and the hydrogen ground-state energy $E_1$.

15. A group of hydrogen atoms are in the same excited state, and photons with at least 1.5-eV energy are required to ionize these atoms. What is the quantum number $n$ for the initial excited state?

16. Find the maximum possible magnitude for the orbital angular momentum of an electron in the $n = 7$ state of hydrogen.

17. Which of the following is not a possible value for the magnitude of the orbital angular momentum in hydrogen: (a) $\sqrt{12}\hbar$; (b) $\sqrt{20}\hbar$; (c) $\sqrt{30}\hbar$; (d) $\sqrt{40}\hbar$; (e) $\sqrt{56}\hbar$?

18. The orbital angular momentum of the electron in a hydrogen atom has magnitude $2.585 \times 10^{-34}$ J·s. Find the minimum possible value for its energy.

19. What is the orbital quantum number for an electron whose orbital angular momentum has magnitude $L = \sqrt{30}\hbar$?

20. A hydrogen atom is in the $6f$ state. What are (a) its energy and (b) the magnitude of its orbital angular momentum?

21. Give a symbolic description for the state of the electron in a hydrogen atom when the total energy is $-1.51$ eV and the orbital angular momentum is $\sqrt{6}\hbar$.

### Section 36.2 Electron Spin

22. Verify the value of the Bohr magneton in Equation 36.14.

23. Theories of quantum gravity predict the existence of a spin-2 particle called the graviton. What would be the magnitude of the graviton's spin angular momentum?

24. Some very short-lived particles known as delta resonances have spin $\frac{3}{2}$. Find (a) the magnitude of their spin angular momentum and (b) the number of possible spin states.

25. What are the possible $j$ values for a hydrogen atom in the $3D$ state?

### Section 36.3 The Exclusion Principle

26. An infinite square well contains nine electrons. Find the energy of the highest-energy electron, expressed in terms of the ground-state energy $E_1$.

27. A quantum harmonic oscillator with frequency $\omega$ contains 21 electrons. What is the energy of the highest-energy electron?

### Section 36.4 Multielectron Atoms and the Periodic Table

28. Use shell notation to characterize the outermost electron of rubidium.

29. Write the full electronic structure of scandium.

30. Write the full electronic structure of bromine.

### Section 36.5 Transitions and Atomic Spectra

31. Show that the wavelength $\lambda$ in nm of a photon whose energy $E$ is given in eV is $\lambda = 1240/E$.

32. The $4f \rightarrow 3p$ transition in sodium produces a spectral line at 567.0 nm. What is the energy difference between these two levels?

33. The $4p \rightarrow 3s$ transition in sodium produces a doublet spectral line at 330.2 and 330.3 nm. What is the energy splitting of the $4p$ level?

## Problems

34. Repeat the calculation in part (b) of Example 36.1 to find the probability that an electron in the hydrogen ground state will be found beyond two Bohr radii.

35. Determine the principal and orbital quantum numbers for a hydrogen atom whose electron has energy $-0.850$ eV and orbital angular momentum of magnitude $\sqrt{12}\hbar$.

36. Find (a) the energy and (b) the magnitude of the orbital angular momentum for an electron in the $5d$ state of hydrogen.

37. Assuming the Moon's orbital angular momentum is quantized, estimate its orbital quantum number $l$.

38. The maximum possible angular momentum for a hydrogen atom in a certain state is $30\sqrt{11}\hbar$. Find (a) the principal quantum number and (b) the energy.

39. A hydrogen atom is in a state with $l = 2$. What are the possible angles its orbital angular momentum vector can make with a given axis?

40. A hydrogen atom has energy $E = -0.850$ eV. Find the maximum possible values for (a) the magnitude of its orbital angular momentum and (b) the component of that angular momentum on a given axis.

41. An electron in hydrogen is in the $5f$ state. What possible values, in units of $\hbar$, could a measurement of the orbital angular momentum component on a given axis yield?

42. Substitute Equation 36.3 into Equation 36.4 and carry out the differentiations to show that you get the unnumbered equation following Equation 36.4.

43. Differentiate the radial probability density for the hydrogen ground state, and set the result to zero to show that the electron is most likely to be found at one Bohr radius.

44. What is the minimum angle between the spin vector and the $z$ axis for (a) a spin-$\frac{3}{2}$ particle and (b) a spin-2 particle?

45. Repeat Exercise 25 for the case where you know only that the principal quantum number is 3; that is, $l$ might have any of its possible values.

46. Draw vector diagrams similar to Fig. 36.10 for spin-orbit coupling of an electron in the $l = 3$ state.

47. A hydrogen atom is in the $4F_{5/2}$ state. What are (a) its energy in units of the ground-state energy, (b) the magnitude of its orbital angular momentum in units of $\hbar$, and (c) the magnitude of its total angular momentum in units of $\hbar$? (d) Which of (b) or (c) is greater? Why?

48. Suppose you put five electrons into an infinite square well. (a) How do the electrons arrange themselves to achieve the lowest total energy? (b) Give an expression for this energy in terms of the electron mass $m$, the well width $L$, and Planck's constant.

49. A harmonic oscillator potential of natural frequency $\omega$ contains eight electrons and is in the state of lowest energy. (a) What is that energy? (b) What would the energy be if the electrons were replaced by spin-1 particles of the same mass?

50. A harmonic oscillator potential of natural frequency $\omega$ contains a number of electrons and is in the state of lowest energy. If that energy is $6.5\hbar\omega$, (a) how many electrons are in the potential well, and (b) what is the energy of the highest-energy electron?

51. Determine the electronic configuration of copper.

52. An electron in a highly excited state of hydrogen ($n_1 \gg 1$) drops into the state $n = n_2$. What is the lowest value of $n_2$ for which the emitted photon will be in the infrared ($\lambda > 700$ nm)?

53. A solid-state laser made from lead–tin selenide has a lasing transition at a wavelength of 30 $\mu$m. If its power output is 2.0 mW, how many lasing transitions occur each second?

54. For hydrogen, fine-structure splitting of the $2p$ state is only about $5 \times 10^{-5}$ eV. What percentage is this difference of the photon energy emitted in the $2p \rightarrow 1s$ transition? *Note:* Your answer shows why it's hard to observe spin-orbit splitting in hydrogen.

55. Find the probability that the electron in the hydrogen ground state will be found in the radial-distance range $r = a_0 \pm 0.1a_0$.

56. A gas contains 1000 hydrogen atoms in their ground states. In how many atoms, on average, will the electrons be within three Bohr radii of the nucleus?

57. What is the most orbital angular momentum that could be added to an atomic electron initially in the $6d$ state without changing its principal quantum number? What would be the new state?

58. (a) Write the alpha-numeric description of a hydrogen atom whose energy is one-ninth of the ground-state energy, whose orbital angular momentum has magnitude $\sqrt{6}\hbar$, and whose spin and orbital angular momenta are as nearly aligned as possible. (b) How many $m_l$ values are there for this state? *Note:* The energies are negative, so this is an excited state.

59. A hydrogen atom is in an $F$ state. (a) Find the possible values for the magnitude of its total angular momentum. (b) For the state with the greatest angular momentum, find the number of possible values for the component of $\vec{J}$ on a given axis.

60. A hydrogen atom is in the $2s$ state. Find the probability that its electron will be found (a) beyond one Bohr radius and (b) beyond 10 Bohr radii.

61. Show that the maximum number of electrons in the $n$th shell of an atom is $2n^2$.

62. Form the radial probability density $P_2(r)$ associated with the $\psi_{2s}$ state of Equation 36.7, and find the most probable radial position for the electron.

63. Substitute the $\psi_2$ wave function in Equation 36.7 into Equation 36.4 to verify that the equation is satisfied and that the energy is given by Equation 36.6 with $n = 2$.

64. (a) Verify Equation 36.8 by considering a single-electron atom with nuclear charge $Ze$ instead of $e$. (b) Calculate the ionization energies for single-electron versions of helium, oxygen, lead, and uranium.

65. A selection rule for the infinite square well allows only those transitions in which $n$ changes by an odd number. Suppose an infinite square well of width 0.200 nm contains an electron in the $n = 4$ state. (a) Draw an energy-level diagram showing all allowed transitions that could occur as this electron drops toward the ground state (include transitions from lower levels that could be reached from $n = 4$). (b) Find all the possible photon energies emitted in these transitions.

66. An ensemble of one-electron square-well systems of width 1.17 nm all have their electrons in highly excited states. They undergo all possible transitions in dropping toward the ground state, obeying the selection rule that $\Delta n$ must be odd. (a) What wavelengths of visible light are emitted? (b) Is there any infrared emission? If so, how many spectral lines fall in the infrared?

67. Use the radial probability density from Equation 36.5, and the normalized ground state hydrogen wave function from Equation 36.3 and Example 36.1, to calculate the average radial distance $r_{av}$ for an electron in this state. *Note:* Because the probability density isn't symmetric, the average radial distance isn't the same as the most probable radial distance shown in Fig 36.4.

68. Follow the procedure in Problem 67 to calculate the average radial distance for an electron in the $2s$ state of hydrogen.

69. The ratio of the magnitudes of the magnetic moment, in units of the Bohr magneton $\mu_B$, to the angular momentum, in units of $\hbar$, is called the g-factor. (a) Show that the classical orbital g-factor for an atomic electron in a circular Bohr orbit is $g_L = 1$. (b) Show that Equation 36.13 gives a g-factor for electron spin of $g_S = 2$.

70. The total magnetic moment of the electron in a hydrogen atom (and most other atoms where spin-orbit coupling of individual electrons is small) is the sum of the orbital and spin magnetic moments $\vec{M} = \vec{M}_L + \vec{M}_S = -\mu_B(\vec{L} + 2\vec{S})/\hbar$ (see Problem 69). When an applied magnetic field is not too large, the Zeeman effect is related to the projection of $\vec{M}$ along the total angular momentum vector $\vec{J} = \vec{L} + \vec{S}$; that is, $M_J = M_J \cos\theta_L + M_S \cos\theta_S$, where $\theta_L$ and $\theta_S$ are the angles that $\vec{L}$ and $\vec{S}$ make with $\vec{J}$ in a vector diagram like Figure 36.10. Use the law of cosines twice to express the cosines in terms of the magnitudes of the vectors and find the g-factor, which is the ratio of the magnitudes of $M_J/\mu_B$ to $J/\hbar$.

71. You're working for a company that makes the red helium–neon lasers widely used in physics experiments. Figure 36.19 shows an energy-level diagram for this laser. An electric current excites helium to the metastable level $E_1$ at 20.61 eV above the ground state. Collisions transfer energy to neon atoms, exciting them to the metastable level $E_2$ at 20.66 eV. The lasing transition drops the atoms to $E_3$, emitting a 632.8-nm photon in the process. You're asked to find the maximum possible efficiency for this laser—that is, the light energy emitted as a percentage of the energy supplied to excite the atoms.

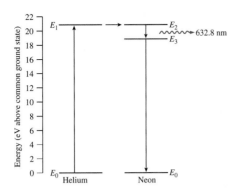

**FIGURE 36.19** Energy-level diagram for the helium–neon laser (Problem 71).

72. You majored in math, obtained a master's degree in mathematics, and now are applying for a teaching position at a community college. As a part of your interview, you must prepare a brief lesson on a topic of your choice. Because you will be teaching a calculus course to pre-engineering and physics majors, you decide to select a topic from atomic physics. You decide that your lesson will be demonstrating how to get the normalization constant, $1/4\sqrt{2\pi a_0^3}$, for the 2s state of hydrogen (Equation 36.7). Use the fact that $P(r)\,dr = 4\pi r^2 \psi_{2s}^2\,dr$ and that the electron must be *somewhere*.

# Answers to Chapter Questions

## Answer to Chapter Opening Question

Only one electron is allowed in a given atomic quantum state, leading to the shell structure of atoms and to chemical properties based on the outermost atomic electrons.

## Answers to GOT IT? Questions

36.1 (c).

36.2 Six.

36.3 $5p \rightarrow 3s$.

# Molecules and Solids

■ What equation determines the structure of a complex molecule, like the aspirin molecule shown here?

In principle, we could apply the Schrödinger equation to all the particles making up a molecule or even a solid, but in practice that's difficult for all but the simplest molecules. Ever-increasing computer power has brought more complex molecules within reach of structural calculations based in the Schrödinger equation, but in many cases it's appropriate to describe molecular and solid structure more qualitatively.

## 37.1 Molecular Bonding

The binding of atoms into molecules involves both electric forces and quantum-mechanical effects associated with the exclusion principle. Although individual atoms are electrically neutral, the distribution of charge within them gives rise to attractive or repulsive forces. When atoms are squeezed closely together, interactions involving the spins of their outermost electrons also result in attractive or repulsive interactions. For atoms with unfilled outer shells, it's energetically favorable for electrons to pair with opposite spins; this causes an attractive interaction. When the outer atomic shells are filled, the exclusion principle forces electrons from the separate atoms into different states as the atoms are brought together. One or more electrons go into higher energy states, raising the overall energy and resulting in a repulsive interaction. Finally, if atoms get very close, the electrical repulsion of the nuclei becomes important. Ultimately, the balance of attractive and repulsive interactions determines the equilibrium configuration of a molecule. In energy terms, we can think of a stable molecule as a minimum-energy configuration of the electrons and nuclei making up two or more atoms (Fig. 37.1). Although such force and energy considerations ultimately govern all molecules, it's convenient to distinguish several molecular binding mechanisms, based on which interactions are most important.

## Ionic Bonding

As we saw in Chapter 36, elements near the left side of the periodic table have few electrons in their outermost shells and correspondingly low ionization energies. In contrast, elements near the right side of the table have nearly filled shells and consequently strong affinities for electrons. When atoms from these different regions of the periodic table come together, relatively little energy is needed to transfer electrons between them. Sodium, for example, has an ionization energy of 5.1 eV, meaning it takes this much energy to make an $Na^+$ ion. Chlorine, at the opposite end of the periodic table, has such a strong electron affinity that the energy of a $Cl^-$ ion is actually 3.8 eV below that of a neutral Cl atom. Thus an expenditure of only 1.3 eV $(5.1\ eV - 3.8\ eV)$ is required to transfer the outermost electron from a sodium to a chlorine atom. The resulting ions are strongly attracted and come together until they reach equilibrium at an internuclear separation of about 0.24 nm. The total energy of the pair is then 4.2 eV below that of neutral chlorine and sodium atoms at large separation (Fig. 37.2). Since it would therefore take 4.2 eV to separate the atoms, this quantity is called the **dissociation energy**.

Because the minimum-energy sodium–chlorine structure consists of ions bound by the electrostatic force, the binding mechanism is termed **ionic bonding**. Ionic bonding generally occurs in crystalline solids. Because the building blocks of an ionically bound substance are electrically charged, each can bind to several of the opposite charge, resulting in a regular crystal pattern (Fig. 37.3). Because the electrostatic force is strong, ionic solids are tightly bound and therefore have high melting points (801°C for NaCl). And because all electrons are bound to individual nuclei, there are no free electrons and therefore ionic solids are electrical insulators.

**FIGURE 37.1** Potential energy of a pair of hydrogen atoms as a function of the distance between their nuclei.

**FIGURE 37.2** Potential-energy curve for $Na^+$ and $Cl^-$ ions, with zero energy corresponding to infinite separation of neutral Na and Cl atoms.

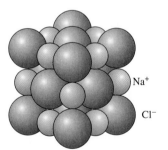

**FIGURE 37.3** A sodium chloride crystal is a regular array of sodium and chlorine atoms, bound by the electrostatic force.

## Covalent Bonding

In an ionic bond, each electron is associated with only one ion. In a **covalent bond**, on the other hand, electrons are shared among atoms. Covalent bonds occur between atoms whose outermost shells aren't full, and whose outer electrons can therefore pair with opposite spins. The simplest example of a covalent bond is the hydrogen molecule, $H_2$. Since each hydrogen atom has a single 1s electron, each could accommodate in its 1s shell a second electron with opposite spin. When two hydrogen atoms join, quantum mechanics predicts a molecular ground state in which both electrons share a single orbital, with the highest probability of finding the electrons between the nuclei (Fig. 37.4). Dissociation energies for covalent bonds are, like those of ionic bonds, on the order of a few electronvolts.

**FIGURE 37.4** Probability density for finding electrons in the ground state of molecular hydrogen ($H_2$).

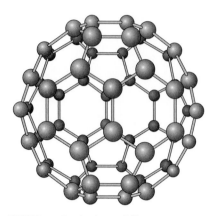

**FIGURE 37.5** Buckminsterfullerene, $C_{60}$, is a symmetric arrangement of 60 carbon atoms. Discovered in the 1980s, $C_{60}$ and related fullerenes hold promise in a wide range of technological applications.

With their outermost molecular orbitals full, covalently bonded molecules often have no room for another electron in their structures. For example, adding a third hydrogen to $H_2$ is impossible because the ground-state orbital already contains two electrons with opposite spins, so the exclusion principle requires that a third electron go into a higher energy state. The energy of that state is higher than that of an $H_2$ molecule and a distant H atom; for this reason $H_3$ isn't a stable molecule. Because their outermost molecular orbitals are full, co-valent molecules interact only weakly, and as a result many common covalent materials—for example, $H_2$, CO, $N_2$, and $H_2O$—are either gases or liquids at ordinary temperatures. In other cases covalent bonds can form crystalline structures. A simple example is dia-mond, a pure-carbon solid formed when each carbon atom bonds covalently to its four nearest neighbors. A more dramatic covalent molecule is buckminsterfullerene, $C_{60}$, a soccer-ball configuration of 60 carbon atoms (Fig. 37.5).

### Hydrogen Bonding

If water consists of covalent molecules, why is it ever a solid? The answer lies in **hydrogen bonds** that form when the tiny, positively charged proton of a hydrogen nucleus nestles close to the negative parts of other molecules. In ice, hydrogen bonds link a proton in one $H_2O$ molecule to the oxygen in another. Because covalent bonding within the water mole-cule leaves the oxygen only slightly negative and the hydrogen only slightly positive, hydrogen bonds are much weaker than ionic or covalent bonds; a typical hydrogen-bond energy is 0.1 eV. Hydrogen bonds are important in determining the overall configuration of complicated molecules. In DNA, for example, covalent bonds join atoms to form long chains; hydrogen bonds then link the chains into the double-helix structure.

### Van der Waals Bonding

In Section 20.5 we mentioned the **van der Waals force** that arises from electrostatic inter-actions between induced dipole moments of otherwise nonpolar molecules. In gases, the van der Waals force causes deviations from the ideal gas law that are most pronounced at high densities. As temperature drops, this weakly attractive force becomes effective in binding molecules into liquids and solids. Liquid and solid oxygen ($O_2$) and nitrogen ($N_2$), for example, are held together by van der Waals bonds.

### Metallic Bonding

In a metal, the outermost atomic electrons aren't bound to individual nuclei, but move throughout the material. The metal forms a crystal lattice of positive ions, bound by this "electron gas." The free electrons give a metal its high electrical and thermal conductivities.

## 37.2 Molecular Energy Levels

In a molecule, electric forces bind electrons and nuclei into a single structure. Like any quantum-mechanical bound system, the energy levels of a molecule are quantized. As in atoms, differences among molecular energy levels are associated with different electronic configurations (Fig. 37.6). But molecules are more complex than atoms, and molecular energy can take additional forms.

In Chapter 18, we found that a complete description of the specific heats of gases required that we consider the rotational and vibrational motion of individual molecules. We hinted at quantum mechanics, pointing out that each of these modes of molecular motion could absorb only certain discrete amounts of energy (see Fig. 18.17). Here, in a quantum-mechanical treatment of molecular energetics, we consider rotational and vibra-tional energy states as well as electronic configuration.

**(a)**

**(b)**

**FIGURE 37.6** Electron probability densities for (a) the ground state and (b) the first excited state of hydrogen. Nuclei are farther apart in the excited state.

### Rotational Energy Levels

If a molecule is rotating, it has angular momentum $L$ whose magnitude, from Equa-tion 11.4, is $L = I\omega$, where $I$ is the rotational inertia and $\omega$ the angular speed. The quantization conditions that we found in Chapter 36 for atomic angular momenta also hold for the angular momentum of molecular rotation, so we have

$$L = \sqrt{l(l + 1)}\hbar \qquad (37.1)$$

where the quantum number $l$ takes on integer values 0, 1, 2, 3, . . . . But then the rotational energy, which from Equation 10.17 is $E_{rot} = \frac{1}{2}I\omega^2$, must also be quantized. Solving the equation $L = I\omega$ for $\omega$ allows us to write the energy as

$$E_{rot} = \frac{1}{2}I\left(\frac{L}{I}\right)^2 = \frac{L^2}{2I}$$

Applying the quantization condition 37.1 for $L$, we then have the quantized rotational energy levels:

$$E_{rot} = \frac{\hbar^2}{2I}l(l + 1) \qquad \text{for } l = 0, 1, 2, 3, \ldots \qquad (37.2)$$

---

**EXAMPLE 37.1** **Molecular Rotation: Computing Molecular Size**

A gas of HCl molecules shows spectral lines that result from transitions between pairs of adjacent rotational energy levels. The energy of the transition increases by 2.63 meV from one spectral line to the next. (a) Use this experimental result to determine the rotational inertia of the HCl molecule. (b) Approximating the more massive chlorine as being essentially fixed, find an expression for the rotational inertia in terms of the hydrogen mass and the separation of the two atomic nuclei. (c) Use the results of parts (a) and (b) to determine the internuclear separation in HCl.

**INTERPRET** We're asked to infer molecular properties from spectroscopic observations. We're given not the wavelength or energy of a given spectral line, but the energy difference associated with adjacent lines.

**DEVELOP** Spectral lines result from photons emitted as a molecule drops from one of the energy levels of Equation 37.2 to the next lower level. So the energy of such a photon is

$$\Delta E_{l \to l-1} = \frac{\hbar^2}{2I}[l(l + 1) - (l - 1)l] = \frac{\hbar^2 l}{I}$$

An adjacent spectral line would result from the transition $l - 1 \to l - 2$, giving $\Delta E_{l-1 \to l-2} = \hbar^2(l - 1)/I$. So our plan is to take the difference between these two transition energies, equate it to the observed energy difference of 2.63 meV between spectral lines, and solve for $I$. For (b) we can treat the molecule as having a fixed center (Cl) with a particle (H) in circular motion. Its rotational inertia is then given by Equation 10.12: $I = mR^2$. Equating the two expressions for $I$ will let us solve for the internuclear separation $R$.

**EVALUATE** (a) The energy difference $\Delta(\Delta E)$ between adjacent transition energies is

$$\Delta(\Delta E) = \frac{\hbar^2 l}{I} - \frac{\hbar^2(l - 1)}{I} = \frac{\hbar^2}{I}$$

Setting this to the observed 2.63-meV spacing and solving for $I$ give $I = 2.65 \times 10^{-47}$ kg·m², where we had to convert meV to J to get the result in SI. (b) We've already shown that $I = mR^2$. (c) Equating our algebraic and numerical expressions for $I$ and solving give the internuclear distance:

$$R = \sqrt{\frac{I}{m}} = \sqrt{\frac{2.65 \times 10^{-47} \text{ kg} \cdot \text{m}^2}{1.67 \times 10^{-27} \text{ kg}}} = 0.126 \text{ nm}$$

where we approximated the hydrogen mass, $m$, with the proton mass listed inside the front cover.

**ASSESS** Our answer makes sense, since it's slightly larger than an isolated hydrogen atom. However, it's approximate because we took the chlorine as perfectly fixed and we also ignored the quantum-mechanical ground-state energy of molecular vibration, which stretches the molecule. Photon energies for low $l$ values are on the order of that 2.63-meV spacing, corresponding to a wavelength of about 0.5 mm. This is in the microwave region of the spectrum, with much lower energy and longer wavelength than we've seen for transitions between atomic energy levels. ∎

---

## Vibrational Energy Levels

The equilibrium configuration of a molecule corresponds to the minimum of the molecular potential-energy curve. In the vicinity of that minimum the curve is well approximated by a parabola (Fig. 37.7). In Chapter 13, we saw that parabolic potential-energy curves result in simple harmonic motion, and in Chapter 35 we used a parabolic potential-energy curve in the Schrödinger equation for the harmonic oscillator. There, we found that quantized vibrational energy levels are given by

$$E_{vib} = \left(n + \frac{1}{2}\right)\hbar\omega \qquad (37.3)$$

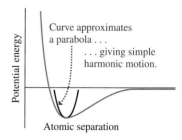

**FIGURE 37.7** Near its minimum, the molecular potential-energy curve approximates a parabola.

where the quantum number $n$ takes on integer values 0, 1, 2, 3, . . . , and where $\omega$ is the natural frequency for classical harmonic oscillations of the molecule. The selection rule for harmonic oscillators limits allowed transitions to those with $\Delta n = \pm 1$, so $\hbar \omega$ is the energy of photons emitted or absorbed in allowed transitions among vibrational energy levels. (Actually, the small-amplitude approximation is often justified for only the lower quantum states, so Equation 37.3 and the selection rule $\Delta n = \pm 1$ may apply to only these states.) For typical diatomic molecules, $\omega$ is on the order of $10^{14}$ s$^{-1}$, in the infrared region of the spectrum. Consequently, study of molecular vibrations involves infrared spectroscopy.

As we found in Chapter 35, the minimum energy of a quantum harmonic oscillator is the ground-state energy $E_0 = \frac{1}{2} \hbar \omega$. Thus a molecule can never have zero vibrational energy, although Equation 37.2 shows that it *can* have zero rotational energy.

---

**EXAMPLE 37.2** **Molecular Energies: Rotational and Vibrational**

An HCl molecule is in its vibrational ground state. Its classical vibration frequency is $f = 8.66 \times 10^{13}$ Hz. If its rotational and vibrational energies are nearly equal, what are its rotational quantum number and angular momentum?

**INTERPRET** We're being asked to compare energies associated with two different processes: vibration and rotation. We're told the vibrational energy state (the ground state, $n = 0$), and we need to find the rotational state with comparable energy.

**DEVELOP** Equation 37.3, $E_{\text{vib}} = \left( n + \frac{1}{2} \right) \hbar \omega$, with $n = 0$ gives the ground-state vibrational energy $E_{\text{vib}} = \frac{1}{2} \hbar \omega = \frac{1}{2} h f$, where we used $\hbar = h/2\pi$ and $\omega = 2\pi f$. Our plan is to equate this to the rotational energy of Equation 37.2, $E_{\text{rot}} = (\hbar^2/2I) l(l+1)$, and solve for the quantity $l(l+1)$. Then we'll use Equation 37.1 to get the angular momentum.

**EVALUATE** Equating the vibrational and rotational energies gives

$$\frac{\hbar^2}{2I} l(l+1) = \frac{1}{2} \hbar \omega = \hbar \pi f$$

where we used $\omega = 2\pi f$. So $l(l+1) = 2\pi f I/\hbar$, and Equation 37.1, $L = \sqrt{l(l+1)} \hbar$, gives

$$L = \sqrt{\frac{2\pi f I}{\hbar}} \hbar = \sqrt{2\pi f I \hbar} = \sqrt{f I h} = 1.23 \times 10^{-33} \text{ J} \cdot \text{s}$$

where we used the rotational inertia $I = 2.65 \times 10^{-47}$ kg·m$^2$ that we found in Example 37.1.

**ASSESS** Our angular momentum is about 12 times $\hbar$; approximating $l(l+1)$ by $l^2$ in Equation 37.1 then shows that $l$ is about 12. So we need a fairly high rotational quantum state to give the same energy as the vibrational ground state. That's consistent with transitions between adjacent rotational states involving microwave photons, while vibrational transitions involve infrared photons. ∎

---

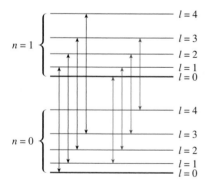

**FIGURE 37.8** Energy-level diagram showing the ground state and first vibrational excited state of a diatomic molecule; also shown are four of the infinitely many rotational states for each $n$.

## Molecular Spectra

A molecule with vibrational quantum number $n$ and rotational quantum number $l$ can undergo transitions obeying the selection rules $\Delta n = \pm 1$ and $\Delta l = \pm 1$. If molecules couldn't rotate, the molecular spectrum would consist of a single line at the classical vibration frequency, corresponding to transitions among adjacent vibrational states. But each vibrational level corresponds to an infinite number of rotational states. The resulting energy-level structure is shown in Fig. 37.8. At typical temperatures, only the ground and first vibrational levels are significantly populated, but with energy distributed among many rotational levels. As a result, molecular spectra show a rich structure, with many lines corresponding to the different transitions of Fig. 37.8. Figure 37.9 is a spectrum of HCl, taken with a high-resolution infrared spectrometer that resolves the individual spectral lines. At lower resolution, the pattern shows up as a broad band, and we often speak of infrared absorption bands in describing the effect of molecules on infrared radiation. For example, absorption bands of atmospheric carbon dioxide limit the escape of infrared radiation from Earth, causing the global warming that we described in Chapter 16. Molecular energy levels are therefore at the heart of today's most global environmental concern.

These lines correspond
to the left-hand arrows
in Fig. 37.8 . . .

. . . these to the
right-hand arrows.

Fine splitting
shows mass
difference
between
$^{35}Cl$ and
$^{37}Cl$.

Absorption

3.3    3.4    3.5    3.6    3.7    3.8

Wavelength ($\mu$m)

←— Increasing energy

**FIGURE 37.9** Absorption spectrum of HCl, showing lines that result from transitions between the $n = 1$ and $n = 0$ vibrational states.

**GOT IT? 37.1**    You meet a scientist who uses microwave technology to study molecular structure. What form of molecular energy is she most concerned with?

## 37.3 Solids

Bonding mechanisms can join relatively few atoms to form a molecule, or many to form a solid. In the lowest energy state, the atoms of a solid are arranged in a regular, repeating pattern; the solid is then **crystalline**. Sometimes solids form without their atoms having the opportunity to achieve a crystal structure; such solids are termed **amorphous**. Glass is a common amorphous solid. Amorphous materials are difficult to analyze due to their inherent randomness, so we concentrate here on crystalline solids.

### Crystal Structure

The hallmark of a crystalline solid is the regular arrangement of atoms. Looking closely shows that a basic pattern repeats throughout the crystal (Fig. 37.10). This basic arrangement is the **unit cell**. Different crystalline materials have different unit cells (Fig. 37.10a, c). Sometimes the same underlying matter may assume different structures, depending on how the solid was formed; this is the case with diamond and graphite, both crystalline forms of carbon.

As with individual molecules, properties like atomic separation in a crystalline solid are determined by the interplay of attractive and repulsive interactions. The situation is complicated, however, because an individual atom experiences forces from many other atoms

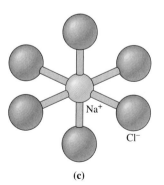

(a)    (b)    (c)

**FIGURE 37.10** (a) The unit cell of cesium chloride has eight chlorine ions surrounding each cesium. (b) A cesium chloride crystal is a periodic array of unit cells. (c) Sodium chloride is different; here each ion is surrounded by only six nearest neighbors of the opposite type.

in the crystal. With ionic bonding, those forces are electrical attraction and repulsion as described by Coulomb's law; that makes ionic crystals most amenable to simple mathematical treatment.

For ionic crystals, we can take individual ions to be point charges. Consider the NaCl structure of Fig. 37.10c. Each sodium ion is surrounded by six nearest chlorine ions, each some distance $r$ away. The potential energy of a singly ionized positive sodium ion in the potential of each negative chlorine ion is $-ke^2/r$. So the contribution to the potential energy of the six nearest chlorines is $-6ke^2/r$, with the minus indicating an attractive interaction. But then there are 12 sodium ions a distance $\sqrt{2}r$ from the sodium; they give rise to a repulsive force and consequently a positive potential energy $+12ke^2/\sqrt{2}r$. At a distance of $\sqrt{3}r$ there are eight more chlorines, giving potential energy $-8ke^2/\sqrt{3}r$. The result is that the electrostatic potential energy of the sodium ion can be written as $U_1 = -\alpha(ke^2/r)$, where $\alpha = 6 - 12/\sqrt{2} + 8/\sqrt{3} - \cdots$; $\alpha$ is called the **Madelung constant**. Many terms in the series are required to compute $\alpha$ accurately, showing that the effect of distant ions is significant in determining the energy of an ion in the crystal. For the NaCl structure, $\alpha$ is approximately 1.748.

As ions are brought closer together, they experience the repulsive effect of the exclusion principle, as we discussed in Section 37.1. This repulsion is described approximately by a potential energy of the form $U_2 = A/r^n$, where $A$ and $n$ are constants. So the total potential energy of an ion in the crystalline solid is

$$U = U_1 + U_2 = -\alpha \frac{ke^2}{r} + \frac{A}{r^n}$$

At equilibrium the potential energy is a minimum (Fig. 37.11), corresponding to zero net force on the ion. Differentiating the potential energy with respect to $r$ and setting $dU/dr$ to zero to find the minimum, we have

$$0 = \frac{\alpha ke^2}{r_0^2} - \frac{nA}{r_0^{n+1}}$$

where $r_0$ designates the equilibrium separation. Solving for $A$ gives $A = \alpha ke^2 r_0^{n-1}/n$, so the potential energy becomes

$$U = -\alpha \frac{ke^2}{r_0} \left[ \frac{r_0}{r} - \frac{1}{n}\left(\frac{r_0}{r}\right)^n \right] \qquad (37.4)$$

The value of $U$ at the equilibrium separation $r_0$ is designated $U_0$ and is called the **ionic cohesive energy**. The magnitude of $U_0$ represents the energy needed to remove an ion entirely from the crystal. The cohesive energy is sometimes given in kcal/mol, in which case its magnitude is the energy per mole needed to break an entire crystal into its constituent ions (see Exercise 23).

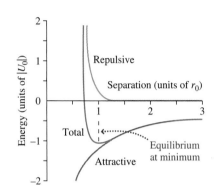

**FIGURE 37.11** Potential-energy function for an ionic crystal, showing separate contributions of the attractive and repulsive terms.

---

**EXAMPLE 37.3** **Potential Energy in a Solid: The NaCl Crystal**

The ionic cohesive energy for NaCl is $-7.84$ eV. The equilibrium separation, which follows from the measured density, is 0.282 nm (see Exercise 22). Use these values and the Madelung constant $\alpha = 1.748$ to find the exponent $n$ in Equation 37.4 for NaCl.

**INTERPRET** Here we're given all but one of the quantities in the expression for a crystal's potential energy, and we're asked to solve for the one unknown, $n$.

**DEVELOP** Equation 37.4 for the potential energy $U$ looks formidable, with $n$ in two places, including an exponent. But we're given the ionic cohesive energy $U_0$, which is the value of $U$ when $r = r_0$. So the two

terms $r_0/r$ become 1, and since $1^n = 1$, the $n$ in the exponent drops out. With $r = r_0$ and $U = U_0$, Equation 37.4 simplifies to

$$U_0 = -\alpha \frac{ke^2}{r_0}\left(1 - \frac{1}{n}\right)$$

Our plan is to solve this equation for $n$.

**EVALUATE** Solving gives

$$n = \left(1 + \frac{U_0 r_0}{\alpha ke^2}\right)^{-1} = 8.22$$

with $k$ the constant in Coulomb's law and $e$ the elementary charge; other quantities are given in the problem statement.

**ASSESS** The large value of this exponent shows that the NaCl crystal is strongly resistant to compression. In Problem 44 you can calculate the associated repulsive force. ∎

## Band Theory

Quantum-mechanical analysis of a solid containing $10^{23}$ atoms or so might seem a hopeless task. But the regularity of a crystalline solid makes that problem, while not easy, at least amenable to mathematical treatment. The physical regularity of the solid is reflected mathematically in the properties of the wave function; specifically, the wave function for a crystalline solid in equilibrium is itself periodic. That's because equivalent points in different unit cells have exactly the same physical properties.

We won't solve the Schrödinger equation for a crystal, or even write the solutions. But we can see what some properties of those solutions must be. Consider two identical atoms, initially widely separated, as they're brought closer together. When the atoms are far apart, they're described by identical wave functions and associated energy-level diagrams; a given electron state, for example, has exactly the same energy in each atom. But as the atoms move closer, their wave functions begin to overlap to form a single wave function that characterizes the entire composite system. Because of the exclusion principle, two electrons that were in identical states in the two widely separated atoms can no longer be in the same state. This effect manifests itself as a separation of what were originally identical energy levels (Fig. 37.12a). As more and more atoms come together, initially identical energy levels split into ever more finely spaced levels (Fig. 37.12b). In a crystalline solid, there are so many atoms that each level splits into an essentially continuous **band** of allowed energies (Fig. 37.12c). **Band gaps** separate the bands arising from distinct single-atom states, as shown in Fig. 37.12c. An electron can have any energy between the top and bottom of a band, but energies in the band gaps are forbidden. The situation is like a single atom, where electrons are allowed only certain discrete energies, except now the discrete levels have broadened into bands.

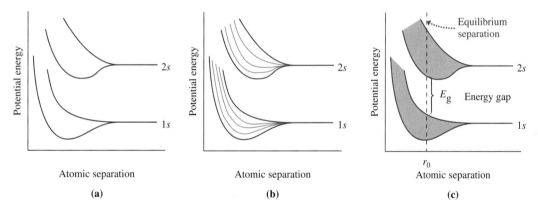

**FIGURE 37.12** (a) Energy levels of the 1s and 2s states as a pair of atoms are brought close together. (b) With five atoms, each level splits into a group of five closely spaced levels. (c) In a crystalline solid, the large number of atoms results in essentially continuous energy bands, separated by gaps.

We're usually interested in the properties of a solid at or near its equilibrium state, designated $r_0$ in Fig. 37.12c. There the solid is characterized by an energy-level diagram in which the energy levels are those of Fig. 37.12c at the value $r = r_0$ (Fig. 37.13).

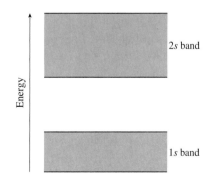

**FIGURE 37.13** Energy-level diagram for the equilibrium separation of Fig. 37.12c.

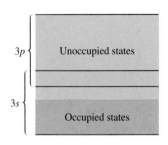

**FIGURE 37.14** Band structure of metallic sodium, with gray indicating filled states and color unfilled states. Bands lower than 3s aren't shown; they correspond to inner electrons, whose levels aren't split significantly.

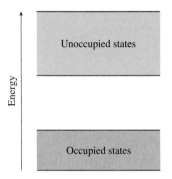

**FIGURE 37.15** Band structure for an insulator.

## Conductors, Insulators, and Semiconductors

Sometimes the splitting and shifting depicted in Fig. 37.12 result in overlapping bands. Figure 37.14 shows the band structure of sodium, in which the 3s and 3p bands overlap. Note that the high-energy band containing electrons—here the 3s/3p band—is not completely full, meaning that energy levels in the upper portion of the band aren't occupied by electrons.

We can determine which of the allowed energy levels of a solid will be occupied in the same way we established the electronic structure of atoms: by placing a given total number of electrons in the lowest possible levels consistent with the exclusion principle. In some materials, like sodium, that filling process results in the highest-energy occupied band being only partially full. But in others, like the material shown in Fig. 37.15, the highest-energy occupied band is completely full.

Figures 37.14 and 37.15 show the essential difference between conductors and insulators. A conductor is a material in which charges are free to move in response to an electric field. Classically, there's no problem with this: Apply an electric field, and if an electron is "free," it will accelerate and gain energy. But quantum mechanically, an electron can gain energy only by jumping into a higher allowed energy level. So there needs to be a higher unoccupied level available.

In sodium, the 3s atomic level contains a single electron, although it has room for two. Put $N$ sodium atoms together to form a crystal, and the 3s band contains only $N$ of the total $2N$ electrons it could hold. So the 3s band isn't full, as Fig. 37.14 shows, and therefore electrons near the top of the filled portion have available unoccupied states with only a little more energy. That makes it easy for an electric field to promote electrons to unoccupied levels. For that reason sodium is an electrical conductor.

In the material of Fig. 37.15, in contrast, one band is completely full and the next higher one empty. An electron in the filled band can't gain energy unless it's enough to jump the band gap. Electric fields of reasonable magnitude can't provide this energy, so the electrons are stuck in the filled bands. That makes the material an insulator.

## Metallic Conductors

We found in Chapter 24 that classical physics can't account for the details of metallic conduction, in particular the temperature dependence of conductivity. Quantum mechanically, the conduction electrons in a metal are like electrons in the three-dimensional box of Section 35.4. They're free to move about inside the metal, but not to leave it. The number of states available to the electrons, per unit energy interval, turns out to increase with energy. You can see the beginnings of this trend in Fig. 35.18, which shows the first few states of the three-dimensional box. We won't do this count; the result, however, is given by

$$g(E) = \left(\frac{2^{7/2}\pi m^{3/2}}{h^3}\right)\sqrt{E} \tag{37.5}$$

where $m$ is the electron mass and $g(E)$ the **density of states**—the number of states per unit volume per unit energy interval centered on the energy $E$.

At absolute zero, electrons fill the lowest available states according to the exclusion principle. The energy of the highest filled level at absolute zero is the **Fermi energy**, $E_F$. At temperature $T = 0$, all states below the Fermi energy are full, and all those above are empty, as shown in Fig. 37.16a.

For $T > 0$, thermal energy promotes some electrons to levels above the Fermi energy, leaving some levels just below $E_F$ vacant (Fig. 37.16b). Now, the Fermi energy in most metals is about 1–10 eV, much higher than the thermal energy at typical temperatures (0.025 eV at room temperature). So the electron distribution changes only slightly—and that means electrons near the Fermi energy carry essentially all the electric current, regardless of temperature. The mean electron speed is therefore quite different from the classical thermal speed (see Problem 48), and that makes the temperature dependence of electrical conductivity in metals very different from the classical prediction.

**FIGURE 37.16** Density of states given by Equation 37.5, with shaded region indicating occupied energy levels. (a) $T = 0$; (b) $T > 0$.

---

**GOT IT? 37.2** Both parts of Fig. 37.16 describe the same piece of metal. How do the shaded areas compare? Why? What's your interpretation of this shaded area?

---

## Semiconductors

In Chapter 24 we gave a classical description of semiconductors—the materials at the heart of our modern electronic world. Here we see how band theory gives a quantum-mechanical explanation of semiconductors.

Our band diagram for an insulator (Fig. 37.15) is strictly correct only at absolute zero. Here the highest occupied band—the **valence band**—is full, and above it the **conduction band** is empty. At temperatures above absolute zero, though, random thermal energy may give an occasional electron enough energy to jump the gap into the conduction band, where it has plenty of nearby states available and can thus respond freely to an electric field. In good insulators, the band gap is many electronvolts and this effect is negligible. But in some materials, notably silicon and germanium, the band gap is on the order of 1 eV (see Table 37.1). At room temperature, thermal excitation promotes enough electrons into the conduction band that these materials conduct electricity, although their conductivity is much lower than in metallic conductors. Such a material is a **semiconductor**. Figure 37.17 compares the band structures for conductors, insulators, and semiconductors.

In Chapter 24 we showed how doping—adding small quantities of impurities—could radically alter the electrical properties of semiconductors. In terms of band theory, a dopant such as phosphorus, with five valence electrons, adds **donor levels** just below the

**TABLE 37.1** Band-Gap Energies for Selected Semiconductors (at 300 K)

| Semiconductor | Band-Gap Energy (eV) |
|---|---|
| Si | 1.14 |
| Ge | 0.67 |
| InAs | 0.35 |
| InP | 1.35 |
| GaP | 2.26 |
| GaAs | 1.43 |
| CdS | 2.42 |
| CdSe | 1.74 |
| ZnO | 3.2 |
| ZnS | 3.6 |

**FIGURE 37.17** Band structures for a conductor, an insulator, and a semiconductor. Gray indicates occupied states.

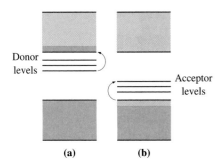

**FIGURE 37.18** Band structures in doped semiconductors: (a) *N*-type; (b) *P*-type.

conduction band (Fig. 37.18a). Thermal energy readily promotes electrons from these levels into the conduction band, greatly increasing the conductivity. This makes an **N-type semiconductor** because its predominant charge carriers are electrons. A dopant like boron, in contrast, creates **acceptor levels** just above the valence band (Fig. 37.18b). Electrons promoted to these levels leave behind **holes** that act as positive charge carriers. The result is a **P-type semiconductor**.

In Chapter 24 we gave a classical explanation of how a junction of *P*- and *N*-type semiconductors conducts electric current in only one direction. From there we went on to describe the operation of the transistor—the semiconductor device at the heart of all modern electronics. We can also understand the *PN* junction in terms of band structure. In Fig. 24.10a we showed how electrons and holes diffuse across a *PN* junction, depleting the junction of charge carriers and making it a poor conductor. Diffusion of electrons also gives the *P*-type side of the junction a net negative charge, and diffusion of the holes makes the *N*-type side positive. This charge separation creates an electric field pointing from *N* to *P*, as shown in Fig. 37.19a. The field opposes further diffusion of charge and thus establishes an equilibrium in which there's no net charge flow across the junction. Because they've moved with the electric field, the electrons that have diffused into the *P*-type region have higher potential energy than those that remained behind in the *N*-type region. (Remember that electrons are negative, so their potential energy increases when they move in the *same* direction as an electric field.) As a result, the band-structure diagram for electrons in the *PN* junction takes the form shown in Fig. 37.19b.

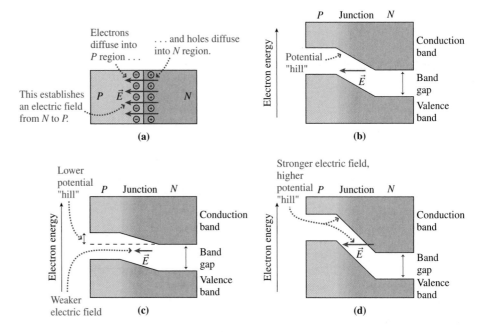

**FIGURE 37.19** (a) Physical picture of an unbiased *PN* junction and (b) the corresponding band structure. (c) Band structure of a forward-biased junction and (d) a reverse-biased junction.

Now suppose we connect a battery to our *PN* junction, with the positive terminal to the *P*-type side of the junction. This condition is called **forward bias**. The effect is to make the *P*-type material less negative, the *N*-type less positive, and thus weaken the electric field and lower the potential "hill" that separates the two regions (Fig. 37.19c). It becomes easier for electrons to move from *N* to *P* and, as we could show with analogous diagrams for holes, it's easier for holes to move from *P* to *N*. So a current flows in the *P*-to-*N* direction, and the forward-biased *PN* junction becomes a good conductor. If, on the other hand, we connect the battery's positive terminal to the *N*-type material, then we strengthen the internal electric field and steepen the potential "hill," making it hard for charges to cross the junction (Fig. 37.19d). Now the *PN* junction is **reverse biased**, and it's a poor conductor.

As electrons and holes pour across a forward-biased junction, many recombine; that is, they drop from the conduction band into the valence band, releasing energy in the process. In light-emitting diodes (LEDs) and diode lasers, this energy appears as photons whose energy is close to that of the band gap. Because $E = hf$, the band gap determines the frequency and, equivalently, the wavelength and color of the emitted light. Development of semiconductor lasers with ever-larger band gaps enabled the evolution from CD to DVD to Blu-ray and HD-DVD discs that we outlined in Chapter 32. Conversely, a material whose band gap corresponds to visible-light photons can absorb light energy, promoting electrons to the conduction band and driving current through an external circuit. Such **photovoltaic cells** (Fig. 37.20) have long been used to generate electricity on spacecraft and in remote terrestrial applications. As their cost continues to drop, they hold promise for large-scale pollution-free electric-power generation.

**FIGURE 37.20** Operation of a photovoltaic cell. Sunlight photons penetrate to the *PN* junction, where they promote electrons to the conduction band and thus produce electron–hole pairs. Energy acquired from solar photons drives current through the metallic contacts and on to an external load.

---

**EXAMPLE 37.4** Semiconductors: Engineering the Band Gap

The lasers used to "read" CDs, DVDs, and Blu-ray discs operate at 780 nm, 650 nm, and 405 nm, respectively. Find the band gap needed for each laser.

**INTERPRET** We're asked to determine the band gap from the laser wavelength. The band gap is the energy difference between the conduction and valence bands, and it's equal to the energy of the photon emitted as electrons "fall" across the gap. So we need to relate photon energy and wavelength.

**DEVELOP** $E = hf$ relates photon energy and frequency; since $f\lambda = c$, $E = hc/\lambda$. Given the wavelengths $\lambda$, we'll use this expression to find the photon energies, which are equal to the band gaps.

**EVALUATE** For the CD,

$$E = \frac{hc}{\lambda} = \frac{(6.63 \times 10^{-34}\,\text{J}\cdot\text{s})(3.00 \times 10^{8}\,\text{m/s})}{(780\,\text{nm})} = 2.55 \times 10^{-19}\,\text{J}$$

or 1.59 eV. Similar calculations give 1.91 eV for DVD and 3.07 eV for Blu-ray.

**ASSESS** The energy gap increases with decreasing wavelength, as we expect because shorter wavelengths correspond to higher frequencies and therefore to higher-energy photons. ∎

## 37.4 Superconductivity

In Chapter 24 we introduced **superconductivity**—the complete loss of electrical resistance in some materials at low temperature. First discovered in mercury in 1911, superconductivity was for decades limited to a few elements and alloys below about 20 K. A breakthrough in 1986 brought a new class of metal-oxide superconductors with superconducting **transition temperatures** of about 100 K; today the highest transition temperatures exceed 160 K. The ultimate goal of a room-temperature superconductor, once thought beyond reach, may yet be achieved.

Superconductors find use in an ever-increasing range of applications, including high-strength electromagnets for MRI scanners, particle accelerators, materials separation, and research; compact, efficient motors for vehicle and marine propulsion; high-$Q$ filters for cell-phone base stations; sensitive magnetic-field sensors for brain-wave imaging and physics research; underground power transmission in crowded cities; and so-called synchronous condensers for optimizing the power factor in AC power transmission (see Section 28.5). Other applications include superconducting electronic devices that promise orders-of-magnitude increases in computer speed, and magnetically levitated vehicles for ground transportation at speeds up to 500 km/h (see the Application on p. 679).

## Superconductivity and Magnetism

The hallmark of a superconductor is zero electrical resistance. Another distinguishing characteristic is the **Meissner effect**, wherein a superconductor excludes magnetic flux from its interior (Fig. 37.21). Figure 37.21c shows why: Currents in the superconductor create their own magnetic field that exactly cancels the field within the material. As we pointed out in Section 27.6, a superconductor's exclusion of magnetic flux makes it perfectly diamagnetic. The magnetic levitation shown in Fig. 27.34 is a manifestation of the Meissner effect, wherein a magnet is supported over a superconductor by mutual repulsion between the magnet and currents in the superconductor.

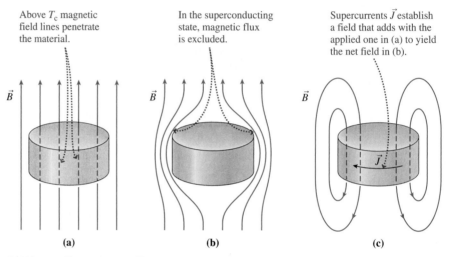

Above $T_c$ magnetic field lines penetrate the material.

In the superconducting state, magnetic flux is excluded.

Supercurrents $\vec{J}$ establish a field that adds with the applied one in (a) to yield the net field in (b).

$\vec{B}$

$\vec{B}$

$\vec{B}$

$\vec{J}$

(a)

(b)

(c)

**FIGURE 37.21** The Meissner effect.

As the strength of an applied magnetic field increases, so do the currents and resulting magnetic field of the superconductor. But beyond a **critical field**, the external magnetic field alters the superconducting state, and the superconductor no longer excludes magnetic flux. In **type I superconductors**, superconductivity ceases abruptly at the critical field (Fig. 37.22a). **Type II superconductors**, in contrast, have upper and lower critical fields, between which superconductivity gradually diminishes (Fig. 37.22b). At the lower critical field the material begins to allow flux penetration, and a regular array of nonsuperconducting regions forms, centered on magnetic field lines. These grow with increasing field, until at the upper critical field the superconducting regions vanish altogether.

Because electric currents generate magnetic fields, the critical field can limit the current-carrying capability of superconductors. Fortunately type II superconductors have high enough upper critical fields to permit substantial currents. Type IIs tend to be alloys or complex mixtures, and include all the high-$T$ superconductors. Critical fields of high-$T$ superconductors are as high as 100 T; however, these materials are brittle ceramics and present engineering challenges to the fabrication of wires and other flexible conductors.

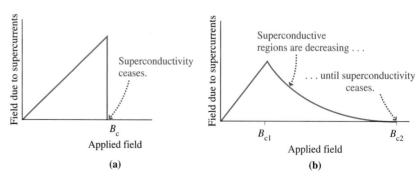

Superconductivity ceases.

$B_c$

Applied field

(a)

Superconductive regions are decreasing . . .

. . . until superconductivity ceases.

$B_{c1}$

$B_{c2}$

Applied field

(b)

**FIGURE 37.22** Responses of (a) type I and (b) type II superconductors to applied magnetic fields. $B_c$ denotes the critical field.

APPLICATION **Maglev!**

Passengers arriving at Shanghai's Pudong International Airport make the 30-km trip to a city metro station in only 7 minutes, "flying" on a

magnetically levitated vehicle—a maglev—at speeds exceeding 400 km/h. The Shanghai system uses conventional electromagnets and electronic feedback circuits to keep the vehicle levitated a mere 1 cm above its guideway. But others are developing maglev systems that rely on superconducting magnets for both levitation and propulsion. Coils in the guideway carry alternating current, alternately pushing and pulling the vehicle's onboard magnet. In effect, vehicle and guideway become a linear electric motor, much like a conventional motor that's been "unwound" to produce straight-line motion. In superconducting systems, any deviation from perfect alignment with the guideway results in induced currents and, correspondingly, magnetic forces that act to keep the maglev vehicle centered in its guideway. Today's superconducting maglevs require onboard refrigeration systems, so development of a room-temperature superconductor would make maglev a much more attractive transportation alternative. The photo shows a Japanese superconducting maglev that has achieved speeds of 450 km/h.

## Theories of Superconductivity

Superconductivity is a purely quantum-mechanical phenomenon; classical physics is totally inadequate to explain its existence. A successful theory of conventional low-$T_c$ superconductors, called the **BCS theory** after its originators, was formulated in 1957 by John Bardeen, Leon Cooper, and John Robert Schrieffer; the trio shared the 1972 Nobel Prize in physics.

In BCS theory, superconductivity results from a quantum-mechanical pairing of electrons that leads to a lower-energy state in which electron pairs move through the crystal lattice with no energy loss to the ions, resulting in zero electrical resistance. The electron pairing involves one electron slightly deforming the ion lattice, with the second electron attracted by the slight positive charge of the deformed lattice (Fig. 37.23). But the paired electrons aren't physically close; typically, a million other electrons, each paired with another distant electron, may lie between the two (Fig. 37.23c). The result of this long-range pairing is coherent motion of the conduction electrons that extends throughout the superconductor. Like well-choreographed dancers, the electrons all move together in a way that precludes energy loss to the ion lattice.

High-temperature superconductors aren't fully understood, although they almost certainly involve quantum-mechanical pairing of charge carriers. The mechanism of the pairing is less clear; one promising candidate involves magnetic interactions, although other mechanisms are under investigation. Superconductivity presents a continuing challenge to both theorists and experimentalists.

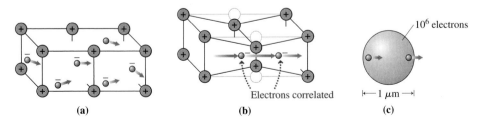

**FIGURE 37.23** Electron pairing in BCS theory. (a) A normal conductor, with uncorrelated electrons. (b) In a superconductor, one electron passing through the lattice deforms it slightly. About $10^{-12}$ s later, a second electron passes through and experiences the potential of the deformed lattice. The two electrons are therefore correlated. (c) Paired electrons are typically 1 $\mu$m apart, with a million others in their vicinity. The coherent motion of all the paired electrons results in superconductivity.

## Big Picture

The big idea here is that quantum mechanics can explain the structure of molecules and solids as well as the atoms treated in Chapter 36. At this level we can't solve the Schrödinger equation for these many-particle systems, but we've argued—using energy and angular momentum quantization and the exclusion principle—for quantum effects in molecules and solids.

## Key Concepts and Equations

Common types of molecular bonding include **ionic, covalent, hydrogen, van der Waals**, and **metallic** bonding. Whatever the bonding mechanism, a stable molecule is at the minimum of its potential-energy curve, shown below for $H_2$.

Molecules exhibit both rotational and vibrational energy, giving a rich structure of quantized energy levels and the resulting spectra.

- Quantization of angular momentum leads to quantized rotational energy levels:
$$E_{rot} = \frac{\hbar^2}{2I} l(l+1), \ l = 0, 1, 2, \ldots.$$

- The vibrational energy levels are those of the harmonic oscillator:
$$E_{vib} = \left(n + \tfrac{1}{2}\right) \hbar \omega, \ n = 0, 1, 2, \ldots.$$

When atoms join to make solids, individual atomic energy levels separate to form **bands**. **Band theory** distinguishes conductors from insulators depending on whether the uppermost occupied band is partially or completely full, respectively. **Semiconductors** are like insulators, but with a much smaller band gap that permits thermal excitation of electrons into the conduction band.

In a metallic conductor, the energy of the highest occupied state at absolute zero is the **Fermi energy**.

Conductor     Insulator     Semiconductor

## Cases and Uses

**Superconductivity** is a quantum-mechanical phenomenon that occurs at low temperatures and admits no classical explanation. Paired electrons move coherently through a superconductor without energy loss to the ion lattice, resulting in zero electrical resistance. Superconductors exclude magnetic fields—the **Meissner effect**—up to a **critical field** that destroys superconductivity, abruptly in **type I superconductors** and gradually in **type II superconductors**.

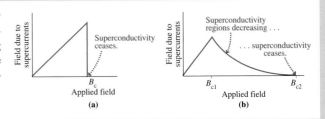

The band structure of doped semiconductors helps explain the one-way conduction of a *PN* junction, a phenomenon at the heart of modern electronics.

# For Thought and Discussion

1. Why is the exclusion principle crucial to the existence of stable molecules?

2. Why do ionically bonded materials have high melting points?

3. Ionic bonds clearly result from electrostatic attraction between ions. In what way do covalent bonds also involve electrostatic attraction?

4. Does it make sense to distinguish individual NaCl molecules in a salt crystal? What about individual $H_2O$ molecules in an ice crystal? Explain.

5. Is it useful to think of the highest-energy electrons "belonging" to individual atoms in an ionically bonded molecule? In a covalently bonded molecule?

6. What are the approximate relative magnitudes of the energies associated with electronic excitation of a molecule, with molecular vibration, and with molecular rotation?

7. Radio astronomers have discovered many complex organic molecules in interstellar space. Why were these discoveries made with radio telescopes and not optical telescopes?

8. In Fig. 18.16, why are the rotational states excited at lower temperatures than the vibrational states?

9. Would you expect solid hydrogen to conduct electricity? Why or why not?

10. The Fermi energy in metals is much higher than the thermal energy at typical temperatures. Why does this make the mean speed of conduction electrons nearly independent of temperature?

11. Why does the size of the band gap determine whether a material is an insulator or a semiconductor?

12. How would you expect the conductivity of an undoped semiconductor to depend on temperature? Why?

13. Name some technological innovations that might result from a room-temperature superconductor.

14. Suppose a room-temperature superconductor were discovered, but it had a very low critical field. In what way would this limit its practical applicability?

15. How do type I and type II superconductors differ?

# Exercises and Problems

## Exercises

### Section 37.2 Molecular Energy Levels

16. Find the energies of the first four rotational states of the HCl molecule described in Example 37.1.

17. Find the wavelength of electromagnetic radiation needed to excite oxygen molecules ($O_2$) to their first rotational excited state. The rotational inertia of an oxygen molecule is $1.95 \times 10^{-46}$ kg·m$^2$.

18. Calculate the wavelength of a photon emitted in the $l = 5$ to $l = 4$ transition of a molecule whose rotational inertia is $1.75 \times 10^{-47}$ kg·m$^2$.

19. Photons of wavelength 1.68 cm excite transitions from the rotational ground state to the first rotational excited state in a gas. What is the rotational inertia of the gas molecules?

20. The classical vibration frequency for diatomic hydrogen ($H_2$) is $1.32 \times 10^{14}$ Hz. Find the spacing between its vibrational energy levels.

21. The energy between adjacent vibrational levels in diatomic nitrogen is 0.293 eV. What is the classical vibration frequency of this molecule?

### Section 37.3 Solids

22. Use the 2.16-g/cm$^3$ density of NaCl to calculate the ionic spacing $r_0$ in the NaCl crystal. *Hint:* Consult Appendix D for atomic weights.

23. Express the 7.84-eV ionic cohesive energy of NaCl in kilocalories per mole of ions.

24. Lithium fluoride, LiF, has the same crystal structure as NaCl and therefore essentially the same Madelung constant $\alpha$. Its ionic cohesive energy is $-10.5$ eV, and the value of $n$ in Equation 37.4 is 6.25. Find the equilibrium ionic separation in LiF.

25. What is the wavelength of light emitted by a gallium phosphide (GaP) light-emitting diode? *Hint:* See Table 37.1.

26. What is the shortest wavelength of light that could be produced by electrons jumping the band gap in a material from Table 37.1? What is the material?

27. Which material in Table 37.1 would provide the longest wavelength of light in a light-emitting diode? What is that wavelength?

28. A common light-emitting diode is made from a combination of gallium, arsenic, and phosphorus (GaAsP) and emits red light at 650 nm. What is its band gap?

## Problems

29. A molecule drops from the $l = 2$ to the $l = 1$ rotational level, emitting a 2.50-meV photon. If the molecule then drops to the rotational ground state, what energy photon will it emit?

30. A molecule absorbs a photon and jumps to the next higher rotational state. If the photon energy is three times what would be needed for a transition from the rotational ground state to the first rotational excited state, between what two levels is the transition?

31. Find an expression for the energy of a photon required for a transition from the $(l-1)$th level to the $l$th level in a molecule with rotational inertia $I$.

32. A molecule of rotational inertia $I$ undergoes a transition from the $l$th rotational level to the $(l-1)$th level. Show that the wavelength of the emitted photon is $\lambda = 4\pi^2 I c/hl$.

33. The rotational spectrum of diatomic oxygen ($O_2$) shows spectral lines spaced 0.356 meV apart in energy. Find the atomic separation in this molecule. *Hint:* See Example 37.1, but remember that the oxygen atoms have equal mass.

34. Use the result given in Problem 56 to find the bond length in carbon monoxide (CO), given that excitation of the first rotational state requires photons of wavelength 2.59 mm.

35. For the HCl molecule of Example 37.2, determine (a) the energy of the vibrational ground state and (b) the energy of a photon emitted in a transition between adjacent vibrational levels. Assume the rotational quantum number doesn't change.

36. Diatomic deuterium ($D_2$) has classical vibration frequency $9.35 \times 10^{13}$ Hz and rotational inertia $9.17 \times 10^{-48}$ kg·m$^2$. Find (a) the energy and (b) the wavelength of a photon emitted in a transition between the $n = 1, l = 1$ state and the $n = 0, l = 2$ state.

37. An oxygen molecule is in its vibrational and rotational ground states. It absorbs a photon of energy 0.19653 eV and jumps to the $n = 1, l = 1$ state. It then drops to the $n = 0, l = 2$ level, emitting a 0.19546-eV photon. Find (a) the classical vibration frequency and (b) the rotational inertia of the molecule.

38. The atomic spacing in diatomic hydrogen ($H_2$) is 74 pm. Find the energy of a photon emitted in a transition from the first rotational excited state to the ground state.

39. What wavelength of infrared radiation is needed to excite a transition between the $n = 0$, $l = 3$ state and the $n = 1$, $l = 2$ state in KCl, for which the rotational inertia is $2.43 \times 10^{-45}$ kg·m$^2$ and the classical vibration frequency is $8.40 \times 10^{12}$ Hz?

40. Find the wavelengths emitted in all possible allowed transitions between the first three rotational states in the $n = 1$ level to any states in the $n = 0$ level in diatomic hydrogen. The rotational inertia of the hydrogen molecule is $4.60 \times 10^{-48}$ kg·m$^2$, and its classical vibration frequency is $3.69 \times 10^{14}$ Hz.

41. Determine the constant $n$ in Equation 37.4 for potassium chloride (KCl), for which $r_0 = 0.315$ nm and $U_0 = -7.21$ eV. The crystal structure is the same as for NaCl.

42. A salt crystal contains $10^{21}$ sodium–chlorine pairs. How much energy would it take to compress the crystal to 90% of its normal size?

43. Lithium chloride, LiCl, has the same structure and therefore the same Madelung constant as NaCl. The equilibrium separation in LiCl is 0.257 nm, and $n = 7$ in Equation 37.4. Find the ionic cohesive energy of the LiCl crystal.

44. (a) Differentiate Equation 37.4 to obtain an expression for the force on an ion in an ionic crystal. (b) Use your result to find the force on an ion in NaCl if the crystal could be compressed to half its equilibrium spacing (see Example 37.3 for relevant parameters). Compare with the electrostatic attraction between the ions at a separation of $\frac{1}{2}r_0$. Your result shows how very "stiff" this ionic crystal is.

45. Equation 37.5 gives the number of states per unit volume per unit energy interval in a metal. Integrating over energy should then give the total number of states per unit volume. Therefore integrating from $E = 0$ to $E = E_F$—over the occupied states, that is—gives the number of conduction electrons per unit volume. Carry out this integration to show that the electron number density is given by

$$ n = \left( \frac{2^{9/2} \pi m^{3/2}}{3h^2} \right) E_F^{3/2} $$

46. The Fermi energy in aluminum is 11.6 eV. Use the result of Problem 45 to find the density of conduction electrons in aluminum.

47. Use the result of Problem 45 to determine the Fermi energy for calcium, which has $4.6 \times 10^{28}$ conduction electrons per cubic meter.

48. The Fermi energy for copper is 7.0 eV. Compare the electron speed associated with this energy with the classical thermal speed for an electron at room temperature (300 K). *Note:* Your result shows how the quantum and classical descriptions of metallic conduction differ.

49. The *Fermi temperature* is defined by equating the thermal energy $kT$ to the Fermi energy, where $k$ is Boltzmann's constant. Calculate the Fermi temperature for silver, for which $E_F = 5.48$ eV, and compare it with room temperature.

50. Photons with energy lower than a semiconductor's band gap are not readily absorbed by the material, so a measurement of light absorption versus wavelength gives the band gap. An absorption spectrum for silicon shows no absorption for wavelengths longer than 1090 nm. Use this information to calculate the band gap in silicon, and verify its entry in Table 37.1.

51. Calculate the median wavelength $\lambda_{median}$ for sunlight, treating the Sun as a 5800-K blackbody (see Equation 34.2b). Use your result to decide whether zinc selenide, with band gap 3.6 eV, would make a good photovoltaic cell. Argue quantitatively.

52. Pure aluminum, which superconducts below 1.20 K, exhibits a critical field of 9.57 mT. Find the maximum current that can be carried in a 30-gauge (0.255-mm-diameter) aluminum superconducting wire without the field from that current exceeding the critical field. *Hint:* Where is the field greatest? Consult Example 26.8.

53. The critical magnetic field in a niobium–titanium superconductor is 15 T. What current is required in a 5000-turn solenoid 75 cm long to produce a field of this strength?

54. The transition from the ground state to the first rotational excited state in diatomic oxygen ($O_2$) requires about 356 $\mu$eV. At what temperature would the thermal energy $kT$ be sufficient to set diatomic oxygen into rotation? Would you ever find diatomic oxygen exhibiting the specific heat of a monatomic gas at normal pressure?

55. The density of rubidium iodide (RbI) is 3.55 g/cm$^3$, and its ionic cohesive energy is $-145$ kcal/mol. Determine (a) the equilibrium separation and (b) the exponent $n$ in Equation 37.4 for RbI.

56. Consider a diatomic molecule consisting of different atoms with masses $m_1$ and $m_2$, separated by a distance $R$. (a) Find the molecule's center of mass. (b) Use your result to show that the rotational inertia about the center of mass is

$$ I = \left( \frac{m_1 m_2}{m_1 + m_2} \right) R^2 $$

The quantity $m_1 m_2 / (m_1 + m_2)$ is called the reduced mass. (c) Use this result to compute a more accurate value for the HCl bond length of Example 42.1.

57. Find the fraction of conduction electrons in a metal at absolute zero that have energies less than half the Fermi energy.

58. The Madelung constant (Section 37.3) is notoriously difficult to calculate because it's the sum of an alternating series of nearly equal terms. But it can be calculated for a hypothetical one-dimensional crystal consisting of a line of alternating positive and negative ions, evenly spaced (Fig. 37.24). Show that the potential energy of an ion in this "crystal" can be written as

$$ U = -\alpha \frac{ke^2}{r_0} $$

where the Madelung constant $\alpha$ has the value 2 ln2.

**FIGURE 37.24** Problem 58

59. The lower-energy states in a covalently bound diatomic molecule can be found approximately from the so-called Morse potential $U(r) = U_0 (e^{2(r - r_0)/a} - e^{-2(r - r_0)/a})$, where $r$ is the atomic separation and $U_0$, $r_0$ and $a$ are constants determined from experimental data. Calculate $dU/dr$ and $d^2U/dr^2$ to show that $U$ has a minimum, and find expressions for (a) $U_{min}$ and (b) the value of the separation $r_{min}$ at which the minimum occurs.

60. (a) Count the number of electron states $N(E)$ with energy $E$ in Equation 35.8 by calculating the volume available to such states in the space with Cartesian coordinate axes $n_x$, $n_y$, $n_z$. *Hint:* Consider each set of positive integers, which is at the corner of a unit cube, and that lies inside a radius $(n_x^2 + n_y^2 + n_z^2)^{1/2}$, and remember that there are two spin values per state. (b) Differentiate $N(E)$ with respect to $E$ to obtain Equation 37.5.

61. Use Equation 37.5 to calculate the average energy of a conduction electron at $T = 0$, in terms of the Fermi energy. *Hint:* Divide the integral of $E\, g(E)\, dE$ by the integral of $g(E)\, dE$, noting that at $T = 0$, $g(E) = 0$ for $E > E_F$.)

62. The current-voltage relation for a *PN* junction at room temperature (as in Figure 37.19) is approximately $I = I_0(e^{eV/kT} - 1)$, where $e$ in the exponent is the elementary charge, $k$ is Boltzmann's constant, $T$ is the absolute temperature, and $V$ is the voltage applied across the junction. Find the DC resistance $R = V/I$ and the AC resistance $r = dV/dI$, and show that both are small in the forward direction $(V > 0)$ and large in the reverse direction $(V < 0)$. *Hint*: At $T = 300$ K, $kT/e = 25.9$ mV, so the magnitude of the exponent is large as long as $|V|$ is appreciable.

63. Your friend the science-fiction writer imagines a universe where everything is opposite our universe. He likes to appeal to the technical mind and he asks for your help with one detail. In this unusual universe, the charge carriers in a conducting material are protons with a density of $10^{28}/m^3$, comparable to electrons in a metal. He asks you to calculate the Fermi energy in this material. What's your answer?

64. You work as a component buyer for a manufacturer of solar-energy systems. A sales rep is in your office trying to sell you ZnS photovoltaic cells that are seemingly very cheap. He claims that the low price is due to their unique manufacturing process and that these cells will provide unparalleled performance in your solar panels. You are suspicious. ZnS has a band gap of 3.6 eV. Are the sales rep's claims valid?

65. You're a manager at a company that makes lasers for CD and DVD players. There is someone to see you, a sales rep with a new semiconductor for you to consider. He looks familiar and you think he's the same one who, just one problem earlier, attempted to sell you a semiconductor for use in solar photovoltaic cells. His new material has a band gap of 2.77 eV. You need a blue-green laser with a wavelength of 447 nm. Do you throw him out of your office a second time, or should you consider purchasing this semiconductor for your laser?

66. Not wanting to take a job for the summer at a fast-food restaurant, you decide to stay on campus and work for your physics professor in her research lab. She's studying molecular nitrogen. The wavelength of electromagnetic radiation needed to excite a transition from the rotational ground state to the first rotational excited state in $N_2$ is 2.47 mm. Your professor asks you to calculate the atomic separation of $N_2$. What do you find?

# Answers to Chapter Questions

## Answer to Chapter Opening Question
The Schrödinger equation.

## Answers to GOT IT? Questions
37.1 Rotational energy.
37.2 The shaded areas are the same; they represent the number of conduction electrons per unit volume.

# 38 Nuclear Physics

■ Changing regions of mental activity are evident in these PET (positron emission tomography) scans of a human brain. What is a PET scan actually "seeing," and why does PET require hospitals to have on-site cyclotrons?

In Chapters 36 and 37 we explored atomic structure and saw how atoms join to form molecules and solids. Now we turn inward, to the atomic nucleus. Since Ernest Rutherford and colleagues discovered the nucleus in 1911, we've known that all the atom's positive charge and nearly all its mass are concentrated in a tiny nuclear region only about $10^{-5}$ of the atom's diameter. By 1920 Rutherford had proposed that nuclei beyond hydrogen contain neutral as well as positive particles, and today we know the nucleus is a composite of positive protons and neutral neutrons—collectively called **nucleons**. As we've seen, the uncertainty principle implies high minimum energies for particles confined in small regions, so we can infer already that the nucleus is a huge energy repository. We'll conclude this chapter with a look at humankind's attempts to harness that energy.

## 38.1 Elements, Isotopes, and Nuclear Structure

We saw in Chapter 36 how the number of electrons determines an atom's shell structure and therefore its chemical behavior. It's the number of protons in the nucleus—the **atomic number**, $Z$—that, in turn, determines the number of electrons in a neutral atom. That means all nuclei with the same $Z$ belong to the same element.

## Isotopes and Nuclear Symbols

Nuclei of the same element can, however, have different numbers of neutrons. That's because neutrons don't affect the nuclear charge and therefore have negligible influence on chemical behavior. Nuclei of the same element with different numbers of neutrons are distinct **isotopes** (Fig. 38.1). We call the total number of nucleons the **mass number**, $A$. Specification of the atomic number $Z$ and mass number $A$ then fully describes a nucleus. Figure 38.1 shows the conventional symbolism used in describing nuclei: the element symbol with a preceding subscript for $Z$ and superscript for $A$. Actually, the atomic number and symbol are redundant. To be helium (He), for example, *means* to have two nuclear protons and therefore $Z = 2$; to be uranium *means* $Z = 92$. Sometimes, therefore, we write helium-4, He-4, or $^4$He to mean the same thing as $^4_2$He.

Elements typically have several naturally occurring isotopes; a few are shown in Fig. 38.1. Most hydrogen has a single proton in its nucleus, but about one in 6500 hydrogen atoms is deuterium ($^2_1$H), whose nucleus contains a proton and a neutron. Most oxygen is $^{16}_8$O, but O-17 and O-18 also occur naturally; their ratios in polar ice cores provide valuable information about past climates. Most uranium is $^{238}_{92}$U, but 0.7% is the U-235 that's used in fission reactors and weapons—hence the great concern about the proliferation of uranium-enrichment facilities to increase the proportion of U-235. Incidentally, the atomic masses listed in the periodic table are averages that reflect the natural abundances of an element's several isotopes. Most elements also have short-lived radioactive isotopes that don't usually occur naturally but can be produced through nuclear reactions; more on these later.

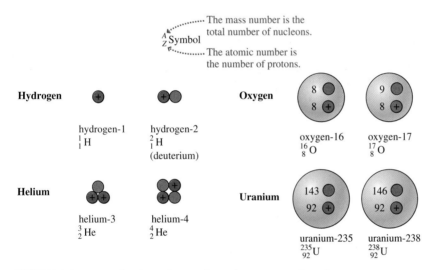

**FIGURE 38.1** Isotopes of a given element have the same number of protons but different numbers of neutrons.

---

**GOT IT? 38.1** Determine the number of protons and neutrons in these nuclei: (a) $^{12}_6$C; (b) $^{15}_8$O; (c) $^{57}_{26}$Fe; (d) $^{239}_{94}$Pu.

---

## The Nuclear Force

Given the electrical repulsion of the protons, there must be another force acting attractively to bind the nuclear constituents. Throughout much of the 20th century, this **nuclear force** was considered fundamental, but we now recognize it as a manifestation of a more fundamental force between the quarks that make up neutrons and protons. We'll explore quarks and their interactions in Chapter 39.

The attractive nuclear force acts between all nucleons—neutrons and protons, protons and protons, neutrons and neutrons. It's very strong at distances of less than a few femtometers ($10^{-15}$ m), but falls approximately exponentially with distance—more rapidly than the inverse-square falloff of the electric force. The attractive nuclear force therefore dominates between two neighboring protons, but electrical repulsion becomes dominant for more widely separated protons. The structure of the nucleus is determined, to a first approximation, by the interplay between the weaker but long-range electric force and the stronger but shorter-range nuclear force.

## Stable Nuclei

Not every combination of protons and neutrons will stick together indefinitely. Too many protons, and electrical repulsion wins out; sooner or later the nucleus decays by emitting a chunk of nuclear material (more details in Section 38.2). In larger nuclei most protons are far apart and therefore experience electrical repulsion more strongly than nuclear attraction (Fig. 38.2). To hold these nuclei together therefore requires more neutrons, which contribute attractive nuclear force but not electrical repulsion. So larger nuclei tend to have a higher ratio of neutrons to protons. Even this effect has its limits, though, and the result is that there are no stable nuclei for $Z > 83$.

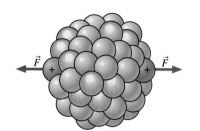

**FIGURE 38.2** Two widely separated protons in a large nucleus experience significant electrical repulsion and negligible nuclear attraction.

Too many neutrons also make a nucleus unstable. That's because the exclusion principle requires extra neutrons to go into higher energy states, making individual particles more likely to escape the nucleus. Furthermore, the neutron itself is an unstable particle; an isolated neutron decays spontaneously into a proton, an electron, and an elusive particle called a neutrino. This decay is suppressed in stable nuclei, but occurs if there are too many neutrons.

The delicate balance between neutrons and protons results in about 400 known stable nuclei, collectively called **nuclides**. Figure 38.3 is a **chart of the nuclides**, showing the stable nuclei, along with many unstable ones, on a chart of atomic number $Z$ versus neutron number $N = A - Z$. The chart shows that lighter nuclei tend to have equal numbers of protons and neutrons, but that heavier nuclei invariably have more neutrons to compensate for the increasing electrical repulsion of their widely separated protons.

## Nuclear Size

Unlike atomic electrons in their widely separated orbitals, nucleons pack tightly into the nucleus. Studies show that most nuclei are spherical, with the **nuclear radius**—defined as the radius at which the density has fallen to half its central value—given approximately by

$$R = R_0 A^{1/3} \qquad (38.1)$$

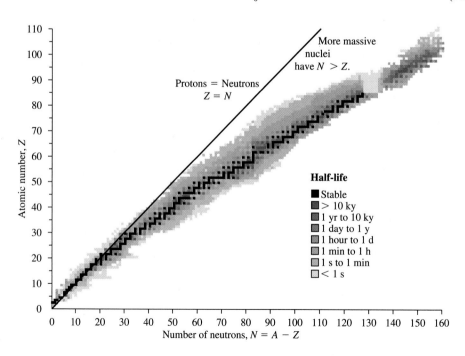

**FIGURE 38.3** A chart of the nuclides, coded by half-life.

where $R_0 = 1.2$ fm and $A$ is the mass number. This cube-root dependence is what we should expect for a tightly packed sphere whose volume is proportional to the number $A$ of its constituent particles, as suggested in Fig. 38.2. This tight packing also suggests that all nuclei have approximately the same density, on the order of $10^{17}$ kg/m³. A teaspoon of nuclear matter has a mass roughly equal to the mass of the Rock of Gibraltar! That absurdly high density reaffirms our picture of the complete atom as mostly empty space with its mass concentrated in a tiny nucleus.

## Nuclear Spin

In Chapter 36 we noted the important role of electron spin in atomic structure. Protons and neutrons are, like electrons, spin-$\frac{1}{2}$ particles. The spins of individual nucleons, combined with any angular momentum associated with their motions within the nucleus, give the nucleus a quantized spin angular momentum $I$ that obeys the same rules we've seen for other quantized angular momenta:

$$I = \sqrt{i(i+1)}\,\hbar \tag{38.2}$$

Here $i$, the nuclear spin quantum number, is a multiple of one-half. The component of $I$ on a given axis is also quantized, just like other angular momenta, according to $I_z = m_i \hbar$, where $m_i$ ranges from $-i$ to $i$ in steps of 1.

The spin quantum number $i$ is an even or odd multiple of one-half depending on whether the number of nucleons is even or odd. That makes even-$A$ nuclei bosons, particles with integer spin that don't obey the exclusion principle. Odd-$A$ nuclei, in contrast, have half-integer spin and are fermions that obey the exclusion principle. This distinction can lead to profound differences in physical behavior between isotopes of the same element. Helium-4, for example, becomes superfluid at low temperatures, meaning it flows without any viscosity. That's possible because helium-4 nuclei are bosons that can all occupy the same quantum state. Similar superfluidity doesn't occur in fermionic helium-3, although at extremely low temperatures He-3 nuclei themselves pair to form spin-1 particles that do make a superfluid.

The angular momentum of the nucleus results in a nuclear magnetic dipole moment, usually expressed in units of the **nuclear magneton**, $\mu_N = e\hbar/2m_p = 5.05 \times 10^{-27}$ J/T, where $m_p$ is the proton mass. The proton itself has a magnetic moment whose component on a given axis takes either of the values $\pm 2.793\,\mu_N = \pm 1.41 \times 10^{-26}$ J/T—a value that's usually listed as "the magnetic moment of the proton" although it's actually the component. Interaction of the nuclear magnetic moment with magnetic fields alters very slightly the energy levels of the atom—although the effect is much smaller than with atomic electrons because the higher proton mass makes for a much smaller magnetic moment. In hydrogen, for example, the proton can have either of two spin orientations relative to the magnetic field due to the electron, and the result is **hyperfine splitting** of the ground state into two levels a mere 5.9 $\mu$eV apart (Fig. 38.4). Transitions between these levels result in a spectral line at a radio wavelength of 21 cm. Radio astronomers use this line to detect interstellar clouds of neutral hydrogen.

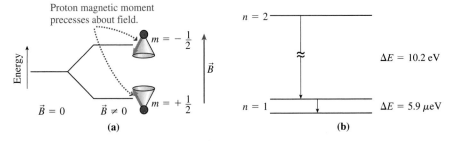

**FIGURE 38.4** (a) A nonzero magnetic field $\vec{B}$ splits the energy level of the spin-$\frac{1}{2}$ proton into two levels. (b) The two possible orientations of the proton in the magnetic field of the electron split the hydrogen ground state into two levels 5.9 $\mu$eV apart.

**Nuclear Magnetic Resonance**

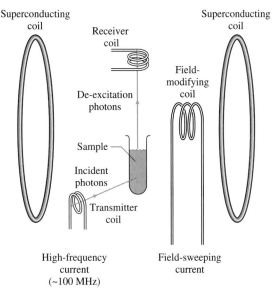

Putting nuclei in an external magnetic field creates two possible energy states, as suggested in Fig. 38.4a, depending on whether the nuclear magnetic moments are more nearly parallel or antiparallel to the field. Applying electromagnetic radiation with the appropriate photon energy will flip nuclei into the higher energy state. But because nuclei also experience magnetic fields from the electrons that surround them, the exact energy required is extremely sensitive to the details of the electron distribution—that is, to the surrounding molecular structure.

**Nuclear magnetic resonance** (NMR) uses this nuclear spin flipping to determine the structure of chemical compounds. In an NMR spectrometer, shown in the figure, the sample under analysis is placed in a uniform magnetic field $B$, usually from superconducting coils. A smaller coil carries AC current at a frequency $f$ corresponding to photon energy $hf$ that would flip the spin of an isolated nucleus in the field $B$. The coil emits electromagnetic waves, and if the nuclei absorb the associated photons, then they flip into their higher states and drop back, emitting radiation of frequency $f$ in the process. A receiver coil detects this radiation.

Because of the extra magnetic effect of the surrounding electrons, nuclei won't generally flip at the exact frequency $f$ and field $B$. So the field is varied until the superposition of the applied field and the

electron-generated field is exactly right. This condition of magnetic resonance then produces the up/down spin flips that generate a signal in the receiver coil. Scanning the field through a range of values detects nuclei in different electron environments, and from this information scientists can deduce the molecular structure.

Nuclear magnetic resonance with protons (H nuclei) is the basis of **magnetic resonance imaging** (MRI), a widely used medical procedure. In MRI, a person is placed inside a large solenoid whose field varies slightly with position. That makes the magnetic resonance frequency a function of position, and thus the resonance signal can be used to localize the protons undergoing magnetic resonance. A computer then uses the resonance information to construct an image. Most of the MRI signal comes from fat and water, making MRI especially good at imaging soft tissue that doesn't show well in X rays. The photo shows an MRI image of a human head and upper torso; soft-tissue structures including the brain are clearly visible.

---

**EXAMPLE 38.1** **Nuclear Spins: Finding the MRI Frequency**

The MRI solenoid we analyzed in Example 26.10 produces a 1.5-T magnetic field. What frequency should be used to drive the transmitter coil in this MRI device?

**INTERPRET** MRI is an implementation of nuclear magnetic resonance using protons (see the Application), so we're being asked for the frequency corresponding to photons that will flip a proton in a 1.5-T magnetic field.

**DEVELOP** We need to calculate the necessary photon energy and then use $E = hf$ to find the corresponding frequency. We've just seen that the proton acts like a magnetic dipole whose component along the field is $\mu_p = \pm 1.41 \times 10^{-26}$ J/T. Equation 26.15 gives the energy of a magnetic dipole: $U = -\vec{\mu} \cdot \vec{B}$. Here we're given the component of the magnetic moment $\vec{\mu}$ along the field, so our two energies become

$U = \pm \mu_p B$, where the signs correspond to the two possible spin orientations. A spin flip changes a proton's energy from $+\mu_p B$ to $-\mu_p B$, so our plan is to find the energy difference between these levels, equate it to the photon energy $hf$, and solve for $f$.

**EVALUATE** We have $E = \mu_p B - (-\mu_p B) = 2\mu_p B$, so

$$f = \frac{E}{h} = \frac{2\mu_p B}{h} = \frac{(2)(1.41 \times 10^{-26} \text{ J/T})(1.5 \text{ T})}{6.63 \times 10^{-34} \text{ J} \cdot \text{s}} = 63.8 \text{ MHz}$$

**ASSESS** This frequency is in the radio region of the electromagnetic spectrum, consistent with the diagram in the Application showing the use of coils and currents, and the approximate transmitter coil frequency of 100 MHz. Our particular answer happens to lie within the frequencies allocated to analog TV channel 3. ∎

## Models of Nuclear Structure

We've seen how the right ratio of neutrons to protons is essential for stable nuclei, and why that ratio increases for larger nuclei. Figure 38.3, the chart of the nuclides, summarizes this information. But take a closer look at that figure: There are more stable nuclei for even values of $Z$, and some—those with the so-called **magic numbers** 2, 8, 20, 28, 50, 82, and 126 protons or neutrons—have many more stable nuclei. Why?

Answering this question and explaining the decay mechanisms and lifetimes of unstable nuclei require a theory of nuclear structure. There still is no complete nuclear theory, analogous to the atomic theory of Chapter 36, that explains all aspects of all nuclei. Our still-imperfect knowledge of the nuclear force, and the tight packing of nucleons, render useless a simple two-particle model like the one we used for hydrogen. Instead, nuclear physicists resort to several models to explain different aspects of nuclear structure. Together, these models provide a good understanding of the nucleus and accurately predict nuclear properties, although not with the precision available in atomic physics.

The **liquid-drop model** provides a reasonable approximation for heavier nuclei, whose many nucleons behave somewhat like the molecules in a drop of liquid. A liquid-drop nucleus can rotate, vibrate, and change shape as long as its volume doesn't change, and the resulting quantized energy levels predict nuclear gamma-ray spectra that are in good agreement with observation. The liquid-drop model also helps explain nuclear fission, as we'll explore in Section 38.4. But it can't account for the dramatic effects of small changes in nucleon number, particularly the role of the magic numbers.

The **nuclear shell model**, advanced in the late 1940s by physicists Maria Goeppert Mayer and J. Hans Jensen, gives the nucleus a shell structure similar to that of atoms. The shells occur because neutrons and protons obey the exclusion principle, and the magic numbers correspond to closed-shell configurations analogous to the electronic structure of inert gases. Closed-shell nucleons are tightly bound, making a magic nucleus particularly stable. Additional nucleons beyond a closed shell stay largely on the outskirts of the nucleus, where they're more readily excited to higher energy levels. Neutrons and protons behave independently in the shell model, and each has its own set of quantum numbers. Closed-shell structure therefore occurs with magic numbers of either protons or neutrons. Some nuclei, like $^{40}_{20}$Ca $(Z = 20, N = 20)$, are "doubly magic" and show exceptional stability.

The **collective model**, advanced by Niels Bohr's son Aage, combines aspects of the liquid-drop and shell models, emphasizing the collective quantum-mechanical behavior of the nucleons. One remarkable prediction of the collective model is that larger, nonmagical nuclei may be more stable if they take nonspherical shapes.

Active areas of nuclear-structure research involve the creation and exploration of exceptionally heavy or neutron-rich nuclei. The creation of elements 115 and 116 in the early 2000s suggests that physicists are approaching a region of longer-lived nuclei dubbed the "island of stability," which may be associated with a new magic neutron number of 184. And experiments in 2005 created silicon-42, whose relative stability implies that its atomic number $Z = 14$ becomes magic in this neutron-bloated $(N = 28)$ species. Until we have a complete nuclear theory, experiments like these will continue to challenge physicists with nuclear surprises.

## 38.2 Radioactivity

In 1896 Henri Becquerel of Paris noticed that a photographic plate stored near uranium compounds became fogged, as though exposed to invisible rays. Becquerel had discovered **radioactivity**, wherein some substances spontaneously emit high-energy particles or photons. Marie and Pierre Curie promptly began a thorough exploration of the phenomenon, for which Marie Curie coined the name "radioactivity." The Curies shared the 1903 Nobel Prize in physics with Becquerel, and Marie Curie won the 1911 Nobel Prize in chemistry for her discovery of polonium and radium.

### Decay Rate and Half-Life

Radioactivity results from the decay of unstable nuclei, a process that occurs at vastly differing rates in different isotopes. The number of decays per unit time is the **activity** of a radioactive sample; the SI unit of activity is the **becquerel** (Bq), equal to one decay per second. An older unit, the **curie** (Ci), is $3.7 \times 10^{10}$ Bq and is approximately the activity of 1 gram of radium-226. For a given isotope, activity is proportional to the number $N$ of nuclei present. $N$ decreases as nuclei decay, so we can write

$$\frac{dN}{dt} = -\lambda N$$

where $\lambda$ is the **decay constant**. As we've seen with discharging capacitors and decaying inductor currents, this equation is a prescription for exponential decay. We solve it the same way, multiplying by $dt/N$ and integrating:

$$\int_{N_0}^{N} \frac{dN}{N} = -\lambda \int_0^t dt$$

where $N_0$ is the initial number of nuclei at time $t = 0$. Evaluating the integrals gives $\ln(N/N_0) = -\lambda t$ or, exponentiating each side and using $e^{\ln x} = x$:

$$N = N_0 e^{-\lambda t} \tag{38.3a}$$

Equation 38.3a shows that the decay constant $\lambda$ is a measure of the exponential decay rate. We can also interpret $\lambda$ as the probability that a given atom will decay in a 1-s time interval. Another convenient measure of exponential decay is the **half-life**, $t_{1/2}$, defined as the time for half the nuclei in a given sample to decay. If we start with $N_0$ nuclei at time $t = 0$, then at a later time $t$ the number of nuclei remaining will be

$$N = N_0 2^{-t/t_{1/2}} \quad \text{(radioactive decay)} \tag{38.3b}$$

You can quickly show that $t_{1/2}$ and $\lambda$ are related by $t_{1/2} = \ln 2/\lambda \approx 0.693/\lambda$ (see Problem 48). Figure 38.5 is a graph of Equation 38.3b. Since activity and number of nuclei are proportional, both decline with the same half-life, as described in Equation 38.3b. Table 38.1 lists some significant radioisotopes and their half-lives.

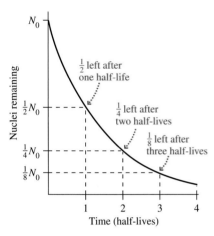

$N_0$

$\frac{1}{2}$ left after one half-life

$\frac{1}{4}$ left after two half-lives

$\frac{1}{8}$ left after three half-lives

Nuclei remaining

$\frac{1}{2}N_0$

$\frac{1}{4}N_0$

$\frac{1}{8}N_0$

1   2   3   4

Time (half-lives)

**FIGURE 38.5** Exponential decay of a radioactive sample.

**TABLE 38.1** Selected Radioisotopes

| Isotope | Half-life | Decay Mode | Comments |
|---|---|---|---|
| Carbon-14 ($^{14}_{6}$C) | 5730 years | $\beta^-$ | Used in radiocarbon dating |
| Iodine-131 ($^{131}_{53}$I) | 8.04 days | $\beta^-$ | Fission product abundant in fallout from nuclear weapons and reactor accidents; damages thyroid gland |
| Oxygen-15 ($^{15}_{8}$O) | 2.03 minutes | $\beta^+$ | Short-lived oxygen isotope produced with cyclotrons and used for medical diagnosis in PET |
| Potassium-40 ($^{40}_{19}$K) | $1.25 \times 10^9$ years | $\beta^-$ | Comprises 0.012% of natural potassium; dominant radiation source within the normal human body; used in radioisotope dating |
| Plutonium-239 ($^{239}_{94}$Pu) | 24,110 years | $\alpha$ | Fissile isotope used in nuclear weapons; produced by neutron capture in $^{238}_{92}$U |
| Radium-226 ($^{226}_{88}$Ra) | 1600 years | $\alpha$ | Highly radioactive isotope discovered by Marie and Pierre Curie; results from decay of $^{238}_{92}$U |
| Radon-222 ($^{222}_{86}$Rn) | 3.82 days | $\alpha$ | Radioactive gas formed naturally in decay of $^{226}_{88}$Ra; seeps into buildings, where it may cause serious radiation exposure |
| Strontium-90 ($^{90}_{38}$Sr) | 29 years | $\beta^-$ | Fission product that behaves chemically like calcium; readily absorbed into bones |
| Tritium ($^{3}_{1}$H) | 12.3 years | $\beta^-$ | Hydrogen isotope used in biological studies and to enhance yields of nuclear weapons |
| Uranium-235 ($^{235}_{92}$U) | $7.04 \times 10^8$ years | $\alpha$ | Fissile isotope comprising 0.72% of natural uranium; used as reactor fuel and in simple nuclear weapons |
| Uranium-238 ($^{238}_{92}$U) | $4.46 \times 10^9$ years | $\alpha$ | Predominant uranium isotope; cannot sustain a chain reaction |

**EXAMPLE 38.2** **Radioactive Decay: Fallout from Chernobyl**

The 1986 nuclear accident at Chernobyl in the Ukraine spread radioactive fallout over Eastern Europe and Scandinavia. Iodine-131 was of great concern, since it's absorbed by the thyroid gland and can cause thyroid cancer. Following the accident, I-131 levels in Romanian milk rose to 2900 Bq per liter. How long did Romanians have to wait before I-131 levels were below their government's 185-Bq/L safety standard?

**INTERPRET** This is a problem about radioactive decay. We're given the initial activity per liter of milk, and we need to find how long it takes for that to decay to the given safety level. Using Table 38.1, we identify I-131's half-life as 8.04 days.

**DEVELOP** Equation 38.3b, $N = N_0 2^{-t/t_{1/2}}$, describes equally well the decline in the number of radioactive nuclei and their activity. The equation shows that after $n$ half-lives, activity drops to $1/2^n$ of its original level. So our plan is to find the number of half-lives $n$ that will lower the milk's I-131 activity from 2900 Bq/L to 185 Bq/L. Mathematically, we want $1/2^n = 185/2900$. We'll solve for $n$ and then use the known half-life to get an actual time.

**EVALUATE** Inverting our expression $1/2^n = 185/2900$ and taking logarithms of both sides give

$$\ln(2^n) = \ln(2900/185).$$

But $\ln(2^n) = n \ln 2$, so

$$n = \frac{\ln(2900/185)}{\ln 2} = 3.97 \text{ half-lives}$$

With $t_{1/2} = 8.04$ days, this amounts to 32 days.

**ASSESS** A quick check shows that our answer is about right: After one half-life the activity has dropped in half, to $2900/2 = 1450$ Bq/L. Another half-life, and it's half of this, or 725 Bq/L. A third, and it's about 360 Bq/L. A fourth, and it's down to about 180 Bq/L. So to reach 185 Bq/L. must take just under 4 half-lives. Romania's 32-day wait time was dependent on both the government safety standard and the initial contamination level. Problem 49 explores the situation in some other affected countries. ∎

✓**TIP** Half-Life and Powers of 2

After $n$ half-lives, activity has dropped by a factor of $1/2^n$. When we estimate activity levels, it's useful to note that $2^{10} = 1024$, or very nearly 1000. Thus activity drops by a factor of about 1000 every 10 half-lives—and therefore by about 1 million in 20 half-lives.

**GOT IT? 38.2** A PET-scan patient is injected with radioactive oxygen-15, whose half-life is 2 min. Approximately what fraction of the original $^{15}$O remains undecayed an hour later?

**APPLICATION** **Radiocarbon Dating**

Carbon-14 formed in the atmosphere is incorporated into a living organism through the food chain.

At death, $^{14}$C uptake ceases.

Cosmic rays
$^{14}CO_2$

(a)

(b)

(c)

(d)

Much later, $^{14}$C activity has decayed considerably.

Archaeologists excavate the long-dead remains. By measuring $^{14}$C activity, they can infer the time since death. Note that the archaeologists, with their active $^{14}$C intake, are more radioactive than their ancient ancestor.

Archaeologists, art historians, geologists, and others use radioactive decay to date ancient objects. For ages up to a few tens of thousands of years, the 5730-year isotope carbon-14 is especially useful. $^{14}$C forms continuously in the atmosphere through reactions of cosmic rays with nitrogen. Living things take in $^{14}$C and maintain a steady concentration through the balance between uptake and radioactive decay. At death, uptake ceases and the level of $^{14}$C begins to drop. Measuring the ratio of $^{14}$C to stable $^{12}$C in a sample of once-living matter and comparing with the ratio found in living material then provides the age (see the figure and Example 38.3).

The cosmic-ray flux at Earth varies with solar activity, and so, therefore, does the $^{14}$C/$^{12}$C ratio. Scientists correct for this effect with data from growth rings in ancient trees, which provide an independent measure of age. Measuring the actual radioactivity takes a fairly large sample, so today the most sophisticated dating is done instead by counting individual C-14 atoms, separating them from ordinary C-12 using a mass spectrometer—a device we described in Example 26.2.

Radiocarbon dating is quite accurate to about 20,000 years and can be used back to about 50,000 years. For longer time spans, up to the billions of years characterizing the ages of rocks, ratios of longer-lived isotopes provide age information. Much knowledge of our own past, and our planet's and our solar system's, comes from radioisotope dating.

**Radioactive Decay: Archaeology**

Archaeologists unearth charcoal from an ancient campfire and find its carbon-14 activity per unit mass to be 7.4% of the activity measured in living wood. Find the charcoal's age.

**INTERPRET** This is a problem about using the decay of carbon-14 to date a once-living material. We want the time it takes for $^{14}$C activity to decline to 7.4% of its original level. From Table 38.1, we identify the half-life of $^{14}$C as 5730 years.

**DEVELOP** Equation 38.3b, $N = N_0 2^{-t/t_{1/2}}$, shows that activity drops by $1/2^n$ in $n$ half-lives, so our plan is to find the number of half-lives that makes the factor $1/2^n$ equal to 0.074. Then we can multiply by the half-life to get the actual time.

**EVALUATE** Solving as we did in Example 38.2 gives

$$n \ln 2 = \ln(1/0.074)$$

which gives $n = 3.76$ half-lives. With $t_{1/2} = 5730$ y, the age is then 21,500 years.

**ASSESS** Again a quick check suffices: One half-life drops activity to 50%; two half-lives to 25%, three to 12.5%, and four to just over 6%. So it must take a little less than four half-lives to get down to 7.4% of the original activity level. ∎

## Types of Radiation

Passing radiation through a magnetic field shows that there are three types: one positively charged, one negative, and one neutral (Fig. 38.6). Early researchers named these alpha, beta, and gamma radiation. Today we know that alpha radiation consists of He-4 nuclei, beta radiation consists of high-energy electrons (or positrons), and gamma rays are high-energy photons. They differ in penetrating power: A sheet of paper can stop alpha particles, several centimeters of matter stop most betas, and uncharged gamma rays can penetrate substantial thicknesses of concrete or lead. Different radioisotopes emit not only different types of radiation but also radiation of different energies.

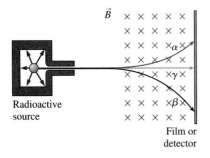

**FIGURE 38.6** The three types of radiation go separate ways in a magnetic field.

## Alpha Decay

Alpha emitters are nuclei with too much positive charge. They shed charge, and mass, by emitting a bundle of two protons and two neutrons—an alpha particle, $^4_2$He. Symbolically,

$$^A_Z X \longrightarrow \, ^{A-4}_{Z-2}Y + \, ^4_2\text{He} \qquad (\text{alpha decay}) \qquad (38.4)$$

Here $X$ is the original or **parent** nucleus, and $Y$ is the **daughter**. Note that the sums of the atomic numbers on both sides of this equation are equal, as are the mass numbers. Energy released in the reaction appears as kinetic energy of the alpha particle. The alpha particle actually emerges with less energy than needed to overcome the nuclear potential barrier, and this provides one of the most direct confirmations of quantum tunneling—which is the only way the alpha particle can escape the nucleus.

## Beta Decay

Beta emitters have too many neutrons, one of which decays into an electron, a proton, and an elusive neutral particle called a neutrino (symbol $\nu$). The electron exits at high energy to form beta radiation, leaving a nucleus with essentially the same mass but its atomic number increased because it has one more unit of positive charge:

$$^A_Z X \longrightarrow \, ^A_{Z+1}Y + e^- + \bar{\nu} \qquad (\text{beta decay}) \qquad (38.5a)$$

In ordinary beta decay the neutrino is, in fact, an antineutrino—hence the bar over its symbol $\nu$.

Beta decay is a manifestation of the weak nuclear force, and in the Sun it produces a steady stream of neutrinos that provide direct information on conditions in the solar core. That's because neutral, nearly massless neutrinos interact only rarely with matter; for example, they pass through the entire Earth with little probability of interaction. We'll see in Chapter 39 how neutrinos nonetheless are opening a new window on distant astrophysical events and the early universe.

A second type of beta decay converts a proton into a neutron, emitting both a positron (an anti-electron, $e^+$) and a neutrino:

$$_Z^A X \longrightarrow {}_{Z-1}^{A}Y + e^+ + \nu \qquad \text{(beta decay, positron emission)} \qquad (38.5b)$$

This reaction occurs in some short-lived isotopes of lighter elements like carbon and oxygen, and gamma rays from the subsequent annihilations of positrons are used in the medical imaging procedure known as positron emission tomography (PET).

A third beta-decay process is **electron capture**, in which a nucleus captures an inner-shell atomic electron, converting a proton to a neutron and ejecting a neutrino:

$$_Z^A X + e^- \longrightarrow {}_{Z-1}^{A}Y + \nu \qquad \text{(electron capture)} \qquad (38.5c)$$

## Gamma Decay

A nucleus in an excited state decays by emitting a photon, just like an atom. But the much higher energy associated with nuclear processes puts such photons in the gamma-ray region of the spectrum. Since the gamma-ray photon is neutral and massless, it doesn't change the type of nucleus; therefore we write

$$_Z^A X^* \longrightarrow {}_Z^A X + \gamma \qquad \text{(gamma decay)} \qquad (38.6)$$

where $X^*$ designates the excited state.

## Decay Series and Artificial Radioactivity

A few radioisotopes, like $^{40}$K and $^{238}$U, have half-lives comparable to Earth's age, so it's not surprising to find these in nature. But we also find shorter-lived species. Some, like cosmic-ray-produced $^{14}$C, result from naturally occurring nuclear reactions. Many others arise in the decay of long-lived isotopes, while some we produce in particle accelerators, nuclear reactors, and nuclear explosions.

Figure 38.7 shows the **decay series** for uranium-238, whose 4.46-billion-year half-life ensures that there's still plenty of it around. The shorter-lived daughter products in this series are present wherever there's uranium. A balance between formation and decay establishes the abundance of each product in the decay series. One of the uranium daughters is radon-222, a radioactive gas that can be a serious health hazard in closed spaces.

In 1930 Marie Curie's daughter Irène and her husband Frédéric Joliot-Curie were the first to induce artificial radioactivity, by bombarding stable isotopes with alpha particles. Today we produce radioisotopes with particle beams, neutrons from nuclear reactors, or by extracting them from the by-products of nuclear fission.

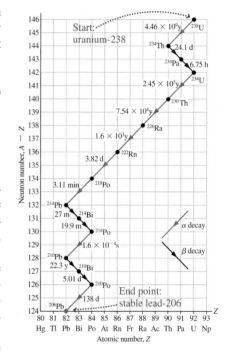

**FIGURE 38.7** The decay of uranium-238 results in a series of shorter-lived nuclei. Times shown are half-lives.

## Uses of Radioactivity

Nuclear radiation has numerous beneficial uses in our technological society. Here we survey just a few:

- **Radioactive Tracers** "Tagging" molecules with radioactive atoms makes it easy to trace their flows through biological and physical systems. Biologists use radioactive tracers routinely to study the uptake and distribution of chemicals. Engineers use radioisotopes to study wear in mechanical parts. Physicians "tag" bone-seeking compounds with radioisotopes to image the skeletal system; the resulting "bone scans" reveal cancer and other diseases.

- **Cancer Treatment** Radiation destroys living cells, especially fast-dividing cancer cells. Early radiation treatment used gamma radiation; today, particle beams deliver radiation with less effect on surrounding tissue. Alternatively, radioisotope "seeds" are embedded directly into a tumor.

- **Food Preservation** High radiation doses destroy bacteria and enzymes that cause food spoilage, providing longer shelf life and a safer food supply. Though controversial, food irradiation is becoming increasingly common.
- **Insect Control** Radiation preferentially damages reproductive cells and can therefore cause sterility. Sterilizing large numbers of pest insects with radiation causes populations to collapse when the sterile insects mate with normal ones. The medfly, a serious pest of citrus crops, has been controlled in this way.
- **Fire Safety** Common smoke detectors contain americium-241, whose alpha radiation ionizes air, allowing it to carry electric current. Smoke particles interfere with the current, triggering the alarm. Exit signs containing radioactive tritium ($^3$H) glow without the need for electricity, providing another measure of fire safety in public buildings.
- **Activation Analysis** Bombarding materials with neutrons or other particles results in excited states or the production of unstable isotopes. Analyzing the resulting radiation helps identify unknown materials. Art historians use this technique to detect forgeries; environmental scientists identify the constituents of pollution; and airport luggage scanners search for the radiation "fingerprint" of chemical explosives.

## Biological Effects of Radiation

Nuclear radiation has sufficiently high energy to ionize or otherwise disrupt biological molecules. Results include cell death, loss of biological functions, and mutations that lead to cancer or to genetic changes in future generations. Many early nuclear scientists, including Marie Curie and her daughter Irène, succumbed to leukemia and other cancers that undoubtedly resulted from radiation exposure.

The energy absorbed in a radiation dose is a rough measure of its biological danger. The SI unit of absorbed dose is the **gray** (Gy), defined as 1 J of energy per kg of absorbing material. A more appropriate measure is the **sievert** (Sv), which is weighted by the biological effectiveness of particular radiation types. Alpha particles, for example, cause more damage per unit energy than do gamma rays, so 1 Gy of alpha radiation is more harmful than 1 Gy of gamma radiation. But 1 Sv of alphas and 1 Sv of gammas cause essentially the same damage.

The biological effects of high radiation doses are well known; exposure to 4 Sv, for example, causes death in about half its human victims. But doses in the 0.1-Sv range and lower are more controversial. There are only a few cases of well-quantified exposures to populations large enough that small effects can be determined accurately. Even less certain are the effects of very low doses, such as the $10^{-5}$-Sv average dose to people living near the 1979 Three Mile Island nuclear accident. On the one hand, biological repair mechanisms may limit damage at low doses. On the other, even low doses may disproportionately affect the young and the unborn. A 2005 study by the U.S. National Academy of Sciences (NAS) suggests that the risk of cancer—the dominant health effect of low-level radiation—should scale linearly with dose. For a one-time dose of 1 mSv, the NAS study estimates a lifetime cancer risk of 1 in 10,000. This compares with a 42% lifetime chance of developing cancer from other causes.

The average U.S. citizen receives about 3.6 mSv of radiation per year, most of it from natural sources (Fig. 38.8). The dominant source, at 55%, is the uranium decay product radon-222, which seeps into buildings from the decay of naturally occurring uranium in the ground and in building materials. Our own bodies account for some 11%, mostly from natural potassium-40. About 19% of our average exposure in the United States comes from artificial sources, mainly medical procedures.

Consumer products—mostly tobacco, drinking water, and building materials—account for about 3%. Less than 1% of our yearly radiation dose is from nuclear power and weapons. Radiation doses vary greatly with location and occupation; for example, residents of high-altitude Denver have greater exposure to cosmic rays, and airline flight crews' cosmic-radiation dose often exceeds the allowed dose for nuclear-plant workers. No matter what your exact dose, though, the risks to your health from radiation exposure pale compared with other risks you knowingly take.

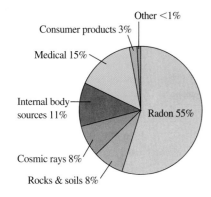

**FIGURE 38.8** Natural (gray) and artificial (color) sources of radiation, as percentages of the U.S. average yearly dose of 3.6 mSv. "Other" includes nuclear power, radioactive waste, and weapons tests.

# 38.3 Binding Energy and Nucleosynthesis

Disassembling a nucleus requires energy to overcome the strong nuclear force. The more tightly bound the nucleus, the higher this **binding energy**. The energies involved in nuclear interactions are high enough that Einstein's mass–energy equivalence is clearly evident, so accounting for energy conservation requires us to consider the rest energy of the particles. Then we can write

$$m_N c^2 + E_b = Z m_p c^2 + (A - Z) m_n c^2 \tag{38.7}$$

where the terms on the left are the rest energy of the nucleus, whose mass is $m_N$, and the binding energy $E_b$. The terms on the right are the rest energies of the $Z$ individual protons and $A - Z$ neutrons that make up the nucleus. So Equation 38.7 shows that we can disassemble a nucleus into its constituent nucleons if we supply additional energy equal to the binding energy. Equivalently, $E_b$ is the energy released if we assemble a nucleus from isolated nucleons.

Equation 38.7 shows that the nuclear mass $m_N$ is *not* the sum of the constituent particles' masses; rather, it's less by the amount $E_b/c^2$. This is clear evidence for mass–energy equivalence. Again, as in Chapter 33, we emphasize that there's nothing uniquely nuclear about this so-called **mass defect**. The mass of a water molecule is also less than the sum of its constituent hydrogen and oxygen atoms—but with chemical binding the effect is so small as to be virtually immeasurable. It's the strength of the nuclear force that makes mass–energy equivalence more obvious in nuclear interactions.

It's convenient to measure nuclear and particle masses in **unified mass units**, u, defined as one-twelfth the mass of a neutral carbon-12 atom. The unified mass unit is very nearly $1.66054 \times 10^{-27}$ kg, slightly less than the mass of the proton or neutron. High-energy physicists, ever cognizant of mass–energy equivalence, often express masses in MeV/$c^2$—a value numerically equal to the rest energy in MeV. Table 38.2 lists selected particle masses in kg, u, and MeV/$c^2$. In practice one often knows atomic rather than nuclear masses, but the difference is generally negligible because the extra mass of the electrons is so small.

**TABLE 38.2**  Selected Masses

|  | Mass (kg) | Mass (u) | Mass (MeV/$c^2$) |
|---|---|---|---|
| Electron | $9.10939 \times 10^{-31}$ | 0.000 548 579 | 0.510 999 |
| Proton | $1.67262 \times 10^{-27}$ | 1.007 276 | 938.272 |
| Neutron | $1.67493 \times 10^{-27}$ | 1.008 665 | 939.566 |
| $^1_1$H atom | $1.67353 \times 10^{-27}$ | 1.007 825 | 938.783 |
| $\alpha$ particle ($^4_2$He nucleus) | $6.64466 \times 10^{-27}$ | 4.001 506 | 3727.38 |
| $^{12}_6$C atom | $1.99265 \times 10^{-26}$ | 12 | 11 177.9 |
| Unified mass unit (u) | $1.66054 \times 10^{-27}$ | 1 | 931.494 |

---

**EXAMPLE 38.4**  **Mass Defect in Helium: Powering the Sun**

Use the appropriate masses from Table 38.2 to find the binding energy of $^4_2$He.

**INTERPRET** This is a question about binding energy—the energy difference between separate constituents of helium-4 and the helium-4 nucleus. We identify the constituent particles from the symbol $^4_2$He: $Z = 2$ protons and $N = A - Z = 2$ neutrons.

**DEVELOP** Equation 38.7 determines the binding energy in terms of the various masses:

$$E_b = Z m_p c^2 + (A - Z) m_n c^2 - m_N c^2$$

**EVALUATE** Using our values for $Z$ and $N = A - Z$, along with the proton, neutron, and alpha-particle (He-4 nucleus) masses from Table 38.2 gives

$$E_b = 2(938.272 \text{ MeV}/c^2)c^2 + 2(939.566 \text{ MeV}/c^2)c^2$$
$$- (3727.38 \text{ MeV}/c^2)c^2 = 28.3 \text{ MeV}$$

**ASSESS** Notice how easy it was to work with mass in units of MeV/$c^2$; the factor $c^2$ canceled and we didn't need to use the speed of light explicitly. The formation of helium through a sequence of nuclear reactions is what powers the Sun, and our 28.3-MeV result is very close to the actual 26.7 MeV released for each He-4 nucleus formed through the solar process. ∎

## The Curve of Binding Energy

Binding energy plays a crucial role in the formation of the elements and in nuclear energy. Figure 38.9 shows the **curve of binding energy**, a plot of binding energy *per nucleon* as a function of mass number $A$. The higher this quantity, the more tightly bound is the nucleus. The broad peak in the vicinity of $A = 60$ shows that nuclei with mass numbers around this value are most tightly bound. That means it's energetically favorable for two lighter nuclei to join through the process of **nuclear fusion**, making a middle-weight nucleus. But heavier nuclei can reach a lower energy state if they split or **fission** into two middle-weight nuclei. We'll discuss fission and fusion later in the chapter.

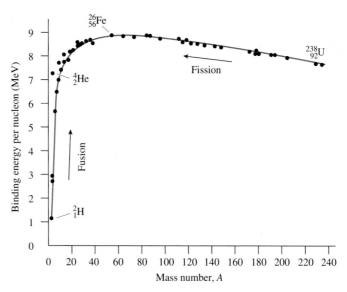

**FIGURE 38.9** The curve of binding energy, showing how fusion and fission can result in the release of nuclear energy.

**GOT IT? 38.3** Rank order these nuclei from the most to the least tightly bound: $^{4}_{2}$He, $^{238}_{92}$U, $^{57}_{26}$Fe, $^{2}_{1}$H, $^{132}_{54}$Xe.

## Nucleosynthesis and the Origin of the Elements

Since it's energetically favorable for light nuclei to fuse together, they'll do so if they have enough energy to overcome their electrical repulsion. This condition held in the high-temperature early universe, particularly from about 1 minute to 30 minutes after the start of the Big Bang. During that time, protons fused to form helium, leaving the universe with approximately its present composition of about 75% hydrogen and 25% helium, with traces of deuterium, lithium, beryllium, and boron. Hundreds of millions of years later the first stars formed, and in the interiors of more massive stars conditions were ripe for a process that fused three helium nuclei to make carbon-12. From there fusion reactions led to the formation of isotopes up to those near the $A = 60$ peak in the curve of binding energy. In fact, the nuclei of essentially all the elements with $A < 60$—including most of the materials in our own bodies—were formed in the interiors of massive stars (Fig. 38.10). Some nuclei with $A > 60$ also formed inside massive stars; others formed in the violent supernova explosions that end such stars' lives. Those explosions spewed fusion-synthesized elements into the interstellar medium where, eons later, these materials may be incorporated into new stars, planets, and living beings.

**FIGURE 38.10** Onionlike structure of a massive star before it goes supernova. Successive stages of fusion reactions produce the elements shown, which accounts for their relative abundance.

## 38.4 Nuclear Fission

Neutrons, first discovered in 1932, make excellent probes of the nucleus because they don't have to overcome electrical repulsion. In 1938 the German chemists Otto Hahn and Fritz Strassmann bombarded uranium with neutrons. They were puzzled to find among the reaction products radioactive versions of the much lighter elements barium and lanthanum. Physicist Lise Meitner and her nephew Otto Frisch interpreted these results to mean that uranium had split or, in their words, **fissioned** (Fig. 38.11). It was the eve of World War II, and the military implications were obvious and ominous: Nuclear fission represented an energy source orders of magnitude more potent than chemical reactions. The race for nuclear weapons was on. With the help of the international physics community, many of whom had fled fascism, the U.S. effort succeeded. A team led by the Italian Enrico Fermi built the first nuclear reactor under the stands of the University of Chicago stadium; it became operational in 1942. Three years later came the first nuclear weapons test, at Trinity site in New Mexico, followed quickly by the nuclear destruction of Hiroshima and Nagasaki.

Although fission can occur spontaneously, it's much more likely when a neutron strikes a heavy nucleus. Figure 38.11 shows a U-235 nucleus absorbing a neutron and becoming, briefly, U-236. The new nucleus undergoes dumbbell-shaped oscillations until electrical repulsion soon tears the dumbbell apart. That's fission.

**Fission products** resulting from this process are usually a pair of unequal-mass middle-weight nuclei. Because they have the neutron-to-proton ratio of the original uranium, they've got too many neutrons for a middle-weight nucleus, so they're intensely radioactive. Typically, two or three neutrons "boil off" almost immediately in the fission process, leaving radioactive isotopes with widely varying half-lives that may be several decades or longer.

We can summarize fission with the nuclear reaction notation we used for radioactive decay. If we skip the intermediate U-236 nucleus, neutron-induced fission in U-235 takes the form

$$\,^{1}_{0}n + \,^{235}_{92}U \longrightarrow X + Y + b\,^{1}_{0}n \qquad \text{(fission)} \qquad (38.8)$$

Here $\,^{1}_{0}n$ is the neutron, with 0 charge and 1 mass unit; $X$ and $Y$ are the fission products; and $b$ is the number of neutrons released immediately. A specific example of Equation 38.8 is $^{235}U$ fission that produces barium and krypton: $\,^{1}_{0}n + \,^{235}_{92}U \longrightarrow \,^{141}_{56}Ba + \,^{92}_{36}Kr + 3\,^{1}_{0}n$. Note how the equation balances: The total charge (subscripts) is the same on both sides, and the mass numbers (superscripts) also agree.

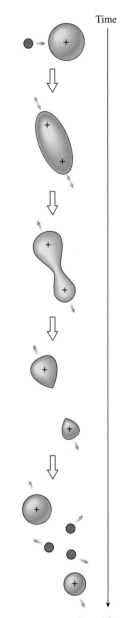

Time

**FIGURE 38.11** Neutron-induced fission of $^{235}U$, showing three neutrons (gray) released in the process.

---

**EXAMPLE 38.5** **Nuclear Fission: Identifying a Fission Product**

Neutron-induced fission of $^{235}U$ yields $^{102}_{42}Mo$, three neutrons, and another fission product. What's that product?

**INTERPRET** We're asked to identify an unknown product of a fission reaction like that of Equation 38.8.

**DEVELOP** We know that charge and mass numbers must both balance, so our plan is to write Equation 38.8 with the information we have and then solve for the unknown's charge and mass. We can then identify it.

**EVALUATE** With the given information, Equation 38.8 becomes $\,^{1}_{0}n + \,^{235}_{92}U \longrightarrow \,^{102}_{42}Mo + \,^{A}_{Z}X + 3\,^{1}_{0}n$, where $X$ is the unknown. We have total mass number 236 on the left, so with 102 mass units in molybdenum and 3 in the neutrons, $A$ must be $236 - 102 - 3 = 131$. Counting the subscripts similarly, we have $Z = 50$. The periodic table shows that the element with atomic number 50 is tin (Sn), so our unknown is $^{131}_{50}Sn$.

**ASSESS** As usual, this particular fission produces two middle-weight products with notably unequal masses. Incidentally, this reaction is significant because Sn-131 quickly undergoes a series of beta decays leading to I-131, an especially dangerous contaminant that we described in Example 38.2. ∎

## Energy from Fission

Fission of a single uranium nucleus releases about 200 MeV of energy. About 165 MeV of this appears as kinetic energy of the fission products, with another 5 MeV in neutron kinetic energy. As they de-excite, the fission products release another 7 MeV in gamma rays, and after a slight delay they emit another 20 MeV in beta particles, gamma rays, and neutrinos.

Spontaneous fission is rare because there's a small energy barrier, associated with unbalanced forces on nucleons at the edge of the nucleus, that normally keeps the nucleus from fissioning. Adding a neutron provides excess energy, which manifests itself in the oscillation and then fission shown in Fig. 38.11.

Many heavy nuclei, including both $^{238}$U and $^{235}$U, can undergo neutron-induced fission and are termed **fissionable**. Fissionable nuclei that will fission with neutrons of *any* energy—especially those with only thermal energy—are called **fissile**. Only three fissile nuclides are known: the uranium isotopes $^{233}$U and $^{235}$U and plutonium-239. The most important of these are $^{235}$U and $^{239}$Pu.

Uranium-235 presently constitutes only about 0.7% of natural uranium; nearly all the rest is $^{238}$U. For most uses, uranium must be enriched in $^{235}$U, to several percent for commercial power reactors and 80% or more for weapons. **Uranium enrichment** is difficult and expensive; since the isotopes $^{235}$U and $^{238}$U are chemically similar, enrichment techniques make use of their very slight mass difference. The technique of choice today involves spinning uranium hexafluoride gas in a sequence of high-speed centrifuges. Enrichment technology is highly sensitive because a nation possessing it can produce weapons-grade uranium.

Plutonium-239, with a 24,110-year half-life, does not occur in nature. It's produced by neutron bombardment of $^{238}$U. The reaction forms $^{239}$U, which undergoes two beta decays to produce first $^{239}$Np and then the fissile $^{239}$Pu:

$$\,_{0}^{1}n + \,_{92}^{238}U \longrightarrow \,_{92}^{239}U$$
$$\,_{92}^{239}U \longrightarrow \,_{93}^{239}Np + e^{-} + \bar{\nu}$$
$$\,_{93}^{239}Np \longrightarrow \,_{94}^{239}Pu + e^{-} + \bar{\nu}$$

Although $^{239}$Pu is produced copiously in nuclear reactors (see Problem 72), **reprocessing** spent reactor fuel to extract plutonium is difficult and dangerous. Contamination with other plutonium isotopes further complicates the process. Like uranium enrichment, plutonium reprocessing is a sensitive technology, and the decision of several European countries and Japan to engage in commercial reprocessing for reactor fuel has made Pu-239 a commercial commodity.

**EXAMPLE 38.6** **Nuclear Fission: Comparing with Coal**

Assuming 200 MeV per fission, estimate the amount of pure $^{235}$U that would provide the same energy as 1 metric ton (1000 kg) of coal.

**INTERPRET** We're asked to compare the energies of nuclear fission and the chemical burning of coal.

**DEVELOP** For coal, we can look up the energy released per unit mass in Appendix C's "Energy Content of Fuels" table. We can then find the number of fission events, at 200 MeV per fission, needed to release the same energy as 1000 kg of coal. Finally, we'll use the mass of a U-235 nucleus to find the corresponding mass of uranium.

**EVALUATE** Appendix C gives an energy content of 29 MJ/Kg for coal, so burning 1000 kg of coal releases 29 GJ of energy. With $1.6 \times 10^{-19}$ J/eV, each 200-MeV fission releases about $3.2 \times 10^{-11}$ J.

Then we need a total of

$$29 \text{ GJ}/3.2 \times 10^{-11} \text{ J/fission} = 9.1 \times 10^{20} \text{ fission events}$$

Each of the $9.1 \times 10^{20}$ U-235 nuclei has a mass of approximately 235 u, so the total mass required is

$$(9.1 \times 10^{20} \text{ nuclei})(235 \text{ u})(1.66 \times 10^{-27} \text{ kg/u}) = 0.35 \text{ g}.$$

**ASSESS** That's about one one-hundredth of an ounce of $^{235}$U, packing as much energy as a ton of coal! Our result shows that U-235 contains more than a million times as much energy as the same amount of coal. That's the reason nuclear power plants are fueled only about once a year, with a truckload or so of nuclear fuel, while coal-burning plants burn many 110-car trainloads of coal each week. It's also the reason for the immense destructive power of nuclear weapons. ∎

## The Chain Reaction

Neutrons induce fission, and fission itself releases more neutrons. This makes possible a **chain reaction**, in which each fission event results in more fission. To sustain a chain reaction, each nucleus that fissions must, on average, cause at least one more fission event;

otherwise, the reaction will fizzle to a halt. In a piece of material that's too small, most neutrons will escape without causing additional fission. For that reason there's a **critical mass** of nuclear fuel necessary to sustain a chain reaction. More than that amount is **supercritical** and results in an exponentially growing chain reaction (Fig. 38.12).

The size of the critical mass depends on the purity of the fissile material, its configuration, and surrounding materials. For plutonium it can be less than 5 kg, and as low as 15 kg for uranium. Those numbers are frighteningly small, and they show why we worry about city-destroying "suitcase bombs."

The **multiplication factor**, $k$, is the average number of neutrons from a fission event that cause additional fission. A critical mass has $k = 1$, and a supercritical mass has $k > 1$. The average time between successive fissions is the **generation time**. In a supercritical mass this can be as short as 10 ns, leading to the entire mass fissioning in about 1 $\mu$s.

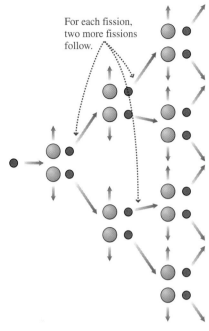

For each fission, two more fissions follow.

**FIGURE 38.12** A supercritical chain reaction with multiplication factor $k = 2$.

## Fission Weapons

A rapidly fissioning supercritical mass is a nuclear explosive. The major technological difficulty in producing a fission weapon is to assemble a supercritical mass so rapidly that the chain reaction consumes enough fissile material before it blows apart. With highly enriched uranium that's not an insurmountable challenge. The crude bomb that destroyed Hiroshima contained about 50 kg of enriched uranium, of which only about 1 kg actually fissioned. So confident were its developers that they never tested this design. Plutonium weapons present a greater challenge; neutrons from spontaneous fission make it more likely that the weapon will "pre-ignite" and blow itself apart.

Construction of a simple fission weapon is distressingly straightforward, but acquisition of weapons-grade fissile material is not. Again, that's why uranium enrichment and plutonium reprocessing technologies are so sensitive. We live in a dangerous and unstable world, and it's going to get more dangerous if fissile materials become widely available.

## Nuclear Power

A **nuclear reactor** uses a controlled fission chain reaction with $k = 1$ to release energy at a steady rate. Since the average number of neutrons emitted in U-235 fission is about 2.5, reactors require that most neutrons don't cause fission. Commercial power reactors limit $k$ in part by keeping the concentration of fissile U-235 low—typically a few percent—so that many neutrons are absorbed by U-238 instead of causing fission. **Control rods** made of neutron-absorbing material provide additional control over $k$; these can be moved into and out of the nuclear fuel to provide precise control of the power level. A small fraction— about 0.65%—of fission-produced neutrons are emitted with delays from about 0.2 s to 1 min, and these **delayed neutrons** allow for relatively slow mechanical control of nuclear reactors. The next example explores this point.

---

**EXAMPLE 38.7**  **Nuclear Fission:**
**Delayed Neutrons and Reactor Control**

A change in operating conditions makes a nuclear reactor slightly supercritical, with $k = 1.001$. Determine the time it would take the reactor power to double (a) if delayed neutrons establish a generation time $\tau = 0.1$ s, and (b) if prompt neutrons—those released immediately—sustain the reaction to give $\tau = 10^{-4}$ s.

**INTERPRET** We're asked to calculate the time until the reaction rate doubles, given the multiplication factor $k$ and two different values for the generation time.

**DEVELOP** A multiplication factor $k = 1.001$ means the rate of fissioning increases by a factor of 1.001 with each generation time; after two generations it will have increased by $k^2$, and so forth. So our plan is to

find the number $n$ of generations that gives $k^n = 2$. Then we can multiply by the two different $\tau$ values to find the actual times: $t = n\tau$.

**EVALUATE** We set $k^n = 2$ and take the logarithm of both sides. With $\ln(k^n) = n \ln k$, we have $n \ln k = \ln 2$, or $n = \ln 2 / \ln k = 693$ with $k = 1.001$. With $\tau = 0.1$ s that gives $t = n\tau = 69.3s$ or just over 1 min, but with $\tau = 10^{-4}$ s it's only 0.07 s.

**ASSESS** With delayed neutrons the doubling time is long enough for the reactor operators and their mechanical controls to take corrective action; with prompt neutrons there isn't time to prevent a serious nuclear accident. Delayed neutrons are crucial to reactor control! Loss of control to a reaction governed by prompt neutrons alone was a key factor in the 1986 Chernobyl nuclear accident. ∎

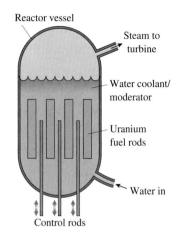

**FIGURE 38.13** A boiling-water reactor, one of two types commonly used in the United States.

High-energy fission neutrons aren't very effective at causing additional fission events, so in most reactor designs they must be slowed to roughly the mean thermal speed. A substance called the **moderator** effects this slowing through elastic collisions between neutrons and the moderator nuclei. In Chapter 9 we found that the maximum energy transfer occurs when colliding particles have equal mass; therefore, the best moderators have low-mass nuclei. The choice of moderator is among the most significant distinguishing features of different reactor designs. Another important choice is the **coolant**, which carries off fission-generated heat.

Power reactors in the United States are **light-water reactors** (LWRs), using ordinary water with the protons of its hydrogen serving as the moderator nuclei. The same water acts as coolant and circulates through a pressure vessel containing uranium fuel rods and control rods. About one-third of the United States' roughly 100 power reactors are **boiling-water reactors** (BWRs), in which water boils in the reactor vessel to make steam that drives a turbine-generator (Fig. 38.13). The remainder are **pressurized-water reactors** (PWRs), in which liquid water under pressure transfers its energy to a secondary loop where water boils to make steam (Fig. 38.14). An advantage of this more complex system is that the steam loop doesn't become radioactive. Both types of light-water reactors have an intrinsic safety feature, in that a loss of coolant also means loss of moderator, and that brings the chain reaction to a halt. But light water has the disadvantage that $_1^1H$ readily absorbs neutrons, and therefore light-water reactor fuel must be enriched in $^{235}U$ in order to sustain the chain reaction. Refueling a LWR is also a big operation: The reactor must be shut down and the lid removed from the pressure vessel—a process that can take a month or longer.

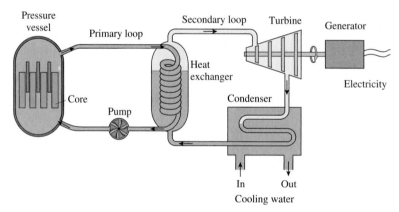

**FIGURE 38.14** A complete power plant using a pressurized-water reactor, the most common type of power reactor in the United States.

The Canadian CANDU design uses heavy water ($_1^2H_2O$, or deuterium oxide) as moderator and coolant. Low neutron absorption means CANDU reactors can operate on natural uranium, eliminating the need for sensitive enrichment technology. And the CANDU design allows continuous refueling, although that increases another proliferation risk by making it easier to extract plutonium.

An older Soviet-era design is the graphite-moderated, water-cooled RBMK reactor. Often built to provide both electric power and plutonium for weapons, this design suffered from the safety defect that loss of coolant not only didn't shut down the chain reaction but could actually accelerate it due to loss of neutron-absorbing hydrogen in the $H_2O$ coolant. The disastrous 1986 Chernobyl accident involved an RBMK reactor. During a test of the emergency cooling system, operators inadvertently put the reactor in an unstable state where an increase in power boiled away more cooling water, resulting in a further increase. The power level soared by a factor of 4000 in 5 seconds, causing a steam explosion that blew the top off the reactor and ignited the flammable graphite moderator. Heavy smoke carried radioactive materials into the atmosphere, resulting in widespread contamination, which we explored in Example 38.2. Today, thousands of square miles surrounding Chernobyl remain officially uninhabitable.

Other reactor designs include gas-cooled reactors that can operate at higher temperatures and therefore greater thermodynamic efficiencies, and **breeder reactors** designed specifically to "breed" plutonium from U-238 and therefore turn most of the nonfissile U-238 into fissile Pu-239. Breeders have no moderator, use liquid sodium coolant, and are critical with fast neutrons alone. Breeders are therefore less stable than so-called thermal reactors using slow neutrons, and widespread adoption of breeder technology entails international trafficking in fissile plutonium.

## Nuclear Waste

We've seen that fission products are highly radioactive because they contain too many neutrons for stable middle-weight nuclei. Because of their high activity, fission products have relatively short half-lives, typically measured in decades. That makes fission-product waste dangerous for centuries to a few millennia. However, neutron absorption in fission reactors also produces plutonium and a host of other **transuranic** isotopes—those heavier than uranium—with much longer lifetimes. It's these substances that mean we'll have to safeguard nuclear waste for tens of thousands of years.

As fission proceeds, the concentration of fission products increases. Before a reactor's $^{235}$U is exhausted, fission products begin absorbing enough neutrons to interfere with the chain reaction. In U.S. LWRs, that requires about one-third of the fuel rods to be replaced annually. Older fuel is also rich in fissile plutonium, and at the end of a fuel rod's 3-year residency in the reactor, more than half the energy generation comes from fissioning plutonium rather than uranium. Figure 38.15 shows the evolution of nuclear fuel in a U.S. LWR.

The disposal of nuclear waste is a thorny issue, mixing political and scientific considerations. To date, there is no operating repository for commercial nuclear waste in the United States, and waste continues to accumulate at nuclear reactor sites. Lest you picture mountains of nuclear garbage, however, remember that factor of $10^7$ difference between nuclear and chemical energy sources. That translates into far less fuel needed for nuclear power plants, and far less waste produced. A 1-GW power reactor produces some 20 tons of nuclear waste annually, while a comparable coal plant produces 1000 tons of carbon dioxide and 30 tons of solid waste *every hour*.

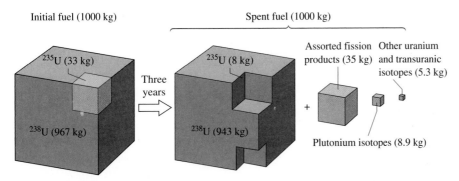

**FIGURE 38.15** Evolution of 1000 kg of 3.3% enriched uranium over its 3-year stay in a light-water reactor.

---

**GOT IT? 38.4** Transportation and mining accidents involving coal are much more frequent than those involving uranium fuel. What's the fundamental reason for this?

---

## Prospects for Nuclear Power

Today, nuclear power supplies some 17% of the world's total electrical energy; in nuclear-intensive France the figure is nearly 80% and in the United States it's 20%. Nuclear power plant construction continues around the world, especially in Asia, largely using advanced versions of light-water reactor designs. However, public opposition and economic considerations have long since halted nuclear power development in the United States, and no new reactors were ordered between 1978 and at least 2005.

Concern over climate change from fossil-fuel combustion has spurred a renewed interest in nuclear power, even among some environmentalists. New reactor designs promise greater reliability, economic viability, and, most important, safety based on intrinsic reactor design rather than complicated safety systems. Most physicists agree that the public's concern over nuclear power is exaggerated. We regularly accept much greater risks from other technologies—for example, some 24,000 premature deaths each year in the United States due to pollution from coal-fired power plants. Comparable estimates for nuclear plants range from about 10 to fewer than 1000 from even the most vigorously antinuclear groups.

Nevertheless, ongoing uncertainties about the risk of catastrophic nuclear accidents, long-term waste storage, terrorism, and weapons proliferation continue to haunt the nuclear power industry. Proliferation, especially, is a very real concern. Although nuclear power and weapons development are different enterprises, they share infrastructure and an educated technological elite that can be put to either purpose. If nuclear power is to advance, it will need to do so under strict international guarantees against diversion of materials and expertise to weapons production.

## 38.5 Nuclear Fusion

The curve of binding energy (Fig. 38.9) shows that fusion of light nuclei provides another approach to nuclear energy production. The curve is steepest at its left end, indicating that the most energy per nucleon comes from the fusion of hydrogen. Indeed, the fusion reactions powering the Sun and many other stars begin with the fusion of hydrogen to form deuterium. Also emitted in the process are a positron, a neutrino, and 0.42 MeV of energy:

$$\mathrm{{}_1^1H + {}_1^1H \longrightarrow {}_1^2H + e^+ + \nu} \qquad (0.42\ \mathrm{MeV}) \qquad (38.9a)$$

Deuterium then fuses with hydrogen to form helium-3 and a gamma ray:

$$\mathrm{{}_1^2H + {}_1^1H \longrightarrow {}_2^3He + \gamma} \qquad (5.49\ \mathrm{MeV}) \qquad (38.9b)$$

Two helium-3 nuclei then react to form helium-4 and a pair of protons ($\mathrm{{}_1^1H}$), releasing 12.86 MeV:

$$\mathrm{{}_2^3He + {}_2^3He \longrightarrow {}_2^4He + 2\,{}_1^1H} \qquad (12.86\ \mathrm{MeV}) \qquad (38.9c)$$

In addition, the positron from reaction 38.9a annihilates with an electron, forming two gamma rays with a total energy of $2mc^2$ or 1.022 MeV. Together, these reactions constitute the **proton–proton cycle**. In the full cycle, reactions 38.9a and b occur twice for each occurrence of reaction 33.9c. The net effect is to convert four protons and two electrons to a single He-4, releasing 26.7 MeV (Fig. 38.16). In massive stars, $\mathrm{{}_2^4He}$ then becomes a building block for still heavier elements, as we discussed earlier.

$$\mathrm{4\,{}_1^1H + 2e^-} \longrightarrow \mathrm{{}_2^4He} + \qquad 26.7\ \mathrm{MeV}$$

**FIGURE 38.16** Net result of the proton–proton cycle of Equations 38.9.

$+$ 26.7 MeV

Reaction 38.9a does not occur readily, and terrestrial fusion research has therefore focused on reactions involving the heavier hydrogen isotopes. Of immediate interest are deuterium–tritium (D-T) and deuterium–deuterium (D-D) reactions, listed below with the energy released in each:

$$\mathrm{{}_1^2H + {}_1^3H \longrightarrow {}_2^4He + {}_0^1n} \qquad (17.6\ \mathrm{MeV};\ \text{D-T reaction}) \qquad (38.10a)$$

$$\mathrm{{}_1^2H + {}_1^2H \longrightarrow {}_2^3He + {}_0^1n} \qquad (3.27\ \mathrm{MeV};\ \text{D-D reaction}) \qquad (38.10b)$$

$$\mathrm{{}_1^2H + {}_1^2H \longrightarrow {}_1^3H + {}_1^1H} \qquad (4.03\ \mathrm{MeV};\ \text{D-D reaction}) \qquad (38.10c)$$

The two outcomes of the D-D reaction have nearly equal probability.

The electrical repulsion between nuclei makes it difficult to get them close enough to fuse. Although quantum tunneling helps, it still takes very high nuclear speeds—corresponding to high temperatures—to initiate fusion. At fusion temperatures, atoms are stripped of their electrons and the fusing material constitutes a plasma. It's necessary somehow to contain this hot plasma. Stars achieve both ends with their immense gravity, which compresses stellar material to fusion temperatures and simultaneously provides confinement. In the Sun's core, for example, the temperature is some 15 MK, and fusing nuclei approach with energies on the order of 1 keV—although even under these conditions the process isn't particularly efficient.

Terrestrial fusion requires still higher temperature, as high-energy particles undergo rapid accelerations that result in the plasma losing energy by radiation. The temperature at which fusion-generated power exceeds radiation loss is the **critical ignition temperature**. For the D-D reactions of Equations 38.10b and c, Fig. 38.17 shows that the ignition temperature is about 600 MK; for D-T it's a lower 50 MK. Net fusion-energy production requires not only high temperature but also confinement for long enough that the fusion energy produced exceeds the energy required to heat the plasma. The heat required depends on the number of nuclei or, on a volume basis, on the number density $n$. However, the rate of fusion-energy production depends on the *square* of the density. That's because doubling $n$ increases both the number of nuclei available to strike other nuclei and the number of nuclei available to be struck; the result is a quadrupling of the fusion rate. The total energy released therefore scales as $n^2 \tau$, where $\tau$ is the **confinement time**. Meanwhile the radiation energy loss depends linearly on $n$, and as a result there's a minimum value of the product $n\tau$ necessary in an energy-producing fusion device. This condition is the **Lawson criterion**, given approximately by

$$n\tau > 10^{22} \text{ s/m}^3 \qquad \text{(Lawson criterion, D-D fusion)}$$
$$n\tau > 10^{20} \text{ s/m}^3 \qquad \text{(Lawson criterion, D-T fusion)}$$

(38.11)

The factor-of-100 difference here shows that D-T fusion will be much easier to achieve.

Fusion technologies use two distinct approaches to the Lawson criterion. **Inertial confinement** strives for very high densities with short confinement times—so short that the particles' inertia alone is sufficient to prevent them from leaving the fusion site during the brief time needed. **Magnetic confinement** holds lower-density plasma in a "magnetic bottle" whose magnetic-field configuration minimizes the chance of escape during a relatively long confinement time. Neither approach has yet produced a sustained energy yield from fusion.

## Inertial Confinement Fusion

Although peaceful fusion devices still elude us, inertial confinement has been used successfully since the 1950s in thermonuclear weapons—often called "hydrogen bombs" to distinguish them from fission explosives (incorrectly called "atomic bombs"). Thermonuclear weapons aren't pure fusion devices, though. They use a fission explosion to achieve the high temperatures needed to ignite fusion, and a clever arrangement for focusing the fission energy on a mixture of lithium deuteride and plutonium-239. The mixture is compressed to fusion temperatures, and fission neutrons convert lithium to helium and tritium. D-T fusion then occurs, providing the device with approximately half its explosive yield. The remainder comes from fission in an outer layer of natural uranium, whose nonfissile U-238 nevertheless fissions under the bombardment of high-energy neutrons. There's essentially no limit to the yield of a thermonuclear weapon, and devices as large as 58 megatons (Mt) TNT equivalent have been tested. That's 5000 times the energy of the fission bomb that destroyed Hiroshima. Today's missile-based thermonuclear weapons range from 40 kt to about 1 Mt, and there are thousands of them in the world's arsenals.

Inertial confinement schemes for controlled fusion focus high-power laser beams on tiny deuterium–tritium targets, producing sequences of miniature thermonuclear explosions. Most advanced is the National Ignition Facility (NIF) at California's Lawrence Livermore National Laboratory. NIF uses 192 laser beams to focus a 1.8-MJ pulse of several nanoseconds duration on the millimeter-diameter D-T target (Fig. 38.18). During

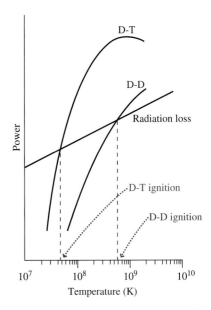

**FIGURE 38.17** Power loss by radiation and power produced by D-D and D-T fusion reactions, as functions of temperature on a log-log plot.

**FIGURE 38.18** Target chamber of the National Ignition Facility is 11 m in diameter and weighs 130 tonnes. Holes are ports for the 192 laser beams that converge on the millimeter-size target.

**FIGURE 38.19** Plasma loss in magnetic confinement. (a) End losses occur when field lines intersect device walls. (b) Curvature of field lines results in cross-field drifts. (c) Instabilities distort the plasma and magnetic field. In (a) and (b) the spiral represents the path of a charged particle.

**FIGURE 38.20** Cutaway diagram of the ITER fusion reactor. D-shaped structures are cross sections of the toroidal plasma chamber.

this brief interval the laser power—energy per time—is about 500 TW, or 1000 times the output of all U.S. electric power plants. The NIF is very much a dual-use facility, aimed as much at simulating nuclear weapons tests as at exploring fusion energy. NIF's first tests of fusion ignition probably won't occur until 2010 or later.

## Magnetic Confinement Fusion

In Chapter 26 we saw how charged particles in highly conducting plasma are essentially "frozen" to the magnetic field lines. Trapping of charged particles on magnetic field lines is the essence of magnetic confinement fusion schemes. The first job of magnetic confinement is to create a magnetic configuration that keeps plasma away from the relatively cool walls of the device. Plasma can escape to the walls in three general ways, as shown in Fig. 38.19.

The most promising magnetic fusion device is the **tokamak**, a Russian invention now used throughout the world. The tokamak has a toroidal configuration whose magnetic field lines never penetrate the walls, eliminating the end loss shown in Fig. 38.19a. Making the machine larger reduces field-line curvature, and with it the cross-field drift of Fig. 38.19b. Additional field components enhance confinement and suppress the instabilities of Fig. 38.19c. After smaller tokamaks paved the way, an international consortium is constructing the ITER device in France for operation in about 2016 (Fig. 38.20). ITER will be the first magnetic fusion system to produce net energy exceeding the energy used for plasma heating. ITER will operate at a plasma temperature higher than 100 MK and should generate 400 MW of fusion power from its 840 cubic meters of D-T plasma. ITER will use deuterium and lithium as fuel, with tritium ($^3_1$H) "bred" right in the reactor by neutron bombardment: $^6_3\text{Li} + ^1_0 n \longrightarrow {}^4_2\text{He} + {}^3_1\text{H}$.

## Prospects for Fusion

When work on controlled fusion began in the 1950s, scientists confidently predicted that limitless fusion energy would be available in a few decades. A half-century later, controlled fusion still appears decades away. But it's a goal worth pursuing: Problem 65 shows that with D-D fusion, a gallon of seawater is equivalent to some 300 gallons of gasoline, and Problem 66 shows that fusion energy resources could last far longer than the Sun will continue to shine!

Once controlled fusion proves scientifically feasible, there will be formidable engineering challenges in the design of a practical fusion power plant. Intense neutron fluxes from D-T fusion degrade materials that form the reaction chamber. Neutron-capture reactions produce radioactive isotopes within the walls, although the associated radioactivity and that of the tritium fuel are much less than the radioactivity of fission waste.

The first practical fusion plants will likely use D-T fusion because its ignition temperature and Lawson criterion are lower than for D-D fusion; the resulting energy will probably run a conventional steam cycle. But the D-D reaction promises cleaner and more efficient power production. Because D-D reaction 38.10c produces charged protons ($^1_1$H) rather than neutral neutrons, there's the possibility of using a magnetohydrodynamic (MHD) generator, in which electromagnetic induction converts charged-particle kinetic energy directly to electricity. MHD generators would bypass the conventional steam cycle and greatly increase the thermodynamic efficiency of a fusion power plant.

There's one caveat to this rosy fusion future: Although fusion itself is relatively clean and produces no greenhouse gases, the availability of unlimited cheap energy would likely spur industrial growth at a level our planet might not tolerate. And ultimately, if fusion energy use grew exponentially, the waste heat from fusion power could itself have climate-changing consequences.

The big idea here is that the tiny but massive atomic nucleus is a repository of vast energy—on the order of $10^7$ times the energy released in chemical reactions. The protons and neutrons that make up the nucleus can take many configurations, with the **atomic number**, $Z$, determining the element and the **mass number**, $A$, determining the particular **isotope**. We write $^A_Z X$ to describe the isotope with mass number $A$ of the element whose atomic number is $Z$ and whose symbol is $X$.

**Helium isotopes**

helium-3   helium-4
$^3_2$He   $^4_2$He

**Stable isotopes** require a delicate balance between protons and neutrons, with near equal numbers for lighter stable nuclei and more neutrons for heavier nuclei. **Unstable isotopes** are **radioactive** and decay by shedding particles. The **curve of binding energy** shows that energy can be released by either **fusion** of lighter nuclei or **fission** of heavier nuclei.

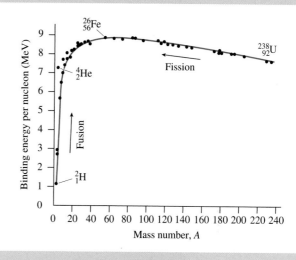

Radioactive isotopes decay with a characteristic **half-life**, $t_{1/2}$: $N = N_0 2^{-t/t_{1/2}}$.

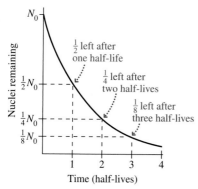

**Alpha decay** emits a helium-4 nucleus:

$$^A_Z X \longrightarrow {}^{A-4}_{Z-2} Y + {}^4_2 \text{He}$$

**Beta decay** emits an electron or a positron and an antineutrino or a neutrino:

$$^A_Z X \longrightarrow {}^A_{Z+1} Y + e^- + \bar{\nu}$$

**Gamma decay** emits a high-energy photon (gamma ray) as an excited nucleus drops to a lower energy state:

$$X^* \longrightarrow X + \gamma$$

Radioactivity is measured in **becquerels**, with 1 Bq equal to one decay per second. **Sieverts** (Sv) measure the biological effects of radiation. Residents of the United States receive an average yearly radiation dose of about 3.6 mSv from both natural and artificial sources.

For fission, the most important isotopes are the **fissile** $^{235}_{92}$U and $^{239}_{94}$Pu, which can fission when struck by low-energy neutrons:

$$^1_0 n + {}^{235}_{92} \text{U} \longrightarrow X + Y + b \, {}^1_0 n.$$

The extra neutrons produced in fission can sustain a **chain reaction** provided there's a **critical mass** of fissile material. Exponentially growing chain reactions power fission weapons, while controlled fission occurs in **nuclear reactors** used for power generation.

**Fusion** powers the Sun and stars but has proved elusive on Earth except in thermonuclear weapons. **Inertial confinement** or **magnetic confinement** fusion may one day provide us with nearly limitless energy.

# For Thought and Discussion

1. Why do nuclei contain neutrons?
2. Why are there no stable nuclei for sufficiently high atomic numbers?
3. Why might future archaeologists have problems dating samples from the second half of the 20th century?
4. Beta decay by positron emission is soon followed by a pair of 511-keV gamma rays. Why?
5. Why would it have been easier to make bombs fueled with uranium-235 a few billion years ago?
6. Why are iodine-131 and strontium-90 particularly dangerous radioisotopes?
7. Which model, liquid-drop or nuclear shell, does a better job explaining (a) nuclear fission and (b) nuclear gamma-ray spectra?
8. On an energy-release-per-unit-mass basis, by approximately what factor do nuclear reactions exceed chemical reactions?
9. Explain and distinguish the roles of the control rods and moderator in a nuclear reactor.
10. Why is a water-moderated reactor intrinsically safer in a loss-of-coolant accident than a graphite-moderated reactor?
11. Is $^{238}U$ fissionable? Is it fissile? Explain the distinction.
12. Why are fission fragments radioactive? Invoke Fig. 38.3 in giving your answer.

# Exercises and Problems

## Exercises

### Section 38.1 Elements, Isotopes, and Nuclear Structure

13. Three of the isotopes of radon $(Z = 86)$ have 125, 134, and 136 neutrons. Write the symbol for each.
14. Write the symbol for the germanium isotope with 44 neutrons.
15. How do (a) the number of nucleons and (b) the nuclear charge compare in the two nuclei $^{35}_{17}Cl$ and $^{35}_{19}K$?
16. Compare the radius of the proton (the $A = 1$ nucleus) with the Bohr radius of the hydrogen atom.
17. A uranium-235 nucleus splits into two roughly equal-size pieces. Find their common radius.

### Section 38.2 Radioactivity

18. How many half-lives will it take for the activity of a radioactive sample to diminish to 10% of its original level?
19. Copper-64 can decay by any of the three beta-decay processes. Write the equation for each decay.
20. Referring to Fig. 38.7, write equations describing the decays of (a) radon-222 and (b) lead-214.
21. Carbon-10 is a positron emitter with a 19.3-s half-life. (a) Write an equation to describe its decay. (b) A $^{10}C$ sample has initial activity 48 MBq. What will be its activity after 1 min, in Bq?
22. A milk sample shows an iodine-131 activity level of 450 pCi/L. What is its activity in Bq/L?
23. Nuclear bomb tests of the 1950s deposited a layer of strontium-90 over Earth's surface. How long will it take from the time of the bomb tests for (a) 99% and (b) 99.9% of this radioactive contaminant to decay?

### Section 38.3 Binding Energy and Nucleosynthesis

24. Find the total binding energy of oxygen-16, given that its nuclear mass is 15.9905 u.
25. Determine the atomic mass of nickel-60, given that its binding energy is very nearly 8.8 MeV/nucleon.
26. Find the nuclear mass of plutonium-239, given that its atomic mass is 239.052157 u.
27. The mass of a lithium-7 nucleus is 7.01435 u. Find the binding energy per nucleon.

### Section 38.4 Nuclear Fission

28. A $^{235}U$ nucleus undergoes neutron-induced fission, resulting in a $^{141}Cs$ nucleus, three neutrons, and another nucleus. What is that nucleus?
29. Neutron-induced fission of $^{235}U$ results in the fission products iodine-139 and yttrium-95. How many neutrons are released?

30. Write a complete equation for the neutron-induced fission of plutonium-239 that results in barium-143, two neutrons, and another nucleus.
31. Assuming 200 MeV per fission, determine the number of fission events occurring each second in a reactor whose thermal power output is 3.2 GW.

### Section 38.5 Nuclear Fusion

32. Verify from Equations 38.9 that the proton–proton cycle yields a net energy of 26.7 MeV.
33. In a magnetic confinement fusion device with confinement time 0.5 s, what density would be required to meet the Lawson criterion for D-T fusion?
34. Inertial confinement schemes generally involve confinement times on the order of 0.1 ns. What is the corresponding density needed to meet the Lawson criterion for D-T fusion?
35. What confinement time is required for the D-T Lawson criterion in the ITER fusion reactor, given its plasma density of $10^{19}$ particles per cubic meter?

## Problems

36. To what size would Earth have to collapse to be at nuclear density?
37. Find the energy needed to flip the spin state of a proton in Earth's magnetic field, whose magnitude is about 30 $\mu$T.
38. An NMR spectrometer is described as a "300-MHz instrument," meaning 300 MHz is the frequency supplied to its transmitter coil to flip the spin states of bare protons. What is the strength of its unperturbed magnetic field?
39. Iron-56 is among the most tightly bound nuclei. Its mass is 55.9206 u. Find the binding energy per nucleon, and check your answer against Fig. 38.9.
40. Find the atomic mass of iridium-193, for which the binding energy is 7.94 MeV/nucleon.
41. In the 4.5-billion-year lifetime of the Earth, how many half-lives have passed of (a) carbon-14, (b) uranium-238, and (c) potassium-40? For each isotope, give the number of atoms remaining today from a sample of $10^6$ atoms at Earth's formation.
42. You measure the activity of a radioactive sample at 2.4 MBq. Thirty minutes later, the activity level is 1.9 MBq. What is the half-life for the radioactive material?
43. If radon infiltration into a house were stopped, what fraction of the initial radon would be left after (a) 1 day, (b) 1 week, (c) 1 month?

44. Nitrogen-13 has a 10-min half-life. A sample of $^{13}$N contains initially $10^5$ atoms. Plot the number of atoms as a function of time from $t = 0$ to $t = 1$ hour. Make your horizontal axis (time) linear, but your vertical axis logarithmic. Why is the curve a straight line? What's the significance of its slope?

45. Thorium-232 is an $\alpha$ emitter with a 14-billion-year half-life. Radium-228 is a $\beta^-$ emitter with a 5.75-year half-life. Actinium-228 is a $\beta^-$ emitter with a 6.13-hour half-life. (a) What is the third daughter in the thorium-232 decay series? (b) Make a chart similar to Fig. 38.7 showing the first three decays in the thorium series.

46. How much cobalt-60 $\left(t_{1/2} = 5.24 \text{ years}\right)$ must be used to make a laboratory source whose activity will exceed 1 GBq for a period of 2 years?

47. Archaeologists unearth a bone and find its carbon-14 content is 34% of that in a living bone. How old is the archaeological find?

48. Show that the decay constant and half-life are related by $t_{1/2} = \ln 2/\lambda \approx 0.693/\lambda$.

49. The table lists reported levels of iodine-131 contamination in milk in three countries affected by the 1986 Chernobyl accident, along with each country's safety guideline. Given that I-131's half-life is 8.04 days, how long did each country have to wait for I-131 levels to decline to the level it considered safe?

| | Level (Bq/L) | |
| Country | Reported | Safety Guideline |
|---|---|---|
| Poland | 2000 | 1000 |
| Austria | 1500 | 370 |
| Germany | 1184 | 500 |

50. How many atoms are in a radioactive sample with activity 12 Bq and half-life 15 days?

51. Analysis of a Moon rock shows that 82% of its initial K-40 has decayed to Ar-40, a process with a half-life of $1.2 \times 10^9$ years. How old is the rock?

52. Neutron activation of the stable nitrogen isotope $^{15}_{7}$N in a bomb-detection system turns it into unstable $^{16}_{7}$N, which decays by beta emission with a 7.13-s half-life. How long after activation will the N-16 activity have dropped by a factor of 1 million?

53. Today, uranium-235 comprises only 0.72% of natural uranium; essentially all the rest is U-238. Use the half-lives given in Table 38.1 to determine the percentage of uranium-235 in natural uranium when Earth formed, about 4.5 billion years ago.

54. Geologists are looking for underground sites that could store nuclear wastes securely for 250,000 years. What fraction of plutonium-239 initially in such waste would remain after that time? See Table 38.1.

55. A sample of oxygen-15 $\left(t_{1/2} = 2.0 \text{ min}\right)$ is produced in a hospital's cyclotron. What should the initial activity concentration be if it takes 3.5 min to get the O-15 to a patient undergoing a PET scan for which an activity of 0.5 mCi/L is necessary?

56. How much $^{235}$U would be needed to fuel the reactor of Exercise 31 for 1 year? *Note:* Your answer is an overestimate because fission of $^{239}$Pu also contributes to the power output.

57. How much uranium-235 would be consumed in a fission bomb with a 20-kt explosive yield?

58. A neutron collides elastically and head-on with a stationary deuteron in a reactor moderated by heavy water. How much of its kinetic energy is transferred to the deuteron? *Hint:* Consult Chapter 9.

59. A buildup of fission products "poisons" a reactor, dropping the multiplication factor to 0.992. How long will it take the reactor power to decrease by half, assuming a generation time of 0.10 s?

60. The total thermal power generated in a nuclear power reactor is 1.5 GW. How much $^{235}$U does it consume in a year?

61. New Hampshire's Seabrook nuclear power plant produces electrical energy at the rate of 1.2 GW, and consumes 1311 kg of $^{235}$U each year. Find (a) its thermal power output and (b) its efficiency, assuming the plant operates continuously.

62. In the dangerous situation of prompt criticality in a fission reactor, the generation time drops to 100 $\mu$s as prompt neutrons alone sustain the chain reaction. If a reactor goes prompt critical with $k = 1.001$, how long does it take for a 100-fold increase in reactor power?

63. How much heavy water (deuterium oxide, $^2$H$_2$O or D$_2$O) would be needed to power a 1000-MW D-D fusion power plant for 1 year?

64. The proton–proton cycle consumes four protons while producing about 27 MeV of energy. (a) At what rate must the Sun consume protons to produce its power output of about $4 \times 10^{26}$ W? (b) The present phase of the Sun's life will end when it has consumed about 10% of its original protons. Estimate how long this phase will last, assuming the Sun's $2 \times 10^{30}$-kg mass was initially 71% hydrogen.

65. About 0.015% of hydrogen nuclei are actually deuterium. (a) How much energy would be released if all the deuterium in a gallon of water underwent fusion? (Use an average of 7.2 MeV per deuteron.) (b) In terms of energy content, to how much gasoline does a gallon of water correspond? Gasoline's energy content is 36 kWh/gal, as listed in Appendix C.

66. The net energy production in the fusion reactions of Equations 38.10b and c, including fusion of the product $^3_2$He and $^3_1$H, is 7.2 MeV per deuteron. About 0.015% of the hydrogen nuclei in seawater are actually deuterium. Estimate how long deuterium in the world's oceans (average depth 3 km) could supply humanity's energy needs at the current consumption rate of about $10^{13}$ W.

67. The atomic masses of uranium-238 and thorium-234 are 238.050784 u and 234.043593 u, respectively. Find the energy released in the alpha decay of $^{238}$U.

68. In the preparation of a transuranic nucleus, $^{209}_{83}$Bi and $^{54}_{24}$Cr combine to form a heavy nucleus plus a neutron. Identify the heavy nucleus.

69. It is possible but difficult to realize the alchemists' dream of synthesizing gold. One reaction involves bombarding mercury-198 with neutrons to produce, for each neutron captured, a gold-197 nucleus and another particle. Write the equation for this reaction.

70. Nickel-65 beta decays by electron emission with decay constant $\lambda = 0.275$ h$^{-1}$. (a) What is the daughter nucleus? (b) In a sample that is initially pure Ni-65, how long will it be before there are twice as many daughter nuclei as parent nuclei?

71. A laser-fusion fuel pellet has mass 5.0 mg and is composed of equal parts (by mass) of deuterium and tritium. (a) If half the deuterium and an equal number of tritium nuclei participate in D-T fusion, how much energy is released? (b) At what rate must pellets be fused in a power plant with 3000-MW thermal power output? (c) What mass of fuel would be needed to run the plant for 1 year? Compare with the $3.6 \times 10^6$ tons of coal needed to fuel a comparable coal-burning power plant.

72. Of the neutrons emitted in each fission event in a light-water reactor, an average of 0.6 neutron is absorbed by $^{238}$U, leading to the formation of $^{239}$Pu. (a) Assuming 200 MeV per fission, how much $^{239}$Pu forms each year in a 30% efficient nuclear power plant whose electric power output is 1.0 GW? (b) With careful design, a fission explosive can be made from 5 kg of $^{239}$Pu. How many potential bombs are produced each year in the power plant of part (a)? *Note:* Extracting the $^{239}$Pu is not an easy job.

73. In the liquid-drop model, the mass of a nucleus with mass number $A$ can be expressed as a quadratic in $Z$: $M(A, Z) = c_1A - c_2Z + (c_2A^{-1} + c_3A^{-1/3})Z^2$, where the $c$'s are constants determined from experimental data. Show that the value of $Z$ that gives the minimum mass (not necessarily an integer) is given by $Z_{min} = (A/2)/[1 + (c_3/c_2)A^{2/3}]$. (A plot of $Z_{min}$ versus $N = A - Z$ gives the line of greatest nuclear stability shown in Figure 38.3.)

74. The probability that a radioactive nucleus will have a lifetime $t$ is the probability that it will survive from time 0 to time $t$ multiplied by the probability that it will decay in the interval from $t$ to $t + dt$. Use this to calculate the average lifetime of a nucleus, and show that it's equal to the inverse of the decay constant in Equation 38.3a.

75. Nucleus A decays into B with decay constant $\lambda_A$, and B decays into a stable product C with decay constant $\lambda_B$. A pure sample starts with an $N_0$ nuclei A at $t = 0$. Find an expression for the total activity of the sample at time $t$. *Hint:* The change in the amount of B in time $dt$ equals the magnitude of the change in A (a positive number since A produces B) plus the change in B (a negative number since B decays). Set up and solve the differential equations for A and B sequentially.

76. (a) Example 38.7 explains that the number of fission events in a chain reaction increases by a factor of $k$ with each generation. Show that the total number of fission events in $n$ generations is $N = (k^{n+1} - 1)/(k - 1)$. (b) In a typical nuclear explosive, $k$ is about 1.5 and the generation time is about 10 ns. Use the result from part (a) to calculate how long it would take for all of the nuclei in a 10-kg mass of $^{235}$U to fission. *Hint:* Sum a series in part (a), and neglect 1 compared to $N$ in part (b).

77. You've decided to give the love of your life a gold ring, but you're keeping the exact nature of the gift a secret. After weeks of incessant questioning, you offer a clue. You say the gift is made mostly of one element and you give the binding energy per nucleon as a hint. Using the data for Au-197, with mass $m = 196.96654$ u, what is your clue?

78. You're taking a history class on Russia and the former Soviet Union. You decide to do your class presentation on the nuclear accident at Chernobyl. You read newspaper accounts of the accident and come across a statement by a Swedish official that I-131 contamination levels in milk would be reduced to a safe amount after 5 days. You decide to verify this claim. The initial activity was 2900 Bq/L and Sweden's safe level is 2000 Bq/L. Is the 5-day figure accurate?

79. A close family member is about to have a brain scan in which technetium-99*, an excited isotope with a 6.01-hour half-life, is used. It just so happens that you're working in the hospital where the procedure will occur. You know the hospital makes the Tc-99* in its own cyclotron and then delivers it to the radiology lab. The isotope will arrive at the lab 3.5 hours after its production, and there must be 10 mg of Tc-99*. The reactor technician says she will produce 15 mg of Tc-99*. Is that adequate?

80. While on your summer vacation, you embark on the great college road trip visiting several national parks in the old West. When you stop at Badlands National Park in South Dakota, you learn about the nearby newly opened Minuteman Missile National Historic Site. You tour the site and are amazed at what you see. Your traveling companion wonders aloud why the missiles had multiple independent warheads instead of one large one. You recall from your course on nuclear weapons and arms reduction that the volume of a nuclear fireball is roughly proportional to the weapon's explosive energy yield. To answer your friend's question, compare the land area subject to a given level of damage for 10 small warheads versus one large one with the same explosive yield. Assume the fireball is a sphere and the ground damage occupies a circular area.

# Answers to Chapter Questions

## Answer to Chapter Opening Question

PET imaging "sees" the gamma rays produced in electron–positron annihilation, using multiple detectors to locate the annihilation site precisely. The positrons themselves are emitted by short-lived radioactive isotopes of elements such as carbon, oxygen, or fluorine. Because the lifetimes are so short (typically minutes), the isotopes must be made on site. Particles accelerated in the cyclotron slam into target materials, causing nuclear reactions that produce the desired isotopes.

## Answers to GOT IT? Questions

38.1 (a) $Z = 6$, $N = 6$; (b) $Z = 8$, $N = 7$; (c) $Z = 26$, $N = 31$; (d) $Z = 94$, $N = 145$.

38.2 About one-billionth.

38.3 $^{57}_{26}$Fe, $^{132}_{54}$Xe, $^{238}_{92}$U, $^{4}_{2}$He, $^{2}_{1}$H.

38.4 There is a much higher concentration of energy in nuclear fuels (by a factor of around $10^7$), which means far less nuclear fuel is mined and transported.

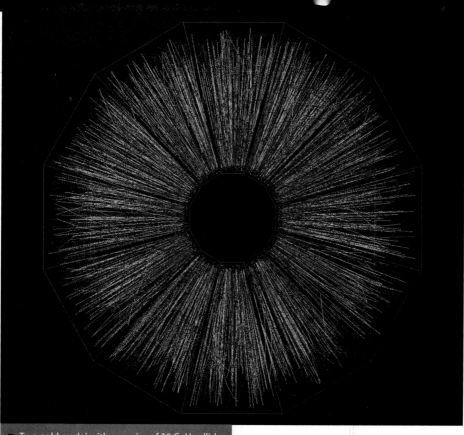

■ Two gold nuclei with energies of 30 GeV collide in the STAR detector of the Relativistic Heavy Ion Collider (RHIC) at Brookhaven National Laboratory, producing a spray of particles. What's this work got to do with the origin of the universe?

The past five chapters have extended the realm of physics to the atomic and molecular scale and then down to the atomic nucleus. Here we go further still, probing the structure of nucleons themselves and trying to make sense of the host of subatomic particles nature reveals. We'll be asking questions about the ultimate nature of matter at the tiniest scales, but in the process we'll find a remarkable connection with questions of the largest scale—questions about the origin and ultimate fate of the entire universe.

## 39.1 Particles and Forces

By 1932 four "elementary" particles of matter were known: the electron, the proton, the neutron, and the neutrino. In addition, there were the positron, antiparticle to the electron, and the photon of electromagnetic radiation. There were also the seemingly fundamental forces—gravity, the electromagnetic force, the nuclear force, and the weak force that manifests itself in beta decay.

In Chapter 34 we saw how the interaction of electromagnetic waves with matter ultimately involves individual photons—the quanta of the electromagnetic field. In the quantum-mechanical view of electromagnetism, the force between two charged particles

(a)

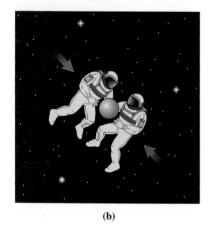

(b)

**FIGURE 39.1** Analogs for particle-mediated forces: (a) repulsive and (b) attractive. Ball represents a photon.

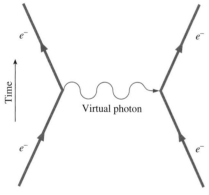

**FIGURE 39.2** A Feynman diagram, showing the interaction of two electrons through the exchange of a virtual photon.

also involves photons, now exchanged between the interacting particles. Imagine two astronauts tossing a ball back and forth (Fig. 39.1a). Catching or throwing the ball, one astronaut gains momentum in a direction away from the other, so the exchange results in a net average repulsive force. If the two astronauts struggle for possession of the ball, then the ball mediates what appears as an attractive interaction (Fig. 39.1b). Figure 39.1 gives classical analogs for the attractive and repulsive electrical interactions involving photon exchange.

We know that photons are emitted when a particle jumps into a lower energy state, with the photon carrying off energy equal to the energy difference between the two states. The process obviously conserves energy. But now we're saying that a single, free electron emits photons that it exchanges with another particle to produce what we call the electromagnetic force. How can that process conserve energy? The energy–time uncertainty relation (Equation 34.16) says that an energy measured in a time $\Delta t$ is necessarily uncertain by an amount $\Delta E \geq \hbar/\Delta t$. The photon exchanged by two particles lasts only a short time, and therefore its energy is uncertain. So we can't really say that energy conservation is violated. A photon created in this way and lasting for only the short time it takes to exchange with another particle is called a **virtual photon**. We never "see" the virtual photon, since it's emitted by one particle and absorbed by the other.

The quantum theory of the electromagnetic interaction is called **quantum electrodynamics** (QED). Although begun by Paul Dirac, it was finally brought to consistent form in 1948 by Richard Feynman, Sin-Itiro Tomonaga, and Julian Schwinger. The fundamental event in QED is the interaction of a photon with an electrically charged particle. Two such events joined by a common virtual photon give the quantum electrodynamical description of the electromagnetic force (Fig. 39.2). The predictions of quantum electrodynamics have been confirmed experimentally to a remarkably high precision, and today QED is our best-verified theory of physical reality.

## Mesons

In 1935 the Japanese physicist Hideki Yukawa proposed that the nuclear force should, like the electromagnetic force, be mediated by exchange of a particle. Yukawa called his hypothetical particle a **meson**. Because the range of the nuclear force is limited, Yukawa argued, the meson should have finite mass. The reason for this connection between mass and range again lies in the energy–time uncertainty relation.

The electromagnetic force falls off as $1/r^2$ and thus has an infinite range. Two particles can be very far apart and still interact electromagnetically. Since photons travel at the finite speed of light, the time $\Delta t$ for a photon interaction can be arbitrarily long. The energy–time uncertainty relation $\Delta E \, \Delta t \geq \hbar$ thus shows that the energy uncertainty $\Delta E$ can be arbitrarily small. Thus, the possible energies for virtual photons must extend downward toward zero—and that can happen only if photons have zero rest energy.

The nuclear force, in contrast, has a finite range of about 1.5 fm. At close to the speed of light, the longest a particle mediating this force would need to exist is for a time $\Delta t$ given by

$$\Delta t = \frac{\Delta x}{c} = \frac{1.5 \times 10^{-15} \text{ m}}{3.0 \times 10^8 \text{ m/s}} = 5.0 \times 10^{-24} \text{ s}$$

Then energy-time uncertainty gives

$$\Delta E \geq \frac{\hbar}{\Delta t} = \frac{1.05 \times 10^{-34} \text{ J} \cdot \text{s}}{5.0 \times 10^{-24} \text{ s}} = 2.1 \times 10^{-11} \text{ J} = 130 \text{ MeV}$$

Yukawa therefore proposed a new particle with mass $130 \text{ MeV}/c^2$, about 250 times that of the electron. Yukawa's prediction was eventually confirmed—but not before nature introduced yet another particle.

## 39.2 Particles and More Particles

In the 1930s the most available source of high-energy particles was cosmic radiation—high-energy protons and other particles of extraterrestrial origin. In 1937 the American physicist Carl Anderson (who had earlier discovered the positron) and his colleagues identified in cosmic rays a particle with a mass 207 times that of the electron. Now called the **muon**, this particle had the same charge and spin as the electron and seemed to behave much like a heavier version of the electron. Two muons were found: the negatively charged $\mu^-$ and its antiparticle $\mu^+$. Although the muon mass was close to Yukawa's prediction, the muon interacted only weakly with nuclei and, therefore, could not be the mediator of the nuclear force.

The real Yukawa particle was discovered 10 years later in 1947, again in cosmic rays, and turned out to have a mass about 270 times that of the electron. This time there were three related particles, now called **pions**: positive $\pi^+$, negative $\pi^-$, and neutral $\pi^0$.

The new particles are all unstable, undergoing decays that ultimately result in well-known stable particles. The negative pion, for example, decays with a mean lifetime of 26 ns to a negative muon and an antineutrino:

$$\pi^- \longrightarrow \mu^- + \bar{\nu}$$

The muon then decays with a 2.2-$\mu$s lifetime to an electron and a neutrino–antineutrino pair:

$$\mu^- \longrightarrow e^- + \nu + \bar{\nu}$$

---

**APPLICATION**    **Particle Detection**

Despite their small size, we can, remarkably, follow the trajectories of individual subatomic particles. Early particle detectors included the **cloud chamber** and the **bubble chamber**, in which particles ionize vapor or liquid, causing visible condensation or bubble formation along the particle tracks. More recent is the **multiwire proportional chamber**, in which crisscrossed wire grids record current pulses from electrons liberated as particles pass through a gas-filled chamber (see the photo, which shows a multiwire chamber at the Stanford Linear Accelerator Center). Analyzing the pulse distribution then reveals particle trajectories. Applying a magnetic field in a particle detector curves the trajectories, enabling scientists to determine the particles' charge-to-mass ratios. **Scintillation detectors** give off light flashes as particles pass through them, and the flash intensity provides a measure of particle energy. **Calorimeters**, consisting of layers of scintillators and energy-absorbing material, analyze the showers of secondary particles produced by a single high-energy particle to determine the original particle's energy. Modern detectors are huge agglomerations of several basic detector types, arranged to extract the maximum information from particle interactions. Computer analysis of detector output allows the identification of events so rare they may occur only once in a million interactions.

---

## Classifying Particles

The availability of increasingly powerful particle accelerators led to an upsurge in particle discoveries. By 1980 there were more than 100 "elementary" particles. Early attempts to classify particles distinguished them by mass, but a more enlightening approach is based on the fundamental forces. We'll now use this approach to outline three particle classes.

**Leptons** are particles that don't experience the strong force. They include the familiar electron, the muon, a more massive particle called the tau, and three types of neutrinos, one associated with each of the charged leptons. The neutrinos were long thought to be massless, but recent experiments show that neutrinos have small nonzero mass and that they "oscillate" among the three types. Each of the leptons has an antiparticle as well. There are thus a total of six lepton–antilepton pairs, and experimental evidence strongly suggests that no others can exist. The leptons all have spin $\frac{1}{2}$ and are therefore **fermions**, which obey the Pauli exclusion principle. Leptons are believed to be true elementary point particles with zero size and no internal structure.

**Hadrons** are particles that do experience the strong force. They fall into two subclasses: mesons and baryons. **Mesons** have integer spin and are therefore **bosons** that don't obey the exclusion principle. Mesons include Yukawa's pions and a host of others; all are unstable. **Baryons** have half-integer spins and are therefore fermions. They include the familiar proton and neutron and similar but more massive particles. The baryons are grouped into pairs, triplets, and higher-multiple groupings of closely related particles. The neutron and proton, for example, form a pair that differ in charge and very slightly in mass. Each baryon has an antiparticle, as do most mesons, but some neutral mesons are their own antiparticles.

The third class of particles comprises the **field particles** or **gauge bosons**, quanta of the different force fields and "carriers" of those forces. These include the familiar photon for the electromagnetic force; three particles called the $W^+$, $W^-$, and $Z$ for the weak force; a particle called the gluon; and a hypothetical graviton that would carry the gravitational force in an as yet incomplete theory of quantum gravity. All the field particles are bosons, carrying spin 1 or, for the graviton, spin 2. You might think Yukawa's meson should be in this category in its role as carrier of the nuclear force. That it doesn't appear here is a hint that the nuclear force isn't really fundamental; as we'll soon see, it's the gluon that plays the more fundamental role.

Table 39.1 lists some of the particles known even before full confirmation of today's elementary particle theories.

**TABLE 39.1** Selected Particles*

| Category/ Particle | Symbol (Particle/ Antiparticle) | Spin | Mass (MeV/c²) | Baryon Number, B | Lepton Number, L | Strangeness, s | Lifetime (s) |
|---|---|---|---|---|---|---|---|
| Field particles | | | | | | | |
| Photon | $\gamma$, $\gamma$ | 1 | 0 | 0 | 0 | 0 | Stable |
| $Z^0$ | $Z^0$, $Z^0$ | 1 | 91,188 | 0 | 0 | 0 | $\sim 10^{-25}$ |
| Leptons | | | | | | | |
| Electron | $e^-$, $e^+$ | $\frac{1}{2}$ | 0.511 | 0 | +1 | 0 | Stable |
| Muon | $\mu^-$, $\mu^+$ | $\frac{1}{2}$ | 105.7 | 0 | +1 | 0 | $2.2 \times 10^{-6}$ |
| Tau | $\tau^-$, $\tau^+$ | $\frac{1}{2}$ | 1777 | 0 | +1 | 0 | $2.9 \times 10^{-13}$ |
| Electron neutrino | $\nu_e$, $\bar{\nu}_e$ | $\frac{1}{2}$ | $<3 \times 10^{-6}$ | 0 | +1 | 0 | Stable |
| Muon neutrino | $\nu_\mu$, $\bar{\nu}_\mu$ | $\frac{1}{2}$ | $<0.19$ | 0 | +1 | 0 | Stable |
| Tau neutrino | $\nu_\tau$, $\bar{\nu}_\tau$ | $\frac{1}{2}$ | $<18.2$ | 0 | +1 | 0 | Stable |
| Hadrons | | | | | | | |
| Mesons | | | | | | | |
| Pion | $\pi^+$, $\pi^-$ | 0 | 139.6 | 0 | 0 | 0 | $2.6 \times 10^{-8}$ |
| Pion | $\pi^0$, $\pi^0$ | 0 | 135.0 | 0 | 0 | 0 | $8.4 \times 10^{-17}$ |
| Eta | $\eta^0$, $\eta^0$ | 0 | 547.8 | 0 | 0 | 0 | $\sim 5 \times 10^{-19}$ |
| Rho | $\rho^0$, $\rho^0$ | 1 | 775.8 | 0 | 0 | 0 | $\sim 4 \times 10^{-24}$ |
| Kaon | $K^+$, $K^-$ | 0 | 493.7 | 0 | 0 | 1 | $1.2 \times 10^{-8}$ |
| Kaon | $K^0$, $K^0$ | 0 | 497.6 | 0 | 0 | 1 | $0.895 \times 10^{-10}$ $5.18 \times 10^{-8}$ † |
| Baryons | | | | | | | |
| Proton | $p$, $\bar{p}$ | $\frac{1}{2}$ | 938.3 | 1 | 0 | 0 | Stable |
| Neutron | $n$, $\bar{n}$ | $\frac{1}{2}$ | 939.6 | 1 | 0 | 0 | 885.7 |
| Lambda | $\Lambda^0$, $\bar{\Lambda}^0$ | $\frac{1}{2}$ | 1115.7 | 1 | 0 | $-1$ | $2.6 \times 10^{-10}$ |
| Sigma | $\Sigma^+$, $\bar{\Sigma}^-$ | $\frac{1}{2}$ | 1189.4 | 1 | 0 | $-1$ | $0.80 \times 10^{-10}$ |
| Sigma | $\Sigma^0$, $\bar{\Sigma}^0$ | $\frac{1}{2}$ | 1192.6 | 1 | 0 | $-1$ | $7.4 \times 10^{-20}$ |
| Sigma | $\Sigma^-$, $\bar{\Sigma}^+$ | $\frac{1}{2}$ | 1197.4 | 1 | 0 | $-1$ | $1.5 \times 10^{-10}$ |
| Omega | $\Omega^-$, $\bar{\Omega}^+$ | $\frac{3}{2}$ | 1672.45 | 1 | 0 | $-3$ | $0.82 \times 10^{-10}$ |

*Does not include any hadrons with $c$, $b$, or $t$ quarks.

†The neutral kaon exists as a quantum-mechanical superposition of states with two different lifetimes.

## Particle Properties and Conservation Laws

Many new particles can be characterized by known properties such as mass, spin, and electric charge. Of these, spin and charge are associated with important conservation laws—conservation of angular momentum and conservation of electric charge. Allowed interactions among particles *must* conserve these quantities. The annihilation of an electron–positron pair, for example, is allowed because the initial particles have no net charge and neither do the resulting photons. Similarly, beta decay of the neutron produces an electron, a proton, and a neutral antineutrino and thus conserves charge:

$$n \longrightarrow p + e^- + \bar{v}_e$$

Here the subscript on the antineutrino indicates that it's an electron antineutrino, as opposed to the muon or tau variety.

Other particle properties appear to be conserved as well. Associated with each baryon or antibaryon is its **baryon number**, assigned the value $+1$ for a baryon and $-1$ for an antibaryon. All experimental evidence to date points to conservation of baryon number: In all observed particle reactions, the sums of the baryon numbers before and after the reaction are equal. An example is, again, beta decay of the neutron: The process starts with a neutron of baryon number $B = 1$ and ends with a proton $(B = 1)$, an electron, and an antineutrino. The last two are leptons, so their baryon numbers are zero, and thus baryon number is conserved. Some theories, which we describe shortly, suggest that baryon number is only approximately conserved. If that is so, then the proton itself is an unstable particle with a mean lifetime in excess of $10^{35}$ years.

Lepton number seems also to be conserved. Again beta decay provides an example: The neutron and proton, being baryons, have lepton number zero, while the resulting electron and *anti*neutrino have lepton numbers $+1$ and $-1$, respectively.

In the late 1950s particles called $K$, $\Lambda$, $\Sigma$, and $\Xi$ were discovered. Strange characteristics of these particles' decays could be explained by introducing a new fundamental property, called **strangeness**, with the new particles having $s = \pm 1$ or $\pm 2$. Strangeness is conserved in strong and electromagnetic interactions, but in weak interactions its value can change. We'll soon see that several other new properties are needed to characterize matter fully.

---

**EXAMPLE 39.1** **Conservation Laws: Evaluating a Particle Interaction**

A pion collides with a proton to produce a neutral kaon and a lambda particle:

$$\pi^- + p \longrightarrow K^0 + \Lambda^0$$

(a) Which of the following are conserved: electric charge, baryon number, lepton number, strangeness? (b) Could another possible result of a pion–proton collision be an electron and a proton?

**INTERPRET** In (a) we're asked to determine whether each quantity listed is the same on both sides of the equation that describes the pion–proton interaction. In (b) we're asked whether a proposed outcome of the interaction is compatible with the conservation laws.

**DEVELOP** We'll need to consult Table 39.1 and sum the various quantities on the left and right sides of the equation to see, in (a), whether they're conserved and, in (b), whether they obey the conservation laws.

**EVALUATE** (a) Table 39.1 shows that all the particles are hadrons, so the lepton number is zero on both sides of the equation. On the left we have a positive and a negative particle and on the right two neutrals, so the electric charge is zero on both sides. The pion is a meson and the proton a baryon, so the baryon number on the left is 1; similarly, the kaon is a meson and the $\Lambda^0$ a baryon, so the baryon number is conserved. Finally, neither pion nor proton is strange, so the total strangeness on the left is zero. Table 39.1 lists the $K^0$ with $s = +1$ and the $\Lambda^0$ with $s = -1$, so strangeness, too, is conserved. For (b), having an electron and a proton as the final state would conserve charge (0), strangeness (0), and baryon number (1). But it wouldn't conserve lepton number, which was originally 0 but would become 1.

**ASSESS** This example shows how conservation laws restrict the possible outcomes of particle interactions. ∎

## Symmetries

Watch a physical process in a mirror, and you expect the image to be a possible physical process; that is, the laws of physics should exhibit **symmetry** with respect to mirror reflection. At the subatomic level, the statement that a process and its mirror image are equally likely is called **conservation of parity**. Mathematically, a system has parity $+1$ if its wave function is unchanged on reflection through the origin—that is, on a coordinate change $x \to -x$, $y \to -y$, $z \to -z$. If the wave function changes sign, then the parity is $-1$. Parity is conserved if its value is unchanged in a particle interaction.

In 1957 theoretical physicists T. D. Lee and C. N. Yang pointed out that parity conservation had not been tested for the weak force. They made the revolutionary suggestion that parity need not be conserved—tantamount to suggesting that nature can distinguish right- from left-handed systems that are otherwise identical. A group led by Chien-Shiung Wu took up the challenge. Wu studied the beta decay of cobalt-60 in a magnetic field that established a left–right symmetry. Her experiments showed a preferential beta emission opposite the field direction—a clear violation of parity conservation (Fig. 39.3).

Although parity might not be conserved, theorists held that a combination of parity reversal ($P$) and charge conjugation ($C$)—changing particles into antiparticles—would result in indistinguishable physical behavior. But in 1964 a violation of this $CP$ conservation was found in a rare decay of the neutral kaon to a pion–antipion pair. The Russian physicist Andrei Sakharov suggested that this asymmetric decay might account for the preponderance of matter over antimatter in today's universe.

It still appears that $CPT$ conservation holds; that is, a combination of mirror reflection, charge conjugation, and reversal of the time coordinate makes a new physical process indistinguishable from the original. There may be a deep philosophical connection here with the direction of time, but the full implications of $CPT$ symmetry and the failure of its individual components aren't fully understood.

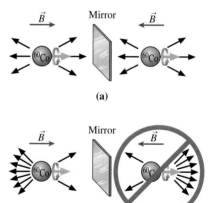

**FIGURE 39.3** Experimental evidence for nonconservation of parity. At left of mirror, a $^{60}$Co nucleus has its spin aligned with a magnetic field. (a) Reflected in the mirror, the spin vector still points to the right, even though the magnetic field is reversed. If the mirror image were equally likely, beta emission (arrows) would occur with equal probability along and opposite the spin direction. (b) Experiment shows that beta emission occurs preferentially opposite the spin direction, so the mirror-image situation at the right in (b) does not occur.

## 39.3 Quarks and the Standard Model

The proliferation of particles distressed physicists used to finding an underlying simplicity in nature. Were all those particles really "elementary," or was there a more fundamental, simpler level? In 1961 physicists Murray Gell-Mann and Yu val Ne'eman independently noticed patterns in the then-known particles. They called their patterns the **Eightfold Way**, after Buddhist principles for right living. An empty spot in one pattern led Gell-Mann to predict a new particle with strangeness $-3$. Experimentalists soon found the particle in bubble-chamber photographs from earlier experiments.

### Quarks

Success of the Eightfold Way convinced physicists that many "elementary" particles weren't really elementary. In 1964 Gell-Mann and his colleague George Zweig independently proposed a set of three particles called **quarks** that combined to form the then-known hadrons. These became known as the **up quark**, the **down quark**, and the **strange quark**. For each there was a corresponding antiquark.

One surprising thing about quarks is that they carry fractional electric charges. The two least massive quarks, the up and the down, carry $+\frac{2}{3}e$ and $-\frac{1}{3}e$, respectively; their antiparticles have the opposite charges. The quarks combine in pairs or triplets to make the two classes of hadrons. Baryons, like the proton and the neutron, consist of three quarks (Fig. 39.4). Mesons contain quark–antiquark pairs (Fig. 39.5). The quarks all have spin $\frac{1}{2}$, which explains why the three-quark baryons all have odd half-integer spin, and why the two-quark mesons have integer spin.

The Pauli exclusion principle precludes three particles having the same quantum numbers, so there must be an additional property distinguishing quarks. Called **color**, this property is a kind of "charge"—not to be confused with electric charge—that can take on any of three values called, whimsically, red, green, and blue. The force binding quarks of different colors is the **strong force**, and the quark theory is known as **quantum chromodynamics (QCD)**. In QCD, **gluons** are the field particles that play the role of photons in quantum electrodynamics, binding particles subject to the strong force. Particles formed from quarks—the mesons and the baryons—are always **colorless**. This is true of mesons because they contain a quark of one color and another of its anticolor. It's true of baryons because they contain three quarks of different colors, which combine to give the baryon zero net color charge. The nuclear force, once thought to be fundamental, is actually a residual manifestation of the strong force, acting between the quarks in colorless particles—in much the same way that the van der Waals force between neutral gas molecules is a "residue" of the stronger electric force among the particles that make up the molecules.

Photons mediate the electromagnetic force between charged particles but are themselves uncharged. In contrast, gluons, like the quarks they bind, carry color charge. There are eight different gluons; six carry combinations like red-antiblue ($R\overline{B}$), green-antired ($G\overline{R}$), and so on; the other two are colorless. Exchange of a colored gluon, unlike photon exchange in quantum electrodynamics, thus changes the colors of the particles involved.

Another surprising aspect of quarks is that the strong force doesn't decrease with separation. For that reason it appears impossible to isolate a single quark (Fig. 39.6). As a result we never see individual free particles with fractional electric charge.

The up, down, and strange quarks soon proved insufficient to account for all the observed particles. Theorist Sheldon Glashow argued for a fourth quark, called the **charmed quark**. Ten years later, following intensive searches, experimental teams at Brookhaven National Laboratory and the Stanford Linear Accelerator Center announced the discovery of a particle that implies the existence of the charmed quark. The charmed and strange quarks form a related pair, similar to the up/down quark pair.

**FIGURE 39.4** Protons and neutrons consist of the quark combinations *uud* and *udd*, respectively.

**FIGURE 39.5** Mesons consist of a quark and an antiquark. The meson here is the $\pi^+$, made from an up quark and a down antiquark.

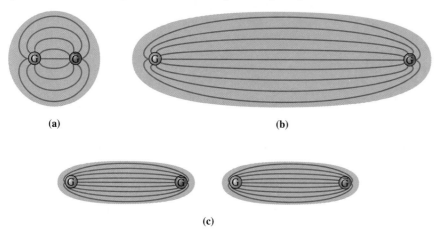

**FIGURE 39.6** Quark confinement in a meson consisting of a green quark G and its antiquark $\overline{G}$. (a) Field lines represent the "color field" that joins the two. (b) The field remains confined as the quarks are moved apart, so the field strength stays essentially constant. (c) Pulling the quarks far apart builds up enough energy to create another quark–antiquark pair, rather than isolating individual parts.

There's still one more quark pair. A 1977 experiment confirmed the existence of the **bottom quark**, and in 1994 experiments at Fermilab gave evidence for the **top quark**. The more exotic quarks are more massive, and therefore, through mass–energy equivalence, it takes more energy to produce particles containing them. This need for higher energy is what drives the push for ever more powerful and expensive particle accelerators. Table 39.2 lists some properties of the six quarks.

**TABLE 39.2** Matter Particles of the Standard Model

| Quark Name | Symbol | Approximate Mass (MeV/$c^2$) | Charge | Corresponding Leptons (Symbol, Mass in (MeV/$c^2$)) |
|---|---|---|---|---|
| Down | $d$ | 6 | $-\frac{1}{3}e$ | Electron ($e$, 0.511), electron neutrino ($\nu_e$) |
| Up | $u$ | 3 | $+\frac{2}{3}e$ | |
| Strange | $s$ | 100 | $-\frac{1}{3}e$ | Muon ($\mu$, 106), muon neutrino ($\nu_\mu$) |
| Charmed | $c$ | 1300 | $+\frac{2}{3}e$ | |
| Bottom | $b$ | 4300 | $-\frac{1}{3}e$ | Tau ($\tau$, 1777), tau neutrino ($\nu_\tau$) |
| Top | $t$ | $1.8 \times 10^5$ | $+\frac{2}{3}e$ | |

---

**EXAMPLE 39.2** **Quarks: Particle Composition and Properties**

Given that the strange quark has strangeness $s = -1$, find the charge and strangeness of the $\Lambda^0$ particle, which has quark composition $uds$.

**INTERPRET** We're asked to find how the charge and strangeness of three individual quarks combine in a composite particle, the $\Lambda^0$.

**DEVELOP** Our plan is to sum the values of the quark's charge and strangeness to get the parameters of the $\Lambda^0$.

**EVALUATE** Table 39.2 gives the charges of the $u$, $d$, and $s$ quarks as $+\frac{2}{3}e$, $-\frac{1}{3}e$, and $-\frac{1}{3}e$, respectively. These sum to zero net charge, so the $\Lambda^0$ is electrically neutral. The up and down quarks aren't strange, so they have $s = 0$. With $s = -1$ for the strange quark, the $\Lambda^0$ must have strangeness $-1$.

**ASSESS** Table 39.1 shows we're right about the strangeness of the $\Lambda^0$, and its superscript 0 implies that this is a neutral particle, as we've found. ∎

---

## The Standard Model

We now have six **flavors** of quarks—up, down, strange, charmed, top, bottom—that seem to be truly elementary constituents of matter. Quarks join to form the hadrons—baryons and mesons. But other particles—namely, leptons and field particles—aren't made from quarks. They, like quarks, seem to be truly indivisible, elementary particles.

In this "zoo" of elementary particles, physicists recognize three distinct "families." The up and the down quarks make the neutron and proton; together with the electron and its related neutrino, they account for the properties of ordinary matter. A second family consists of the strange and charmed quarks, the electron-like muon, and the muon neutrino. The quarks of this family are more massive than the up and down quarks, and the muon is more massive than the electron. More massive still are the particles of the third family, consisting of the top and bottom quarks, the electron-like tau particle, and the tau neutrino. Table 39.2 shows these three families from which all known matter is constructed.

You may be expecting that future editions of this book will tell of additional quarks, and thus of additional families of matter. Whether such additional families exist was an open question until physicists at the Large Electron Positron Collider in Geneva, Switzerland, examined more than half a million particle-decay events and concluded that the number of different types of neutrinos that can exist is $2.99 \pm 0.06$. Since there's presumably a neutrino type for each family, this result seems to preclude the existence of additional families.

The theory that currently describes elementary particles and their interactions is called the **standard model**. In addition to the particles shown in Table 39.2, the standard model includes the photons that mediate the electromagnetic interaction, the gluons of the strong force, the $W$ and $Z$ particles that mediate the weak force, and an as yet undetected particle called the Higgs boson, believed responsible for other particles' masses (Table 39.3). The standard model is successful in explaining the phenomena of particle physics, but it leaves many fundamental questions unanswered. Why, for example, do the quarks and leptons have the particular masses they do? Why are there only three families of elementary particles? Why are leptons and quarks distinct? Are these particles really elementary, or are there even smaller structures at hitherto unexplored scales? Continuing theoretical work and experiments at ever-higher energies may someday answer these questions.

**TABLE 39.3** Field Particles and Forces

| Particle | Mass (GeV/$c^2$) | Electric and Color Charges* | Force Mediated | Range | Approximate Strength at 1 fm (Relative to Strong Force) |
|---|---|---|---|---|---|
| Graviton | 0 | 0, 0 | Gravity | Infinite | $10^{-38}$ |
| $W^\pm$ | 80.2 | $\pm1, 0$ | Weak | $<2.4 \times 10^{-18}$ m | $10^{-13}$ |
| $Z^0$ | 91.2 | 0, 0 | | | |
| Photon, $\gamma$ | 0 | 0, 0 | Electromagnetic | Infinite | $10^{-2}$ |
| Gluon, $g$ (8 varieties) | 0 | 0, 6 color–anticolor combinations, 2 colorless | Strong | Infinite$^\dagger$ | 1 |
| Higgs boson, $H^0$ | 40–1000? | 0, 0 | The Higgs boson is an as yet undetected particle needed to account for the masses of the other particles. | | |

*Color is a quark property analogous to, but more complicated than, electric charge.
$^\dagger$The nuclear force is the residual strong force between colorless particles and has a range of about 1 fm ($10^{-15}$ m).

# 39.4 Unification

We introduced the fundamental forces of nature in Chapter 4: gravity, the electroweak force, and the strong force. It was not always that simple, though. In Chapters 20–29 we studied the electric and magnetic forces, first separately but then with the realization that they fall under the single umbrella of electromagnetism. The unification of electricity and magnetism was a major step forward in our understanding of physical reality. Physicists continue to strive for further unification, with the ultimate hope that someday all the forces will be understood as a manifestation of a single common interaction.

## Electroweak Unification

In the 1960s and early 1970s, a century after Maxwell formalized the unification of electromagnetism, physicists Steven Weinberg, Abdus Salam, and Sheldon Glashow proposed that the electromagnetic force and the weak force are really aspects of the same thing. Their theory predicted the existence of the particles $W^+$, $Z^0$, and $W^-$, the "carriers" of the unified electroweak force. In 1983 a huge international consortium headed by Carlo Rubbia discovered the $W$ and $Z$ particles, using advances in accelerator technology developed by Simon van der Meer (Fig. 39.7). That discovery confirmed the electroweak unification, and Rubbia and van der Meer joined a long list of physicists who had won the Nobel Prize for contributions to our understanding of the structure of matter.

**FIGURE 39.7** Particles tracks from the decay of a $Z$ particle help confirm electroweak unification.

**FIGURE 39.8** Japan's Super Kamiokande experiment consists of 50,000 tons of pure water in an underground chamber, surrounded by some 10,000 photomultiplier tubes to detect flashes from rare nuclear reactions including neutrino interactions and hypothetical proton decays.

## Further Unification

Electroweak unification led to the present situation in physics, with the strong force, the electroweak force, and gravity comprising the **fundamental forces** that describe all interactions of matter. A further step, the **grand unification theories** (GUTs), attempts to merge the electroweak and strong forces. Some versions of GUT predict that the proton should decay on the very long timescale of some $10^{36}$ years. We can't wait that long, but we can put $10^{34}$ protons together in the form of tens of thousands of tons of water and watch for proton decay (Fig. 39.8). Such experiments have not found the predicted decays, but another GUT prediction—that neutrinos have mass—was verified, also in the device of Fig. 39.8. Many physicists believe that some form of grand unification will soon be achieved.

Even grand unification would still leave two forces, one of them gravity. Attempts to reconcile our current theory of gravity—Einstein's general theory of relativity—with quantum mechanics have so far made little progress. Yet such a reconciliation is a necessary prerequisite for a final unification of all known forces. A possible candidate is **string theory**, which pictures elementary particles as vibration modes on stringlike structures that may be as short as $10^{-35}$ m (Fig. 39.9). String theory is set not in the four-dimensional spacetime to which we're accustomed, but in a spacetime with 10 or more dimensions. The extra dimensions are "compactified" in a way that makes them undetectable in normal interactions. To some physicists, string theories hold the promise of a "theory of everything," explaining all our observations about the behavior of the universe. To others they're another in a long line of unsuccessful attempts at a comprehensive explanation of physical phenomena. Only further research will tell.

**FIGURE 39.9** (a) A particle decay involving point particles in the standard model. (b) A similar decay in string theory. Note that there's no single point in spacetime at which the decay takes place.

(a)          (b)

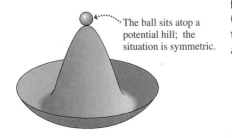

The ball sits atop a potential hill; the situation is symmetric.

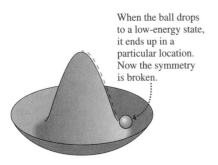

When the ball drops to a low-energy state, it ends up in a particular location. Now the symmetry is broken.

**FIGURE 39.10** A mechanical analogy for symmetry breaking, showing a ball subject to a hat-shaped potential-energy curve.

## Symmetry Breaking

Unification theories predict that phenomena that appear distinct under normal conditions will be seen as one at sufficiently high energies. The observed unification represents a kind of symmetry that's "broken" as the energy level drops. Figure 39.10 shows a mechanical analogy for such **symmetry breaking**. With high energy, a ball sits atop a potential "hill," and the situation is symmetric. But when the ball drops to a low-energy state, it ends up at a particular angular position, and the symmetry is broken. Analogously, at energies above 100 GeV, what we call the electromagnetic and weak forces are one and the same. But at lower energies the symmetry is broken, and we see two distinct forces. Particle accelerators now being planned will exceed the energy of electroweak symmetry breaking, allowing us to explore that interaction in its fundamental simplicity. But the energy at which symmetry breaking occurs increases to some $10^{15}$ GeV as we move from electroweak to grand unification, making it unlikely that we'll achieve that energy in the foreseeable future. And the energy at which gravity, too, would join a single unified force is an even more remote $10^{19}$ GeV.

## APPLICATION    Particle Accelerators

Aerial view showing the location of the 4.3-km-diameter Large Hadron Collider on the Swiss–French border. The accelerator itself is in a tunnel 50–175 m below ground. The device accelerates protons to 7 TeV. The smaller ring accelerates the protons to 450 GeV before they're injected into the main accelerator.

Most particles are more massive than the proton, and some, like the weak-force mediators $W^{\pm}$ and $Z$, are extremely massive. Since the more massive particles are all unstable, discovering them involves first creating them—and that requires energy of at least $mc^2$, with $m$ the particle mass. That energy requirement, along with the hint of new phenomena such as force unification, drives particle physicists' seemingly insatiable desire for particle accelerators of ever-higher energies.

The earliest accelerators were electrostatic devices that established large potential differences between conducting electrodes, and they used the associated electric field to accelerate charged particles. But such accelerators are limited to maximum energies of about 20 MeV because of the difficulties of handling high voltages. We saw in Chapter 26 how this problem is cleverly circumvented in the cyclotron, an accelerator that uses a magnetic field to keep particles in circular orbits so they can gain energy on each orbit from a modest electric field. But cyclotrons work for only nonrelativistic particles, for which the cyclotron frequency is independent of particle energy. Today's high-energy experiments call for ultrarelativistic particles, whose speeds differ only minutely from the speed of light. As a result, today's accelerators are primarily variations on the **synchrotron**, a device in which the magnetic field increases with the particle energy to maintain particles in a circular orbit of fixed radius. An alternative to the synchrotron is the **linear accelerator**, the largest of which is the 3-km-long Stanford Linear Accelerator.

A head-on collision between two cars is much more damaging than a collision of a moving car with a stationary one, since in the former case all the energy goes into damaging the cars while in the latter a great deal of energy goes into accelerating the initially stationary "target" car. For the same reason head-on collisions of high-energy particles make much more energy available for creation of new particles. As a result, most of the highest-energy accelerators today are colliders, with particle beams accelerated in opposite directions and brought to collide inside elaborate detectors. The largest accelerator today is the Large Hadron Collider (LHC) at CERN, the European Laboratory for Particle Physics, at Geneva, Switzerland. With a 27-km circumference, LHC collides proton beams at energies up to 14 TeV. Both LHC and the Relativistic Heavy Ion Collider at Brookhaven National Laboratory in the United States create conditions that existed when the universe was only $10^{-12}$ s old; more on this in the next section.

## 39.5 The Evolving Universe

We come at the end to the broadest possible questions about physical reality: How did the universe begin? What is its overall structure? What will its future bring? Remarkably, these cosmological questions are closely linked with the questions of particle physics.

### Expansion of the Universe

Early in the 20th century, astronomers argued about the nature of certain fuzzy patches visible in telescope photographs. Many thought they were gas clouds scattered among the visible stars, but others made a more radical proposal: that some of these "nebulae" were gravitationally bound systems containing billions of stars and that they were almost inconceivably distant.

In the 1920s, the opening of the 2.5-m telescope at California's Mt. Wilson Observatory finally resolved the issue. There, astronomer Edwin Hubble proved that some nebulae were indeed distant galaxies like our own Milky Way, each containing $10^9$ to $10^{13}$ stars. Today cosmologists think of galaxies as individual "point particles" whose distribution traces the overall structure of the universe.

Hubble continued to study the galaxies throughout the 1920s, and by analyzing their spectra he made a remarkable discovery: Spectral lines from distant galaxies are shifted toward the red, with the amount of shift dependent on the distance to the galaxies. The most reasonable and widely held explanation is that the redshift is caused by the Doppler effect (Section 14.8). Then the implication of Hubble's work is that the distant galaxies are receding from us at speeds proportional to their distances. This result is known as **Hubble's law**:

$$v = H_0 d \qquad (39.1)$$

where $v$ is the recession speed, $d$ the distance, and $H_0$ the **Hubble constant**, whose value is now known to be very nearly 22 kilometers per second per million light years of distance. Astronomers now use the Hubble relation to find the distances to remote galaxies, measuring their redshifts and using Equation 39.1 to infer their distances.

It may sound like Hubble's law puts us right in the center of things, in grotesque violation of modern science's view that Earth and its inhabitants don't occupy a favored position in the universe. But actually the inhabitants of any other galaxy would observe the same thing: All the distant galaxies would be receding from them at speeds proportional to their distances. As long as the universe is infinite in extent, none can claim to be at the center. And if it's not infinite, then Einstein's general theory of relativity gives it a closed-curve shape that still has no center.

Hubble's law implies that the universe is expanding as the galaxies move apart. Extrapolating backward in time suggests there was a time when all the galaxies were together. Thus the **Hubble expansion** suggests that the universe had a definite beginning, in the form of a colossal explosion that flung matter into an expansion that continues today. Based on additional evidence that we'll describe shortly, scientists are quite certain that the universe began with such a **Big Bang**. Some of that evidence comes from the Hubble Space Telescope, named to honor Hubble's pioneering studies (Fig. 39.11).

**FIGURE 39.11** A portion of the Hubble Deep Field, a Hubble Space Telescope image showing distant galaxies whose redshifts provide information about cosmic expansion.

---

**EXAMPLE 39.3** **Hubble's Law: Calculating the Age of the Universe**

Using $H_0 = 22$ km/s/Mly and assuming the expansion rate has been constant, find out how long the universe has been expanding.

**INTERPRET** We're given the Hubble constant $H_0$ and asked to extrapolate back in time until all the galaxies were together.

**DEVELOP** If a galaxy has been moving with constant speed for time $t$, then its distance from us today is $d = vt$. But Hubble's law gives its speed in terms of distance: $v = H_0 d$. So our plan is to substitute $H_0 d$ for $v$ and solve for the time $t$.

**EVALUATE** We have $t = d/v = d/H_0 d = 1/H_0$. To evaluate this expression we need to convert from the mixed units used for the Hubble constant:

$$t = \frac{1}{H_0} = \frac{1}{(22 \text{ km/s/Mly})/[3.0 \times 10^5 \text{ (km/s)/(ly/y)}]} = 13.6 \text{ Gy}$$

**ASSESS** Our calculation shows that the universe is about 13.6 billion years old—on the assumption that the expansion rate hasn't changed. We'll soon see that this assumption isn't quite correct, but nevertheless our result is a good measure of the age of the universe. Note that we used $c = 3.0 \times 10^5$ km/s in converting from km/s to ly/y so our answer would come out in years. ∎

---

## The Cosmic Background Radiation

In 1965 Arno Penzias and Robert Wilson at Bell Laboratories found a faint "noise" of microwave radiation in a satellite communications antenna they were testing. The noise seemed to come from all directions. Theorists at Princeton identified Penzias and Wilson's "noise" as radiation dating to a much earlier era in the universe. This **cosmic microwave background radiation** is the strongest evidence yet for the Big Bang.

The Big Bang theory suggests that the universe started very hot and then cooled as it expanded, doing work against its own gravitation. At first it was so hot that any atoms that formed would be dashed apart by collisions at the high thermal energy prevailing. Thus, in its early times the universe was populated by individual charged particles. These interacted readily with electromagnetic radiation, making the universe opaque. But by about 400,000 years the temperature had dropped to several thousand kelvins, and at that point atoms of hydrogen and helium could form. Since neutral atoms interact much more weakly with electromagnetic radiation, the universe became transparent, and photons emitted as the atoms formed could travel throughout the universe with little chance of being subsequently absorbed. Those photons became the cosmic background radiation, permeating the entire universe.

Expansion of the universe would cool not only matter but also radiation; the Big Bang scenario suggests that radiation from the time the atoms formed should now be cooled to that of a blackbody at 2.7 K—a radiation distribution that peaks in the microwave region of the electromagnetic spectrum. Analysis of the microwave background shows that it has indeed precisely that blackbody character and temperature (Fig. 39.12). Thus, the observed cosmic microwave background is a direct reflection of conditions some 400,000 years after the universe began.

The cosmic microwave background (CMB) is remarkably uniform, but not perfectly so. Today, satellite instruments record minute spatial variations in the microwave background, and analysis of these variations reveals a wealth of information about the early universe (Fig. 39.13). For example, the CMB variations are consistent with a universe whose overall geometry is flat, and the variations themselves represent the "seeds" of galaxies, galaxy clusters, and even larger structures that dominate the universe today. Analysis of CMB variations also provides our most accurate value for the Hubble constant, a value that independently corroborates galaxy redshift observations.

**FIGURE 39.12** Spectrum of the cosmic microwave background matches perfectly that of a blackbody at 2.726 K.

## The Earliest Times

The cosmic microwave background shows us the universe as it was some 400,000 years after the Big Bang. Nuclear physics takes us back even further, to the time from about 1 second to 30 minutes when the lightest nuclei were forming. The first composite nuclei were the simplest: deuterium, consisting of a proton and a neutron. The rate of deuterium formation is critically sensitive to the expansion rate of the universe. Measurements of deuterium abundance, based on spectral lines from interstellar deuterium, therefore provide direct evidence for conditions early in the Big Bang.

Evidence for still earlier times comes from particle physics, which predicts the interactions and particle populations under the hot, high-energy conditions that existed a fraction of a second after the Big Bang. In 2005, experiments using the Relativistic Heavy Ion Collider (RHIC) at Brookhaven National Laboratory produced a so-called quark–gluon plasma similar to the state of the universe just microseconds after the Big Bang. Thus RHIC and other high-energy particle accelerators can act as "time machines," allowing us to study conditions that prevailed in the very early universe. Figure 39.14 summarizes our understanding of the universe's evolution, showing that the phenomena of particle physics and unification are inextricably tied with cosmic expansion.

**FIGURE 39.13** All-sky map of the cosmic microwave background shows minute spatial variations in the radiation intensity. Data were taken with the Wilkinson Microwave Anisotropy Probe (WMAP) satellite.

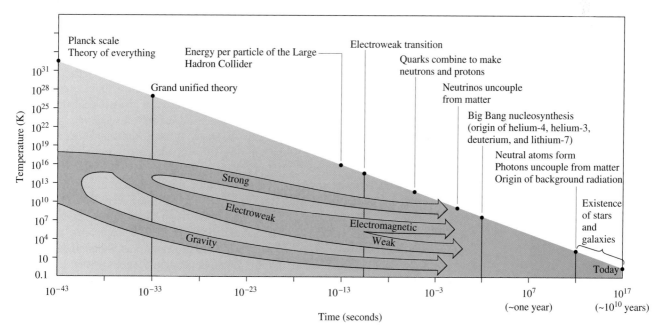

**FIGURE 39.14** Evolution of the universe from the earliest times to the present. Note the highly logarithmic scales.

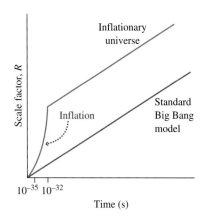

**FIGURE 39.15** Expansion in the standard and inflationary Big Bang theories. The scale factor $R$ measures the amount of expansion.

## The Inflationary Universe

The original Big Bang theory has difficulty explaining several features of the observed universe. Why, for example, do we find only matter but not antimatter? Why does the universe seem homogeneous and in thermodynamic equilibrium on the largest scales, when the most distant regions are so far apart that light could not have traveled between them in the time since the beginning? And why does the overall geometry of the universe appear flat, when general relativity allows for curved space?

The solution to these conundrums is **inflation**, an idea first advanced by Alan Guth of MIT. Guth's theory holds that the universe underwent a period of exponential expansion beginning at about $10^{-35}$ s and lasting until about $10^{-32}$ s (Fig. 39.15). The expansion was the result of a delay in the symmetry breaking that made the fundamental forces appear distinct. Because of the tremendous expansion, now-distant locations would once have been close enough to reach the thermodynamic equilibrium that we now observe. Furthermore, the inflationary expansion would "flatten out" any overall curvature, giving us the flat universe we see today.

## Dark Matter, Dark Energy, and the Future of the Universe

Will the universe expand forever, or will the expansion eventually stop and reverse? That's like asking whether a spacecraft will escape Earth forever or ultimately return, and the answer is the same: If the kinetic energy exceeds the magnitude of the (negative) potential energy, then expansion will proceed forever. Thus a single parameter—designated $\Omega$ (capital Greek omega)—determines the fate of the universe:

$$\Omega = \frac{|\text{potential energy of the universe}|}{\text{kinetic energy of the universe}}$$

where $\Omega > 1$ implies eventual collapse, and $\Omega < 1$ means continued expansion. An alternative way of expressing the same dichotomy is in terms of the average density. A simple Newtonian calculation based on kinetic and gravitational potential energies of particles in an expanding universe gives $\Omega$ in terms of the average density $\rho$ and the Hubble constant: $\Omega = 8\pi G\rho/3H_0^2$—a result that turns out to be identical to the correct general relativistic calculation. Setting $\Omega = 1$ gives the **critical density** that divides eternal expansion from eventual collapse:

$$\rho_c = \frac{3H_0^2}{8\pi G} \quad \text{(critical density)} \tag{39.2}$$

Analysis of the cosmic microwave background fluctuations strongly suggests that the actual universe is at the critical density, with $\Omega = 1$. But when we total the visible matter in the universe, we come up with far less than the critical density. Elementary particle theories corroborate this observational result, suggesting that there can be only enough ordinary matter—protons, neutrons, nuclei, and atoms—to make up about 4% of the critical density. Furthermore, the motions of stars in galaxies suggest the presence of a great deal more mass than is visible. All this implies the existence of **dark matter** whose composition is unknown and which can't be the sort of matter—made mostly from quarks—with which we're familiar.

Another approach to the cosmic density is to study the most distant galaxies, whose light has taken so long to reach us that we're seeing them as they were in the early universe. One might expect that cosmic expansion was faster in earlier times, and slowed as the galaxies did work against their mutual gravitational attraction. Comparing the recession speeds of ancient, distant galaxies with the speeds of nearer ones should then give the rate of cosmic deceleration—which, in turn, should depend on cosmic density. But observations in 1998 gave a surprising and unexpected result: The cosmic expansion is actually accelerating!

Cosmic acceleration implies a kind of "antigravity" operating on the largest scales. Ironically, Einstein had proposed just such a phenomenon in his original formulation of general relativity. Einstein needed this **cosmological constant** in his theory in order to keep the universe static—which, in 1916, astronomers thought it was. When Hubble then showed that the universe is expanding, Einstein dropped the cosmological constant and called it "the greatest blunder of my life." Now it appears that Einstein had the right idea in the first place.

The source of the cosmic acceleration is **dark energy**, which may be just another name for Einstein's cosmological constant or may be a different phenomenon with the same "antigravity" effect. At this point we don't know quite what it is. But we do know how much of it there is: Dark energy is fully 73% of the "stuff" that makes up our universe. Another 23% is dark matter. That means only 4% of the universe is in the form of familiar matter made from baryons and thus ultimately from quarks. These numbers come from a confluence of recent observations of the cosmic microwave background, distant supernovae, and surveys of distant galaxies; together, they tightly constrain the relative amounts of matter and dark energy (Fig. 39.16). Here, at the end of a long physics course, it's sobering to realize that much of our universe consists of "stuff" about which we know so very little!

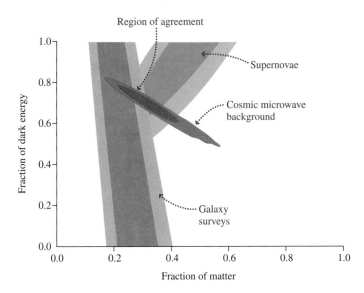

**FIGURE 39.16** Constraints on the relative abundances of dark energy and of matter in the universe, from three independent sets of observations. All three converge to agreement on approximately 73% dark energy and 27% matter; the latter includes 23% dark matter and 4% ordinary matter.

## Understanding the Universe

With this chapter's brief survey, we've reached the limits of present understanding of the universe. On the way we've seen that particle physics and cosmology are inextricably linked. To understand our universe, we need to understand all its aspects from the largest to the smallest—and that means understanding all the forces, from gravity to the weak force; all the physical laws, from Newton's and Maxwell's to the laws of quantum mechanics; and the nature of the elementary particles. In this text we've touched on all these topics, and we hope we've given you a foundation for further understanding and appreciation of the richness and diversity of the physical universe.

## Big Picture

The big idea here is that the structure of ordinary matter and its interactions can be explained in terms of a handful of particles: the **quarks**, which make up the familiar proton and neutron and a host of other **hadrons; leptons**, including the familiar electrons and the elusive neutrinos; and the **gauge bosons**, which mediate the fundamental forces and include the photon for the electromagnetic force. The **standard model** describes the elementary particles and their interactions; it sheds light not only on the structure of matter we see today but on the early universe as well. But there's a humbling caveat: Only about 4% of the universe consists of familiar matter. The rest is **dark energy** and **dark matter**, about which we know very little.

## Key Concepts and Equations

**Quarks**, which come in six flavors and three colors, join in threes to form **baryons**, including protons and neutrons, and in twos to form **mesons**. Quarks carry fractional electric charges, but because the **strong force** between quarks doesn't decrease with distance, it appears impossible to have an isolated quark.

Proton

Neutron

Meson

**Leptons** are the other class of elementary matter particles, and include the electron and its more massive cousins, the muon and tau, and the three types of neutrinos.

Exchange of **gauge bosons** explains the forces between particles.

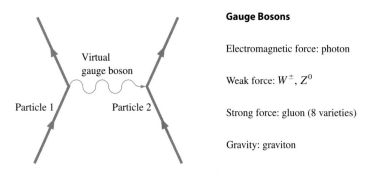

Virtual gauge boson

Particle 1    Particle 2

**Gauge Bosons**

Electromagnetic force: photon

Weak force: $W^{\pm}$, $Z^0$

Strong force: gluon (8 varieties)

Gravity: graviton

The three **fundamental forces**—strong, electroweak, and gravity—are believed to be manifestations of the same fundamental interactions that appear unified at high enough energies.

## Cases and Uses

Our knowledge of particles and their interactions combines with general relativity and with observations of the **Hubble expansion** and the **cosmic microwave background radiation** to give a picture of the origin and structure of the universe. The universe began some 14 billion years ago in a hot **Big Bang**. The simplest nuclei formed within the first half-hour, and the first atoms at about 400,000 years; at that point the universe became transparent and the cosmic microwave background formed. We've learned recently that the universe is overall flat and that its expansion is accelerating under the influence of mysterious **dark energy**. There's still much that we don't know!

# For Thought and Discussion

1. Why did Yukawa conclude that the particle mediating the strong force should have nonzero mass?
2. How can we detect the tracks of individual particles?
3. How are baryons fundamentally different from leptons?
4. What coordinates are changed under the inversion processes $P$ and $T$?
5. Why are we unlikely to observe an isolated quark?
6. Describe the relation between the strong force and the nuclear force.
7. What's the role of gluons?
8. Classify (a) mesons and (b) baryons as fermions or bosons, and relate your classification to the particles' quark compositions.
9. Name the fundamental force involved in (a) binding of a proton and a neutron to make a deuterium nucleus; (b) decay of a neutron to a proton, an electron, and a neutrino; (c) binding of an electron and a proton to make a hydrogen atom.
10. What forces are unified in the electroweak theory?
11. What forces would be unified by GUTs?
12. Why do we need higher-energy particle accelerators to explore fully the standard model?
13. How can Hubble's law hold without the universe having a center?
14. Is it possible for a charged particle to be its own antiparticle?
15. Describe the origin of the cosmic microwave background.
16. Explain how particle accelerators can help us understand the early universe.

# Exercises and Problems

## Exercises

### Section 39.1 Particles and Forces

17. How long could a virtual photon of 633-nm red laser light exist without violating conservation of energy?
18. Some scientists have speculated on a possible "fifth force," with a range of about 100 m. Following Yukawa's reasoning, what would be the mass of the field particle mediating such a force?

### Section 39.2 Particles and More Particles

19. Write the equation for the decay of the positive pion to a muon and a neutrino, being sure to label the type of neutrino. *Hint:* The positive muon is an antiparticle.
20. Use Table 39.1 to find the total strangeness before and after the decay $\Lambda^0 \rightarrow \pi^- + p$, and use your answer to determine the force involved in this reaction.
21. The $\eta^0$ particle is a neutral nonstrange meson that can decay to a positive, a negative, and a neutral pion. Write the reaction for this decay, and verify that it conserves charge, baryon number, and strangeness.
22. Are either or both of these decay schemes possible for the tau particle: (a) $\tau^- \rightarrow e^- + \overline{v}_e + v_\tau$, (b) $\tau^- \rightarrow \pi^- + \pi^0 + v_\tau$?
23. Is the interaction $p + p \rightarrow p + \pi^+$ allowed? If not, what conservation law does it violate?

### Section 39.3 Quarks and the Standard Model

24. Determine the quark composition of the $\pi^-$.
25. The Eightfold Way led Gell-Mann to predict a baryon with strangeness $-3$. Determine this particle's quark composition.
26. The $\Sigma^+$ and $\Sigma^-$ have quark compositions *uus* and *dds*, respectively. Are the $\Sigma^+$ and $\Sigma^-$ each other's antiparticles? If not, give the quark compositions of their antiparticles.

### Section 39.4 Unification

27. Estimate the volume of the 50,000 tons of water used in the Super Kamiokande experiment shown in Fig. 39.8.
28. Estimate the temperature in a gas of particles such that the thermal energy $kT$ is high enough to make electromagnetism and the weak force appear as a single phenomenon.
29. Repeat Exercise 28 for the $10^{15}$-GeV energy of grand unification.

### Section 39.5 The Evolving Universe

30. Express the Hubble constant in SI units.
31. Find the distance to a galaxy that is receding from us at $2 \times 10^4$ km/s.

32. What is the recession speed of a galaxy 300 Mly from Earth?
33. What would be the age of the universe, assuming constant expansion rate, if the Hubble constant were 25 km/s/Mly?

## Problems

34. The mass of the photon is assumed to be zero, but experiments put only an upper limit of $5 \times 10^{-63}$ kg on the photon mass. What would the range of the electromagnetic force be if the photon mass were actually at this upper limit?
35. Which of the reactions (a) $\Lambda^0 \rightarrow \pi^+ + \pi^-$ and (b) $K^0 \rightarrow \pi^+ + \pi^-$ is not possible, and why not?
36. Both the neutral kaon and the neutral $\rho$ meson can decay to a pion–antipion pair. Which of these decays is mediated by the weak force? How can you tell?
37. Grand unification theories suggest that the decay $p \rightarrow \pi^0 + e^+$ may be possible, in which case all matter may eventually become radiation. Are (a) baryon number and (b) electric charge conserved in this hypothetical proton decay?
38. Consider systems described by wave functions that are proportional to the terms (a) $xy^2z$, (b) $x^2yz$, and (c) $xyz$, where $x$, $y$, and $z$ are the spatial coordinates. Which pairs of these systems could be transformed into each other under an interaction obeying conservation of parity?
39. The $J/\psi$ particle is an uncharmed meson that nevertheless includes charmed quarks. Determine its quark composition.
40. List all the possible quark triplets formed from any combination of the up, down, and charmed quarks, along with the charge of each.
41. The Tevatron at Fermilab accelerates protons to energy of 1 TeV. (a) How much is this in joules? (b) How far would a 1-g mass have to fall in Earth's gravitational field to gain this much energy?
42. (a) What is the relativistic factor $\gamma$ for a 7-TeV proton in the Large Hadron Collider? (b) Find an accurate value for the proton's speed.
43. How long, as measured in the lab frame, does it take a 7-TeV proton to circle the 27-km circumference of the Large Hadron Collider?
44. Estimate the critical density of the universe.
45. Estimate the diameter to which the Sun would have to be expanded for its average density to be the critical density.
46. A baryon called the neutral lambda particle has mass 1116 MeV/$c^2$. Find the minimum speed necessary for the particles in a proton–antiproton collider to produce lambda–antilambda pairs.

47. A so-called muonic atom is a hydrogen atom with the electron replaced by a muon; the muon's mass is 207 times the electron's. Find (a) the size and (b) the ground-state energy of a muonic atom.

48. (a) By what factor must the magnetic field in a proton synchrotron be increased as the proton energy increases by a factor of 10? Assume the protons are highly relativistic, so $\gamma \gg 1$. (b) By what factor must the diameter of the accelerator be increased to raise the energy by a factor of 10 without changing the magnetic field?

49. A galaxy's hydrogen-$\beta$ spectral line, normally at 486.1 nm, appears at 495.4 nm. (a) Use the Doppler shift of Chapter 14 to find the galaxy's recession speed, and (b) infer the distance to the galaxy. Are you justified in using Chapter 14's nonrelativistic Doppler formulas? Use $H_0 = 22$ km/s/Mly.

50. At the time the cosmic microwave background radiation originated, the temperature of the universe was about 3000 K. What were (a) the median wavelength (Equation 34.2b) of the newly formed radiation and (b) the corresponding photon energy?

51. Many particles are far too short-lived for their lifetimes to be measured directly. Instead, tables of particle properties often list "width," measured in energy units and indicating the width of distribution of measured rest energies for the particle. For example, the $Z^0$ has a mass of 91.18 GeV and a width of 2.5 GeV. Use the energy–time uncertainty relation to estimate the corresponding lifetime.

52. A mix of particles starts with equal numbers of the three types of sigma particles listed in Table 39.1. Find the relative portion of each after (a) $5 \times 10^{-20}$ s and (b) $5 \times 10^{-10}$ s. Give your answer in a reference frame in which the particles are at rest.

53. You pick up an older astronomy book and read that the Hubble constant is 17 km/s/Mly. You believe astronomers now think it's larger, which implies that the universe is younger than previously thought. Curious, you calculate the age of the universe for that older value of $H_0$. Looking on the Internet, you find the current number for $H_0$ is about 22 km/s/Mly. Compare the ages associated with your two values for $H_0$.

54. A friend believes the universe is very young, perhaps less than a million years old. Based on Hubble's law, how would you argue that the universe is older? What would the Hubble constant be for a 1-million-year-old universe?

55. You're contemplating cosmology. The universe may continue to expand for a very, very long time. You wonder if there will be people around when the universe is 60 Gy old. If so, what will they measure for the Hubble constant?

56. You're studying the history of cosmology. You see a photo of Arno Penzias and Robert Wilson standing in front of the antenna with which they discovered the cosmic background radiation. Your roommate comments on the strange shape of the antenna, and you explain it was a microwave antenna. "Why microwaves?" your roommate asks. Answer by finding the median wavelength for a blackbody at the appropriate temperature (see Section 34.2).

# Answer to Chapter Question

## Answer to Chapter Opening Question

RHIC creates conditions similar to those in the first microseconds of the universe, providing physicists a direct look at the behavior of matter in these early times.

# Modern Physics

**Modern physics**, developed since the year 1900, contrasts with the **classical physics** that came before. Modern physics is essential in understanding physical reality at the atomic scale, at very low temperatures, at very high relative velocities, in regions of very strong gravity, and in the evolution and large-scale structure of the universe.

The two big ideas in modern physics are **relativity** and **quantum mechanics**. Relativity is based on a simple principle but alters our commonsense notions of space and time, matter and energy, and the nature of gravity. Quantum mechanics replaces Newtonian determinism with a statistical description in which matter and energy exhibit both wave-like and particle-like behaviors.

Einstein's **special theory of relativity** is based in the statement that the laws of physics are the same for all observers in uniform motion. Therefore Maxwell's prediction that there should be electromagnetic waves propagating at the speed of light $c$ is valid in all uniformly moving reference frames. So measures of space and time cannot be absolute, but depend on one's reference frame.

The time between two events is shortest in a reference frame where events occur at the same place; here that's the reference frame of clock $C$.

An object's length is longest in a reference frame in which it's at rest.

**General relativity** is Einstein's theory of gravity, which explains gravity as the geometric curvature of spacetime. General relativity is at the foundations of modern astrophysics and cosmology, describing phenomena from black holes to the overall structure of the universe.

Energy $E$, momentum $p$, and mass $m$ in relativity are related by $E^2 = p^2c^2 + (mc^2)^2$.

For an object at rest with respect to an observer, this gives $E = mc^2$, showing the relativistic equivalence of mass and energy.

**Quantum physics** arose from attempts to explain several phenomena observed around the turn of the 20th century. These include blackbody radiation from hot objects, the photoelectric effect, and the existence and spectra of atoms.

Violet        Increasing wavelength →        Red

Hydrogen spectrum

Wave–particle duality is at the heart of quantum physics. The energy in electromagnetic radiation of frequency $f$ is concentrated in particle-like "bundles" called photons. Thus electromagnetic energy is quantized, with each photon carrying energy $E = hf$, where $h = 6.63 \times 10^{-34}$ J·s is Planck's constant. Conversely, matter exhibits wave-like behavior. The de Broglie wavelength associated with a particle of momentum $p$ is $\lambda = h/p$.

Quantization of atomic angular momentum leads to the Bohr model for the atom, with quantized energy levels that help explain atomic spectra.

Hydrogen energy levels

**Quantum mechanics** describes phenomena at the atomic scale. The **Schrödinger equation** gives the **wave function**, $\psi$, for a particle of mass $m$ with potential energy $U$ and total energy $E$:

$$-\frac{\hbar^2}{2m}\frac{d^2\psi(x)}{dx^2} + U(x)\psi = E\psi(x)$$

$\psi^2$ is the **probability density** for finding the particle. Application of the Schrödinger equation to bound systems results in quantized energy levels.

Infinite square well        Harmonic oscillator

Application of the Schrödinger equation to atoms and molecules explains atomic and molecular structure, the organization of the chemical elements in the periodic table, and the behavior of crystalline solids.

Wave–particle duality is closely related to the Heisenberg uncertainty principle, which states that it's impossible to measure a particle's position and momentum simultaneously with perfect accuracy. Rather, the uncertainties $\Delta x$ and $\Delta p$ in position and momentum must obey the inequality

$$\Delta x \, \Delta p \geq \hbar \text{ (uncertainly principle), where } \hbar = h/2\pi$$

**Nuclear physics** plunges into the heart of the atom, and shows that larger nuclei require more neutrons than protons in order for the strong nuclear force to overcome the repulsive electric interaction between protons. Nuclear physics describes such phenomena as radioactivity and the production of energy by nuclear fission and fusion.

Applying the principles of physics from the subatomic scale of quantum and particle physics to the largest scales described by general relativity gives us our modern-day picture of the origin and evolution of the universe.

The cosmic microwave background radiation, pictured here in an image from the WMAP satellite.

**Part Six challenge problem**

Derive Equation 39.2 for the critical density of the universe on the assumption that the universe is spherically symmetric, is homogeneous, and obeys Newton's laws on the large scale.

# Mathematics

## A-1  Algebra and Trigonometry

### Quadratic Formula

If $ax^2 + bx + c = 0$, then $x = \dfrac{-b \pm \sqrt{b^2 - 4ac}}{2a}$.

### Circumference, Area, Volume

Where $\pi \approx 3.14159 \ldots$

| | |
|---|---|
| circumference of circle | $2\pi r$ |
| area of circle | $\pi r^2$ |
| surface area of sphere | $4\pi r^2$ |
| volume of sphere | $\frac{4}{3}\pi r^3$ |
| area of triangle | $\frac{1}{2}bh$ |
| volume of cylinder | $\pi r^2 l$ |

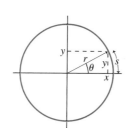

### Trigonometry

definition of angle (in radians): $\theta = \dfrac{s}{r}$

$2\pi$ radians in complete circle

1 radian $\approx 57.3°$

### Trigonometric Functions

$\sin\theta = \dfrac{y}{r}$

$\cos\theta = \dfrac{x}{r}$

$\tan\theta = \dfrac{\sin\theta}{\cos\theta} = \dfrac{y}{x}$

Values at Selected Angles

| $\theta \rightarrow$ | 0 | $\dfrac{\pi}{6}$ (30°) | $\dfrac{\pi}{4}$ (45°) | $\dfrac{\pi}{3}$ (60°) | $\dfrac{\pi}{2}$ (90°) |
|---|---|---|---|---|---|
| $\sin\theta$ | 0 | $\dfrac{1}{2}$ | $\dfrac{\sqrt{2}}{2}$ | $\dfrac{\sqrt{3}}{2}$ | 1 |
| $\cos\theta$ | 1 | $\dfrac{\sqrt{3}}{2}$ | $\dfrac{\sqrt{2}}{2}$ | $\dfrac{1}{2}$ | 0 |
| $\tan\theta$ | 0 | $\dfrac{\sqrt{3}}{3}$ | 1 | $\sqrt{3}$ | $\infty$ |

## Graphs of Trigonometric Functions

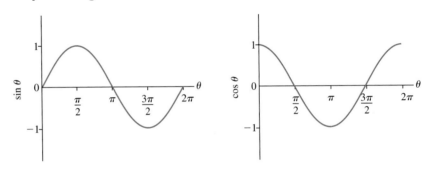

## Trigonometric Identities

$$\sin(-\theta) = -\sin\theta$$

$$\cos(-\theta) = \cos\theta$$

$$\sin\left(\theta \pm \frac{\pi}{2}\right) = \pm\cos\theta$$

$$\cos\left(\theta \pm \frac{\pi}{2}\right) = \mp\sin\theta$$

$$\sin^2\theta + \cos^2\theta = 1$$

$$\sin 2\theta = 2\sin\theta\cos\theta$$

$$\cos 2\theta = \cos^2\theta - \sin^2\theta = 1 - 2\sin^2\theta = 2\cos^2\theta - 1$$

$$\sin(\alpha \pm \beta) = \sin\alpha\cos\beta \pm \cos\alpha\sin\beta$$

$$\cos(\alpha \pm \beta) = \cos\alpha\cos\beta \mp \sin\alpha\sin\beta$$

$$\sin\alpha \pm \sin\beta = 2\sin\left[\tfrac{1}{2}(\alpha \pm \beta)\right]\cos\left[\tfrac{1}{2}(\alpha \mp \beta)\right]$$

$$\cos\alpha + \cos\beta = 2\cos\left[\tfrac{1}{2}(\alpha + \beta)\right]\cos\left[\tfrac{1}{2}(\alpha - \beta)\right]$$

$$\cos\alpha - \cos\beta = -2\sin\left[\tfrac{1}{2}(\alpha + \beta)\right]\sin\left[\tfrac{1}{2}(\alpha - \beta)\right]$$

## Laws of Cosines and Sines

Where $A$, $B$, $C$ are the sides of an arbitrary triangle and $\alpha$, $\beta$, $\gamma$ the angles opposite those sides:

*Law of cosines*

$$C^2 = A^2 + B^2 - 2AB\cos\gamma$$

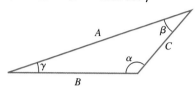

*Law of sines*

$$\frac{\sin\alpha}{A} = \frac{\sin\beta}{B} = \frac{\sin\gamma}{C}$$

## Exponentials and Logarithms

$$e^{\ln x} = x, \quad \ln e^x = x \quad e = 2.71828\ldots$$

$$a^x = e^{x\ln a} \qquad \ln(xy) = \ln x + \ln y$$

$$a^x a^y = a^{x+y} \qquad \ln\left(\frac{x}{y}\right) = \ln x - \ln y$$

$$(a^x)^y = a^{xy} \qquad \ln\left(\frac{1}{x}\right) = -\ln x$$

$$\log x \equiv \log_{10} x = \ln(10)\ln x \simeq 2.3\ln x$$

## Approximations

For $|x| \ll 1$, the following expressions provide good approximations to common functions:

$$e^x \simeq 1 + x$$

$$\sin x \simeq x$$

$$\cos x \simeq 1 - \tfrac{1}{2}x^2$$

$$\ln(1 + x) \simeq x$$

$$(1 + x)^p \simeq 1 + px \quad \text{(binomial approximation)}$$

Expressions that don't have the forms shown may often be put in the appropriate form. For example:

$$\frac{1}{\sqrt{a^2 + y^2}} = \frac{1}{a\sqrt{1 + \dfrac{y^2}{a^2}}} = \frac{1}{a}\left(1 + \frac{y^2}{a^2}\right)^{-1/2} \simeq \frac{1}{a}\left(1 - \frac{y^2}{2a^2}\right) \quad \text{for } y^2/a^2 \ll 1, \text{ or } y^2 \ll a^2.$$

## Vector Algebra

### Vector Products

$$\vec{A} \cdot \vec{B} = AB\cos\theta$$

$|\vec{A} \times \vec{B}| = AB\sin\theta$, with direction of $\vec{A} \times \vec{B}$ given by the right-hand rule:

## Unit Vector Notation

An arbitrary vector $\vec{A}$ may be written in terms of its components $A_x$, $A_y$, $A_z$ and the unit vectors $\hat{\imath}, \hat{\jmath}, \hat{k}$ that have length 1 and lie along the $x$, $y$, $z$ axes:

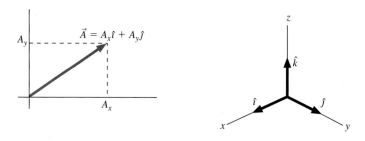

In unit vector notation, vector products become

$$\vec{A} \cdot \vec{B} = A_x B_x + A_y B_y + A_z B_z$$

$$\vec{A} \times \vec{B} = (A_y B_z - A_z B_y)\hat{\imath} + (A_z B_x - A_x B_z)\hat{\jmath} + (A_x B_y - A_y B_x)\hat{k}$$

## Vector Identities

$$\vec{A} \cdot \vec{B} = \vec{B} \cdot \vec{A}$$

$$\vec{A} \times \vec{B} = -\vec{B} \times \vec{A}$$

$$\vec{A} \cdot (\vec{B} \times \vec{C}) = \vec{B} \cdot (\vec{C} \times \vec{A}) = \vec{C} \cdot (\vec{A} \times \vec{B})$$

$$\vec{A} \times (\vec{B} \times \vec{C}) = (\vec{A} \cdot \vec{C})\vec{B} - (\vec{A} \cdot \vec{B})\vec{C}$$

# A-2 Calculus

## Derivatives

### Definition of the Derivative

If $y$ is a function of $x$, then the **derivative of $y$ with respect to $x$** is the ratio of the change $\Delta y$ in $y$ to the corresponding change $\Delta x$ in $x$, in the limit of arbitrarily small $\Delta x$:

$$\frac{dy}{dx} = \lim_{\Delta x \to 0} \frac{\Delta y}{\Delta x}$$

Algebraically, the derivative is the rate of change of $y$ with respect to $x$; geometrically, it is the slope of the $y$ versus $x$ graph—that is, of the tangent line to the graph at a given point:

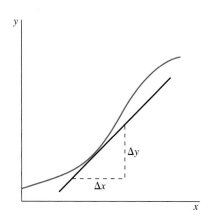

## Derivatives of Common Functions

$$\frac{da}{dx} = 0 \quad (a \text{ is a constant})$$

$$\frac{dx^n}{dx} = nx^{n-1} \quad (n \text{ need not be an integer})$$

$$\frac{d}{dx}\sin x = \cos x$$

$$\frac{d}{dx}\cos x = -\sin x$$

$$\frac{d}{dx}\tan x = \frac{1}{\cos^2 x}$$

$$\frac{de^x}{dx} = e^x$$

$$\frac{d}{dx}\ln x = \frac{1}{x}$$

## Derivatives of Sums, Products, and Functions of Functions

**1. Derivative of a constant times a function**

$$\frac{d}{dx}[af(x)] = a\frac{df}{dx} \quad (a \text{ is a constant})$$

**2. Derivative of a sum**

$$\frac{d}{dx}[f(x) + g(x)] = \frac{df}{dx} + \frac{dg}{dx}$$

**3. Derivative of a product**

$$\frac{d}{dx}[f(x)g(x)] = g\frac{df}{dx} + f\frac{dg}{dx}$$

*Examples*

$$\frac{d}{dx}(x^2\cos x) = \cos x\frac{dx^2}{dx} + x^2\frac{d}{dx}\cos x = 2x\cos x - x^2\sin x$$

$$\frac{d}{dx}(x\ln x) = \ln x\frac{dx}{dx} + x\frac{d}{dx}\ln x = (\ln x)(1) + x\left(\frac{1}{x}\right) = \ln x + 1$$

**4. Derivative of a quotient**

$$\frac{d}{dx}\left[\frac{f(x)}{g(x)}\right] = \frac{1}{g^2}\left(g\frac{df}{dx} - f\frac{dg}{dx}\right)$$

*Example*

$$\frac{d}{dx}\left(\frac{\sin x}{x^2}\right) = \frac{1}{x^4}\left(x^2\frac{d}{dx}\sin x - \sin x\frac{dx^2}{dx}\right) = \frac{\cos x}{x^2} - \frac{2\sin x}{x^3}$$

**5. Chain rule for derivatives**
If $f$ is a function of $u$ and $u$ is a function of $x$, then

$$\frac{df}{dx} = \frac{df}{du}\frac{du}{dx}$$

*Examples*

**a.** Evaluate $\dfrac{d}{dx}\sin(x^2)$. Here $u = x^2$ and $f(u) = \sin u$, so

$$\frac{d}{dx}\sin(x^2) = \frac{d}{du}\sin u\frac{du}{dx} = (\cos u)\frac{dx^2}{dx} = 2x\cos(x^2)$$

**b.** $\dfrac{d}{dt}\sin \omega t = \dfrac{d}{d\,\omega t}\sin \omega t\,\dfrac{d}{dt}\omega t = \omega\cos \omega t \quad (\omega \text{ is a constant})$

**c.** Evaluate $\dfrac{d}{dx}\sin^2 5x$. Here $u = \sin 5x$ and $f(u) = u^2$, so

$$\frac{d}{dx}\sin^2 5x = \frac{d}{du}u^2\frac{du}{dx} = 2u\frac{du}{dx} = 2\sin 5x\frac{d}{dx}\sin 5x$$

$$= (2)(\sin 5x)(5)(\cos 5x) = 10\sin 5x\cos 5x = 5\sin 2x$$

## Second Derivative

The second derivative of $y$ with respect to $x$ is defined as the derivative of the derivative:

$$\frac{d^2y}{dx^2} = \frac{d}{dx}\left(\frac{dy}{dx}\right)$$

*Example*

If $y = ax^3$, then $dy/dx = 3ax^2$, so

$$\frac{d^2y}{dx^2} = \frac{d}{dx}3ax^2 = 6ax$$

## Partial Derivatives

When a function depends on more than one variable, then the partial derivatives of that function are the derivatives with respect to each variable, taken with all other variables held constant. If $f$ is a function of $x$ and $y$, then the partial derivatives are written

$$\frac{\partial f}{\partial x} \quad\text{and}\quad \frac{\partial f}{\partial y}$$

*Example*

If $f(x, y) = x^3\sin y$, then

$$\frac{\partial f}{\partial x} = 3x^2\sin y \quad\text{and}\quad \frac{\partial f}{\partial y} = x^3\cos y$$

## Integrals

## Indefinite Integrals

Integration is the inverse of differentiation. The **indefinite integral**, $\int f(x)\,dx$, is defined as a function whose derivative is $f(x)$:

$$\frac{d}{dx}\left[\int f(x)\,dx\right] = f(x)$$

If $A(x)$ is an indefinite integral of $f(x)$, then because the derivative of a constant is zero, the function $A(x) + C$ is also an indefinite integral of $f(x)$, where $C$ is any constant. Inverting the derivatives of common functions listed in the preceding section gives the integrals that follow (a more extensive table appears at the end of this appendix).

$$\int a\,dx = ax + C \qquad\qquad \int \cos x\,dx = \sin x + C$$

$$\int x^n\,dx = \frac{x^{n+1}}{n+1} + C, \quad n \neq -1 \qquad \int e^x\,dx = e^x + C$$

$$\int \sin x\,dx = -\cos x + C \qquad\qquad \int x^{-1}\,dx = \ln x + C$$

## Definite Integrals

In physics we're most often interested in the **definite integral**, defined as the sum of a large number of very small quantities, in the limit as the number of quantities grows arbitrarily large and the size of each arbitrarily small:

$$\int_{x_1}^{x_2} f(x)\,dx \equiv \lim_{\substack{\Delta x \to 0 \\ N \to \infty}} \sum_{i=1}^{N} f(x_i)\,\Delta x$$

where the terms in the sum are evaluated at values $x_i$ between the limits of integration $x_1$ and $x_2$; in the limit $\Delta x \to 0$, the sum is over all values of $x$ in the interval.

The key to evaluating the definite integral is provided by the **fundamental theorem of calculus**. The theorem states that, if $A(x)$ is an *indefinite* integral of $f(x)$, then the *definite integral* is given by

$$\int_{x_1}^{x_2} f(x)\,dx = A(x_2) - A(x_1) \equiv A(x)\Big|_{x_1}^{x_2}$$

Geometrically, the definite integral is the area under the graph of $f(x)$ between the limits $x_1$ and $x_2$.

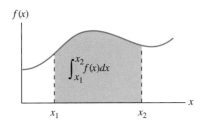

## Evaluating Integrals

The first step in evaluating an integral is to express all varying quantities within the integral in terms of a single variable; Chapter 9 outlines a general strategy for setting up an integral. Once you've set up an integral, you can evaluate it yourself or look it up in tables. Two common techniques can help you evaluate integrals or convert them to forms listed in tables:

1. **Change of variables**

   An unfamiliar integral can often be put into familiar form by defining a new variable. For example, it is not obvious how to integrate the expression

   $$\int \frac{x\,dx}{\sqrt{a^2 + x^2}}$$

   where $a$ is a constant. But let $z = a^2 + x^2$. Then

   $$\frac{dz}{dx} = \frac{da^2}{dx} + \frac{dx^2}{dx} = 0 + 2x = 2x$$

   so $dz = 2x\,dx$. Then the quantity $x\,dx$ in our unfamiliar integral is just $\frac{1}{2}dz$, while the quantity $\sqrt{a^2 + x^2}$ is just $z^{1/2}$. So the integral becomes

   $$\int \frac{1}{2} z^{-1/2}\,dz = \frac{\frac{1}{2}z^{1/2}}{1/2} = \sqrt{z}$$

   where we have used the standard form for the integral of a power of the independent variable. Substituting back $z = a^2 + x^2$ gives

   $$\int \frac{x\,dx}{\sqrt{a^2 + x^2}} = \sqrt{a^2 + x^2}$$

## 2. Integration by parts

The quantity $\int u\,dv$ is the area under the curve of $u$ as a function of $v$ between specified limits. In the figure, that area can also be expressed as the area of the rectangle shown minus the area under the curve of $v$ as a function of $u$. Mathematically, this relation among areas may be expressed as a relation among integrals:

$$\int u\,dv = uv - \int v\,du \qquad \text{(integration by parts)}$$

This expression may often be used to transform complicated integrals into simpler ones.

*Example*

Evaluate $\int x\cos x\,dx$. Here let $u = x$, so $du = dx$. Then $dv = \cos x\,dx$, so we have $v = \int dv = \int \cos x\,dx = \sin x$. Integrating by parts then gives

$$\int x\cos x\,dx = (x)(\sin x) - \int \sin x\,dx = x\sin x + \cos x$$

where the $+$ sign arises because $\int \sin x\,dx = -\cos x$.

## Table of Integrals

More extensive tables are available in many mathematical and scientific handbooks; see, for example, **Handbook of Chemistry and Physics** (Chemical Rubber Co.) or Dwight, **Tables of Integrals and Other Mathematical Data** (Macmillan). Some math software, including *Mathematica* and *Maple*, can also evaluate integrals symbolically. Wolfram Research provides *Mathematica*-based integration at *http://integrals.wolfram.com*.

In the expressions below, $a$ and $b$ are constants. An arbitrary constant of integration may be added to the right-hand side.

$$\int e^{ax}\,dx = \frac{e^{ax}}{a}$$

$$\int \frac{dx}{x^2 + a^2} = \frac{1}{a}\tan^{-1}\left(\frac{x}{a}\right)$$

$$\int \sin ax\,dx = -\frac{\cos ax}{a}$$

$$\int \frac{x\,dx}{\sqrt{a^2 - x^2}} = -\sqrt{a^2 - x^2}$$

$$\int \cos ax\,dx = \frac{\sin ax}{a}$$

$$\int \frac{x\,dx}{\sqrt{x^2 \pm a^2}} = \sqrt{x^2 \pm a^2}$$

$$\int \tan ax\,dx = -\frac{1}{a}\ln(\cos ax)$$

$$\int \frac{dx}{(x^2 \pm a^2)^{3/2}} = \frac{\pm x}{a^2\sqrt{x^2 \pm a^2}}$$

$$\int \sin^2 ax\,dx = \frac{x}{2} - \frac{\sin 2ax}{4a}$$

$$\int xe^{ax}\,dx = \frac{e^{ax}}{a^2}(ax - 1)$$

$$\int \cos^2 ax\,dx = \frac{x}{2} + \frac{\sin 2ax}{4a}$$

$$\int x^2 e^{ax}\,dx = \frac{x^2 e^{ax}}{a} - \frac{2}{a}\left[\frac{e^{ax}}{a^2}(ax - 1)\right]$$

$$\int x\sin ax\,dx = \frac{1}{a^2}\sin ax - \frac{1}{a}x\cos ax$$

$$\int \frac{dx}{a + bx} = \frac{1}{b}\ln(a + bx)$$

$$\int x\cos ax\,dx = \frac{1}{a^2}\cos ax + \frac{1}{a}x\sin ax$$

$$\int \frac{dx}{(a + bx)^2} = -\frac{1}{b(a + bx)}$$

$$\int \frac{dx}{\sqrt{a^2 - x^2}} = \sin^{-1}\left(\frac{x}{a}\right)$$

$$\int \ln ax\,dx = x\ln ax - x$$

$$\int \frac{dx}{\sqrt{x^2 \pm a^2}} = \ln\left(x + \sqrt{x^2 \pm a^2}\right)$$

# The International System of Units (SI)

This material is from the U.S. edition of the English translation of the seventh edition of "Le Système International d'Unités (SI)," the definitive publication in the French language issued in 1991 by the International Bureau of Weights and Measures (BIPM). The year the definition was adopted is given in parentheses.

**length (meter):** The meter is the length of the path traveled by light in vacuum during a time interval of 1/299 792 458 of a second. (1983)

**mass (kilogram):** The kilogram is equal to the mass of the international prototype of the kilogram. (1889)

**time (second):** The second is the duration of 9 192 631 770 periods of the radiation corresponding to the transition between the two hyperfine levels of the ground state of the cesium-133 atom. (1967)

**electric current (ampere):** The ampere is that constant current which, if maintained in two straight parallel conductors of infinite length, of negligible circular cross section, and placed 1 meter apart in vacuum, would produce between these conductors a force equal to $2 \times 10^{-7}$ newton per meter of length. (1948)

**temperature (kelvin):** The kelvin, unit of thermodynamic temperature, is the fraction 1/273.16 of the thermodynamic temperature of the triple point of water. (1967)

**amount of substance (mole):** The mole is the amount of substance of a system that contains as many elementary entities as there are atoms in 0.012 kilogram of carbon-12. (1971)

**luminous intensity (candela):** The candela is the luminous intensity, in a given direction, of a source that emits monochromatic radiation of frequency $540 \times 10^{12}$ hertz and that has a radiant intensity in that direction of $\left(1/683\right)$ watt per steradian. (1979)

## SI Base and Supplementary Units

| Quantity | SI Unit Name | SI Unit Symbol |
|---|---|---|
| **Base Unit** | | |
| Length | meter | m |
| Mass | kilogram | kg |
| Time | second | s |
| Electric current | ampere | A |
| Thermodynamic temperature | kelvin | K |
| Amount of substance | mole | mol |
| Luminous intensity | candela | cd |
| **Supplementary Units** | | |
| Plane angle | radian | rad |
| Solid angle | steradian | sr |

## SI Prefixes

| Factor | Prefix | Symbol |
|---|---|---|
| $10^{24}$ | yotta | Y |
| $10^{21}$ | zetta | Z |
| $10^{18}$ | exa | E |
| $10^{15}$ | peta | P |
| $10^{12}$ | tera | T |
| $10^{9}$ | giga | G |
| $10^{6}$ | mega | M |
| $10^{3}$ | kilo | k |
| $10^{2}$ | hecto | h |
| $10^{1}$ | deka | da |
| $10^{0}$ | — | — |
| $10^{-1}$ | deci | d |
| $10^{-2}$ | centi | c |
| $10^{-3}$ | milli | m |
| $10^{-6}$ | micro | $\mu$ |
| $10^{-9}$ | nano | n |
| $10^{-12}$ | pico | p |
| $10^{-15}$ | femto | f |
| $10^{-18}$ | atto | a |
| $10^{-21}$ | zepto | z |
| $10^{-24}$ | yocto | y |

Some SI Derived Units with Special Names

| Quantity | Name | Symbol | Expression in Terms of Other Units | Expression in Terms of SI Base Units |
|---|---|---|---|---|
| Frequency | hertz | Hz | | $s^{-1}$ |
| Force | newton | N | | $m \cdot kg \cdot s^{-2}$ |
| Pressure, stress | pascal | Pa | $N/m^2$ | $m^{-1} \cdot kg \cdot s^{-2}$ |
| Energy, work, heat | joule | J | $N \cdot m$ | $m^2 \cdot kg \cdot s^{-2}$ |
| Power | watt | W | $J/s$ | $m^2 \cdot kg \cdot s^{-3}$ |
| Electric charge | coulomb | C | | $s \cdot A$ |
| Electric potential, potential difference, electromotive force | volt | V | $J/C$ | $m^2 \cdot kg \cdot s^{-3} \cdot A^{-1}$ |
| Capacitance | farad | F | $C/V$ | $m^{-2} \cdot kg^{-1} \cdot s^4 \cdot A^2$ |
| Electric resistance | ohm | $\Omega$ | $V/A$ | $m^2 \cdot kg \cdot s^{-3} \cdot A^{-2}$ |
| Magnetic flux | weber | Wb | $V \cdot s$ | $m^2 \cdot kg \cdot s^{-2} \cdot A^{-1}$ |
| Magnetic field | tesla | T | $Wb/m^2$ | $kg \cdot s^{-2} \cdot A^{-1}$ |
| Inductance | henry | H | $Wb/A$ | $m^2 \cdot kg \cdot s^{-2} \cdot A^{-2}$ |
| Radioactivity | becquerel | Bq | 1 decay/s | $s^{-1}$ |
| Absorbed radiation dose | gray | Gy | J/kg, 100 rad | $m^2 \cdot s^{-2}$ |
| Radiation dose equivalent | sievert | Sv | J/kg, 100 rem | $m^2 \cdot s^{-2}$ |

# Conversion Factors

The listings below give the SI equivalents of non-SI units. To convert from the units shown to SI, multiply by the factor given; to convert the other way, divide. For conversions within the SI system, see the table of SI prefixes in Appendix B, Chapter 1, or the inside front cover. Conversions that are not exact by definition are given to, at most, four significant figures.

## Length

1 inch (in) = 0.0254 m

1 foot (ft) = 0.3048 m

1 yard (yd) = 0.9144 m

1 mile (mi) = 1609.34 m

1 nautical mile = 1852 m

1 angstrom (Å) = $10^{-10}$ m

1 light year (ly) = $9.46 \times 10^{15}$ m

1 astronomical unit (AU) = $1.5 \times 10^{11}$ m

1 parsec = $3.09 \times 10^{16}$ m

1 fermi = $10^{-15}$ m = 1 fm

## Mass

1 slug = 14.59 kg

1 metric ton (tonne; t) = 1000 kg

1 unified mass unit (u) = $1.661 \times 10^{-27}$ kg

Force units in the English system are sometimes used (incorrectly) for mass. The units given below are actually equal to the number of kilograms multiplied by $g$, the acceleration of gravity.

1 pound (lb) = weight of 0.454 kg

1 ton = 2000 lb = weight of 908 kg

1 ounce (oz) = weight of 0.02835 kg

## Time

1 minute (min) = 60 s

1 hour (h) = 60 min = 3600 s

1 day (d) = 24 h = 86 400 s

1 year (y) = 365.2422 d$^*$ = $3.156 \times 10^7$ s

## Area

1 hectare (ha) = $10^4$ m$^2$

1 square inch (in$^2$) = $6.452 \times 10^{-4}$ m$^2$

1 square foot (ft$^2$) = $9.290 \times 10^{-2}$ m$^2$

1 acre = 4047 m$^2$

1 barn = $10^{-28}$ m$^2$

1 shed = $10^{-30}$ m$^2$

## Volume

1 liter (L) = 1000 cm$^3$ = $10^{-3}$ m$^3$

1 cubic foot (ft$^3$) = $2.832 \times 10^{-2}$ m$^3$

1 cubic inch (in$^3$) = $1.639 \times 10^{-5}$ m$^3$

1 fluid ounce = 1/128 gal = $2.957 \times 10^{-5}$ m$^3$

1 barrel = 42 gal = 0.1590 m$^3$

1 gallon (U.S.; gal) = $3.785 \times 10^{-3}$ m$^3$

1 gallon (British) = $4.546 \times 10^{-3}$ m$^3$

## Angle, Phase

1 degree (°) = $\pi/180$ rad = $1.745 \times 10^{-2}$ rad

1 revolution (rev) = 360° = $2\pi$ rad

1 cycle = 360° = $2\pi$ rad

* The length of the year changes very slowly with changes in Earth's orbital period.

## Speed, Velocity

$1 \text{ km/h} = (1/3.6) \text{ m/s} = 0.2778 \text{ m/s}$      $1 \text{ ft/s} = 0.3048 \text{ m/s}$

$1 \text{ mi/h (mph)} = 0.4470 \text{ m/s}$      $1 \text{ ly/y} = 3.00 \times 10^8 \text{ m/s}$

## Angular Speed, Angular Velocity, Frequency, and Angular Frequency

$1 \text{ rev/s} = 2\pi \text{ rad/s} = 6.283 \text{ rad/s } (\text{s}^{-1})$      $1 \text{ rev/min (rpm)} = 0.1047 \text{ rad/s } (\text{s}^{-1})$

$1 \text{ Hz} = 1 \text{ cycle/s} = 2\pi \text{ s}^{-1}$

## Force

$1 \text{ dyne} = 10^{-5} \text{ N}$      $1 \text{ pound (lb)} = 4.448 \text{ N}$

## Pressure

$1 \text{ dyne/cm}^2 = 0.10 \text{ Pa}$      $1 \text{ lb/in}^2 \text{ (psi)} = 6.895 \times 10^3 \text{ Pa}$

$1 \text{ atmosphere (atm)} = 1.013 \times 10^5 \text{ Pa}$      $1 \text{ in H}_2\text{O (60°F)} = 248.8 \text{ Pa}$

$1 \text{ torr} = 1 \text{ mm Hg at } 0°\text{C} = 133.3 \text{ Pa}$      $1 \text{ in Hg (60°F)} = 3.377 \times 10^3 \text{ Pa}$

$1 \text{ bar} = 10^5 \text{ Pa} = 0.987 \text{ atm}$

## Energy, Work, Heat

$1 \text{ erg} = 10^{-7} \text{ J}$      $1 \text{ Btu}^* = 1.054 \times 10^3 \text{ J}$

$1 \text{ calorie}^* \text{ (cal)} = 4.184 \text{ J}$      $1 \text{ kWh} = 3.6 \times 10^6 \text{ J}$

$1 \text{ electronvolt (eV)} = 1.602 \times 10^{-19} \text{ J}$      $1 \text{ megaton (explosive yield; Mt)}$

$1 \text{ foot-pound (ft} \cdot \text{lb)} = 1.356 \text{ J}$      $= 4.18 \times 10^{15} \text{ J}$

## Power

$1 \text{ erg/s} = 10^{-7} \text{ W}$      $1 \text{ Btu/h (Btuh)} = 0.293 \text{ W}$

$1 \text{ horsepower (hp)} = 746 \text{ W}$      $1 \text{ ft} \cdot \text{lb/s} = 1.356 \text{ W}$

## Magnetic Field

$1 \text{ gauss (G)} = 10^{-4} \text{ T}$      $1 \text{ gamma } (\gamma) = 10^{-9} \text{ T}$

## Radiation

$1 \text{ curie (ci)} = 3.7 \times 10^{10} \text{ Bq}$      $1 \text{ rad} = 10^{-2} \text{ Gy}$

     $1 \text{ rem} = 10^{-2} \text{ Sv}$

### Energy Content of Fuels

| Energy Source | Energy Content |
|---|---|
| Coal | $29 \text{ MJ/kg} = 7300 \text{ kWh/ton} = 25 \times 10^6 \text{ Btu/ton}$ |
| Oil | $43 \text{ MJ/kg} = 39 \text{ kWh/gal} = 1.3 \times 10^5 \text{ Btu/gal}$ |
| Gasoline | $44 \text{ MJ/kg} = 36 \text{ kWh/gal} = 1.2 \times 10^5 \text{ Btu/gal}$ |
| Natural gas | $55 \text{ MJ/kg} = 30 \text{ kWh/100 ft}^3 = 1000 \text{ Btu/ft}^3$ |
| Uranium (fission) | |
|     Normal abundance | $5.8 \times 10^{11} \text{ J/kg} = 1.6 \times 10^5 \text{ kWh/kg}$ |
|     Pure U-235 | $8.2 \times 10^{13} \text{ J/kg} = 2.3 \times 10^7 \text{ kWh/kg}$ |
| Hydrogen (fusion) | |
|     Normal abundance | $7 \times 10^{11} \text{ J/kg} = 3.0 \times 10^4 \text{ kWh/kg}$ |
|     Pure deuterium | $3.3 \times 10^{14} \text{ J/kg} = 9.2 \times 10^7 \text{ kWh/kg}$ |
|     Water | $1.2 \times 10^{10} \text{ J/kg} = 1.3 \times 10^4 \text{ kWh/gal} = 340 \text{ gal gasoline/gal}$ |
| $H_2O$ | |
|     100% conversion, matter | |
|       to energy | $9.0 \times 10^{16} \text{ J/kg} = 931 \text{ MeV/u} = 2.5 \times 10^{10} \text{ kWh/kg}$ |

\* Values based on the thermochemical calorie; other definitions vary slightly.

# The Elements

The atomic weights of stable elements reflect the abundances of different isotopes; values given here apply to elements as they exist naturally on Earth. For stable elements, parentheses express uncertainties in the last decimal place given. For elements with no stable isotopes (indicated in **boldface**), at most three isotopes are given; for elements 99 and beyond, only the longest-lived isotope is given. (Exceptions are the unstable elements thorium, protactinium, and uranium, for which atomic weights reflect natural abundances of long-lived isotopes.) See also the periodic table inside the back cover.

| Atomic Number | Names | Symbol | Atomic Weight |
|---|---|---|---|
| 1 | Hydrogen | H | 1.00794 (7) |
| 2 | Helium | He | 4.002602 (2) |
| 3 | Lithium | Li | 6.941 (2) |
| 4 | Beryllium | Be | 9.012182 (3) |
| 5 | Boron | B | 10.811 (5) |
| 6 | Carbon | C | 12.011 (1) |
| 7 | Nitrogen | N | 14.00674 (7) |
| 8 | Oxygen | O | 15.9994 (3) |
| 9 | Fluorine | F | 18.9984032 (9) |
| 10 | Neon | Ne | 20.1797 (6) |
| 11 | Sodium (Natrium) | Na | 22.989768 (6) |
| 12 | Magnesium | Mg | 24.3050 (6) |
| 13 | Aluminum | Al | 26.981539 (5) |
| 14 | Silicon | Si | 28.0855 (3) |
| 15 | Phosphorus | P | 30.973762 (4) |
| 16 | Sulfur | S | 32.066 (6) |
| 17 | Chlorine | Cl | 35.4527 (9) |
| 18 | Argon | Ar | 39.948 (1) |
| 19 | Potassium (Kalium) | K | 39.0983 (1) |
| 20 | Calcium | Ca | 40.078 (4) |
| 21 | Scandium | Sc | 44.955910 (9) |
| 22 | Titanium | Ti | 47.88 (3) |
| 23 | Vanadium | V | 50.9415 (1) |
| 24 | Chromium | Cr | 51.9961 (6) |
| 25 | Manganese | Mn | 54.93805 (1) |
| 26 | Iron | Fe | 55.847 (3) |
| 27 | Cobalt | Co | 58.93320 (1) |
| 28 | Nickel | Ni | 58.69 (1) |
| 29 | Copper | Cu | 63.546 (3) |
| 30 | Zinc | Zn | 65.39 (2) |
| 31 | Gallium | Ga | 69.723 (1) |
| 32 | Germanium | Ge | 72.61 (2) |
| 33 | Arsenic | As | 74.92159 (2) |

*cont'd.*

| Atomic Number | Names | Symbol | Atomic Weight |
|---|---|---|---|
| 34 | Selenium | Se | 78.96 (3) |
| 35 | Bromine | Br | 79.904 (1) |
| 36 | Krypton | Kr | 83.80 (1) |
| 37 | Rubidium | Rb | 85.4678 (3) |
| 38 | Strontium | Sr | 87.62 (1) |
| 39 | Yttrium | Y | 88.90585 (2) |
| 40 | Zirconium | Zr | 91.224 (2) |
| 41 | Niobium | Nb | 92.90638 (2) |
| 42 | Molybdenum | Mo | 95.94 (1) |
| **43** | **Technetium** | **Tc** | **97, 98, 99** |
| 44 | Ruthenium | Ru | 101.07 (2) |
| 45 | Rhodium | Rh | 102.90550 (3) |
| 46 | Palladium | Pd | 106.42 (1) |
| 47 | Silver | Ag | 107.8682 (2) |
| 48 | Cadmium | Cd | 112.411 (8) |
| 49 | Indium | In | 114.82 (1) |
| 50 | Tin | Sn | 118.710 (7) |
| 51 | Antimony (Stibium) | Sb | 121.75 (3) |
| 52 | Tellurium | Te | 127.60 (3) |
| 53 | Iodine | I | 126.90447 (3) |
| 54 | Xenon | Xe | 131.29 (2) |
| 55 | Cesium | Cs | 132.90543 (5) |
| 56 | Barium | Ba | 137.327 (7) |
| 57 | Lanthanum | La | 138.9055 (2) |
| 58 | Cerium | Ce | 140.115 (4) |
| 59 | Praseodymium | Pr | 140.90765 (3) |
| 60 | Neodymium | Nd | 144.24 (3) |
| **61** | **Promethium** | **Pm** | **145, 147** |
| 62 | Samarium | Sm | 150.36 (3) |
| 63 | Europium | Eu | 151.965 (9) |
| 64 | Gadolinium | Gd | 157.25 (3) |
| 65 | Terbium | Tb | 158.92534 (3) |
| 66 | Dysprosium | Dy | 162.50 (3) |
| 67 | Holmium | Ho | 164.93032 (3) |
| 68 | Erbium | Er | 167.26 (3) |
| 69 | Thulium | Tm | 168.93421 (3) |
| 70 | Ytterbium | Yb | 173.04 (3) |
| 71 | Lutetium | Lu | 174.967 (1) |
| 72 | Hafnium | Hf | 178.49 (2) |
| 73 | Tantalum | Ta | 180.9479 (1) |
| 74 | Tungsten (Wolfram) | W | 183.85 (3) |
| 75 | Rhenium | Re | 186.207 (1) |
| 76 | Osmium | Os | 190.2 (1) |
| 77 | Iridium | Ir | 192.22 (3) |
| 78 | Platinum | Pt | 195.08 (3) |
| 79 | Gold | Au | 196.96654 (3) |
| 80 | Mercury | Hg | 200.59 (3) |
| 81 | Thallium | Tl | 204.3833 (2) |
| 82 | Lead | Pb | 207.2 (1) |
| 83 | Bismuth | Bi | 208.98037 (3) |

*cont'd.*

| Atomic Number | Names | Symbol | Atomic Weight |
|---|---|---|---|
| 84 | Polonium | Po | 209, 210 |
| 85 | Astatine | At | 210, 211 |
| 86 | Radon | Rn | 211, 220, 222 |
| 87 | Francium | Fr | 223 |
| 88 | Radium | Ra | 223, 224, 226 |
| 89 | Actinium | Ac | 227 |
| 90 | Thorium | Th | 232.0381 (1) |
| 91 | Protactinium | Pa | 231.03588 (2) |
| 92 | Uranium | U | 238.0289 (1) |
| 93 | Neptunium | Np | 237, 239 |
| 94 | Plutonium | Pu | 239, 242, 244 |
| 95 | Americium | Am | 241, 243 |
| 96 | Curium | Cm | 245, 247, 248 |
| 97 | Berkelium | Bk | 247, 249 |
| 98 | Californium | Cf | 249, 250, 251 |
| 99 | Einsteinium | Es | 252 |
| 100 | Fermium | Fm | 257 |
| 101 | Mendelevium | Md | 258 |
| 102 | Nobelium | No | 259 |
| 103 | Lawrencium | Lr | 262 |
| 104 | Rutherfordium | Rf | 263 |
| 105 | Dubnium | Db | 268 |
| 106 | Seaborgium | Sg | 266 |
| 107 | Bohrium | Bh | 272 |
| 108 | Hassium | Hs | 277 |
| 109 | Meitnerium | Mt | 276 |
| 110 | Darmstadtium | Ds | 281 |
| 111 | Roentgenium | Rg | 280 |
| 112 | — | — | 285 |
| 113 | — | — | 284 |
| 114 | — | — | 289 |
| 115 | — | — | 288 |
| 116 | — | — | 292 |

# Astrophysical Data

## Sun, Planets, Principal Satellites

| Body | Mass ($10^{24}$ kg) | Mean Radius ($10^6$ m Except as Noted) | Surface Gravity (m/s$^2$) | Escape Speed (km/s) | Sidereal Rotation Period[*] (days) | Mean Distance from Central Body[†] ($10^6$ km) | Orbital Period | Orbital Speed (km/s) |
|---|---|---|---|---|---|---|---|---|
| **Sun** | $1.99 \times 10^6$ | 696 | 274 | 618 | 36 at poles 27 at equator | $2.6 \times 10^{11}$ | 200 My | 250 |
| **Mercury** | 0.330 | 2.44 | 3.70 | 4.25 | 58.6 | 57.6 | 88.0 d | 48 |
| **Venus** | 4.87 | 6.05 | 8.87 | 10.4 | −243 | 108 | 225 d | 35 |
| **Earth** | 5.97 | 6.37 | 9.81 | 11.2 | 0.997 | 150 | 365.3 d | 30 |
| Moon | 0.0735 | 1.74 | 1.62 | 2.38 | 27.3 | 0.385 | 27.3 d | 1.0 |
| **Mars** | 0.642 | 3.38 | 3.74 | 5.03 | 1.03 | 228 | 1.88 y | 24.1 |
| Phobos | $9.6 \times 10^{-9}$ | 9–13 km | 0.001 | 0.008 | 0.32 | $9.4 \times 10^{-3}$ | 0.32 d | 2.1 |
| Deimos | $2 \times 10^{-9}$ | 5–8 km | 0.001 | 0.005 | 1.3 | $23 \times 10^{-3}$ | 1.3 d | 1.3 |
| **Jupiter** | $1.90 \times 10^3$ | 69.1 | 26.5 | 60.6 | 0.414 | 778 | 11.9 y | 13.0 |
| Io | 0.0888 | 1.82 | 1.8 | 2.6 | 1.77 | 0.422 | 1.77 d | 17 |
| Europa | 0.479 | 1.57 | 1.3 | 2.0 | 3.55 | 0.671 | 3.55 d | 14 |
| Ganymede | 0.148 | 2.63 | 1.4 | 2.7 | 7.15 | 1.07 | 7.15 d | 11 |
| Callisto | 0.107 | 2.40 | 1.2 | 2.4 | 16.7 | 1.88 | 16.7 d | 8.2 |
| and 13 smaller satellites | | | | | | | | |
| **Saturn** | 569 | 56.8 | 11.8 | 36.6 | 0.438 | $1.43 \times 10^3$ | 29.5 y | 9.65 |
| Tethys | 0.0007 | 0.53 | 0.2 | 0.4 | 1.89 | 0.294 | 1.89 d | 11.3 |
| Dione | 0.00015 | 0.56 | 0.3 | 0.6 | 2.74 | 0.377 | 2.74 d | 10.0 |
| Rhea | 0.0025 | 0.77 | 0.3 | 0.5 | 4.52 | 0.527 | 4.52 d | 8.5 |
| Titan | 0.135 | 2.58 | 1.4 | 2.6 | 15.9 | 1.22 | 15.9 d | 5.6 |
| and 12 smaller satellites | | | | | | | | |
| **Uranus** | 86.6 | 25.0 | 9.23 | 21.5 | −0.65 | $2.87 \times 10^3$ | 84.1 y | 6.79 |
| Ariel | 0.0013 | 0.58 | 0.3 | 0.4 | 2.52 | 0.19 | 2.52 d | 5.5 |
| Umbriel | 0.0013 | 0.59 | 0.3 | 0.4 | 4.14 | 0.27 | 4.14 d | 4.7 |
| Titania | 0.0018 | 0.81 | 0.2 | 0.5 | 8.70 | 0.44 | 8.70 d | 3.7 |
| Oberon | 0.0017 | 0.78 | 0.2 | 0.5 | 13.5 | 0.58 | 13.5 d | 3.1 |
| and 11 smaller satellites | | | | | | | | |
| **Neptune** | 103 | 24.0 | 11.9 | 23.9 | 0.768 | $4.50 \times 10^3$ | 165 y | 5.43 |
| Triton | 0.134 | 1.9 | 2.5 | 3.1 | 5.88 | 0.354 | 5.88 d | 4.4 |
| and 7 smaller satellites | | | | | | | | |
| **Pluto** | 0.015 | 1.2 | 0.4 | 1.2 | −6.39 | $5.91 \times 10^3$ | 249 y | 4.7 |
| Charon | 0.001 | 0.6 | | | −6.39 | 0.02 | 6.39 d | 0.2 |
| and 2 smaller satellites | | | | | | | | |

[*]Negative rotation period indicates retrograde motion, in opposite sense from orbital motion. Periods are sidereal, meaning the time for the body to return to the same orientation relative to the distant stars rather than the Sun.

[†]Central body is galactic center for Sun, Sun for planets, and planet for satellites.

# Answers to Odd-Numbered Exercises and Problems

**A-17**

## Chapter 1
11. 100,000 bigger
13. 0.108 782 775 7 ns
15. $10^8$
17. 0.62 rad
19. 28 g
21. $10^6$
23. 8.59 $m^2$/L
25. 3.6 km/h
27. 57.3°
29. $7.4 \times 10^6$ $m/s^2$
31. $4 \times 10^6$
33. 41 m
35. $\approx 5.20$
37. $2 \times 10^6$ cows
39. About 0.1%
41. $9 \times 10^{26}$
43. 0.025 cm
45. 7500
47. (a) $\approx \pm 5\%$  (b) $\pm 1\%$  (c) $\pm 0.5\%$
49. 42.195 km
51. 30 tuners in Chicago
53. \$8.12/lb., \$10.06/bag including shipping
55. About $^1/_4$ kg

## Chapter 2
13. 10.16 m/s
15. (a) 24 km north  (b) 9.6 km/h north
   (c) .16 km/h south  (d) 0  (e) 0
17. 7.84 m/s
19. 1 m/s = 2.24 mi/h
21. (a) 2.0 m/s  (b) 0  (c) $-5.0$ m/s
   (d) 1.2 m/s  (e) 0.17 m/s
23. (a) $b - 2ct$  (b) 8.4 s after launch
25. 0.354 $m/s^2$
27. While falling, $-9.82$ $m/s^2$. While
   stopping, $+84.0$ $m/s^2$.
29. 17.4 $m/s^2$
31. $v = \dfrac{dx}{dt} = \dfrac{d}{dt}\left(x_0 + v_0 t + \dfrac{1}{2}at^2\right) = v_0 + at$
33. (a) 46 $m/s^2$  (b) 61 s
35. $-26.9$ $ft/s^2$
37. 12 km/s
39. 94.9 m
41. (a) 123 m  (b) 39.2 m/s, 44.1 m
   (c) 9.8 m/s, 118 m  (d) 19.6 m/s, 103 m
43. 10.5 m/s
45. 48 mi/h
47. (a) 84 km/h  (b) 55 h
49. 2.6 h later; 1800 km from New York
51. $\bar{v} = bt^3$
53. (a) 28.1 m/s  (b) 22.5 $m/s^2$
   (c) 9.38 m/s  (d) 11.25 $m/s^2$

55. 55%
57. (a) 0.014 s  (b) 0.51 m
59. 886 m
61. (a) 24.8 m/s  (b) 179 m
63. $4.6 \times 10^{-3}$ $m/s^2$
65. 10.59 m/s
67. 273 m
69. (a) Diver A: $-7.88$ m/s; Diver B:
   $-7.67$ m/s  (b) Diver A hits first about
   0.162 s before Diver B.
71. 3.85 s; 6.24 m/s
73. 5.0 s; 17 km/h.
75. (a) $\bar{v} = L/t = \dfrac{1}{2}(v_1 + v_2)$

   (b) $\bar{v} = \dfrac{L}{t} = \dfrac{2v_1 v_2}{v_1 + v_2}$
77. 70.7%
79. $-0.309$ $m/s^2$
81. (a) $v(t) = v_0 + a_0 t + \dfrac{1}{2}bt^2$

   (b) $x(t) = x_0 + v_0 t + \dfrac{1}{2}a_0 t^2 + \dfrac{1}{6}bt^3$
83. (a) 9.1 m/s  (b) 9.1 m/s  (c) 0.18 s
85. 15 drops/s
87. Walking $\approx$933 hours, or 39 days;
   bicycling $\approx$180 hours, or 7.5 days;
   car $\approx$51 hours, or 2.1 days

## Chapter 3
17. 266m, 34° N of W
19. 702 km, 21.3° W of N
21. $C = 5.0$ m; 127° clockwise from $\vec{A}$
23. 36.4 m; 20.9°
25. $\vec{a} = 3.0\hat{i} + 15\hat{j}$ $m/s^2$
27. (a) 5.4 mi at 32° E of N
   (b) 15 mi/h at 32° E of N
29. Southwest
31. 0°
33. $(259\ km/h)\hat{i} + (65.9\ km/h)\hat{j}$
35. 196 km/h
37. 6.0 $m/s^2$ at 53° below $x$ axis
39. 0.728 s for both
41. 1.54 m
43. $3.44 \times 10^6$ m/s
45. 60.7 mi/h
47. 12 h
49. The two possible solutions are $\theta_B = 117°$,
   or 243°.
51. 0.869 $\mu$m/s
53. 5.70 $m/s^2$
55. (a) 0.32 cm/s  (b) $3.36 \times 10^{-2}$ $cm/s^2$
   (c) 90°

57. They are equal.
59. 0.5 $m/s^2$
61. 5.7 m/s
63. 8.32 m/s; 61°
65. 30° = 5.10 s; 60° = 8.84 s
69. 70.4 km/h; the motorcyclist was probably
   speeding.
71. 31.2° or 65.7°
73. 89 m/s
75. 90 km
77. 7.2 m/s at 77° to horizontal
79. 38° below the $x$ axis
81. $\sqrt{2gh + gR^2/2h}$
83. The rocket will produce a change in
   velocity corresponding to a 24.2° change
   in angle. However, the displacement is
   only 1008 km and is not enough to avoid
   collision, unless the asteroid "coasts" for
   additional time after the rocket ceases
   firing.
85. Including air resistance, the optimum
   angle is less than 45 degrees. The density
   of air is less and thus the air resistance is
   less, when the atmospheric pressure is
   lower (higher altitude, as in Denver) or
   the temperature is higher.

## Chapter 4
13. 3.75 MN
15. $1.53 \times 10^3$ kg
17. Quadruples
19. 30.7 kN
21. 7.72 cm
23. Venus
25. (a) 3.33 N  (b) 12.0 oz
27. 620 N
29. $-1.76$ kN
31. (a) 4.41 MN  (b) 4.91 MN
33. Apparent weight is 55% of actual
   weight
35. $1.3 \times 10^{-23}$ m
37. 15.9 cm
39. 5.77N at 72.2° to the $x$ axis
41. (a) 725 N  (b) 725 N  (c) 725 N
   (d) 851 N  (e) 599 N
43. 0.531 s
45. 6 N
47. (a) 5.26 kN  (b) 1.08 kN  (c) 494 N
   (d) 589 N
49. 830 g
51. 4.29 cm

53. (a) 42.1 kN (b) 55.9 kN (c) 55.9 kN
(d) 42.1 kN (e) 28.4 kN
55. $n \le 9$
59. 900 N
61. 14.9 kN
63. 0.30 MN
65. (a) $a = 1.0$ m/s² upward
(b) 756 N
67. $F_{net} = \dfrac{ma}{(1 - u^2/c^2)^{3/2}}$
69. 20.5 g

## Chapter 5
13. $(4.02 \text{ N})\hat{\imath} + (1.66 \text{ N})\hat{\jmath}$
15. 9.40°
17. 880 N
19. Right-hand mass 2.5 times left-hand mass
21. (a) 3.86 m/s² (b) 533 N
25. 132 m
27. 493 km/h
29. 0.18
31. 0.12
33. 43 cm
35. 98 N in horizontal string, 139 N in the other string
37. (a) $T = m_2 g$ (b) $\tau = 2\pi \sqrt{\dfrac{m_1 R}{m_2 g}}$
39. (a) Horizontal: 357 N; vertical: 441 N
(b) 39°
41. 2.89 m/s
43. 342 N
45. 11.4 m/s²
47. 2.68 cm
49. 107 m
51. 59.3 mi/h
55. 0.46
59. (a) lower at equator, where a net downward force is needed to keep you accelerating with Earth's rotation
(b) 0.34%
63. (a) 0.6 $mg$ (b) 37°
65. $v_y = -(mg/b)(1 - e^{-bt/m})$
67. $T_0 = T_s \cos\theta_s$, $mg = 2T_s \sin\theta_s$,
$d^2y/dx^2 = (m/LT_0)(1 + (dy/dx)^2)^{1/2}$
69. 11.5 m/s, considerably slower than a bullet
71. 7.64 km

## Chapter 6
13. 494 J
15. 9.6 MJ
17. 1.1 $mgh$
19. (a) 86.9 u² (b) 20.3 u²
21. 1.9 m
23. (a) 1.0 J (b) 3.0 J
25. 29.9 cm
27. $3.7 \times 10^{-13}$ J
29. 2.25 kJ
31. (a) 23.5 J (b) 17.7 m/s
33. 97 W
35. (a) 60 kW (b) 1 kW (c) 42 W
37. 9.4 MJ
39. 430 W

41. 22 s
43. (a) 400 J (b) 31 kg
45. 4.4 MN
47. (a) all three are equal to 1
(b) all three are equal to 0
(c) $\vec{A} \cdot \vec{B} = A_x B_x + A_y B_y + A_z B_z$
49. 65.8°, 49.6°, 115°
51. (a) 360 J (b) 350 J (c) 357.5 J
(d) 359.375 J
53. (a) 33 J (b) 60 J (c) 78 J
57. 22.9 m
59. 220 J
61. 0.634 kg/s, 10.1 gal/min
63. 2.1 MJ
65. 7.7°
67. (a) 450 W (b) 8.0 kJ
69. 42 kJ
$\dfrac{1}{3}(18W/s^2)\big[(20s)^3 - (10s)^3\big] = 42$ kJ
75. (a) 33 J (b) 167 J
77. 135 J
79. 8.7 MJ, not efficient
81. The floor exerts an average force on the leg of 2.82 kN, or 36 times its weight.

## Chapter 7
11. (a) $-2\,\mu mgL$, (b) $-\sqrt{2}\,\mu mgL$
13. (a) 60.4 kJ (b) 109 kJ (c) $U_1 = 0$
15. $-2.24$ MJ
17. 842 m
19. 55 cm
21. 50 m/s (180 km/h)
23. 84 cm
25. (a) 4.85 m/s (b) 7.00 m/s (c) 11 m
27. (a) $-2$ N (b) 0 (c) 8 N (d) 1 N
(e) $-4$ N (f) 0
29. (a) $4.2 \times 10^{13}$ J (b) 11 h
31. $U = mgy$
33. (a) 1.07 J (b) 1.12 J
35. 778 J, 4.9%
37. (a) $U(x_2) - U(x_1) = A\left(\dfrac{1}{x_2} - \dfrac{1}{x_1}\right)$
39. 95.3 m
41. 4.73 m/s
43. (a) 16 m/s (b) 29%
45. 12.9 km/h
47. 1.59 m
49. 4.22 m and 0.488 m
51. 2.82 Å
53. 20.1 m/s, 30.4 m/s
55. 1.39 m
57. 62.5 cm from left end of frictional zone
59. 2.87 m
61. 14 m
63. $v = \sqrt{2K/m} = \sqrt{2(U_0 - U/m)}$
$= 2(a/3m)^{1/2} x^{3/4}$
65. (a) Yes, 1.60 J (b) No, 80.9 cm
67. $2.14 \times 10^{-30}$ J $\cdot$ m

69. $P = \dfrac{W}{t} = \dfrac{\frac{1}{2}mv^2 + mgd}{2\dfrac{d}{\sqrt{2g(h-d)}}}$
$= \dfrac{\frac{1}{2}m(v^2 + 2gd)\sqrt{2g(h-d)}}{2d}$
$= \dfrac{m(v^2 + 2gd)\sqrt{2g(h-d)}}{4d}$

## Chapter 8
13. $2.73 \times 10^{-3}$ m/s²; acceleration due to Earth's gravity at $3.85 \times 10^8$ m
15. (a) 3.70 m/s² (b) 1.35 m/s²
17. 46 nN
19. $\approx 1690$ km
21. 1.87 y
23. 109 min
25. $1.77 \times 10^{32}$ J
27. 1670 km
29. 58 MJ
31. 7.7 Mm
33. $8.85 \times 10^5$ m
35. $0.414 R_p$
39. (a) $4.56 \times 10^6$ m/s, or about 1.5% the speed of light (b) 8.77 s
41. $(T_A/T_B) = (r_A/r_B)^{3/2} = 2.83$
43. 616 km/s
45. $R_E/99 = 64$ km; underestimate
47. $\sqrt{2}$
49. (a) 11.2 km/s (b) 9.74 km/s
(c) no
51. $v = \sqrt{2GM\left(\dfrac{3}{R} + \dfrac{1}{r}\right)}$
53. 15 km/s, 23 km/s
55. 7.96 km/s
57. (a) 0.601 AU (b) $5.3 \times 10^{11}$ J
(c) 38.4 km/s
63. The ratio is about 0.46 for the tide-producing force, but about 180 for the gravitational force.
65. R $<$ 2.9 km
67. The distance from the center of the Earth $3.47 \times 10^8$ m. The distance from the surface of the Earth is $3.40 \times 10^8$ m

## Chapter 9
13. 2m
15. $\sqrt{3}\,L/6$ along any altitude
17. 2.1 m from the stern
19. $-(0.6 \text{ m/s})\hat{\imath} - (0.8 \text{ m/s})\hat{\jmath}$
21. 8.4 km/h
23. 1.23 J
25. $K_{int} = 3K_0$, $K_{cm} = K_0$
29. 3875 kg
31. 46 m/s
33. $v_{1f} = -11$ Mm/s; $v_{2f} = 6.9$ Mm/s
35. 0.115$a$ above the vertex of the missing trangle
37. 72.2 cm
39. A distance 1/4$h$ from the base along the axis of symmetry

41. (a) 99 cm   (b) 3.9 m/s

43. (a) $a_{x0} = \dfrac{v_0}{M}\left(\dfrac{dm}{dt}\right)$   (b) $v_c = v_0$

45. $-17.4°$

47. $-(46.7 \text{ m/s})\hat{\imath} + (67.7 \text{ m/s})\hat{\jmath}$

49. If the two masses are equal, 60°.

51. 3.0 mN

53. 35.2 m/s

55. 91.5 cm/s

59. 120°

61. $m_1 = (3 \pm \sqrt{8})m_2 = 5.83m_2$
   or $(5.83)^{-1}m_2$

63. Start with Equation 9.14 and solve for $v_{1f}$: $v_{1f} = v_{2f} - v_{1i} + v_{2i}$. Plug this into Equation 9.12a: $m_1 v_{1i} + m_2 v_{2i} = m_1(v_{2f} - v_{1i} + v_{2i}) + m_2 v_{2f}$. Solve for $v_{2f}$.

65. 18.6%

67. $v = 658$ m/s, $\theta_x = 11.5°$

71. 7.95 s

73. $\dfrac{K_{2f}}{K_{1i}} = \dfrac{4(m_2/m_1)}{(1 + m_2/m_1)^2}$

75. Along the bisector, a distance $\dfrac{4R}{3\theta}\sin\left(\dfrac{\theta}{2}\right)$ from the center

77. $v_{\text{PrimitiveEarth}} = 32$ km/s, $v_{\text{Earth}} = 30$ km/s, so about an 8% change

79. $v_c = 38.7$ mi/hr, $v_t = 52.3$ mi/hr. Both are speeding.

81. $1.83 \times 10^{13}$ J

# Chapter 10

15. (a) 463 m/s   (b) $(463 \text{ m/s})\cos\theta$

17. 45.8 m/s

19. (a) 12.1 min   (b) $2.17 \times 10^4$ rev

21. 160 N·m

23. (a) 146 N   (b) 155 N

25. 0.072 N·m

27. $6.14 \times 10^2$ kg·m²

29. (a) 36.9 kg   (b) If not all the mass of the wheel is concentrated at the rim, the total mass is greater than this minimum.

31. (a) $9.69 \times 10^{37}$ kg·m²
   (b) $2.59 \times 10^{19}$ N·m

33. (a) $1.17 \times 10^{-3}$ kg·m²   (b) 1.24 N·m

35. (a) 446 J   (b) 139 W

37. $\approx 0.089\%$

39. (a) 160 MJ   (b) 16 MW

41. 1/3

43. (a) 65.7 rpm   (b) 3.65 s

45. (a) $21.7$ s$^{-1}$, 207 rpm
   (b) $34.7$ s$^{-1}$, 331 rpm

47. $-1.30$ s$^{-2}$

49. 0.147 N·m

53. (a) 15.6 kg·m²   (b) 0.303 s

55. (a) 7.21 h   (b) $\cong 1950$ rev

57. 0.36

59. $2.09$ s$^{-1}$

61. 7.03 m/s

63. 17.1%

65. $0.498\,MR^2$

67. 32.9 m

69. (a) $M = \frac{2}{3}\rho_0 \pi R^2 w$;   (b) $I = \frac{3}{5}MR^2$

71. The object is in the $x$-$y$ plane. Add these,

$$I_y = \int x^2\, dm \qquad I_x = \int y^2\, dm \qquad x^2 + y^2 = z^2$$

But $I_x + I_y = \displaystyle\int (x^2 + y^2)\, dm$

$$I_x + I_y = \int z^2\, dm = I_z$$

73. $\frac{3}{10}MR^2$

75. 184 N

77. $F > 10.6 \times 10^3$ N

# Chapter 11

13. 3.55 rad/s² in the southeast direction

15. 18.8° west of north

17. (a) $-\hat{k}$   (b) $\hat{k}$   (c) $-\hat{\jmath} + \hat{\imath}$

19. (a) $8.1\hat{k}$ N·m   (b) $14.7\hat{k}$ N·m

21. 4.14 J·s

23. 2.31 J·s

25. 17.4 rpm

27. 2.5 d

29. 9.2 rpm

31. $-120°$ along the $x$ axis

35. 36.9 J·s

37. $2.70 \times 10^5$ J·s, direction out of the plane of Fig. 11.13

39. 0.206 kg·m²

41. 0.630, or 63% of a full circle relative to the ground

43. (a) $0.537$ s$^{-1}$   (b) 6.44 m/s   (c) 207 N

45. 3.14 rpm

47. 21.9 g

49. Sun's rotation 2.8%; Jupiter's orbital motion 60%.

51. (a) 142 rpm   (b) 26.5% lost

53. (a) $\omega = \dfrac{2\omega_0}{7}$   (b) $t = \dfrac{2\omega_0 R}{7\mu g}$

55. (a) $\tau = \dfrac{MgL\cos\theta}{2}$   (b) 48.2°

59. No, the acceleration is 0.6 rad/s²

61. 0.93 kg m²/s

# Chapter 12

17. (a) $0 = \Sigma F_x = -F_1 + F_2\sin\phi + F_3$
      $0 = \Sigma F_y = -F_2\cos\phi + F_4$
      $0 = (\Sigma\tau_z)_O$
      $\quad = -L_2 F_2 - L_1 F_3\sin\phi + L_1 F_4\cos\phi$
   (b) $0 = (\Sigma\tau_z)_P = -L_1 F_1\sin\phi + (L_1 - L_2)F_2$

19. (a) $\tau_A = \dfrac{1}{2}mgL$
   (b) $\tau_B = 0$
   (c) $\tau_C = \dfrac{1}{4}mgL$

21. 0.20 m from the left end

23. 0.718 m on the opposite side of $C$ from the worker

25. (a) $x = 47$ m   (b) unstable

27. $n_L = -1.17 \times 10^4$ N, which is downward and must be exerted by the bolt

29. 0.866

31. (a) 1298 N   (b) 1941 N

33. 50.2 kN

35. 6.05 kN

37. 1.17 m

39. 87 kg

41. 63.4°, unstable

43. The block slides before it tips.

45. 74.3 kg

47. 0.366 $mgs$

49. $m \le m_L\dfrac{2\mu_s - \tan\theta}{2(\tan\theta - \mu_s)}$

51. $F_{app} = Mg\dfrac{\sin\theta}{1 + \cos\theta} = Mg\tan\left(\dfrac{\theta}{2}\right)$

53. Block will slide before it tips over if $\mu_s < \tan\alpha = 0.5$

57. $\mu = \dfrac{\sin 2\theta}{3 + \cos 2\theta}$

59. 836 N

61. 28°

63. (a) $0.438\dfrac{GmM}{R_E^{\,2}}$, 192° from North

   (b) $0.0356\dfrac{GmM}{R_E}$

65. 4170 N, tension

# Chapter 13

17. 2.27 ms

19. (a) $x(t) = A\cos(\omega t + \phi)$
       $= (10 \text{ cm})\cos\left[(10\,\pi\text{s}^{-1})t\right]$
   (b) $x(t) = (2.5 \text{ cm})\sin\left[(5 \text{ s}^{-1})t\right]$

21. (a) 0.842 Hz   (b) 1.19 s
   (c) 1/32 m/s   (d) 1.40 N

23. 0.589 Hz, 1.70 s

25. (a) $\approx 4.29$ s$^{-1}$   (b) 0.918 N/m
   (c) 0.817 m

27. (a) $2.21$ s$^{-1}$   (b) 2.84 s   (c) 0.632 m

29. 1.21 s

31. 1.64 s

33. Seven oscillations in the $x$ direction are completed for four in the $y$ direction.

35. 20.0 cm

37. 0.248 s

39. 64.8 km/h

41. 0.694 s

45. (a) $T = 2\pi\sqrt{\dfrac{L}{|\vec{g} - \vec{a}_0|}}$

   (b) $T = 2\pi\sqrt{\dfrac{L}{g + \frac{1}{2}g}} = 2\pi\sqrt{\dfrac{2L}{3g}}$

   (c) $T = 2\pi\sqrt{\dfrac{L}{g - \frac{1}{2}g}} = 2\pi\sqrt{\dfrac{2L}{g}}$

   (d) $T = \infty$

47. 0.342 s

49. $R = \sqrt{\dfrac{2\kappa}{k}}$

51. 5.04 g

53. 0.147%

61. $T = 2\pi\sqrt{\dfrac{m}{k}} = 2\pi\sqrt{\dfrac{m}{2mga}} = \dfrac{2\pi}{\sqrt{2ga}}$

65. 1.53 s

67. (a) $E_2 = \dfrac{1}{4}E_1$;   (b) $a_{2max} = \dfrac{1}{4}a_{1max}$

69. 27.16°

71. $\dfrac{d^2x}{dt^2} = -\dfrac{10}{7}gax$, therefore $T = 2\pi\sqrt{\dfrac{7}{10ga}}$

73. 0.540 Hz, −0.68 m/s, −0.109 rad

79. 2.1 m/s$^2$

## Chapter 14

17. (a) 0.192s   (b) 6.54 cm
19. (a) 300 m   (b) 1.58 m   (c) 3 cm
    (d) 7.5 $\mu$m   (e) 500 nm   (f) 3 Å (See Appendix C on units.)
21. (a) $3.02 \times 10^7\,s^{-1}$   (b) $2.03 \times 10^4\,m^{-1}$
    (c) $1.49 \times 10^3$ m/s
23. (a) 1.3 cm   (b) 9.11 cm   (c) 0.203 s$^{-1}$
    (d) 44.9 cm/s   (e) In the negative $x$ direction
25. 250 m/s
27. (a) 7.6 N
29. 9.9 W
31. 343 m/s
33. 421 m/s
35. 942 Hz; in normal air, the frequency would be 686 Hz
37. 5.4 m
39. (a) 280 Hz   (b) 70 Hz   (c) 210 Hz
41. 567 Hz
43. 93.3 Hz
45. The galaxy is receding with speed $\approx 2.38 \times 10^4$ km/s
47. 30.4 m/s
49. $4\pi^2 A^2 F/\lambda$
51. $\approx 100$ W
55. $v = \sqrt{kL(L - L_0)/m}$
57. 10 m
59. $L_0 = \dfrac{5}{7}L_1$
61. 190 m/s
65. 6.3 m
69. 25 m/s
71. 7.25 km
73. 41 m/s
75. 15.5 kHz
77. 60.033 Hz
79. 31.7 dB

## Chapter 15

17. $10^{-14}$
19. 1 torr = 133 Pa; 1 inch of mercury = 3.39 kPa
21. 101.3 kN
23. 21 N
25. 4.96 km
27. 890 Pa gauge
29. 93 cm higher in the eye
31. $2.53 \times 10^3$ kg/m$^3$, too small for the specific gravity of the diamond
33. (a) no   (b) yes
37. (a) $1.8 \times 10^4$ m$^3$/s   (b) 1.5 m/s

39. 1.75 m/s
41. 831 cm$^2$
43. (a) 624 Pa gauge   (b) 1.25 kPa
45. 3.6 mm
47. $8.11 \times 10^3$ kg
49. 59 g
51. 27 m
53. (a) 49 kg   (b) 2500 kg
55. 14.4 kPa gauge
57. 14.3 m
59. (a) 1.5 m/s   (b) 0.473 L/s
61. 70%
63. (a) $(1.63\%)p_{atm}$   (b) 16.8 cm
65. 14.6 kg
67. (a) 1.18 MW   (b) 3.98 MW   (c) 981
69. $t = \dfrac{A_0}{A_1}\sqrt{\dfrac{2h}{g}}$
71. (a) $\rho(h) = \dfrac{\rho_0}{h_0 g}e^{-h/h_0}$   (b) 5.7 km
75. 2.79 TN · m
77. yes, diluted 1:1 with water
79. $p = p_0 - \rho g h$ use
    $\rho g = 62.2$ lb/ft$^3$, $h = 33$ ft, $p_0 = 18$ lb/in$^2$
    and convert ft to in, $p = 3.8$ lb/in$^2$

## Chapter 16

15. 20° C
17. −40
19. 102.4°F
21. 32 kJ
23. 96.9 W
25. (a) 168 J/K   (b) 480 J/kg·K
27. 0.293 W
29. 55 kW
31. −4.17 kW
33. $R_{air} = 5.56$ per in; $R_{conc} = 0.143$ per in;
    $R_{fg} = 3.45$ per in; $R_{glass} = 0.167$ to 0.2 per in;
    $R_{sf} = 5$ per in; $R_{pine} = 1.28$ per in
35. 2240 W
37. 22 mm$^2$
39. (a) 138 kPa   (b) 33.4 kPa   (c) 233 kPa
41. −10.0°C
43. 0.36 kg
45. (a) 23 kJ   (b) 337 kJ   (c) 65 kJ
47. 2.51 min
49. (a) 560 g   (b) 0.27 K/s
51. 1.8 kg
53. 9.22 K
55. 197 g
57. −14.2 W
59. 6.47 gal
61. 23°C
63. (a) $1.15 \times 10^3$ K   (b) 700 K
65. 24°C
67. 1151 K
69. (a) $204   (b) $23.76
71. 40.7 K
73. 7.5 K
75. 1.5 h
77. 2.9 J

## Chapter 17

19. $3.17 \times 10^{23}$
21. (a) $1.27V_2$   (b) $2V_2$

23. $2.7 \times 10^7$
25. $H_2$
27. Lead
29. 143 kJ
31. 120 mol/m$^3$
33. 0.985 L
35. 263°C.
37. $2.69 \times 10^{25}$ m$^{-3}$, about $2.5 \times 10^{10}$ times greater
39. (a) 235 mol   (b) 5.65 m$^3$
41. (a) 1.27 atm   (b) 0.0268 mol
    (c) 0.786 atm
43. $1.25 \times 10^{10}$ kg
45. (a) $3.26 \times 10^{19}$ J to melt at 0°C
    (b) $1.80 \times 10^{13}$ W
47. 11.8 d
49. 3.55 MJ
51. 1.51 h
53. 5.0 kg
55. 0.798 kg
57. 79 g
59. 1.2 kg ice, 0.80 kg water, all at 0°C
61. ethyl alcohol
63. 59.2 L
65. 8.35 h
69. 1.00034 L
71. 36.5 d, fast
73. Yes. On Venus, one mole would occupy 0.67 L, so a 1 L container is big enough.
75. No; it would take 50 minutes.

## Chapter 18

15. 29 kJ
17. Increase by 250 J
19. 13.9 kW
21. $2p_1V_1$
23. (a) 1.33   (b) 215 J
25. $V = 0.18V_0$
27. 57.7%
29. (a) 200 K   (b) 120 K
31. (a) 3.3 kJ   (b) 0.392 mol
33. (a) 1.49 mm   (b) 10.7 $\mu$J
35. 1.35
37. (a) 300 kPa   (b) −243 J
39. 440°C
41. (a) 810 K   (b) 25.8 atm
43. (a) 300 K, 1.5 kJ   (b) 336 K 0 J
    (c) 326 K, 429 J
45. (a) 39.9 kPa   (b) 83.3 kPa   (c) 80.2 kJ
47. −928 J
49. (a) 211 J   (b) 12.9 L
51. 91.6 J
53. 335 K
55. (a) 598 J   (b) 2500 J flows in
57. 79%
59. 25 m
61. (a) 218 J   (b) 2.71 kJ
65. (a) 15.2°C   (b) 15.5 kJ

67. $e = 1 - \dfrac{r^{1-\gamma}(r_c^{\gamma} - 1)}{\gamma(r_c - 1)}$

69. $W = nRT \ln\left[\dfrac{V_2 - nb}{V_1 - nb}\right] + an^2\left[\dfrac{1}{V_2} - \dfrac{1}{V_1}\right]$

71. $-9.8$ K/km

73. 1.4 atm, $V = 8.67$ L

75. 130 J

## Chapter 19

15. 99.95%

17. (a) 39%   (b) 550 J   (c) 190°C

19. 53.3 kJ

21. 275 J/K

23. 21.9 kg

25. (a) 420 J   (b) 47.2%   (c) 290 K
    (d) 9.24 kW

27. (a) 1.75GW   (b) 43%   (c) 505 K

29. $2 \times 10^7$ kg/s, slightly more than the Mississippi's flow

31. (a) 8.53   (b) $1.10 \times 10^3$ kg

33. (a) 69%   (b) 967 K

35. (a) 4.3   (b) maximum COP $= 11$

37. 2.83

39. (a) 4.72   (b) 4.24 kW   (c) heat pump 54¢/h, oil $1.30/h,

41. (a) 17.4%   (b) 83.3%.

43. $1.35 \times 10^8$ J/K

47. 11.7 J/k $\cdot$ s

49. 74.9 kJ

51. 163 J/K

53. (a) 35%   (b) 5.88 h

55. (a) $e_{\text{otto}} = 1 - r^{1-\gamma}$, $e_{\text{carnot}} = 1 - (1/3)r^{1-\gamma}$

57. 61%

59. (a) $Th = T_{h0}e^{-\rho_0 t}[mc(T_{h0} - T_c)]$
    (b) $b = 0$ at $t = \dfrac{mc(T_{h0} - T_c)}{P_0} \ln\left(\dfrac{T_{h0}}{T_c}\right)$

61. $-0.150$ J/k; cooling produces a more ordered state

63. (a) $-368$ kJ/K   (b) 490 kJ/K
    (c) $-65.3$ kJ/K

65. No, 5.9 h

## Chapter 20

15. $1.56 \times 10^{20}$

17. $2.3 \times 10^{39}$

19. 5.08 m

21. (a) $\hat{j}$   (b) $-\hat{i}$
    (c) $(\hat{i} + 3\hat{j})/\sqrt{10} \simeq 0.32\hat{i} + 0.95\hat{j}$

23. $3.81 \times 10^9$ N/C

25. (a) 2.2 MN/C   (b) 77 N

27. $5.15 \times 10^{11}$ N/C

29. 39 pm

31. $-6.50$ $\mu$C/m

33. $3.3 \times 10^{-12}$ kg

35. (a) 1.35 cm   (b) reverse direction, accelerates and exits field region at $3.8 \times 10^5$ m/s

37. $1.04 \times 10^{-11}$

39. $-0.18\hat{i} + 0.64\hat{j}$ nN

41. $x = 5.45a$

43. $1.6\hat{i} - 0.33\hat{j}$ N

45. (a) Magnitude $4kqQ/a^2$; (b) directed toward the negative charge for positive $Q$, and vice versa

47. 4.14 cm

49. (a) $\vec{E} = \dfrac{2kqy}{(a^2 + y^2)^{3/2}}\hat{j}$   (b) $y = \pm a/\sqrt{2}$

53. (a) $\vec{E} = 4kq\hat{j}[4y(a^2 + 4y^2)^{-3/2} + (2y - \sqrt{3}a)^{-2}]$ for $y > \sqrt{3}a/2$

55. $6.94 \times 10^{-2}$ m$^3$

57. 2.8 Mm/s

59. $-14$ $\mu$C/m

61. $1.29 \times 10^{-30}$ C·m

63. (a) $\vec{\tau} = -(2kQqa/x^2)\hat{k}$
    (b, c) $\vec{F} = (2kQqa/x^3)\hat{j}$, i.e., parallel to the dipole moment

65. 7.0 cm, 0.54 $\mu$C

67. (a) 2.5 $\mu$C/m   (b) $3.0 \times 10^5$ N/C
    (c) 1.8 N/C

73. $y = \dfrac{a}{\sqrt{2}}$

75. $-\hat{i}k\lambda_0(1.5 - 2\ln 2)/L$

77. 900 MN force, not negligible

79. 19.6 kN/C, upward

## Chapter 21

21. $Q_C = -Q_B = 2Q_A$

23. 650 kN/C

25. $\pm 1.48$ kN·m$^2$/C

27. (a) $-q/\epsilon_0$   (b) $-2q/\epsilon_0$   (c) 0   (d) 0

29. $4.9 \times 10^4$ N·m$^2$/C

31. (a) 1.21 MN/C   (b) 2.02 MN/C
    (c) 504 kN/C

33. Line symmetry

35. 49 kN/C

37. (a) $5.14 \times 10^6$ N/C   (b) 34.0 N/C

39. (a) 2.0 MN/C   (b) 7.2 kN/C

41. (a) $\rho = 0$   (b) $\sigma = 4.0$ mC/m$^2$
    (c) Other charges would destroy the symmetry, making $\sigma$ nonuniform.

43. 1.8 MN/C

45. $\pm E_0 \displaystyle\int_0^a y\,dy = \pm\dfrac{1}{2}E_0 a^2$

47. (a) 0   (b) 3.53 kN/C   (c) 1.42 $\mu$C

49. (a) $8kQ/R^2$, inward   (b) $kQ/4R^2$, inward
    (c) a would not change, b would become 0

51. (a) 3.6 MN/C   (b) 3.8 MN/C   (c) a would not change, b would nearly double

53. (a) 20.2 kN/C   (b) 1.68 kN/C

55. 6.3 $\mu$C/m$^3$

57. (a) $E = \rho x/\epsilon_0$   (b) $E = \rho d/2\epsilon_0$

59. 18 N/C

61. (b) $-Q$

65. (a) $\pi\rho^0 a^3$   (b) $E(r) = \rho^0 r^2/4\epsilon_0 a$

67. $5\rho^0/3R^2$

69. $\vec{E} = \dfrac{\rho_0 R}{\epsilon_0}(e - 2)v/r$

71. $\vec{F} = m\vec{g} = -GmM_E\vec{r}/R_E^3$

## Chapter 22

17. $1.92 \times 10^{-17}$ J

21. 27.9 J

23. $-E_0 y$

25. 53 nC

27. (a) 442 kV   (b) 9.2 Mm/s

31. (a) 4 V   (b) $E_x = 1$ V/m, $E_y = -12$ V/m, $E_z = 3$ V/m

33. 3 kV

35. 5.6 kV/m

37. 80 mV

39. $9.37 \times 10^7$ m/s

43. 6.1 $\mu$C

45. $v = \sqrt{2keQ/mR}$

47. $V(R) - V(0) = -kQ/2R$

49. 21.0 V

51. $3\sqrt{3}kq/a$

53. $kq\{[(x - a)^2 + y^2]^{-1/2} + [(x + a)^2 + y^2]^{-1/2}\}$

55. (a) 905 V   (b) 452 V

57. 16.2 cm, 809 nC

59. (a) $V(x, y) = -E_0(x + y)$
    $= -150(x + y)$ V/m
    (b) 150 V

63. $E = V_0/R$, radially otward

65. (a) 43 kV   (b) 1.7 MV/m   (c) 540 V
    (d) 0

67. (a) $-2.17$ kV   (b) no change

69. (a) $x = -3$ m, 0 m, 1 m
    (b) $\vec{E} = (3x^2 + 4x - 3)\hat{i}$
    (c) $x = -1.87$ m, 0.535 m

73. $4.07 \times 10^7$ m/s, $6.96 \times 10^5$ m/s

75. $\dfrac{kq}{2a}\ln\left(\dfrac{\sqrt{2} + 1}{\sqrt{2} - 1}\right) \simeq \dfrac{0.881kq}{a}$

77. $V(x) = -\dfrac{k\lambda_0}{L^2}x\left[x\ln\left\{\dfrac{x - (L/2)}{x + (L/2)}\right\} + L\right]$

79. Yes

## Chapter 23

15. $U = \dfrac{kq^2(2\sqrt{2} + 1)}{2a\sqrt{2}}$

17. $v = \sqrt{\dfrac{2kq^2}{ma}}$

19. (a) 2.0 MV/m   (b) 9.9 kV   (c) 5.5 mJ

21. (a) 0.74 $\mu$C   (b) 40 kV

23. $\pm 14$ C

27. 1.91 m$^2$

29. 70 nF

31. Equal

33. (a) 6.0 $\mu$F,   (b) 0.55 $\mu$F,   (c) 0.83 $\mu$F, 1.3 $\mu$F, 1.5 $\mu$F, 2.2 $\mu$F, 2.8 $\mu$F, 3.7 $\mu$F

35. 1 km$^3$

37. $9 \times 10^{30}$ J/m$^3$

39. $\dfrac{1}{2}kQ^2\left(\dfrac{1}{a} - \dfrac{1}{b}\right)$

41. $C = \dfrac{2\pi\epsilon_0 L}{\ln(b/a)}$

45. 1 $\mu$F stores 15 times as much energy

47. (a) 30 $\mu$F   (b) 0.1 $\mu$F   (c) 0.01 $\mu$F

49. (a) 5.0 kW   (b) 250 $\mu$F   (c) 0.50 W

51. (a) Increases by factor of 2.5   (b) drops to 40% of its original value

53. (a) 2 series pairs in parallel or two parallel pairs in series   (b) 4 in series

55. 102 $\mu$J

57. 75 V

59. (a) 3.5 mm   (b) 87 kV
61. $\pm 1\%$
63. $kQ^2/10R$
65. 2.22 GeV
67. 7.75 J
71. 1.41 fm
73. (b) $U = C_0 V_0^2 (\kappa + 1)/4$

   (c) $F_x = \dfrac{1}{2} \dfrac{C_0 V_0^2 (\kappa - 1)}{L}$

75. $U/L = \pi \rho^2 R^4/8\epsilon_0$

77. (a) $U = \dfrac{Q^2 x}{2A\epsilon_0}$   (b) $F = \dfrac{Q^2}{2A\epsilon_0}$ The force is
   the charge on one plate times the electric
   field due to the other plate. It is not the
   charge on one plate times the field due to
   both plates!

79. $u = \dfrac{\lambda^2}{8\pi^2 \epsilon_0 r^2}$, $U = \dfrac{Q^2 \ln(b/a)}{4\pi\epsilon_0 L}$

81. 266 pF, within specs

## Chapter 24
15. $2.9 \times 10^5$ C
17. 7.65 MA/m$^2$
19. 0.17 V/m
21. $2.2 \times 10^{-6}$ $\Omega \cdot$m
23. 25 $\Omega$
25. 2.34 mA
27. 25 mA
29. 1.38 kW
31. 160 $\mu$A
33. 240 $\Omega$
35. 48 k$\Omega$
37. (a) 0.48 pA   (b) $1.5 \times 10^4$
39. 0.31 mA
41. 0.723 mm/s, 3.26 mm/s, $3.62 \times 10^7$ m/s
43. (a) $4.29 \times 10^6$ A/m$^2$   (b) 3.37 A
45. $v_{Fe}/v_{Ag} = 6.1$
47. $d_{Al} = 1.26 d_{Cu}$
49. (a) 2.07 cm   (b) 2.60 cm   (c) aluminum
51. 34 m$\Omega$
53. Resistor with more power has $\sqrt{2}$ time
   greater diameter
55. 0.625 A
57. 2.9 A
59. (a) 1.11 kW   (b) 13.0 $\Omega$
61. 0.171 mm/s
65. 5 $\Omega$
67. $5.59 \times 10^{-5}$ $\Omega \cdot$m
69. 50 min

## Chapter 25
17. 9 V
19. 216 kJ
21. 229 k$\Omega$
23. 0.02 $\Omega$
25. (a) $3 + 2/3 = 11/3$, $2 + 3/4 = 11/4$,
   $1 + 6/5 = 11/5$ $\Omega$
   (b) $3 \times 3/(3 + 3) = 3/2$,
   $2 \times 4(2 + 4) = 4/3$,
   $1 \times 5/(1 + 5) = 5/6$ $\Omega$
   (c) $1 + 2 + 3 = 6$ $\Omega$
   (d) $(1^{-1} + 2^{-1} + 3^{-1})^{-1} = 6/11$ $\Omega$
27. 2.07 A, 0.214 A, 1.86 A

29. 0
31. 0.662% lower
37. $\mathcal{E} R_2/(R_1 + R_2)$
39. 1.5 A
41. 30 A
43. 14.4 W
45. (a) 162 $\Omega$   (b) 125 mW
47. (a) $\dfrac{R_1 \mathcal{E}}{R + 2R_1}$   (b) $\dfrac{1}{2}\mathcal{E}$
49. 2.45 W
51. $\dfrac{7}{5} R$
53. $-4.77$ mA, downward
55. 48.4 V, 57.3 V, 59.9 V
57. (a) 0.992 A   (b) 0.83%
59. (a) 0.35 s   (b) 0.17 s
61. 3.4 $\mu$F
63. (a) $I_1 = 25$ mA, $I_2 = 0$, $V_C = 0$
   (b) $I_1 = I_2 = 10$ mA, $V_C = 60$ V
   (c) $I_1 = 0$, $I_2 = 10$ mA, $V_C = 60$ V
   (d) $I_1 = I_2 = 0$, $V_C = 0$
65. (a) 4.51 V   (b) 35.2 $\Omega$
67. 83 $\mu$s
71. 1.84 V, 20.8 $\Omega$
75. 8.5 nA, downward
77. (a) as capacitance in parallel with
   a resistance   (b) 595 s
79. $I_2(t) = \dfrac{\mathcal{E}}{4R}\left[2 - e^{-\frac{2}{RC}(1 + \sqrt{2})t} - e^{-\frac{2}{RC}(1 - \sqrt{2})t}\right]$
81. 3.3 k$\Omega$ in parallel with 4.7 k$\Omega$, this
   combination in series with 1.5 k$\Omega$

## Chapter 26
17. (a) 1.6 mT   (b) 2.3 mT
19. (a) $2.0 \times 10^{-14}$ N   (b) $1.0 \times 10^{-14}$ N
   (c) 0
21. 400 km/s
23. 358 ns
25. $r_{proton} = 43 r_{electron}$
27. 1.3 $\mu$s
29. (a) 48.9 mT   (b) 0.734 N/m
31. 0.12 T
33. (a) 40.8 $\mu$T   (b) 5.09 nT
35. 4.02 A
37. 0.62 G
39. 482 mT
41. 24 A
45. (a) $-1.1\hat{i} + 1.5\hat{j} + 1.7\hat{k}$ mN
47. 40.1° or 140°
51. (a) 70.6 $\mu$m   (b) 444 $\mu$m
53. (a) 5.88 A   (b) upward from A to B
55. 377 mT
57. $1.97 \times 10^{-25}$ J $= 1.23$ $\mu$e V
59. 732
61. 23° west of magnetic north
63. 18.2 mN
65. (a) $\mu_0 J_0 r^2/3R$   (b) $\mu_0 J_0 R^2/3r$
67. (a) $-\mu_0 J_s \hat{j}$   (b) 0
69. (a) $2.27 \times 10^3$ m$^{-1}$   (b) 3.31 kW
73. (a) 8.0 $\mu$T   (b) 4.0 $\mu$T   (c) 0
75. (a) $B\pi R^2 \mu_0 I/2w$   (b) $B \approx \mu_0 I/2\pi r$

77. (a) $\dfrac{1}{3J_0}$   (b) $B = \mu_0 J_0 R^2/6r$

   (c) $B = \dfrac{1}{2}\mu_0 J_0 r(1 - 2r/3R)$

79. (a) $\mu = IL^2/4\pi N$   (b) $N = 1$
81. 3.8 mm
83. (a) $b = a$   (b) $B_0 = 8\ \mu_0 I/5^{3/2}a$
85. $1.25 \times 10^{-5}$ N
87. $\mu_0 I\sqrt{2}/\pi a$
89. $F = 3\ \mu_0 \mu I/8\sqrt{2}a^2$, attractive
91. $3.8 \times 10^9$ A

## Chapter 27
17. (a) $6.28 \times 10^{-4}$ Wb   (b) $6.91 \times 10^{-3}$ Wb
   (c) 2.51 mA   (d) counterclockwise
19. 199
21. 0.4 H
23. 1350
25. 130 $\Omega$
27. 100 mA
29. 0.510 $\mu$J
31. $9.9 \times 10^8$ J/m$^3$
33. 1.08 T/ms
35. $B(t) = (bR/A)t^3/3$
37. $B^2 \ell^3 v/R$
39. $16\ B_0 x_0^3/3$
41. (a) 3 s   (b) clockwise
43. (a) $-60$ mA   (b) $-32.4$ mA
45. 20.6 V
47. (a) CCW   (b) $(Blv)^2/R$
49. (c) $v_{terminal} = \mathcal{E}/Bl$
51. $-(8.40 \times 10^{-18}$ N$)\hat{j} + (6.45 \times 10^{-16}$ N$)\hat{i}$
55. 11.0 s
57. (a) 0.11 H   (b) 20 mA
59. (a) 40 A/s   (b) 9.23 A/s
61. (a) 2.18 A   (b) 1.68 A
63. 1.77 ms
65. (a) 5.7 MJ   (b) 31 m$\Omega$   (c) 39.3s
67. Smaller by factor $1/4n^2 R^2$
69. (a) $P = I_0^2 Re^{-2Rt/L}$
71. (b) $v_t = mgR/w^2 B^2$
   (c) counterclockwise
75. $L = \mu_0 N^2 (A - (A^2 - R^2)^{1/2})$
77. $I_2 = (\mathcal{E}_0/(R_1 + R_2)) \exp(-R_p t/L)$, where
   $R_p = R_1 R_2/(R_1 + R_2)$
79. No, it stores 80 J
81. $\mathcal{E} = \dfrac{1}{2}B_x l_\omega^2 \approx 19$ mV; not practical

## Chapter 28
15. 8.91 V
17. (a) 350 mA   (b) 1.50 kHz
19. 45 mA
21. (a) 804 $\Omega$   (b) 48.2 $\Omega$   (c) 2.41 $\Omega$
23. 1.47 $\mu$F
25. 8.23 kHz
27. 4.9 pF to 42 pF
29. 67 nH
31. 3.5 k$\Omega$
33. 5.0 mA, 10 mA, 20 mA
35. 0.39 A
37. (a) 522   (b) 14 A
39. (a) 79.6 Hz   (b) 2 H   (c) $X_L = 4X_C$

41. (a) 10 mA   (b) 14 V   (c) 7.7 mA
43. 0.32 A
45. 0.199 $\mu$H
47. (a) $1/\sqrt{2}$   (b) 1/2   (c) $-1/\sqrt{2}$   (d) 1/2
49. 50.4
51. 6.20 $\Omega$
53. (a) above   (b) current lags by approximately 49
55. (a) 0.333   (b) 4 W
57. (a) 5.5%   (b) 9.1%   (c) large
59. 3.73 mF
61. 2.73 V
63. 1.63 kHz
65. 36.2 nF, 11.5 nF
69. $R = 400\ \Omega$, $L = 67$ mH, $C = 0.094\ \mu$F
73. $V_p/\sqrt{3}$
75. 101 V, $-190$ V, 134 V
79. Yes, $I_{peak} = 1.45$ A

## Chapter 29

13. 1.3 nA
15. $-\hat{k}$
19. 7.5 km
21. 2.57 s
23. 5000 km
25. $x$ direction
27. 11.7%
29. 12 GW/m$^2$
31. $2.7 \times 10^{-10}$ W/m$^2$
33. 3.18 $\mu$W/m$^2$
35. (a) $7.20 \times 10^{11}$ V/m·s   (b) increasing
37. 15 kV/m
39. 1.1 pT
41. 90.5%
43. 18.8%
47. 1.1%
49. $4 \times 10^{10}$
51. $E_p = 40$ V/m, $B_p = 0.13\ \mu$T
53. (a) 2.5 J   (b) $8.33 \times 10^{-9}$ kg·m/s
55. 6.0 mPa
57. 110 km$^2$
59. $1.05 \times 10^{16}$ W
61. 211 mPa
63. Long, with cylindrical symmetry
67. $n = 6$
69. 2.75 m
71. about $10^{16}$ y

## Chapter 30

13. (a) three   (b) along the reverse of the entry path
15. 240° CCW (or 120° CW)
17. 126 nm
19. 1.57
21. 67.5°
23. (a) 49.8°   (b) 42.2°   (c) 22.4°
25. (a) 61.3°   (b) 80.9°   (c) there is none
27. 1.07°
29. (a) 2   (b) 210°
33. (a) 40 cm   (b) 32.5 cm   (c) 25.0
35. 9.48°
37. 1.30 m
39. 42.2°
41. $\sqrt{2}$

43. $2.62 \times 10^8$ m/s
47. 36.5° to 38.1°
49. $v = c \sin \theta_c$
51. $1.96 \times 10^8$ m/s
53. 2.72 m
57. $n = 1.17$
61. 50.9°
67. 375 nm, yes
69. $\theta_c = 37°$, no

## Chapter 31

17. Mirror surface 55° to horizontal
19. (a) Image height is $^1/_4$ object height
    (b) inverted
21. (a) At the center of curvature of the mirror (twice its focal length)   (b) real
23. $M = -2$ (i.e., a real, inverted image)
25. 21.1 cm
27. 27 cm
29. 40 cm
31. $-0.857$ mm
33. $(25\ \text{cm})^{-1} + (-55\ \text{cm})^{-1} = 2.18$ diopters
35. $-1.25$ diopters, the minus sign signifying a diverging lens
37. $-200$
39. (a) 24.3 cm behind the mirror   (b) 29 mm
    (c) virtual
41. 18 cm in front of the mirror
43. 7.42 cm
45. (a) 2.0 m   (b) 3.0 m in front of the mirror
47. 12 cm
49. (a) Real, inverted, 7.7 cm high
    (b) virtual, upright, 7.7 cm high
51. 29 cm or 41 cm from the candle
53. 11.1 cm
55. Real image, 109 cm on other side of lens
57. $L' = -67.9$ cm
59. 2
61. $n = 2$
63. 1.72
65. Real image, magnified 2.74 times
67. 3.33 diopters, converging lens
69. $20.8' \approx \dfrac{1}{3}°$
71. 72 cm
83. 0.257 mm
85. 45 mm

## Chapter 32

11. 1.65 cm
13. 420 nm
15. 4
17. (a) 4.8°, 9.7°   (b) 2.9°, 6.8°
19. (a) 2$^{nd}$   (b) 1$^{st}$
21. 103 nm
23. 424 nm, 594 nm
25. the top 4.5 cm
27. 29.3°
29. 0.0162
31. 36 cm
33. 1.3 m
35. 94.1°
37. (a) $m_{max} = 38$   (b) $m_{max} = 3$

39. The fourth bright 550-nm fringe ($m = 4$) coincides with the sixth dark 400-nm fringe ($m = 5$).
41. 5.33°
43. (a) 500 nm, 560 nm   (b) same as (a)
    (c) $m = 2$, no overlap
45. 0.01 nm
47. 77.2 nm
49. Echelle grating has 60% greater resolving power
51. 3.28 Å
53. 200 nm
55. 2.16 cm
57. Next integer greater than $nN + (1/2)(n-1)$
59. 545 nm
61. 83.8 cm
63. (a, b) objects separated by 22.8 cm can be resolved; no to newspaper headlines, yes to billboards   (c) objects separated by 4.6 cm (1.8 in) can be resolved; yes to banner headlines and billboards
65. 6.9 km
67. 484 nm
69. $n - 1 = m\lambda_{vac}/2L$
73. $y_1 = 17.7$ cm, $y_2 = 44.7$ cm
77. No, $\Delta\lambda = 6.33$ pm.
79. 210 nm

## Chapter 33

13. (a) 4.50 h   (b) 4.56 h   (c) 4.62 h
15. 33.1 ly
17. 40.4 m
19. $0.14c$
21. (a) 2   (b) 2.53
23. 0.141
25. (a) 2.1 MeV   (b) 1.6 MeV
29. (a) $0.857c$   (b) 9.69 min
31. $c/\sqrt{2}$
33. Twin A: 83 years; twin B: 40 years
35. $0.96c$
39. B before A by $3.47 \times 10^5\ y$
41. In Problem 39, the events are simultaneous in a frame moving from A to B at $v/c = 0.5$; in Problem 40, there is no frame in which they are simultaneous.
45. 0.936
47. (a)  12.5 y   (b) 7.50 y
51. (a) 4.2 ly   (b) B $-$ 2.4 y (i.e., B occurs earlier)
53. (a) 0.758   (b) $5.82 \times 10^{-19}$ kg·m/s
55. 25 h
57. (a) 0.256 eV   (b) 128 keV   (c) 3.11 MeV
63. $0.866c$
65. $0.95c$

## Chapter 34

15. 16
17. $\lambda_{peak} = 10\ \mu$m; $\lambda_{medium} = 14.3\ \mu$m
19. (a) 500 nm, yellow-green   (b) 710 nm, deep red near the limit of the visible spectrum
21. 191 nm; ultraviolet
23. $P_{blue} = 1.44P_{red}$

25. $\lambda = 122$ nm, 103 nm, 97.2nm
27. 91.2 nm; ultraviolet
29. (a) $3.7 \times 10^{-63}$ m   (b) 73 nm
31. $v_p = 5.46 \times 10^{-4} v_e$
33. $6.3 \times 10^7$ m/s
35. 126 nm
37. 23 keV
39. $R_{200\,nm} = 0.056 R_{500\,nm}$
41. (a) $4.39 \times 10^3$ K   (b) 0.474
43. (a) $1.7 \times 10^{28}$ s$^{-1}$   (b) $3.2 \times 10^{15}$ s$^{-1}$
    (c) $1.3 \times 10^{18}$ s$^{-1}$
45. (a) $1.12 \times 10^{15}$ Hz   (b) 2.80 eV
47. Copper, nickel, silicon
49. 443 nm
51. (a) 154 pm   (b) 222 eV
53. No
55. (a) 26.4 cm   (b) 4.70 $\mu$eV
57. $n_1 = 229$
61. (a) 27 pm   (b) 40.8 eV
63. 1.62 km/s
65. 2.90 km/s
67. 1.16 ps
71. $\dfrac{1}{2} m_e c^2 \left[ (\gamma - 1) + \sqrt{(\gamma - 1)(\gamma + 3)} \right]$
79. $5.49 \times 10^{-2}$ eV

## Chapter 35

11. (a) At $x = 0$   (b) At $x = \pm 0.589a$
13. 0.0625
15. $n = 5$
17. $6.03 \times 10^{-22}$ J
19. $h = 312$ J·s, $4.71 \times 10^{-35}$ times the actual value of Planck's constant
21. $6.79 \times 10^{26}$
23. 0.23 MeV
25. $2.66 \times 10^{14}$ Hz
27. $3 \times 10^{-32}$ J
29. $9.87 \times 10^{-11}$ J = 616 MeV
31. $0.968 b^{-2.5}$
35. (a) $3.48 \times 10^{-19}$ J = 2.18 eV
    (b) 571 nm
37. 21.3 $\mu$m
39. 2.20 nm, 2.75 nm, 4.71 nm, 4.12 nm, 6.59 nm, 11.0 nm
41. (a) $\Psi_{n\,odd}(x') = \sqrt{\dfrac{2}{L}} \cos\left( \dfrac{n\pi x'}{L} \right);$

    $\Psi_{n\,even}(x') = \sqrt{\dfrac{2}{L}} \sin\left( \dfrac{n\pi x'}{L} \right)$

    (b) $E_n = n^2 h^2 / 8mL^2$
43. 0.759 nm
45. $8.36 \times 10^{31}$
47. $n > 99$
49. (a) 0.48   (b) 0.091   (c) 0.20   (d) 0.26
    (e) 0.25
53. $n = 3$
55. (a) 4   (b) 2
59. (a) no bound states   (b) one bound state, at $\mathcal{E} = 6.44$   (c) two bound states, at $\mathcal{E} = 7.53$, $\mathcal{E} = 29.3$
63. $B/A = (k_1 - k_2)/(k_1 + k_2)$, $C/A = 2k_1/(k_1 + k_2)$

65. 2.38 nm, yes
67. $2 \times 10^{13}$ K

## Chapter 36

15. $\sqrt{13.6/1.5} = 3$
17. d.
19. $\ell = 5$
21. $3d$
23. $\sqrt{6}\hbar$
25. $\dfrac{3}{2}, \dfrac{5}{2}$
27. $\dfrac{21}{2}\hbar\omega$
29. $1s^2 2s^2 2p^6 3s^2 3p^6 3d^1 4s^2$
33. 1.14 meV
35. $n = 4$, $\ell = 3$
37. about $3 \times 10^{68}$
39. 90°, 65.9°, 114°, 35.3°, 145°
41. $0, \pm 1, \pm 2, \pm 3$
45. $\dfrac{1}{2}, \dfrac{1}{2}, \dfrac{3}{2}, \dfrac{3}{2}$, and $\dfrac{5}{2}$
47. (a) $E_4 = E_1/16$   (b) $L = \sqrt{12}\hbar$.
    (c) $J = \sqrt{35}\hbar/2$   (d) The magnitude of $L$ is greater than $J$.
49. (a) $16\hbar\omega$   (b) $\hbar\omega/2$
51. $1s^2 2s^2 2p^6 3s^2 3p^6 3d^{10} 4s^1$
53. $3.02 \times 10^{17}$
55. 0.11
57. $3.03\hbar$
59. (a) $\ell \pm \dfrac{1}{2} = \dfrac{5}{2}$ or $\dfrac{7}{2}$   (b) $8m_j$-values
67. $r_{av} = 1.5 a_0$
71. (a) 18.7 eV   (b) 9.5%

## Chapter 37

17. 3.48 mm
19. $9.40 \times 10^{-46}$ kg·m$^2$
21. $7.07 \times 10^{13}$ Hz
23. 181 kcal/mol
25. 548.7 nm
27. InAs, 3.54 $\mu$m
29. 1.25 meV
31. $\ell \hbar^2 / I$
33. 0.121 nm
35. (a) 0.179 eV   (b) 0.359 eV
37. (a) $4.72 \times 10^{13}$ Hz
    (b) $1.95 \times 10^{-46}$ kg·m$^2$
39. 35.8 $\mu$m
41. 10.2
43. $-8.40$ eV
47. 4.7 eV
49. $6.35 \times 10^4$ K, over 200 times room temperature
51. 709 nm, zinc selenide wouldn't make a good photovoltaic cell
53. 1.79 kA
55. (a) 0.368 nm   (b) 12.3
57. 35.4%
59. (a) $U_{min} = 0$   (b) $r_{min} = r_0$
61. $E_{av} = (3/5) E_F$
63. 2.5 meV
65. 2.78 eV. Yes, it works okay.

## Chapter 38

13. $^{211}_{86}$Ra, $^{220}_{86}$Ra, $^{222}_{86}$Ra
15. (a) same   (b) $Q_K - Q_{Cl} + 2e$
17. 5.9 fm
19. $^{64}_{29}$Cu $\to$ $^{64}_{30}$Zn $+ \beta^- + \bar{\nu}$ (40%)
    $^{64}_{29}$Cu $\to$ $^{64}_{28}$Ni $+ \beta^+ + \nu$ (19%)
    $^{64}_{29}$Cu $+ e^- \to$ $^{64}_{28}$Ni $+ \nu$ (41%)
21. (a) $^{10}_{6}$C $\longrightarrow$ $^{10}_{5}$B $+ e^+ + \bar{\nu}_e$   (b) 5.56 MBq
23. (a) 193 y   (b) 289 y
25. 59.95 u
27. 5.61 MeV/nucleon
29. 2
31. $1.0 \times 10^{20}$ fission/s
33. $2 \times 10^{20}$ m$^{-3}$
35. 10 s
37. $5.3 \times 10^{-12}$ eV
39. 8.80 meV/nucleaon
41. (a) $7.85 \times 10^5$, essentially 0   (b) 1.009, $5.0 \times 10^5$   (c) 3.6, $8.2 \times 10^4$
43. (a) 0.834   (b) 0.281   (c) 0.432%
45. (a) $^{228}_{90}$Th
47. $8.92 \times 10^3$ y
49. 8.04 d for Poland, 16.2 d for Austria, 10.0 d for Germany
51. $3.0 \times 10^9$ y
53. 23%
55. 1.68 mCi/L
57. 1.02 kg
59. 8.63 s
61. (a) 3.41 GW   (b) 35.2%
63. 454 kg
65. (a) $4.4 \times 10^{10}$ J   (b) 340 gal
67. 4.3 MeV
69. $^1_0$n $+ ^{198}_{80}$Hg $\longrightarrow$ $^{197}_{79}$Au $+ ^A_Z$X
71. (a) 1.1 GJ   (b) 2.8 pellets/s   (c) 450 kg
75. $\lambda_A N_A + \lambda_B N_B = \lambda_A N_0 [ (\lambda_A - 2\lambda_B) \exp(-\lambda_A t) + \lambda_B \exp(-\lambda_B t) ]/(\lambda_A - \lambda_B)$
77. 1568 Mev
79. 10 mg left, barely enough!

## Chapter 39

17. $3 \times 10^{-16}$ s
19. $\pi^+ \to \mu^+ + \nu_\mu$
21. $\eta \to \pi^+ \pi^- \pi^0$, conserves charge $(1 - 1 + 0 = 0)$
23. No, the conservation of baryon number
25. $sss$
27. $4.54 \times 10^4$ m$^3$, about $1.2 \times 10^7$ gallons
29. $10^{28}$ K
31. 1.33 Gly
33. 12 Gy
35. a, violates conservation of baryon number
37. (a) No   (b) yes
39. $c\bar{c}$
41. (a) 0.16 $\mu$J   (b) 16.3 $\mu$m
43. 90 $\mu$s
45. 313 ly
47. (a) 256 fm   (b) $-2.81$ keV
49. (a) $5.74 \times 10^3$ km/s   (b) 383 Mly
51. $2.6 \times 10^{-25}$ s
53. 18 Gy, 13.6 Gy
55. 5 km/s/Mly

# Credits

## Chapter 24
Page 398: Danial Sambraus/Photo Researchers. Page 410, Figure 24.16: Erik Borg/Addison Wesley Longman. Page 410, Figure 24.17: William Ferguson.

## Chapter 25
Page 415: Tom Pantages. Page 419, Figure 25.7: Erik Borg/Addison Wesley Longman.

## Chapter 26
Page 436: M. Aschwanden et al. (LMSAL), TRACE, NASA. Page 437, Figure 26.1: Stuart Field. Page 440: Ivan Massar. Page 446, Figure 26.22: NASA. Page 448: John Wilson White/AWL. Page 452, Figure 26.30: University Corporation for Atmospheric Research UCAR. Page 456, Figure 26.37: Richard Megna/Fundamental Photographs.

## Chapter 27
Page 464: Charles E. Rotkin/Corbis. Page 479, Figure 27.28: NASA. Page 483, Figure 27.34: Milt and Joan Mann/Cameramann International, Ltd.

## Chapter 28
Page 489: Lester Lefkowitz/Corbis. Page 497, Figure 28.13: Erik Borg/Addison Wesley. Page 501, Figure 28.21 TL: Norbert Schaefer/Corbis. Page 501, Figure 28.21 TM: Taxi/Getty Images. Page 501, Figure 28.21 TR: Jonathan Nourak/PhotoEdit. Page 501, Figure 28.21 ML: Jonathan Nourak/PhotoEdit. Page 501, Figure 28.21 MR: D2 Productions/Omni-Photo Communications. Page 501, Figure 28.21 BL: Jonathan Nourak/PhotoEdit. Page 501, Figure 28.21 BR: Michael Mathers/PeterArnold.

## Chapter 29
Page 508 T: PhotoDisc/Getty Images. Page 508 BL: PhotoDisc Red/Getty Images. Page 508 BR: Michael Keller/Corbis. Page 518, Figure 29.8: Richard Megna/Fundamental Photographs. Page 523, 29.14: Illustration: Babakin Space Center, courtesy of The Planetary Society.

## Chapter 30
Page 529: Craig Tuttle/Corbis. Page 530: Courtesy of Schott Corporation. Page 535, 30.10: Richard Megna/Fundamental Photographs. Page 536: Spencer Grant/Photo Researchers. Page 537, Figure 30.16: Courtesy of R. Giovanelli and H. R. Gillett/CSIRO National Measurement Laboratory, Australia. Page 539, Figure 30.18: Library of Congress.

## Chapter 31
Page 542: Yoav Levy/Phototake. Page 543, Figure 31.2: Martin Bough/Fundamental Photographs. Page 544, Figure 31.4: Courtesy of the Space Telescope Institute/NASA. Page 545, Figure 31.7: Erik Borg/Addison Wesley Longman. Page 545, Figure 31.9: Stewart Weir/Corbis. Page 547, Figure 31.11: Perkin Elmer. Page 551, Figure 31.21: Erik Borg/Addison Wesley Longman. Page 555: Chris Barry/Phototake. Page 558, Figure 31.34: Giant Magellan Telescope, Carnegie Observatories.

## Chapter 32
Page 563: Chris Collins/Corbis. Page 565: Andrew Syred/Photo Researchers. Page 565, Figure 32.3: M. Cagnet et al. Atlas of Optical Phenomena Springer-Cerlag, 1962. Page 568, Figure 32.8: Chris Jones. Page 571, Figure 32.13: Richard Megna/Fundamental Photographs. Page 571, Figure 32.14: Courtesy of Jay M. Pasachoff. Page 572, 32.16: Erik Borg/Addison Wesley. Page 573, Figure 32.17: NASA/Ames Research Center/Ligo Project. Page 576, Figure 32.23: Courtesy of Cagnet Atlas. Page 576, Figure 32.24: Courtesy of Cagnet Atlas. Page 577, 32.27: Chris Jones. Page 579: Courtesy of the Blu-Ray Disc Association. Page 580: Chris Jones.

## Chapter 33
Page 585: Delft University of Technology/Photo Researchers, Inc. Page 586: AP/Wide World. Page 589, Figure 33.4: Einstein Archives, Hebrew University, Jerusalem. Page 604, Figure 33.20: Space Telescope Science Institute/Photo Researchers.

## Chapter 34
Page 609: Courtesy Wolfgang Ketterle, MIT. Page 620: Andrew Syred/Photo Researchers. Page 621, Figure 34.13: Millie LeBlanc/Courtesy of The Educational Development Center.

## Chapter 35
Page 629: IBM Corporation. Page 630, Figure 35.2: Courtesy of Elisha Huggins, Dartmouth. Page 639: Courtesy of IBM.

## Chapter 36
Page 646: Chip Clark. Page 654, Figure 36.14: Courtesy Wolfgang Ketterle, MIT.

## Chapter 37
Page 666: Alfred Pasieka/Photo Researchers. Page 679: Koichi Kamoshida/Liaison/Getty Images.

## Chapter 38
Page 684: WDCN/Univ. College London/Photo Researchers. Page 688: David Job/Getty Images. Page 703, Figure 38.18: Lawrence Livermore National Laboratory/Photo Researchers.

## Chapter 39
Page 709: Brookhaven National Laboratory. Page 717, Figure 39.7: Patrice Loiez, CERN/Photo Researchers. Page 718, Figure 39.8: ICRR Institute for Cosmic Ray Research. Page 719: CERN. Page 720, Figure 39.11: NASA Headquarters. Page 721, Figure 39.13: NASA Headquarters.

# A

Aberration
  chromatic, 537, 554, 558
  of lenses, 554, 556
  of mirrors, 543–544
  of starlight, 588, 589
ABS brakes, 78
Absolute motion, 587, 590
Absolute zero, 263, 264
Absorption spectra, 616, 658, 660
AC circuits. *See* Alternating current circuits
Accelerated reference frames, 53, 61–62, 604
Acceleration
  angular, 157–158, 174–175
  average, 19, 20, 34
  centripetal, 43
  as change in direction, 34–35
  in circular motion, 42–44, 73–76
  constant, 20–23
    gravitational, 23–25, 38
    in two dimensions, 36–38
  definition of, 19
  force and, 51–53
  of gravity, 119–120
    near Earth's surface, 23–25, 38
  instantaneous, 19, 20, 34
  in one dimension, 19–23
  radial, 44, 157
  tangential, 44, 157
  in two dimensions, 34–35
    constant, 36–38
  as vector, 34–35
  with zero velocity, 20
Accelerators, particle, 594, 719
Acceptor levels, 676
Accuracy, in calculations, 6–7
Acousto-optic modulator (AOM), 570
Actinide series, 658
Action-at-a-distance forces, 54, 126, 127, 333
Activation analysis, 694
Activity, of radioisotope, 689–690
Adiabat, 298
Adiabatic exponent, 298–299, 301–302
Adiabatic free expansion, 319–320
Adiabatic processes, 297–299, 300
  entropy change in, 319–320
Aerodynamics, 252–253
Air pollution, 299
  electrostatic precipitators and, 379
Air resistance, 23, 38, 80
Allowed transitions, 659, 670
Alpha particles, 692, 694
Alternating current, 489–490
Alternating current circuits. *See also* Electric circuits
  circuit elements in, 490–494
  *LC* circuit, 495–497
  of radio transmitter, 520
  phasor diagrams for, 493, 498, 499
  power in, 500
  power supplies in, 501–502
  *RLC* circuit, 497–499
  transformers in, 500–501

Ammeter, 415, 426
Amorphous solids, 671
Ampere (A), 3, 398, 445, 737
Ampère's law
  for changing currents, 509–511
  compared to Faraday's law, 509
  in magnetic field calculations, 453–457
  for plane wave, 514–515
  for steady currents, 451–457, 509, 510
Amperian loop, 452, 453
Amplifier, transistor as, 404
Amplitude
  of AC signal, 489, 490, 493
  of oscillatory motion, 205
    damped, 214–215
    driven, 216
    peak-to-peak, 205
    simple harmonic, 208
  of wave, 223
    electromagnetic, 516
Anderson, Carl, 710
Angle, units for, 3, 739
Angle of incidence, 531
Angle of reflection, 531
Angular acceleration, 157–158, 174–175
Angular displacement, 156
  of torsional oscillator, 210
Angular frequency
  of AC signal, 490, 493
  of simple harmonic motion, 206–207
  of wave motion, 224–225
Angular magnification, 556, 557
Angular momentum, 177–181.
        *See also* Spin
  in Bohr atom, 617, 620, 650
  conservation of, 178–180, 713
  definition of, 177
  precession and, 180–181
  in quantum mechanics
    atomic, 650–654, 656
    molecular, 668–669
    nuclear, 687–688
    selection rules and, 659, 670
Angular speed, 156–157
Angular velocity, 156–157, 174–175
Antennas, 520, 523
Antilock braking systems (ABS), 78
Antinodes, of standing wave, 234
Antiparticles, 601, 641, 711, 712, 714
Antireflection coatings, 572
AOM (acousto-optic modulator), 570
Aphelion, 123
Apparent weight, 61–62
Apparent weightlessness, 56, 58
Approximation, 7
Archimedes' principle, 247–248
Area, 729, 739
Aristotelian view of motion, 49–50,
    51, 121
Astronomy. *See also* Telescopes
  beyond visible spectrum, 578, 654, 687
Atmosphere (atm), 244
Atmospheric pressure, 244, 245

Atom(s). *See also* Elements; Hydrogen atom
  Bohr model, 617–619, 620, 624, 650
  earliest theories of, 610
  as magnetic dipole, 449, 483
  multielectron, 655–658
Atomic fountain clock, 25
Atomic number, 655, 684
Atomic spectra, 616–619, 658–659
Atomic weights, 741–743
Audio systems
  *RC* circuits in, 428–429
  speakers in, 494, 499
Automobile. *See* Car
Average acceleration, 19, 20, 34
Average angular velocity, 156
Average speed, 14–15, 16
Average velocity, 15–16, 18–19, 34
  with constant acceleration, 20–21
Axis of rotation, 156
Axis of symmetry, 356

# B

Back emf, 475, 476
Ballistic pendulum, 145
Balmer, Johann, 616
Band gaps, 673, 675, 677
Band theory, 673–674, 675–677
Bandwidth, 536
Bardeen, John, 679
Barometer, 245
Baryon number, 713
Baryons, 712, 713, 715
Baseball, aerodynamics of, 253
Battery
  internal resistance of, 419–420
  as source of emf, 416
BCS theory, 679
Beam splitter, 572
Beats, 230–231
Becquerel (Bq), 690
Becquerel, Henri, 689
Bernoulli, Daniel, 250
Bernoulli effect, 251–252
Bernoulli's equation, 250–251
Beta decay, 692–693, 713, 714
Bias, 676
Bicycle, 181
Big Bang, 720–722
  magnetic monopoles and, 447
  nucleosynthesis and, 696
Binding energy, nuclear, 695–696
Binnig, Gerd, 639
Biot, Jean Baptiste, 442
Biot–Savart law, 443–445, 451
Black holes, 654
Blackbody, 270, 610
Blackbody radiation, 610–612
  from Big Bang, 721
  vs. discrete spectra, 616
  Rayleigh–Jeans law for, 611–612, 624
Blu-ray disc, 579, 677
Bohr, Aage, 689